# RISK AND RELIABILITY IN GEOTECHNICAL ENGINEERING

# RISK AND RELIABILITY IN GEOTECHNICAL ENGINEERING

EDITED BY
## KOK-KWANG PHOON
## JIANYE CHING

CRC Press
Taylor & Francis Group
Boca Raton  London  New York

CRC Press is an imprint of the
Taylor & Francis Group, an **informa** business

A SPON BOOK

The pictures on the book cover were contributed by Lina Ximena Garzón from her PhD thesis "Physical Modeling of Soil Spatial Variability," Universidad de Los Andes, Bogotá, Colombia.

MATLAB® is a trademark of The MathWorks, Inc. and is used with permission. The MathWorks does not warrant the accuracy of the text or exercises in this book. This book's use or discussion of MATLAB® software or related products does not constitute endorsement or sponsorship by The MathWorks of a particular pedagogical approach or particular use of the MATLAB® software.

CRC Press
Taylor & Francis Group
6000 Broken Sound Parkway NW, Suite 300
Boca Raton, FL 33487-2742

© 2015 by Taylor & Francis Group, LLC
CRC Press is an imprint of Taylor & Francis Group, an Informa business

No claim to original U.S. Government works

ISBN-13: 978-1-4822-2721-5 (hbk)
ISBN-13: 978-1-138-89286-6 (pbk)

---

**Library of Congress Cataloging-in-Publication Data**

---

Risk and reliability in geotechnical engineering / [edited by] Kok-Kwang Phoon and Jianye Ching.
   pages cm
  Includes bibliographical references and index.
  ISBN 978-1-4822-2721-5 (alk. paper)
  1. Geotechnical engineering. 2. Reliability. 3. Rock mechanics. 4. Soil mechanics. I. Phoon, Kok-Kwang, editor. II. Ching, Jianye, editor.

TA706.R48 2014
624.1'51--dc23
                                       2014023520

---

Visit the Taylor & Francis Web site at
http://www.taylorandfrancis.com

and the CRC Press Web site at
http://www.crcpress.com

# Contents

## 6 Polynomial chaos expansions and stochastic finite-element methods     265

BRUNO SUDRET

**PART III**
**Design**          337

8  **LRFD calibration of simple limit state functions in geotechnical
   soil-structure design**     339

  RICHARD J. BATHURST

## PART IV
## Risk and decision

## 11  Practical risk assessment for embankments, dams, and slopes

LUIS ALTAREJOS-GARCÍA, FRANCISCO SILVA-TULLA, IGNACIO ESCUDER-BUENO, AND ADRIÁN MORALES-TORRES

# Preface

*Risk and Reliability in Geotechnical Engineering* was originally conceived as an update to our 2008 Taylor & Francis book *Reliability-Based Design in Geotechnical Engineering— Computations and Applications*. However, R&D in this domain has gained pace over the past 6 years and it is the collective opinion of the contributors that it is timely to write a new book. One important milestone that took place after 2008 is the recognition by the broader structural community that geotechnical reliability is fundamentally distinctive from structural reliability in various important aspects. A new Annex D on "Reliability of Geotechnical Structures" will be included in the third edition of ISO2394 (2015) to emphasize the need to be sensitive to the practical needs of geotechnical engineering in the general application reliability principles. In other words, one expects more attention to be paid on realism in research and the practice of geotechnical reliability.

There is little doubt that the evaluation of the subsurface condition (including soil/rock properties) is one key aspect that distinguishes geotechnical engineering from structural engineering practice. Soils and rocks are naturally occurring geomaterials that cannot be produced according to factory specifications. Dealing with variability is the norm, not the exception, in geotechnical engineering practice. Site investigation is mandated by building regulations in many countries, in part because it is important to appreciate variable subsurface conditions in geotechnical design. Given the diversity of site conditions and the associated diversity of local practices that evolved to deal with these site-specific conditions, it is also possible for geotechnical variability to be comparable or even larger than the variability in the loadings. This variability is already considered in existing practice, albeit implicitly, through the application of a sizeable factor of safety and other risk mitigating measures (e.g., observational approach, integrity test, load test). Information is clearly collected prior to the design (site investigation) and during the construction process (quality assurance). It is very useful to quantify the value of this information for design and risk mitigation in a unified way. At present, reliability-based designs can be viewed as a simplified form of risk-based design where different consequences of failure are implicitly covered by the adoption of different target reliability indices. Explicit risk management methodologies are required for large geotechnical systems where soil and loading conditions are too varied to be conveniently slotted into a few reliability classes (typically three) and an associated simple discrete tier of target reliability indices. Some of these issues are examined in this book in the context of realistic geotechnical examples covering piles, slopes, retaining structures, dams, embankments, and soil liquefaction.

This book focuses on making these important reliability and risk methodologies more accessible to practitioners and researchers by presenting those soil statistics which are necessary inputs, by explaining how calculations can be carried out using simple tools, and by presenting illustrative or actual examples showcasing the benefits and limitations of these analyses. In short, this book adheres closely to the educational theme that has made the previous 2008 book a success. The reader will find "need to know" information for a non-specialist to calculate and interpret the reliability index and risk of geotechnical structures in a realistic and robust way. It will suit engineers, researchers, and students who are interested in the practical outcomes of reliability and risk analyses without going into the intricacies of the underlying mathematical theories.

MATLAB® and Simulink® are registered trademarks of The MathWorks, Inc. For product information, please contact:

The MathWorks, Inc.
3 Apple Hill Drive
Natick, MA 01760-2098 USA
Tel: 508 647 7000
Fax: 508-647-7001
E-mail: info@mathworks.com
Web: www.mathworks.com

# Acknowledgments

This book is a collaborative project among some of the top practitioners in our georisk community. It owes its success to all the contributors who have invested significant efforts to make the chapters useful and accessible to the nonspecialists.

The editors are grateful for the patience exercised by everyone in meeting our deadlines and responding to our review comments. We would also like to thank our senior editor from Taylor & Francis, Tony Moore, for helping us to steer this book project to fruition.

The pictures on the book cover were contributed by Lina Ximena Garzón from her PhD thesis "Physical Modeling of Soil Spatial Variability," Universidad de Los Andes, Bogotá, Colombia.

We welcome any constructive comments and suggestions from the readers. Please submit your views to our email addresses below.

Email of Kok-Kwang Phoon: kkphoon@nus.edu.sg

Email of Jianye Ching: jyching@gmail.com

# Editors

**Kok-Kwang Phoon** is a distinguished professor and head of the Department of Civil and Environmental Engineering, National University of Singapore (NUS). He is a professional engineer in Singapore and past president of the Geotechnical Society of Singapore. His main research interests include statistical characterization of geotechnical parameters and reliability-based design in geotechnical engineering. He is the recipient of numerous research awards, including the ASCE Norman Medal in 2005, the NUS Outstanding Researcher Award in 2010, and the John Booker Medal in 2014. He is the founding editor of *Georisk* and chair of TC304 (Engineering Practice of Risk Assessment and Management) in the International Society for Soil Mechanics and Geotechnical Engineering (ISSMGE). He was former chair of ASCE Geo-Institute Risk Assessment and Management Committee. He is fellow of ASCE and fellow of the Academy of Engineering Singapore.

**Jianye Ching** is a professor in the Department of Civil Engineering, National Taiwan University (NTU). His main research interests include geotechnical reliability analysis and reliability-based design, basic uncertainties in soil properties, random fields and spatial variability, and geotechnical design codes. He is the secretary of TC304 (risk) in the International Society for Soil Mechanics and Geotechnical Engineering (ISSMGE). He is the recipient of the Outstanding Research Award and the Wu-Da-Yu Memorial Award from the National Science Council of Taiwan, Republic of China.

# Contributors

Ashraf Ahmed is a lecturer in the Department of Civil Engineering, Aswan University, Egypt. He earned his PhD in 2012 from the University of Nantes, France. His main research interests include risk analysis, geotechnical reliability analysis and design, and spatial variability of soil properties. He is the recipient of the 2012 best paper award from the journal *Georisk* (Assessment and Management of Risk for Engineered Systems and Geohazards). Email: Ashraf.Ahmef@gmail.com

Luis Altarejos-García graduated at Universidad Politécnica de Valencia (UPV), Spain, with a PhD in civil engineering. He is a university professor at Universidad Politécnica de Cartagena (UPCT) and partner at iPresas Risk Analysis, a spin-off company from the UPV that provides consultancy services for enhanced governance, safety and security assessments, and investment prioritization in critical infrastructures. He is a certified project manager professional (PMP) from the Project Management Institute (PMI) and a registered professional engineer in Spain. He has been a member of the Technical Committee on Computational Aspects of Analysis and Design of Dams from the Spanish National Committee on Large Dams (SPANCOLD) since 2009. He has worked worldwide as a consultant in several projects for public institutions, hydropower companies, and mining groups, mainly in water management infrastructures, such as canals, pipes, spillways, and dams. He has also been a consultant for the World Bank in dam safety projects. Email: luis.altarejos@upct.es

Gregory B. Baecher is a Glenn L. Martin Institute Professor of Engineering at the University of Maryland. He earned a BSCE from University of California, Berkeley and a PhD from the Massachusetts Institute of Technology. His principal area of work is engineering risk management. He coauthored with J. T. Christian in *Reliability and Statistics in Geotechnical Engineering* (Wiley, 2003), and with D.N.D. Hartford *Risk and Uncertainty in Dam Safety* (Thos. Telford, 2004). He is a recipient of the ASCE Middlebrooks and State-of-the-Art Awards, and a member of the US National Academy of Engineering. Email: gbaecher@mac.com

Richard J. Bathurst graduated from Queen's University, Ontario, Canada, in 1985 with a PhD in geotechnical engineering. He is currently professor of civil engineering and research director at the GeoEngineering Centre at Queen's-RMC in Kingston, Ontario, Canada. He holds adjunct positions at the University of Waterloo and Edith Cowan University in Australia. His primary areas of research are reinforced soil wall and slope technologies with specific emphasis on the development of static and seismic design methods, reliability analysis, and LRFD calibration. He has authored or coauthored more than 300 papers and 20 book chapters. Dr. Bathurst is president of the Canadian Geotechnical Society, past-president of the International Geosynthetics Society, past-president of the North American

Geosynthetics Society and an elected fellow of the Engineering Institute of Canada and Canadian Academy of Engineering. Dr. Bathurst is the editor of the technical journal *Geosynthetics International*. Email: bathurst-r@rmc.ca

**Zijun Cao** is currently an associate professor in the School of Water Resources and Hydropower Engineering, Wuhan University, China. He earned his PhD in geotechnical engineering from City University of Hong Kong in 2012. His main research areas include probabilistic site characterization with particular interests in the quantification of uncertainties in soil properties, efficient probabilistic analysis, and risk assessment of slope stability, practical reliability-based design of geotechnical structures. Email: zijuncao@whu.edu.cn

**John T. Christian** delivered the Terzaghi Lecture in 2003 and is the current chairman of the Civil Engineering Section of the National Academy of Engineering (NAE). He has served as the editor-in-chief of the journal *Geotechnical and Geoenvironmental Engineering*, and now serves as its ombudsman. He is a pioneer in the use of computer methods and coauthored the first general, user-friendly computer program for analysis of slope stability with circular and noncircular failure surfaces. His distinguished career includes chairing the NAE's review of Boston's "Big Dig" project, service on NAE committees on New Orleans Regional Hurricane Projects and the Louisiana Coastal Restoration Project, probabilistic studies of the stability of an earth dam and of the seismic behavior of mine waste slopes, as well as numerous publications. He is the 1996 winner of the ASCE Thomas A. Middlebrooks Award. Email: jtchrist36@comcast.net

**J. Michael Duncan** is a professor emeritus of civil engineering at Virginia Tech. He has been a consultant on geotechnical engineering projects since 1965, and has supervised 45 PhD students at University of California, Berkeley and Virginia Tech. He has authored more than 300 geotechnical engineering publications, including the textbook *Soil Strength and Slope Stability*, coauthored with Stephen G. Wright. He is a member of the National Academy of Engineering and a distinguished member of American Society of Civil Engineers. He has been a member of the Panama Canal Geotechnical Advisory Board since 1986. Email: jmd@vt.edu

**Ignacio Escuder-Bueno** graduated from Universitat Politècnica de València (UPV), Spain, earned his Master of Science in civil engineering from the University of Wisconsin-Milwaukee (UWM), and a PhD in civil engineering from UPV. He is a university professor at UPV as well as promoter and founder associate of iPresas, a technology-based UPV spin-off company. He is a registered professional engineer in Spain, and has worked internationally as a consultant in numerous works related to safety studies, risk analysis, or design projects concerning more than 70 dams. He is chairman of the International Committee on Dams Computational Aspects of the International Commission on Large Dams (ICOLD) since 2011, secretary general of ICOLD European Club since 2010, and full member of the Spanish National Committee on Large Dams (SPANCOLD) since 2007. He has promoted and chaired the International Week on Risk Analysis, Dam Safety, Dam Security, and Critical Infrastructure Management (2005, 2008, and 2011). Email: iescuder@hma.upv.es

**Robert B. Gilbert** is the Brunswick Abernathy professor in civil, architectural and environmental engineering at the University of Texas at Austin. He joined the faculty in 1993. Prior to that, he earned a BS (1987), MS (1988), and PhD (1993) in civil engineering from the University of Illinois at Urbana-Champaign. He also practiced with Golder Associates Inc. as a geotechnical engineer from 1988 to 1993. His expertise is the assessment, evaluation, and management of risk in civil engineering. Applications include building foundations, slopes, pipelines, dams and levees, energy production systems, landfills, and groundwater and soil remediation systems. Email: Bob_Gilbert@mail.utexas.edu

**Mahdi Habibi** is currently a PhD candidate in the Department of Civil, Architectural, and Environmental Engineering at the University of Texas at Austin. He earned a BS (2003) and MS (2006) in civil engineering from the University of Tehran, Iran. He worked for three years as a practicing geotechnical engineer with Pars Geometry Consultant Co. in Tehran. Email: mhabibi@utexas.edu

**Charng Hsein Juang** is currently the Glenn Professor of the Glenn Department of Civil Engineering at Clemson University, South Carolina, and a fellow of American Society of Civil Engineers (ASCE). He also holds a chair professorship at several overseas universities: Tongji University, Shanghai China; Wuhan University, Hubei China; and National Central University of Taiwan. Dr. Juang earned his BS and MS from Cheng Kung University and his PhD from Purdue University. He has authored more than 200 refereed papers, including over 150 journal papers, with a Google Scholar h-index at 29 and a citation count of over 2600. Dr. Juang has received five best paper awards from various civil engineering societies including the TK Hsieh Award from the Institution of Civil Engineers. His latest research interests focus on liquefaction, excavation, slope stability, reliability, and robust geotechnical design. Dr. Juang is an editor-in-chief for *Engineering Geology*, an international research journal. Email: hsein@clemson.edu

**Sara Khoshnevisan** is currently a PhD student at Clemson University, South Carolina, majoring in geotechnical engineering, having earned her BS in mining engineering from Tehran University, Iran, and her MS in civil engineering from Clemson University, South Carolina. Khoshnevisan is a student member of the American Society of Civil Engineers and a member of Phi Kappa Phi. Her research interests focus on liquefaction, supported excavation, reliability, and robust geotechnical design. Email: khoshnevisan.sara@gmail.com

**Dian-Qing Li** is a professor in the State Key Laboratory of Water Resources and Hydropower Engineering Science, Wuhan University, China. His main research interests include risk and reliability in geotechnical engineering, risk and uncertainty in dam safety, embankment dams, and slopes. He is a member of TC304 (risk) in the International Society for Soil Mechanics and Geotechnical Engineering (ISSMGE) and a member of the American Society of Civil Engineers Technical Committee on Risk Assessment and Management. Currently, he is an editorial board member (EBM) of *Georisk*, and an EBM of ASCE-ASME journal, *Risk and Uncertainty in Engineering System*. He is the recipient of the 2012 National Science Fund for Distinguished Young Scholars, China, and the 2013 Young and Middle-Aged Leading Scientists, Engineers, and Innovators, Ministry of Science and Technology, China. Email: dianqing@whu.edu.cn

**Bak-Kong Low** earned his BS and MS in civil engineering from the Massachusetts Institute of Technology, and PhD from the University of California at Berkeley. He is a fellow of the American Society of Civil Engineers, and a registered professional engineer of Malaysia. He currently teaches at the School of Civil and Environmental Engineering at the Nanyang Technological University (NTU) in Singapore. Apart from his long teaching and research career at NTU, he had done research while on sabbaticals at Hong Kong University of Science and Technology (Sept.–Dec. 1996), University of Texas at Austin (Jan.–April 1997) and Norwegian Geotechnical Institute (May–Aug. 2006). His research interests and publications in soil and rock engineering can be found at http://alum.mit.edu/www/bklow. Email: bklow@alum.mit.edu

**Adrián Morales-Torres** is a civil engineering graduate and earned a master degree in hydraulic engineering and environment from the Universidad Politécnica de Valencia (UPV), Spain. In 2010, he received the most outstanding graduate award by the Civil Engineers Association in Valencia. Currently, he is completing his PhD in civil engineering about risk-informed decision making in dam safety management at UPV. Since 2009, he has worked as researcher in UPV and has participated in different works related to dam safety management and risk analysis for more than 30 dams in Spain. In addition, he has been part of the SUFRI European project (Sustainable strategies of urban flood risk management with non-structural measures to cope with the residual risk, 2009–2011) and he is currently developing a decision support tool for the E²STORMED European project (Improvement of energy efficiency in the water cycle by the use of innovative storm water management in smart Mediterranean cities, 2013–2015). Email: admotor@upvnet.upv.es

**Trevor L.L. Orr** is an associate professor at Trinity College Dublin (TCD). His main research interests include geotechnical design and design codes, geotechnical risk and reliability, tunneling, and underground structures. He has been closely involved in Eurocode 7, the European standard for geotechnical design, since the initial work started in 1981. He is currently a member of CEN's Sub-Committee 7 for the development of Eurocode 7 and is a convenor of its Evolution Group 3: Model Solutions. He is a coauthor of two books on Eurocode 7, including the *Designers' Guide to Eurocode* 7 published by the Institution of Civil Engineers (ICE), London, and has presented many papers and invited lectures on Eurocode 7. He was awarded the Crampton Prize for his paper on Eurocode 7, published in the ICE journal *Geotechnical Engineering* in 2012. He is an editor of the ICE journal *Geotechnical Engineering* (2015–2017). Email: torr@tcd.ie

**Iason Papaioannou** is a postdoctoral research associate and lecturer at the Engineering Risk Analysis Group of the Technische Universität München (TUM), Germany. He is co-founder and partner at Eracons, a company specialized in consulting and software development for reliability assessment of engineering structures. His research interests include random fields and spatial variability, reliability assessment, and Bayesian analysis of numerical models with emphasis on applications to geotechnical structures. He studied civil engineering at the National Technical University of Athens (NTUA) and computational mechanics at the Technische Universität München (TUM). He earned his doctoral degree from TUM in 2012. His thesis was awarded the Bauer prize for the best PhD thesis of the Faculty of Civil, Geo, and Environmental Engineering of TUM. He is currently secretary of the Geotechnical Safety Network (GEOSNet). Email: iason.papaioannou@tum.de

**Francisco Silva-Tulla** completed his engineering degree at the University of Illinois, Champaign–Urbana and graduate studies at MIT. Dr. Silva has 43 years of professional experience including 39 years as a consulting engineer, participating in engineering and environmental projects, and studies in roles ranging from complete responsibility (program or project management, design, plans and specifications, budget control, construction supervision, quality control, and long-term surveillance) to specialized services (supervision of work performed by other engineering firms; guidance and leadership to groups of consultants; quality assurance; value engineering; expert witness and litigation support; dam and levee safety, risk assessments, soil and groundwater exploration, sampling, analysis, and characterization; and environmental assessments and impact studies). Dr. Silva's experience includes work for the mining, construction, chemical, petrochemical, petroleum, power, transportation, manufacturing, and waste disposal industries as well as U.S. and foreign government agencies. A large part of his professional experience relates to earth structures (especially embankments, dams, levees, dikes, slopes, landfills, and excavations) and the safety of constructed facilities (including probabilistic risk assessments and dam safety). Dr. Silva also has corporate management experience at the operating group level in consulting organizations with up to US$700M annual revenues. Email: silva@alum.mit.cdu

**Matthew D. Sleep** is an assistant professor of civil engineering at the Oregon Institute of Technology. Prior to Oregon Tech, Matthew earned his PhD at Virginia Tech researching slope stability, levees, transient seepage, and reliability. Dr. Sleep has worked for the United States Army Corps of Engineers and private consulting. He currently teaches and continues research on reliability and transient seepage. Email: matthew.sleep@oit.edu

**Abdul-Hamid Soubra** is a professor of geotechnical engineering in the Department of Civil Engineering, University of Nantes, France. His main research interests include reliability analysis of geotechnical structures, spatial variability of soil properties, and the stability problems in geotechnical engineering. He serves as an associate editor of international journals (e.g., *Journal of Geotechnical and Geoenvironmental Engineering. ASCE* and *Georisk: Assessment and Management of Risk for Engineered Systems and Geohazards*) and is a member of the Risk Assessment and Management (GI-RAM) of the American Society of Civil Engineers. He is the recipient of the 2012 best paper award from *Georisk*. Email: Abed.Soubra@univ-nantes.fr

**Daniel Straub** is professor for engineering risk and reliability analysis at Technische Universität München (TUM), Germany. His interest is in developing physically based stochastic models and methods for the decision support of infrastructure, environmental, and general engineering systems, with a particular focus on Bayesian techniques. Daniel earned his Dipl.-Ing. degree in civil engineering in 2000 and his PhD in 2004 from ETH Zürich, Switzerland, and consequently was a postdoc and adjunct faculty at UC Berkeley, before joining TUM in 2008. He has also been frequently active as a consultant to the industry on reliability and risk assessments for structures, infrastructures, and the oil and gas industry. His awards include the ETH Silbermedaille and the Early Achievement Research Award of IASSAR (International Association for Structural Safety and Reliability), and he is an honorary professor at the University of Aberdeen, United Kingdom. Email: straub@tum.de

**Bruno Sudret** has been a professor of risk, safety, and uncertainty quantification at the Department of Civil, Environmental and Geomatic Engineering of the Swiss Federal Institute of Technology (ETH Zürich, Switzerland) since 2012. He has been working in probabilistic engineering mechanics and uncertainty quantification methods for engineering systems since 2000, first as a postdoctoral fellow at the University of California at Berkeley, then as a senior researcher at EDF R&D (France). From 2008 to 2011 he worked as the director of research and strategy at Phimeca Engineering (France), a consulting company specialized in structural reliability. As a young researcher he received the "Jean Mandel" prize awarded by the French Association of Mechanics in 2005. He is a member of the Joint Committee on Structural Safety (JCSS) and of the Board of Directors of the International Civil Engineering Risk and Reliability Association (CERRA). He also serves on the editorial board of the ASCE-ASME *Journal of Risk and Uncertainty in Engineering Systems*. Email: sudret@ibk.baug.ethz.ch

**Xiao-Song Tang** is currently working as a lecturer at Wuhan University, China. He earned his Bachelor of Engineering in 2008 from Sichuan University (Chengdu, China) and his PhD from Wuhan University (Wuhan, China) in 2014. From 2012 to 2013, he studied as an exchange PhD student at the National University of Singapore (NUS) supervised by Professor Kok-Kwang Phoon. His main research interests include uncertainty modeling of correlated geotechnical parameters using copulas and reliability analysis of geotechnical structures. Email: xstang@whu.edu.cn

**Yu Wang** is an associate professor in the Department of Architecture and Civil Engineering, City University of Hong Kong. He earned his PhD in geotechnical engineering from Cornell University, New York. His main research interests include geotechnical risk and reliability (e.g., probabilistic characterization of geotechnical properties, reliability-based design in geotechnical engineering, and probabilistic slope stability analysis), seismic risk assessment of critical civil infrastructure systems (e.g., water supply systems), soil–structure interaction, and geotechnical laboratory and in situ testing. Dr. Wang was the president of the American Society of Civil Engineers (ASCE)–Hong Kong Section in 2012–2013. He is also a member of several international technical committees (TCs), including an ASCE Geo-Institute TC on risk and two ISSMGE (International Society of Soil Mechanics and Geotechnical Engineering) TCs on risk and in-situ testing, respectively. He is the recipient of the inaugural Wilson Tang Best Paper Award in 2012. Email: yuwang@cityu.edu.hk

**Jie Zhang** is an associate professor in the Department of Geotechnical Engineering at Tongji University, China. His teaching and research encompass geotechnical reliability, Bayesian updating of geotechnical systems, and civil engineering risk assessment. Dr. Zhang is a corresponding member of TC304 in the International Society for Soil Mechanics and Geotechnical Engineering (ISSMGE), and the secretary of the Young Scientists Committee of Risk and Insurance Research Branch, China Society of Civil Engineering. He is a recipient of the Natural Science Award from the Minister of Science and Technology of China in 2011, and the Outstanding Reviewer Award from Computers and Geotechnics in 2013. Email: cezhangjie@gmail.com

**Limin Zhang** is professor of geotechnical engineering and associate director of the geotechnical centrifuge facility at the Hong Kong University of Science and Technology. His research areas include slopes and embankment dams, geotechnical risk assessment, pile foundations, multiphase flows, and centrifuge modeling. He is chair of Geotechnical Safety Network (GEOSNet), vice chair of the International Press-In Association, editor-in-chief of the international journal *Georisk*, associate editor of the American Society of Civil Engineers' *Journal of Geotechnical and Geoenvironmental Engineering*, and editorial board member of six other journals. He received the Mao Yi-Sheng Soil Mechanics and Geotechnical Engineering Award and the Overseas Young Investigator Award from the National Natural Science Foundation of China. He has published over 150 international journal papers. Email: cezhangl@ust.hk

# Part 1

# Properties

# Chapter 1

# Constructing multivariate distributions for soil parameters

*Jianye Ching and Kok-Kwang Phoon*

## 1.1 INTRODUCTION

Reliability-based design (RBD) is known to provide a rational basis for incorporating uncertainties in the design environment explicitly into the geotechnical design (e.g., pile length). However, one recurring criticism of RBD is that there is no particular reason to use it because it seems to produce designs comparable to the existing practice. In particular, the link between potential reduction of uncertainties resulting from collection of more information and how this reduction could translate to actual savings in design dimensions has remained a vague theoretical possibility so far. More information can be collected using two approaches. One approach is to take field measurements/samples at more locations, that is, increase the amount of data for a given test type. The second approach is to conduct more test types, for example, supplement standard penetration test (SPT) with cone penetration test (CPT). The former approach increases information quantitatively, while the latter approach increases information qualitatively. This distinction is important, particularly pertaining to reduction of bias in the estimation of design parameters such as the undrained shear strength. The first approach may be effective in improving precision, but is usually not effective in reducing estimation bias. Both approaches are typically carried out simultaneously in practice to varying degrees, depending on the needs of the project and economics.

While it is theoretically correct that reduction in uncertainties will translate to design savings, critical questions of paramount importance to practice such as "how much reduction in pile length?" and "is it worth my time/money to collect more site information?" cannot be answered theoretically. These critical questions can be answered only by applying RBD to actual design problems where the amount of site information can be varied systematically. In reality, site investigation information always appears in a multivariate form. For instance, when borehole samples are drawn, SPT-N values are usually available; moreover, the information regarding unit weight, plasticity index (PI), liquid limit (LL), and water content can quickly be obtained through laboratory tests. Many of these test indices may be simultaneously correlated to a design parameter such as the undrained shear strength $s_u$. With the bivariate correlations at hand, it seems straightforward to update the first two moments [mean and coefficient of variation (COV)] of $s_u$ conditioning on a single-test index (e.g., SPT-N); however, it is not obvious as how to conduct the same analysis conditioning on multivariate test indices [e.g., SPT-N and overconsolidation ratio (OCR) simultaneously]. An option is to discard all test indices but retain the most relevant and/or most accurate test index. However, it is uneconomical to eliminate costly information in the updating process because the abandoned test indices may further reduce the uncertainties. Even test indices that are weakly correlated to the design parameter may reduce uncertainties substantially if a number of test indices are available and their effects can be combined. Another common approach is to simply take the average of $s_u$ estimates from different tests. This approach is intuitive, but is purely empirical with no assurance that it will be effective in all cases. Ching et al. (2010) have demonstrated that even a more sophisticated variance-weighted averaging approach is less effective than a rigorous Bayesian updating approach. In addition, it is a good geotechnical practice to cross-validate interpretation of soil properties from different sources of information, given the significant assumptions and empiricism underlying most bivariate correlations. In our opinion, combining multivariate information in a rational and systematic way is a major useful application of probability theory, because it is impossible even for an expert probabilist to combine multiple sources of uncertain information that are dependent on each other in some complex way using pure physical intuition or engineering judgment.

The purpose of constructing multivariate distributions for soil parameters is to emulate the information content of a real site as realistically as possible. Ching et al. (2014a) called data simulated with the intent of replicating the multidimensional correlation structure underlying a basket of laboratory and field tests as "virtual site" data. It is currently not possible to emulate every aspect of a real site. One hopes to construct a virtual site that would at least reproduce realistic data within a scope of interest. In this chapter, the scope is to reproduce the information content arising from a typical basket of laboratory and field tests conducted in a clay site for the purpose of estimating important design parameters such as the undrained shear strength and the preconsolidation stress. The critical feature here is the consistent and realistic coupling of different test data, which are achieved using a multivariate normal distribution. Data from different tests will be coupled (or correlated in the context of a multivariate normal distribution), because they measure the same mass of soil, although they could be possessing different aspects of soil behavior under different boundary conditions and over different volumes. The current virtual site does not model spatial variability. This rather limited virtual site model will be expanded to include more tests and to improve realism (such as spatial variability) in the future. The purpose of developing a virtual site is not to replace the actual site investigation. It is meant to serve as a tool for engineers to explore design savings accrued from conducting better and/or more tests in the context of RBD.

The idea of simulating a "virtual site" is not new. For example, Jaksa et al. (2003, 2005) and Goldsworthy et al. (2007) used three-dimensional random fields and Monte Carlo simulation to simulate the spatially variable elastic modulus of a "virtual" site. Each spatially variable realization constitutes a plausible full-information scenario. Site investigation is then carried out numerically by sampling the continuous random field at discrete locations. The site investigation data so obtained constitute the typical partial-information scenario commonly encountered in practice. The scope of these studies was to quantify the discrepancy between the design based on partial spatial information (information at discrete points typically measured in a site investigation) and the design based on complete spatial information (an ideal state where all subsurface information have been mapped/characterized as continuous functions of spatial coordinates without measurement errors). The focus of these studies is to evaluate how increasing the number of measurement points for a single test will affect design decisions. Our "virtual site" examines the complementary aspect of how increasing the number of tests at a given depth will affect design decisions. Spatial variability is not considered in this chapter.

The basic goal of this chapter is to explain how a useful multivariate non-normal probability model can be constructed from actual geotechnical data. Standard undergraduate texts on probability and statistics present univariate probability models, which are not adequate for geotechnical data. The construction method is explained by building "need-to-know" theoretical tools incrementally: (1) single normal random variable, (2) bivariate normal vector as two normal random variables coupled by a correlation coefficient, (3) multivariate normal vector as a generalization of the bivariate case coupled by a correlation *matrix*, (4) single non-normal random variable as a nonlinear transform of the normal random variable, and (5) multivariate non-normal vector as a component-by-component nonlinear transform of the multivariate normal case. No prior knowledge of probability and statistics is required, but the reader may need to read standard texts for details. The emphasis in this chapter is on how to use the theoretical tools to produce useful results in practice. In other words, given a table of measured numbers (multivariate data), how would an engineer (1) identify a reasonable probability model ("goodness-of-fit" problem) from data, (2) estimate the model parameters (e.g., mean, COV) from data, (3) simulate "virtual site" data from the probability model, and (4) draw useful engineering conclusions from the probability model? The sample size of geotechnical data is typically small. Statistical uncertainties are ubiquitous and play a significant role in practice. Complete multivariate data are also rarely available. These aspects and other important limitations are comprehensively discussed to ensure that the engineer is fully informed of the practical limits of statistical inference in geotechnical engineering.

## 1.2 NORMAL RANDOM VARIABLE

### 1.2.1 Random data

Random data can be viewed as a list of numbers taking a range of values and assuming a different frequency of occurrences when plotted in the form of a histogram. Random data can be modeled as a random variable following a cumulative distribution function (CDF). The CDF can be presented in the form of its derivative for continuous variables. This derivative is called the probability density function (PDF).

It is crucial to distinguish between a random variable and a list of *measured* values, say a list of undrained shear strength ($s_u$) values obtained by performing unconfined compression test on undisturbed samples. The former is a mathematical model. The latter is reality—what you measure in practice. There are two challenges in linking what you measure to a random variable.

First, the number of data points in a list of measurements (called sample size) must be *finite*. It is relatively easy to simulate a finite list of values if the random variable is defined. For example, if the undrained shear strength is normally distributed with a mean of 100 kPa and a standard deviation of 20 kPa, we can obtain, say, 30 values using the MATLAB® function normrnd(100, 20, 30, 1). You can perform simulation using Data > Data Analysis > Random Number Generation in EXCEL as well. It is important to note that the theoretical properties such as the mean of a random variable can be obtained only from an infinite sample (called a population). The arithmetic average obtained from a finite sample is called the "sample mean." In this chapter, the term "mean" is associated with a random variable while the term "sample mean" is associated with a finite sample. The same terminology applies to other properties also. It is possible to simulate different finite samples. Given the random nature of the data, the sample mean computed from one sample will be different from the sample mean computed from another sample. This phenomenon is called "statistical uncertainty" and it is crucial to appreciate that all quantities estimated from a finite sample will be subjected to this fundamental limitation. The upshot is that no theoretical properties can be estimated with perfect precision. A point estimate will be implicitly associated with a statistical error. It is arguably more accurate to report an estimate of a theoretical property in the form of a confidence interval. An alternate method is to report the *p*-value associated with a null hypothesis for a theoretical property. These concepts will be made specific in the following sections, but it suffices to appreciate that the above statistical tools are needed because the sample size is finite in practice. In geotechnical engineering, our sample sizes are typically small and statistical uncertainty cannot be ignored. Simulation is an important tool to study statistical uncertainty.

Second, it is important to appreciate that a list of "random" looking measurements does not necessarily follow a random variable model. From the authors' experience in statistical modeling of geotechnical engineering data, we have found this model adequate in the sense of producing meaningful and useful results for practice. This chapter presupposes that a random variable model is adequate for geotechnical engineering data. If one accepts this leap of faith, the obvious follow-up question is which CDF would be appropriate. In view of the finite sample size, this "goodness-of-fit" question cannot be resolved with certainty. Some standard "goodness-of-fit" tests would be presented below, but one should be mindful that it is not sufficient to find a good fit for a list of measurements (say a column of numbers in EXCEL). Geotechnical engineering data are multivariate in nature, for example, they may measure several properties such as the undrained shear strength, natural water content, Atterberg limits, and preconsolidation pressure from the same undisturbed sample. While there is a wide choice of probability models to fit a single column of number (univariate data), there is only one practical choice to fit multiple columns (multivariate data). This choice involves a column-by-column nonlinear transformation of a multivariate *normal* probability model. Because of this restriction, it is more convenient to choose a univariate probability model that is a transformation of the standard normal model. The Johnson system of distributions is generated by such a transformation and it is useful to start testing goodness of fit using this system of distributions.

The above "need to know" concepts are explained and illustrated using simulated data in the sections below. Simulated data are "perfect" in the sense that they are theoretically derived from a fully defined random variable. Hence, in contrast to actual data, there is no question that a random variable model works! In addition, it is useful to compare statistics computed from a finite sample size with the theoretical answers, which are also known since the random variable is fully defined.

## 1.2.2 Normal random variable

The normal distribution is also called the Gaussian distribution. Symbolically, "$Y \sim N(\mu, \sigma^2)$" means that Y is normally distributed with mean, $\mu$, and standard deviation, $\sigma$. The normal distribution is the most important distribution in characterizing a physical parameter that can take a range of values with a different likelihood of occurrences. Its importance will be apparent in the context of non-normal multivariate distributions discussed in Section 1.6.

The concepts discussed below are illustrated using normally distributed undrained shear strength values with mean of 100 kPa and standard deviation of 20 kPa, unless stated otherwise. These values were simulated using the MATLAB function normrnd. The reader can reproduce the data by initializing the pseudorandom sequence using randn('state', 13).

### 1.2.2.1 Probability density function

The PDF for the normal distribution is

$$f(y) = \frac{1}{\sqrt{2\pi} \cdot \sigma} \exp\left[\frac{-(y - \mu)^2}{2 \cdot \sigma^2}\right] \tag{1.1}$$

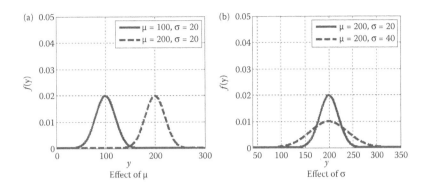

*Figure 1.1* PDFs of normal random variables.

where $f(y)$ can be evaluated using EXCEL function NORMDIST$(y, \mu, \sigma, 0)$ or MATLAB function normpdf$(y, \mu, \sigma)$. Figure 1.1 shows some examples of the normal PDFs. It is clear that the center of the PDF is equal to the mean value, $\mu$. The spread of the PDF is controlled by the standard deviation, $\sigma$. The larger the spread, the larger the value of $\sigma$. The normal distribution is a two-parameter probability model. It is widely used in practice, in part, because the parameters $(\mu, \sigma)$ can be estimated easily and quite accurately from the sample sizes commonly encountered in geotechnical engineering practice, say between 30 and 100 data points.

The COV is defined as

$$\mathrm{COV} = \frac{\sigma}{\mu} \tag{1.2}$$

The COV is widely reported in the geotechnical engineering literature, because it is dimensionless. However, one should not jump to the conclusion that the COV is a *constant*, because of the normalization in Equation 1.2. It is possible that the standard deviation is a constant, in which case, the COV must decrease with the mean value by definition as shown in Figure 1.2. It is important to develop a physical sense of COV based on the guidelines

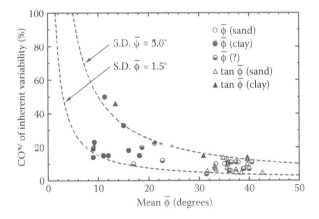

*Figure 1.2* COV of the effective stress friction angle ($\bar{\phi}$) decreases with the mean value. (From Phoon, K.K. and Kulhawy, F.H. 1999. *Canadian Geotechnical Journal*, 36(4), 612–624, reproduced with permission of the *Canadian Geotechnical Journal*.)

*Table 1.1* Ranges of soil property variability for reliability calibration

| Geotechnical parameter | Property variability | COV (%) |
|---|---|---|
| Undrained shear strength | Low[a] | 10 30 |
| | Medium[b] | 30–50 |
| | High[c] | 50–70 |
| Effective stress friction angle | Low[a] | 5–10 |
| | Medium[b] | 10–15 |
| | High[c] | 15–20 |
| Horizontal stress coefficient | Low[a] | 30–50 |
| | Medium[b] | 50–70 |
| | High[c] | 70–90 |

Source: Phoon, K.K. and Kulhawy, F.H. 2008. Serviceability limit state reliability-based design. Chapter 9, *Reliability-Based Design in Geotechnical Engineering: Computations and Applications*. Taylor & Francis, 344–384.

[a] Typical of good-quality direct laboratory or field measurements.
[b] Typical of indirect correlations with good field data, except for the SPT.
[c] Typical of indirect correlations with SPT field data and with strictly empirical correlations.

for low, medium, and high variabilities of some geotechnical properties given in Table 1.1. These guidelines were developed from an extensive compilation of soil statistics (Phoon and Kulhawy 1999).

There is a special normal distribution—the standard normal distribution $N(0, 1^2)$. Its PDF can be calculated using

$$\varphi(x) = \frac{1}{\sqrt{2\pi}} \exp\left(\frac{-x^2}{2}\right) \tag{1.3}$$

where $\varphi(x)$ can be evaluated using EXCEL function NORMSDIST($x$, 0) or MATLAB function normpdf(x). The COV is not defined when the mean is zero. It is noteworthy that a general normal random variable (Y) with mean, $\mu$, and standard deviation, $\sigma$ can be obtained from X using a simple linear transform

$$Y = \mu + \sigma \cdot X \tag{1.4}$$

It is a normal practice to divide the full range of values occurring in a data set into bins as shown in Figure 1.3. The number of data points falling into each bin can be plotted as a familiar histogram. The data set in Figure 1.3 is simulated from a normal distribution with a mean of 100 kPa and a standard deviation of 20 kPa. The sample size n = 1000. The main advantage of the PDF is that it can be compared with an empirical histogram visually (after some appropriate scaling). The solid line in Figure 1.3 is the theoretical PDF. However, there are disadvantages too. For example, the number of bins must be selected in a sensible way. Figure 1.4 shows the effect of bin numbers. Figure 1.4a (five bins) loses all the details, whereas Figure 1.4b (100 bins) is rather noisy. It is also important to emphasize that a histogram constructed from perfect normal data (i.e., data simulated from a normal distribution) will not follow a normal PDF unless the sample size is sufficiently large. Figure 1.5 shows how simulated data from the normal distribution will compare with the theoretical PDF for sample sizes of 30, 50, and 100. Contrary to popular belief, it is actually common to see non-normal looking histograms (e.g., nonsymmetrical histograms) even when the data are simulated from a normal distribution for small sample sizes! This is a manifestation of statistical uncertainty. The Kolmogorov–Smirnov (K–S) test described in Section

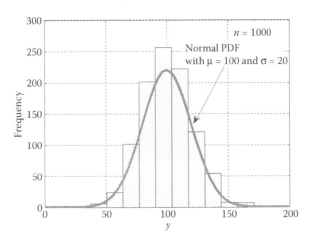

Figure 1.3 Histogram of Y. The solid line is the normal PDF with μ = 100 kPa and σ = 20 kPa.

1.2.3 provides a more robust method of assessing goodness of fit to a normal distribution, although it is less visually appealing.

### 1.2.2.2 Cumulative distribution function

The cumulative distribution function (CDF) of Y is defined as the probability that a random variable Y is less than or equal to a specific numerical value y. In this chapter, an uppercase symbol denotes a random variable and a lowercase symbol denotes an ordinary variable (it is helpful to imagine this ordinary variable taking a specific constant number, say 100 kPa). This distinction is critical. Symbolically, the CDF is denoted by F(y). Basically, F(y) is the integral of the PDF (or PDF is the derivative of CDF). For the normal distribution, its CDF is

$$F(y) = \int_{-\infty}^{y} \frac{1}{\sqrt{2\pi} \cdot \sigma} \exp\left[\frac{-(t - \mu)^2}{2 \cdot \sigma^2}\right] \cdot dt \tag{1.5}$$

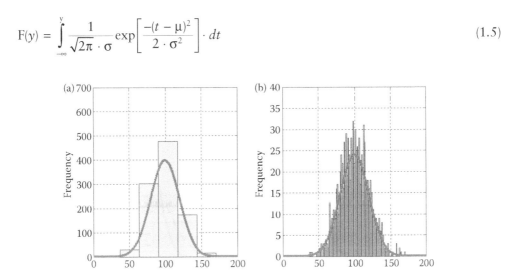

Figure 1.4 Histograms of Y with two different number of bins.

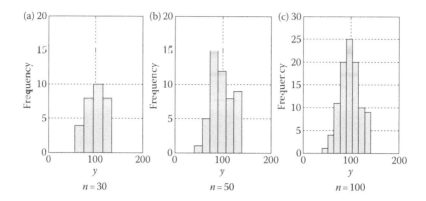

*Figure 1.5* Histograms of Y with three different sample sizes (*n*).

The concept of a CDF and its associated S-shaped curve should be familiar to all geo-technical engineers. It is identical to the grain size distribution, which plots the percent-age finer by weight versus grain diameter. F(*y*) can be evaluated using EXCEL function NORMDIST(*y*, μ, σ, 1) or MATLAB function normcdf(*y*, μ, σ). When μ increases from 100 (stiff clay) to 200 kPa (very stiff clay), the curve shifts to the right (Figure 1.6a). When σ increases from 20 (low variability with COV = 20%) to 40 kPa (medium variability with COV = 40%), the curve becomes less steep (Figure 1.6b).

For standard normal X, the CDF is denoted by Φ(*x*):

$$\Phi(x) = \int_{-\infty}^{x} \frac{1}{\sqrt{2\pi}} \exp\left(\frac{-t^2}{2}\right) \cdot dt \qquad (1.6)$$

Φ(*x*) can be evaluated using EXCEL function NORMSDIST(*x*, 1) or MATLAB function normcdf(*x*). It is noteworthy that F(*y*) can be computed from Φ by appropriate shifting and scaling

$$F(y) = P(Y \leq y) = P(\mu + \sigma X \leq y) = P\left(X \leq \frac{y-\mu}{\sigma}\right) = \Phi\left(\frac{y-\mu}{\sigma}\right) \qquad (1.7)$$

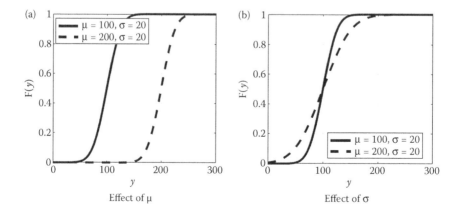

*Figure 1.6* CDFs of normal random variables.

There is an important engineering concept related to the CDF. This concept is the percentile. The probability of finding values smaller than the η-percentile = $y_\eta$ is η, that is,

$$\Phi\left(\frac{y_\eta - \mu}{\sigma}\right) = \eta \tag{1.8}$$

and hence

$$y_\eta = \mu + \sigma \times \Phi^{-1}(\eta) \tag{1.9}$$

$\Phi^{-1}(\eta)$ can be evaluated using EXCEL function NORMSINV(η) or MATLAB function norminv(η). It is clear from Equation 1.9 that the percentile is simply the inverse CDF. The concept of a percentile is widely used to define a characteristic value in design codes. For example, Eurocode 7 (BS EN1997-1:2004) mentions in Section 2.4.5.2 on "Characteristic values of geotechnical parameters" Clause (11) that "If statistical methods are used, the characteristic value should be derived such that the calculated probability of a worse value governing the occurrence of the limit state under consideration is not greater than 5%" (p. 28). For the normal distribution, this 5% percentile or characteristic value will be given by

$$y_\eta = \mu + \sigma \times \Phi^{-1}(0.05) = \mu - 1.64 \times \sigma = \mu(1 - 1.64 \times \text{COV}) \tag{1.10}$$

Equation 1.10 is not applicable for non-normal distribution.

It is possible to estimate the CDF of Y, given the data points of Y

$$F(y) = P(Y \leq y) \approx \frac{1}{n}\sum_{k=1}^{n} I(Y^{(k)} \leq y) \equiv F_n(y) \tag{1.11}$$

where $Y^{(k)}$ is the $k$th data point of Y; $n$ is the total number of data; and I(.) is the indicator function: it is unity if the enclosed statement is true and is zero otherwise. The CDF, $F_n(y)$, estimated by Equation 1.11 from a finite sample size ($n$) is called the "empirical" CDF. The analytical CDF shown in Equation 1.5 or 1.6 is called a "theoretical" CDF. There are several estimators for the CDF. Equation 1.11 is called the Kaplan–Meier estimator. Another rather common estimator is the median estimator

$$F_n(y) = \frac{1}{n + 0.4}\left(\sum_{k=1}^{n}\left[I(Y^{(k)} \leq y)\right] - 0.3\right) \tag{1.12}$$

Figure 1.7a shows two empirical CDFs (ECDFs) from two different simulated data sets of size $n - 10$. The theoretical normal CDF with a mean of 100 kPa and a standard deviation of 20 kPa is plotted as a solid line. It is clear that the ECDFs are different and are scattered around the theoretical line. This is how statistical uncertainty manifests itself in a CDF plot. Figure 1.7b is the same as Figure 1.7a, except that the sample size increases to 100. The scatter is now smaller. It can be proved theoretically that statistical uncertainty decreases with the sample size.

The CDF is less visually appealing compared to the PDF because an engineer cannot see the frequency of occurrences over a given range of value as readily as what is shown in a histogram/PDF. Nonetheless, it is important to note that the ECDF can be constructed

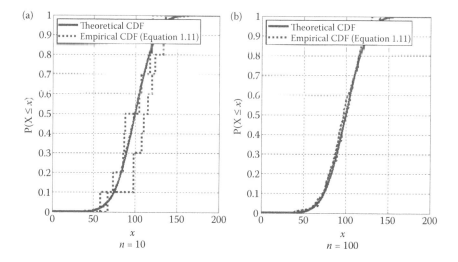

*Figure 1.7* Theoretical and ECDFs of Y.

from data almost uniquely. Different estimators do produce different curves, but the effect is rather minor (Kimball 1960). In contrast, the histogram can vary significantly depending on the number of bins as shown in Figure 1.4. It is noteworthy that goodness-of-fit tests are based on the CDF.

### 1.2.3 Estimation of normal parameters

There are two parameters in the normal distribution model: $\mu$ and $\sigma$. There are various ways of estimating these two parameters based on the observed data. Here, we review point estimators for $\mu$ and $\sigma$ using the (a) method of moments, (b) percentile method, and (3) maximum likelihood method.

The calculation procedure and behavior of each estimator are illustrated using simulated normal data of size $n = 10$, $\mu = 100$ kPa, and $\sigma = 20$ kPa. We illustrate all three methods using the same simulated set [initialized by randn('state', 13) before executing normrnd(100, 20, 10, 1)]. The sample values are shown in Table 1.2.

*Table 1.2* Sample values of Y ($n = 10$)

| Simulated data $Y^{(k)}$ | Index | Sorted Y data k | Rank |
|---|---|---|---|
| 124.06 | 1 | 71.43 | 1 |
| 113.35 | 2 | 74.40 | 2 |
| 134.75 | 3 | 81.15 | 3 |
| 90.52 | 4 | 89.81 | 4 |
| 133.54 | 5 | 90.52 | 5 |
| 74.40 | 6 | 105.41 | 6 |
| 81.15 | 7 | 113.35 | 7 |
| 105.41 | 8 | 124.06 | 8 |
| 71.43 | 9 | 133.54 | 9 |
| 89.81 | 10 | 134.75 | 10 |

### 1.2.3.1 Method of moments

The model parameters $\mu$ and $\sigma$ can be estimated by the sample mean ($m$) and sample standard deviation ($s$):

$$\mu \approx \frac{1}{n}\sum_{k=1}^{n} Y^{(k)} \equiv m \qquad \sigma \approx \sqrt{\frac{1}{n-1}\sum_{k=1}^{n}\left(Y^{(k)} - m\right)^{2}} \equiv s \qquad (1.13)$$

The sample mean and sample standard deviation obtained from simulated data are 101.84 and 23.82, respectively. They are different from the theoretical values of $\mu = 100$ kPa and $\sigma = 20$ kPa, because of the statistical uncertainty. We can obtain a comprehensive view of statistical uncertainty by simulating 1000 different sets of values. Each set consists of 10 numbers. Equation 1.13 is applied to each set, thus producing 1000 sample means and 1000 sample standard deviations. Figure 1.8 shows the empirical histograms for $(m-\mu)/(s/n^{0.5})$ and $(n-1)s^{2}/\sigma^{2}$. The sample mean m can be as small as 81.43 and as large as 120.36, although $\mu = 100$ kPa. Note that $s^{2}$ is not normally distributed, although the underlying data are normally distributed. It can be proved theoretically that $(n-1)s^{2}/\sigma^{2}$ is $\chi$-squared distributed. These "sampling distributions" are discussed in more detail below.

### 1.2.3.2 Percentile method

The percentile has been defined in Equation 1.9. The median of Y is the 0.5-percentile of Y. For the normal distribution, the mean value $\mu$ is identical to the median (0.5-percentile) by substituting $\Phi^{-1}(0.5) = 0$ in Equation 1.9. As a result, one can estimate $\mu$ by estimating the 0.5-percentile. A simple illustration of the sample value of 0.5-percentile is given. Table 1.2 shows the sorted Y data, from the smallest to the largest. The third column contains the simulated data sorted in an ascending order. The fourth column contains the rank. The smallest number is rank 1. The second smallest number is rank 2 and so forth. The sample size $n = 10$; hence, the sample median can be taken as the average of the Y values with ranks 5 and 6, namely sample median = $(90.52 + 105.41)/2 = 97.97$. It is not the same as the actual median (=100) because of the statistical uncertainty.

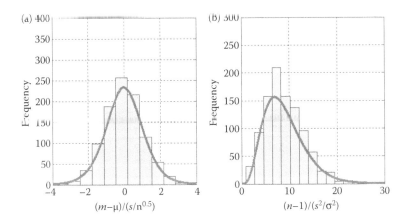

Figure 1.8 Histograms of $(m-\mu)/(s/n^{0.5})$ and $(n-1)s^{2}/\sigma^{2}$ and the sampling distributions.

For the standard deviation $\sigma$, a percentile method is proposed by Donoho and Johnstone (1994): $\sigma$ can be estimated as the sample median of the absolute values of the deviations from the sample mean m, divided by 0.6745. Namely,

$$\sigma \approx \frac{\text{Sample median of } \left|Y^{(1)} - m\right|, \left|Y^{(2)} - m\right|, \ldots, \left|Y^{(k)} - m\right|}{0.6745} \tag{1.14}$$

For normal distribution, the sample mean in Equation 1.14 can also be estimated as the sample median. Table 1.3 shows the $|Y-m|$ data and the sorted data ($m = 97.97$ is the sample median of Y). The sample median of the sorted $|Y-m|$ data is $(16.82 + 23.57)/2 = 20.19$. The percentile estimate for $\sigma$ is therefore $20.19/0.6745 = 29.94$.

The percentile method of estimating $\mu$ and $\sigma$ is more robust than the method of moments, as outliers will not affect the point estimates significantly. For example, if the maximum value is recorded wrongly by a factor of 10, the sample median value remains the same (=97.97), but the sample mean ($m$) is now 223.12. The actual mean is 100. The estimate from Equation 1.14 still remains the same (=29.94), but the sample standard deviation ($s$) is now 295.61. Robust statistics is an important practical topic. The reader can refer to advanced texts for more details.

### 1.2.3.3 Maximum likelihood method

The principle of maximum likelihood says that the best estimates for $\mu$ and $\sigma$ are those that maximize the likelihood function. For *independent* observations ($Y^{(1)}, \ldots, Y^{(n)}$), the likelihood function is the joint density of ($Y^{(1)}, \ldots, Y^{(n)}$)

$$f(Y^{(1)}, Y^{(2)}, \ldots, Y^{(n)} | \mu, \sigma) = \prod_{k=1}^{n} f(Y^{(k)} | \mu, \sigma) = \prod_{k=1}^{n} \frac{1}{\sqrt{2\pi} \cdot \sigma} \exp\left[\frac{-(Y^{(k)} - \mu)^2}{2 \cdot \sigma^2}\right] \tag{1.15}$$

Equivalently, one can find the maximum likelihood estimates for $\mu$ and $\sigma$ by maximizing the logarithm of the likelihood function

$$\ln[f(Y^{(1)}, Y^{(2)}, \ldots, Y^{(n)} | \mu, \sigma)] = \sum_{k=1}^{n} \left[-0.5\ln(2\pi) - \ln(\sigma) - \frac{(Y^{(k)} - \mu)^2}{2 \cdot \sigma^2}\right] \tag{1.16}$$

Table 1.3 Sample values of $|Y-m|$ ($m$ = sample median = 97.97)

| $|Y^{(k)}-m|$ | Index k | Sorted $|Y-m|$ data | Rank |
|---|---|---|---|
| 26.10 | 1 | 7.45 | 1 |
| 15.39 | 2 | 7.45 | 2 |
| 36.78 | 3 | 8.15 | 3 |
| 7.45 | 4 | 15.39 | 4 |
| 35.58 | 5 | 16.82 | 5 |
| 23.57 | 6 | 23.57 | 6 |
| 16.82 | 7 | 26.10 | 7 |
| 7.45 | 8 | 26.54 | 8 |
| 26.54 | 9 | 35.58 | 9 |
| 8.15 | 10 | 36.78 | 10 |

It is easy to verify that the best estimates are

$$\mu_{MLE} = \frac{1}{n}\sum_{k=1}^{n} Y^{(k)} = 101.84 \quad \sigma_{MLE} = \sqrt{\frac{1}{n}\sum_{k=1}^{n}\left(Y^{(k)} - \mu_{MLE}\right)^2} = 22.60 \qquad (1.17)$$

Note that $m = \mu_{MLE}$ and $(n-1)s^2 = n\sigma_{MLE}^2$. This relation is not true for other distribution types. In fact, in many cases, there are no closed-form solutions to Equation 1.17 and optimization is needed.

### 1.2.3.4 Normal probability plot

The method of moments, percentile method, and maximum likelihood method provide only point estimates of $\mu$ and $\sigma$. They do not address the more basic question as to whether Y is normally distributed in the first place. The normal probability plot compares the ECDF $F_n(y)$ of the data with the theoretical normal CDF, $F(y)$. The concept is relatively simple as explained below. Let us say you suspect Y to be normally distributed with the unknown mean, $\mu$, and standard deviation, $\sigma$. Then the CDF of Y must be

$$F(y) = \Phi\left(\frac{y - \mu}{\sigma}\right) \qquad (1.18)$$

and

$$\Phi^{-1}\left[F(y)\right] = \frac{y - \mu}{\sigma} \qquad (1.19)$$

If Y is indeed normal, it is clear from Equation 1.19 that a plot with y values on the vertical axis and $\Phi^{-1}[F(y)]$ on the horizontal axis will produce a straight line with the vertical intercept, $\mu$, and gradient, $\sigma$. In summary, linearity of the plot implies that the normal hypothesis is reasonable. Equation 1.19 also provides another method to estimate $\mu$ and $\sigma$!

Table 1.4 illustrates how this method can be applied to simulated normal data of size = 10, $\mu = 100$ kPa, and $\sigma = 20$ kPa. Again, we initialized by randn('state', 13) before executing normrnd(100, 20, 10, 1). The first column contains the simulated data sorted in an ascending order. The second column contains the rank. The third column computes the ECDF using Equation 1.12. The fourth column is obtained by applying the inverse of the standard normal CDF to the third column. Finally, the normal probability plot is obtained by drawing the first column on the vertical axis and the fourth column on the horizontal axis. Linear regression can be used to fit a straight line to the normal probability plot as shown in Figure 1.9 to get estimates for $\mu$ and $\sigma$ from the vertical intercept and gradient, respectively.

### 1.2.3.5 Statistical uncertainties in the $\mu$ and $\sigma$ estimators

#### 1.2.3.5.1 Sampling distributions and confidence intervals

Let us denote the sample mean and sample standard deviation in Equation 1.13 by m and s, respectively. For the sample mean m, its sampling distribution is a normal distribution with mean $= \mu$ and standard deviation $= \sigma/n^{0.5}$, where n is the number of data points.

*Table 1.4* Sorted Y data and the ECDF

| Sorted Y data | Rank k | ECDF $F_n(y)$ from Equation 1.12 | $\Phi^{-1}[F_n(y)]$ |
|---|---|---|---|
| 71.43 | 1 | 0.067 | −1.50 |
| 74.40 | 2 | 0.163 | −0.98 |
| 81.15 | 3 | 0.260 | −0.64 |
| 89.81 | 4 | 0.356 | −0.37 |
| 90.52 | 5 | 0.452 | −0.12 |
| 105.41 | 6 | 0.548 | 0.12 |
| 113.35 | 7 | 0.644 | 0.37 |
| 124.06 | 8 | 0.740 | 0.64 |
| 133.54 | 9 | 0.837 | 0.98 |
| 134.75 | 10 | 0.933 | 1.50 |

If the actual standard deviation $\sigma$ is known, the standardized $(m - \mu)/(\sigma/n^{0.5})$ is distributed as the standard normal distribution. One can then establish the 95% confidence interval of $\mu$ by

$$m + x_{0.025} \cdot \sigma/n^{0.5} \leq \mu \leq m + x_{0.975} \cdot \sigma/n^{0.5} \tag{1.20}$$

where $x_{0.025} = -1.96$ and $x_{0.975} = 1.96$ are, respectively, the 0.025- and 0.975-percentiles of the standard normal distribution. The confidence interval is more informative than simply reporting a point estimate, because statistical uncertainty is quantified.

However, it is usually the case that the actual standard deviation $\sigma$ is unknown. In this case, if Y is indeed normally distributed (which may not be true), the standardized $(m - \mu)/(s/n^{0.5})$ is distributed as the Student's t-distribution with $n-1$ degrees of freedom (DOF). An empirical example of this t-distribution with 9 DOFs is shown in Figure 1.8a. One can then establish the 95% confidence interval of $\mu$ by

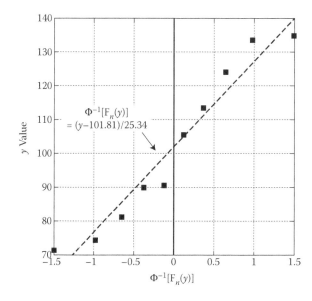

*Figure 1.9* Normality plot: y versus $\Phi^{-1}[F_n(y)]$.

$$m + t_{0.025} \cdot s/n^{0.5} \leq \mu \leq m + t_{0.975} \cdot s/n^{0.5} \tag{1.21}$$

where $t_{0.025}$ and $t_{0.975}$ are, respectively, the 0.025 and 0.975 percentiles of the Student's $t$-distribution with $n - 1$ DOF.

For the sample standard deviation, $s$, if Y is indeed normally distributed (again, which may not be true), the standardized $(n - 1)s^2/\sigma^2$ is distributed as the $\chi$-squared distribution with $(n - 1)$ DOF. An empirical example of this $\chi$-squared distribution with 9 DOFs is shown in Figure 1.8b. One can then establish the 95% confidence interval of $\sigma^2$ by

$$(n - 1) \cdot s^2/\chi_{0.975}^2 \leq \sigma^2 \leq (n - 1) \cdot s^2/\chi_{0.025}^2 \tag{1.22}$$

where $\chi_{0.025}^2$ and $\chi_{0.975}^2$ are, respectively, the 0.025 and 0.975 percentiles of the $\chi$-squared distribution with $(n - 1)$ DOF.

### 1.2.3.5.2 Bootstrapping

Equations 1.20 through 1.22 are based on the strong assumption that the data are normally distributed. In practice, we do not know the distribution of the data. We can test for normality using the K–S test described below, but it would be convenient to obtain confidence intervals for $\mu$ and $\sigma$ without making an assumption on the distribution. The nonparametric bootstrapping (Efron and Tibshirani 1993) is a general framework of obtaining approximate samples from the sampling distribution of any statistics. Let the statistics of interest be denoted by $g(Y^{(1)}, Y^{(2)}, ..., Y^{(n)})$. For the sample mean, m, $g(Y^{(1)}, Y^{(2)}, ..., Y^{(n)}) = (Y^{(1)} + Y^{(2)} + ... + Y^{(n)})/n$. The steps for bootstrapping are as follows:

1. Resampling $(Y^{(1)}, Y^{(2)}, ..., Y^{(n)})$ with replacement. Denote the resampled Y by $(Y'^{(1)}, Y'^{(2)}, ..., Y'^{(n)})$. It is noteworthy that after the resampling, there may be repetitive values in $(Y'^{(1)}, Y'^{(2)}, ..., Y'^{(n)})$, because they are resampled with replacement.
2. Evaluate $g(Y'^{(1)}, Y'^{(2)}, ..., Y'^{(n)})$. This is a resampled g value.
3. Repeat steps 1 and 2 to obtain B resampled g value. Note that B is distinctive from $n$.

The B resampled g values can be viewed as approximate realizations of the sampling distribution of g.

Again, we initialized by randn('state', 13) before executing normrnd(100, 20, 10, 1). The sample mean, $m = 101.84$ and the sample standard deviation, $s = 23.82$. These are the point estimates for $\mu$ and $\sigma$. However, it is not clear how large the statistical uncertainties are. Figure 1.10 shows the histograms of B = 1000 resampled $m$ values and $s$ values (B = 1000) based on the bootstrapping procedure. The 95% confidence intervals of $\mu$ and $\sigma$ can be estimated as the interval bounded by the 0.025 and 0.975 sample percentiles of the resampled values. This confidence interval is called the 95% bootstrap confidence intervals, and this method is called the percentile method (Efron 1981). For $\mu$, the 95% bootstrap confidence interval is [88.37, 115.58]. This can be compared to the 95% analytical confidence interval [84.80, 118.88] based on Equation 1.21. For $\sigma$, the 95% bootstrap confidence interval is [15.51, 27.90]. This can be compared to the 95% analytical confidence interval [16.38, 43.49] based on Equation 1.22. The difference between the bootstrap and analytical confidence intervals is due to the small sample $n = 10$. The problem of "insufficient coverage" for bootstrap confidence intervals of $\sigma$ was discussed in Schenker (1985): the probability for the bootstrap confidence interval to cover the actual value of $\sigma$ is lower than expected. This problem may occur when the sample size (n) is small. The bootstrap method is based on the assumption that the discrete samples

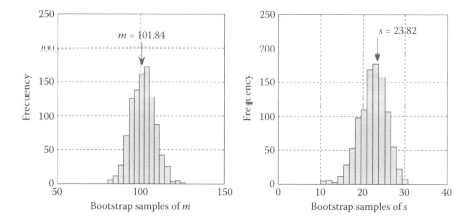

*Figure 1.10* Bootstrap samples of m and s. The arrows indicate the point estimates.

$(Y^{(1)}, Y^{(2)}, ..., Y^{(n)})$ represent the true population, and this assumption breaks down when $n$ is small, leading to the problem of insufficient coverage. Efron (1987) proposed an improved bootstrap method consisting of bias corrections and acceleration adjustments (BCa) to mitigate this problem. An easy-to-follow guide for the BCa method can be found in Carpenter and Bithell (2000). Using the BCa method, the 95% bootstrap confidence interval for $\sigma$ becomes [18.65, 29.91] for our case with $n = 10$. This confidence interval is closer to the analytical one [16.38, 43.49] based on Equation 1.22. However, the difference is still large.

To investigate the sample size under which the problem of insufficient coverage will be minimal, the probability for the 95% bootstrap confidence interval to cover the actual value of $\sigma$ is simulated. This is done by the following steps:

1. Simulate $(Y^{(1)}, Y^{(2)}, ..., Y^{(n)})$ from a normal distribution with $\mu = 100$ and $\sigma = 20$.
2. Construct the 95% bootstrap confidence interval of $\mu$ (or $\sigma$) based on $(Y^{(1)}, Y^{(2)}, ..., Y^{(n)})$.
3. See whether the 95% bootstrap confidence interval covers the actual value of $\mu = 100$ (or the actual value of $\sigma = 20$).

Do this for 1000 independent realizations of $(Y^{(1)}, Y^{(2)}, ..., Y^{(n)})$. The coverage probability is simply the ratio of successful coverage among the 1000 realizations. The bootstrap confidence interval works properly if the coverage probability is close to 95%.

Table 1.5 shows the coverage probabilities for the 95% bootstrap confidence intervals of $\mu$ and $\sigma$. The numbers in the parentheses are for the BCa method. It is clear that the coverage

*Table 1.5* Coverage probabilities for the 95% bootstrap confidence intervals of $\mu$ and $\sigma$ (numbers in parentheses are for the BCa method)

| Sample size n | $\mu$ | $\sigma$ |
|---|---|---|
| $n = 10$ | 0.91 (0.91) | 0.79 (0.85) |
| $n = 20$ | 0.92 (0.92) | 0.86 (0.90) |
| $n = 50$ | 0.93 (0.93) | 0.89 (0.92) |
| $n = 100$ | 0.93 (0.94) | 0.93 (0.94) |
| $n = 1000$ | 0.95 (0.95) | 0.94 (0.95) |

probability for $\sigma$ is noticeably less than 0.95 when $n$ is less than 100. The improvement brought about by the BCa method for the 95% bootstrap confidence intervals of $\sigma$ for a small sample size ($n \leq 50$) is evident. In general, the coverage probability is close to 0.95 for $n \geq 100$ for both $\mu$ and $\sigma$. It is therefore recommended that the sample size $n$ should be $\geq 100$ for the bootstrap confidence intervals to work properly.

### 1.2.3.5.3 Goodness-of-fit test (K–S test)

The normal probability plot is a good visual tool to judge whether the normal distribution provides a satisfactory fit to the data. In this section, the K–S test (Conover 1999) is introduced to characterize the goodness of fit for the normal distribution formally using the framework of hypothesis testing. The null hypothesis $H_0$ for the K–S test is

$$H_0 : Y \text{ is normally distributed} \tag{1.23}$$

Namely, $F(y) = \Phi[(y - \mu)/\sigma]$. Under this null hypothesis, the following statistics $D_n$ is asymptotically distributed as the Kolmogorov distribution:

$$D_n = \sqrt{n} \cdot \sup_y \left| F_n(y) - F(y) \right| = \sqrt{n} \cdot \sup_y \left| F_n(y) - \Phi[(y - \mu)/\sigma] \right| \tag{1.24}$$

where "sup" denotes the supremum (the least upper bound); $n$ is the sample size of the data points. One can see that if the null hypothesis $H_0$ is true, $D_n$ should be small, because $F_n(y)$ will be close to $F(y)$ under $H_0$. As a result, $H_0$ can be rejected if $D_n$ is large. Consider the following criterion of rejecting/accepting $H_0$:

$$\text{Reject } H_0 \text{ if } D_n > c \quad \text{Do not reject } H_0 \text{ if } D_n \leq c \tag{1.25}$$

where $c$ is called the critical value. It is a prescribed threshold for $D_n$. The critical value $c$ is typically chosen such that the probability of committing Type I error [probability of rejecting a true $H_0$, namely $P(D_n > c)$] is equal to a small number $\alpha$ (e.g., $\alpha = 0.05$). The $\alpha$ value is called the significance level of the test. The threshold c is in fact the $(1 - \alpha)$ percentile of the Kolmogorov distribution and can be found in textbooks.

In MATLAB, the command $[h, p, d_n] = \text{kstest}(\mathbf{X}, [], \alpha)$ is for the K–S test for the standard normal distribution. The inputs include the vector $\mathbf{X}$ that contains the data ($X^{(1)}, X^{(2)}, ..., X^{(n)})^T$ (the superscript 'T' means the matrix transpose) and $\alpha$. The outputs include $h$ ($h = 1$ means $H_0$ is rejected), $p$ ($p$-value), and $d_n$ (the realization of $D_n$). To implement the standard normal K–S test, one needs to first convert the data ($X^{(1)}, X^{(2)}, ..., X^{(n)}$) into their standardized form:

$$X^{(k)} = \frac{Y^{(k)} - m}{s} \tag{1.26}$$

The $p$-value is defined to be $P(D_n > d_n)$. The null hypothesis $H_0$ is rejected if $p < \alpha$. It can be seen that the p-value quantifies how strong the rejection is: a small $p$-value indicates strong rejection. The K–S test with $\alpha = 0.05$ on the 10 samples of Y gives $h = 0$ ($H_0$ is not rejected) and $p = 0.835$. Therefore, the normal distribution hypothesis is not rejected at a significance level of 0.05. This is expected as the Y samples are simulated from a normal distribution. However, if one repeats this procedure with different simulated samples say 100 times, one expects five simulated samples to be rejected *on the average*, despite the fact that all samples belong to a normal distribution. This is another interpretation of the significance level α. In the presence of statistical uncertainty, one must expect to make a wrong conclusion (reject null hypothesis although it is true) on some occasions. One can also make a wrong conclusion by failing to reject the null hypothesis although it is false. This is known as Type II error. We do not elaborate further on hypothesis testing, because details are available in standard texts.

### 1.2.4 Simulation of a normal random variable

#### 1.2.4.1 Simulating standard uniform random variable U

The standard uniform distribution refers to the uniform distribution that is defined on [0, 1]. Both MATLAB and EXCEL have built-in function or add-in for simulating the standard uniform random variable, denoted by U. These are in fact "pseudo" random number generators that generate deterministic numbers that mimic the behaviors of random numbers. In MATLAB, the function rand($n$, 1) will generate an ($n \times 1$) vector that contains $n$-independent and identically distributed (iid) samples of U. In EXCEL, the "data analysis" add-in can generate samples of U.

#### 1.2.4.2 Simulating standard normal random variable X

Let us denote the standard normal random variable by X. It is first noted that $\Phi(X)$ is a standard uniform random variable. This is due to the fact that the CDF of $\Phi(X)$ has the following expression:

$$P\left[\Phi(X) \leq u\right] = P\left[X \leq \Phi^{-1}(u)\right] = \Phi\left[\Phi^{-1}(u)\right] = u \qquad (1.27)$$

Since $F(u) = u$ is the CDF of the standard uniform distribution, $\Phi(X)$ is indeed a standard uniform random variable. Namely,

$$U = \Phi(X) \qquad (1.28)$$

where U is a standard uniform random variable. In other words,

$$X = \Phi^{-1}(U) \qquad (1.29)$$

where $\Phi^{-1}$ is the inverse of the standard normal CDF. As a result, the following steps can be used to simulate X:

1. Simulate U using MATLAB function 'rand' or EXCEL data analysis add-in.
2. Let $X = \Phi^{-1}(U)$. The $\Phi^{-1}$ function can be evaluated using MATLAB function norminv(U), or using EXCEL NORMSINV(U).

It is worthwhile to point out that Equation 1.29 can be generalized to simulate any random variable by replacing $\Phi^{-1}$ with the inverse CDF of the random variable. However, the inverse CDF is usually not available in closed form and alternate methods have been developed to circumvent this computational difficulty. For example, the normal random variable can be simulated using the following Box–Muller method (Box and Muller 1958):

1. Simulate two independent standard uniform random variables ($U_1$, $U_2$).
2. Simulate two independent standard normal random variables ($X_1$, $X_2$)

$$X_1 = \sqrt{-2\ln(U_1)} \cdot \cos(2\pi U_2) \quad X_2 = \sqrt{-2\ln(U_1)} \cdot \sin(2\pi U_2) \tag{1.30}$$

The Marsaglia–Bray method circumvents the evaluation of trigonometric functions in Equation 1.30 (Marsaglia and Bray 1964):

1. Simulate two independent standard uniform random variables ($U_1$, $U_2$). Compute

$$V_1 = 2U_1 - 1 \quad V_2 = 2U_2 - 1 \tag{1.31}$$

2. Compute

$$R = \sqrt{V_1^2 + V_2^2} \tag{1.32}$$

   If $|R| \geq 1$ or $R = 0$, repeat step 1.
3. Simulate two independent standard normal random variables ($X_1$, $X_2$)

$$X_1 = V_1 \sqrt{\frac{-2\ln(R^2)}{R^2}} \quad X_2 = V_2 \sqrt{\frac{-2\ln(R^2)}{R^2}} \tag{1.33}$$

### 1.2.4.3 Simulating normal random variable Y

The normal random variable Y with mean, $\mu$, and standard deviation, $\sigma$, can be simulated using

$$Y = \mu + \sigma \cdot X \tag{1.34}$$

One can also simulate a table of iid Y samples containing $n$ rows and $d$ columns using the MATLAB function normrnd ($\mu$, $\sigma$, $n$, $d$), without the need of going through the above details.

## 1.3 BIVARIATE NORMAL VECTOR

## 1.3.1 Bivariate data

What are bivariate data? It is instructive to distinguish between univariate and bivariate data. One may be tempted to say that if univariate data refer to one column of data (e.g., undrained shear strength), then bivariate data simply refer to two columns of data (e.g., undrained shear strength and undrained modulus). However, there is more than one method of collecting two columns of data. One method is to extract the undrained shear strength and undrained modulus from the same stress–strain curve. The undrained shear strength ($s_u$) is the ultimate stress. The undrained modulus ($E_u$) can be defined as the secant Young's modulus at a stress level equal to one-third of the ultimate stress. The crux here is that $s_u$ and $E_u$ are properties of a single soil sample. It is meaningful to take the ratio $E_u/s_u$ to obtain the rigidity index, which is another soil property. If you repeat this procedure for different soil samples, you will obtain different pairs of values that can be entered as two columns of numbers in EXCEL. However, it is crucial to know that each *row* refers to one soil sample. Such a database is called bivariate data. Another method is to extract $s_u$ from one soil sample and $E_u$ from a different soil sample. In other words, the pair of values $(s_u, E_u)$ is extracted from two different stress–strain curves or perhaps, more commonly, one value is measured in the laboratory and the other value is measured in the field away from the sampling location. This procedure will also produce two columns of numbers in EXCEL. Such a database is called two sets of univariate data. The critical difference is that each pair of values in a bivariate data could be related in some way, because they are merely different aspects of the same stress–strain curve. For univariate data, there is no relation between $s_u$ and $E_u$ even if they appear in the same row, because they refer to properties from two different soil samples. In fact, one can randomly shuffle the $s_u$ values (or $E_u$ values) in one column and nothing changes because the $s_u$ value and the $E_u$ value appearing in a single row are still associated with different soil samples. In summary, row has no meaning for two sets of univariate data.

In contrast, it is critical to enter the pair of values $(s_u, E_u)$ in the correct row for bivariate data. It is more accurate to visualize bivariate data as the coordinates of a point in a scatter plot than two columns of numbers. Consider the pair $(s_u = 40 \text{ kPa}, E_u = 8 \text{ MPa})$ and the pair $(s_u = 200 \text{ kPa}, E_u = 20 \text{ MPa})$. When the above pairs of values are plotted as points, say with $E_u$ as the vertical axis and $s_u$ as the horizontal axis, it is clear that the points will be shifted drastically when one column is shuffled creating two new pairs of values: $(s_u = 200 \text{ kPa}, E_u = 8 \text{ MPa})$ and $(s_u = 40 \text{ kPa}, E_u = 20 \text{ MPa})$. One would also be changing the physical property of the soil in a fundamental way in terms of the rigidity index. The original rigidity indices are 100 and 200. The new rigidity indices are 40 and 500! Clearly, minor errors in entering values are less significant than entering values in the incorrect row for bivariate data.

The scatter plot provides a graphical overview of an important concept called "dependency" between two variables. An actual example is shown in Figure 1.11. The $a$ and $n$ parameters control a nonlinear equation, called the van Genuchten model, which describes how the volumetric water content of an unsaturated soil varies with the matrix suction. It is quite clear that a downward trend exists, that is, a small value of $a$ is associated with a large value of $n$ and vice versa. If you shuffle the columns containing these values, you may produce a pair of values where both $a$ and $n$ are large. Such a data point does not exist in practice and there are good physical reasons to explain this. In fact, the tight clustering of data around a downward trend, which is a "signature" profile of the data in Figure 1.11, will be lost when coordinates are randomized in the shuffling process. Hence, in contrast to shuffling having no effect on two columns of univariate data, shuffling effectively destroys two columns of bivariate data and the underlying physical basis.

It is easy to appreciate the concept of dependency in a visual way using a scatter plot. However, it is less straightforward to characterize dependency quantitatively. Mathematically, you would need a bivariate probability model, which can be expressed as a two-dimensional PDF or CDF. The most common model is the bivariate normal model that is discussed in detail below. Dependency is succinctly characterized by a single number called the Pearson or product–moment correlation coefficient in this model. Nonetheless, it is important to note that this model is not unique. Here, we do not mean that the marginal distribution of each component (e.g., $s_u$) is non-normal. We will show below that it is always possible to transform a non-normal component into a normal component. Nonetheless, two normal components do not necessarily constitute a bivariate normal vector. It suffices to state here that a random vector is *bivariate normal* (vis-à-vis univariate normal) if and only if all linear

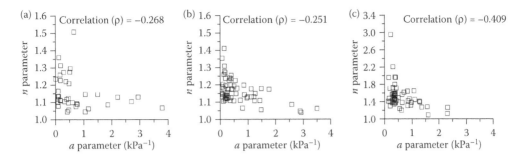

*Figure 1.11* Correlation between van Genuchten parameters for (a) sandy clay loam, (b) loam, and (c) loamy sand. (From Phoon, K.K., Santoso, A., and Quek, S.T. 2010. *Journal of Geotechnical and Geoenvironmental Engineering, ASCE*, 136(3), 445–455, reproduced with permission of the American Society of Civil Engineers.)

sums of the components result in a normal random variable. This condition may not be satisfied by actual bivariate data. There are alternate bivariate probability models. Some of them are discussed in Chapter 2.

The bivariate model is a special case of the multivariate probability model. Nonetheless, it is worthwhile covering this special case, because geotechnical engineering is replete with correlations between two soil parameters, for example, between undrained shear strength and PI, between OCR and cone tip resistance, and many others (Kulhawy and Mayne 1990). In addition, there are many curve-fitting procedures that involve two parameters. The most well-known example is the cohesion and friction angles produced when a non-linear failure envelope is fitted to a linear Mohr–Coulomb envelope. The cohesion is the vertical intercept and the friction angle is the gradient of this linear envelope. A soil–water characteristic curve can be fitted to a nonlinear van Genuchten containing two parameters as shown in Figure 1.11. A third example is a two-parameter hyperbolic model for fitting a load-displacement curve produced by a pile load test (Dithinde et al. 2011). Aside from the practical significance of a bivariate model, it is a good introduction to the more general multivariate model.

We use simulated bivariate standard normal data below to clarify the concept of the correlation coefficient and how to estimate this index of dependency from data. To obtain the simulated data, we initialize the pseudorandom sequence using randn('state', 13). Next, we obtain two columns of independent standard normal data using $Z = \text{normrnd}(0, 1, n, 2)$, where $n$ is the sample size. The correlation matrix is entered as $C = [1\ \delta_{12}; \delta_{12}\ 1]$, where $\delta_{12}$ is the Pearson correlation coefficient. The upper triangle Cholesky factor is computed as $U = \text{chol}(C)$ and two columns of correlated standard normal data are obtained from $X^T = Z^T \times U$, where the superscript 'T' means the matrix transpose. Details on simulation of correlated data are given in Section 1.3.4.

## 1.3.2 Bivariate normal distribution

The bivariate normal distribution can be used to model two jointly distributed normal random variables. Let us denote the vector $[Y_1\ Y_2]^T$ by $Y$. This normal random vector is described by a mean vector $\mu = [\mu_1\ \mu_2]^T$ and a covariance matrix $C$:

$$C = \begin{bmatrix} \text{COV}(Y_1, Y_1) & \text{COV}(Y_1, Y_2) \\ \text{COV}(Y_2, Y_1) & \text{COV}(Y_2, Y_2) \end{bmatrix} = \begin{bmatrix} \sigma_1^2 & \delta_{12}\sigma_1\sigma_2 \\ \delta_{12}\sigma_1\sigma_2 & \sigma_2^2 \end{bmatrix} \tag{1.35}$$

where COV(,) denotes the covariance; $\sigma_i$ is the standard deviation of $Y_i$; and $\delta_{ij}$ is the Pearson (product–moment) correlation coefficient between $Y_i$ and $Y_j$. The mean and standard deviation have been defined in the preceding discussion on a single normal random variable. The only new concept required to describe a bivariate normal random vector is the covariance (or correlation coefficient).

The PDF for the bivariate normal distribution is

$$f(y) = \frac{1}{\sqrt{2\pi}^2 \cdot \sqrt{|C|}} \exp\left[\frac{-(y - \mu)^T C^{-1}(y - \mu)}{2}\right]$$ 

(1.36)

where $|C|$ is the determinant of C matrix; the superscript '–1' symbolically denotes the matrix inverse.

### 1.3.2.1 Bivariate standard normal

Consider the case where X is standard normal ($\mu = 0$ and $\sigma = 1$). Hence, the two random variables ($X_1$, $X_2$) are both standard normal, and their bivariate normal distribution is simplified into

$$f(x) = \frac{1}{\sqrt{2\pi}^2 \cdot \sqrt{|C|}} \exp\left(\frac{-x^T C^{-1} x}{2}\right)$$

(1.37)

where

$$C = \begin{bmatrix} 1 & \delta_{12} \\ \delta_{12} & 1 \end{bmatrix}$$

(1.38)

### 1.3.2.2 Correlation coefficient

The Pearson product–moment correlation coefficient is defined as

$$\delta_{ij} = \frac{COV(X_i, X_j)}{\sigma_i \cdot \sigma_j}$$

(1.39)

$\delta_{ij}$ quantifies the degree of linear correlation between $X_i$ and $X_j$. It is equal to 1 for perfectly positive linear correlation, –1 for perfectly negative linear correlation, and 0 for uncorrelated $X_i$ and $X_j$. Figure 1.12 shows the contour plots for two bivariate standard normal distributions with $\delta_{12}$ equal to 0.9 and 0 (random sequence initiated using randn['state', 13]). The random samples from the distributions are also shown. For $\delta_{12} = 0.9$, a strong positive linear correlation exists between $X_1$ and $X_2$. For $\delta_{12} = 0$, $X_1$ and $X_2$ are uncorrelated.

## 1.3.3 Estimation of $\delta_{12}$

### 1.3.3.1 Method of moments

Given samples of ($X_1$, $X_2$) that are bivariate standard normal, it is possible to estimate $\delta_{12}$. The most common method is the method of moments:

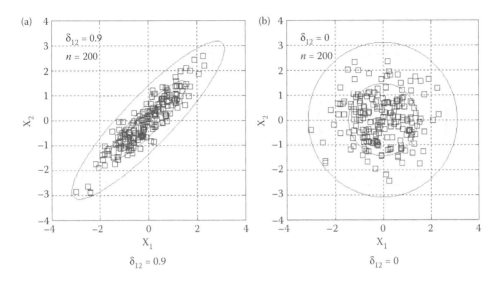

*Figure 1.12* Contour plots for the bivariate standard normal distribution of $(X_1, X_2)$.

$$\delta_{12} \approx \frac{1/(n-1)\sum_{k=1}^{n}\left(X_1^{(k)}-m_1\right)\cdot\left(X_2^{(k)}-m_2\right)}{\sqrt{1/(n-1)\sum_{k=1}^{n}\left(X_1^{(k)}-m_1\right)^2 \times 1/(n-1)\sum_{k=1}^{n}\left(X_2^{(k)}-m_2\right)^2}} \tag{1.40}$$

where the superscript $(k)$ is the sample index; $m_i$ is the sample mean of $X_i$. Note that the denominator is simply the product of the sample standard deviation of $X_1$ and the sample standard deviation of $X_2$. The MATLAB function corr($X_1$, $X_2$,'type', 'Pearson') and the EXCEL command CORREL($X_1$, $X_2$) can be used to find the $\delta_{12}$ estimate in Equation 1.40. Here, $X_1 = \left(X_1^{(1)},..., X_1^{(n)}\right)^T$ is an $(n \times 1)$ vector that contains all samples of $X_1$.

### 1.3.3.2 Maximum likelihood method

The maximum likelihood method can also be employed:

$$\mu_{\text{MLE}}, \sigma_{\text{MLE}}, \delta_{12,\text{MLE}} \approx \underset{\mu_1,\mu_2,\sigma_1,\sigma_2,\delta_{12}}{\arg\max} \prod_{k=1}^{n} \frac{1}{\sqrt{2\pi}^2 \cdot \sqrt{|C|}} \exp\left[-0.5 \times \left(X^{(k)}-\mu\right)^T C^{-1}\left(X^{(k)}-\mu\right)\right]$$

$$- \underset{\mu_1,\mu_2,\sigma_1,\sigma_2,\delta_{12}}{\arg\max} \sum_{k=1}^{n}\left[\ln(|C|)\ \left(X^{(k)}-\mu\right)^T C^{-1}\left(X^{(k)}-\mu\right)\right] \tag{1.41}$$

where

$$X^{(k)} = \begin{bmatrix} X_1^{(k)} \\ X_2^{(k)} \end{bmatrix} \qquad \mu = \begin{bmatrix} \mu_1 \\ \mu_2 \end{bmatrix} \qquad C = \begin{bmatrix} \sigma_1^2 & \delta_{12}\sigma_1\sigma_2 \\ \delta_{12}\sigma_1\sigma_2 & \sigma_2^2 \end{bmatrix} \tag{1.42}$$

### 1.3.3.3 Rank correlation method

The Spearman rank correlation between $X_1$ and $X_2$, denoted by $\rho_{12}$, is the Pearson correlation between the ranks of $X_1$ and $X_2$. Namely, each X samples are converted into its ranks. For instance, if there are five $(X_1, X_2)$ samples (random sequence initiated using randn['state', 13]):

$$
\begin{aligned}
\mathbf{X}^T &= \begin{bmatrix} X_1^{(1)} & X_1^{(2)} & X_1^{(3)} & X_1^{(4)} & X_1^{(5)} \\ X_2^{(1)} & X_2^{(2)} & X_2^{(3)} & X_2^{(4)} & X_2^{(5)} \end{bmatrix} \\
&= \begin{bmatrix} 1.20 & 0.67 & 1.74 & -0.47 & 1.68 \\ -1.28 & -0.94 & 0.27 & -1.43 & -0.51 \end{bmatrix}
\end{aligned}
\tag{1.43}
$$

Then the ranks are

$$
\mathbf{X}_r^T = \begin{bmatrix} 3 & 2 & 5 & 1 & 4 \\ 2 & 3 & 5 & 1 & 4 \end{bmatrix}
\tag{1.44}
$$

The Spearman rank correlation $\rho_{12}$ is simply the Pearson correlation between the two rows of ranks in Equation 1.44. The Spearman rank correlation $\rho_{12}$ quantifies how well the relationship between two variables can be described as a monotonic function. For $\rho_{12} = 1$, the relationship between the two variables is perfectly positively monotonic (not necessarily linear), and for $\rho_{12} = -1$, the relationship is perfectly negatively monotonic. In MATLAB, $\rho_{12}$ can be estimated using the function corr($\mathbf{X}_1$, $\mathbf{X}_2$,'type', 'Spearman').

In general, the Spearman correlation $\rho_{12}$ is not the same as the Pearson correlation $\delta_{12}$. Table 1.6 shows an example where X and Y are perfectly monotonically correlated. In fact, $Y = X^3$. The Pearson correlation $\delta = 0.906$ is computed by corr($\mathbf{X}$, $\mathbf{Y}$, 'type', 'Pearson'), whereas the Spearman correlation $\rho = 1$ is computed by the MATLAB function corr($\mathbf{X}$, $\mathbf{Y}$, 'type', 'Spearman'). They are not equal. The Pearson correlation $\delta$ is not exactly 1 because it quantifies "linear correlation," and in this case, the relationship between X and Y is nonlinear. The Spearman correlation $\rho$ is exactly 1 because the ranks of X and Y are identical (see the fourth and fifth columns of Table 1.6). As a result, the Spearman correlation $\rho$ quantifies how well the relationship between (X, Y) can be described as a monotonic function.

It is well known that the Pearson product–moment correlation $\delta_{12}$ between the bivariate normal random variables $(X_1, X_2)$ is approximately equal to its Spearman rank correlation $\rho_{12}$. It is therefore possible to estimate $\delta_{12}$ using the rank correlation $\rho_{12}$.

Table 1.6 Values of (X, Y) data points

| Index k | X | Y | X rank | Y rank |
|---------|---|---|--------|--------|
| 1 | 0 | 0 | 1 | 1 |
| 2 | 1 | 1 | 2 | 2 |
| 3 | 2 | 8 | 3 | 3 |
| 4 | 3 | 27 | 4 | 4 |
| 5 | 4 | 64 | 5 | 5 |
| 6 | 5 | 125 | 6 | 6 |

### *1.3.3.4 Statistical uncertainties in the $\delta_{12}$ estimate*

The bootstrapping can be used to assess the sampling distribution of $\delta_{12}$. The steps for bootstrapping are as follows:

1. Resampling the sample index with replacement. For the data in Equation 1.43, there are in total five data points ($n = 5$). As a result, the sample *index* includes 1, 2, 3, 4, 5. Resampling these indices with replacement can be done by implementing the MATLAB function randsample($n$, $n$, 1). Note that after the resampling, there may be repetitive indices.
2. Suppose the resampled indices are (5, 3, 1, 1, 2). The resampled samples are therefore

$$
\begin{aligned}
\mathbf{X'}^{\mathrm{T}} &= \begin{bmatrix} X_1^{(5)} & X_1^{(3)} & X_1^{(1)} & X_1^{(1)} & X_1^{(2)} \\ X_2^{(5)} & X_2^{(3)} & X_2^{(1)} & X_2^{(1)} & X_2^{(2)} \end{bmatrix} \\
&= \begin{bmatrix} 1.68 & 1.74 & 1.20 & 1.20 & 0.67 \\ -0.51 & 0.27 & -1.28 & -1.28 & -0.94 \end{bmatrix}
\end{aligned}
\tag{1.45}
$$

3. Estimate the sample Pearson correlation between the resampled ($X_1$, $X_2$) using one of the methods presented above. This is a resampled $\delta_{12}$ value.
4. Repeat steps 1 and 2 to obtain B resampled $\delta_{12}$ value. Note that B is not the same as $n$.

The B resampled $\delta_{12}$ values can be viewed as approximate realizations of the sampling distribution of $\delta_{12}$.

Now, consider the ($X_1$, $X_2$) data in the left plot in Figure 1.12. The actual value of $\delta_{12}$ is 0.9. The sample value of $\delta_{12}$ can be obtained from the methods of moments (Equation 1.40). For $n = 50$, the estimated value of $\delta_{12}$ is 0.937. For $n = 200$, the estimated value of $\delta_{12}$ is 0.912. However, these are point estimates of $\delta_{12}$, and it is not clear how large the statistical uncertainties are. Figure 1.13 shows the histograms of 10,000 resampled $\delta_{12}$ values for $n = 50$ and $n = 200$ based on the bootstrapping procedure (bootstrap sample size B = 10,000). The 95% bootstrap confidence interval of $\delta_{12}$ can be estimated as the interval bounded by the 0.025 and 0.975 sample percentiles of the resampled $\delta_{12}$ values. This is called the percentile method (Efron 1981). It is clear that the confidence interval is narrower when the number of data points gets larger.

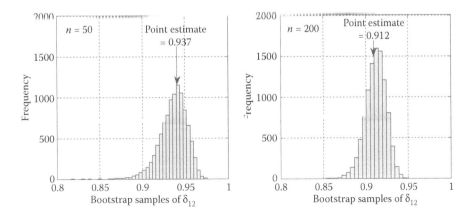

*Figure 1.13* Bootstrap samples of $\delta_{12}$. The arrows indicate the point estimates.

It is essential to verify whether the problem of "insufficient coverage" exists for the boot-strap confidence intervals of $\delta_{12}$. The coverage probability refers to the probability for the bootstrap confidence interval to cover the actual value of $\delta_{12}$. To investigate the sample size under which the problem of insufficient coverage will be minimal, the probability for the 95% bootstrap confidence interval of $\delta_{12}$ to cover the actual value of $\delta_{12}$ is simulated. This is done by the following steps:

1. Simulate $(X_1, X_2)$ data of sample size = $n$ from a bivariate standard normal distribution with a chosen $\delta_{12}$.
2. Construct the 95% bootstrap confidence interval of $\delta_{12}$ based on the simulated $(X_1, X_2)$ data.
3. See whether the 95% bootstrap confidence interval covers the actual of $\delta_{12}$.

Do this for 1000 independent realizations of $(X_1, X_2)$ data. The coverage probability is simply the ratio of successful coverage among the 1000 realizations. The bootstrap confidence interval works properly if the coverage probability is close to 95%.

Table 1.7 shows the coverage probabilities for the 95% bootstrap confidence intervals of $\delta_{12}$ for three chosen values of $\delta_{12}$ (0, 0.5, and 0.9). The numbers in the parentheses are for the BCa method. It is clear that the coverage probability for $\delta_{12}$ is close to 0.95 when $n \geq 50$. The improvement brought about by the BCa method for the 95% bootstrap confidence intervals of $\delta_{12}$ is insignificant. It is therefore recommended that the percentile method is sufficient and that the sample size $n$ should be $\geq 50$ for the bootstrap confidence intervals to work properly.

### 1.3.3.5 Goodness-of-fit test (the line test)

The normality of a column of data can be checked using the K–S test (see Equations 1.23 through 1.26). However, the bivariate normality of two columns of data requires a separate check. The line test (Hald 1952; Kowalski 1970) is a reasonably simple test for bivariate normality. The line test is based on the fact that if $(X_1, X_2)$ are bivariate standard normal, the variable Q

$$ Q = \frac{1}{1 - \delta_{12}^2} \left[ \left( \frac{X_1 - \mu_1}{\sigma_1} \right)^2 - 2\delta_{12} \left( \frac{X_1 - \mu_1}{\sigma_1} \right) \left( \frac{X_2 - \mu_2}{\sigma_2} \right) + \left( \frac{X_2 - \mu_2}{\sigma_2} \right)^2 \right] \tag{1.46} $$

is $\chi$-square distribution with 2 DOF. In reality, the value of $\mu$, $\sigma$, and $\delta$ is not available. Only the sample versions are available. Hence, $(\mu_1, \mu_2)$ are replaced by $(m_1, m_2)$, $(\sigma_1, \sigma_2)$ are replaced by $(s_1, s_2)$, and $\delta_{12}$ is replaced by its sample estimate (say using Equation 1.40). The CDF for the 2-DOF $\chi$-square distribution is

Table 1.7 Coverage probabilities for the 95% bootstrap confidence intervals of $\delta_{12}$ (Numbers in parentheses are for the BCa method)

| | $\delta_{12} = 0$ | $\delta_{12} = 0.5$ | $\delta_{12} = 0.9$ |
|---|---|---|---|
| $n = 10$ | 0.91 (0.94) | 0.92 (0.94) | 0.90 (0.93) |
| $n = 20$ | 0.92 (0.93) | 0.92 (0.93) | 0.92 (0.92) |
| $n = 50$ | 0.94 (0.95) | 0.94 (0.94) | 0.95 (0.95) |
| $n = 100$ | 0.95 (0.95) | 0.94 (0.94) | 0.95 (0.95) |
| $n = 1000$ | 0.95 (0.94) | 0.94 (0.94) | 0.94 (0.94) |

$$F(q) = 1 - \exp\left(-\frac{q}{2}\right) \tag{1.47}$$

and

$$-2\ln\left[1 - F(q)\right] = q \tag{1.48}$$

If Q is indeed 2-DOF $\chi$-square, it is clear from Equation 1.48 that a plot with $q$ values on the vertical axis and $-2\ln[1 - F(q)]$ on the horizontal axis will produce a 1:1 line. As a result, if the actual $q$ versus $-2\ln[1 - F_n(q)]$ relationship lies close to the 1:1 line, the bivariate normal hypothesis is reasonable. $F_n(q)$ is the ECDF of Q, which can be estimated using Equations 1.11 or 1.12 according to the Q data points.

Table 1.8 illustrates how the line test can be applied to the first 10 simulated bivariate standard normal data in Figure 1.12a. The first column contains the simulated Q data sorted in an ascending order. The second column contains the rank. The third column computes the ECDF $F_n(q)$ using Equation 1.12. The fourth column is obtained by applying the logarithm of $1 - F_n(q)$ in the third column. Finally, the probability plot is obtained by drawing the first column on the $y$-axis and the fourth column on the $x$-axis. Figure 1.14 shows the resulting $q$ versus $-2\ln[1 - F_n(q)]$ relationship. If it is indeed close to the 1:1 line, we can conclude that there is no strong evidence to reject the bivariate standard normal model.

### 1.3.4 Simulation of bivariate standard normal random variables

Given the Pearson correlation $\delta_{12}$, random samples of bivariate standard normal $(X_1, X_2)$ can be readily simulated by the following steps:

1. Simulate independent standard normal random variables $(Z_1, Z_2)$. The details have been described in Section 1.2.4.

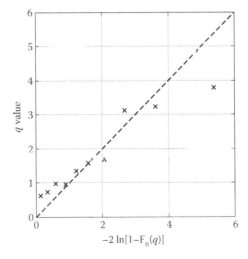

*Figure 1.14* Line test: $q$ versus $-2\ln[1 - F_n(q)]$ plot. The dashed line is the 1:1 line.

*Table 1.8* Sorted Q data and the ECDF

| Sorted Q data | Rank k | ECDF $F_n(q)$ from Equation 1.12 | |
|---|---|---|---|
| 0.62 | 1 | 0.067 | 0.14 |
| 0.72 | 2 | 0.164 | 0.36 |
| 0.96 | 3 | 0.260 | 0.60 |
| 0.97 | 4 | 0.356 | 0.88 |
| 1.33 | 5 | 0.452 | 1.20 |
| 1.57 | 6 | 0.548 | 1.59 |
| 1.69 | 7 | 0.644 | 2.07 |
| 3.12 | 8 | 0.740 | 2.70 |
| 3.23 | 9 | 0.837 | 3.62 |
| 3.78 | 10 | 0.933 | 5.40 |

2. Let

$$X_1 = Z_1$$
$$X_2 = \delta_{12} \cdot Z_1 + \sqrt{1 - \delta_{12}^2} \cdot Z_2 \tag{1.49}$$

It is easy to verify that both $X_1$ and $X_2$ are still standard normal. Unlike $Z_1$ and $Z_2$ being independent, $X_1$ and $X_2$ are correlated with Pearson correlation $= \delta_{12}$.

Another way of writing Equation 1.49 is

$$X^T = Z^T \times U \tag{1.50}$$

where

$$X = \begin{bmatrix} X_1 \\ X_2 \end{bmatrix} \qquad Z = \begin{bmatrix} Z_1 \\ Z_2 \end{bmatrix} \qquad U = \begin{bmatrix} 1 & \delta_{12} \\ 0 & \sqrt{1 - \delta_{12}^2} \end{bmatrix} \tag{1.51}$$

Note that

$$U^T \times U = C = \begin{bmatrix} 1 & \delta_{12} \\ \delta_{12} & 1 \end{bmatrix} \tag{1.52}$$

The **U** matrix is called the upper Cholesky factor of **C**. The Cholesky factor can be viewed as the "square-root" of a matrix. In MATLAB, the function 'chol' can be used to compute **U**: **U** = chol(**C**). Figure 1.12 shows the simulated samples of $(X_1, X_2)$ with mean values = 0, standard deviations = 1, and with the Pearson correlation = $\delta_{12}$ ($\delta_{12} = 0.9$ and 0.0).

## 1.4 MULTIVARIATE NORMAL VECTOR

### 1.4.1 Multivariate data

Multivariate data can be understood as a generalization of bivariate data to higher dimensions. We recall that bivariate data occupy two columns in EXCEL. Multivariate data occupy multiple columns. Bivariate data can be visualized as a point in a scatter plot. Multivariate data can also be visualized as a point in higher-dimensional space, but it is hard to visualize scatter beyond our familiar three-dimensional space. We have emphasized that the two values in any given row are linked for bivariate data. For the case of bivariate normal data, we have discussed the characterization of this linkage or dependency using the Pearson correlation coefficient, a number lying between −1 and 1. If there are three columns of data, it is easy to envisage the computation of three distinct correlation coefficients: one between columns 1 and 2, one between columns 1 and 3, and one between columns 2 and 3. The astute reader may generalize the Pearson correlation coefficient involving the product of two columns to a higher-order coefficient involving the product of three columns. Although these higher-order product–moment coefficients can be defined mathematically, they are not used practically for two reasons. First, statistical uncertainty increases with the order of the product–moment coefficient. In other words, computing a coefficient involving three columns will incur more statistical uncertainty than the one involving two columns. Second, the only practical multivariate probability model is the multivariate normal model. This model requires only the computation of all bivariate Pearson correlation coefficients. It does not require higher-order product–moment coefficients. In fact, all higher-order product–moment coefficients can be computed from Pearson correlation coefficients using closed-form equations for the *special case* of multivariate normal data (Isserlis 1918).

Multivariate information is usually gathered in a typical site investigation. For instance, when undisturbed samples are extracted for odometer and triaxial tests, SPT and/or piezocone test (CPTU) may be conducted in close proximity. Moreover, index properties such as the unit weight, natural water content, plastic limit, LL, and liquidity index (LI) are commonly determined from relatively simple laboratory tests on disturbed samples. It is generally known that data from these varied sources are not independent if they are measured in close physical proximity. The definition of "close" is related to the spatial variability of the site. These data sources are typically correlated to a design parameter, for example, the undrained shear strength ($s_u$). These correlations can be exploited to reduce the COV of the design parameter. The impact on RBD is obvious.

When multivariate geotechnical data exist in sufficient amount, it is of significant practical usefulness to construct a multivariate probability distribution function, which is usually based on the multivariate normal distribution (Ching and Phoon 2012, 2013; Ching et al. 2014b). The applications include: (a) deriving the mean and COV of any parameter given the information contained in a subset with possibly more than one parameter, and (b) evaluating if new strong bivariate correlations can be found either among the original components or some derived components. For the former, it is likely for the COV of a design parameter, say the undrained shear strength ($s_u$), to reduce when other parameters, say the normalized cone tip resistance and OCR, have been measured. This aspect is significant for RBD. In fact, COV reduction can be viewed as a measure of the value of information and may eventually underpin a defensible "information-sensitive" framework for justifying the investment to measure an additional parameter. In addition, the ability to predict the existence of new correlations not included as part of the model calibration provides a stronger scientific underpinning to correlation studies in geotechnical engineering. The reason is that these predictions can be falsified by taking new observations, which is the cornerstone of the scientific method. In other words, it is a lot harder to develop multivariate models, but if they stand the test of time, they are usually more robust than bivariate models. There is an oft-expressed adage, "don't confuse me with more data!", that nicely expressed the challenge of combining varied data sources.

## 1.4.2 Multivariate normal distribution

The multivariate normal distribution can be easily understood as a generalization of the bivariate normal distribution. Let us denote the $d \times 1$ vector $[Y_1\ Y_2\ ...\ Y_d]^T$ by $Y$. It has a mean vector $\mu = [\mu_1\ \mu_2\ ...\ \mu_d]^T$ and a covariance matrix $C$:

$$C = \begin{bmatrix} \sigma_1^2 & \delta_{12}\sigma_1\sigma_2 & \cdots & \delta_{1d}\sigma_1\sigma_d \\ & \sigma_2^2 & \cdots & \delta_{2d}\sigma_2\sigma_d \\ & & & \vdots \\ \text{symmetric} & & & \sigma_d^2 \end{bmatrix} \tag{1.53}$$

The PDF for the multivariate normal distribution is

$$f(y) = \frac{1}{\sqrt{2\pi}^d \cdot \sqrt{|C|}} \exp\left[ \frac{-(y-\mu)^T C^{-1}(y-\mu)}{2} \right] \tag{1.54}$$

If $[X_1\ X_2\ ...\ X_d]^T$ are multivariate standard normal, the multivariate standard normal distribution is simplified into

$$f(x) = \frac{1}{\sqrt{2\pi}^d \cdot \sqrt{|C|}} \exp\left( \frac{-x^T C^{-1} x}{2} \right) \tag{1.55}$$

where $C$ is the correlation matrix:

$$C = \begin{bmatrix} 1 & \delta_{12} & \cdots & \delta_{1d} \\ & 1 & \cdots & \delta_{2d} \\ & & & \vdots \\ \text{sym.} & & & 1 \end{bmatrix} \tag{1.56}$$

It is important to note that $C$ must satisfy a matrix property called positive definiteness. A positive-definite matrix is like a positive number. You can find the square root of a positive number. In a similar way, you can find the Cholesky factor of a positive-definite matrix. This requirement was not discussed in the bivariate case, because it is automatically satisfied when $\delta_{12}^2 < 1$. However, for the multivariate case, $C$ may not be positive-definite even if all $\delta_{ij}$ lie between $-1$ and $1$. This is a critical difference between the bivariate and the multivariate model.

## 1.4.3 Estimation of correlation matrix C

The correlation coefficients $\delta_{ij}$ in the matrix $C$ can be estimated using two methods:

a. Full multivariate manner based on a full multivariate dataset $(X_1, X_2, ..., X_d)$:

$$
\mathbf{C} \approx \begin{bmatrix} s_1^{-1} & & \\ & \ddots & \\ & & s_d^{-1} \end{bmatrix} \times \frac{1}{n-1} \sum_{k=1}^{n} \left( \begin{bmatrix} X_1^{(k)} - m_1 \\ \vdots \\ X_d^{(k)} - m_d \end{bmatrix} \times \begin{bmatrix} X_1^{(k)} - m_1 \\ \vdots \\ X_d^{(k)} - m_d \end{bmatrix}^{\mathrm{T}} \right) \times \begin{bmatrix} s_1^{-1} & & \\ & \ddots & \\ & & s_d^{-1} \end{bmatrix} \quad (1.57)
$$

where $m_i$ and $s_i$ are the sample mean and sample standard deviation for $X_i$. Note that the full multivariate dataset $(X_1, X_2, \ldots, X_d)$ is required for this method. This method guarantees that the resulting $\mathbf{C}$ is at least semipositive-definite. The issue of positive definiteness will be discussed later.

b. Entry-by-entry bivariate manner based on a bivariate dataset $(X_i, X_j)$:

$$
\delta_{ij} \approx \frac{1/(n_{ij}-1) \sum_{k=1}^{n_{ij}} \left( X_i^{(k)} - m_i \right) \cdot \left( X_j^{(k)} - m_j \right)}{\sqrt{1/(n_{ij}-1) \sum_{k=1}^{n_{ij}} \left( X_i^{(k)} - m_i \right)^2 \times 1/(n_{ij}-1) \sum_{k=1}^{n_{ij}} \left( X_j^{(k)} - m_j \right)^2}} \quad (1.58)
$$

where $n_{ij}$ is the number of the bivariate $(X_i, X_j)$ data points. The benefit of this method is that the full multivariate dataset $(X_1, X_2, \ldots, X_d)$ is not required. Only *all possible* bivariate datasets $(X_i, X_j)$ are needed.

To illustrate the first method, we simulate a full multivariate standard normal dataset $(X_1, X_2, \ldots, X_d)$ with $d = 3$ and sample size $n = 10$ using the procedure described in Section 1.4.4: the random seed is initiated by the MATLAB function randn('state', 13). Three columns of independent standard normal data are simulated using $\mathbf{Z} = \text{normrnd}(0, 1, n, 3)$. The upper triangle Cholesky matrix is computed as $\mathbf{U} = \text{chol}(\mathbf{C})$, where $\mathbf{C}$ is the correlation matrix of $(X_1, X_2, X_3)$. In the current case,

$$
\mathbf{C} = \begin{bmatrix} 1 & -0.57 & 0.59 \\ -0.57 & 1 & 0.05 \\ 0.59 & 0.05 & 1 \end{bmatrix} \quad (1.59)
$$

Then, three columns of X data are obtained from $\mathbf{X}^{\mathrm{T}} = \mathbf{Z}^{\mathrm{T}} \times \mathbf{U}$. The simulated data points are shown in Table 1.9. The data in Table 1.9 has full multivariate information because $(X_1, X_2, X_3)$ are simultaneously known for each case. The first method implements Equation 1.57 to estimate $\mathbf{C}$. The resulting $\mathbf{C}$ estimate is

$$
\mathbf{C} \approx \begin{bmatrix} 1 & -0.713 & 0.824 \\ -0.713 & 1 & -0.445 \\ 0.824 & -0.445 & 1 \end{bmatrix} \quad (1.60)
$$

The MATLAB function $\mathbf{C} = \text{corr}(\mathbf{X})$ ($\mathbf{X}$ is the $10 \times 3$ matrix shown in Table 1.9) will give the same result as above.

To illustrate the second method, we simulate three bivariate standard normal datasets for the pairs of $(X_1, X_2)$, $(X_1, X_3)$, and $(X_2, X_3)$ with $n_{12} = 10$, $n_{13} = 11$, and $n_{23} = 9$. Each pair is simulated independently of the other two pairs. Again, the random seed is initiated by the MATLAB function randn('state', 13) before simulation is carried out. We obtain two columns of independent standard normal data using $\mathbf{Z} = \text{normrnd}(0, 1, n_{12}, 2)$. The correlation

*Table 1.9* Simulated full multivariate X data with $d = 3$ and $n = 10$

| k | $X_1$ | $X_2$ | $X_3$ |
|---|---|---|---|
| 1 | 1.10 | 0.63 | 0.75 |
| 2 | 0.67 | −0.69 | 0.40 |
| 3 | 1.74 | 0.27 | 0.94 |
| 4 | 0.47 | 0.19 | −0.31 |
| 5 | 1.68 | −1.58 | 1.17 |
| 6 | −1.28 | 0.17 | −1.08 |
| 7 | −0.94 | 1.49 | −0.51 |
| 8 | 0.27 | 0.79 | 1.02 |
| 9 | −1.43 | 1.41 | −1.92 |
| 10 | −0.51 | 1.51 | 0.66 |

matrix is entered as $C = [1\ \delta_{12};\ \delta_{12}\ 1] = [1\ -0.570;\ -0.570\ 1]$. The upper triangle Cholesky matrix is computed as $U = \text{chol}(C)$ and two columns of correlated standard normal data of $(X_1, X_2)$ are obtained from $X^T = Z^T \times U$. The resulting dataset is called Bivariate dataset #1. This procedure is subsequently repeated for $n_{13} = 11$ and $C = [1\ \delta_{13};\ \delta_{13}\ 1]$, and for $n_{23} = 9$ and $C = [1\ \delta_{23};\ \delta_{23}\ 1]$. The resulting datasets are called Bivariate datasets #2 and #3. These three bivariate datasets are shown in Table 1.10. The data in Table 1.10 do not have full multivariate information because $(X_1, X_2, X_3)$ are not simultaneously known. For Bivariate dataset #1, only $(X_1, X_2)$ are simultaneously known. This mimics the reality in geotechnical literature when bivariate correlation datasets (e.g., simultaneously known OCR and $s_u$) are abundant, but full multivariate datasets are rare (e.g., simultaneously known OCR, $s_u$, and $S_t$). The second method applies Equation 1.58 to estimate (1) $\delta_{12}$ using Bivariate dataset #1, (2) $\delta_{13}$ using Bivariate dataset #2, and (3) $\delta_{23}$ using Bivariate dataset #3. The resulting C estimate is

$$C \approx \begin{bmatrix} 1 & -0.713 & 0.706 \\ -0.713 & 1 & -0.214 \\ 0.706 & -0.214 & 1 \end{bmatrix} \tag{1.61}$$

### 1.4.3.1 Positive definiteness of the correlation matrix C

Let us consider a case with $d = 3$: there are three random variables $X_1$, $X_2$, and $X_3$, and there are three correlation coefficients $\delta_{12}$, $\delta_{13}$, and $\delta_{23}$. It is not possible for $\delta_{12}$, $\delta_{13}$, and $\delta_{23}$ to take arbitrary values between −1 and 1. Consider the following correlation matrix:

$$C = \begin{bmatrix} 1 & 0.1 & 0.8 \\ 0.1 & 1 & 1.0 \\ 0.8 & 1.0 & 1 \end{bmatrix} \tag{1.62}$$

Note that $\delta_{12} = 0.1$ and $\delta_{13} = 0.8$. In other words, $X_1$ and $X_2$ are poorly correlated, whereas $X_1$ and $X_3$ are highly correlated. It is obvious that $\delta_{23} = 1.0$ is absurd. If $\delta_{23}$ were indeed 1.0, $X_2$ and $X_3$ would have been perfectly correlated, then $X_1$ should have been highly correlated to $X_2$. This contradicts the fact that $X_1$ and $X_2$ are poorly correlated. In fact, $\delta_{23}$ can only

*Table 1.10* Simulated bivariate X data with $n_{12} = 10$, $n_{13} = 11$, and $n_{23} = 9$

| k | Bivariate dataset #1 | | Bivariate dataset #2 | | Bivariate dataset #3 | |
|---|---|---|---|---|---|---|
| | $X_1$ | $X_2$ | $X_1$ | $X_3$ | $X_2$ | $X_3$ |
| 1 | 1.20 | −0.63 | −0.59 | −1.30 | −0.16 | 1.72 |
| 2 | 0.67 | −0.69 | 0.28 | 1.22 | −0.53 | −0.35 |
| 3 | 1.74 | 0.27 | −1.22 | −1.49 | 0.30 | −1.94 |
| 4 | −0.47 | 0.19 | 0.023 | −0.83 | −0.11 | −0.0038 |
| 5 | 1.68 | −1.58 | 0.81 | 2.08 | −1.87 | 0.020 |
| 6 | −1.28 | 0.17 | −0.0093 | −1.43 | 0.64 | −0.94 |
| 7 | −0.94 | 1.49 | −0.77 | −0.87 | −2.99 | 0.60 |
| 8 | 0.27 | 0.79 | 0.49 | 0.87 | −0.39 | −1.38 |
| 9 | −1.43 | 1.41 | −2.16 | −2.12 | 1.34 | 0.59 |
| 10 | −0.51 | 1.51 | 0.41 | −1.07 | | |
| 11 | | | −0.78 | 0.12 | | |

take values $<0.677$. This restriction is related to the concept of matrix positive definiteness. The eigenspectrum of **C** contains only positive values if and only if **C** is positive-definite. Namely, the **C** matrix in Equation 1.62 is not positive-definite. Indeed, it has a negative eigenvalue of −0.2089. Positive definiteness can be guaranteed only if the correlation matrix **C** is estimated from a full multivariate dataset $(X_1, X_2, ..., X_d)$ as shown in Table 1.9 (i.e., using Equation 1.57) and if $n \gg d$. The **C** matrix estimated using the entry-by-entry bivariate method in Equation 1.58 is not guaranteed to be positive-definite. Examples of producing nonpositive definite **C** based on actual data are shown in Section 1.7.3.

To illustrate the absurdity of the **C** matrix in Equation 1.62, consider a random variable $Y = X_1 + X_2 - X_3$. It is common practice to encounter this linear sum, usually in the context of a first-order Taylor series expansion of a nonlinear function. The variance of Y is equal to

$$\text{Var}(Y) = \sigma_1^2 + \sigma_2^2 + \sigma_3^2 + 2\delta_{12}\sigma_1\sigma_2 - 2\delta_{13}\sigma_1\sigma_3 - 2\delta_{23}\sigma_2\sigma_3 = 3 + 2\delta_{12} - 2\delta_{13} - 2\delta_{23} = -0.4$$

(1.63)

where $\sigma_i = 1$ is the standard deviation of $X_i$. Note that the variance of any random variable is positive by definition. The nonpositive definite **C** matrix in Equation 1.62 can produce a negative variance as shown in Equation 1.63. Hence, positive definiteness is not an academic concept that we can safely ignore in practice, notwithstanding the rather abstract nature of this concept.

### 1.4.3.2 Goodness-of-fit test

Multivariate normality requires separate checks. For example, if the scatter plot of $X_i$ versus $X_j$ shows a distinct nonlinear trend, then the multivariate normal distribution assumption is not suitable. There are numerous formal tests for multivariate normality in the statistics literature, but the state of practice is less established than formal tests for univariate normality (e.g., K–S test). The first method is the generalization of the line test in Section 1.3.3. This method is applicable to nonstandard multivariate normal distribution with an arbitrary dimension ($d$) and is based on the fact that the Mahalanobis distance $Q_d$ between a multivariate normal vector $(X_1, X_2, ..., X_d)$ and its mean vector follows the $\chi$-square distribution with $m$ DOF. The Mahalanobis distance is defined as

$$Q_d = (\mathbf{X} - \mathbf{\mu})^T \mathbf{C}^{-1} (\mathbf{X} - \mathbf{\mu}) = \begin{bmatrix} X_1 - \mu_1 \\ X_2 - \mu_2 \\ \vdots \\ X_d - \mu_d \end{bmatrix}^T \begin{bmatrix} \sigma_1^2 & \delta_{12}\sigma_1\sigma_2 & \cdots & \delta_{1d}\sigma_1\sigma_d \\ & \sigma_2^2 & & \delta_{2d}\sigma_2\sigma_d \\ & & \ddots & \vdots \\ & & & \sigma_d^2 \end{bmatrix}^{-1} \begin{bmatrix} X_1 - \mu_1 \\ X_2 - \mu_2 \\ \vdots \\ X_d - \mu_d \end{bmatrix} \quad (1.64)$$

where $\mu_i$ and $\sigma_i$ are the mean and standard deviation of $X_i$, and the matrix to be inverted is the covariance matrix. In reality, the mean $\mu_i$ is replaced by sample mean $m_i$, and the covariance matrix is replaced by the sample covariance matrix:

$$\hat{\mathbf{C}} = \frac{1}{n-1} \sum_{k=1}^{n} \left( \begin{bmatrix} X_1^{(k)} - m_1 \\ \vdots \\ X_d^{(k)} - m_d \end{bmatrix} \times \begin{bmatrix} X_1^{(k)} - m_1 \\ \vdots \\ X_d^{(k)} - m_d \end{bmatrix}^T \right) \quad (1.65)$$

For the dataset in Table 1.9, $(m_1, m_2, m_3) = (0.092, 0.29, 0.072)$, and the sample covariance matrix is

$$\hat{\mathbf{C}} = \begin{bmatrix} 1.42 & -0.88 & 0.99 \\ & 1.08 & -0.47 \\ \text{sym.} & & 1.02 \end{bmatrix} \quad (1.66)$$

Therefore, the Mahalanobis distance for the $k$th sample in Table 1.9 can be determined as

$$Q_d^{(k)} = \begin{bmatrix} X_1^{(k)} - 0.092 \\ X_2^{(k)} - 0.29 \\ X_3^{(k)} - 0.072 \end{bmatrix}^T \begin{bmatrix} 1.42 & -0.88 & 0.99 \\ & 1.08 & -0.47 \\ \text{sym.} & & 1.02 \end{bmatrix}^{-1} \begin{bmatrix} X_1^{(k)} - 0.092 \\ X_2^{(k)} - 0.29 \\ X_3^{(k)} - 0.072 \end{bmatrix} \quad (1.67)$$

Table 1.11 shows the calculated Mahalanobis distances for the dataset in Table 1.9.

In contrast to Equation 1.46, the CDF for $Q_d$ does not have an elegant analytical form. However, the following equation still holds:

$$F_{\chi_d^2}^{-1} \left[ F_{\chi_d^2}(q_d) \right] = q_d \quad (1.68)$$

where $F_{\chi_d^2}$ is the $\chi$-square CDF with $d$ DOFs. If we replace the $F_{\chi_d^2}$ inside the square bracket by the ECDF, we get

$$F_{\chi_d^2}^{-1} \left[ F_n(q_d) \right] \approx q_d \quad (1.69)$$

The ECDF can be estimated using Equations 1.11 or 1.12, whereas the $F_{\chi_d^2}^{-1}(p)$ can be evaluated using MATLAB function chi2inv($p$, $d$). If $Q_d$ is indeed $d$-DOF $\chi$-square, it is clear

Table 1.11 Mahalanobis distances for the dataset in Table 1.9

| $k$ | $X_1 - m_1$ | $X_2 - m_2$ | $X_3 - m_3$ | $Q_d$ |
|---|---|---|---|---|
| 1 | 1.11 | −0.92 | 0.28 | 1.62 |
| 2 | 0.58 | −0.98 | 0.33 | 1.00 |
| 3 | 1.65 | −0.026 | 0.87 | 4.87 |
| 4 | −0.57 | −0.11 | −0.38 | 0.69 |
| 5 | 1.59 | −1.87 | 1.09 | 3.42 |
| 6 | −1.37 | −0.12 | −1.15 | 3.14 |
| 7 | −1.03 | 1.20 | −0.59 | 1.33 |
| 8 | 0.18 | 0.50 | 0.95 | 2.23 |
| 9 | −1.52 | 1.12 | −1.99 | 5.02 |
| 10 | −0.60 | 1.21 | 0.59 | 3.67 |

from Equation 1.69 that a plot with $q_d$ values on the vertical axis and $F_{\chi_d^2}^{-1}[1 - F_n(q_d)]$ on the horizontal axis will produce a 1:1 line. As a result, if the actual $q_d$ versus $F_{\chi_d^2}^{-1}[1 - F_n(q_d)]$ relationship lies close to the 1:1 line, we can conclude that there is no strong evidence to reject the multivariate normal model.

Table 1.12 illustrates how the line test can be applied to simulated multivariate standard normal data in Table 1.9. The first column contains the simulated $Q_d$ data sorted in ascending order. The second column contains the rank. The third column computes the ECDF $F_n(q_d)$ using Equation 1.12. The fourth column is $F_{\chi_d^2}^{-1}[1 - F_n(q_d)]$, where $F_n(q_d)$ is the data in the third column. Finally, the probability plot is obtained by drawing the first column on the $y$-axis and the fourth column on the $x$-axis. Figure 1.15 shows the resulting $q_d$ versus $F_{\chi_d^2}^{-1}[1 - F_n(q_d)]$, relationship. It is indeed close to the 1:1 line, indicating the multivariate normal model is reasonable.

## 1.4.4 Simulation of multivariate standard normal random vector X

Given the correlation matrix C, random samples of multivariate standard normal $(X_1, X_2, ..., X_d)$ can be readily simulated by the following steps:

1. Simulate independent standard normal random vector $Z = [Z_1, Z_2, ..., Z_d]^T$.

Table 1.12 Sorted $Q_d$ data and the EDF

| Sorted $Q_d$ data | Rank $k$ | ECDF $F_n(q_d)$ from Equation 1.12 | $F_{\chi_d^2}^{-1}[1 - F_n(q_d)]$ |
|---|---|---|---|
| 0.69 | 1 | 0.067 | 0.44 |
| 1.00 | 2 | 0.163 | 0.85 |
| 1.33 | 3 | 0.260 | 1.25 |
| 1.62 | 4 | 0.356 | 1.67 |
| 2.23 | 5 | 0.452 | 2.12 |
| 3.14 | 6 | 0.548 | 2.63 |
| 3.42 | 7 | 0.644 | 3.24 |
| 3.67 | 8 | 0.740 | 4.02 |
| 4.87 | 9 | 0.837 | 5.12 |
| 5.02 | 10 | 0.933 | 7.15 |

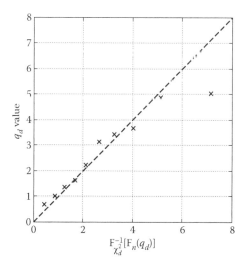

*Figure 1.15* Mahalanobis distance test: $q_d$ versus $F_{\chi_d^2}^{-1}[1 - F_n(q_d)]$ plot. The dashed line is the 1:1 line.

2. Let **U** matrix be the Cholesky decomposition of **C**:

$$\mathbf{U}^T \times \mathbf{U} = \mathbf{C} \tag{1.70}$$

In MATLAB, $\mathbf{U} = \text{chol}(\mathbf{C})$.

3. Let

$$\mathbf{X}^T = \mathbf{Z}^T \times \mathbf{U} \tag{1.71}$$

It is noteworthy that the above steps break down if **C** is not positive-definite, because the Cholesky decomposition **U** will contain complex numbers if **C** is not positive-definite. This is analogous to producing a complex number if you take the square root of a negative number.

### 1.4.5 Conditional normal and updating

Using a multivariate normal distribution, it is possible to update the marginal distribution of any one parameter or the multivariate distribution of any group of parameters given information from other parameters. This updated distribution is called the conditional distribution. To illustrate the conditioning, let us consider the following example with four soil parameters: $s_u$ (undrained shear strength), $\sigma'_p$ (preconsolidation stress), N (SPT-N value), and $q_t - \sigma_v$ (net cone tip resistance). These four parameters are correlated.

To begin with, let us assume they are normally distributed with mean values and COVs given in Table 1.13. As a result,

$$
\begin{aligned}
s_u &= \mu_1 + \sigma_1 \times X_1 \\
\sigma'_p &= \mu_2 + \sigma_2 \times X_2 \\
N &= \mu_3 + \sigma_3 \times X_3 \\
q_t - \sigma_v &= \mu_4 + \sigma_4 \times X_4
\end{aligned}
\tag{1.72}
$$

*Table 1.13* Mean, COV, and standard deviation for $s_u$, $\sigma'_p$ N, and $q_t - \sigma_v$

| Variable | | $\mu$ | COV | $\sigma$ |
|---|---|---|---|---|
| Undrained shear strength | $s_u$ | 200 kPa | 0.2 | 40 kPa |
| Preconsolidation stress | $\sigma'_p$ | 800 kPa | 0.2 | 160 kPa |
| SPT blowcount | SPT-N value | 20 | 0.3 | 6 |
| Net cone tip resistance | $q_t - \sigma_v$ | 2500 kPa | 0.3 | 750 kPa |

where $\mu$'s and $\sigma$'s are the mean values and standard deviations, respectively, listed in Table 1.13; $(X_1, X_2, X_3, X_4)$ are individually standard normal. The mean values in Table 1.13 are chosen such that $s_u/\sigma'_p \approx 0.25$, $(q_t - \sigma_v)/s_u \approx 12.5$, and $(\sigma'_p/P_a)/N \approx 0.4$ ($P_a = 101.3$ kPa is the atmosphere pressure). These values are typical (e.g., Kulhawy and Mayne 1990). Let us further assume the correlation matrix C for $(X_1, X_2, X_3, X_4)$ is

$$C = \begin{bmatrix} 1 & 0.9 & 0.5 & 0.7 \\ 0.9 & 1 & 0.4 & 0.6 \\ 0.5 & 0.4 & 1 & 0.4 \\ 0.7 & 0.6 & 0.4 & 1 \end{bmatrix} \tag{1.73}$$

Suppose site investigation yields the following information at a certain depth in a clay layer: N = 10 and $q_t - \sigma_v = 2000$ kPa. On the basis of this information, the purpose is to update the marginal distributions of $s_u$ and $\sigma'_p$ for the clay at the same depth. The unconditional distribution for $s_u$ is normal with mean = 200 kPa and COV = 0.2. The updated (conditional) distribution for $s_u$ is expected to be different.

The effect of updating (or conditioning) can be explained by simulated data. A large amount ($n = 2 \times 10^6$) of $(s_u, \sigma'_p, N, q_t - \sigma_v)$ data are simulated. This can be done by first simulating $Z = \text{normrnd}(0, 1, n, 4)$. Then, the Cholesky factor is computed as $U = \text{chol}(C)$. Finally, $X^T = Z^T \times U$. The first column in $X$ contains the $X_1$ samples; so, $s_u = \mu_1 + \sigma_1 \times X_1$ will yield $n = 2 \times 10^6$ samples of $s_u$. The same procedure will yield $n = 2 \times 10^6$ samples of $\sigma'_p$ N, and $q_t - \sigma_v$. These samples are plotted as light-gray crosses in Figure 1.16. It is clear that $(s_u, N)$ are positively correlated, and so are $(s_u, q_t - \sigma_v)$. These are the unconditional samples

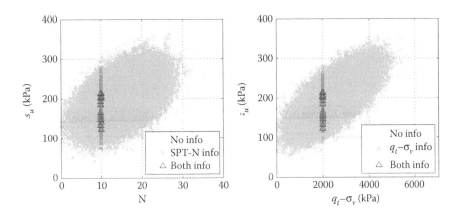

*Figure 1.16* Illustration of conditioning using simulated $(s_u, N, q_t - \sigma_v)$ samples.

(absence of other sources of information). Among the unconditional $(s_u, N)$ samples, there are some samples satisfying $N \approx 10$ $(9.9 < N < 10.1)$. These samples are marked as dark gray crosses in the left plot. Similarly, the $(s_u, q_t - \sigma_v)$ samples satisfying $q_t - \sigma_v \approx 2000$ kPa $(1990 < q_t - \sigma_v < 2010)$ are marked in dark gray in the right plot. Finally, the samples satisfying $N \approx 10$ and $q_t - \sigma_v \approx 2000$ kPa are marked as black triangles in both plots. These are the conditional samples of $s_u$ given the information $N \approx 10$ and $q_t - \sigma_v \approx 2000$ kPa.

Figure 1.17 shows the histograms of the unconditional and conditional $s_u$ samples. There are two distinct features: (a) the mean value of $s_u$ changes after the conditioning, and (b) the COV of $s_u$ is smaller after the conditioning. This illustrates the effect of conditioning: given the information that $N \approx 10$ and $q_t - \sigma_v \approx 2000$ kPa, the probability distribution can be updated. The updated (conditional) distribution has a mean value different from the unconditional mean and has a COV that is less than the unconditional COV. The reduction in COV deserves more attention: it means that the magnitude of uncertainty is reduced in the presence of additional sources of information. Engineers can appreciate this effect intuitively, but conditioning allows this effect to be quantified consistently. The effect of conditioning on the distribution of $\sigma_p'$ is similar.

The above example is intended to illustrate the effect of conditioning without mathematics. In fact, the conditional distribution of $(s_u, \sigma_p')$ can be obtained analytically using Bayesian analysis. The tedious simulation involving 2 million samples can be avoided. The information $N = 10$ and $q_t - \sigma_v = 2000$ kPa is equivalent to

$$
\begin{aligned}
X_3 &= \left(N - \mu_3\right)/\sigma_3 = \left(10 - 15\right)/6 = -1.11 \\
X_4 &= \left[\left(q_t - \sigma_v\right) - \mu_4\right]/\sigma_4 = \left(2000 - 2500\right)/750 = -0.67
\end{aligned}
\tag{1.74}
$$

Let $\mathbf{X}$ be partitioned in $\mathbf{X}^{[1]}$ and $\mathbf{X}^{[2]}$, where $\mathbf{X}^{[2]} = [X_3, X_4]^T = [-1.1\ -0.67]^T$ are known values, and $\mathbf{X}^{[1]} = [X_1, X_2]^T$ are unknown random variables:

$$
\mathbf{X} = \begin{bmatrix} \mathbf{X}^{[1]} \\ \mathbf{X}^{[2]} \end{bmatrix}
\tag{1.75}
$$

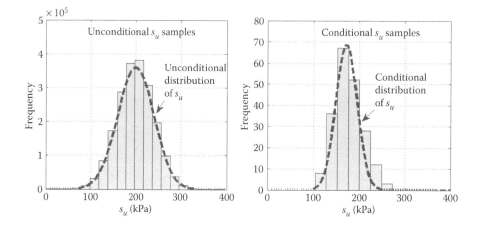

Figure 1.17 Unconditional and conditional PDFs of $s_u$.

The correlation matrix C can be partitioned as

$$C = \begin{bmatrix} 1 & 0.9 & | & 0.5 & 0.7 \\ 0.9 & 1 & | & 0.4 & 0.6 \\ \hline 0.5 & 0.4 & | & 1 & 0.4 \\ 0.7 & 0.6 & | & 0.4 & 1 \end{bmatrix} = \begin{bmatrix} C^{[11]} & C^{[12]} \\ C^{[21]} & C^{[22]} \end{bmatrix} \tag{1.76}$$

It can be proved that the conditional distribution of $X^{[1]}$ given $X^{[2]}$ is a multivariate normal distribution with the following updated mean vector and covariance matrix:

$$\mu_{update}^{[1]} = C^{[12]} \times \left(C^{[22]}\right)^{-1} \times X^{[2]} = \begin{bmatrix} 0.5 & 0.7 \\ 0.4 & 0.6 \end{bmatrix} \times \begin{bmatrix} 1 & 0.4 \\ 0.4 & 1 \end{bmatrix}^{-1} \times \begin{bmatrix} -1.111 \\ -0.667 \end{bmatrix} = \begin{bmatrix} -0.69 \\ -0.56 \end{bmatrix}$$

$$C_{update}^{[11]} = C^{[11]} - C^{[12]} \times \left(C^{[22]}\right)^{-1} \times C^{[21]} \tag{1.77}$$

$$= \begin{bmatrix} 1 & 0.9 \\ 0.9 & 1 \end{bmatrix} - \begin{bmatrix} 0.5 & 0.7 \\ 0.4 & 0.6 \end{bmatrix} \times \begin{bmatrix} 1 & 0.4 \\ 0.4 & 1 \end{bmatrix}^{-1} \times \begin{bmatrix} 0.5 & 0.4 \\ 0.6 & 0.7 \end{bmatrix} = \begin{bmatrix} 0.45 & 0.44 \\ 0.44 & 0.61 \end{bmatrix}$$

The conditional mean value for $X_1$ is $-0.69$ and the conditional variance is $0.45$. The conditional mean and variance of $s_u$ can be calculated using the relationship $s_u = 200 + 40X_1$: the conditional mean value for $s_u = 200 + 40 \times (-0.69) = 172.49$ kPa, and the conditional variance for $s_u = 40^2 \times 0.45 = 723.81$ kPa². The conditional COV is therefore $723.81^{0.5}/172.49 = 0.16$. This conditional distribution for $s_u$ is plotted on the right plot in Figure 1.17, showing a good agreement with the histogram of the conditional $s_u$ samples. The conditional mean, variance, and COV for $\sigma'_p$ can also be calculated in a similar way.

The conditional bivariate distribution for $(s_u, \sigma'_p)$ can also be obtained. Recall that $(s_u, \sigma'_p)$ are related to $(X_1, X_2)$ through the following equation:

$$\begin{bmatrix} s_u \\ \sigma'_p \end{bmatrix} = \begin{bmatrix} 200 \\ 2500 \end{bmatrix} + \begin{bmatrix} 40 & 0 \\ 0 & 750 \end{bmatrix} \times \begin{bmatrix} X_1 \\ X_2 \end{bmatrix} \tag{1.78}$$

The conditional mean vector for $(s_u, \sigma'_p)$ is therefore

$$\mu_{update} = \begin{bmatrix} \mu_{s_u,update} \\ \mu_{\sigma'_p,update} \end{bmatrix} = \begin{bmatrix} 200 \\ 2500 \end{bmatrix} + \begin{bmatrix} 40 & 0 \\ 0 & 750 \end{bmatrix} \times \mu_{update}^{[1]} = \begin{bmatrix} 172.48 \\ 710.26 \end{bmatrix} \tag{1.79}$$

The conditional covariance matrix for $(s_u, \sigma'_p)$ is therefore

$$C_{update} = \begin{bmatrix} 40 & 0 \\ 0 & 750 \end{bmatrix} \times C_{update}^{[11]} \times \begin{bmatrix} 40 & 0 \\ 0 & 750 \end{bmatrix} = \begin{bmatrix} 723.81 & 2803.81 \\ 2803.81 & 15603.81 \end{bmatrix} \tag{1.80}$$

This conditional bivariate distribution can be evaluated using Equation 1.36. Its contours are plotted in Figure 1.18. The left plot shows the contours for the unconditional PDF for $(s_u, \sigma'_p)$, whereas the right plot shows the contours for the conditional PDF for $(s_u, \sigma'_p)$. Shown together with the contours are the unconditional samples of $(s_u, \sigma'_p)$ (left plot) and the conditional samples of $(s_u, \sigma'_p)$ (right plot).

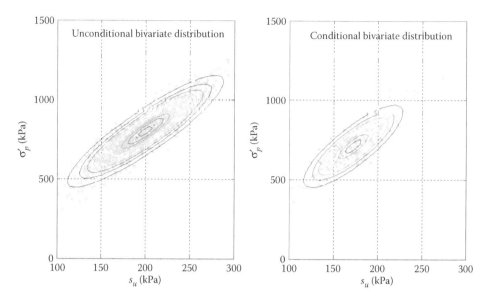

*Figure 1.18* Contours for the unconditional and conditional bivariate PDF for $(s_u, \sigma'_p)$.

## 1.5 NON-NORMAL RANDOM VARIABLE

### 1.5.1 Non-normal data

Most engineers refer to "non-normal" data as a single column of data (univariate data) that do not exhibit the classical bell-shaped histogram. We have demonstrated in Figure 1.5 that even simulated normal data will not produce a bell shape when the sample size is small. Hence, one should not be hasty to conclude that a column of data is not normally distributed based on visual inspection of the histogram alone. We have emphasized that the normal distribution is very useful, because it can be fitted by two simple parameters (mean and standard deviation) and it can be extended to a multivariate distribution with reasonable data requirement in the form of a correlation matrix. In view of its practicality and analytical tractability, one should abandon only the normal assumption in the presence of strong evidence.

There are two different sources of evidence that may compel one to reject the normal hypothesis. One is statistical evidence based on goodness-of-fit test as discussed in Section 1.2.3. We note in passing that goodness-of-fit tests are applied to the entire data set. In actuality, we are more interested in deviations from normality at the probability tails. The reason is that it is below average strength that will lead to failure. The typical acceptable probability of failure of civil engineering infrastructures is one in a thousand. If strength is the only random variable governing the probability of failure, then we are interested only in values that are approximately three standard deviations below the mean strength. We do not discuss this nonnormality in the probability tails below, because it is usually difficult to gather sufficient data to characterize this tail behavior with sufficient confidence.

The second reason for rejecting normality is violation of physics. A normal random variable is unbounded, that is, it can take all values on the real line. A soil parameter such as the undrained shear strength $(s_u)$ can only take positive values. The OCR is larger than 1 by definition. The ratio between the interface friction angle and the soil friction angle is always <1. A bounded random variable is nonnormal. This source of nonnormality can affect the probability of realizing small values significantly (behavior of lower probability tail). For example, if $s_u$ is normally distributed with a mean = 100 kPa and a standard deviation = 40 kPa, the probability of realizing values lower than 25 kPa is 0.03. If $s_u$ is log-normally distributed with the same mean and standard deviation, the probability is 0.0003—two orders of magnitude lower than the corresponding probability under the normal assumption. The moral of the story here is that it is important to include a lower bound in a probability model when it exists. It is not recommended to retain the normal assumption out of theoretical simplicity or expediency. A less clear-cut situation arises when one appeals to engineering judgment or experience to impose a lower bound. For example, one can venture to stake a lower bound of 10 kPa for $s_u$. The presence of very soft clay (classified by Terzaghi et al. [1996] as $s_u < 12.5$ kPa) is easy to detect, because it cannot be easily retained in a conventional sampler. Nonetheless, there is a possibility for small lenses of very soft clay to exist, because site investigation is typically too limited to eliminate this possibility with 100% confidence. One can appeal to engineering judgment say based on extensive knowledge of sites with similar geology elsewhere, but the reason for adopting $s_u > 10$ kPa is admittedly weaker than $s_u > 0$ kPa.

It is useful to briefly review the concepts of censored and truncated data that are distinct from the concept of bounded nonnormal random variables. The SPT blow count value (N-value) lies between 1 and 100. The lower bound is the minimum blow count while the upper bound is imposed to avoid damage to the sampler. When an SPT-N value of 100 is recorded, we know that the actual value is *at least* 100. This type of "censored data" is commonly produced by field tests, because there are physical limits to the strength that can be measured without damaging or without exceeding the sensitivity of the test equipment. Another example is a proof load test that is meant to assure an engineer that the capacity of a pile, soil nail, or ground anchor is larger than a certain value. This type of "censored data" appears naturally when the intention is to avoid testing a component to ultimate failure.

In an SPT record, we know the number of data points with a value of 100. There are examples where the number of data points exceeding a threshold is unknown. For example, a contractor provides a record of pile head settlement under working load. The reported settlements are <25 mm. The contractor did not record the number of piles with pile head settlement exceeding 25 mm, because these piles are deemed defective and new piles are automatically driven nearby to compensate for these defective piles. We do not know the number of piles settling more than 25 mm under working load. The SPT record is called right-censored data. The pile settlement record is called right-truncated data.

It is clear from the above brief digression that nonnormality is only one aspect of statistical characterization of actual data. We focus on characterizing histograms using nonnormal probability models below. Many nonnormal models can be found in standard statistical texts, but only the Johnson system of distributions is reviewed below. The reason is that the Johnson system of distributions can be transformed into a standard normal distribution using closed-form equations. This computational advantage is quite significant, because the only practical method to model multivariate non-normal data is to transform each component *individually* into standard normal data and to link these transformed standard normal components using a multivariate normal distribution. We have highlighted the usefulness of multivariate models compared to univariate models. We have noted that a multivariate model is a more natural fit to geotechnical engineering data, because a number of laboratory and field tests are typically conducted in a site investigation program. With these observations in mind, it is obvious that fitting each parameter to a probability distribution is only an intermediate goal and cannot be conducted without regard to the more stringent theoretical and practical constraints imposed by the multivariate model, which is the *final goal* in the characterization of geotechnical engineering data.

### 1.5.2 Non-normal distribution

It is quite unlikely for real data points to follow the normal distribution. As explained above, this is partly because most soil parameters are non-negative (e.g., soil shear strengths, effective stresses, moduli, etc.). Consider the normalized preconsolidation stress ($\sigma'_p/P_a$; $P_a$ is one atmosphere pressure) data points in the Clay/10/7490 database (Ching and Phoon 2014a). This database contains 2028 data points for the normalized preconsolidation stress. Its histogram is shown in the left plot in Figure 1.19. The histogram obviously does not resemble a normal distribution. In fact, even when the natural logarithm is applied, the histogram of $\ln(\sigma'_p/P_a)$ still exhibits a certain degree of asymmetry, which is not consistent with the normal model. In this example, there are sufficient data points for one to suspect that these departures from normality are not caused by statistical uncertainties. We cover the Johnson system of distributions below, starting with the lognormal distribution, which is the most well-known member in the geotechnical engineering literature.

#### 1.5.2.1 Lognormal and shifted lognormal distributions

Let Y be a nonnegative soil parameter. It is clear that Y cannot be normal. The simplest distribution model for Y is the lognormal distribution. If Y is lognormal, $\ln(Y)$ is normal with mean $= \lambda$ and standard deviation $= \xi$:

$$\lambda = \ln\left(\frac{\mu}{\sqrt{1 + \text{COV}^2}}\right) \quad \xi = \sqrt{\ln(1 + \text{COV}^2)} \tag{1.81}$$

where $\mu$ and COV are the mean value and COV of Y. As a result, the relationship between the standard normal X and Y is as follows:

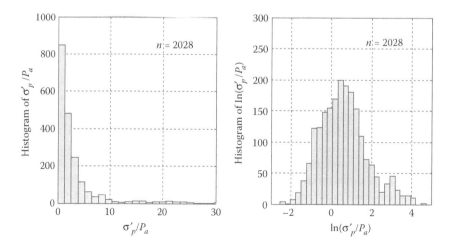

*Figure 1.19* Histograms for $\sigma'_p/P_a$ and $\ln(\sigma'_p/P_a)$.

$$\lambda + \xi \cdot X = \ln(Y) \quad \text{or} \quad Y = \exp(\lambda + \xi \cdot X) \tag{1.82}$$

It is clear that Y cannot be negative, because the smallest possible value of the exponential function is 0. It can be shown that the PDF of Y can be written as

$$f(y) = \frac{1}{\sqrt{2\pi} \cdot \xi \cdot y} \exp\left(\frac{-[\ln(y) - \lambda]^2}{2 \cdot \xi^2}\right) \quad y > 0 \tag{1.83}$$

The lognormal distribution has zero as its lower bound. The shifted lognormal distribution generalizes the lognormal distribution to account for nonzero lower bounds. If Y is shifted lognormally, the relationship between X and Y is

$$\frac{X - b_X^*}{a_X} = \ln(Y - b_Y) \tag{1.84}$$

where $a_X$, $b_X^*$, and $b_Y$ are the parameters for the shifted lognormal distribution. The parameter $b_Y$ is the lower bound of Y and it is typically determined by physics. The remaining parameters can be determined by the method of moments:

$$\frac{1}{a_X^2} = \ln\left[1 + \frac{\mu^2 \cdot COV^2}{(\mu - b_Y)^2}\right] \quad b_X^* = \frac{1}{2a_X} - a_X \ln(\mu - b_Y) \tag{1.85}$$

It is clear that when $b_Y = 0$, the shifted lognormal distribution reduces to lognormal distribution with $a_X = 1/\xi$ and $b_X^* = -\lambda/\xi$. The notation in Equation 1.84 are chosen to be different from those in Equation 1.82 to accommodate the other members of the Johnson system of distributions.

### 1.5.2.2 Johnson system of distributions

Phoon (2008) and Phoon and Ching (2013) highlighted that the shifted lognormal distribution is a member of a more general Johnson system, which can be expressed in the following form following the notations presented by Slifker and Shapiro (1980):

$$\frac{X - b_X}{a_X} = \kappa\left(\frac{Y - b_Y}{a_Y}\right) = \kappa(Y_n) \tag{1.86}$$

where $Y_n = (Y - b_Y)/a_Y$ is the normalized Y. The SU member is unbounded and is defined by

$$\kappa(Y_n) = \sinh^{-1}(Y_n) = \ln\left(Y_n + \sqrt{1 + Y_n^2}\right) \tag{1.87}$$

The SB member is bounded between $[b_Y, a_Y + b_Y]$ and is defined by

$$\kappa(Y_n) = \ln\left(\frac{Y_n}{1 - Y_n}\right) \tag{1.88}$$

The SL member (which is the shifted lognormal member) is bounded from *below* by $b_Y$ and is defined by

$$\kappa(Y_n) = \ln(Y_n) \tag{1.89}$$

Clearly, Equations 1.84 and 1.89 are identical if $b_X^* = b_X - a_X \ln(a_Y)$. Their PDFs are as follows:

$$f(y) = \begin{cases} a_X/a_Y \cdot \exp\left(-0.5\left[b_X + a_X \sinh^{-1}(y_n)\right]^2\right) / \sqrt{2\pi(1 + y_n^2)} & \text{for SU} \\ a_X/a_Y \cdot \exp\left(-0.5\left(b_X + a_X \ln\left[y_n/(1 - y_n)\right]\right)^2\right) / \left[\sqrt{2\pi} \cdot y_n(1 - y_n)\right] & \text{for SB} \\ a_X \cdot \exp\left(-0.5\left[b_X^* + a_X \ln(y - b_y)\right]^2\right) / \left[\sqrt{2\pi} \cdot (y - b_y)\right] & \text{for SL} \end{cases} \tag{1.90}$$

Figure 1.20 shows some distributions in the Johnson system—the Johnson system can generate distributions with a wide range of shapes. For each of the SU, SB, and SL distributions, a baseline case is plotted. Then, the effect of each of the parameters ($a_X$, $b_X$, $a_Y$, $b_Y$) is shown in the figure.

### 1.5.3 Selection and parameter estimation for the Johnson distribution

Slifker and Shapiro (1980) proposed an elegant selection and parameter estimation approach for the Johnson distribution using percentiles:

1. Choose a number $z > 0$. We assume $z = 0.7$ in this chapter, as recommended in Slifker and Shapiro (1980).
2. Compute the percentiles corresponding to $-3z$, $-z$, $z$, $3z$ using the standard CDF. Hence, for $z = 0.7$, the four percentiles are $p_a = \Phi(-2.1) = 0.018$, $p_b = \Phi(-0.7) = 0.242$, $p_c = \Phi(0.7) = 0.758$, and $p_d = \Phi(2.1) = 0.982$. A useful rule of thumb is to ensure that the sample size is larger than $10/p_a$. If this rule is not satisfied, the value of $z$ should be reduced. For $z = 0.7$, $10/p_a = 556$, and hence, a sample size of about 550 is required for this choice of $z$. The typical sample size for geotechnical engineering data is in the order of 100 or less. Further research is needed to look into fitting data from more realistic sample sizes to the Johnson system.
3. Compute the values of Y corresponding to these four percentiles. Formally, $y_a = \mathrm{F}^{-1}(p_a)$, $y_b = \mathrm{F}^{-1}(p_b)$, $y_c = \mathrm{F}^{-1}(p_c)$, and $y_d = \mathrm{F}^{-1}(p_d)$. However, this is not practical because the CDF of Y, F(y), is unknown. Fortunately, $y_a$, $y_b$, $y_c$, and $y_d$ can be obtained directly from data without the knowledge of F(y) using sample percentiles: $y_i = p_i$ sample percentile. In MATLAB command, $y_i = \mathrm{prctile}(\mathbf{Y}, 100 * p_i)$, where the $\mathbf{Y}$ vector contains all Y data points.
4. Three parameters are computed from $y_a$, $y_b$, $y_c$, and $y_d$: $m = y_d - y_c$, $n = y_b - y_a$, and $p = y_c - y_b$.
5. Finally, identify the Johnson member as SU if $mn/p^2 > 1$, SB if $mn/p^2 < 1$, and SL if $mn/p^2 = 1$.

Once the distribution type has been identified (SU, SB, or SL), the distribution parameters can be computed as follows:

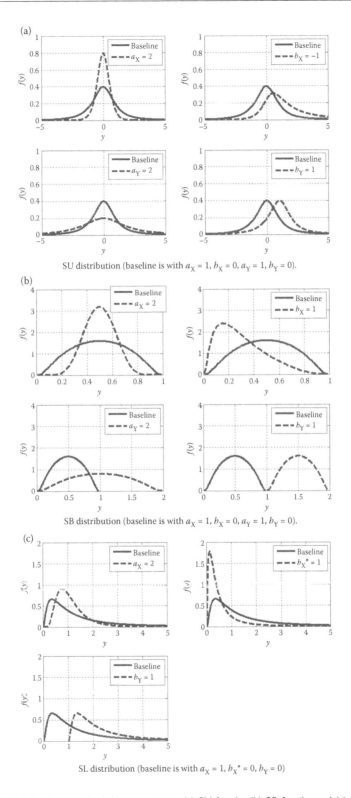

Figure 1.20 Some distributions in the Johnson system: (a) SU family; (b) SB family; and (c) SL family.

For SU,

$$
\begin{aligned}
a_X &= 2z/\cosh^{-1}[0.5(m/p + n/p)] \qquad\qquad a_X > 0\\
b_X &= a_X \sinh^{-1}\{(n/p - m/p)/[2(D - 1)^{0.5}]\}\\
a_Y &= 2p(D - 1)^{0.5}/[(m/p + n/p - 2)(m/p + n/p + 2)^{0.5}] \quad a_Y > 0\\
b_Y &= (y_c + y_b)/2 + p(n/p - m/p)/[2(m/p + n/p - 2)]
\end{aligned}
\tag{1.91}
$$

For SB,

$$
\begin{aligned}
a_X &= z/\cosh^{-1}\{0.5[(1 + p/m)(1 + p/n)]^{0.5}\} \qquad\qquad a_X > 0\\
b_X &= a_X \sinh^{-1}\{(p/n - p/m)[(1 + p/m)(1 + p/n) - 4]^{0.5}/[2(D^{-1} - 1)]\}\\
a_Y &= p\{[(1 + p/m)(1 + p/n) - 2]^2 - 4\}^{0.5}/(D^{-1} - 1) \qquad a_Y > 0\\
b_Y &= (y_c + y_b)/2 - a_y/2 + p(p/n - p/m)/[2(D^{-1} - 1)]
\end{aligned}
\tag{1.92}
$$

For SL,

$$
\begin{aligned}
a_X &= 2z/\ln(m/p)\\
b_X^* &= a_X \ln\{(m/p - 1)/[p(m/p)^{0.5}]\}\\
b_Y &= (y_c + y_b)/2 - 0.5p(m/p + 1)/(m/p - 1)
\end{aligned}
\tag{1.93}
$$

in which $D = mn/p^2$.

As an example, consider the Johnson SU distribution with $a_X = 1$, $b_X = -1$, $a_Y = 1$, and $b_Y = 0$. By initiating the random at randn('state', 13), random samples of Y can be simulated using the procedure introduced in Section 1.5.4. The sample size $n = 1000$. The histogram of the simulated Y data is shown in Figure 1.21a, together with the underlying PDF. The ECDF can be computed by Equation 1.12 and is shown in Figure 1.21b. The percentiles $(y_a, y_b, y_c, y_d)$ can be readily identified using MATLAB command $y_i = \text{prctile}(Y, 100 * p_i)$. Graphically, $y_a$ is the location on the horizontal axis such that the ECDF is equal to $p_a$. The resulting $(y_a, y_b, y_c, y_d)$ are (−1.352, 0.285, 2.633, 10.773). As a result, $m = y_d - y_c = 8.140$, $n = y_b - y_a = 1.637$, $p = y_c - y_b = 2.347$, and $mn/p^2 = 2.419$. In this example, the SU family is correctly identified from the data. In addition, the SU parameters can be identified using Equation 1.91: $a_X = 1.027$, $b_X = -1.019$, $a_Y = 1.040$, and $b_Y = -0.042$. These values are reasonably close to the actual values $a_X = 1$, $b_X = -1$, $a_Y = 1$, and $b_Y = 0$.

### 1.5.3.1 Probability plot and the goodness-of-fit test (K–S test)

#### 1.5.3.1.1 Converting a Johnson random variable into standard normal

Given the simulated Y data, it is desirable to construct a probability plot similar to Figure 1.9 to check whether the Johnson SU distribution fits well. However, it is more convenient to plot the "normal" probability plot for the X data converted from the Y data. In general, this conversion has the following form:

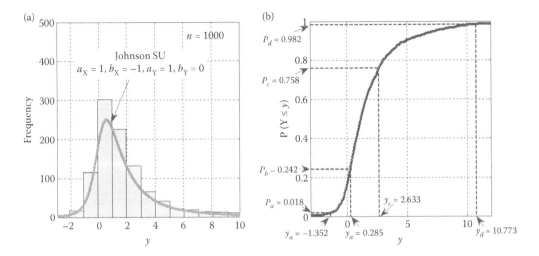

*Figure 1.21* (a) Histograms for the simulated Y; (b) ECDF of Y, and the four percentiles.

$$X = \Phi^{-1}\left[F_Y\left(Y\right)\right] \tag{1.94}$$

where $F_Y$ is the CDF of Y. With the family type chosen and the parameters $(a_X, b_X, a_Y, b_Y)$ identified, this conversion has the following convenient analytical form:

$$X = \begin{cases} b_X + a_X \times \ln\left(\dfrac{Y - b_Y}{a_Y} + \sqrt{1 + \left(\dfrac{Y - b_Y}{a_Y}\right)^2}\right) & \text{SU} \\[3mm] b_X + a_X \times \ln\left[\dfrac{\left(Y - b_Y\right)/a_Y}{1 - \left(Y - b_Y\right)/a_Y}\right] & \text{SB} \\[3mm] b_X^* + a_X \times \ln\left(Y - b_Y\right) & \text{SL} \end{cases} \tag{1.95}$$

Figure 1.22a shows the histogram of the X data that was converted from the aforementioned simulated Y data using Equation 1.95. The conversion is based on the model parameters: $a_X = 1.027$, $b_X = -1.019$, $a_Y = 1.040$, and $b_Y = -0.042$. The standard normal PDF is also plotted for comparison. Visually, the X converted from the Y data seems to fit a standard normal model reasonably well.

### 1.5.3.1.2 Normal probability plot and K–S test

The normal probability plot for the converted X data can be obtained using the procedure shown in Figure 1.9. Figure 1.22b shows the normal probability plots for the converted X data. The p-value for the K–S test can be computed by using MATLAB function $[h, p] = \text{kstest}(X)$. Note that **X** must be taken as the input to this function (not **Y**), because the

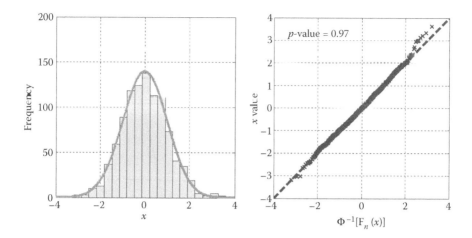

Figure 1.22 (a) Histogram for the X data converted from the simulated Y data; (b) x versus $\Phi^{-1}[F(x)]$ plot for the X data.

MATLAB kstest function assumes the target distribution is standard normal. The resulting $p$-value is shown in Figure 1.22b. It is clear that the X data converted from the Y data has a $p$-value larger than 0.05 ($p$-value = 0.97). Hence, the standard normal distribution hypothesis cannot be rejected at a significant level of 0.05. Namely, the Johnson SU hypothesis for the Y data cannot be rejected.

### 1.5.4 Simulation of the Johnson random variable

Given the family type (SU, SB, or SL) and the four parameters ($a_X$, $b_X$, $a_Y$, $b_Y$), random samples of the Johnson random variable Y can be simulated by the following steps:

1. Simulate a standard normal random variable X.
2. Simulate the Johnson random variable Y using the inverse transform of Equation 1.94:

$$Y = F_Y^{-1}\left[\Phi(X)\right] \tag{1.96}$$

With the family type chosen and the parameters ($a_X$, $b_X$, $a_Y$, $b_Y$) identified, this transformation has the following analytical form:

$$Y = \begin{cases} b_Y + a_Y \times \sinh\left(\dfrac{X - b_X}{a_X}\right) & \text{SU} \\[2ex] \dfrac{b_Y + (a_Y + b_Y) \times \exp\left[(X - b_X)/a_X\right]}{1 + \exp\left[(X - b_X)/a_X\right]} & \text{SB} \\[2ex] b_Y + \exp\left(\dfrac{X - b_X^*}{a_X}\right) & \text{SL} \end{cases} \tag{1.97}$$

### 1.5.5 Some practical observations

#### 1.5.5.1 Choice of z

Previously, we have chosen $z = 0.7$, which is recommended in Slifker and Shapiro (1980). It turns out that this choice is robust. Figure 1.23 shows the variation (with respect to $z$) of the identified distribution type and $(a_X, b_X, a_Y, b_Y)$ for *one realization* of the simulated Y data. For plotting on a numerical scale, SU is indexed as "1," SB is indexed as "2," and SL is indexed as "3." The values for the actual Y distribution (namely, type $= 1$, $a_X = 1$, $b_X = -1$, $a_Y = 1$, and $b_Y = 0$) are plotted as dashed lines. The effect of sample size is illustrated in the subplots using $n = 30$, 100, and 1000. When $n = 30$, the identified type is incorrect regardless of the choice of $z$. This false identification is less severe when $n = 100$. However, we still see few false identifications when $z$ is near 0.4. There is no false identification when $n = 1000$. The identified $(a_X, b_X, a_Y, b_Y)$ for $n = 30$ are very different from the actual values because the type has been identified wrongly. The effect of these parameters on the PDF is dependent on the probability model (Figure 1.20). The identified $(a_X, b_X, a_Y, b_Y)$ for $n = 100$ are close to the actual values for $z = 0.5$–0.8. The identified $(a_X, b_X, a_Y, b_Y)$ for $n = 1000$ are close to the actual values for a broad range of $z$. In general, $z = 0.7$ seems to be a robust choice for $n = 100$ and 1000. Note that the subplot for $n = 30$ and possibly $n = 100$ will change from realization to realization due to statistical uncertainty.

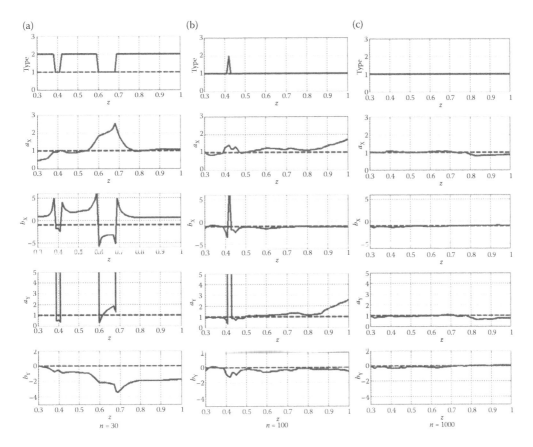

*Figure 1.23* Variations of the identified type and $(a_X, b_X, a_Y, b_Y)$ for *one realization* of the simulated Y data with (a) $n = 30$; (b) $n = 100$; and (c) $n = 1000$.

### 1.5.5.2 Parameter estimation under prescribed lower and/or upper bound

There are soil parameters with clear lower and/or upper bounds. Many parameters cannot be negative. It is very rare for the undrained shear strength of a clay to be <10 kPa. By definition, the OCR and sensitivity ($S_t$) are bounded from below by 1. The Poisson ratio is bounded between 0 and 0.5. The percentile procedure described in Section 1.5.3 cannot accommodate these bounds directly. This fact is demonstrated by the $S_t$ data points in the Clay/10/7490 database (Ching and Phoon 2014a). The number of data points $n = 1591$. The histogram of this $Y = \ln(S_t)$ dataset is shown in Figure 1.24a. There is a physical lower bound of zero, because $S_t > 1$. However, the percentile procedure gives the SU type, which is unbounded from below. This fitted SU distribution is plotted in Figure 1.24a. It can be seen that the PDF is not zero for $y < 0$. To reinforce the physical lower bound at $y = 0$, one method is to bypass the percentile procedure and choose the SL distribution (SL is bounded from below). The lower bound of Y (namely $b_Y$) is prescribed to be $-0.01$ (not 0). This is to avoid the numerical problem that may occur when we have data points with $S_t = 1$ (if $b_Y = 0$, $S_t = 1$ implies that $Y = 0$ and that $\ln(Y + b_Y) = \ln(0) = $ negative infinity). The method of maximum likelihood is used to estimate $a_X$ and $b_X^*$ With $b_y = -0.01$ being prescribed, the PDF for the SL distribution is (see Equation 1.90)

$$f(y) = a_X \cdot \exp\left(-0.5\left[b_X^* + a_X \ln(y + 0.01)\right]^2\right)\Big/\left[\sqrt{2\pi} \cdot (y + 0.01)\right] \tag{1.98}$$

The maximum likelihood estimates for $a_X$ and $b_X^*$ can be found by maximizing the logarithm of the likelihood function:

$$\ln\left[f\left(Y^{(1)}, Y^{(2)}, \ldots, Y^{(n)} \mid a_X, b_X^*\right)\right]$$
$$= \sum_{k=1}^{n}\left(\ln(a_X) - \ln(Y^{(k)} + 0.01) - 0.5\left[b_X^* + a_X \ln(Y^{(k)} + 0.01)\right]^2\right) \tag{1.99}$$

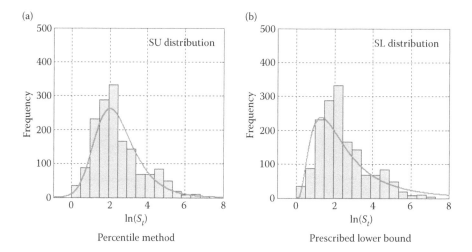

*Figure 1.24* Histograms of $\ln(S_t)$ and the identified Johnson distributions using (a) the percentile method and (b) the maximum likelihood with a prescribed lower bound $= -0.01$.

The resulting best estimates are $a_X = 1.42$ and $b_X^* = -1.08$. The fitted SL distribution is plotted in Figure 1.24b for comparison. Currently, there is no fully satisfactory method for dealing with bounds *and* retaining the simplicity of the Johnson system for multivariate nonnormal analysis discussed in the next section.

## 1.6 MULTIVARIATE NON-NORMAL RANDOM VECTOR

### 1.6.1 Multivariate non-normal data

Section 1.4 presents the treatment of multivariate *normal* data. Section 1.5 presents the treatment of *univariate* non-normal data. We are now in a position to deal with *multivariate non-normal* data, which are commonly encountered in geotechnical engineering. In this section, we will only present a multivariate non-normal probability model that is constructed from the multivariate normal probability model. This construction method involves transforming each normal component into a non-normal component *in isolation of other components* (also called a "memoryless" approach in the random process literature). It will be apparent to the reader after reading Section 1.6 that this construction method is chosen to accommodate data limitations. The key data limitation in practice is sample size. The size of a typical dataset in geotechnical engineering is only sufficient to estimate the marginal distributions underlying each component (univariate information) and the correlation matrix (bivariate information on all possible pairs of components). It will be illustrated in Section 1.7 that actual data limitations could be even more severe. We may not have a complete multivariate dataset to estimate every entry in the correlation matrix. Section 1.7 presents a pragmatic approach to fill in this data gap.

In summary, the key constraint in the construction of a multivariate non-normal probability model is insufficient data. Because of insufficient data, it is *theoretically* not possible to establish a multivariate non-normal model *uniquely*. The construction method presented in this section is able to fit available data (in the form of marginal distributions and a correlation matrix), but it is by no means the only available method. For example, one could replace the backbone multivariate normal distribution by a multivariate *t* distribution (Arslan 2004). You could adopt other methods based on completely different principles, such as the skew-normal distribution (Azzalini and Capitanio 2003), but most methods are difficult to apply in practice. As an engineer, it is important to appreciate two practical constraints that are intrinsic to the construction method based on a multivariate normal distribution. First, it is possible that the multivariate non-normal probability model constructed using this method cannot fit an actual dataset. For example, the correlation matrix for the physical variables cannot be reproduced by a valid correlation matrix for the backbone multivariate normal distribution (Arwade 2005). Another example is the rank (or Spearman) correlation matrix for the physical variables that is significantly different from the product–moment (or Pearson) correlation matrix for the backbone multivariate normal distribution. Second, this model cannot accommodate more data. For example, if one has sufficient data at hand to estimate both marginal and bivariate distributions or higher order moments beyond correlation coefficients, it is not possible to introduce these additional sources of information into the construction method, because it only requires marginal distributions and a correlation matrix as inputs only—no more and no less.

The construction of a multivariate non-normal probability model using the CDF transform approach is illustrated using simulated data below. The most critical weakness of this approach is its inability to produce a correlation coefficient spanning the full range between −1 and 1 for a pair of physical variables. This weakness is elucidated at the end of this section.

### 1.6.2 CDF transform approach

Let $(Y_1, Y_2, ..., Y_d)$ denote multivariate non-normally distributed random variables. One well known CDF transform approach for constructing a valid multivariate distribution for these random variables is

1. Define

$$X_i = \Phi^{-1}\left[F_i(Y_i)\right] \tag{1.100}$$

where $\Phi^{-1}(\cdot)$ = inverse standard normal CDF and $F_i(\cdot)$ = CDF of $Y_i$. By definition, $(X_1, X_2, ..., X_d)$ are *individually* standard normal random variables. That is, the histogram of any component, $X_i$, will look normal (bell-shaped).

2. Assume $(X_1, X_2, ..., X_d)$ follows a multivariate standard normal distribution as defined by Equation 1.55. It is crucial to note here that *collectively* $(X_1, X_2, ..., X_d)$ does not necessarily follow a multivariate standard normal distribution even if each component is standard normal. For example, if the scatter plot of $X_i$ versus $X_j$ shows a distinct nonlinear trend, then the multivariate normal distribution assumption is incorrect. You can apply the Mahalanobis distance test in Section 1.4.3 as well. The entries in the correlation matrix $C$ in Equation 1.56 are the Pearson moment–product correlations among $(X_1, X_2, ..., X_d)$. Recall that for multivariate standard normal $(X_1, X_2, ..., X_d)$, the Pearson and Spearman (rank) correlations are nearly identical. Together with the fact that the rank correlation between $(X_i, X_j)$ is identical to that between $(Y_i, Y_j)$, the entries in $C$ are nearly the same as the rank correlations among $(Y_1, Y_2, ..., Y_d)$.

### 1.6.3 Estimation of the marginal distribution of Y

Consider a simulated multivariate dataset of $(Y_1, Y_2, Y_3)$ shown in Figure 1.25 ($n = 1000$). These 1000 data points have full multivariate information: each data point has known $(Y_1, Y_2, Y_3)$ values. In contrast, incomplete multivariate information will contain data points such as $(Y_1, Y_2, ?)$, $(Y_1, ?, Y_3)$, and $(?, Y_2, Y_3)$. The question marks denote unknown values. The treatment of incomplete multivariate information is presented in Section 1.7. The multivariate data points are simulated using the procedure discussed in Section 1.6.5, with the

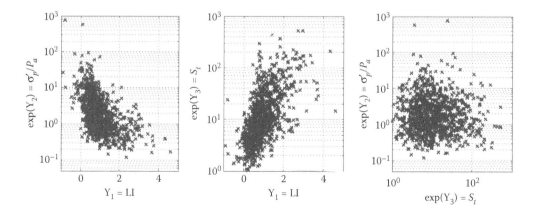

*Figure 1.25* Simulated multivariate datasets for non-normal $(Y_1, Y_2, Y_3)$.

*Table 1.14* Types and parameters for the underlying distributions of $(Y_1, Y_2, Y_3)$

| Random variable | Soil parameter | Distribution type | Distribution parameters | | | |
| --- | --- | --- | --- | --- | --- | --- |
| | | | $a_X$ | $b_X$ | $a_Y$ | $b_Y$ |
| $Y_1$ | LI | SU | 1.434 | −1.068 | 0.629 | 0.358 |
| $Y_2$ | $\ln(\sigma'_p/P_a)$ | SU | 4.495 | −9.572 | 1.199 | −4.481 |
| $Y_3$ | $\ln(S_t)$ | SU | 2.453 | −2.233 | 1.888 | 0.343 |

random seed initialized by randn('state', 13). The underlying marginal distributions for $(Y_1, Y_2, Y_3)$ are Johnson SU, and the underlying C matrix is

$$C = \begin{bmatrix} 1 & -0.57 & 0.59 \\ -0.57 & 1 & 0.05 \\ 0.59 & 0.05 & 1 \end{bmatrix} \qquad (1.101)$$

The statistics of these underlying distributions are given in Table 1.14. $(Y_1, Y_2, Y_3)$ are realistic clay parameters: $Y_1$ represents LI, $Y_2$ represents the logarithm of the normalized preconsolidation stress $\ln(\sigma'_p/P_a)$ ($P_a$ is one atmosphere pressure), and $Y_3$ represents the logarithm of sensitivity $\ln(S_t)$. The scatters and trends in Figure 1.25 are similar to those observed in the Clay/10/7490 database (Ching and Phoon 2014a). Hence, although the data are simulated, they are realistic rather than mathematically contrived with no relation to geotechnical engineering data. Ching et al. (2014a) called data simulated from a realistic geotechnical engineering context as "virtual site" data. It is significant that nonlinear correlation trends are observed among the simulated data points. Namely, $(Y_1, Y_2, Y_3)$ are multivariate non-normal. In fact, $(Y_1, Y_2, Y_3)$ are individually non-normal. Figure 1.26 shows the histograms of the simulated $(Y_1, Y_2, Y_3)$ data.

Given the simulated $(Y_1, Y_2, Y_3)$ data, the procedure described in Section 1.5.3 is used to identify the types and parameters for the Johnson distribution. The identified types and parameters are given in the parentheses in Table 1.15. This table should be compared to Table 1.14. Note that Table 1.14 is the exact solution while Table 1.15 is the estimated

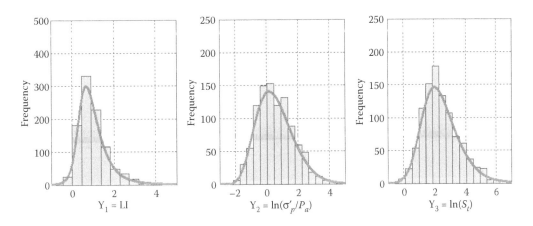

*Figure 1.26* Histograms of the simulated non-normal $(Y_1, Y_2, Y_3)$ data. The solid lines are the fitted Johnson distributions.

*Table 1.15* Types and parameters for the identified distributions of $(Y_1, Y_2, Y_3)$

| Random variable | Soil parameter | Distribution type | Distribution parameters | | | |
|---|---|---|---|---|---|---|
| | | | $a_X$ | $b_X$ | $a_Y$ | $b_Y$ |
| $Y_1$ | LI | SU | 1.491 | −1.114 | 0.650 | 0.328 |
| $Y_2$ | $\ln(\sigma_p'/P_a)$ | SB | 2.506 | 2.938 | 16.177 | 3.372 |
| $Y_3$ | $\ln(S_t)$ | SU | 3.046 | −2.965 | 2.270 | −0.285 |

solution based on 1000 data points. The differences arise from statistical uncertainty. The fitted distributions are shown in Figure 1.26—the fits are fairly satisfactory, albeit the identification is not perfect (e.g., $Y_2$ is incorrectly identified as Johnson SB). For $Y_1$ and $Y_3$, the identified type is SU. Again, these observations can be fully explained using statistical uncertainty.

The rule of the game in this section is to assume that the exact solution in the form of Table 1.14 is not available. Only the numerical values of the data points are made available to the engineer, say in the form of an EXCEL spreadsheet containing 1000 rows and three columns. Following the spirit of this game, the following equation can be used to transform $(Y_1, Y_3)$ into standard normal $(X_1, X_3)$ (see Equation 1.95):

$$X_i = b_{Xi} + a_{Xi} \times \ln\left(\frac{Y_i - b_{Yi}}{a_{Yi}} + \sqrt{1 + \left(\frac{Y_i - b_{Yi}}{a_{Yi}}\right)^2}\right)$$ (1.102)

where $(a_X, b_X, a_Y, b_Y)$ can be found in Table 1.15. For $Y_2$, the identified type is SB. Hence, the following equation can be used to transform $Y_2$ into standard normal $X_2$ (see Equation 1.95):

$$X_2 = b_{X2} + a_{X2} \times \ln\left[\frac{(Y_2 - b_{Y2})/a_{Y2}}{1 - (Y_2 - b_{Y2})/a_{Y2}}\right]$$ (1.103)

The model parameters $a_{X2}$, $b_{X2}$, $a_{Y2}$, and $b_{Y2}$ are taken from Table 1.15.

Figure 1.27 shows the histograms of the transformed $(X_1, X_2, X_3)$ data. They are similar to standard normal. The $p$-values based on the standard normal K–S test are shown in the figure. The large $p$-values suggest that the standard normal distribution cannot be rejected for the converted $(X_1, X_2, X_3)$ data. That is to say, the fitted Johnson distributions cannot be rejected for the simulated $(Y_1, Y_2, Y_3)$ data.

### 1.6.4 Estimation of the correlation matrix C

The transformed $(X_1, X_2, X_3)$ data points are shown in Figure 1.28. The converted $(X_1, X_2, X_3)$ data do not exhibit any nonlinear trend. The line test discussed in Section 1.4.3 (see Equation 1.64) is carried out on the converted $(X_1, X_2, X_3)$ data. The result for the line test is shown in Figure 1.29, suggesting that the multivariate normal hypothesis is suitable for the converted $(X_1, X_2, X_3)$ data. The correlation matrix C can be estimated using Equation 1.57, because we have full multivariate data in this simulated example. The estimated C is

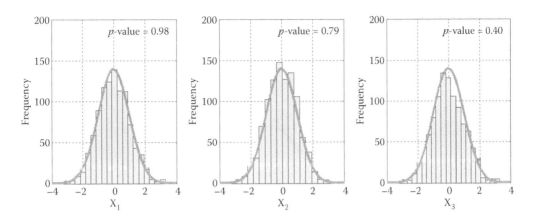

*Figure 1.27* Histograms of the converted $(X_1, X_2, X_3)$ data. Solid lines are the standard normal PDF.

$$\hat{C} = \begin{bmatrix} 1 & -0.599 & 0.645 \\ -0.599 & 1 & -0.049 \\ 0.645 & -0.049 & 1 \end{bmatrix} \tag{1.104}$$

As discussed earlier, the Pearson product–moment correlations among $(X_1, X_2, X_3)$ are nearly the same as the Spearman rank correlations among $(Y_1, Y_2, Y_3)$. Hence, the **C** matrix can be also estimated as the Spearman rank correlations among $(Y_1, Y_2, Y_3)$. This can be easily done in MATLAB using **C** = corr(**Y**, 'type', 'Spearman'). The resulting **C** matrix is

$$\hat{C} = \begin{bmatrix} 1 & -0.571 & 0.630 \\ -0.571 & 1 & -0.045 \\ 0.630 & -0.045 & 1 \end{bmatrix} \tag{1.105}$$

When only bivariate data are available, Equation 1.58 should be used in an entry-by-entry manner. This entry-by-entry approach will be demonstrated in Section 1.7 for the Clay/10/7490 database.

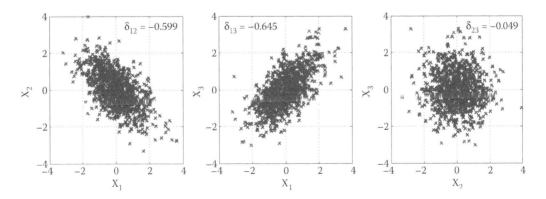

*Figure 1.28* Transformed $(X_1, X_2, X_3)$ data points.

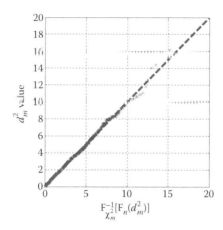

*Figure 1.29* Line test for the transformed $(X_1, X_2, X_3)$ data.

## 1.6.5 Simulation

Given the identified types and parameters in Table 1.15 and the estimated correlation matrix **C** in Equation 1.104, random samples of $(Y_1, Y_2, Y_3)$ can be readily simulated by the following steps:

1. Simulate independent standard normal random vector $\mathbf{Z} = [Z_1, Z_2, Z_3]^T$.
2. Let **U** matrix be the Cholesky decomposition of **C**:

$$\mathbf{U}^T \times \mathbf{U} = \mathbf{C} \tag{1.106}$$

In MATLAB, $\mathbf{U} = \text{chol}(\mathbf{C})$.
3. Let

$$\mathbf{X}^T = \mathbf{Z}^T \times \mathbf{U} \tag{1.107}$$

4. Finally, $X_i$ can be transformed into $Y_i$ using Equations 1.96 or 1.97.

The simulated $(Y_1, Y_2, Y_3)$ data behave fairly similarly to those in Figure 1.25.

## 1.6.6 Some practical observations

Can we simulate $(Y_1, Y_2, Y_3)$ without considering the correlations among $(Y_1, Y_2, Y_3)$? This can be done by simulating each $Y_i$ separately. However, it is wrong to ignore such correlations, namely, set $\delta_{12} = \delta_{13} = \delta_{23} = 0$ in violation of nonzero correlations exhibited by the actual data. Figure 1.30 shows the simulated $(Y_1, Y_2, Y_3)$ data without considering correlations. Figure 1.30 should be compared to Figure 1.25. It is clear that the correlations shown in Figure 1.25 are not observed in Figure 1.30.

Another extreme scenario is to simulate $(Y_1, Y_2, Y_3)$ with $\delta_{12} = \delta_{13} = \delta_{23} = 1$. Figure 1.31 shows the simulated $(Y_1, Y_2, Y_3)$ data. It is clear that deterministic correlations exist among $(Y_1, Y_2, Y_3)$. This is also grossly inconsistent with the data scatter shown in Figure 1.25 (actual data). Note that $\delta_{12}, \delta_{13}$, and $\delta_{23}$ are the Pearson product–moment correlations among $(X_1, X_2, X_3)$. They are not the Pearson correlations among $(Y_1, Y_2, Y_3)$. The Pearson correlations among the $(Y_1, Y_2, Y_3)$ data shown in Figure 1.31 are, surprisingly, not 1. In fact, they

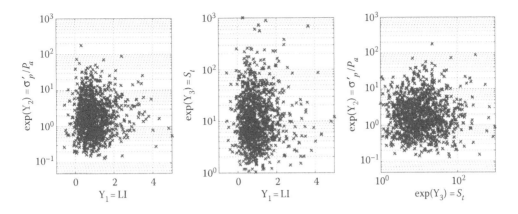

*Figure 1.30* Simulated multivariate dataset of $(Y_1, Y_2, Y_3)$ by forcing $\delta_{12} = \delta_{13} = \delta_{23} = 0$.

are <1. For instance, the Pearson correlation between $(Y_1, Y_2)$ is only 0.979. This is because the Pearson correlation is a measure of "linear" correlation. Even if $Y_1$ is related to $Y_2$ in a deterministic way, the Pearson correlation is <1 if this relationship is not linear.

It is worthwhile to point out that the Pearson correlation between $(Y_1, Y_2)$ cannot exceed 0.979. The reason is that the correlation coefficient in nonnormal space is monotonically related to the correlation coefficient in the normal space in the CDF transform approach and it is not possible for the Pearson correlation coefficient between $(X_1, X_2)$ to exceed 1. One may rightfully wonder as an engineer if this limitation is inconsequential to practice. It is rare for physical variables to produce correlation coefficients near to 1 in practice and, in any case, a correlation coefficient of 0.979 is as good as 1 in the presence of statistical uncertainty. Unfortunately, this theoretical limitation associated with the CDF transform approach can be practically important as illustrated below.

First, we state without proof here that the exact relation between these correlation coefficients is given by the integral equation below:

$$\delta_{Y_1 Y_2} = \frac{\int_{-\infty}^{\infty} \int_{-\infty}^{\infty} F_1^{-1}[\Phi(x_1)] \times F_2^{-1}[\Phi(x_2)] \times f(x_1, x_2; \delta_{12}) dx_1 dx_2 - \mu_{Y_1} \times \mu_{Y_2}}{\sigma_{Y_1} \times \sigma_{Y_2}} \qquad (1.108)$$

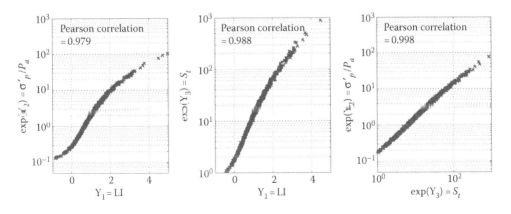

*Figure 1.31* Simulated multivariate dataset of $(Y_1, Y_2, Y_3)$ by forcing $\delta_{12} = \delta_{13} = \delta_{23} = 1$.

where $\delta_{Y1Y2}$ is the Pearson correlation between $(Y_1, Y_2)$; $\delta_{12}$ is the Pearson correlation between $(X_1, X_2)$; and $f(x_1, x_2; \delta_{12})$ is the bivariate standard normal PDF defined in Equation 1.37:

$$f\left(x_1, x_2, \delta_{12}\right) = \exp\left( 0.5\begin{bmatrix} x_1 \\ x_2 \end{bmatrix}^t \begin{bmatrix} 1 & \delta_{12} \\ \delta_{12} & 1 \end{bmatrix}^{-1} \begin{bmatrix} x_1 \\ x_2 \end{bmatrix} \right) \Big/ \left(2\pi \times \sqrt{1 - \delta_{12}^2}\right) \tag{1.109}$$

and $\mu_{Y_i}$ and $\sigma_{Y_i}$ are the mean and standard deviation of $Y_i$, respectively:

$$\mu_{Y_i} = \int_{-\infty}^{\infty} F_i^{-1}\left[\Phi(x)\right] \times \varphi(x)dx \quad \sigma_{Y_i} = \int_{-\infty}^{\infty} \left(F_i^{-1}\left[\Phi(x)\right] - \mu_{Y_i}\right)^2 \times \varphi(x)dx \tag{1.110}$$

$\varphi(x)$ is the univariate standard normal PDF (Equation 1.3).

This relation between $\delta_{Y1Y2}$ and $\delta_{12}$ is plotted on the left plot in Figure 1.32. When evaluating the CDF transform $F_1^{-1}[\Phi(x_1)]$, the parameters $(a_{X1}, b_{X1}, a_{Y1}, b_{Y1})$ in Table 1.15 are used. The same approach is followed for evaluating $F_2^{-1}[\Phi(x_2)]$. It is clear that $\delta_{Y1Y2}$ and $\delta_{12}$ are not identical. Moreover, even though $\delta_{12}$ spans the full range $-1.0$–$1.0$, $\delta_{Y1Y2}$ only spans a nonfull range of $-0.91$–$0.98$. This gets even worse if we reduce $a_{X1}$ and $a_{X2}$ in Table 1.15 by a factor of 2 ($a_{X1} = 1.491/2 = 0.754$; $a_{X2} = 2.506/2 = 1.253$). By doing so, the standard deviations of $(Y_1, Y_2)$ will increase. The resulting relation between $\delta_{Y1Y2}$ and $\delta_{12}$ is plotted on the right plot in Figure 1.32. It is seen that the lower bound for $\delta_{Y1Y2}$ is now only $-0.4$! The inability of the CDF transform approach to reproduce strong negative correlations among physical variables is arguably the most critical weakness of this approach.

## 1.7 REAL EXAMPLE

In this section, the construction of multivariate probability distributions of soil parameters will be demonstrated using the Clay/10/7490 database compiled by Ching and Phoon

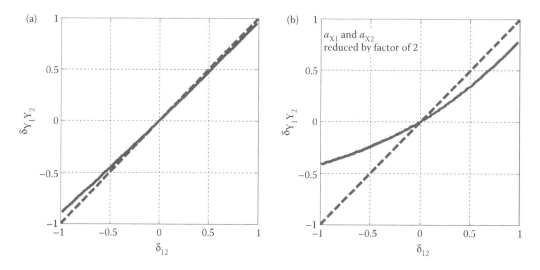

*Figure 1.32* Relation between $\delta_{Y1Y2}$ and $\delta_{12}$ (a) for parameters given in Table 1.15; and (b) $a_{X1}$ and $a_{X2}$ reduced by a factor of 2. The dashed lines are the 1:1 lines.

(2014a). In addition, we generalize the concept of conditioning discussed in Section 1.4.5 from the normal case to the non-normal (Johnson distribution) case (Section 1.7.3).

Several valuable multivariate databases have been compiled by the authors so far (Ching and Phoon 2012, 2013, 2014a; Ching et al. 2014b). These databases are labeled in accordance to the template: (soil type)/(number of parameters of interest)/(number of data points). The Clay/10/7490 database is the largest database compiled so far in terms of the number of data points and the number of parameters of interest (equal to dimension of multivariate probability distribution function).

### 1.7.1 Clay/10/7490 database

The Clay/10/7490 database compiles data from 251 studies. The number of data points associated with each study varies from 1 to 419 with an average of 30 data points per study. The clay properties cover a wide range of OCR (but mostly 1 ~ 10), a wide range of sensitivity $(S_t)$ (sites with $S_t = 1$ ~ tens or hundreds are fairly typical), and a wide range of PI (but mostly 8 ~ 100). Details are reported by Ching and Phoon (2014a).

Ten dimensionless clay parameters are compiled in this database. These parameters can be divided into the following three broad categories:

1. Index properties:
   - LL
   - PI
   - LI
2. Stresses and strengths:
   - Normalized vertical effective stress $(\sigma'_v/P_a)$ ($P_a$ is one atmosphere pressure = 101.3 kPa)
   - Normalized preconsolidation stress $(\sigma'_p/P_a)$
   - Normalized undrained shear strength $(s_u/\sigma'_v)$ [All $s_u$ values are converted into the "mobilized" $s_u$ values, denoted by $s_u$(mob), which is the *in situ* undrained shear strength mobilized in embankment and slope failures (Mesri and Huvaj 2007)]
   - Sensitivity $(S_t = s_u/s_u^{rc})$ ($s_u^{rc}$ is the remolded undrained shear strength)
3. CPTU parameters:
   - Pore pressure ratio $B_q = (u_2 - u_0)/(q_t - \sigma_v)$ ($u_2$ is the pore pressure behind the cone; $u_0$ is the hydrostatic pore pressure; $q_t$ is the corrected cone tip resistance; and $\sigma_v$ is the total effective stress)
   - Normalized cone tip resistance $(q_t - \sigma_v)/\sigma'_v$
   - Normalized effective cone tip resistance $(q_t - u_2)/\sigma'_v$

It is clear that other dimensionless parameters of interest, such as the OCR, can be derived from the above 10 dimensionless parameters. The basic statistics of these parameters (10 basic parameters together with OCR) are listed in Table 1.16 to provide a feel of the range and distribution of the parameters. The number of data points ($n$) for each parameter is shown in the second column. The statistics are the mean value, COV, minimum value (min), and maximum value (max).

To keep notation concise, the physical random variables (or natural logarithm transform) are denoted by

1. $Y_1 = \ln(\text{LL})$ ⎫
2. $Y_2 = \ln(\text{PI})$ ⎬ Index properties
3. $Y_3 = \text{LI}$ ⎭

*Table 1.16* Statistics of the 10 geotechnical parameters in Clay/10/7490

| Parameters | n | Mean | COV | Min | Max |
|---|---|---|---|---|---|
| LL | 3822 | 67.7 | 0.00 | 10.1 | 515 |
| PI | 4265 | 39.7 | 1.08 | 1.9 | 363 |
| LI | 3661 | 1.01 | 0.78 | −0.75 | 6.45 |
| $\sigma'_v/P_a$ | 3370 | 1.80 | 1.47 | 1.13E-3 | 38.74 |
| $\sigma'_p/P_a$ | 2028 | 4.37 | 2.31 | 0.094 | 193.30 |
| $s_u/\sigma'_v$ | 3538 | 0.51 | 1.25 | 3.68E-3 | 7.78 |
| $S_t$ | 1589 | 35.0 | 2.88 | 1 | 1467 |
| $B_q$ | 1016 | 0.58 | 0.35 | 0.01 | 1.17 |
| $(q_t-\sigma_v)/\sigma'_v$ | 862 | 8.90 | 1.17 | 0.48 | 95.98 |
| $(q_t-u_2)/\sigma'_v$ | 668 | 5.34 | 1.37 | 0.61 | 108.20 |
| OCR | 3531 | 3.85 | 1.56 | 1.0 | 60.23 |

Source: Ching, J. and Phoon, K.K. 2014b. *Canadian Geotechnical Journal*, 51(6), 686–704, reproduced with permission of the NRC Research Press.

4. $Y_4 = \ln(\sigma'_v/P_a)$
5. $Y_5 = \ln(\sigma'_p/P_a)$
6. $Y_6 = \ln(s_u/\sigma'_v)$    } Stresses and strengths
7. $Y_7 = \ln(S_t)$
8. $Y_8 = B_q$
9. $Y_9 = \ln[(q_t-\sigma_v)/\sigma'_v]$    } CPTU parameters
10. $Y_{10} = \ln[(q_t-u_2)/\sigma'_v]$

The natural logarithm transform is applied to eight non-negative parameters. For the parameters that can be potentially negative (LI and $B_q$), no transform is needed and $Y_i$ simply denotes the physical parameter itself. There are data points where two or more of these parameters are simultaneously known. For instance, a disturbed clay sample is extracted to determine LI, and an undisturbed clay sample is extracted at a nearby borehole at the same depth to determine $\sigma'_p$. In this study, we describe $Y_3 = $ LI and $Y_5 = \ln(\sigma'_p/P_a)$ as being "simultaneously" known in this sense.

In principle, the construction of a multivariate distribution for $(Y_1, Y_2, ..., Y_{10})$ will require multivariate data points. This means a single set of values of $(Y_1, Y_2, ..., Y_{10})$ is obtained from the same soil sample or more practically, from soil samples extracted from adjacent boreholes at comparable depths. The criterion is that a single set of values must be measured from samples that are strongly correlated in the spatial sense. The database will be a multivariate database if all 10 tests were conducted in each cited reference. It is evident that such multivariate data are not available in geotechnical engineering. Nonetheless, it is not uncommon to have *bivariate* data measured in close spatial proximity, for example, both $Y_3$ and $Y_5$ [LI and $\ln(\sigma'_p/P_a)$] are measured in two adjacent boreholes at comparable depths. One would expect these $Y_3$ and $Y_5$ profiles to be correlated. Since there are 10 basic parameters, it is possible to form $^{10}C_2 = 45$ distinct pairs of parameters. The off-diagonal numbers in Table 1.17 show the numbers of data points associated with each pair of parameters. The numbers in the leading diagonal of Table 1.17 show the numbers of data points associated with a single parameter, which are identical to the numbers shown in the second column of Table 1.16.

*Table 1.17* Numbers of data points with bivariate information

|  | $Y_1$ | $Y_2$ | $Y_3$ | $Y_4$ | $Y_5$ | $Y_6$ | $Y_7$ | $Y_8$ | $Y_9$ | $Y_{10}$ | OCR |
|---|---|---|---|---|---|---|---|---|---|---|---|
| $Y_1$ | 3822 | 3822 | 3412 | 2084 | 1362 | 1835 | 1184 | 680 | 618 | 541 | 1475 |
| $Y_2$ |  | 4265 | 3424 | 2169 | 1433 | 2173 | 1203 | 688 | 626 | 549 | 1745 |
| $Y_3$ | Index properties |  | 3661 | 1999 | 1314 | 1709 | 1279 | 660 | 598 | 521 | 1388 |
| $Y_4$ |  |  |  | 3370 | 1944 | 2419 | 853 | 965 | 862 | 668 | 1959 |
| $Y_5$ |  |  |  |  | 2028 | 1423 | 554 | 780 | 691 | 543 | 1984 |
| $Y_6$ |  |  |  |  |  | 3532 | 715 | 595 | 533 | 525 | 2120 |
| $Y_7$ |  |  | Stresses and strengths |  |  |  | 1589 | 240 | 230 | 190 | 586 |
| $Y_8$ |  |  |  |  |  |  |  | 1016 | 862 | 668 | 832 |
| $Y_9$ |  | Symmetry |  |  |  |  |  |  | 862 | 590 | 692 |
| $Y_{10}$ |  |  |  |  |  |  |  | CPTU parameters |  | 668 | 544 |
| OCR |  |  |  |  |  |  |  |  |  |  | 3531 |

Source: Ching, J. and Phoon, K.K. 2014b. *Canadian Geotechnical Journal*, 51(6), 686–704, reproduced with permission of the NRC Research Press.

## 1.7.2 Construction of multivariate distribution

The multivariate non-normal distribution for $(Y_1, Y_2, ..., Y_{10})$ is constructed using the approach discussed in Section 1.6: (1) Fit a Johnson distribution to each component $(Y_i)$, (2) convert $Y_i$ into standard normal $X_i$, and (3) compute the correlation matrix for $(X_1, X_2, ..., X_{10})$. The key challenge is to compute a valid *positive-definite* correlation matrix in step (3). As discussed in Section 1.4.3, the correlation matrix is guaranteed to be at least semipositive definite if it is estimated from multivariate data (Equation 1.57). However, multivariate data are rare in geotechnical engineering. The practical approach is to estimate each entry in the correlation separately from *bivariate* data (Equation 1.58). The downside is that such a piecemeal approach will not guarantee a positive-definite correlation matrix. This rather abstract theoretical property cannot be dismissed without running the risk of producing completely absurd answers such as a negative variance as shown in Section 1.4.3. This section demonstrates how a correlation matrix can be estimated from actual bivariate data while preserving the critical property of positive definiteness.

### 1.7.2.1 Fit a Johnson distribution to each component ($Y_i$)

Each component ($Y_i$) can be fitted to a Johnson distribution using the procedures introduced in Section 1.6.3. The distribution type and parameters are summarized in Table 1.18.

### 1.7.2.2 Convert $Y_i$ into standard normal $X_i$

It is easy to transform the soil parameters into standard normal random variables ($X_1, X_2, ..., X_{10}$) using Equation 1.94 (or Equation 1.95) and the distribution type/parameters from Table 1.18. The normality of this transformed data can be checked using a probability plot or the K–S test. Using MATLAB function kstest, the *p*-values associated with the K–S test for ($X_1, X_2, ..., X_{10}$) can be computed, as listed in Table 1.18 (the right-most column). There are five *p*-values <0.05 ($X_1, X_2, X_3, X_6, X_7$), indicating that there is sufficient evidence to reject the null hypothesis that $X_i$ is a standard normal random variable at a level of significance of 5%. However, the Johnson distribution is still adopted in the ensuing analysis because of its analytical elegance (discussed later).

*Table 1.18* Distribution type and distribution parameters for $(Y_1, Y_2, ..., Y_{10})$

| Random variable | Soil parameter or its log transform | Distribution type | Distribution parameters | | | | |
|---|---|---|---|---|---|---|---|
| | | | $a_X$ | $b_X$ | $a_Y$ | $b_Y$ | p-Value |
| $Y_1$ | $\ln(LL)$ | SU | 1.636 | −1.166 | 0.616 | 3.479 | 5.7e-07 |
| $Y_2$ | $\ln(PI)$ | SU | 1.433 | −0.265 | 0.918 | 3.178 | 3.0e-05 |
| $Y_3$ | $LI$ | SU | 1.434 | −1.068 | 0.629 | 0.358 | 1.2e-07 |
| $Y_4$ | $\ln(\sigma'_v/P_a)$ | SB | 3.150 | 0.256 | 14.458 | −7.010 | 0.40 |
| $Y_5$ | $\ln(\sigma'_p/P_a)$ | SB | 4.600 | 21.548 | 576.785 | −4.793 | 0.16 |
| $Y_6$ | $\ln(s_u/\sigma'_v)$ | SU | 2.039 | −0.517 | 1.427 | −1.461 | 2.9e-09 |
| $Y_7$ | $\ln(S_t)$ | SU | 2.393 | −2.080 | 1.885 | 0.461 | 7.1e-14 |
| $Y_8$ | $B_q$ | SU | 2.676 | 0.161 | 0.513 | 0.615 | 0.31 |
| $Y_9$ | $\ln[(q_t - \sigma_v)/\sigma'_v]$ | SU | 1.340 | −0.572 | 0.659 | 1.476 | 0.53 |
| $Y_{10}$ | $\ln[(q_t-u_2)/\sigma'_v]$ | SU | 2.134 | −1.102 | 1.154 | 0.657 | 0.57 |

Source: Ching, J. and Phoon, K.K. 2014b. *Canadian Geotechnical Journal*, 51(6), 686–704, reproduced with permission of the NRC Research Press.

### 1.7.2.3 Compute the correlation matrix for $(X_1, X_2, ..., X_{10})$

Figure 1.33 presents the bivariate correlation structure underlying the 10 soil parameters *after* they have been transformed into standard normal random variables using Equation 1.95. As mentioned previously, there are 45 possible bivariate correlations for a database containing $d = 10$ parameters. The simplest method to quantify the bivariate correlation between $X_i$ and $X_j$ is to compute the Pearson correlation coefficients $\delta_{ij}$ using Equation 1.58. Here, a simplified version is used:

$$\delta_{ij} \approx \frac{1}{n_{ij}} \sum_{k=1}^{n_{ij}} X_i^{(k)} \cdot X_j^{(k)} \tag{1.111}$$

where $n_{ij}$ is the total number of bivariate $(X_i, X_j)$ data points. This simplified version is based on the fact that the mean and standard deviation of $X_i$ are equal to 0 and 1, respectively, because $X_i$ is standard normal. Therefore,

$$\delta_{ij} = \frac{\text{COV}(X_i, X_j)}{\sigma_i \cdot \sigma_j} = \frac{E(X_i X_j) - \mu_i \mu_j}{\sigma_i \cdot \sigma_j} = E(X_i X_j) \tag{1.112}$$

It is useful to recollect that this additional step of converting a non-normal variable Y into a standard normal variable X is necessary if one were to exploit the multivariate normal distribution to couple individual components together in a consistent way. The bivariate correlation structure presented in Figure 1.33 is *sufficient* to fully characterize a multivariate probability distribution only if the multivariate normal hypothesis is true.

### 1.7.2.4 Problem of nonpositive definiteness

The bootstrapping technique (Efron and Tibshirani 1993) introduced in Section 1.3.3 is applied to obtain 1000 bootstrap samples of $\delta_{ij}$. Figure 1.34 shows the histograms of the 1000 $\delta_{ij}$ estimates for the $X_1 - X_2$ and $X_7 - X_9$ pairs, namely $\delta_{12}$ and $\delta_{79}$. The 90% confidence

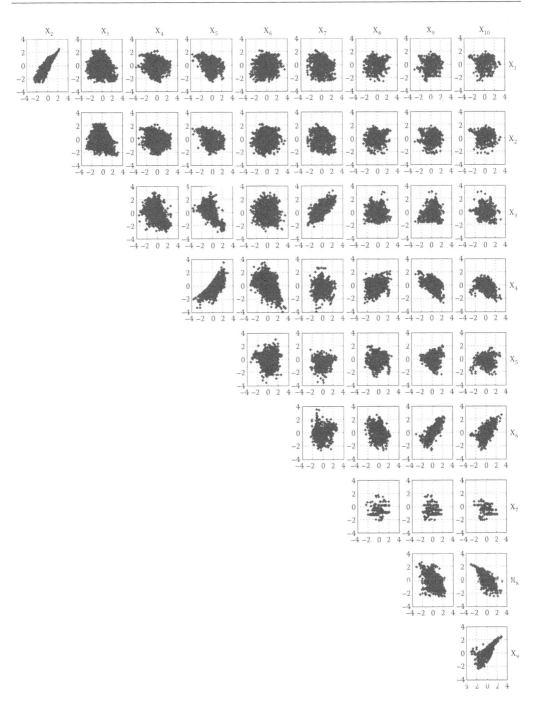

*Figure 1.33* Scatter plots between $X_i$ and $X_j$ (horizontal axis = column variable, vertical axis = row variable). (From Ching, J. and Phoon, K.K. 2014b. *Canadian Geotechnical Journal*, 51(6), 686–704, reproduced with permission of the NRC Research Press.)

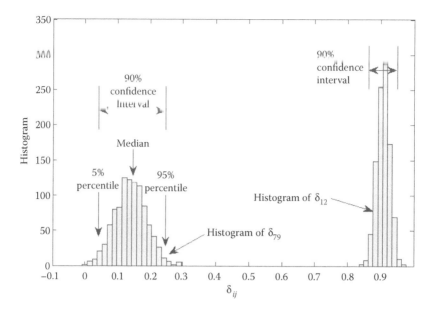

*Figure 1.34* Histograms of the bootstrap $\delta_{ij}$ samples. (From Ching, J. and Phoon, K.K. 2014b. *Canadian Geotechnical Journal*, 51(6), 686–704, reproduced with permission of the NRC Research Press.)

interval for $\delta_{12}$ is significantly narrower than that for $\delta_{79}$ probably because there are more data points in the $X_1 - X_2$ pair ($n_{12} = 3822$) than in the $X_7 - X_9$ pair ($n_{79} = 230$). On the basis of $\delta_{ij}$ samples, the 90% confidence intervals (5% and 95% percentiles) and the median of $\delta_{ij}$ can be identified. Table 1.19 shows the 90% confidence intervals (range bounded by the 5% and 95% percentiles) and the median of the $\delta_{ij}$ estimates (median values are in the parentheses).

The correlation matrix C presented in Table 1.19, which is formed from the median values of $\delta_{ij}$, may not be positive-definite, because $\delta_{ij}$ is estimated *independently* of other entries in Table 1.19. There is one negative eigenvalue in the C matrix based on the median values of $\delta_{ij}$.

The choice of using the median value as a point estimate of each entry in C is a matter of convenience. The median value is not the only possible value of $\delta_{ij}$, given the statistical uncertainty shown in Figure 1.34. Ching and Phoon (2014b) formed a positive-definite C matrix using the following steps:

1. For each bivariate correlation, a bootstrap sample of $\delta_{ij}$ is obtained. There are $d(d - 1)/2 = 45$ possible bivariate correlations; so, this step is conducted for 45 times.
2. Determine whether the resulting C matrix is positive-definite by checking its eigenvalues. If C is positive-definite, it is accepted (and the $\delta_{ij}$ values are accepted). Otherwise, C is rejected.
3. The above steps are repeated until 1000 C matrices are accepted. The ratio of the acceptance is about 16% in this example.
4. The final matrix is obtained by averaging the 1000 accepted C matrices. This matrix is shown in Table 1.20. Table 1.20 happens to be similar to Table 1.19 in this example. However, Table 1.20 is positive-definite, because it can be mathematically proved that the average of positive-definite matrices is also positive-definite.

*Table 1.19* Statistics of the estimated $\delta_{ij}$ in the form of 90% confidence interval (Median)

|   |   | $X_1$ | $X_2$ | $X_3$ | $X_4$ | $X_5$ | $X_6$ | $X_7$ | $X_8$ | $X_9$ | $X_{10}$ |
|---|---|---|---|---|---|---|---|---|---|---|---|
| | $X_1$ | 1.00 | 0.88~0.94 (0.91) | −0.28~−0.22 (−0.25) | −0.27~−0.21 (−0.24) | −0.35~−0.27 (−0.31) | 0.06~0.14 (0.10) | −0.26~−0.15 (−0.20) | 0.04~0.15 (0.09) | 0.04~0.14 (0.09) | 0.01~0.13 (0.07) |
| | $X_2$ | | 1.00 | −0.35~−0.29 (−0.32) | −0.24~−0.18 (−0.21) | −0.31~−0.24 (−0.27) | 0.01~0.07 (0.04) | −0.30~−0.19 (−0.24) | 0.06~0.16 (0.11) | −0.04~0.05 (0.00) | −0.06~0.05 (−0.01) |
| | $X_3$ | Index properties | | 1.00 | −0.53~−0.45 (−0.49) | −0.64~−0.53 (−0.58) | −0.05~0.03 (−0.01) | 0.55~0.65 (0.60) | −0.11~0.00 (−0.06) | 0.00~0.13 (0.07) | −0.12~0.01 (−0.05) |
| | $X_4$ | | | | 1.00 | 0.68~0.76 (0.72) | −0.59~−0.50 (−0.54) | −0.02~0.09 (0.03) | 0.15~0.25 (0.20) | −0.43~−0.33 (−0.38) | −0.37~−0.26 (−0.31) |
| $C =$ | $X_5$ | | | | | 1.00 | −0.03~−0.05 (0.01) | 0.00~0.12 (0.06) | −0.09~0.02 (−0.03) | 0.06~0.18 (0.12) | −0.03~0.11 (0.03) |
| | $X_6$ | | | | | | 1.00 | 0.09~−0.21 (0.15) | −0.30~−0.18 (−0.24) | 0.65~0.79 (0.72) | 0.55~0.70 (0.62) |
| | $X_7$ | | | Stresses and strengths | | | | 1.00 | 0.11~−0.27 (0.19) | 0.05~0.22 (0.14) | −0.17~0.01 (−0.08) |
| | $X_8$ | | | | | | | | 1.00 | −0.51~−0.38 (−0.44) | −0.70~−0.56 (−0.63) |
| | $X_9$ | | Symmetry | | | | | | | 1.00 | 0.67~0.82 (0.74) |
| | $X_{10}$ | | | | | | | | CPTU parameters | | 1.00 |

Source: Ching, J. and Phoon, K.K. 2014b. *Canadian Geotechnical Journal*, 51(6), 686–704, reproduced with permission of the NRC Research Press.

## 1.7.3  Conditioning: Bayesian analysis

On the basis of multivariate probability distribution constructed in Section 1.7.2, it is possible to update the marginal distribution of any one parameter or even the multivariate distribution of any group of parameters given information from other parameters covered by the probability distribution. One example of the former is to update the distribution of $Y_6$ based on $Y_9$ measurements. One example of the latter is to update the bivariate distribution of $(Y_5, Y_6)$ based on measurements from $(Y_3, Y_4, Y_8, Y_{10})$. Detailed calculations for this example are presented below. Ching and Phoon (2014b) covered another example where the bivariate distribution of $(Y_5, Y_6)$ is updated based on measurements from $(Y_4, Y_8, Y_9)$. The theory has been covered in Section 1.4.5. The estimation of a design property from multiple data sources is one important practical outcome of a site investigation program. The "conditioning" procedure described in this section can be viewed as a rationalization of simpler empirical procedures widely adopted in practice such as averaging estimates from different tests or choosing the most conservative estimate produced by all tests.

Consider an example involving updating the normalized preconsolidation stress $(Y_5)$ and the normalized undrained shear strength $(Y_6)$ at a given depth based on data from other sources measured at the same depth: LI $(Y_3)$, the normalized effective vertical stress $(Y_4)$, the pore pressure ratio $(Y_8)$, and the normalized effective cone tip resistance $(Y_{10})$. This

*Table 1.20* Average of 1000 positive-definite matrices obtained from bootstrap samples

|  |  | $X_1$ | $X_2$ | $X_3$ | $X_4$ | $X_5$ | $X_6$ | $X_7$ | $X_8$ | $X_9$ | $X_{10}$ |
|---|---|---|---|---|---|---|---|---|---|---|---|
|  | $X_1$ | 1.00 | 0.91 | −0.25 | −0.24 | −0.30 | 0.10 | −0.21 | 0.09 | 0.09 | 0.07 |
|  | $X_2$ |  | 1.00 | −0.32 | −0.21 | −0.27 | 0.04 | −0.25 | 0.11 | 0.00 | −0.01 |
|  | $X_3$ | Index properties | | 1.00 | −0.49 | −0.57 | 0.01 | 0.59 | −0.05 | 0.06 | −0.05 |
| $C =$ | $X_4$ |  |  |  | 1.00 | 0.72 | −0.50 | 0.00 | 0.20 | −0.38 | −0.32 |
|  | $X_5$ |  |  |  |  | 1.00 | 0.01 | 0.06 | −0.03 | 0.11 | 0.04 |
|  | $X_6$ |  |  |  |  |  | 1.00 | 0.18 | −0.24 | 0.73 | 0.63 |
|  | $X_7$ |  |  | Stresses and strengths | | | | 1.00 | 0.18 | 0.15 | −0.08 |
|  | $X_8$ |  |  |  |  |  |  |  | 1.00 | −0.45 | −0.63 |
|  | $X_9$ |  | Symmetry |  |  |  |  |  |  | 1.00 | 0.74 |
|  | $X_{10}$ |  |  |  |  |  |  |  | CPTU parameters | | 1.00 |

Source: Ching, J. and Phoon, K.K. 2014b. *Canadian Geotechnical Journal*, 51(6), 686–704, reproduced with permission of the NRC Research Press.

is a realistic example as $Y_8$ and $Y_{10}$ are routinely measured simultaneously in piezocone surroundings.

The solution requires the distribution of a six-dimensional random vector. In standard normal space, this random vector can be partitioned into two subvectors:

$$\mathbf{X} = \begin{bmatrix} X_5 \\ X_6 \\ X_3 \\ X_4 \\ X_8 \\ X_{10} \end{bmatrix} = \begin{bmatrix} \mathbf{X}^{[1]} \\ \mathbf{X}^{[2]} \end{bmatrix} \qquad (1.113)$$

The correlation matrix for the six-dimensional random vector (which is simply a submatrix extracted from Table 1.20) is partitioned as shown below to make explicit the correlation matrix for each subvector ($C^{[11]}$ and $C^{[22]}$) and the cross-correlation matrix ($C^{[12]}$ = transpose of $C^{[21]}$):

$$\mathbf{C} = \begin{matrix} & \begin{matrix} X_5 & X_6 & X_3 & X_4 & X_8 & X_{10} \end{matrix} \\ \begin{matrix} X_5 \\ X_6 \\ X_3 \\ X_4 \\ X_8 \\ X_{10} \end{matrix} & \left( \begin{array}{cc|cccc} 1 & 0.01 & -0.57 & -0.72 & -0.03 & 0.04 \\ 0.01 & 1 & 0.01 & -0.50 & -0.24 & 0.63 \\ \hline -0.57 & 0.01 & 1 & -0.49 & -0.05 & -0.05 \\ -0.72 & -0.50 & -0.49 & 1 & 0.20 & -0.32 \\ -0.03 & -0.24 & -0.05 & 0.20 & 1 & -0.63 \\ 0.04 & 0.63 & -0.05 & -0.32 & -0.63 & 1 \end{array} \right) \end{matrix} = \begin{bmatrix} \mathbf{C}^{[11]} & \mathbf{C}^{[12]} \\ \mathbf{C}^{[21]} & \mathbf{C}^{[22]} \end{bmatrix} \qquad (1.114)$$

The conditional distribution of $\mathbf{X}^{[1]}$ given $\mathbf{X}^{[2]}$ is a *bivariate* normal distribution with the following mean vector and covariance matrix:

$$\begin{aligned} \mu_{\text{update}}^{[1]} &= \mathbf{C}^{[12]} \left( \mathbf{C}^{[22]} \right)^{-1} \mathbf{X}^{[2]} = \begin{bmatrix} -0.228 & 0.691 & -0.035 & 0.230 \\ -0.152 & -0.419 & 0.233 & 0.634 \end{bmatrix} \mathbf{X}^{[2]} \\[2mm] \mathbf{C}_{\text{update}}^{[11]} &= \mathbf{C}^{[11]} - \mathbf{C}^{[12]} \left( \mathbf{C}^{[22]} \right)^{-1} \mathbf{C}^{[21]} = \begin{bmatrix} 0.359 & 0.204 \\ 0.204 & 0.451 \end{bmatrix} \end{aligned} \qquad (1.115)$$

To illustrate the practical usefulness of a conditional distribution, the following values measured at a depth = 3.16 m are extracted from a field study in Berthierville (Canada) reported by Rochelle et al. (1988):

$$Y_3 = LI = 1.793$$

$$Y_4 = \ln(\sigma'_v/P_a) = \ln(0.402) = -0.912$$

$$Y_8 = B_q = 0.504$$

$$Y_{10} = \ln[(q_t - u_2)/\sigma'_v] = \ln(3.541) = 1.265$$

Using Equation 1.95 and the probability model parameters $(a_X, b_X, a_Y, b_Y)$ given in Table 1.18, the values in standard normal space can be computed as $X_3 = 1.174$, $X_4 = -0.737$, $X_8 = -0.411$, and $X_{10} = -0.024$. Hence, the conditional mean vector is

$$\mu^{[1]}_{update} = \begin{bmatrix} -0.228 & 0.691 & -0.035 & 0.230 \\ -0.152 & -0.419 & 0.233 & 0.634 \end{bmatrix} \begin{bmatrix} 1.174 \\ -0.737 \\ -0.411 \\ -0.024 \end{bmatrix} = \begin{bmatrix} -0.769 \\ 0.0189 \end{bmatrix} \tag{1.116}$$

In other words, the updated version of $X_5$ is a normal random variable with mean = −0.769 and standard deviation = $(0.359)^{1/2} = 0.599$. The updated version of $X_6$ is a normal random variable with mean = 0.0189 and standard deviation = $(0.451)^{1/2} = 0.671$. Note that the updated versions of $X_5$ and $X_6$ are no longer standard normal random variables. For brevity, we drop the subscript and denote the updated *nonstandard* normal random variable produced by conditioning as X′. The symbol X denotes a standard normal random variable. The updated physical random variable corresponding to X′ is Y′ and its probability distribution can be deduced from Equation 1.86:

$$\kappa\left(\frac{Y' - b_Y}{a_Y}\right) = \frac{X' - b_X}{a_X} = \frac{(\sigma'_X X + \mu'_X) - b_X}{a_X} = \frac{X - (b_X - \mu'_X)/\sigma'_X}{a_X/\sigma'_X} \tag{1.117}$$

where $\mu'_X$ and $\sigma'_X$ are the updated mean and standard deviation of X′. Comparing Equations 1.86 and 1.117, it is clear that Y′ remains a Johnson random variable with the distribution type unchanged, whereas the parameters are updated into $(a_X/\sigma'_X, (b_X - \mu'_X)/\sigma'_X, a_Y, b_Y)$. In general, the distribution function for Y′ (conditioned) is not the same as the distribution function for Y (unconditioned). Additional efforts are needed to obtain the distribution function for Y′. For a Johnson distribution, both Y and Y′ follow the same $\kappa(\cdot)$ function. The only difference is the numerical values of the model parameters and these updated model parameters for Y′ can be calculated in closed form. This is a significant practical advantage and explains why a Johnson distribution is recommended in Section 1.5.

On the basis of the above observation, it is clear that the updated version of $Y_5$ is an SU distribution with $a_X = 4.600/0.599 = 7.682$, $b_X = (21.548 + 0.763)/0.599 = 37.267$, $a_Y = 576.785$, and $b_Y = -4.793$. The unconditioned mean and standard deviation of $Y_5$ are 0.61 and 1.17, respectively. The conditioned mean and standard deviation of $Y_5$ are −0.28

and 0.59, respectively. The actual measured value of $Y_5$ is $-0.727$ ($\sigma'_p/P_a = 0.483$ (Rochelle et al. 1988).

The updated version of $Y_6$ is an SU distribution with $a_X = 2.039/0.671 = 3.038$, $b_X = (-0.51/-0.028)/0.671 = -0.799$, $a_Y = 1.427$, and $b_Y = 1.461$. The unconditional mean and standard deviation of $Y_6$ are $-1.05$ and $0.82$, respectively. The conditioned mean and standard deviation of $Y_6$ are $-1.06$ and $0.51$, respectively. The actual measured value of $Y_6$ is $-1.196$ ($s_u/\sigma'_v = 0.302$) (Rochelle et al. 1988). The unconditioned and conditioned distributions for $Y_5$ and $Y_6$ are shown in Figure 1.35.

It is also possible to plot the bivariate distribution of $(Y_5, Y_6)$. Note that $(X_5, X_6)$ are bivariate normal with mean equal to $\mu^{[1]}$ and covariance matrix equal to $C^{[11]}$. The bivariate normal PDF of $(X_5, X_6)$ is

$$f_X(x_5, x_6) = \frac{1}{2\pi\sqrt{|C^{[11]}|}} \times \exp\left(-\frac{1}{2}\left[\begin{Bmatrix} x_5 \\ x_6 \end{Bmatrix} - \mu^{[1]}\right]^T \left(C^{[11]}\right)^{-1} \left[\begin{Bmatrix} x_5 \\ x_6 \end{Bmatrix} - \mu^{[1]}\right]\right) \quad (1.118)$$

According to Equation 1.95, we have

$$x_5 = b_{X5} + a_{X5} \times \sinh^{-1}\left[(y_5 - b_{Y5})/a_{Y5}\right] \quad x_6 = b_{X6} + a_{X6} \times \sinh^{-1}\left[(y_6 - b_{Y6})/a_{Y6}\right] \quad (1.119)$$

because both $(Y_5, Y_6)$ are Johnson SU. The derivative of $x$ with respect to $y$ is

$$\frac{dx_5}{dy_5} = \frac{a_{X5}}{a_{Y5}\sqrt{1 + [(y_5 - b_{Y5})/a_{Y5}]^2}} \quad \frac{dx_6}{dy_6} = \frac{a_{X6}}{a_{Y6}\sqrt{1 + [(y_6 - b_{Y6})/a_{Y6}]^2}} \quad (1.120)$$

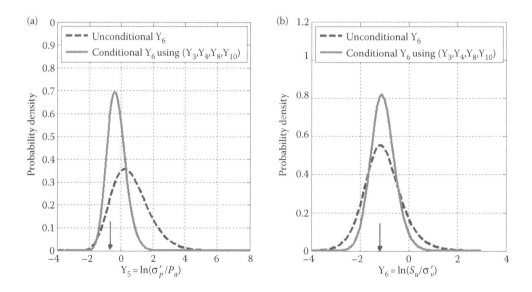

Figure 1.35 Conditional probability density distribution based on $Y_3 = 1.793$, $Y_4 = -0.912$, $Y_8 = 0.504$, and $Y_{10} = 1.265$ for (a) normalized preconsolidation stress ($Y_5$) and (b) normalized undrained shear strength ($Y_6$). (The arrow indicates actual measured value reported by Rochelle, P.L., Zebdi, M., Leroueil, S., Tavenas, F., and Virely, D. 1988. *Proceedings of the 1st International Symposium on Penetration Testing*, Orlando, 2, 831–841.)

Hence, the bivariate PDF of $(Y_5, Y_6)$ is

$$f_Y(y_5, y_6) = f_X\left[x_5 = b_{X5} + a_{X5} \times \sinh^{-1}\left(\frac{y_5 - b_{Y5}}{a_{Y5}}\right), x_6 = b_{X6} + a_{X6} \times \sinh^{-1}\left(\frac{y_6 - b_{Y6}}{a_{Y6}}\right)\right]$$
$$\times \left|\frac{dx_5}{dy_5} \times \frac{dx_6}{dy_6}\right|$$

$$(1.121)$$

Before the conditioning,

$$\mu^{[1]} = \begin{Bmatrix} 0 \\ 0 \end{Bmatrix} \qquad C^{[11]} = \begin{bmatrix} 1 & 0.01 \\ 0.01 & 1 \end{bmatrix} \qquad (1.122)$$

The updated mean vector and covariance matrix have been computed in Equations 1.115 and 1.116:

$$\mu^{[1]} = \mu_{update}^{[1]} = \begin{bmatrix} -0.769 \\ 0.0189 \end{bmatrix} \quad C^{[11]} = C_{update}^{[11]} = \begin{bmatrix} 0.359 & 0.204 \\ 0.204 & 0.451 \end{bmatrix} \qquad (1.123)$$

The unconditioned and conditioned bivariate PDFs for $(Y_5, Y_6)$ are shown in Figure 1.36. The cross-symbol in the figure indicates the actual measured values of $(Y_5, Y_6)$. The reduction in the scatter is even more pronounced in Figure 1.36 than in Figure 1.35.

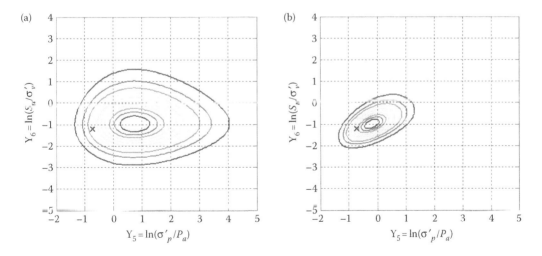

*Figure 1.36* Unconditional and conditional bivariate PDFs of $(Y_5, Y_6)$, based on the information $Y_3 = 1.793$, $Y_4 = -0.912$, $Y_8 = 0.504$, and $Y_{10} = 1.265$; (a) unconditional PDF and (b) conditional PDF. (The cross indicates actual measured values of $(Y_5, Y_6)$ reported by Rochelle, P.L., Zebdi, M., Leroueil, S., Tavenas, F., and Virely, D. 1988. *Proceedings of the 1st International Symposium on Penetration Testing*, Orlando, 2, 831–841.)

## 1.8 FUTURE CHALLENGES

This chapter discusses how to construct and how to identify a virtual site from limited geotechnical data within the framework of a multivariate non-normal probability model. The practical usefulness of a virtual site in reducing the uncertainties in the estimation of design parameters is illustrated. The effect of reducing these parametric uncertainties on RBD is discussed elsewhere (Ching et al. 2014a).

Ideally, a virtual site should emulate every aspect of a real site. The realism of the virtual site presented in this chapter can be improved in at least six ways. First, spatial variability should be included as it is a fundamental feature of geotechnical data. The mathematical framework in the form of a random vector field exists, but very few studies have been conducted so far on how to identify this field statistically from limited data. Hence, the challenge here is statistical, not theoretical.

Second, different tests are coupled using the standard concept of product–moment correlation. This concept is strictly applicable only to linear relationships. Currently, one attempts to linearize nonlinear relationships using a CDF transform, but there is no guarantee that this method will work for all real cases. Copulas can accommodate more general dependencies, but they appear to be of limited use beyond modeling bivariate data.

Third, the proposed multivariate non-normal probability model is constructed from an underlying multivariate normal probability model. The K–S test is routinely used to verify goodness of fit to a normal distribution. It cannot be extended to verify multivariate normality. Goodness-of-fit tests for multivariate normality do exist, but they are more difficult to apply and no "gold standard" has arisen in the statistics literature.

Fourth, genuine multivariate data are rare in geotechnical engineering. It is not trivial to construct a valid positive-definite correlation matrix from bivariate data, which are commonly available in a site investigation program. A partial expedient solution is offered in this chapter, but it could be possible to develop a more rigorous and directed approach in the future.

Fifth, it is important to realize that fitting each parameter to a marginal distribution is only an intermediate goal and cannot be conducted without regard to the more stringent constraints imposed by the multivariate model, which is the *final goal* in the characterization of geotechnical engineering data. If the multivariate nonnormal probability model is constructed from an underlying multivariate normal probability model as proposed in this chapter, then the non-normal marginal distribution should preferably be transformed from a normal distribution using the closed-form equation. The Johnson system of distributions is elegant in this sense. However, distribution identification using percentiles is not robust for a small sample size, say <30 data points. In addition, its model parameters are not related to physical bounds in a direct way. Physical lower/upper bounds are important to avoid absurd realizations such as negative shear strengths or parameters such as SPT-N, OCR, sensitivity, and so on taking values <1.

Sixth, the basic framework of constructing a multivariate non-normal probability model by transforming each component in a multivariate normal probability model individually is not general and may not work for actual data. In addition, this framework cannot accommodate more information beyond marginal distributions and a correlation matrix.

## LIST OF SYMBOLS

| | |
|---|---|
| $\Phi(\cdot)$ | Cumulative distribution function for standard normal |
| $\varphi(\cdot)$ | Probability density function for standard normal |

| | |
|---|---|
| $\phi$ | Friction angle |
| $\mu$ | Mean value |
| $\lambda$ | Mean value of $\ln(Y)$ |
| $\mu$ | Mean vector $(\mu_1, \mu_2, ..., \mu_d)^T$ |
| $\delta$ | Pearson product–moment correlation coefficient |
| $\alpha$ | Significance level of the test |
| $\rho$ | Spearman rank correlation coefficient |
| $\sigma$ | Standard deviation |
| $\xi$ | Standard deviation of $\ln(Y)$ |
| $\kappa$ | The transformation between X and Y for the Johnson distribution |
| $\sigma'_v$ | Vertical effective stress |
| $\Phi^{-1}(.)$ | Inverse cumulative distribution function for standard normal |
| $\chi^2_\eta$ | $\eta$-Percentile of $\chi$-squared distribution |
| $\sigma^2$ | Variance |
| $\mu_i$ | Mean value of $X_i$ or $Y_i$ |
| $\sigma_i$ | Standard deviation of $X_i$ or $Y_i$ |
| $\delta_{ij}$ | Pearson product–moment correlation coefficient between $X_i$ and $X_j$ |
| $\rho_{ij}$ | Spearman rank correlation coefficient between $X_i$ and $X_j$ |
| $\mu_{MLE}$ | Maximum likelihood estimate for $\mu$ |
| $\sigma_{MLE}$ | Maximum likelihood estimate for $\sigma$ |
| $\delta_{MLE}$ | Maximum likelihood estimate for $\delta$ |
| $\sigma'_p$ | Preconsolidation stress |
| $(q_t - u_2)/\sigma'_v$ | Effective cone tip resistance |
| $(q_t - \sigma_v)/\sigma'_v$ | Normalized cone tip resistance normalized |
| $(u_2 - u_0)/\sigma'_v$ | Normalized excess pore pressure |
| $\mu_{update}$ | Updated (conditional) mean vector |
| $|C|$ | Determinant of C matrix |
| $a_X, b_X, b_X^*, a_Y, b_Y$ | Parameters for the Johnson distribution |
| B | Bootstrap sample size |
| $B_q$ | Pore pressure ratio $= (u_2 - u_0)/(q_t - \sigma_v)$ |
| C | Correlation matrix (or covariance matrix) |
| $C^{[ij]}$ | $(i, j)$ Partition of C matrix |
| $C^{-1}$ | Inverse of C matrix |
| CDF | Cumulative distribution function |
| COV(.,.) | Covariance |
| COV | Coefficient of variation |
| CPTU | Piezocone penetration test |
| $C_{update}$ | Updated (conditional) covariance matrix |
| $d$ | Dimension of X (or dimension of Y) |
| $E_u$ | Undrained modulus |
| F(.) | Cumulative distribution function (CDF) |
| $f(.)$ | Probability density function (PDF) |
| $\Gamma_{\chi^2_d}$ | CDF of the $\chi$-square distribution with d degrees of freedom |
| $F_n(\cdot)$ | Empirical cumulative distribution function |
| $H_0$ | Null hypothesis |
| $H_1$ | Alternative hypothesis |
| $I(\cdot)$ | Indicator function |
| iid | Independent identically distributed |
| LL | Liquid limit |
| $m$ | Sample mean |

| | |
|---|---|
| $m_i$ | Sample mean of $X_i$ or $Y_i$ |
| $N(\mu, \sigma^2)$ | Normal distribution with mean $= \mu$ and variance $= \sigma^2$ |
| $n$ | Sample size |
| N | SPT-N value |
| $n_{ij}$ | Number of the bivariate $(X_i, X_j)$ data points |
| $N_{kF}$ | Cone factor $= (q_t - u_2)/s_u$ |
| $N_{kT}$ | Cone factor $= (q_t - \sigma_v)/s_u$ |
| OCR | Overconsolidation ratio |
| $P(\cdot)$ | Probability |
| $p$ | $p$-Value in hypothesis testing |
| $P_a$ | One atmosphere pressure ($=101.3$ kPa) |
| PDF | Probability density function |
| PI | Plasticity index |
| $Q_d$ | Mahalanobis distance |
| $q_t$ | Corrected cone tip resistance |
| $s$ | Sample standard deviation |
| $s_i$ | Sample standard deviation of $X_i$ or $Y_i$ |
| SPT | Standard penetration test |
| $S_t$ | Sensitivity |
| $s_u(\text{mob})$ | *In situ* undrained shear strength mobilized in embankment and slope |
| SU, SB, SL | Three members for the Johnson distribution |
| $s_u$ | Undrained shear strength |
| $s_u^{re}$ | Remolded undrained shear strength |
| $t_\eta$ | $\eta$-Percentile of Student's $t$-distribution |
| U | Uniform $[0, 1]$ random variable |
| U | Upper triangle Cholesky factor of $\mathbf{C}$ matrix |
| $u_0$ | Hydrostatic pore pressure |
| $u_2$ | Pore pressure behind the cone |
| $x_\eta$ | $\eta$-Percentile of standard normal distribution |
| $X^{(k)}$ | kth sample of X |
| $\mathbf{X}$ | (Correlated) standard normal vector $(X_1, X_2, ..., X_d)^T$ |
| X | (Standard) normal random variable |
| $\mathbf{X}^{[i]}$ | ith partition vector of $\mathbf{X}$ |
| $X_i$ | ith component in $\mathbf{X} = (X_1, X_2, ..., X_d)^T$ |
| $X_n$ | Normalized X $[=(X - b_X)/a_X]$ |
| $X_r$ | Rank of X |
| $y_\eta$ | $\eta$-Percentile of Y |
| $Y^{(k)}$ | kth sample of Y |
| $\mathbf{Y}$ | Random vector of $(Y_1, Y_2, ..., Y_d)^T$ |
| Y | Uncertain soil parameter |
| $Y_i$ | ith component in $\mathbf{Y} = (Y_1, Y_2, ..., Y_d)^T$ |
| $Y_n$ | Normalized Y $[=(Y - b_Y)/a_Y]$ |
| Z | Uncorrelated standard normal vector $(Z_1, Z_2, ..., Z_d)^T$ |
| $\sigma_v$ | Total vertical stress |

## REFERENCES

Arslan, O. 2004. Family of multivariate generalized t distributions. *Journal of Multivariate Analysis*, 89(2), 329–333.

Arwade, S.R. 2005. Translation vectors with non-identically distributed components. *Probabilistic Engineering Mechanics*, 20, 158–167.

Azzalini, A. and Capitanio, A. 2003. Distributions generated by perturbation of symmetry with emphasis on a multivariate skew distribution. *Journal of Royal Statistical Society*, series B, 65, 367–389.

Box, G.E.P. and Muller, M.E. 1958. A note on the generation of random normal deviates. *Annals of Mathematical Statistics*, 29, 610–611.

Carpenter, J. and Bithell, J. 2000. Bootstrap confidence intervals: When, which, what? A practical guide for medical statisticians. *Statistics in Medicine*, 19, 1141–1164.

Ching, J. and Phoon, K.K. 2012. Modeling parameters of structured clays as a multivariate normal distribution. *Canadian Geotechnical Journal*, 49(5), 522–545.

Ching, J. and Phoon, K.K. 2013. Multivariate distribution for undrained shear strengths under various test procedures. *Canadian Geotechnical Journal*, 50(9), 907–923.

Ching, J. and Phoon, K.K. 2014a. Transformations and correlations among some parameters of clays—The database. *Canadian Geotechnical Journal*, 51(6), 663–685.

Ching, J. and Phoon, K.K. 2014b. Correlations among some clay parameters—The multivariate distribution. *Canadian Geotechnical Journal*, 51(6), 686–704.

Ching, J., Phoon, K.K., and Chen, C.H. 2014b. Modeling CPTU parameters of clays as a multivariate normal distribution. *Canadian Geotechnical Journal*, 51(1), 77–91.

Ching, J., Phoon, K.K., and Chen, Y.C. 2010. Reducing shear strength uncertainties in clays by multivariate correlations. *Canadian Geotechnical Journal*, 47(1), 16–33.

Ching, J., Phoon, K.K., and Yu, J.W. 2014a. Linking site investigation efforts to final design savings with simplified reliability-based design methods. *Journal of Geotechnical and Geoenvironmental Engineering*, ASCE, 140(3), 04013032.

Conover, W.J. 1999. *Practical Nonparametric Statistics*. 3rd edition. John Wiley & Sons, Inc., New York.

Dithinde, M., Phoon, K.K., De Wet, M., and Retief, J.V. 2011. Characterisation of model uncertainty in the static pile design formula. *Journal of Geotechnical and Geoenvironmental Engineering*, ASCE, 137(1), 70–85.

Donoho, D.L. and Johnstone, I.M. 1994. Ideal spatial adaptation by wavelet shrinkage. *Biometrika*, 81, 425–455.

Efron, B. 1981. Nonparametric standard errors and confidence intervals. *Canadian Journal of Statistics*, 9, 139–172.

Efron, B. 1987. Better bootstrap confidence intervals. *Journal of the American Statistical Association*, 82, 316–331.

Efron, B. and Tibshirani, R. 1993. *An Introduction to the Bootstrap*. Chapman and Hall/CRC Press, Boca Raton, FL.

Goldsworthy, J.S., Jaksa, M.B., Fenton, G.A., Griffiths, D.V., Kaggwa W.S., and Poulos, H.G. 2007. Measuring the risk of geotechnical site investigations. *Proceedings of Geo-Denver 2007*, Denver.

Hald, A. 1952. *Statistical Theory with Engineering Applications*. John Wiley and Sons, New York.

Isserlis, L. 1918. On a formula for the product–moment coefficient of any order of a normal frequency distribution in any number of variables. *Biometrika*, 12(1/2), 134–139.

Jaksa, M.B., Goldsworthy, J.S., Fenton, G.A., Kaggwa, W.S., Griffiths, D.V., Kuo, Y.L., and Poulos, H.G. 2005. Towards reliable and effective site investigations. *Géotechnique*, 55(2), 109–121.

Jaksa, M.B., Kaggwa, W.S., Fenton, G.A., and Poulos, H.G. 2003. A framework for quantifying the reliability of geotechnical investigations. *Applications of Statistics and Probability in Civil Engineering*, ICASP9, Millpress, San Francisco, Rotterdam, 2, 1285–1291.

Kimball, B.F. 1960. On the choice of plotting positions on probability paper. *Journal of the American Statistical Association*, 55, 546–560.

Kowalski, C.J. 1970. The performance of some rough tests for bivariate normality before and after coordinate transformations to normality. *Technometrics*, 12(3), 517–544.

Kulhawy, F.H. and Mayne, P.W. 1990. *Manual on Estimating Soil Properties for Foundation Design*. Report EL-6800, Electric Power Research Institute, Palo Alto.

Marsaglia, G. and Bray, T.A. 1964. A convenient method for generating normal variables. *SIAM Review*, 6, 260–264.

Mesri, G. and Huvaj, N. 2007. Shear strength mobilized in undrained failure of soft clay and silt deposits. *Advances in Measurement and Modeling of Soil Behavior (GSP 173)*, Ed. D.J. DeGroot et al., ASCE, Reston, VA, 1–22.

Phoon, K.K. 2008. Numerical recipes for reliability analysis—A primer. *Chapter 1, Reliability-Based Design in Geotechnical Engineering: Computations and Applications*. Taylor & Francis, New York, 1–75.

Phoon, K.K. and Ching, J. 2013. Multivariate model for soil parameters based on Johnson distributions. *Foundation Engineering in the Face of Uncertainty, Geotechnical Special Publication* honoring Professor F.H. Kulhawy, Reston, VA, 337–353.

Phoon, K.K. and Kulhawy, F.H. 1999. Characterization of geotechnical variability. *Canadian Geotechnical Journal*, 36(4), 612–624.

Phoon, K.K. and Kulhawy, F.H. 2008. Serviceability limit state reliability-based design. Chapter 9, *Reliability-Based Design in Geotechnical Engineering: Computations and Applications*. Taylor & Francis, New York, 344–384.

Phoon, K.K., Santoso, A., and Quek, S.T. 2010. Probabilistic analysis of soil water characteristic curves. *Journal of Geotechnical and Geoenvironmental Engineering*, ASCE, 136(3), 445–455.

Rochelle, P.L., Zebdi, M., Leroueil, S., Tavenas, F., and Virely, D. 1988. Piezocone tests in sensitive clays of eastern Canada. *Proceedings of the 1st International Symposium on Penetration Testing*, Orlando, 2, 831–841.

Schenker, N. 1985. Qualms about bootstrap confidence intervals. *Journal of the American Statistical Association*, 80, 360–361.

Slifker, J.F. and Shapiro, S.S. 1980. The Johnson system: Selection and parameter estimation. *Technometrics*, 22(2), 239–246.

Terzaghi, K., Peck, R.B., and Mesri, G. 1996. *Soil Mechanics in Engineering Practice*. 3rd edition. John Wiley and Sons, New York.

# Chapter 2

# Modeling and simulation of bivariate distribution of shear strength parameters using copulas

*Dian-Qing Li and Xiao-Song Tang*

## 2.1 INTRODUCTION

It is well known that the shear strength parameters [cohesion ($c$) and friction angle ($\phi$)] are important parameters for evaluating deformation and stability of geotechnical structures, such as slope stability, bearing capacity of foundations, and earth pressure of retaining walls. As far as the reliability analysis of these geotechnical structures is concerned, the shear strength parameters are typically treated as uncertain parameters (Griffiths et al. 2011; Cherubini 2000; Abd Alghaffar and Dymiotis-Wellington 2007). Furthermore, it is widely accepted that $c$ and $\phi$ are negatively correlated in the literature (e.g., Low 2007; Li et al. 2011; Tang et al. 2012, 2013a). To evaluate the reliability of geotechnical structures exactly, the joint cumulative distribution function (CDF) or probability density function (PDF) of shear strength parameters should be known. It is concluded by the previous studies (e.g., Low 2007; Li et al. 2011; Tang et al. 2012, 2013a) that the negative correlation between cohesion and friction angle has a significant effect on geotechnical reliability and ignoring such a correlation would lead to an overestimate of the probability of failure.

Many researchers studied the geotechnical reliability considering the dependence between the shear strength parameters. In these studies, one or more of the three fundamental assumptions have been made. First, the marginal distributions of the shear strength parameters are normal distribution or have been transformed into normal distribution. Second, the shear strength parameters have the same type of marginal distributions. Third, the dependence structure between shear strength parameters is characterized by a Gaussian copula. With regard to the third assumption, the well-known examples include the commonly used Nataf model (Nataf 1962) and translation approach (Lebrun and Dutfoy 2009a, b; Li et al. 2012b; 2013b,c). In geotechnical practice, however, the shear strength parameters do not always follow normal distribution and the same marginal distributions. Furthermore, the Gaussian copula may not be adequate for characterizing the dependence structure between $c$ and $\phi$ as demonstrated by Tang et al. (2013a). Hence, it is of practical interest to develop a more general and flexible approach for modeling the bivariate distribution of shear strength parameters associated with geotechnical reliability problems.

To overcome the aforementioned three shortcomings, the past couple of years have witnessed a growing interest in applying copulas for modeling the joint probability distribution of multivariate data, particularly bivariate data (e.g., McNeil et al. 2005; Nelsen 2006). Copulas are functions that join multivariate distribution functions to their one-dimensional marginal distribution functions. There are many copulas in literature such as Gaussian, $t$, Frank, Clayton, Gumbel, and Plackett copulas. Each copula has its own dependence structure. Copulas provide a fairly general way for constructing multivariate distributions that satisfy some nonparametric measure of dependence and prescribed marginal distributions. The copula theory has been extensively used for financial and hydrological applications (e.g., McNeil et al. 2005; Genest and Favre 2007; Salvadori and De Michele 2007). Recently, the copula theory has been applied to geotechnical engineering. For example, Uzielli and Mayne (2011, 2012) investigated the dependence between load-displacement model parameters underlying vertically loaded shallow footings on sands using copula. Li et al. (2012a, 2013a) constructed the bivariate distribution of hyperbolic curve-fitting parameters underlying load-settlement curves of piles using copulas. Tang et al. (2013a) investigated the impact of copula selection on geotechnical reliability. Wu (2013a) proposed a copula-based sampling method for probabilistic slope stability analysis. Wu (2013b) further employed the Gaussian and Frank copulas to model the trivariate distribution among cohesion, friction angle, and unit weight of soils. It is evident that the applications of copulas to geotechnical engineering are scarce, especially in geotechnical reliability analysis (Dutfoy and Lebrun 2009; Tang et al. 2013b, c). The potential applications of copulas for geotechnical reliability analyses should be further explored.

This chapter aims to develop a copula-based approach for modeling and simulating the bivariate distribution of shear strength parameters. In this approach, the aforementioned three limitations are removed. This chapter is organized as follows. In Section 2.2, the copula theory is briefly introduced. Thereafter, the procedure for step-by-step modeling of the bivariate distribution of shear strength parameters is illustrated using measured data. In Section 2.4, the simulation algorithms for copulas and bivariate distribution are presented in detail. The effect of copulas on the probability of retaining wall overturning failure is investigated in Section 2.5.

## 2.2 COPULA THEORY

As mentioned in the introduction, one objective of this chapter is to model and simulate the bivariate distribution of shear strength parameters using copulas. To facilitate the understanding of subsequent copula applications, the copula theory is first introduced. The definition of copulas is presented in Section 2.2.1. Then, two commonly used dependence measures, namely Pearson linear correlation coefficient and Kendall rank correlation coefficient, are explained in Section 2.2.2. The adopted four bivariate copulas in this chapter are provided in Section 2.2.3.

### 2.2.1 Definition of copulas

The word *copula* originated from a Latin word for "link" or "tie" that connects different things. To define it (e.g., Nelsen 2006): Copulas are functions that join or couple multivariate distribution functions to their one-dimensional marginal distribution functions. Alternatively, copulas are multivariate distribution functions whose one-dimensional marginal distributions are uniform in the interval of [0, 1]. Since Sklar's theorem is the foundation of many applications of the copula theory, such a theorem is introduced first.

**Sklar's theorem** (e.g., Nelsen 2006). Let $F(x_1, x_2, ..., x_n)$ be the joint CDF of a random vector $\mathbf{X} = [X_1, X_2, ..., X_n]$. Its marginal CDFs are $F_1(x_1)$, $F_2(x_2)$, ..., $F_n(x_n)$. The concept of a marginal CDF has been introduced in Chapter 1. A joint CDF is defined as the probability that $\mathbf{X}$ is less than or equal to a specific numerical vector $[x_1, x_2, ..., x_n]$:

$$F(x_1, x_2, ..., x_n) = P(X_1 \leq x_1, X_2 \leq x_2, ..., X_n \leq x_n) \quad (2.1)$$

where $P(.)$ denotes the probability. Then there exists an $n$-dimensional copula $C$ such that for all real $[x_1, x_2, ..., x_n]$,

$$F(x_1, x_2, \ldots, x_n) = C(F_1(x_1), F_2(x_2), \ldots, F_n(x_n); \theta) \tag{2.2}$$

where $\theta$ is a vector of copula parameters describing the dependence among $X_1, X_2, \ldots, X_n$. The dimension of this vector varies with copula types from $1 \times 1$ to $1 \times 0.5n(n-1)$. For the Gaussian and $t$ copulas, the number of copula parameters is $0.5n(n-1)$, which is the same as the number of the correlation coefficient pairs in a multivariate normal distribution discussed in Chapter 1. If $C$ is an $n$-dimensional copula and $F_1(x_1), F_2(x_2), \ldots, F_n(x_n)$ are marginal CDFs, then the function $F(x_1, x_2, \ldots, x_n)$ defined by Equation 2.2 is a joint CDF with marginal distributions $F_1(x_1), F_2(x_2), \ldots, F_n(x_n)$.

Sklar's theorem essentially states that the joint CDF of $\mathbf{X}$ can be expressed in terms of a copula function and its marginal CDFs. In other words, fitting a joint CDF to measured data using copulas involves two steps: (1) determining the best-fit marginal distributions for all individual random variables, and (2) identifying the copula that provides the best fit to the measured dependence structure from a set of candidate copulas. The above two steps can be carried out separately, allowing different marginal distributions and dependence structures to be incorporated into a multivariate distribution. This advantage underlying the copula approach is critical since the processes of determining the marginal distributions and the correlation structure in a multivariate distribution are decoupled.

The joint PDF of $\mathbf{X}$ can be derived from the joint CDF of $\mathbf{X}$. Basically, the joint PDF is the derivative of the joint CDF (or the joint CDF is the integration function of the joint PDF). By taking derivatives of Equation 2.2, the joint PDF of $\mathbf{X}$, $f(x_1, x_2, \ldots, x_n)$, can be obtained as (e.g., McNeil et al. 2005)

$$
\begin{aligned}
f(x_1, x_2, \ldots, x_n) &= \frac{\partial^n C(F_1(x_1), F_2(x_2), \ldots, F_n(x_n); \theta)}{\partial F_1(x_1) \ldots \partial F_n(x_n)} \prod_{i=1}^{n} \frac{\partial F_i(x_i)}{\partial x_i} \\
&= D(F_1(x_1), F_2(x_2), \ldots, F_n(x_n); \theta) \prod_{i=1}^{n} f_i(x_i)
\end{aligned}
\tag{2.3}
$$

where $D = \partial^n C(F_1(x_1), F_2(x_2), \ldots, F_n(x_n); \theta) / \partial F_1(x_1) \ldots \partial F_n(x_n)$ is a copula density function, which is the derivative of a copula function $C$. $f_i(x_i) = \partial F_i(x_i)/\partial x_i$ is the marginal PDF of $X_i$ for $i = 1, 2, \ldots, n$. By introducing $u_i = F_i(x_i)$, the copula function $C(F_1(x_1), F_2(x_2), \ldots, F_n(x_n); \theta)$ and the copula density function $D(F_1(x_1), F_2(x_2), \ldots, F_n(x_n); \theta)$ can be rewritten as $C(u_1, u_2, \ldots, u_n; \theta)$ and $D(u_1, u_2, \ldots, u_n; \theta)$, respectively. Note that $U_i = F_i(X_i)$ is a standard uniform random variable. This is because the CDF of $U_i$ can be expressed as

$$P(U_i \le u_i) = P(F_i(X_i) \le u_i) = P(X_i \le F_i^{-1}(u_i)) = F_i(F_i^{-1}(u_i)) = u_i \tag{2.4}$$

It is clear from Equation 2.4 that the CDF of $U_i$ is equal to $u_i$. Thus, $U_i$ is indeed a standard uniform random variable that is bounded on the interval of $[0, 1]$.

Since the focus of this chapter is the modeling of the bivariate distribution of shear strength parameters, the copula theory involving two variables (i.e., $n = 2$) is introduced in detail from hereon. According to Sklar's theorem, the bivariate CDF of two random variables $X_1$ and $X_2$, $F(x_1, x_2)$, can be given by

$$F(x_1, x_2) = C(F_1(x_1), F_2(x_2); \theta) = C(u_1, u_2; \theta) \tag{2.5}$$

where $C(u_1, u_2; \theta)$ is a bivariate copula function in which $\theta$ is a copula parameter measuring the dependence between $X_1$ and $X_2$. It is worthwhile to point out that the vector $\theta$ in Equation 2.2 is reduced to a single parameter $\theta$ when $n = 2$. In other words, all bivariate copulas have only one copula parameter $\theta$. By taking derivatives of Equation 2.5, the bivariate PDF of $X_1$ and $X_2$, $f(x_1, x_2)$, is obtained as

$$f(x_1, x_2) = D(F_1(x_1), F_2(x_2); \theta)f_1(x_1)f_2(x_2) = D(u_1, u_2; \theta)f_1(x_1)f_2(x_2) \tag{2.6}$$

where $D(u_1, u_2; \theta)$ is a bivariate copula density function, which is given by

$$D(u_1, u_2; \theta) = \partial^2 C(u_1, u_2; \theta)/\partial u_1 \partial u_2 \tag{2.7}$$

Theoretically, the joint CDF and PDF of $X_1$ and $X_2$ can be determined by Equations 2.5 and 2.6 if the marginal CDFs of $X_1$ and $X_2$, and the copula function are known. For example, when both $X_1$ and $X_2$ are standard normal random variables (the CDF and PDF of a standard normal random variable are given in Chapter 1), substituting the CDFs of $X_1$ and $X_2$ into the Gaussian copula in Table 2.1 leads to the well-known bivariate standard normal distribution. It is noted that this is not the case for other copulas. If the CDFs of $X_1$ and $X_2$ are substituted into the Plackett copula, a bivariate standard normal distribution will not be expected. In other words, the dependence structure underlying a bivariate standard normal distribution is uniquely characterized by a Gaussian copula (Lebrun and Dutfoy 2009a, b). Hence, the Pearson correlation coefficient $\rho$ (see Chapter 1) underlying the bivariate standard normal distribution is incidentally equal to the copula parameter $\theta$ underlying the Gaussian copula.

Figure 2.1 shows the contour plots for the bivariate distributions of two standard normal random variables $X_1$ and $X_2$ using a Gaussian copula with $\theta$ equal to $-0.5$, $0$, and $0.5$. Note that these copula parameters $\theta$ are obtained from the Kendall rank correlation coefficients $\tau$ (see Section 2.2.2) equal to $-1/3$, $0$, and $1/3$, respectively. It is clear that there is a strong negative correlation between $X_1$ and $X_2$ when $\theta = -0.5$ ($\tau = -1/3$), whereas a strong positive correlation exists between $X_1$ and $X_2$ when $\theta = 0.5$ ($\tau = 1/3$). For $\theta = 0$ ($\tau = 0$), $X_1$ and $X_2$ are uncorrelated. It is well known that an important step in copula modeling is the determination of copula parameters $\theta$. Since various copulas have their own parameters, it is desirable to have a common dependence measure such as the Kendall rank correlation coefficient adopted above to obtain the copula parameters $\theta$ from the measured data. In the next section, two commonly used dependence measures are introduced for this purpose.

### 2.2.2 Dependence measures

This section focuses on two kinds of dependence measure: (1) the usual Pearson linear correlation coefficient (Pearson's rho, $\rho$), and (2) the Kendall rank correlation coefficient (Kendall's tau, $\tau$). Their relations to the copula parameter $\theta$ are highlighted.

#### 2.2.2.1 Pearson's rho

The Pearson linear correlation coefficient has been defined in Chapter 1. In this chapter, the Pearson linear correlation coefficient and its relation to the copula parameter $\theta$ are further explained within the framework of the copula theory. Pearson's rho, which is also called a

Table 2.1 Summary of the adopted bivariate copula functions and their parameter domains

| Copula | Copula function, $C(u_1, u_2; \theta)$ | Copula density function, $D(u_1, u_2; \theta)$ | Generator function, $\varphi_\theta(t)$ | Range of $\theta$ |
|---|---|---|---|---|
| Gaussian | $\Phi_\theta(\Phi^{-1}(u_1),\ \Phi^{-1}(u_2))$ | $\dfrac{1}{\sqrt{1-\theta^2}} \exp\left[-\dfrac{\varsigma_1^2\theta^2 - 2\theta\varsigma_1\varsigma_2 + \varsigma_2^2\theta^2}{2(1-\theta^2)}\right],\ \begin{aligned}\varsigma_1 &= \Phi^{-1}(u_1)\\ \varsigma_2 &= \Phi^{-1}(u_2)\end{aligned}$ | — | $[-1, 1]$ |
| Plackett | $\dfrac{S - \sqrt{S^2 - 4u_1u_2\theta(\theta-1)}}{2(\theta-1)},$ $S = 1 + (\theta-1)(u_1 + u_2)$ | $\dfrac{\theta[1 + (\theta-1)(u_1 + u_2 - 2u_1u_2)]}{\{[1 + (\theta-1)(u_1 + u_2)]^2 - 4u_1u_2\theta(\theta-1)\}^{(3/2)}}$ | — | $(0, \infty)\backslash\{1\}$ |
| Frank | $-\dfrac{1}{\theta}\ln\left[1 + \dfrac{(e^{-\theta u_1}-1)(e^{-\theta u_2}-1)}{e^{-\theta}-1}\right]$ | $\dfrac{-\theta(e^{-\theta}-1)e^{-\theta(u_1+u_2)}}{\left[(e^{-\theta}-1) + (e^{-\theta u_1}-1)(e^{-\theta u_2}-1)\right]^2}$ | $-\ln\dfrac{e^{-\theta t}-1}{e^{-\theta}-1}$ | $(-\infty, \infty)\backslash\{0\}$ |
| No.16 | $\dfrac{1}{2}\left(S + \sqrt{S^2 + 4\theta}\right),$ $S = u_1 + u_2 - 1 - \theta\left(\dfrac{1}{u_1} + \dfrac{1}{u_2} - 1\right)$ | $\dfrac{1}{2}\left(1 + \dfrac{\theta}{u_1^2}\right)\left(1 + \dfrac{\theta}{u_2^2}\right)S^{-(1/2)}\left\{-S^{-1}\left[u_1 + u_2 - 1 - \theta\left(\dfrac{1}{u_1} + \dfrac{1}{u_2} - 1\right)\right]^2 + 1\right\},$ $S = \left[u_1 + u_2 - 1 - \theta\left(\dfrac{1}{u_1} + \dfrac{1}{u_2} - 1\right)\right]^2 + 4\theta$ | $\left(\dfrac{\theta}{t} + 1\right)(1 - t)$ | $[0, \infty)$ |

Note: The symbol "—" denotes that the generator function is not available; $\Phi^{-1}$ is the inverse standard normal distribution function with Pearson linear correlation coefficient $\theta$; and $\Phi_\theta$ is the bivariate standard normal distribution function with Pearson linear correlation coefficient $\theta$

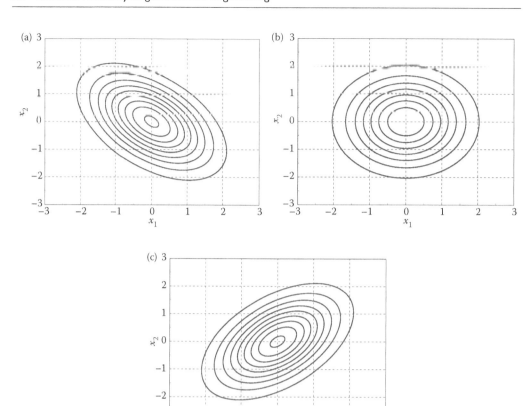

*Figure 2.1* Contour plots for the bivariate distributions of $X_1$ and $X_2$ using a Gaussian copula. (a) $\theta = -0.5$, $\tau = -1/3$, (b) $\theta = 0$, $\tau = 0$, and (c) $\theta = 0.5$, $\tau = 1/3$.

product–moment correlation coefficient, measures the degree of linear dependence between $X_1$ and $X_2$ as (e.g., Mari and Kozt 2001)

$$\rho = \frac{\text{Cov}(X_1, X_2)}{\sigma_1 \sigma_2} \tag{2.8}$$

where $\text{Cov}(X_1, X_2)$ is the covariance between $X_1$ and $X_2$; $\sigma_1$ and $\sigma_2$ are the standard deviations of $X_1$ and $X_2$, respectively. The effect of Pearson's rho on the shape of the bivariate standard normal distributions can be found in Figure 1.12 of Chapter 1. By definition of the covariance, Pearson's rho can be further expressed as

$$\rho = \int_{-\infty}^{+\infty} \int_{-\infty}^{+\infty} \left( \frac{x_1 - \mu_1}{\sigma_1} \right) \left( \frac{x_2 - \mu_2}{\sigma_2} \right) f(x_1, x_2) dx_1 dx_2 \tag{2.9}$$

where $\mu_1$ and $\mu_2$ are the means of $X_1$ and $X_2$, respectively. Substituting Equation 2.6 into Equation 2.9, the integral relation between $\rho$ and $\theta$ can be obtained as

$$\rho = \int\limits_{-\infty}^{+\infty} \int\limits_{-\infty}^{+\infty} \left(\frac{x_1 - \mu_1}{\sigma_1}\right)\left(\frac{x_2 - \mu_2}{\sigma_2}\right) D(F_1(x_1), F_2(x_2); \theta) f_1(x_1) f_2(x_2) dx_1 dx_2 \qquad (2.10)$$

It is evident from Equation 2.10 that the determination of copula parameters using Pearson's rho depends on both the correlation coefficient and the marginal distributions. For given marginal distributions $F_1(x_1)$ and $F_2(x_2)$ of $X_1$ and $X_2$, and correlation coefficient $\rho$ between $X_1$ and $X_2$, the preceding integral equation can be solved iteratively to obtain $\theta$. For a Gaussian copula, Equation 2.10 is reduced to Equation 1.108 of Chapter 1. The observation that the absolute values of Pearson's rho between non-normal variables are smaller than those between the transformed standard normal variables (see Figure 1.32 in Chapter 1) can be readily explained by Equation 2.10. In practical application, Pearson's rho is often estimated from the measured data. The sample version of Pearson's rho is expressed as (e.g., Mari and Kozt 2001)

$$\rho = \frac{\sum_{i=1}^{N} \left(x_{1i} - \overline{x_1}\right)\left(x_{2i} - \overline{x_2}\right)}{\sqrt{\sum_{i=1}^{N} \left(x_{1i} - \overline{x_1}\right)^2}\sqrt{\sum_{i=1}^{N} \left(x_{2i} - \overline{x_2}\right)^2}} \qquad (2.11)$$

where $(x_{1i}, x_{2i})$ denotes a pair of $X_1$ and $X_2$ values; $N$ is the sample size; and $\overline{x_1}$ and $\overline{x_2}$ are the sample means of $X_1$ and $X_2$, respectively. It is noted that this equation is the same as Equation 1.40 of Chapter 1. Since Pearson's rho is a measure of linear dependence, it is valid only when the joint CDF is Gaussian. Therefore, it is not adequate for characterizing a nonlinear dependence between two random variables. Furthermore, Pearson's rho is invariant only under strictly monotonic linear transformations; thus, the dependence measure needs to be evaluated for each nonlinear transformation. If random variables are not jointly Gaussian or some nonlinear transformations are used, Pearson's rho is not an effective dependence measure. Sometimes, it may produce misleading results (Tang et al. 2013a).

### 2.2.2.2  Kendall's tau

To remove the aforementioned limitations underlying Pearson's rho, rank correlation can be used. The commonly used rank correlation includes Spearman rank correlation coefficient and Kendall rank correlation coefficient. The former has been introduced in Chapter 1. In this chapter, the latter is adopted to determine the copula parameters due to its intrinsic relation to a copula function. As a rank correlation, Kendall's tau depends only on the ranks underlying the sample for each variable of interest rather than the actual numerical values. The idea of a "rank" is quite simple. Consider a list of 10 numbers, containing realizations of a random variable $X_1$. The smallest number is assigned a rank = 1, the second smallest number is assigned a rank = 2, and so forth. The list of 10 numbers $(x_{1i})$ is replaced by a list of ranks $(r_{1i})$, which is a permutation of integers from 1 to 10. Numerical examples to illustrate a rank are referred to Equations 1.43 and 1.44 in Chapter 1.

Unlike Pearson's rho, Kendall's tau measures the degree of concordance between $X_1$ and $X_2$. The concept of concordance is straightforward: $X_1$ and $X_2$ are concordant if "large" values of $X_1$ tend to be with "large" values of $X_2$ or "small" values of $X_1$ tend to be with "small" values of $X_2$. Let $(X_1', X_2')$ be an independent copy of $(X_1, X_2)$. Then, a formulation of concordance is as follows: $(X_1, X_2)$ and $(X_1', X_2')$ are concordant if $(X_1 - X_1')(X_2 - X_2') > 0$

and discordant if $(X_1 - X_1')(X_2 - X_2') < 0$. In mathematics, Kendall's tau is defined as the probability of concordance minus the probability of discordance between $X_1$ and $X_2$ (e.g., Nelsen 2006)

$$\tau = P[(X_1 - X_1')(X_2 - X_2') > 0] - P[(X_1 - X_1')(X_2 - X_2') < 0] \tag{2.12}$$

In Equation 2.12, the first term on the right-hand side is the probability of concordance and the second term is the probability of discordance. Like Pearson's rho, Kendall's tau also produces correlation coefficients between –1 and 1. However, Kendall's tau does not assume that the relationship between two random variables is linear. Thus, it is invariant with respect to strictly monotonic linear and nonlinear transformations, which allows for a unique dependence measure for all transformed variables. Kendall's tau can be further expressed in terms of a copula function $C(u_1, u_2; \theta)$ as (e.g., Nelsen 2006)

$$\tau = 4 \int_0^1 \int_0^1 C(u_1, u_2; \theta) dC(u_1, u_2; \theta) - 1 \tag{2.13}$$

The derivation of Equation 2.13 from Equation 2.12 can be found in Nelsen (2006). It is clear from Equation 2.13 that the determination of copula parameters using Kendall's tau depends only on the correlation coefficient. It is independent of the marginal distributions. This is a critical difference between Pearson's rho and Kendall's tau as demonstrated by Li et al. (2012a). For a given correlation coefficient $\tau$ between $X_1$ and $X_2$, the preceding integral equation can be solved iteratively to find $\theta$. The sample version of Kendall's tau is given by

$$\tau = \frac{\sum_{i<j} \text{sign}[(x_{1i} - x_{1j})(x_{2i} - x_{2j})]}{0.5N(N-1)} \tag{2.14}$$

where $N$ is the sample size; sign(.) is calculated by

$$\text{sign} = \begin{cases} 1 & (x_{1i} - x_{1j})(x_{2i} - x_{2j}) \geq 0 \text{ (concordant)} \\ -1 & (x_{1i} - x_{1j})(x_{2i} - x_{2j}) < 0 \text{ (discordant)} \end{cases} \quad i, j = 1, 2, ..., N \tag{2.15}$$

The numerator in Equation 2.14 is the difference between the number of concordant pairs and the number of discordant pairs, whereas the denominator denotes the total number of observation pairs for a sample size of $N$. Consider the $(X_1, X_2)$ samples in Equation 1.43 of Chapter 1. The total number of pairs for $N = 5$ is $0.5 \times 5 \times (5-1) = 10$. The number of concordant pairs is 9 and 1 is for the number of discordant pairs. Thus, Kendall's tau is computed as $\tau = (9-1)/10 = 0.8$. This value can also be obtained using MATLAB® function corr($X_1$, $X_2$, 'type', 'Kendall').

Generally, Kendall's tau is not the same as Pearson's rho. For the $(X, Y)$ data points in Table 1.6 of Chapter 1, the Pearson linear correlation coefficient is computed as $\rho = 0.906$ using corr($X$, $Y$, 'type', 'Pearson'), whereas the Kendall rank correlation coefficient $\tau = 1$ is obtained by the MATLAB function corr($X$, $Y$, 'type', 'Kendall'). They are not the same; this is because Pearson's rho measures the degree of linear dependence between $X_1$ and $X_2$ and, in this example, $X_1$ and $X_2$ are nonlinear. On the contrary, Kendall's tau measures the concordance between $X_1$ and $X_2$. It is clear that the relationship $Y = X^3$ is perfectly concordant and, thus, a Kendall rank correlation coefficient of 1 is expected.

### 2.2.3 Four selected copulas

There are many copulas in the literature belonging to various classes of copula families. The difference among various copulas is mainly characterized by dependency characteristics such as symmetry, tail dependence, and range of correlation coefficients (e.g., Nelsen 2006). Owing to the strongly negative correlation between $c$ and $\phi$, the copulas allowing a wide range of negative correlation coefficients are selected to fit the dependence structure between $c$ and $\phi$. A literature review reveals that the Gaussian copula, Plackett copula, Frank copula, and No.16 copula (e.g., Nelsen 2006) are appropriate for describing such dependence structure between $c$ and $\phi$. The aforementioned four copulas along with the copula parameter, $\theta$, are summarized in Table 2.1. Among the four copulas, the Gaussian copula is an elliptical copula. The Plackett copula is a member of the Plackett copula family. The Frank and No.16 copulas are Archimedean copulas. All the four copulas can describe negative dependences, and the values of the Kendall rank correlation coefficients between $c$ and $\phi$ can approach $-1$. For a more general rule of choosing copulas, the interested readers are referred to Nelsen (2006) and McNeil et al. (2005).

Figure 2.2 shows the contour plots for the bivariate distributions of two standard normal variables $X_1$ and $X_2$ using Plackett, Frank, and No.16 copulas. The bivariate distributions of $X_1$ and $X_2$ using a Gaussian copula have been presented in Figure 2.1. These bivariate

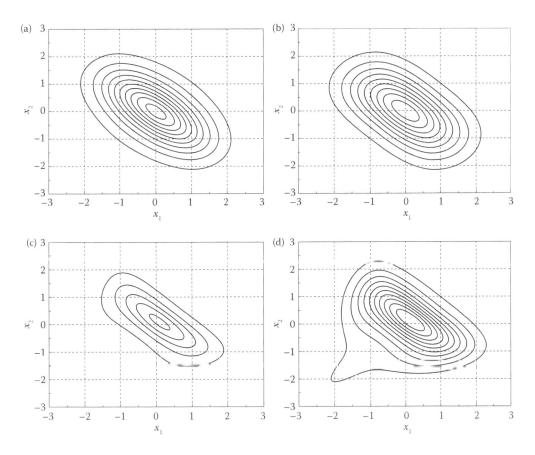

*Figure 2.2* Contour plots for the bivariate distributions of $X_1$ and $X_2$ using Plackett, Frank and No.16 copulas. (a) Plackett copula ($\theta = 0.213$, $\tau = -1/3$), (b) Frank copula ($\theta = -3.306$, $\tau = -1/3$), (c) No.16 copula ($\theta = 0.035$, $\tau = -1/3$), and (d) No.16 copula ($\theta = 0.067$, $\theta = -1/5$).

distributions are constructed by substituting the CDFs of $X_1$ and $X_2$ into the corresponding copula functions in Table 2.1. The copula parameters θ are obtained from Kendall rank correlation coefficients τ using Equation 2.13. It is clear that the bivariate distributions constructed from different copulas are different, indicating that each copula has its own dependence structure. It is well known that the dependence structure underlying a copula has a significant impact on reliability (Tang et al. 2013a,b,c). To represent a dependence structure, the concepts of symmetry and tail dependence can be used. These two concepts are introduced below.

There are two forms of symmetry for a copula function. One is the exchangeability. The other is the radial symmetry. For the former, a copula function $C(u_1, u_2; \theta)$ is said to be symmetric when $C(u_1, u_2; \theta) = C(u_2, u_1; \theta)$ for all $(u_1, u_2)$ in $[0, 1]^2$. Alternatively, $C(u_1, u_2; \theta)$ is symmetric when $u_1$ and $u_2$ are exchangeable (e.g., Nelsen 2006). The exchangeability can only guarantee that a copula function is symmetrical with respect to the 45° diagonal line of a unit square (i.e., a domain defined by $[0, 1]^2$). It is clear from Table 2.1 that $u_1$ and $u_2$ are exchangeable for the Gaussian, Plackett, Frank, and No.16 copulas. Therefore, the Gaussian, Plackett, Frank, and No.16 copulas are symmetric copulas. If the variables $X_1$ and $X_2$ have the same marginal CDFs, the bivariate distributions of $X_1$ and $X_2$ using the symmetric copulas are also symmetric. For example, the bivariate distributions of $X_1$ and $X_2$ using the Gaussian, Plackett, Frank, and No.16 copulas in Figures 2.1 and 2.2 are symmetrical with respect to the 45° diagonal line of a square (i.e., a domain defined by $[-3, 3]^2$ in this example).

Besides the changeability, the other symmetry is the radial symmetry. A copula function $C(u_1, u_2; \theta)$ is said to be radially symmetric when $C(u_1, u_2; \theta) = u_1 + u_2 - 1 + C(1 - u_1, 1 - u_2; \theta)$ for all $(u_1, u_2)$ in $[0, 1]^2$ (e.g., Nelsen 2006). The radial symmetry implies that a copula function is symmetrical with respect to the center (i.e., a point defined by $[0.5, 0.5]$) of a unit square. With the above requirement, it can be concluded from Table 2.1 that the Gaussian, Plackett, and Frank copulas are radially symmetric. On the other hand, the No.16 copula is not radially symmetric. The reason is that $C(u_1, u_2; \theta)$ is not identical to $u_1 + u_2 - 1 + C(1 - u_1, 1 - u_2; \theta)$ for the No.16 copula. If $X_1$ and $X_2$ have the same marginal CDFs, the bivariate distributions of $X_1$ and $X_2$ using the radially symmetric copulas are also radially symmetric. For example, the bivariate distributions of $X_1$ and $X_2$ using the Gaussian, Plackett, and Frank copulas in Figures 2.1 and 2.2 are symmetrical with respect to the center (i.e., a point defined by $[0, 0]$ in this example) of a square. On the contrary, the bivariate distributions of $X_1$ and $X_2$ using the No.16 copula in Figure 2.2 are not symmetrical with respect to the center of a square. The common feature of radial symmetry underlying the Gaussian, Plackett, and Frank copulas will lead to relatively similar reliability results as demonstrated in Section 2.5.

Tail dependence, as its name says, measures the dependence between $X_1$ and $X_2$ in the upper-right quadrant or in the lower-left quadrant of a bivariate distribution. Recall that $F_1(x_1)$ and $F_2(x_2)$ are the marginal CDFs of two continuous random variables $X_1$ and $X_2$, respectively. Let $F_1^{-1}(x_1)$ and $F_2^{-1}(x_2)$ be the inverse CDFs of $X_1$ and $X_2$, respectively. Then, the coefficient of upper tail dependence, $\lambda_U$, is defined as the limit (if it exists) of the conditional probability that $X_2$ is greater than the $100t$-th percentile of $F_2$ given that $X_1$ is greater than the $100t$-th percentile of $F_1$ as $t$ approaches 1 (e.g., Nelsen 2006)

$$\lambda_U = \lim_{t \to 1^-} P[X_2 > F_2^{-1}(t) | X_1 > F_1^{-1}(t)] \tag{2.16}$$

where $F_1^{-1}(t)$ and $F_2^{-1}(t)$ are the $100t$-th percentiles of $F_1$ and $F_2$, respectively. If $\lambda_U \in (0,1]$, $X_1$ and $X_2$ are said to be asymptotically dependent the upper tail; if $\lambda_U = 0$, $X_1$ and $X_2$ are said to be asymptotically independent the upper tail. Similarly, the coefficient of lower tail dependence, $\lambda_L$, is defined as the limit (if it exists) of the conditional probability that $X_2$ is less than or equal to the $100t$-th percentile of $F_2$ given that $X_1$ is less than or equal to the $100t$-th percentile of $F_1$ as $t$ approaches 0 (e.g., Nelsen 2006)

$$\lambda_L = \lim_{t \to 0^+} P[X_2 \le F_2^{-1}(t) | X_1 \le F_1^{-1}(t)] \tag{2.17}$$

If $\lambda_L \in (0,1]$, $X_1$ and $X_2$ are said to be asymptotically dependent at the lower tail; if $\lambda_L = 0$, $X_1$ and $X_2$ are said to be asymptotically independent at the lower tail.

It is noted that although the tail dependence is associated with the bivariate distribution of $X_1$ and $X_2$, the two parameters $\lambda_U$ and $\lambda_L$ are nonparametric and depend only on the copula function of $X_1$ and $X_2$. Specifically, $\lambda_U$ can be expressed in terms of a copula function as

$$
\begin{aligned}
\lambda_U &= \lim_{t \to 1^-} P[X_2 > F_2^{-1}(t) | X_1 > F_1^{-1}(t)] \\
&= \lim_{t \to 1^-} \frac{P[X_1 > F_1^{-1}(t), X_2 > F_2^{-1}(t)]}{P[X_1 > F_1^{-1}(t)]} \\
&= \lim_{t \to 1^-} \frac{1 - P[X_1 \le F_1^{-1}(t)] - P[X_2 \le F_2^{-1}(t)] + P[X_1 \le F_1^{-1}(t), X_2 \le F_2^{-1}(t)]}{1 - P[X_1 \le F_1^{-1}(t)]} \\
&= \lim_{t \to 1^-} \frac{1 - 2t + C(t, t; \theta)}{1 - t}
\end{aligned}
\tag{2.18}
$$

Similarly, $\lambda_L$ can be expressed in terms of a copula function as

$$
\begin{aligned}
\lambda_L &= \lim_{t \to 0^+} P[X_2 \le F_2^{-1}(t) | X_1 \le F_1^{-1}(t)] \\
&= \lim_{t \to 0^+} \frac{P[X_1 \le F_1^{-1}(t), X_2 \le F_2^{-1}(t)]}{P[X_1 \le F_1^{-1}(t)]} \\
&= \lim_{t \to 0^+} \frac{C(t, t; \theta)}{t}
\end{aligned}
\tag{2.19}
$$

Equations 2.18 and 2.19 clearly state that the tail dependence of a bivariate distribution can be directly obtained from its copula function. As mentioned in the literature (e.g., Nelsen 2006), the Gaussian, Plackett, and Frank copulas do not have tail dependence. That is, $\lambda_U = 0$ and $\lambda_L = 0$ for the Gaussian, Plackett, and Frank copulas. Unlike the aforementioned three copulas, the No.16 copula has only lower tail dependence and the coefficient of lower tail dependence $\lambda_L$ is a constant of 0.5 for the whole range of the $\theta$ parameter. The lower tail dependence underlying the No.16 copula can be observed in Figure 2.2d for a relatively small correlation coefficient $\tau = -1/2$. It is noted that although $\lambda_L$ is a constant, the effect of this lower tail dependence is reduced and the No.16 copula becomes approximately radially symmetric when the negative correlation becomes strong (e.g., Nelsen 2006). This phenomenon can be observed by comparing Figure 2.2d ($\tau = -1/2$) with Figure 2.2c ($\tau = -1/3$). In general, the lower tail dependence of the No.16 copula has an important effect on reliability (see Section 2.5), which should be noted in practical applications.

It is known that a key step in copula modeling is the determination of copula parameters. In the previous applications, the copula parameters $\theta$ underlying the four copulas are obtained from the Kendall rank correlation coefficients $\tau$ using Equation 2.13. This is a dual integral equation. Solving this equation needs great efforts. In the following equations, some analytical or more concise relationships between $\tau$ and $\theta$ are presented. For example, if $C(u_1, u_2; \theta)$ is a Gaussian copula, then there is an explicit expression between $\tau$ and $\theta$:

$$\tau = \frac{2\arcsin(\theta)}{\pi} \tag{2.20}$$

Hence, the copula parameter $\theta$ underlying the Gaussian copula can be readily determined once the Kendall's tau $\tau$ between $X_1$ and $X_2$ is known. This $\theta$ can also be obtained using MATLAB function copulaparam('Gaussian', tau, 'type', 'Kendall').

For Archimedean copulas such as the Frank and No. 16 copulas, the double integral relationship between $\tau$ and $\theta$ in Equation 2.13 can be further simplified to a single integral (e.g., Nelsen 2006)

$$\tau = 1 + 4\int_0^1 \frac{\varphi_\theta(t)}{\varphi_\theta'(t)}\,dt \tag{2.21}$$

where $\varphi_\theta(t)$ is the generator function of an Archimedean copula and $\varphi_\theta'(t)$ is the first derivative of $\varphi_\theta(t)$ with respect to $t$. For the Frank copula, the MATLAB function copulaparam('Frank', tau, 'type', 'Kendall') can also be adopted to determine the copula parameter $\theta$.

It is noted that an Archimedean copula is associated with a generator function. In other words, an Archimedean copula is uniquely determined by its generator function. Table 2.1 also lists the generator functions for the Frank and No.16 copulas. Recollect that $F(x_1, x_2)$ is the joint distribution function of two continuous random variables $X_1$ and $X_2$. $F_1(x_1)$ and $F_2(x_2)$ are the marginal distribution functions of $X_1$ and $X_2$, respectively. Then, a generator function $\varphi_\theta$ is defined as a function that can make $F(x_1, x_2)$ factor into a sum of functions of $F_1(x_1)$ and $F_2(x_2)$ (e.g., Nelsen 2006)

$$\varphi_\theta(F(x_1,\ x_2)) = \varphi_\theta(F_1(x_1)) + \varphi_\theta(F_2(x_2)) \tag{2.22}$$

For copulas, Equation 2.22 can be rewritten as

$$\varphi_\theta(C(u_1, u_2; \theta)) = \varphi_\theta(u_1) + \varphi_\theta(u_2) \tag{2.23}$$

If $\varphi_\theta$ is a convex-decreasing function from $[0, 1]$ to $[0, \infty)$ such that $\varphi_\theta(0) = \infty$ and $\varphi_\theta(1) = 0$ and the inverse $\varphi_\theta^{-1}$ is convex decreasing on $[0, \infty)$, then a bivariate Archimedean copula can be constructed by inversing Equation 2.23 as

$$C(u_1,\ u_2;\ \theta) = \varphi_\theta^{-1}[\varphi_\theta(u_1) + \varphi_\theta(u_2)] \tag{2.24}$$

It is worthwhile to point out that there are many functions satisfying the requirements as a generator function. As a consequence, many Archimedean copulas can be constructed from these functions as demonstrated by Nelsen (2006).

## 2.3 MODELING BIVARIATE DISTRIBUTION OF SHEAR STRENGTH PARAMETERS

The previous section presents the basic idea behind the copula theory. In this section, the copula approach is employed to model the bivariate distribution of shear strength parameters.

This approach includes the following three steps: (1) the measured cohesions and friction angles of soils are collected, (2) the best-fit marginal distributions are determined, and (3) the copula that provides the best fit to measured dependence structure is identified. To decouple steps (2) and (3), Kendall's tau is adopted to determine the copula parameters underlying the four candidate copulas (Gaussian, Plackett, Frank, and No.16 copulas). From hereon, $X_1$ and $X_2$ denote the cohesion $c$ and friction angle $\phi$, respectively.

### 2.3.1 Measured data of cohesion and friction angle

To construct the bivariate distribution of shear strength parameters, some measured data of cohesion and friction angle reported in the literature (Li et al. 2000; Wu et al. 2005; Zhang et al. 2013) are collected. Tables 2.2 through 2.4 summarize the measured cohesions and friction angles of soils from Xiaolangdi Hydropower Station, silty clay in Taiyuan area, and several typical strata in the Hangzhou area of China, respectively. On the basis of these measured data, the coefficients of variation (COV) (see Equation 1.2 of Chapter 1), Pearson's rho (Equation 2.11), and Kendall's tau (Equation 2.14) can be obtained, which are also listed in these tables. Significant variations underlying the measured shear strength parameters can be observed. Furthermore, both Pearson's rho and Kendall's tau indicate that there exists a strongly negative correlation between cohesion and friction angle. These observations coincide with the observation reported in Tang et al. (2012, 2013a). Since the data in Table 2.2 have the largest sample sizes, they are further employed to construct the bivariate distributions of shear strength parameters. These data are obtained from various tests, namely (1) consolidated-drained test (referred to as CD hereafter), (2) consolidated-undrained test (CU), and (3) unconsolidated-undrained test (UU). The sample sizes (N) are 63, 64, and 61, respectively. The resulting Pearson's rho are −0.544, −0.702, and −0.623 for CD, CU, and UU datasets, respectively. The corresponding Kendall's tau are −0.384, −0.544, and −0.447.

### 2.3.2 Identification of best-fit marginal distributions

To fit the marginal distributions of $c$ and $\phi$, four candidate distributions, namely Normal truncated below zero (referred to as TruncNormal hereafter), Lognormal, Gumbel truncated below zero (referred to as TruncGumbel hereafter), and Weibull distributions are examined (Lumb 1970; Baecher and Christian 2003). These four distributions can guarantee that the simulated shear strength data are positive, which satisfy the requirements of positive $c$ and $\phi$. Table 2.5 summarizes the PDFs $f(x; p, q)$ and domains of distribution parameters $(p, q)$ associated with the four candidate distributions. The relationships between $(p, q)$ and $(\mu, \sigma)$ for the four distributions are also provided in Table 2.5.

There are many goodness-of-fit test methods for identifying the best-fit marginal distributions underlying measured data. For example, Section 1.2.3 of Chapter 1 introduced the commonly used Kolmogorov–Smirnov (K–S) test. In this chapter, both the Akaike Information Criterion (AIC) (Akaike 1974) and Bayesian Information Criterion (BIC) (Schwarz 1978) are adopted to identify the best-fit marginal distributions. A marginal distribution resulting in the smallest AIC and BIC values is considered to be the best-fit marginal distribution. The AIC and BIC are, respectively, defined as

$$\text{AIC} = -2\sum_{i=1}^{N} \ln f(x_i; p, q) + 2k_1 \tag{2.25}$$

*Table 2.2* Measured shear strength parameters from Xiaolangdi hydropower station in China

| Test number | CD (43 data) | | CU (61 data) | | UU (81 data) | |
|---|---|---|---|---|---|---|
| | c (kPa) | φ (°) | c (kPa) | φ (°) | c (kPa) | ψ (°) |
| 1 | 75.11 | 20.78 | 128.14 | 18.00 | 209.17 | 5.41 |
| 2 | 45.72 | 25.45 | 41.70 | 25.95 | 185.73 | 7.28 |
| 3 | 57.40 | 22.22 | 66.73 | 20.93 | 109.50 | 12.01 |
| 4 | 81.43 | 23.30 | 56.83 | 21.47 | 123.57 | 13.97 |
| 5 | 67.32 | 24.78 | 48.73 | 23.55 | 165.68 | 11.48 |
| 6 | 56.15 | 20.06 | 74.51 | 16.22 | 218.74 | 7.65 |
| 7 | 92.19 | 18.77 | 97.56 | 15.84 | 173.55 | 7.60 |
| 8 | 65.97 | 24.76 | 67.22 | 26.22 | 201.81 | 13.71 |
| 9 | 64.56 | 18.49 | 75.62 | 17.04 | 225.11 | 10.98 |
| 10 | 20.71 | 22.95 | 72.62 | 16.59 | 212.94 | 10.55 |
| 11 | 27.43 | 24.75 | 31.89 | 25.69 | 220.78 | 9.14 |
| 12 | 80.83 | 23.66 | 9.19 | 31.15 | 210.08 | 5.42 |
| 13 | 49.37 | 17.02 | 68.34 | 16.59 | 201.31 | 9.63 |
| 14 | 36.39 | 23.16 | 56.06 | 21.59 | 131.66 | 13.45 |
| 15 | 75.74 | 20.07 | 39.54 | 23.48 | 202.82 | 8.85 |
| 16 | 81.04 | 16.63 | 99.66 | 14.04 | 206.09 | 9.13 |
| 17 | 71.61 | 21.03 | 36.53 | 23.47 | 218.46 | 7.10 |
| 18 | 69.81 | 23.62 | 50.77 | 24.94 | 222.71 | 8.85 |
| 19 | 53.50 | 22.70 | 103.66 | 17.53 | 179.08 | 9.25 |
| 20 | 88.02 | 20.43 | 87.45 | 17.04 | 285.78 | 4.97 |
| 21 | 93.48 | 21.41 | 78.70 | 20.00 | 170.47 | 12.07 |
| 22 | 74.30 | 22.81 | 92.11 | 20.10 | 141.35 | 11.71 |
| 23 | 65.03 | 22.14 | 48.40 | 22.26 | 181.53 | 11.96 |
| 24 | 62.27 | 22.28 | 68.61 | 19.63 | 208.19 | 15.52 |
| 25 | 56.54 | 22.13 | 51.01 | 23.62 | 253.06 | 5.69 |
| 26 | 77.68 | 22.87 | 87.31 | 21.24 | 217.38 | 6.63 |
| 27 | 71.40 | 24.00 | 88.02 | 21.21 | 187.52 | 6.30 |
| 28 | 86.09 | 18.70 | 65.90 | 18.17 | 220.22 | 8.83 |
| 29 | 58.35 | 22.33 | 85.61 | 21.25 | 139.16 | 10.13 |
| 30 | 91.49 | 21.37 | 35.76 | 23.46 | 100.28 | 12.86 |
| 31 | 91.52 | 21.33 | 86.07 | 19.44 | 178.51 | 4.30 |
| 32 | 18.91 | 22.75 | 68.96 | 20.95 | 134.02 | 8.79 |
| 33 | 81.19 | 19.92 | 100.39 | 19.06 | 216.67 | 8.66 |
| 34 | 64.93 | 19.20 | 46.24 | 24.05 | 168.59 | 10.30 |
| 35 | 59.37 | 22.68 | 81.10 | 19.65 | 206.83 | 10.92 |
| 36 | 105.58 | 21.56 | 82.09 | 23.72 | 160.58 | 14.76 |
| 37 | 56.38 | 26.14 | 59.70 | 27.36 | 149.78 | 13.89 |
| 38 | 51.75 | 23.44 | 30.08 | 24.95 | 164.39 | 12.58 |
| 39 | 87.80 | 22.05 | 52.36 | 22.91 | 245.28 | 7.92 |
| 40 | 83.94 | 22.47 | 61.27 | 22.84 | 73.25 | 8.90 |
| 41 | 112.53 | 20.58 | 142.73 | 19.78 | 82.55 | 12.78 |
| 42 | 61.65 | 22.50 | 109.88 | 17.80 | 161.54 | 10.71 |
| 43 | 85.44 | 20.65 | 63.16 | 21.42 | 161.68 | 7.38 |

Table 2.2 (Continued) Measured shear strength parameters from Xiaolangdi hydropower station in China

| Test number | CD (63 data) | | CU (64 data) | | UU (61 data) | |
|---|---|---|---|---|---|---|
| | c (kPa) | φ (°) | c (kPa) | φ (°) | c (kPa) | φ (°) |
| 44 | 38.82 | 27.34 | 88.95 | 18.88 | 257.40 | 10.29 |
| 45 | 70.72 | 21.14 | 23.10 | 27.96 | 231.41 | 11.01 |
| 46 | 91.14 | 23.63 | 57.29 | 22.55 | 343.89 | 7.89 |
| 47 | 63.85 | 22.83 | 65.76 | 22.14 | 101.11 | 12.53 |
| 48 | 87.13 | 20.28 | 85.31 | 21.43 | 107.15 | 13.87 |
| 49 | 79.26 | 21.19 | 27.91 | 23.75 | 90.23 | 15.89 |
| 50 | 46.37 | 23.09 | 51.68 | 24.09 | 190.91 | 13.00 |
| 51 | 80.48 | 22.56 | 4.62 | 27.82 | 181.85 | 12.35 |
| 52 | 81.03 | 23.77 | 93.75 | 21.06 | 186.31 | 16.67 |
| 53 | 33.87 | 27.03 | 113.64 | 22.73 | 76.08 | 18.20 |
| 54 | 51.75 | 25.60 | 9.54 | 25.70 | 206.40 | 9.23 |
| 55 | 71.57 | 23.74 | 11.37 | 27.76 | 188.57 | 9.44 |
| 56 | 101.74 | 21.44 | 98.96 | 22.68 | 104.63 | 14.97 |
| 57 | 30.67 | 27.31 | 86.69 | 20.93 | 115.33 | 14.63 |
| 58 | 48.77 | 26.81 | 53.96 | 21.57 | 89.45 | 23.06 |
| 59 | 52.46 | 26.18 | 54.10 | 24.87 | 141.47 | 22.09 |
| 60 | 43.74 | 23.29 | 57.28 | 21.85 | 95.56 | 17.11 |
| 61 | 72.41 | 22.62 | 93.57 | 24.06 | 62.19 | 20.35 |
| 62 | 37.68 | 30.32 | 64.46 | 22.77 | – | – |
| 63 | 14.57 | 27.20 | 81.35 | 23.85 | – | – |
| 64 | – | – | 33.41 | 25.56 | – | – |
| COV | 0.33 | 0.11 | 0.43 | 0.15 | 0.32 | 0.36 |
| ρ | –0.544 | | –0.702 | | –0.623 | |
| τ | –0.384 | | –0.544 | | –0.447 | |

Source: After Zhang L, Tang XS, Li DQ. Journal of Civil Engineering and Management 2013; 30(2): 11–17 (in Chinese).

Note: CD, Consolidated-drained triaxial compression test; CU, consolidated-undrained triaxial compression test; UU, unconsolidated-undrained triaxial compression test.

and

$$\text{BIC} = -2\sum_{i=1}^{N} \ln f(x_i; p, q) + k_1 \ln N \qquad (2.26)$$

where $\sum_{i=1}^{N} \ln f(x_i, p, q)$ is the logarithm of the likelihood function for a specified distribution; $k_1$ is the number of distribution parameters. For the four selected distributions, all of them are two-parameter distributions. Therefore, $k_1 = 2$ is used in Equations 2.25 and 2.26. Specifically, identification of the best-fit marginal distributions using AIC and BIC consists of the following four steps:

1. Compute the sample mean μ and sample standard deviation σ of the measured shear strength parameters using Equation 1.13 of Chapter 1.

*Table 2.3* Measured shear strength parameters for silty clay in Taiyuan, China

| Test number | Q (15 data) | | CQ (15 data) | | UU (15 data) | | CU (15 data) | |
|---|---|---|---|---|---|---|---|---|
| | c (kPa) | φ (°) | c (kPa) | φ (°) | c (kPa) | φ (°) | c (kPa) | φ (°) |
| I | 32.2 | 8.2 | 35.1 | 6.2 | 25.3 | 16.0 | 40.1 | 7.1 |
| 2 | 40.2 | 10.5 | 60.3 | 16.3 | 58.4 | 6.5 | 60.2 | 8.2 |
| 3 | 30.1 | 15.2 | 32.6 | 15.2 | 43.5 | 9.6 | 60.4 | 18.1 |
| 4 | 35.6 | 12.8 | 55.6 | 22.6 | 52.1 | 10.7 | 70.6 | 11.5 |
| 5 | 49.3 | 8.1 | 75.2 | 9.1 | 60.0 | 6.5 | 80.5 | 8.3 |
| 6 | 23.5 | 13.6 | 29.8 | 15.4 | 40.3 | 6.9 | 45.1 | 10.1 |
| 7 | 17.6 | 24.5 | 25.7 | 35.6 | 25.7 | 16.0 | 45.3 | 20.6 |
| 8 | 40.1 | 15.1 | 45.7 | 25.1 | 35.6 | 7.5 | 36.3 | 15.7 |
| 9 | 19.8 | 16.8 | 22.6 | 16.2 | 20.2 | 12.1 | 28.3 | 19.3 |
| 10 | 20.6 | 18.1 | 35.3 | 18.2 | 33.3 | 9.2 | 53.2 | 19.5 |
| 11 | 16.3 | 20.3 | 30.4 | 30.8 | 21.6 | 15.8 | 16.1 | 26.4 |
| 12 | 23.7 | 23.1 | 41.5 | 31.7 | 24.1 | 14.9 | 30.4 | 22.5 |
| 13 | 19.1 | 26.2 | 31.2 | 25.4 | 49.6 | 15.7 | 63.1 | 25.1 |
| 14 | 30.5 | 16.2 | 51.2 | 23.5 | 49.1 | 13.5 | 53.5 | 19.1 |
| 15 | 10.2 | 25.0 | 19.5 | 30.4 | 28.7 | 15.8 | 30.1 | 23.7 |
| COV | 0.40 | 0.35 | 0.39 | 0.40 | 0.36 | 0.33 | 0.37 | 0.38 |
| ρ | −0.813 | | −0.367 | | −0.591 | | −0.504 | |
| τ | −0.676 | | −0.257 | | −0.425 | | −0.333 | |

Source:  After Li XY et al. *Chinese Journal of Geotechnical Engineering* 2000; 22(6): 668–672 (in Chinese).

Note:  Q, Quick direct shear test; CQ, consolidated-quick direct shear test; UU, unconsolidated-undrained triaxial compression test; CU, consolidated-undrained triaxial compression test.

2. Estimate the distribution parameters $p$ and $q$ underlying the four candidate distributions based on the relationships between $(p, q)$ and $(\mu, \sigma)$ as shown in Table 2.5.
3. Evaluate the AIC and BIC values associated with the four candidate distributions using Equations 2.25 and 2.26, respectively.
4. The marginal distribution resulting in the smallest AIC and BIC values is identified to be the best-fit marginal distribution to the measured data of shear strength parameters.

As an example, consider the cohesion ($c$) data of CU dataset, the sample mean and sample standard deviation are computed as $\mu = 66.483$ and $\sigma = 28.778$, respectively. The distribution parameters underlying the Lognormal distribution are then obtained as $p = 4.111$ and $q = 0.414$. Substituting the observed data $\{c_i, i = 1, 2, ..., N\}$ into $f(x; p, q)$ of the Lognormal distribution leads to the logarithm of the likelihood function equal to −337.146. Finally, the AIC and BIC for the Lognormal distribution are obtained as AIC $= -2 \times (-337.146) + 2 \times 2 = 678.292$ and BIC $= -2 \times (-337.146) + 2 \times \ln(64) = 682.611$. Following the above procedure, Table 2.6 shows the AIC and BIC values associated with the four marginal distributions for various datasets of ($c$, φ). Note that both the AIC and BIC values indicate that the Weibull and Lognormal distributions are the best-fit distributions underlying the CD and UU datasets for $c$ and φ, respectively. For the CU dataset, the best-fit distributions for both $c$ and φ are the TruncNormal distributions. To further examine the capabilities of the identified marginal distributions to fit the measured data, Figure 2.3 shows the PDFs of the four candidate marginal distributions along with the histograms of the measured data. It is evident that the

Table 2.4 Measured shear strength parameters for several typical strata in the Hangzhou area of China

| Stratum number | Test number | Q (10 data) | | CQ (10 data) | | UU (10 data) | | CU (10 data) | |
|---|---|---|---|---|---|---|---|---|---|
| | | c (kPa) | φ (°) | c (kPa) | φ (°) | c (kPa) | φ (°) | c (kPa) | φ (°) |
| III | 1 | 4.7 | 8.2 | 6.3 | 13.9 | 12.9 | 4.1 | 22.3 | 8.5 |
| | 2 | 8.3 | 14.7 | 10.7 | 16.2 | 18.5 | 9.5 | 15.3 | 13.3 |
| | 3 | 6.7 | 17.2 | 8.5 | 19.7 | 9.0 | 12.4 | 21.5 | 16.5 |
| | 4 | 9.4 | 12.8 | 11.9 | 14.5 | 19.3 | 8.3 | 26.4 | 12.5 |
| | 5 | 14.4 | 9.1 | 15.4 | 13.1 | 22.3 | 5.8 | 36.2 | 8.1 |
| | 6 | 18.8 | 6.1 | 21.8 | 11.4 | 25.1 | 5.5 | 30.8 | 6.8 |
| | 7 | 8.5 | 15.5 | 9.1 | 20.5 | 8.6 | 10.1 | 13.7 | 12.3 |
| | 8 | 11.7 | 13.8 | 13.3 | 9.2 | 28.1 | 6.4 | 28.8 | 8.6 |
| | 9 | 6.6 | 21.6 | 8.0 | 18.7 | 15.6 | 9.2 | 22.3 | 14.1 |
| | 10 | 13.9 | 8.3 | 12.5 | 16.4 | 18.2 | 6.8 | 34.1 | 10.3 |
| | COV | 0.42 | 0.38 | 0.38 | 0.24 | 0.36 | 0.32 | 0.30 | 0.28 |
| | ρ | −0.628 | | −0.614 | | −0.582 | | −0.609 | |
| | τ | −0.511 | | −0.422 | | −0.422 | | −0.450 | |
| IV | 1 | 12.2 | 10.8 | 14.3 | 14.0 | 16.6 | 5.2 | 30.6 | 9.3 |
| | 2 | 15.7 | 12.3 | 24.7 | 12.9 | 24.7 | 4.4 | 34.8 | 6.6 |
| | 3 | 19.1 | 15.9 | 18.7 | 14.2 | 30.2 | 6.9 | 37.1 | 10.8 |
| | 4 | 14.3 | 8.1 | 20.1 | 13.6 | 21.8 | 5.8 | 27.0 | 12.2 |
| | 5 | 21.3 | 18.4 | 12.5 | 23.5 | 26.7 | 14.2 | 19.2 | 15.9 |
| | 6 | 15.6 | 16.4 | 13.8 | 17.5 | 28.4 | 10.4 | 29.9 | 13.3 |
| | 7 | 8.2 | 11.7 | 21.2 | 13.4 | 23.5 | 9.7 | 35.3 | 9.1 |
| | 8 | 9.0 | 7.2 | 25.4 | 8.8 | 19.2 | 7.6 | 37.7 | 6.8 |
| | 9 | 15.5 | 9.7 | 19.6 | 12.0 | 21.1 | 5.5 | 29.4 | 6.2 |
| | 10 | 20.3 | 10.4 | 22.5 | 11.3 | 29.8 | 8.3 | 40.6 | 10.8 |
| | COV | 0.29 | 0.31 | 0.23 | 0.28 | 0.19 | 0.38 | 0.19 | 0.31 |
| | ρ | 0.586 | | − 0.820 | | 0.438 | | − 0.578 | |
| | τ | 0.422 | | − 0.778 | | 0.289 | | − 0.225 | |
| V | 1 | 10.5 | 18.2 | 27.0 | 22.5 | 23.3 | 16.6 | 41.7 | 15.6 |
| | 2 | 36.1 | 8.9 | 32.0 | 15.9 | 56.1 | 8.0 | 67.2 | 9.7 |
| | 3 | 28.8 | 12.3 | 53.5 | 18.6 | 62.0 | 10.2 | 45.5 | 12.4 |
| | 4 | 25.3 | 14.7 | 51.8 | 16.3 | 42.6 | 12.5 | 71.6 | 13.0 |
| | 5 | 23.4 | 20.0 | 20.3 | 22.7 | 36.0 | 18.2 | 33.6 | 18.2 |
| | 6 | 31.7 | 11.5 | 45.5 | 16.2 | 35.3 | 11.5 | 45.1 | 14.7 |
| | 7 | 18.4 | 26.8 | 14.5 | 25.1 | 25.4 | 15.2 | 16.0 | 22.3 |
| | 8 | 24.2 | 18.1 | 22.5 | 21.8 | 30.9 | 14.3 | 44.8 | 15.9 |
| | 9 | 13.6 | 29.3 | 12.7 | 34.7 | 28.6 | 22.7 | 35.9 | 25.1 |
| | 10 | 10.0 | 24.6 | 22.5 | 21.4 | 35.2 | 10.1 | 42.3 | 17.9 |
| | COV | 0.10 | 0.37 | 0.50 | 0.26 | 0.34 | 0.32 | 0.36 | 0.28 |
| | ρ | −0.816 | | −0.757 | | −0.662 | | −0.777 | |
| | τ | −0.778 | | −0.629 | | −0.467 | | −0.733 | |
| VI | 1 | 37.0 | 10.4 | 46.7 | 11.5 | 66.8 | 6.1 | 80.6 | 6.7 |
| | 2 | 41.9 | 13.6 | 32.2 | 14.7 | 71.2 | 6.7 | 89.1 | 9.4 |
| | 3 | 33.4 | 19.0 | 42.3 | 16.2 | 62.4 | 8.4 | 75.5 | 11.5 |
| | 4 | 26.5 | 16.8 | 40.7 | 19.5 | 50.1 | 9.5 | 63.5 | 15.2 |

continued

*Table 2.4* (Continued) Measured shear strength parameters for several typical stratums in the Hangzhou area of China

| Stratum number | Test number | Q (10 data) | | CQ (10 data) | | UU (10 data) | | CU (10 data) | |
|---|---|---|---|---|---|---|---|---|---|
| | | c (kPa) | $\phi$ (°) | c (kPa) | $\phi$ (°) | c (kPa) | $\phi$ (°) | c (kPa) | $\phi$ (°) |
| | 5 | 54.1 | 13.6 | 68.5 | 14.5 | 81.5 | 8.1 | 85.7 | 8.2 |
| | 6 | 37.6 | 15.2 | 36.3 | 22.7 | 53.4 | 11.4 | 68.0 | 12.4 |
| | 7 | 29.4 | 25.1 | 31.7 | 28.0 | 69.2 | 13.4 | 73.4 | 19.0 |
| | 8 | 44.1 | 9.6 | 57.9 | 12.9 | 78.8 | 7.1 | 84.9 | 6.4 |
| | 9 | 28.8 | 21.7 | 26.6 | 23.4 | 62.3 | 13.8 | 65.1 | 17.7 |
| | 10 | 49.6 | 8.4 | 64.1 | 9.8 | 80.8 | 5.9 | 92.6 | 8.3 |
| | COV | 0.24 | 0.35 | 0.32 | 0.34 | 0.16 | 0.32 | 0.13 | 0.39 |
| | $\rho$ | −0.720 | | −0.723 | | −0.485 | | −0.765 | |
| | $\tau$ | −0.584 | | −0.644 | | −0.378 | | −0.467 | |

Source: After Wu CF, Zhu XR, Liu XM. *Chinese Journal of Geotechnical Engineering* 2005; 27(1): 94–99 (in Chinese).

Note: Q, Quick direct shear test; CQ, consolidated-quick direct shear test; UU, unconsolidated-undrained triaxial compression test; CU, consolidated-undrained triaxial compression test.

*Table 2.5* PDFs and domains of distribution parameters associated with the selected four distributions

| Distribution | $f(x; p, q)$ | $\mu$ and $\sigma^2$ | Range of p | Range of q |
|---|---|---|---|---|
| Trunc Normal | $\dfrac{1}{q\sqrt{2\pi}}\exp\left[-\dfrac{1}{2}\left(\dfrac{x-p}{q}\right)^2\right]\bigg/\left[1-\Phi\left(-\dfrac{p}{q}\right)\right]$ | $\mu = p,\ \sigma^2 = q^2$ | $(-\infty, \infty)$ | $(0, \infty)$ |
| Log normal | $\dfrac{1}{qx\sqrt{2\pi}}\exp\left[-\dfrac{1}{2}\left(\dfrac{\ln x-p}{q}\right)^2\right]$ | $\mu = \exp(p + 0.5q^2),$ $\sigma^2 = [\exp(q^2) - 1]\exp(2p + q^2)$ | $(-\infty, \infty)$ | $(0, \infty)$ |
| Trunc Gumbel | $\dfrac{q\exp\{-q(x-p) - \exp[-q(x-p)]\}}{\{1 - \exp[-\exp(pq)]\}}$ | $\mu = p + 0.5772/q,\ \sigma^2 = \pi^2/(6q^2)$ | $(-\infty, \infty)$ | $(0, \infty)$ |
| Weibull | $\dfrac{q}{p}\left(\dfrac{x}{p}\right)^{q-1}\exp\left[-\left(\dfrac{x}{p}\right)^q\right]$ | $\mu = p\Gamma(1 + 1/q),$ $\sigma^2 = p^2[\Gamma(1 + 2/q) - \Gamma^2(1 + 1/q)]$ | $(0, \infty)$ | $(0, \infty)$ |

Note: $\Phi$ denotes the standard normal distribution function; $\Gamma$ is the gamma function.

*Table 2.6* AIC and BIC values for the TruncNormal, Lognormal, TruncGumbel, and Weibull distributions

| Dataset | Parameter | TruncNormal [AIC, BIC] | Lognormal [AIC, BIC] | TruncGumbel [AIC, BIC] | Weibull [AIC, BIC] |
|---|---|---|---|---|---|
| CD | c | 569.36, 573.65 | 599.31, 603.60 | 599.53, 603.81 | **569.12, 573.41** |
| | $\phi$ | 300.88, 305.17 | **300.26, 304.55** | 314.60, 318.89 | 315.93, 320.21 |
| CU | c | **613.31, 617.63** | 678.29, 682.61 | 632.61, 636.93 | 617.04, 621.35 |
| | $\phi$ | **340.35, 344.67** | 342.19, 346.50 | 357.50, 361.82 | 345.03, 349.35 |
| UU | c | 667.96, 672.18 | 677.13, 681.35 | 680.20, 684.43 | **667.49, 671.71** |
| | $\phi$ | 345.39, 349.62 | **339.06, 343.29** | 339.17, 343.39 | 344.46, 348.68 |

Note: The AIC and BIC values are bold if the corresponding distribution is preferred.

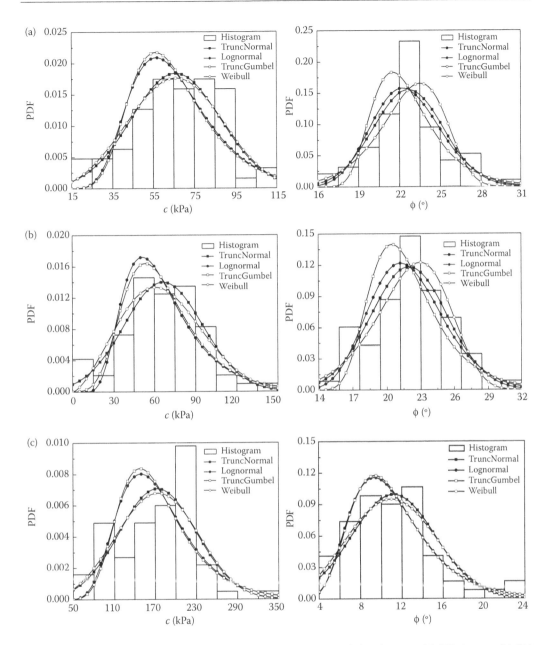

*Figure 2.3* Histograms of measured data and PDFs of fitted marginal distributions. (a) CD dataset, (b) CU dataset, and (c) UU dataset.

identified best-fit marginal distributions using AIC and BIC provide better fits to the properties of measured shear strength parameters than the other candidate marginal distributions.

## 2.3.3 Identification of best-fit copula

After determining the best-fit marginal distributions of $c$ and $\phi$, the next step is to identify the copula that provides the best fit to the measured dependence structure between $c$ and $\phi$.

As stated previously, the Gaussian, Plackett, Frank, and No.16 copulas are selected as candidate copulas to fit the measured dependence structure between $c$ and $\phi$. Similarly, the AIC and BIC can be used to identify the best-fit copula underlying the measured data. A copula producing the smallest AIC and BIC values is considered to be the best-fit copula. The corresponding AIC and BIC are, respectively, defined as

$$\text{AIC} = -2\sum_{i=1}^{N} \ln D(u_{1i}, u_{2i}; \theta) + 2k_2 \tag{2.27}$$

and

$$\text{BIC} = -2\sum_{i=1}^{N} \ln D(u_{1i}, u_{2i}; \theta) + k_2 \ln N \tag{2.28}$$

where $D(u_{1i}, u_{2i}; \theta)$ is the copula density function shown in Table 2.1; $\sum_{i=1}^{N} \ln D(u_{1i}, u_{2i}; \theta)$ is the logarithm of the likelihood function for a specified copula; $k_2$ is the number of copula parameters; and $\{(u_{1i}, u_{2i}), i = 1, 2, ..., N\}$ are the empirical distribution values of measured $(c, \phi)$, which are defined as

$$\begin{cases} u_{1i} = \dfrac{\text{rank}(c_i)}{N+1} \\ u_{2i} = \dfrac{\text{rank}(\phi_i)}{N+1} \end{cases} \quad i = 1, 2, ..., N \tag{2.29}$$

in which $\text{rank}(c_i)$ [or $\text{rank}(\phi_i)$] denotes the rank of $c_i$ (or $\phi_i$) among the list $\{c_1, ..., c_N\}$ (or $\{\phi_1, ..., \phi_N\}$) in an ascending order. Note that Equation 2.29 is another estimator of the empirical CDF, which is similar to the empirical CDFs in Equations 1.11 and 1.12 of Chapter 1. Hence, $(u_{1i}, u_{2i})$ are realizations of standard uniform variables. For the four selected copulas, all of them are single-parameter copulas. Therefore, $k_2 = 1$ is used in Equations 2.27 and 2.28. Similarly, identification of the best-fit copula using AIC and BIC includes the following five steps:

1. Compute Kendall's tau, $\tau$, of the measured data of shear strength parameters using Equations 2.14 and 2.15. The corresponding results are listed in Table 2.2.
2. Estimate the copula parameters $\theta$ underlying the four candidate copulas using Equation 2.13 or Equation 2.20 for the Gaussian copula or Equation 2.21 for the Frank and No.16 copulas based on the measured $\tau$ in Step 1. The results are listed in Table 2.7.
3. Calculate the empirical distributions $\mathbf{U} = (U_1, U_2)$ of measured $(c, \phi)$ using Equation 2.29. Figure 2.4 shows the scatter plots of $U_1$ and $U_2$. There is a strong negative dependence between $U_1$ and $U_2$. Furthermore, the samples of $U_1$ and $U_2$ for the three datasets are basically symmetrical with respect to the 135° diagonal line of a unit square, which graphically demonstrates that the selected four copulas are capable of capturing the dependence structure underlying the measured data.
4. Evaluate the AIC and BIC values for the four candidate copulas using Equations 2.27 and 2.28, respectively.
5. The copula producing the smallest AIC and BIC values is identified to be the best-fit copula to the measured data.

*Table 2.7* Copula parameters, AIC, and BIC values for the Gaussian, Plackett, Frank, and No.16 copulas

| Dataset | Gaussian [θ, AIC, BIC] | Plackett [θ, AIC, BIC] | Frank [θ, AIC, BIC] | No.16 [θ, AIC, BIC] |
|---|---|---|---|---|
| CD | −0.567, −19.63, −17.49 | 0.164, −18.00, −15.85 | −3.946, **−20.18, −18.04** | 0.027, −15.31, −13.17 |
| CU | −0.754, −43.50, −41.35 | 0.068, −43.00, −40.84 | −6.590, **−44.10, −41.95** | 0.011, −32.39, −30.23 |
| UU | −0.645, −26.40, −24.29 | 0.118, **−27.38, −25.27** | −4.841, −27.23, −25.11 | 0.020, −23.28, −21.17 |

Note: The AIC and BIC values are bold if the corresponding copula is preferred.

Table 2.7 summarizes the AIC and BIC values associated with the four copulas for various datasets of $(c, \phi)$. Note that both the AIC and BIC values indicate that the Plackett copula is the best-fit copula for the UU dataset. For the CD and CU datasets, the Frank copula is the best-fit copula. All of them are better fits to measured data than the Gaussian copula based on AIC and BIC. These results indicate that the Gaussian copula may not provide the best fit to the dependence between $c$ and $\phi$. In short, the marginal distributions and copula to construct the bivariate distribution of shear strength parameters should be determined based on the measured data in some sense such as the AIC and BIC. It is recommended that the marginal distributions of $c$ and $\phi$ and the copula for characterizing the dependence structure between $c$ and $\phi$ must be selected carefully. This is because they have a significant impact

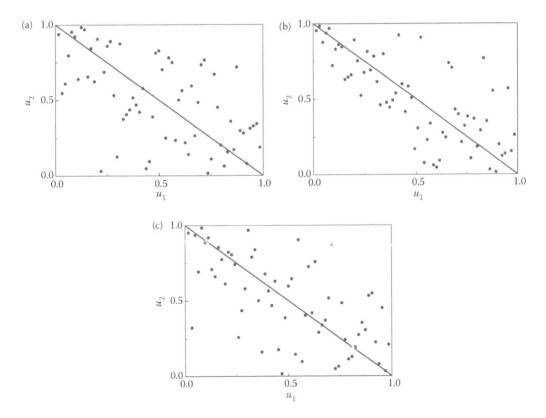

*Figure 2.4* Scatter plots of $u_1$ and $u_2$ for measured cohesions and friction angles. (a) CD dataset, (b) CU dataset, and (c) UU dataset.

on geotechnical reliability as illustrated in Section 2.5. To facilitate the applications of the copula theory for nonspecialists such as geotechnical engineers, the MATLAB codes for identifying the best-fit marginal distributions and copulas using AIC and BIC are appended.

## 2.4 SIMULATING BIVARIATE DISTRIBUTION OF SHEAR STRENGTH PARAMETERS

With the identified best-fit marginal distributions and copula, the bivariate distribution of shear strength parameters is readily determined using Equations 2.5 and 2.6. In this section, the simulation of the bivariate distribution of shear strength parameters is presented. Section 2.4.1 summarizes the simulation algorithms of copulas and bivariate distribution. A comparison between the simulated data and measured data underlying the bivariate distribution of shear strength parameters is presented in Section 2.4.2.

### 2.4.1 Algorithms for simulating bivariate distribution

The algorithms for generating $m$ samples $\mathbf{U}_{m \times 2} = [U_1, U_2]$ from a specified copula and $\mathbf{X}_{m \times 2} = [X_1, X_2]$ from a bivariate distribution are presented below. All the algorithms require samples of two independent standard normal vectors. Each vector contains $m$ standard normal random variables of zero mean and unit variance.

#### 2.4.1.1 Gaussian copula

1. Simulate two independent standard normal vectors $\mathbf{Z}_{m \times 2} = [Z_1, Z_2]$. This can be obtained using MATLAB: `z = randn (m,2)` and MATLAB function `randn('state',1)` is used to fix the initial seed. If the sample size $m$ is small, the MATLAB command `z*inv(chol(cov(z)))` is further adopted to eliminate the sampling correlations underlying the simulated $\mathbf{Z}_{m \times 2}$.
2. Find the Cholesky decomposition $\mathbf{Q}$ of $\mathbf{R}$, in which $\mathbf{R} = [1\ \theta; \theta\ 1]$ is a correlation matrix containing the Gaussian copula parameters $\theta$ and $\mathbf{Q} = [1\ \theta; 0\ (1-\theta^2)^{0.5}]$ is an upper triangular matrix. This can be done by applying the MATLAB function `chol` as below: $\mathbf{Q} = \text{chol}(\mathbf{R})$ (see Section 1.3.4 of Chapter 1).
3. Set $\mathbf{Y} = \mathbf{ZQ}$. Then two correlated standard normal vectors $\mathbf{Y}_{m \times 2} = [Y_1, Y_2]$ are obtained.
4. Set $\mathbf{U} = \Phi(\mathbf{Y})$, in which $\Phi$ is the CDF of a standard normal variable. This can be obtained using MATLAB: $\mathbf{U} = \text{normcdf} (\mathbf{Y})$. Then two correlated standard uniform vectors are obtained as $\mathbf{U}_{m \times 2} = [U_1, U_2]$ belonging to the Gaussian copula.

#### 2.4.1.2 Plackett copula

1. Simulate two independent standard normal vectors $\mathbf{Z}_{m \times 2} = [Z_1, Z_2]$. This can be obtained using MATLAB: `z = randn (m,2)` and MATLAB function `randn('state',1)` is used to fix the initial seed. If the sample size $m$ is small, the MATLAB command `z*inv(chol(cov(z)))` is further adopted to eliminate the sampling correlations underlying the simulated $\mathbf{Z}_{m \times 2}$.
2. Set $\mathbf{V} = \Phi(\mathbf{Z})$. Then two independent standard uniform vectors $\mathbf{V}_{m \times 2} = [V_1, V_2]$ are obtained. This can be realized from MATLAB using $\mathbf{V} = \text{normcdf} (\mathbf{Z})$.
3. Set $a = V_2(1 - V_2)$, $b = \theta + a(\theta - 1)^2$, $c = 2a(V_1\theta^2 + 1 - V_1) + \theta(1 - 2a)$, and $d = \sqrt{\theta}\sqrt{\theta + 4aV_1(1 - V_1)(1 - \theta)^2}$ (e.g., Nelsen 2006).

4. Set $U_1 = V_1$ and $U_2 = [c - (1 - 2V_2)d]/2b$. Then, two correlated standard uniform vectors are obtained as $\mathbf{U}_{m\times2} = [U_1, U_2]$ belonging to the Plackett copula (e.g., Nelsen 2006).

### 2.4.1.3 Frank and No.16 copulas

1. Simulate two independent standard normal vectors $\mathbf{Z}_{m\times2} = [Z_1, Z_2]$. This can be obtained using MATLAB: z = randn (m,2) and MATLAB function randn('state',1) is used to fix the initial seed. If the sample size $m$ is small, the MATLAB command z*inv(chol(cov(z))) is further adopted to eliminate the sampling correlations underlying the simulated $\mathbf{Z}_{m\times2}$.
2. Set $\mathbf{V} = \Phi(\mathbf{Z})$. Then two independent standard uniform vectors $\mathbf{V}_{m\times2} = [V_1, V_2]$ are obtained. This can be realized from MATLAB using $\mathbf{V} = $ normcdf $(\mathbf{Z})$.
3. Set $U_1 = V_1$.
4. Set $V_2 = C_2(U_2|U_1)$ in which $C_2(U_2|U_1)$ is the conditional distribution of $U_2$ given the values of $U_1$. It can be calculated by (e.g., Nelsen 2006)

$$C_2(u_2|u_1) = \frac{\varphi^{-1(1)}(\varphi(u_1) + \varphi(u_2))}{\varphi^{-1(1)}(\varphi(u_1))} \tag{2.30}$$

in which $\varphi(\cdot)$ is the generator function of an Archimedean copula. The generator functions for the Frank and No.16 copulas are listed in Table 2.1. Then $U_2$ is determined by solving the equation $V_2 = C_2(U_2|U_1)$ using the bisection method. Two correlated standard uniform vectors are obtained as $\mathbf{U}_{m\times2} = [U_1, U_2]$ belonging to the Frank copula or No.16 copula (e.g., Nelsen 2006).

After simulating the correlated standard uniform samples $\mathbf{U}_{m\times2} = [U_1, U_2]$ from the four copulas, the physical samples of shear strength parameters $\mathbf{X}_{m\times2} = [X_1, X_2] = [c, \phi]$ can be easily obtained using the usual CDF transform method. Set $U_1 = F_1(X_1)$ and $U_2 = F_2(X_2)$, then $\mathbf{X}_{m\times2} = [X_1, X_2] = [F_1^{-1}(U_1), F_2^{-1}(U_2)]$ in which $F_1^{-1}(.)$ and $F_2^{-1}(.)$ are the inverse CDFs of $X_1$ and $X_2$, respectively (e.g., Ang and Tang 1984). The inverse CDFs for the four distributions are summarized in Table 2.8. It can be seen that $\mathbf{U}_{m\times2}$ depends on the correlation between shear strength parameters only, whereas $\mathbf{X}_{m\times2}$ relies on both the correlation and marginal distributions underlying the shear strength parameters.

## 2.4.2 Simulation of copulas and bivariate distribution

The copulas are simulated through obtaining their correlated standard uniform samples $\mathbf{U}_{m\times2} = [U_1, U_2]$ using the simulation algorithms in Section 2.4.1. The samples of $U_1$ and $U_2$

Table 2.8 Transformations of U to X for the selected four distributions

| Distribution | $X = F^{-1}(U; p, q)$ |
| --- | --- |
| TruncNormal | $X = p + q\Phi^{-1}\{U[1 - \Phi(-p/q)] + \Phi(-p/q)\}$ |
| Lognormal | $X = \exp[q\Phi^{-1}(U) + p]$ |
| TruncGumbel | $X = p - \dfrac{1}{q}\ln\left\{-\ln\left[U\left(1 - \exp(-\exp(pq))\right) + \exp(-\exp(pq))\right]\right\}$ |
| Weibull | $X = p[-\ln(1 - U)]^{(1/q)}$ |

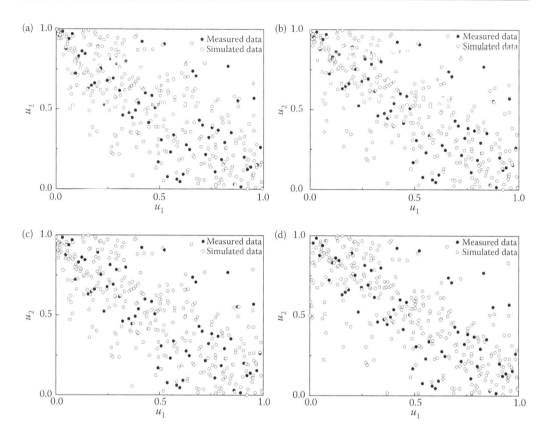

*Figure 2.5* Scatter plots of measured and simulated $u_1$ and $u_2$ for CU dataset. (a) Gaussian copula, (b) Plackett copula, (c) Frank copula, and (d) No.16 copula.

for measured datasets of shear strength parameters are given by Equation 2.29. To examine the fitness of the identified copulas to dependence structures underlying the measured data, 300 samples (i.e., $m = 300$) are drawn from the four selected copulas. As an example, Figure 2.5 shows the simulated (300 samples) and measured (64 samples) data of $U_1$ and $U_2$ for CU dataset. Note that all the copulas selected can capture the symmetry of the measured data about the 135° diagonal line of a unit square. From visual inspection, it is very difficult to distinguish the best-fit copula among them due to the limited number of measured data. Generally, the Frank copula, having been identified as the best-fit copula using the AIC and BIC criteria, can match the measured data adequately.

Similarly, the bivariate distribution of shear strength parameters is simulated by obtaining its physical samples $\mathbf{X}_{m \times 2} = [c, \phi]$ using the algorithms presented previously. In this simulation, the best-fit marginal distributions (TruncNormal distribution for both $c$ and $\phi$ in CU dataset) are adopted to convert $\mathbf{U}_{m \times 2}$ into $\mathbf{X}_{m \times 2}$. To further examine the fitness of the constructed bivariate distributions using various copulas, Figure 2.6 compares the simulated (300 samples) and measured (64 samples) data of $c$ and $\phi$ for CU dataset. The bivariate distribution constructed using the Frank copula agrees well with the measured data in comparison with the other three copulas. Compared with the results in Figure 2.5 that are affected by the copula goodness of fit alone, the results in Figure 2.6 are affected by both

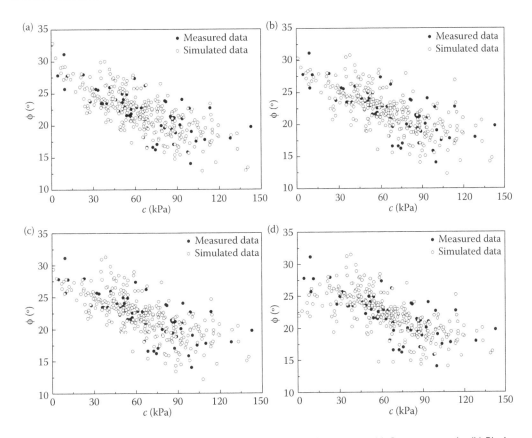

*Figure 2.6* Scatter plots of measured and simulated $c$ and $\phi$ for CU dataset. (a) Gaussian copula, (b) Plackett copula, (c) Frank copula, and (d) No.16 copula.

marginal distributions and copula goodness of fits. Similarly, the MATLAB codes for simulating copulas and bivariate distribution are appended.

On the basis of simulated samples $\mathbf{U}_{m \times 2} = [U_1, U_2]$, Pearson's rho, $\rho$, and Kendall's tau, $\tau$, between $U_1$ and $U_2$ can be calculated by Equations 2.11 and 2.14, respectively. Applying the similar approach, the $\rho$ and $\tau$ between $X_1$ and $X_2$ can be obtained using the simulated samples $\mathbf{X}_{m \times 2} = [X_1, X_2]$. The results are summarized in Table 2.9. For comparison, the $\rho$ and $\tau$ of CU dataset are also listed. Note that the $\rho$ between $X_1$ and $X_2$ are significantly different from those between $U_1$ and $U_2$. The reason is that a nonlinear monotonic transformation underlying the CDF transformation is employed to transform $\mathbf{U}$ into $\mathbf{X}$, and Pearson's rho is not invariant under such nonlinear transformations as mentioned previously. Unlike Pearson's rho, the $\tau$ between $X_1$ and $X_2$ are the same as those between $U_1$ and $U_2$ as shown in the third and fourth lines of Table 2.9. This is because Kendall's tau is invariant under monotone transformations. In addition, the $\tau$ between $X_1$ and $X_2$ associated with various copulas agree well with the prescribed $\tau$ between the measured data and the maximum relative error in $\tau$ is only 2%, while the resulting $\rho$ between $X_1$ and $X_2$ may differ considerably from the prescribed $\rho$ between the measured data. The reason is that there is no non-Gaussian simulation technique available to date that can match rank and product–moment correlation simultaneously (Grigoriu 1998; Phoon et al. 2004).

*Table 2.9* Comparison of simulation results associated with different copulas for CU dataset

| | Gaussian | Plackett | Frank | No.16 |
|---|---|---|---|---|
| $\tau$ Between measured $c$ and $\phi$ | | $-0.544$ | | |
| $\tau$ Between simulated $U_1$ and $U_2$ | $-0.546$ | $-0.551$ | $-0.550$ | $-0.555$ |
| $\tau$ Between simulated $X_1$ and $X_2$ | $-0.546$ | $-0.551$ | $-0.550$ | $-0.555$ |
| $\rho$ Between measured $c$ and $\phi$ | | $-0.702$ | | |
| $\rho$ Between simulated $U_1$ and $U_2$ | $-0.745$ | $-0.741$ | $-0.754$ | $-0.757$ |
| $\rho$ Between simulated $X_1$ and $X_2$ | $-0.749$ | $-0.719$ | $-0.710$ | $-0.684$ |

## 2.5 IMPACT OF COPULA SELECTION ON RETAINING WALL RELIABILITY

A copula-based approach for modeling and simulating the bivariate distribution of shear strength parameters was presented in the previous sections. It is concluded that the Gaussian copula may not provide the best fit to the measured dependence structure between shear strength parameters. However, in geotechnical practice, the Gaussian copula is often adopted to characterize the dependence structure among multiple geotechnical parameters in terms of Nataf transformation or the translation approach (Phoon et al. 2010; Dithinde et al. 2011; Ching and Phoon 2012, 2013). In particular, Chapter 1 also uses the Gaussian copula for constructing the multivariate distribution of multiple geotechnical parameters.

There are three practical reasons to stick to the Gaussian copula: (1) Reliability-based design in geotechnical engineering often involves multiple geotechnical parameters. The Gaussian copula is one of the relatively few copulas that have practical $n$-dimensional generalizations; (2) small sample size is a real feature of geotechnical data. On the basis of the limited data, only marginal distributions and covariance underlying geotechnical parameters can be determined, which are the only required inputs of the Gaussian copula. In addition, there is large statistical uncertainty in identifying a best-fit copula using limited data even for the bivariate case; and (3) complete multivariate data (i.e., multiple geotechnical parameters for a single depth at a site known simultaneously) are not available in geotechnical practice. With the incomplete multivariate data (i.e., multiple geotechnical parameters for a single depth at a site known partially) at hand, the Gaussian copula is the only copula that can be constructed using bivariate data as demonstrated in Section 1.7 of Chapter 1.

Although the application of copulas is associated with the above limitations, there are still three cases in which the copula approach can be applied: (1) There is one pair or multiple independent pairs of correlated geotechnical parameters in a specific geotechnical problem such as the shear strength parameters studied in this chapter; (2) the sample size of the geotechnical data is sufficiently large enough to accurately identify the best-fit copula; and (3) complete multivariate data are available in geotechnical engineering. In this case, the multivariate $t$ copula as well as the multivariate Gaussian copula can be adopted to model the multivariate distribution of the multiple geotechnical parameters. When the above conditions are satisfied, it is desirable to try other copulas (not just the Gaussian copula) since the copula type will influence the geotechnical reliability significantly (e.g., Tang et al. 2013a). With this in mind, this section presents some investigations to show the effect of copula selection on reliability. The reliability of a retaining wall with overturning failure mode is studied.

### 2.5.1 Retaining wall example

A semigravity retaining wall (Low 2005) shown in Figure 2.7 is employed to investigate the reliability of a retaining wall. Generally, three failure modes should be considered in the design of a semigravity retaining wall: (1) overturning of the wall about its toe, (2) sliding along its base, and (3) bearing capacity failure of the foundation soil. The overturning failure mode is examined below because the relationship between the strength parameters and factor of safety associated with this failure mode is strongly nonlinear. The existing deterministic approach evaluates a lumped factor of safety against overturning failure about the wall's toe as (Low 2005)

$$FS = \frac{M_{\text{resisting}}}{M_{\text{overturning}}} = \frac{W_1 \times Arm_1 + W_2 \times Arm_2}{P_a \times Arm_a} \tag{2.31}$$

where $M_{\text{resisting}}$ and $M_{\text{overturning}}$ denote the actual resisting moments and overturning moments, respectively; $W_1$ and $W_2$ are the component weights of the retaining wall, with horizontal lever distances $Arm_1$ and $Arm_2$, respectively, measured from the toe of the wall; and $P_a$ is the active earth thrust with a vertical lever distance $Arm_a$. In this chapter, Rankine's theory is used to compute $P_a$, which is based on the assumption that the back of the retaining wall is frictionless. For backfill with cohesion $c$ and internal friction angle $\phi$, $P_a$ is given by

$$P_a = \frac{1}{2}\gamma_{\text{soil}}H^2 K_a - 2cH\sqrt{K_a} + \frac{2c^2}{\gamma_{\text{soil}}} \tag{2.32}$$

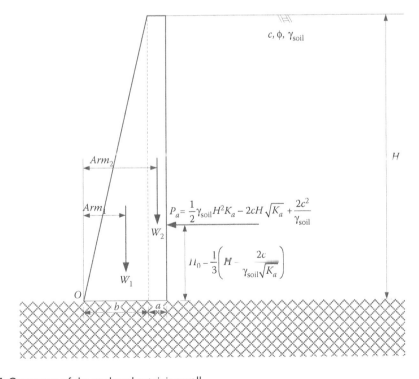

Figure 2.7 Geometry of the analyzed retaining wall.

in which $K_a$ is the coefficient of active earth pressure; $\gamma_{soil}$ is the unit weight of the backfill; and $H$ is the height of the wall. According to Rankine's theory, $K_a$ is expressed as

$$K_a = \tan^2\left(\frac{\pi}{4} - \frac{\phi}{2}\right) \tag{2.33}$$

It is assumed that the back of the retaining wall is fully drained, that is, the water table is below the base of the wall and it has no effect on the stability of the wall. The active earth thrust $P_a$ will act at a height of $H_0$ ($Arm_a = H_0$) above the base of the wall with a horizontal direction. In Figure 2.7, the following equations can be established, for a wall with a vertical back:

$$W_1 = \frac{1}{2}\gamma_{wall}bH, \ Arm_1 = \frac{2}{3}b, \ W_2 = \gamma_{wall}aH, \ Arm_2 = b + \frac{a}{2} \tag{2.34}$$

in which $\gamma_{wall}$ is the unit weight of the retaining wall concrete.

It is well known that the shear strength parameters $c$ and $\phi$ of the retained soil have a significant influence on the probability of overturning failure (Abd Alghaffar and Dymiotis-Wellington 2007). Therefore, both $c$ and $\phi$ are treated as random variables. Following Jimenez-Rodriguez et al. (2006), a Lognormal distribution is adopted to model the distributions of $c$ and $\phi$. The other five parameters, namely $H$, $a$, $b$, $\gamma_{soil}$, and $\gamma_{wall}$, are assumed to be constants so that the correlation between $c$ and $\phi$ can be effectively explored without interference from the other random variables. The mean and COV of $c$ are assumed to be 12 kPa and 0.4, respectively. The mean and COV of $\phi$ are assumed to be 20° and 0.2, respectively. The deterministic parameters are listed as follows: $H = 6$ m, $a = 0.4$ m, $b = 1.4$ m, $\gamma_{soil} = 18$ kN/m³, and $\gamma_{wall} = 24$ kN/m³. In this example, keeping the aforementioned advantages and weaknesses of Pearson's rho in mind, Pearson's rho is adopted to determine the copula parameters $\theta$ to be consistent with the geotechnical engineering practice. On the basis of $\rho$ in Tables 2.2 through 2.4, a $\rho = -0.5$ is used to account for the effect of correlation on the probability of retaining wall overturning failure.

## 2.5.2 Probability of failure using direct integration

In the current reliability literature, many reliability methods such as the first-order reliability method (FORM), second-order reliability method (SORM), and Monte Carlo simulation (MCS), are available for determining the probability of failure. To remove the errors resulting from linearization at the design point underlying the FORM and SORM, and statistical errors due to sampling sizes underlying the MCS, a direct integration method is adopted to determine the probability of failure. In this way, the effect of different copulas on the probability of failure can be identified accurately.

The following performance function is adopted for the retaining wall example:

$$g(c, \phi) = FS(c, \phi) - 1 \tag{2.35}$$

in which $FS(c, \phi)$ is determined by Equation 2.31. It should be noted that the performance function of the retaining wall example is a cubic equation with respect to cohesion $c$.

The probability of failure, $p_f$, is given by the following double integral:

$$p_f = \iint_{FS\leq1} f(c,\phi)dcd\phi$$

(2.36)

where $f(c, \phi)$ is the joint PDF of $c$ and $\phi$. Applying Equation 2.6, Equation 2.36 can be further expressed as

$$p_f = \iint_{FS\leq1} f_1(c)f_2(\phi)D(F_1(c),\ F_2(\phi);\ \theta)dcd\phi$$

(2.37)

It is evident from Equation 2.37 that the double integral could be time-consuming. For this reason, the first derivative of a copula function is employed, which is given by

$$M(u_1,u_2;\theta) = \partial C(u_1,u_2;\theta)/\partial u_2$$

(2.38)

By substituting Equation 2.38 into Equation 2.37, the double integral in Equation 2.37 can be reduced to a single integral:

$$p_f = \int_{FS\leq1} f_2(\phi)M(F_1(c),\ F_2(\phi);\ \theta)d\phi$$

(2.39)

For further derivation, the expression of $c$ in terms of $\phi$ should be available, which can be obtained based on the limit state function $g(c, \phi) = FS(c, \phi) - 1 = 0$. For the retaining wall example, by applying Shengjin's formulas (Fan 1989) to solve the cubic equation with respect to $c$, $c$ can be expressed as

$$c = 0.5\gamma_{soil}\sqrt{K_a}\left[H - \sqrt[3]{\frac{6(W_1 \times Arm_1 + W_2 \times Arm_2)}{\gamma_{soil}K_a}}\right]$$

(2.40)

Substituting Equation 2.40 into Equation 2.39, one can obtain the probability of retaining wall failure:

$$p_f = \int_0^{\phi_0} f_2(\phi)M\left(F_1\left(0.5\gamma_{soil}\sqrt{K_a}\left[H - \sqrt[3]{\frac{6(W_1 \times Arm_1 + W_2 \times Arm_2)}{\gamma_{soil}K_a}}\right]\right),\ F_2(\phi);\ \theta\right)d\phi \quad (2.41)$$

in which $\phi_0$ is calculated by

$$\phi_0 = \frac{\pi}{2} - 2\arctan\left(\sqrt{\frac{6(W_1 \times Arm_1 + W_2 \times Arm_2)}{\gamma_{soil}H^3}}\right)$$

(2.42)

Note that the probability of failure shown in Equation 2.41 is easily solved when the first derivative of a copula is available. For convenience, Table 2.10 summarizes the first

*Table 2.10* First derivatives of the selected four copulas

| Copula | First derivative of $C(u_1, u_2; \theta)$ with respect to $u_2$, $M(u_1, u_2; \theta)$ |
|---|---|
| Gaussian | $\Phi\left(\dfrac{\Phi^{-1}(u_1) - \theta\Phi^{-1}(u_2)}{\sqrt{1 - \theta^2}}\right)$ |
| Plackett | $\dfrac{1}{2} - \dfrac{1 + (\theta - 1)u_2 - (\theta + 1)u_1}{2\{[1 + (\theta - 1)(u_1 + u_2)]^2 - 4u_1 u_2 \theta(\theta - 1)\}^{(1/2)}}$ |
| Frank | $\dfrac{e^{-\theta u_2}(e^{-\theta u_1} - 1)}{(e^{-\theta} - 1) + (e^{-\theta u_1} - 1)(e^{-\theta u_2} - 1)}$ |
| No.16 | $\dfrac{1}{2}\left(1 + \dfrac{\theta}{u_2^2}\right)\left[1 + S(S^2 + 4\theta)^{-\frac{1}{2}}\right], \; S = u_1 + u_2 - 1 - \theta\left(\dfrac{1}{u_1} + \dfrac{1}{u_2} - 1\right)$ |

derivatives of the four copulas considered. When the copula parameters $\theta$ are known, the probabilities of failure for the retaining wall can be efficiently evaluated using Equation 2.41.

### 2.5.3 Nominal factor of safety for retaining wall stability

It is clear that a mean factor of safety computed by substituting mean values for the random variables in Equation 2.31 cannot account for the COVs of shear strength parameters and the correlation between cohesion and friction angle. To take into consideration these statistics approximately, nominal factors of safety involving cautious estimates of the shear strength parameters are introduced. Following the Eurocode 7 practice (Orr 2000), a 5% fractile value of the factors of safety shown in Equation 2.31 is defined as the nominal factor of safety. In this chapter, the nominal factor of safety, $FS_n$, is obtained from simulations as illustrated below using the Plackett copula (Tang et al. 2013a). The algorithm for simulating the nominal factor of safety consists of the following three steps:

1. Simulate two correlated standard uniform vectors $\mathbf{U}_{m \times 2} = [U_1, U_2]$ belonging to the Plackett copula using the algorithms in Section 2.4.1. Note that a sample size of $m = 10^6$ is adopted for the simulation.
2. Let $\mathbf{X}_{m \times 2} = (c, \phi) = (F_1^{-1}(U_1), F_2^{-1}(U_2))$ in which $F_1^{-1}(.)$ and $F_2^{-1}(.)$ are the inverse CDFs of $c$ and $\phi$, respectively. In this example, $c$ and $\phi$ are Lognormal variables.
3. Substitute $\mathbf{X}_{m \times 2} = (c, \phi)$ into Equation 2.31. The vector of the factor of safety $\mathbf{FS}_{m \times 1}$ for the retaining wall is obtained. The 5% fractile value of the simulated factors of safety is obtained using the MATLAB function `quantile(FS,0.05)`, which is taken as $FS_n$.

It is evident from the above simulation procedures that $FS_n$ is a function of deterministic parameters (e.g., geometrical parameters $a$ and $b$) and statistical parameters (e.g., COV and $\rho$ of shear strength parameters). In the parametric studies presented below, the variation of $p_f$ with various deterministic and statistical parameters is studied by plotting against $FS_n$ rather than the mean factor of safety. There are two reasons for this more complicated choice. First, a single horizontal axis based on $FS_n$ can be applied in all parameter studies (deterministic and statistical parameters), thus providing a unified and concise presentation of the results. On the other hand, the mean factor of safety is just a function of deterministic parameters. If $p_f$ is plotted against the mean factor of safety, only the variations of $p_f$ with deterministic parameters can be studied. Second, $FS_n$ is closer to the nominal factor of safety computed in practice, because engineers will use lower bound strength values, rather than mean values.

### 2.5.4 Reliability results produced by different copulas

The impact of copula selection on probability of failure is systematically studied based on three factors: (1) geometrical parameters $(a, b)$ (Figure 2.7), (2) COV scaling factor, $\lambda$ defined as $COV_c = 0.4/\lambda$ and $COV_\phi = 0.2/\lambda$, and (3) $\rho$. In the parametric studies shown in Figure 2.8, each factor is varied over a range of values shown in the figure caption while the other parameters remain unchanged. To facilitate comparisons between the four subplots in Figure 2.8, changes in each factor are presented in a uniform way as changes in the nominal factor of safety. In other words, the horizontal axes are identical, although changes in the nominal factor of safety are caused by different factors in each subplot.

#### 2.5.4.1 Effect of geometrical parameters on probability of failure

Applying the direct integration method in Section 2.5.2, the probability of failure of the retaining wall is obtained. The nominal factor of safety is calculated by the steps presented

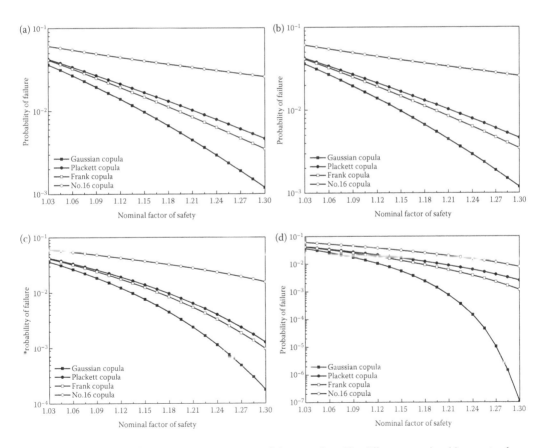

*Figure 2.8* Probabilities of retaining wall overturning failure produced by different copulas. (a) *a* varying from 0.40 m to 0.58 m, (b) *b* varying from 1.40 m to 1.64 m, (c) $\lambda$ varying from 1.00 to 1.95, and (d) $\rho$ varying from −0.50 to −0.88.

in Section 2.5.3. Figures 2.8a and b show the probabilities of failure on log scale produced by different copulas for various values of $FS_n$ as a result of varying $a$ and $b$, respectively. It is clear that the probabilities of failure produced by different copulas differ considerably. The Gaussian copula results in the smallest probability of failure among the four selected copulas. On the other hand, the No.16 copula leads to the largest probability of failure. Therefore, the Gaussian copula, commonly used for modeling the bivariate distribution of shear strength parameters, will significantly overestimate the retaining wall reliability. As $a$ or $b$ increases, $FS_n$ increases, while the probability of failure decreases. It is interesting to note that the probabilities of failure for equal values of $FS_n$ as a result of varying $a$ and $b$ are the same, indicating the same sensitivity of $a$ and $b$ to the safety of the retaining wall.

Table 2.11 summarizes the relative differences in probabilities of failure associated with different copulas and nominal factors of safety. In Table 2.11, the values are calculated by $p_f/p_{f\text{Gaussian}}$ in which $p_f$ is the probability of failure produced by the Plackett, Frank, and No.16 copulas while $p_{f\text{Gaussian}}$ is the probability of failure produced by the Gaussian copula. It is evident that the differences in probabilities of failure produced by different copulas are significant, especially for $FS_n = 1.30$. For example, the ratios $p_f/p_{f\text{Gaussian}}$ associated with $FS_n = 1.30$ are 3.89, 2.92, and 21.68 for the Plackett, Frank, and No.16 copulas, respectively. These results further indicate that the probabilities of failure of the retaining wall can differ fairly significantly. Thus, it is important to identify the best-fit copula underlying the measured data for shear strength parameters. Otherwise, the misuse of copula may cause unacceptable errors in the probability of failure. The differences in probabilities of failure increase with increasing $FS_n$ or decreasing probability of failure, especially for the No.16 copula. For example, the ratio $p_f/p_{f\text{Gaussian}}$ associated with the No.16 copula increases from 1.68 to 21.68 when $FS_n$ increases from 1.03 to 1.30.

### 2.5.4.2 Effect of COV of shear strength parameters on probability of failure

Figure 2.8c shows the probabilities of failure on a log scale produced by different copulas for various values of $FS_n$ as a result of varying the COV scaling factor, $\lambda$. Like the results shown in Figures 2.8a and b, the probabilities of failure produced by different copulas differ considerably. Again, the Gaussian copula produces the smallest probability of failure among the selected four copulas. As the $COV_c$ and $COV_\phi$ decrease (or $\lambda$ increases), $FS_n$ increases and the probability of failure decreases. However, the probability of failure is more sensitive to $FS_n$ than those shown in Figures 2.8a and b. For instance, when $FS_n$ ranges from 1.03 to 1.30, the probabilities of failure for the Gaussian copula are within [3.63E-02, 1.77E-04] as shown in Figure 2.8c, which is significantly wider than [3.63E-02, 1.19E-03] as shown in Figures 2.8a and b. The relative differences in probabilities of failure produced by different copulas and nominal factors of safety are also listed in Table 2.11. When the same $FS_n$ is adopted, the differences in probabilities of failure associated with different copulas are significant. For the Plackett, Frank, and No.16 copulas, the ratios $p_f/p_{f\text{Gaussian}}$ associated with $FS_n = 1.30$ are 7.17, 5.47, and 87.45, respectively. The probability of failure for the Gaussian copula is 87.45 times smaller than that for the No.16 copula. In addition, for a specified copula, the differences in probabilities of failure increase as the $FS_n$ increases.

### 2.5.4.3 Effect of correlation between cohesion and friction angle on probability of failure

Figure 2.8d shows the probabilities of failure on log scale produced by different copulas for various values of $FS_n$ as a result of varying $\rho$. The results are qualitatively the same as

Modeling and simulation of bivariate distribution of shear strength parameters   109

Table 2.11 Comparison of probabilities of failure associated with different copulas and nominal factors of safety

| Copula | $a = [0.40, 0.58\ m]$ | | | $b = [1.40, 1.64\ m]$ | | | $\lambda = [1.00, 1.95]$ | | | $\rho = [-0.50, -0.88]$ | | |
|---|---|---|---|---|---|---|---|---|---|---|---|---|
| | $FS_n = 1.03$ | $FS_n = 1.17$ | $FS_n = 1.30$ | $FS_n = 1.03$ | $FS_n = 1.17$ | $FS_n = 1.30$ | $FS_n = 1.03$ | $FS_n = 1.17$ | $FS_n = 1.30$ | $FS_n = 1.03$ | $FS_n = 1.17$ | $FS_n = 1.30$ |
| Gaussian | 1 | 1 | 1 | 1 | 1 | 1 | 1 | 1 | 1 | 1 | 1 | 1 |
| Plackett | 1.17 | 1.84 | 3.89 | 1.17 | 1.84 | 3.89 | 1.17 | 1.98 | 7.17 | 1.17 | 3.78 | $2.45 \times 10^4$ |
| Frank | 1.14 | 1.60 | 2.92 | 1.14 | 1.60 | 2.92 | 1.14 | 1.75 | 5.47 | 1.14 | 2.96 | $1.13 \times 10^4$ |
| No.16 | 1.68 | 4.77 | 21.68 | 1.68 | 4.77 | 21.68 | 1.68 | 5.88 | 87.45 | 1.68 | 7.39 | $7.54 \times 10^4$ |

Note: The value is calculated by $p_f / p_{f\text{Gaussian}}$.

those shown in Figures 2.8a through c, but the probabilities of failure associated with the Gaussian copula can be several orders of magnitude smaller than those associated with the other three copulas, especially for a strong negative correlation between cohesion and friction angle. Furthermore, the probabilities of failure associated with the Gaussian copula are very sensitive to the change of $FS_n$. When $FS_n$ varies from 1.03 to 1.17, the probability of failure associated with the Gaussian copula decreases from 3.63E 02 to 1.02E-07 (more than four orders of magnitude!). The relative differences in probabilities of failure associated with different copulas and nominal factors of safety are listed in the last three columns of Table 2.11. The same conclusions as those drawn from the other three cases can also be made. However, the ratios $p_f/p_{f\text{Gaussian}}$ associated with $FS_n = 1.30$ are significantly larger than those for the other three cases. For instance, the ratios $p_f/p_{f\text{Gaussian}}$ associated with $FS_n = 1.30$ are 2.45E04, 1.13E04, and 7.54E04 for the Plackett, Frank, and No.16 copulas, respectively.

## 2.5.5 Discussions

It can be concluded from the above results that the probabilities of overturning failure for a retaining wall associated with the four selected copulas differ greatly, especially when small COVs or large correlation coefficients underlying shear strength parameters are used. In this section, some discussions are presented to explain such observations following two ways: (1) a comparison among simulated samples of cohesion and friction angle associated with various copulas is carried out, and (2) relative locations between the limit state surfaces and the joint PDF isolines of cohesion and friction angle are investigated.

Figure 2.9 shows the simulated samples of cohesion and friction angle from the selected four copulas for $\rho = -0.5$, $COV_c = 0.4$, and $COV_\phi = 0.2$. The sample size is 1000. The contour lines of constants $FS = 1.0$, 1.4, and 2.0 for a representative retaining wall with $H = 6$ m, $a = 0.4$ m, $b = 1.4$ m, $\gamma_{\text{soil}} = 18$ kN/m$^3$, and $\gamma_{\text{wall}} = 24$ kN/m$^3$ are also plotted in Figure 2.9. Again, the numbers ($N$) of samples falling in the regions associated with $FS \leq 1.0$, $1.0 < FS \leq 1.4$, $1.4 < FS \leq 2.0$, and $FS > 2.0$ are shown in the corresponding regions. Note that different copulas characterize different dependence structures between cohesion and friction angle although the same marginal distributions and correlation coefficient of shear strength parameters are followed. For example, the numbers of samples falling in the aforementioned four regions for the Gaussian copula are 38, 290, 323, and 349. They are 63, 197, 393, and 347 for the No.16 copula. It is noted that the region associated with $FS \leq 1.0$ is of more significance to practice because the probability of failure is basically derived from this region. Thus, geotechnical engineers often pay more attention to the differences in this region associated with different copulas. The numbers of samples falling in this region are 38, 47, 47, and 63 for the Gaussian, Plackett, Frank, and No.16 copulas, respectively. It is evident from these results that the No.16 copula leads to the largest probability of failure while the Gaussian copula results in the smallest probability of failure.

To make a better comparison between the Gaussian copula and the other three copulas, Figure 2.10 shows the joint PDF isolines of shear strength parameters associated with the four copulas selected. The joint PDF isoline associated with the Gaussian copula is plotted using a dashed line. For illustration, a typical PDF isoline value of 0.001 is used. This makes the PDF isoline envelope nearly the whole domain where simulated samples may appear. Note that the shape of such an isoline is similar to that of the scatter plots of simulated samples. It is evident that the joint PDFs of the shear strength parameters associated with different copulas differ considerably, especially between the Gaussian and No.16 copulas. Such a difference further leads to the difference in probability of failure between Gaussian

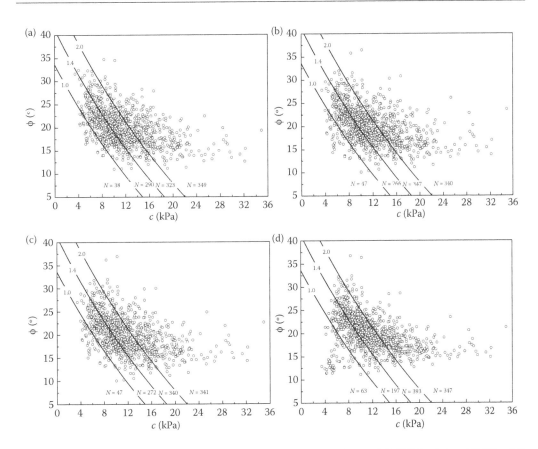

*Figure 2.9* Scatter plots of $c$ and $\phi$ generated by different copulas with $\rho = -0.5$, $COV_c = 0.4$ and $COV_\phi = 0.2$. (a) Gaussian copula, (b) Plackett copula, (c) Frank copula, and (d) No.16 copula.

copula and the other three copulas, as shown in Figure 2.8. To investigate the effect of relative locations between the limit state surfaces and the joint PDF isolines on reliability, three limit state surfaces are considered. They are associated with three retaining walls defined by (1) $a = 0.4$ m, $b = 1.4$ m, and $FS_n = 1.03$, (2) $a = 0.58$ m, $b = 1.4$ m, and $FS_n = 1.30$, and (3) $a = 0.4$ m, $b = 1.64$ m, and $FS_n = 1.30$. The respective limit state surfaces are referred to as "Limit state I", "Limit state II", and "Limit state III". It can be seen that the shift of $FS_n$ from 1.03 to 1.30 is due to the shift of $a$ from 0.4 to 0.58 m or $b$ from 1.4 to 1.64 m. Similarly, the shift of $FS_n$ from 1.03 to 1.30 is caused by the shift of limit state surfaces from I to II or from I to III. Consequently, the probability of failure decreases when $FS_n$ increases from 1.03 to 1.30 although the joint PDF of the shear strength parameters remains unchanged.

Figure 2.11 shows the joint PDF isolines of the shear strength parameters associated with different copulas for $\rho = -0.5$, $COV_c = 0.2$, and $COV_\phi = 0.1$. Compared with Figure 2.10, the boundary of the joint PDF isoline becomes smaller as the COVs of $c$ and $\phi$ decrease. In other words, the probability content of the joint PDF over the same failure set of the retaining wall decreases. As a result, the probabilities of failure of the retaining wall will thus decrease although the limit state surfaces do not change. Similarly, Figure 2.12 shows the joint PDF isolines of the shear strength parameters associated with different copulas for $\rho = -0.8$, $COV_c = 0.4$, and $COV_\phi = 0.2$. In comparison with Figure 2.10, as the negative correlation

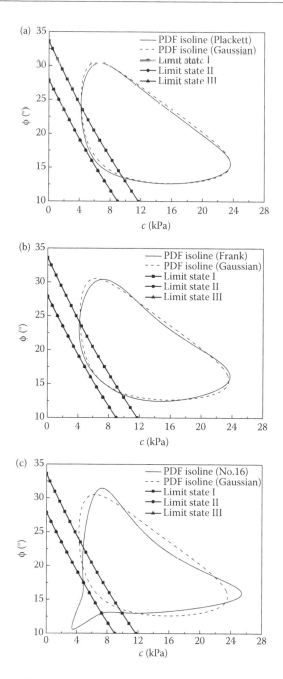

*Figure 2.10* Comparison of PDF isolines of shear strength parameters associated with different copulas for $\rho = -0.5$, $COV_c = 0.4$ and $COV_\phi = 0.2$. (a) Plackett copula versus Gaussian copula, (b) Frank copula versus Gaussian copula, and (c) No.16 copula versus Gaussian copula.

between cohesion and friction angle becomes stronger, the PDF isoline becomes narrower, and the difference among the joint PDFs associated with different copulas becomes more significant. Like the results shown in Figure 2.11, the probabilities of failure will become smaller although the limit state surfaces remain unchanged.

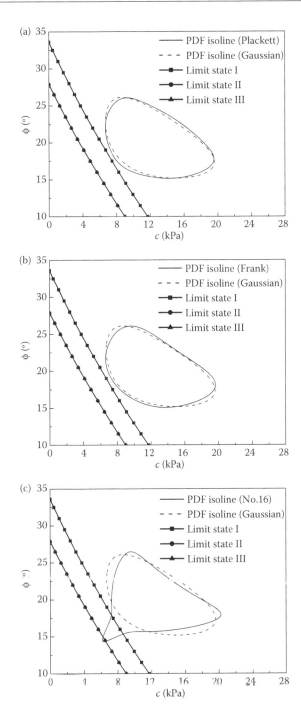

*Figure 2.11* Comparison of PDF isolines of shear strength parameters associated with different copulas for $\rho = -0.5, COV_c = 0.2$ and $COV_\phi = 0.1$. (a) Plackett copula versus Gaussian copula, (b) Frank copula versus Gaussian copula, and (c) No.16 copula versus Gaussian copula.

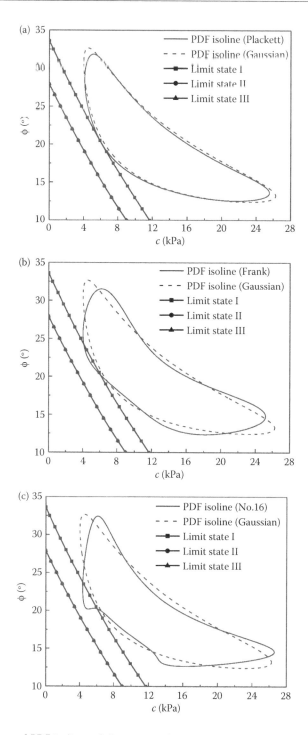

*Figure 2.12* Comparison of PDF isolines of shear strength parameters associated with different copulas for $\rho = -0.8$, $COV_c = 0.4$ and $COV_\phi = 0.2$. (a) Plackett copula versus Gaussian copula, (b) Frank copula versus Gaussian copula, and (c) No. 16 copula versus Gaussian copula.

## 2.6 SUMMARY AND CONCLUSIONS

This chapter has employed the copula approach to model and simulate the bivariate distribution of shear strength parameters. Some measured data of shear strength parameters reported in the literature are collected. Four copulas, namely Gaussian, Plackett, Frank, and No.16 copulas, are selected to construct the joint probability distribution of cohesion and friction angle. The impact of copulas on retaining wall reliability is investigated. Several conclusions can be drawn:

1. Copulas can effectively decouple the determinations of marginal distributions and joint probability distributions of shear strength parameters. Applying the copula theory, it is not necessary that the shear strength parameters follow the same marginal distributions or must share a dependence structure characterized by the Gaussian copula. In short, the copula theory provides a general and convenient approach for modeling and simulating the bivariate distributions of shear strength parameters, which should be highlighted in practical geotechnical applications.
2. The copula selection has a significant impact on geotechnical reliability. The failure probabilities produced by different copulas differ considerably. This difference increases with decreasing probability of failure or increasing nominal factor of safety. Significant difference in probability of failure could be observed for relatively small COVs of shear strength parameters or a strongly negative correlation between cohesion and friction angle.
3. The Gaussian copula may not capture the dependence structure between cohesion and friction angle properly. When the Gaussian copula is applied to reliability analyses, the probabilities of failure for the geotechnical structures may be underestimated significantly. Thus, it is of importance to select the appropriate copula to characterize the dependence structure underlying the shear strength parameters when enough data are available.
4. The applications of copulas to geotechnical engineering are mostly concerned with bivariate data. One reason for this is that relatively few copulas have practical $n$-dimensional generalizations. One well-known example is the elliptical copulas such as the Gaussian and $t$ copulas. They can be easily extended to model and simulate the multivariate distributions of multiple geotechnical parameters. However, the multivariate $t$ copula is applicable only when complete multivariate data are available. With incomplete multivariate data at hand, only the multivariate Gaussian copula can be adopted.

## ACKNOWLEDGMENTS

This work was supported by the National Science Fund for Distinguished Young Scholars (Project No. 51225903) and the National Basic Research Program of China (973 Program) (Project No. 2011CB013506) and the National Natural Science Foundation of China (Project No. 51329901).

## APPENDIX 2A: MATLAB® CODES

This appendix presents MATLAB codes stored in M-files for modeling and simulating the bivariate distribution of shear strength parameters using copulas: (1) MATLAB codes for identification of the best-fit marginal distributions and copula underlying measured data, (2) MATLAB codes for simulation of copulas and bivariate distribution, and (3) MATLAB codes for definition of subfunctions used in (1) and (2).

## (1) Identification of best-fit margins and copula using AIC and BIC

```
% Identification of best-fit margins and copula using AIC and BIC
% Filename: Identification.m
% © (2013) Dian-Qing Li and Xiao-Song Tang
clc;
format short;
% Read observed cohesions and friction angles stored in XLS-files, data
data = xlsread('data.xls');
%
% Select best-fit margin among TruncNormal,Lognormal,TruncGumbel and
% Weibull distributions
% Rows,columns,mean,standard deviation of data, rows,cols,mu,sigma
[rows cols] = size(data);
mu = mean(data);
sigma = std(data);
% Number of distribution parameters, Parameter_Margin
Parameter_Margin = 2;
%
% Calculate the AIC and BIC values for variable X1
PDF_TruncNormal_X1 = normpdf(data(:,1),mu(1),sigma(1))/(1-normcdf(-mu(1)/
sigma(1)));
AIC_TruncNormal_X1 = -2*sum(log(PDF_TruncNormal_X1))+2*Parameter_Margin;
BIC_TruncNormal_X1 = -2*sum(log(PDF_TruncNormal_X1))+log(rows)*Parameter_
Margin;
%
sLn = sqrt(log(1+(sigma(1)/mu(1))^2)); mLn = log(mu(1))-sLn^2/2;
PDF_Lognormal_X1 = lognpdf(data(:,1),mLn,sLn);
AIC_Lognormal_X1 = -2*sum(log(PDF_Lognormal_X1))+2*Parameter_Margin;
BIC_Lognormal_X1 = -2*sum(log(PDF_Lognormal_X1))+log(rows)*Parameter_
Margin;
%
a = 1.282549808/sigma(1); b = mu(1)-(0.5772156649/a);
PDF_TruncGumbel_X1 = a*exp(-a*(data(:,1)-b)-exp(-a*(data(:,1)-b)))/
(1-exp(-exp(a*b)));
AIC_TruncGumbel_X1 = -2*sum(log(PDF_TruncGumbel_X1))+2*Parameter_Margin;
BIC_TruncGumbel_X1 = -2*sum(log(PDF_TruncGumbel_X1))+log(rows)*Parameter_
Margin;
%
k = k_solveWeibull(mu(1),sigma(1)); u = mu(1)/gamma(1+1/k);
PDF_Weibull_X1 = k/u*(data(:,1)/u).^(k-1).*exp(-(data(:,1)/u).^k);
AIC_Weibull_X1 = -2*sum(log(PDF_Weibull_X1))+2*Parameter_Margin;
BIC_Weibull_X1 = -2*sum(log(PDF_Weibull_X1))+log(rows)*Parameter_Margin;
%
AIC_X1 = [AIC_TruncNormal_X1 AIC_Lognormal_X1 AIC_TruncGumbel_X1 AIC_
Weibull_X1];
display(AIC_X1);
[AIC_min_X1 Index_X1] = min(AIC_X1,[],2);
if Index_X1 == 1
    display('The best-fit margin for X1 using AIC is TruncNormal
distribution');
```

```
elseif Index_X1 == 2
   display('The best-fit margin for X1 using AIC is Lognormal
distribution');
elseif Index_X1 == 3
   display('The best-fit margin for X1 using AIC is TruncGumbel
distribution');
elseif Index_X1 == 4
   display('The best-fit margin for X1 using AIC is Weibull
distribution');
end
%
BIC_X1 = [BIC_TruncNormal_X1 BIC_Lognormal_X1 BIC_TruncGumbel_X1
BIC_Weibull_X1];
display(BIC_X1);
[BIC_min_X1 Index_X1] = min(BIC_X1,[],2);
if Index_X1 == 1
   display('The best-fit margin for X1 using BIC is TruncNormal
distribution');
elseif Index_X1 == 2
   display('The best-fit margin for X1 using BIC is Lognormal
distribution');
elseif Index_X1 == 3
   display('The best-fit margin for X1 using BIC is TruncGumbel
distribution');
elseif Index_X1 == 4
   display('The best-fit margin for X1 using BIC is Weibull
distribution');
end
%
% Calculate the AIC and BIC values for variable X2
PDF_TruncNormal_X2 = normpdf(data(:,2),mu(2),sigma(2))/(1-normcdf(-mu(2)/
sigma(2)));
AIC_TruncNormal_X2 = -2*sum(log(PDF_TruncNormal_X2)) + 2*Parameter_
Margin;
BIC_TruncNormal_X2 = -2*sum(log(PDF_TruncNormal_X2)) + log(rows)*Parameter_
Margin;
%
sLn = sqrt(log(1 + (sigma(2)/mu(2))^2)); mLn = log(mu(2))-sLn^2/2;
PDF_Lognormal_X2 = lognpdf(data(:,2),mLn,sLn);
AIC_Lognormal_X2 = -2*sum(log(PDF_Lognormal_X2)) + 2*Parameter_Margin;
BIC_Lognormal_X2 = -2*sum(log(PDF_Lognormal_X2)) + log(rows)*Parameter_
Margin;
%
a = 1.282549808/sigma(2); b = mu(2) - (0.5772156649/a);
PDF_TruncGumbel_X2 = a*exp(-a*(data(:,2)-b)-exp(-a*(data(:,2)-b)))/
(1-exp(-exp(a*b)));
AIC_TruncGumbel_X2 = -2*sum(log(PDF_TruncGumbel_X2)) + 2*Parameter_Margin;
BIC_TruncGumbel_X2 = -2*sum(log(PDF_TruncGumbel_X2)) + log(rows)*Parameter_
Margin;
%
k = k_solveWeibull(mu(2),sigma(2)); u = mu(2)/gamma(1 + 1/k);
PDF_Weibull_X2 = k/u*(data(:,2)/u).^(k-1).*exp(-(data(:,2)/u).^k);
AIC_Weibull_X2 = -2*sum(log(PDF_Weibull_X2)) + 2*Parameter_Margin;
BIC_Weibull_X2 = -2*sum(log(PDF_Weibull_X2)) + log(rows)*Parameter_Margin;
%
```

```
AIC_X2 = [AIC_TruncNormal_X2 AIC_Lognormal_X2 AIC_TruncGumbel_X2 AIC_
Weibull_X2];
display(AIC_X2);
[AIC_min_X2 Index_X2] = min(AIC_X2,[],2);
if Index_X2 == 1
    display('The best-fit margin for X2 using AIC is TruncNormal
distribution');
elseif Index_X2 == 2
    display('The best-fit margin for X2 using AIC is Lognormal
distribution');
elseif Index_X2 == 3
    display('The best-fit margin for X2 using AIC is TruncGumbel
distribution');
elseif Index_X2 == 4
    display('The best-fit margin for X2 using AIC is Weibull
distribution');
end
%
BIC_X2 = [BIC_TruncNormal_X2 BIC_Lognormal_X2 BIC_TruncGumbel_X2 BIC_
Weibull_X2];
display(BIC_X2);
[BIC_min_X2 Index_X2] = min(BIC_X2,[],2);
if Index_X2 == 1
    display('The best-fit margin for X2 using BIC is TruncNormal
distribution');
elseif Index_X2 == 2
    display('The best-fit margin for X2 using BIC is Lognormal
distribution');
elseif Index_X2 == 3
    display('The best-fit margin for X2 using BIC is TruncGumbel
distribution');
elseif Index_X2 == 4
    display('The best-fit margin for X2 using BIC is Weibull
distribution');
end
%
% Select best-fit copula among Gaussian,Plackett,Frank and No.16 copulas
% Empirical distributions of data, U
[datasort,dataindex] = sort(data);
Ranks_data = data;
for m = 1:cols
    Ranks_data(dataindex(:,m),m) = 1:rows;
end
U = Ranks_data/(rows + 1);
%
% Kendall's tau of data, Kendall_data
Kendall_data = corr(data,'type','Kendall');
%
% Estimation of copula parameters
rho_Gaussian = sin(pi/2*Kendall_data);
theta_Plackett = theta_estimationPlackett_Kendall(Kendall_data(1,2));
theta_Frank = copulaparam('Frank',Kendall_data(1,2),'type','Kendall');
theta_16 = theta_estimation16_Kendall(Kendall_data(1,2));
Copula_parameter = [rho_Gaussian(1,2) theta_Plackett theta_Frank
theta_16];
```

```
display(Copula_parameter);
%
% Number of copula parameters, Parameter_Copula
Parameter_Copula = 1;
cPlackett = @(u,v,theta)((1 + (theta-1)*(u + v))^2-4*u*v*theta*(theta-1))
^(-3/2)*theta*(1 + (theta-1)*(u + v-2*u*v));
c16 = @(u,v,theta)0.5*(1 + theta/u^2)*(1 + theta/v^2)*((u + v-1-theta*
(1/u + 1/v-1))^2 + 4*theta)^(-0.5)*(-((u + v-1-theta*(1/u + 1/v-1))^2 +
4*theta)^(-1)*(u + v-1-theta*(1/u + 1/v-1))^2 + 1);
%
% Calculate the AIC and BIC values for the four candidate copulas
PDF_Gaussian = copulapdf('Gaussian',U,rho_Gaussian(1,2));
AIC_Gaussian = -2*sum(log(PDF_Gaussian)) + 2*Parameter_Copula;
BIC_Gaussian = -2*sum(log(PDF_Gaussian)) + log(rows)*Parameter_Copula;
%
PDF_Plackett = zeros(rows,1);
for m = 1:rows
    PDF_Plackett(m) = cPlackett(U(m,1),U(m,2),theta_Plackett);
end
AIC_Plackett = -2*sum(log(PDF_Plackett)) + 2*Parameter_Copula;
BIC_Plackett = -2*sum(log(PDF_Plackett)) + log(rows)*Parameter_Copula;
%
PDF_Frank = copulapdf('Frank',U,theta_Frank);
AIC_Frank = -2*sum(log(PDF_Frank)) + 2*Parameter_Copula;
BIC_Frank = -2*sum(log(PDF_Frank)) + log(rows)*Parameter_Copula;
%
PDF_16 = zeros(rows,1);
for m = 1:rows
    PDF_16(m) = c16(U(m,1),U(m,2),theta_16);
end
AIC_16 = -2*sum(log(PDF_16)) + 2*Parameter_Copula;
BIC_16 = -2*sum(log(PDF_16)) + log(rows)*Parameter_Copula;
%
AIC_Copula = [AIC_Gaussian AIC_Plackett AIC_Frank AIC_16];
display(AIC_Copula);
[AIC_min_Copula Index_Copula] = min(AIC_Copula,[],2);
if Index_Copula == 1
    display('The best-fit copula using AIC is Gaussian copula');
elseif Index_Copula == 2
    display('The best-fit copula using AIC is Plackett copula');
elseif Index_Copula == 3
    display('The best-fit copula using AIC is Frank copula');
elseif Index_Copula == 4
    display('The best-fit copula using AIC is No.16 copula');
end
%
BIC_Copula = [BIC_Gaussian BIC_Plackett BIC_Frank BIC_16];
display(BIC_Copula);
[BIC_min_Copula Index_Copula] = min(BIC_Copula,[],2);
if Index_Copula == 1
    display('The best-fit copula using BIC is Gaussian copula');
elseif Index_Copula == 2
    display('The best-fit copula using BIC is Plackett copula');
elseif Index_Copula == 3
    display('The best-fit copula using BIC is Frank copula');
```

```
elseif Index_Copula == 4
    display('The best-fit copula using BIC is No.16 copula');
end
```

## (2) Simulation of copulas and bivariate distribution

```
% Simulation of copulas and bivariate distribution
% Filename: CopulaSim.m
% © (2013) Dian-Qing Li and Xiao-Song Tang
clc;
format short;
% Read observed cohesions and friction angles stored in XLS-files, data
data = xlsread('data.xls');
% Number of realizations, N
N = 300;
% Type of Copula, Type_Copula
% Options: Gaussian,Plackett,Frank,No.16
Type_Copula = 'No.16';
% Marginal distributions of data, distri
% Options: TruncNormal,Lognormal,TruncGumbel,Weibull
distri(1,1) = {'TruncNormal'}; distri(1,2) = {'TruncNormal'};
% Seeds fixed at 1
randn('state',1);
%
% Rows,columns,mean,standard deviation of data, rows,cols,mu,sigma
[rows cols] = size(data);
mu = mean(data);
sigma = std(data);
%
% Pearson's linear correlation coefficient of data, Pearson_data
Pearson_data = corr(data);
display(Pearson_data);
%
% Kendall's tau of data, Kendall_data
Kendall_data = corr(data,'type','Kendall');
display(Kendall_data);
%
% Estimation of copula parameters
switch Type_Copula
    case 'Gaussian'
        rho = sin(pi/2*Kendall_data);
        display(rho);
    case 'Plackett'
        theta = theta_estimationPlackett_Kendall(Kendall_data(1,2));
        display(theta);
    case 'Frank'
        theta = copulaparam('Frank',Kendall_data(1,2),'type','Kendall');
        display(theta);
    case 'No.16'
        theta = theta_estimation16_Kendall(Kendall_data(1,2));
        display(theta);
end
%
% Simulation of correlated uniformly distributed variables from copulas,
% U
z = randn(N,2);
```

```
zz = z*inv(chol(cov(z)));
switch Type_Copula
    case 'Gaussian'
        Q = chol(rho);
        Y = zz*Q;
        U = normcdf(Y);
    case 'Plackett'
        V = normcdf(zz);
        U = zeros(size(V));
        U(:,1) = V(:,1);
        a = V(:,2).*(1-V(:,2));
        b = theta + a.*(theta-1)^2;
        c = 2*a.*(U(:,1).*theta^2+1-U(:,1)) + theta.*(1-2.*a);
        d = sqrt(theta).*sqrt(theta+4.*a.*U(:,1).*(1-U(:,1)).*(1-theta)^2);
        U(:,2) = (c-(1-2.*V(:,2)).*d)./(2.*b);
    case 'Frank'
        V = normcdf(zz);
        U = zeros(size(V));
        U(:,1) = V(:,1);
        U(:,2) = -1/theta.*log(1+V(:,2).*(1-exp(-theta))./(V(:,2).*(exp(-
theta.*U(:,1))-1)-exp(-theta.*U(:,1)))));
    case 'No.16'
        V = normcdf(zz);
        U = zeros(size(V));
        U(:,1) = V(:,1);
        for m = 1:N
            U(m,2) = U2_solve16(U(m,1),V(m,2),theta);
        end
end
%
% Mean of U, mu_U
mu_U = mean(U);
display(mu_U);
%
% Pearson's linear correlation coefficient of U, Pearson_U
Pearson_U = corr(U);
display(Pearson_U);
%
% Kendall's tau of U, Kendall_U
Kendall_U = corr(U,'type','Kendall');
display(Kendall_U);
%
% Simulation of physical variables from constructed joint probability
% distribution, X
X = zeros(size(U));
for m = 1:2
    X(:,m) = X_value(U(:,m),mu(m),sigma(m),distri{1,m});
end
%
% Pearson's linear correlation coefficient of X, Pearson_X
Pearson_X = corr(X);
display(Pearson_X);
%
% Kendall's tau of X, Kendall_X
Kendall_X = corr(X,'type','Kendall');
```

```
display(Kendall_X);
%
% Scatter plots of U
figure;
scatter(U(:,1),U(:,2),'.','b');
if(Kendall_data(1,2)>0)
    legend('Simulated data for U','Location','SouthEast');
else
    legend('Simulated data for U','Location','SouthWest');
end
axis([-0.05 1.05 -0.05 1.05]);
box on
xlabel('{\it U}_1','fontname','Times New Roman');
ylabel('{\it U}_2','fontname','Times New Roman');
%
% Scatter plots of X
figure;
scatter(X(:,1),X(:,2),'.','b');
if(Kendall_data(1,2)>0)
    legend('Simulated data for X','Location','SouthEast');
else
    legend('Simulated data for X','Location','NorthEast');
end
xmin=min(data(:,1))-0.1*(max(data(:,1))-min(data(:,1)));
xmax=max(data(:,1))+0.4*(max(data(:,1))-min(data(:,1)));
ymin=min(data(:,2))-0.1*(max(data(:,2))-min(data(:,2)));
ymax=max(data(:,2))+0.4*(max(data(:,2))-min(data(:,2)));
axis([xmin xmax ymin ymax]);
box on
xlabel('{\it c}','fontname','Times New Roman');
ylabel('{\it fai}','fontname','Times New Roman');
```

### (3) Definition of subfunctions used in the above programs

```
% Definition of No.16 copula's dependence parameter estimation function,
% theta_estimation16_Kendall
function y=theta_estimation16_Kendall(tau)
a=1e-6;
b=1e4;
eps=1e-6;
fx=@(theta,tau)1+4*quadgk(@(t)(theta./t+1).*(1-t)./(theta./t.^2.*(t-1)-
(theta./t+1)),0,1)-tau;
fa=fx(a,tau);
fb=fx(b,tau);
if fa*fb>0
    disp('error:[a b] interval no solution, please re-enter a and b');
    return
end
while b-a>eps
    c=0.5*(a+b);
    fc=fx(c,tau);
    if(fc*fa<0)
        b=c;
        fb=fc;
```

```matlab
    else
        a = c;
        fa = fc;
    end
end
y = 0.5*(a+b);
return
%
% Definition of Plackett copula's dependence parameter estimation
% function, theta_estimationPlackett_Kendall
function y = thcta_estimationPlackett_Kendall(tau)
a = 1e-6;
b = 1e4;
eps = 1e-6;
fx = @(theta,tau)4*dblquad(@(u,v)(1+(theta-1).*(u+v)-sqrt((1+(theta-
1).*(u+v)).^2-4.*u.*v.*theta.*(theta-1)))./2./(theta-1).*((1+(theta-
1).*(u+v)).^2-4.*u.*v.*theta.*(theta-1)).^(-3/2).*theta.*(1+(theta-
1).*(u+v-2.*u.*v)),0,1,0,1)-1-tau;
fa = fx(a,tau);
fb = fx(b,tau);
if fa*fb > 0
    disp('error:[a b] interval no solution, please re-enter a and b');
    return
end
while b-a > eps
    c = 0.5*(a+b);
    fc = fx(c,tau);
    if(fc*fa < 0)
        b = c;
        fb = fc;
    else
        a = c;
        fa = fc;
    end
end
y = 0.5*(a+b);
return
%
% Definition of NO.16 copula's U2 solving function, U2_solve16
function y = U2_solve16(U1,V2,theta)
a = 1e-6;
b = 0.999999;
eps = 1e-6;
fa = V2-0.5*(1+theta/U1^2)*(1+((U1+a-1-theta*(1/U1+1/a-
1))^2+4*theta)^(-0.5)*(U1+a-1-theta*(1/U1+1/a-1)));
fb = V2-0.5*(1+theta/U1^2)*(1+((U1+b-1-theta*(1/U1+1/b-
1))^2+4*theta)^(-0.5)*(U1+b-1-theta*(1/U1+1/b-1)));
if fa*fb > 0
    disp('error:[a b] interval no solution, please re-enter a and b');
    return
end
while b-a > eps
    c = 0.5*(a+b);
    fc = V2-0.5*(1+theta/U1^2)*(1+((U1+c-1-theta*(1/U1+1/c-
    1))^2+4*theta)^(-0.5)*(U1+c-1-theta*(1/U1+1/c-1)));
```

```
    if(fc*fa < 0)
        b = c;
        fb = fc;
    else
        a = c;
        fa = fc;
    end
end
y = 0.5*(a + b);
return
%
% Definition of U to X function, X_value
function X = X_value(U,mu,sigma,distri)
switch distri
    case 'TruncNormal'
        X = mu + sigma*norminv((1-normcdf(-mu/sigma))*U + normcdf(-mu/sigma));
    case 'Lognormal'
        X = exp(norminv(U)*sqrt(log(1 + sigma^2/(mu^2))) + log(mu) -
0.5*log(1 + sigma^2/(mu^2)));
    case 'TruncGumbel'
        a = 1.282549808/sigma;
        b = mu - (0.5772156649/a);
        X = b-log(-log(U*(1-exp(-exp(a*b))) + exp(-exp(a*b))))/a;
    case 'Weibull'
        k = k_solveWeibull(mu,sigma);
        u = mu/gamma(1 + 1/k);
        X = u*(-log(1-U)).^(1/k);
end
return
%
% Definition of Weibull distribution parameter k solving function,
% k_solveWeibull
function y = k_solveWeibull(mu,sigma)
a = 0.02;
b = 40;
eps = 1e-6;
fa = 1/sqrt(gamma(1 + 2/a)/gamma(1 + 1/a)^2-1) -mu/sigma;
fb = 1/sqrt(gamma(1 + 2/b)/gamma(1 + 1/b)^2-1) -mu/sigma;
if fa*fb > 0
    disp('error:[a b] interval no solution, please re-enter a and b');
    return
end
while b-a > eps
    c = 0.5*(a + b);
    fc = 1/sqrt(gamma(1 + 2/c)/gamma(1 + 1/c)^2-1) -mu/sigma;
    if fc*fa < 0;
        b = c;
        fb = fc;
    else
        a = c;
        fa = fc;
    end
end
y = 0.5*(a + b);
return
```

## LIST OF SYMBOLS

| | |
|---|---|
| $\Phi(.)$ | Standard normal distribution function |
| $\Phi^{-1}(.)$ | Inverse standard normal distribution function |
| $\Phi_{\theta}(.,.)$ | Bivariate standard normal distribution function with Pearson's rho $\theta$ |
| $\varphi_{\theta}(.)$ | Generator function |
| $\phi$ | Friction angle |
| $\theta$ | Copula parameter |
| $\theta$ | Vector of copula parameters |
| $\lambda$ | COV scaling factor |
| $\lambda_{U}$ | Coefficient of upper tail dependence |
| $\lambda_{L}$ | Coefficient of lower tail dependence |
| $\Gamma(.)$ | $\gamma$-Function |
| $\mu$ | Mean value |
| $\mu_{i}$ | Mean value of $X_i$ |
| $\sigma$ | Standard deviation |
| $\sigma^2$ | Variance |
| $\sigma_{i}$ | Standard deviation of $X_i$ |
| $\gamma_{soil}$ | Unit weight of the backfill |
| $\gamma_{wall}$ | Unit weight of the retaining wall concrete |
| $\rho$ | Pearson linear correlation coefficient (Pearson's rho) |
| $\tau$ | Kendall rank correlation coefficient (Kendall's tau) |
| AIC | Akaike information criterion |
| $Arm_a$ | Vertical lever distance of $P_a$ |
| $a, b, H$ | Geometrical parameters of the retaining wall |
| $Arm_1$ and $Arm_2$ | Horizontal lever distances of $W_1$ and $W_2$ |
| BIC | Bayesian information criterion |
| $c$ | Cohesion |
| CDF | Cumulative distribution function |
| CQ | Consolidated-quick direct shear test |
| CD | Consolidated-drained triaxial compression test |
| CU | Consolidated-undrained triaxial compression test |
| $Cov(.,.)$ | Covariance |
| COV | Coefficient of variation |
| $COV_c$ | Coefficient of variation of $c$ |
| $COV_{\phi}$ | Coefficient of variation of $\phi$ |
| $C(u_1, u_2, ..., u_n; \theta)$ | $n$-Dimensional copula function |
| $C(u_1, u_2; \theta)$ | Bivariate copula function |
| $C_2(U_2|U_1)$ | Conditional distribution of $U_2$ given the values of $U_1$ |
| $D(u_1, u_2, ..., u_n; \theta)$ | $n$-Dimensional copula density function |
| $D(u_1, u_2; \theta)$ | Bivariate copula density function |
| $FS$ | Mean factor of safety |
| $FS_n$ | Nominal factor of safety |
| $F(x_1, x_2, ..., x_n)$ | Joint CDF of $[X_1, X_2, ..., X_n]$ |
| $f(x_1, x_2, ..., x_n)$ | Joint PDF of $[X_1, X_2, ..., X_n]$ |
| $F(x_1, x_2)$ | Bivariate CDF of $X_1$ and $X_2$ |
| $f(x_1, x_2)$ | Bivariate PDF of $X_1$ and $X_2$ |
| $f(c, \phi)$ | Bivariate PDF of $c$ and $\phi$ |
| $F_i(.)$ | Marginal CDF of $X_i$ |
| $F_1(c)$ | Marginal CDF of $c$ |

| | |
|---|---|
| $F_2(\phi)$ | Marginal CDF of $\phi$ |
| $f_i(.)$ | Marginal PDF of $X_i$ |
| $f_1(c)$ | Marginal PDF of $c$ |
| $f_2(\phi)$ | Marginal PDF of $\phi$ |
| $f(x; p, q)$ | Marginal PDF of $X$ with distribution parameters $p$ and $q$ |
| $F_i^{-1}(.)$ | Inverse CDF of $X_i$ |
| $g(.)$ | Performance function |
| $k_1$ | Number of distribution parameters |
| $k_2$ | Number of copula parameters |
| $K_a$ | Coefficient of active earth pressure |
| $m$ | Sample size of simulated data |
| $M_{\text{resisting}}$ | Resisting moments |
| $M_{\text{overturning}}$ | Overturning moments |
| $M(u_1, u_2; \theta)$ | First derivative of $C(u_1, u_2; \theta)$ with respect to $u_2$ |
| $n$ | Dimension of $\mathbf{X}$ |
| $N$ | Sample size of measured data |
| PDF | Probability density function |
| $P(.)$ | Probability |
| $P_a$ | Active earth thrust |
| $p_f$ | Probability of failure |
| $p, q$ | Marginal distribution parameters |
| $p_{f\text{Gaussian}}$ | Probability of failure produced by Gaussian copula |
| Q | Quick direct shear test |
| $\mathbf{Q}$ | Upper triangular Cholesky factor of $\mathbf{R}$ matrix |
| $\mathbf{R}$ | Correlation matrix underlying the Gaussian copula |
| sign(.) | Indicator function |
| UU | Unconsolidated-undrained triaxial compression test |
| $\mathbf{U}$ | Correlated standard uniform vector $[U_1, U_2, ..., U_n]$ |
| $U_i$ | $i$th component in $\mathbf{U} = [U_1, U_2, ..., U_n]$ |
| $U$ | Standard uniform variable |
| $\mathbf{V}$ | Uncorrelated standard uniform vector $[V_1, V_2, ..., V_n]$ |
| $W_1, W_2$ | Component weights of the retaining wall |
| $\bar{x}_1, \bar{x}_2$ | Sample means of $X_1$ and $X_2$ |
| $\mathbf{X}$ | Random vector of $[X_1, X_2, ..., X_n]$ |
| $X_i$ | $i$th component in $\mathbf{X} = [X_1, X_2, ..., X_n]$ |
| $X$ | Uncertain soil parameter |
| $\mathbf{Y}$ | Correlated standard normal vector $[Y_1, Y_2, ..., Y_n]$ |
| $\mathbf{Z}$ | Uncorrelated standard normal vector $[Z_1, Z_2, ..., Z_n]$ |

## REFERENCES

Abd Alghaffar MA, Dymiotis-Wellington C. Time-variant reliability of retaining walls and calibration of partial factors. *Structure and Infrastructure Engineering* 2007; 3(3): 187–198.

Akaike H. A new look at the statistical model identification. *IEEE Transactions on Automatic Control* 1974; 19(6): 716–723.

Ang AH-S, Tang WH. *Probability Concepts in Engineering Planning and Design, Vol. II: Decision, Risk, and Reliability.* John Wiley and Sons: New York, 1984.

Baecher GB, Christian JT. *Reliability and Statistics in Geotechnical Engineering.* John Wiley and Sons: New York, 2003.

Cherubini C. Reliability evaluation of shallow foundation bearing capacity on c', ϕ' soils. *Canadian Geotechnical Journal* 2000; 37(1): 264–269.

Ching J, Phoon KK. Modeling parameters of structured clays as a multivariate normal distribution. *Canadian Geotechnical Journal* 2012; 49(5): 522–545.

Ching J, Phoon KK. Multivariate distribution for undrained shear strengths under various test procedures. *Canadian Geotechnical Journal* 2013; 50(9): 907–923.

Dithinde M, Phoon KK, De Wet M, Retief JV. Characterization of model uncertainty in the static pile design formula. *Journal of Geotechnical and Geoenvironmental Engineering* 2011; 137(1): 70–85.

Dutfoy A, Lebrun R. Practical approach to dependence modelling using copulas. *Proceedings of the Institution of Mechanical Engineers, Part O: Journal of Risk and Reliability* 2009; 223(4): 347–361.

Fan SJ. A new extracting formula and a new distinguishing means on the one variable cubic equation. *Natural Science Journal of Hainan Teachers College* 1989; 2(2): 91–98.

Genest C, Favre AC. Everything you always wanted to know about copula modeling but were afraid to ask. *Journal of Hydrologic Engineering* 2007; 12(4): 347–368.

Griffiths DV, Huang JS, Fenton GA. Probabilistic infinite slope analysis. *Computers and Geotechnics* 2011; 38(4): 577–584.

Grigoriu MD. Simulation of non-Gaussian translation processes. *Journal of Engineering Mechanics* 1998; 124(2): 121–126.

Jimenez-Rodriguez R, Sitar N, Chacon J. System reliability approach to rock slope stability. *International Journal of Rock Mechanics and Mining Sciences* 2006; 43(6): 847–859.

Lebrun R, Dutfoy A. Do Rosenblatt and Nataf isoprobabilistic transformations really differ? *Probabilistic Engineering Mechanics* 2009a; 24(4): 577–584.

Lebrun R, Dutfoy A. A generalization of the Nataf transformation to distributions with elliptical copula. *Probabilistic Engineering Mechanics* 2009b; 24(2): 172–178.

Li DQ, Chen YF, Lu WB, Zhou CB. Stochastic response surface method for reliability analysis of rock slopes involving correlated non-normal variables. *Computers and Geotechnics* 2011; 38(1): 58–68.

Li DQ, Jiang SH, Wu SB, Zhou CB, Zhang LM. Modeling multivariate distributions using Monte Carlo simulation for structural reliability analysis with complex performance function. *Proceedings of the Institution of Mechanical Engineers, Part O: Journal of Risk and Reliability* 2013c; 227(2): 109–118.

Li DQ, Phoon KK, Wu SB, Chen YF, Zhou CB. Impact of translation approach for modelling correlated non-normal variables on parallel system reliability. *Structure and Infrastructure Engineering* 2013b; 9(10): 969–982.

Li DQ, Tang XS, Phoon KK, Chen YF, Zhou CB. Bivariate simulation using copula and its application to probabilistic pile settlement analysis. *International Journal for Numerical and Analytical Methods in Geomechanics* 2013a; 37(6): 597–617.

Li DQ, Tang XS, Zhou CB, Phoon KK. Uncertainty analysis of correlated non-normal geotechnical parameters using Gaussian copula. *Science China Technological Sciences* 2012a; 55(11): 3081–3089.

Li DQ, Wu SB, Zhou CB, Phoon KK. Performance of translation approach for modeling correlated nonnormal variables. *Structural Safety* 2012b; 39:52–61.

Li XY, Xie KH, Bai XH, Shi MY. The spatial probabilistic characteristics of strength indexes for Taiyuan silty clay. *Chinese Journal of Geotechnical Engineering* 2000; 22(6): 668–672 (in Chinese)

Low BK. Reliability-based design applied to retaining walls. *Geotechnique* 2005; 55(1): 63–75.

Low BK. Reliability analysis of rock slopes involving correlated nonnormals. *International Journal of Rock Mechanics and Mining Sciences* 2007; 44(6): 922–935.

Lumb P. Safety factors and the probability distribution of soil strength. *Canadian Geotechnical Journal* 1970; 7(3): 225–242.

Mari DD, Kozt S. *Correlation and Dependence*. Imperial College Press: UK, 2001.

McNeil AJ, Frey R, Embrechts P. *Quantitative Risk Management: Concepts, Techniques and Tools*. Princeton University Press: Princeton, 2005.

Nataf A. Détermination des distributions de probabilité dont les marges sont données. *Comptes Rendus de l'Académie des Sciences* 1962; 225:42–43.

Nelsen RB. *An Introduction to Copulas*. 2nd edition, Springer: New York, 2006.

Orr TLL. Selection of characteristic value and partial factors in geotechnical designs to Eurocode 7. *Computers and Geotechnics* 2000; 26(3–4): 263–279.

Phoon KK, Quek ST, Huang HW. Simulation of non-Gaussian processes using fractile correlation *Probabilistic Engineering Mechanics* 2004; 19(4): 287–292.

Phoon KK, Santoso A, Quek ST. Probabilistic analysis of soil–water characteristic curves. *Journal of Geotechnical and Geoenvironmental Engineering* 2010; 136(3): 445–455.

Salvadori G, De Michele C. On the use of copulas in hydrology: Theory and practice. *Journal of Hydrological Engineering* 2007; 12(4): 369–380.

Schwarz G. Estimating the dimension of a model. *Annals of Statistics* 1978; 6(2): 461–464.

Tang XS, Li DQ, Chen YF, Zhou CB, Zhang LM. Improved knowledge-based clustered partitioning approach and its application to slope reliability analysis. *Computers and Geotechnics* 2012; 45:34–43.

Tang XS, Li DQ, Rong G, Phoon KK, Zhou CB. Impact of copula selection on geotechnical reliability under incomplete probability information. *Computers and Geotechnics* 2013a; 49:264–278.

Tang XS, Li DQ, Zhou CB, Phoon KK, Zhang LM. Impact of copulas for modeling bivariate distributions on system reliability. *Structural Safety* 2013c; 44:80–90.

Tang XS, Li DQ, Zhou CB, Zhang LM. Bivariate distribution models using copulas for reliability analysis. *Proceedings of the Institution of Mechanical Engineers, Part O: Journal of Risk and Reliability* 2013b; 227(5): 499–512.

Uzielli M, Mayne PW. Serviceability limit state CPT-based design for vertically loaded shallow footings on sand. *Geomechanics and Geoengineering: An International Journal* 2011; 6(2): 91–107.

Uzielli M, Mayne PW. Load-displacement uncertainty of vertically loaded shallow footings on sands and effects on probabilistic settlement. *Georisk* 2012; 6(1): 50–69.

Wu CF, Zhu XR, Liu XM. Studies on variability of shear strength indexes for several typical stratums in Hangzhou area. *Chinese Journal of Geotechnical Engineering* 2005; 27(1): 94–99 (in Chinese).

Wu XZ. Probabilistic slope stability analysis by a copula-based sampling method. *Computational Geosciences* 2013a; 17(5): 739–755.

Wu XZ. Trivariate analysis of soil ranking-correlated characteristics and its application to probabilistic stability assessments in geotechnical engineering problems. *Soils and Foundations* 2013b; 53(4): 540–556.

Zhang L, Tang XS, Li DQ. Bivariate distribution model of soil shear strength parameter using copula. *Journal of Civil Engineering and Management* 2013; 30(2): 11–17 (in Chinese).

# Part II

# Methods

# Chapter 3

# Evaluating reliability in geotechnical engineering

*J. Michael Duncan and Matthew D. Sleep*

## 3.1 PURPOSE OF RELIABILITY ANALYSIS

The factors of safety used in conventional geotechnical practice are based on experience, which is logical. However, it is common to use the same factor of safety value for a given type of application, such as $F = 1.3$ for short-term slope stability or $F = 3.0$ for erosion and piping, without regard to the degree of uncertainty involved in the calculation in a particular instance. Through regulation or tradition, the same safety factor value is often applied to conditions that involve widely varying degrees of uncertainty. This is not logical.

Reliability calculations provide a means of evaluating the combined effects of uncertainties, and a means of distinguishing between conditions where uncertainties are particularly high or low. In spite of the fact that it has potential value, the reliability theory has not been used much in routine geotechnical practice. There are two reasons for this. First, the reliability theory involves terms and concepts that are not familiar to many geotechnical engineers. Second, it is commonly perceived that using the reliability theory would require more data, time, and effort than are available in most circumstances.

"Reliability" as it is used in the reliability theory is the probability of an event occurring or the probability of a "positive outcome." Reliability is the complement of probability of failure. Thus, if there is a 0.5% probability that the factor of safety could be less than 1.0, the reliability (the probability that the factor of safety is greater than 1.0) is 99.5%.

Evaluating reliability affords a means of assessing the degree of uncertainty involved in geotechnical engineering calculations. Christian et al. (1994), Tang et al. (1999), and others have presented examples of reliability use in geotechnical engineering, and explained the underlying theories. The purpose of this chapter is to show that reliability concepts can be applied in simple ways, without more data, time, or effort than are commonly available in geotechnical engineering practice. Working with the same quantity and types of data, and the same types of engineering judgments that are used in conventional analyses, it is possible to make useful evaluations of reliability. Although evaluations of reliability are inevitably approximate, they provide a useful complement to conventional factors of safety.

The results of simple reliability analyses, of the type described in this chapter, will be neither more nor less accurate than conventional deterministic analyses that use the same types of data, judgments, and approximations. While neither deterministic nor reliability analyses are precise, they both have value, and each enhances the value of the other.

It is not advocated here that the factor of safety analyses be abandoned in favor of reliability analyses. Instead, it is suggested that factor of safety and reliability be used together, as complementary measures of acceptable design. The simple types of reliability analyses described in this chapter require only modest extra effort as compared to that required to evaluate factors of safety, but they will add considerable value to the results of the analyses.

## 3.2 PROBABILITY OF FAILURE AND RISK

This chapter focuses on methods of evaluating the probability of failure for geotechnical structures such as retaining walls, slopes, levees, and dams. Probabilities of failure are associated with particular modes of failure, such as retaining wall sliding, overturning, or bearing capacity failure.

Risk is related to the probability of failure, and also includes the consequences of failure—what will be the consequences if failure occurs? For example, would the consequences of retaining wall sliding be simply an unattractive but still stable wall, or would sliding require expensive remediation, or, in the extreme, could it pose a threat to life? Evaluating risk involves evaluating both the probability of failure and the consequences should failure occur. Where loadings recur over time, for example, river stage variation as shown in Figure 3.1, risk can be stated on an annual basis. A study of the probability of levee failure due to overtopping and underseepage could be stated as "the annualized risk of flood damage is X dollars per year for this section of levee," or "the annualized risk to human life is Y lives lost per year for this section of levee."

## 3.3 LANGUAGE OF STATISTICS AND PROBABILITY

### 3.3.1 Variables

When tests are performed to measure values of a physical property, for example, the undrained shear strength of saturated clay, it is commonly found that a different value is measured in each test. In this context physical properties of soil or rock are appropriately called

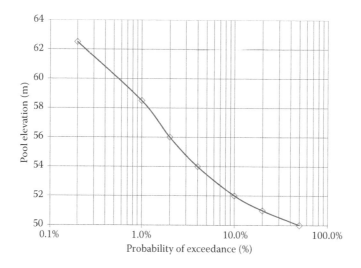

*Figure 3.1* Flood frequency curve for a reservoir.

"variables," even though a single value of the variable, such as the average, may be used in calculations.

While the average value of a variable is useful, and sufficient for many purposes, other characteristics of scattered measured values are also important, and are needed for calculations that reflect the level of uncertainty in the measured values. The standard deviation and the coefficient of variation (COV), defined below, are useful measures of the degree of scatter in a set of values of a variable.

### 3.3.2 Correlated and uncorrelated variables

The reliability methods as presented in this chapter (Taylor Series, Point Estimate, Simplified Hasofer Lind, and Monte Carlo simulation) consider only "uncorrelated" variables, for two reasons:

1. Determining correlation coefficients for geotechnical parameters is difficult and uncertain, and
2. Incorporating correlation coefficients complicates reliability analyses.

Methods for incorporating correlation coefficients into reliability analyses can be found in Baecher and Christian (2003), Harr (1987), Low (1996), Low and Tang (1997, 2004), and Low and Phoon (2002).

### 3.3.3 Standard deviation

Standard deviation is a measure of dispersion, or scatter, in values of a variable.

The standard deviation ($\sigma$) is defined mathematically as the square root of the average of the squared values of the difference between each of the measured values and the average, as expressed in

$$\sigma = \sqrt{\frac{\sum_{i=1}^{n}(x_i - \bar{x})^2}{n-1}} \tag{3.1}$$

where
   $\sigma$ is the "sample standard deviation" of $n$ measured values of $x$
   $\sum_{i=1}^{n}$ is the sum of values from 1 to $n$
   $x_i$ is the $i^{\text{th}}$ measured value
   $\bar{x}$ is the average of the measured values
   $n$ is the number of values

A slightly different standard deviation is the "population standard deviation" which is calculated using Equation 3.2

$$\sigma = \sqrt{\frac{\sum_{i=1}^{n}(x_i - \bar{x})^2}{n}} \tag{3.2}$$

Baecher and Christian (2003) indicate that using Equation 3.1 corrects for a statistical bias that results from the fact that the average is estimated from the same data that are used to calculate the standard deviation.

For use in characterizing the uncertainties in soil and rock properties, the difference between these definitions is not significant, and for practical purposes they can be used interchangeably. The standard deviation of $x$ has the same units as $x$.

### EXAMPLE

Table 3.1 contains 25 measured values of the shear strength measured on samples of soft saturated clay. The values of $s_u$ vary from 10 to 18 kPa, and the average is 14.36 kPa. The standard deviation, calculated using Equation 3.1 (STDEV in Excel) is 2.14 kPa. The standard deviation calculated using Equation 3.2 is 2.13 kPa.

*Table 3.1* Undrained shear strength data

| (a) Sorted by test | | (b) Sorted by $s_u$ value | | (c) Sorted into bins | | |
| --- | --- | --- | --- | --- | --- | --- |
| Test number | Measured $s_u$ (kPa) | Test number | Measured $s_u$ (kPa) | Interval or "bin"* | No. of values in bin | Relative frequency |
| 1 | 10 | 21 | 10 | 10–10.9 kPa | 2 | 0.08 |
| 2 | 15 | 24 | 10 | | | |
| 3 | 13 | 12 | 11 | 11–11.9 kPa | 1 | 0.04 |
| 4 | 14 | 22 | 12 | 12–12.9 kPa | 2 | 0.08 |
| 5 | 16 | 20 | 12 | | | |
| 6 | 14 | 2 | 13 | 13–13.9 kPa | 3 | 0.12 |
| 7 | 15 | 25 | 13 | | | |
| 8 | 18 | 18 | 13 | | | |
| 9 | 14 | 3 | 14 | 14–14.9 kPa | 5 | 0.20 |
| 10 | 17 | 5 | 14 | | | |
| 11 | 14 | 8 | 14 | | | |
| 12 | 16 | 10 | 14 | | | |
| 13 | 11 | 13 | 14 | | | |
| 14 | 14 | 6 | 15 | 15–15.9 kPa | 5 | 0.20 |
| 15 | 18 | 17 | 15 | | | |
| 16 | 15 | 1 | 15 | | | |
| 17 | 17 | 19 | 15 | | | |
| 18 | 15 | 15 | 15 | | | |
| 19 | 13 | 23 | 16 | 16–16.9 kPa | 3 | 0.12 |
| 20 | 15 | 4 | 16 | | | |
| 21 | 12 | 11 | 16 | | | |
| 22 | 12 | 9 | 17 | 17–17.9 kPa | 2 | 0.08 |
| 23 | 16 | 16 | 17 | | | |
| 24 | 10 | 7 | 18 | 18–18.9 kPa | 2 | 0.08 |
| 25 | 13 | 14 | 18 | | | |
| Average | 14.36 kPa | | | | | |
| Std. Dev. | 2.14 kPa | | | | | |
| COV | 15% | | | | | |

*Data intervals, or ranges of values, have several different names. They are called "groups" or "classes" in statistics, and "bins" in Microsoft Excel.

### 3.3.4 Coefficient of variation

While the standard deviation is a useful indicator of the amount of scatter in a variable, the degree of dispersion is easier to see in context if it is expressed in terms of the COV, which is the ratio of the standard deviation divided by the average ($\sigma/\bar{x}$).

$$COV = \frac{\sigma}{\bar{x}} \qquad (3.3)$$

where COV is the coefficient of variation.

The COV is a dimensionless measure of the amount of scatter, and is usually expressed as a percentage. For the measured values of undrained shear strength listed in Table 3.1, the COV = $(2.14/14.36) \times 100\% = 15\%$.

### 3.3.5 Histograms and relative frequency diagrams

Histograms and relative frequency diagrams are graphical representations of a series of measured values in the form of bar charts, as shown in Figure 3.2.

The range of measured values, from the lowest to the highest, is shown on the horizontal axis. In order to display the data as vertical bars, the horizontal axis is divided into "bins." The height of each bar in the histogram indicates the number of values in the bin. The height of each bar in the relative frequency diagram indicates the number of values in each bin divided by the total number, 25 in this case.

Although the width of the bins is arbitrary, the bin width affects the appearance of the histogram. The width of the bins used to plot Figure 3.2 is 1 kPa. The numbers of values in the bins, and the values of relative frequency, are shown in Table 3.1(c).

### 3.3.6 Probability and probability theory

The *Random House Dictionary of the English Language* defines probability as "the relative possibility that an event will occur" and probability theory as "the theory of analyzing and making statements concerning the probability of occurrence of uncertain events."

### 3.3.7 Probability density function

Probability density functions (PDFs) are continuous distributions that indicate the probability of occurrence of any value of the variable, within the range covered by the distribution. Some PDFs extend from minus infinity to plus infinity; others cover finite ranges of values. PDFs provide a means of making generalizations about distributions of variables, and can be thought of as hypotheses regarding how the values would be found to be distributed if a very large number of values were available.

The two most commonly assumed distributions are the normal distribution and the log-normal distribution, which are shown in Figure 3.3. Both of these are theoretical distributions and can be calculated based solely on the values of the mean and standard deviation of a variable. The distributions shown in Figure 3.3 were calculated for an average value of $s_u = 14.36$ kPa, and a standard deviation of 2.14 kPa, corresponding to the undrained shear strength data in Table 3.1 and Figure 3.2.

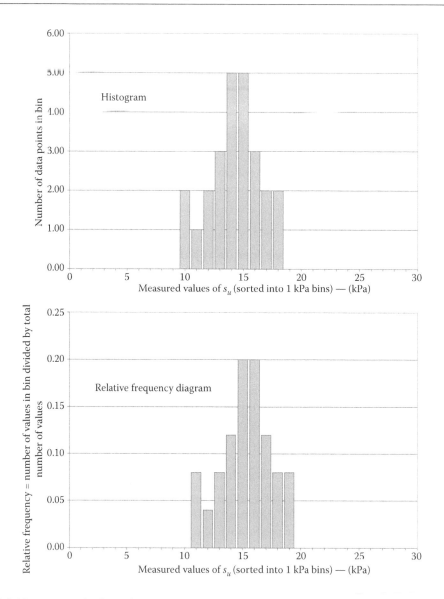

*Figure 3.2* Histogram and relative frequency diagram of undrained shear strength data shown in Table 3.1.

### 3.3.8 Normal and lognormal distributions

The normal distribution is a symmetrical bell-shaped curve, with the peak at the average value. The normal probability distribution is given by Equation 3.4

$$\text{Normal } p(x) = \frac{1}{\sigma\sqrt{2\pi}} \exp\left[ -\frac{1}{2}\left( \frac{x - \bar{x}}{\sigma} \right)^2 \right] \tag{3.4}$$

where
    $p(x)$ is the probability of a particular value of $x$

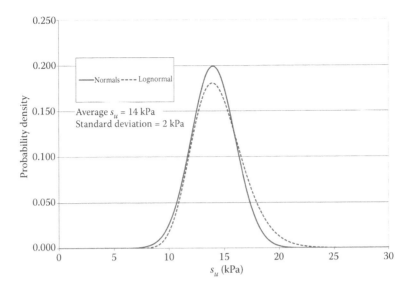

*Figure 3.3* Normal and lognormal distributions, or PDFs.

$x$ is the value of the variable
$\sigma$ is the standard deviation
$\bar{x}$ is the average value of $x$

Equation 3.4 gives the probability for a single value of $x$. The normal distribution curve shown in Figure 3.3 was developed by calculating probabilities of values of $s_u$ ranging from 0 to 30 kPa, plotting these values against the value of $s_u$, and drawing a smooth curve through the values.

The Excel function NORMDIST was used to calculate the normal distribution curve shown in Figure 3.3. This Excel function has the form:

**NORMDIST** $(x, \bar{x}, \sigma, \textbf{FALSE})$

This function was evaluated many times, for different values of $x$ and the same values of $\bar{x}$ and $\sigma$, to develop the normal distribution curve shown in Figure 3.3.

The fourth argument of the NORMDIST function, FALSE, prompts Excel to calculate the value of the PDF. If the final argument is TRUE, NORMDIST calculates the value of the cumulative density function (CDF), discussed below.

The width of the bell-shaped curve is governed by the value of the standard deviation. Figure 3.4a shows normal distributions for $s_u = 14.36$ kPa, and the values of standard deviation are 1, 2.14, and 3.5 kPa. As the standard deviation and the width of the bell-shaped curve increase, the peak probability decreases. The areas under all three of the curves in Figure 3.4a are equal to unity, consistent with the fact that the total of all probabilities for any distribution is always equal to unity.

A normal distribution of the measured values in Table 3.1 (average $s_u = 14.36$ kPa, standard deviation = 2.14 kPa) is shown with the relative frequency diagram of measured values in Figure 3.6. The theoretical normal distribution can be thought of as a generalized

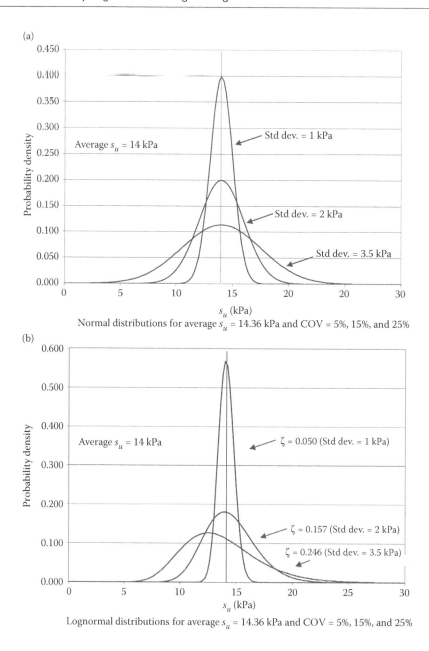

(a)

*Probability density*

Average $s_u$ = 14 kPa

Std dev. = 1 kPa

Std dev. = 2 kPa

Std dev. = 3.5 kPa

$s_u$ (kPa)

Normal distributions for average $s_u$ = 14.36 kPa and COV = 5%, 15%, and 25%

(b)

*Probability density*

Average $s_u$ = 14 kPa

$\zeta$ = 0.050 (Std dev. = 1 kPa)

$\zeta$ = 0.157 (Std dev. = 2 kPa)

$\zeta$ = 0.246 (Std dev. = 3.5 kPa)

$s_u$ (kPa)

Lognormal distributions for average $s_u$ = 14.36 kPa and COV = 5%, 15%, and 25%

*Figure 3.4* Normal and lognormal PDFs.

hypothesis regarding the variability of the data that extends beyond the results of the 25 tests that were performed.

There is some theoretical justification for the normal distribution. The "central limit theorem" indicates that the sum of a large number of distributions approaches the normal distribution as the number of distributions approaches infinity. However, use of the normal or any other theoretical distribution to form generalizations about distributions of data is really an assumption about the distribution. Such assumptions are made partly because they are thought to be reasonable approximations of reality, and partly because they are convenient for use in calculations of probability and reliability.

(a) Normal distributions for average $s_u = 14.36$ kPa and COV = 5%, 15%, and 25%
(b) Lognormal distribution for average $s_u = 14.36$ kPa and COV = 5%, 15%, and 25%

### 3.3.9 Lognormal distribution

The lognormal distribution is an unsymmetrical bell-shaped curve, with the peak to the left of the average value. The shape of the lognormal distribution is given by Equation 3.5:

$$\text{Lognormal } p(x) = \frac{1}{\varsigma x \sqrt{2\pi}} \exp\left[-\frac{1}{2}\left(\frac{\ln x - \lambda}{\varsigma}\right)^2\right] \tag{3.5}$$

where
 $p(x)$ is the lognormal probability of a particular value of $x$
 $x$ is the variable
 $\varsigma$ (zeta) is the standard deviation of the natural logarithm of $x$

$$\varsigma = \sqrt{\ln\left(1 + COV^2\right)} \tag{3.6}$$

COV is the coefficient of variation, defined by Equation 3.3
 $\lambda$ (lambda) is the average value of the natural logarithm of $x$

$$\lambda = \ln(\bar{x}) - \frac{1}{2}\varsigma^2 \tag{3.7}$$

The lognormal distribution is based on the assumption that the logarithm of the variable is normally distributed. The PDF does not extend to the left of zero (to negative values), whereas the normal distribution extends infinitely far to the left and right of the average value. Not extending to the left of zero seems more reasonable for quantities, such as shear strength or factor of safety, which cannot be negative. Even though the normal distribution extends to negative values, that part of the curve usually has negligible significance, and the implication that negative values are possible is not a reason to reject use of the normal distribution if it is otherwise reasonable and convenient.

The width of the bell-shaped lognormal curve is governed by the value of $\varsigma$, the standard deviation of the logarithm of x. $\varsigma$ is approximately equal to the COV. Figure 3.4b shows lognormal distributions for average $s_u = 14.36$ kPa, and values of $\varsigma = 0.050$, 0.149, and 0.246, corresponding to COV = 0.050, 0.150, and 0.250, (values of standard deviation = 1, 2.14, and 3.5 kPa).

As in the case of the normal distributions, the areas under all three of the lognormal curves in Figure 3.4b are equal to unity, consistent with the fact that the total of all probabilities is unity for any distribution.

A lognormal distribution of the measured variables in Table 3.1 (average $s_u = 14.36$ kPa, standard deviation = 2.14 kPa) is shown with the relative frequency diagram of measured values in Figure 3.5. The use of the lognormal distribution to represent these data would involve an assumption that the data follow a lognormal distribution. Judging solely by the degree of conformance between the shape of the relative frequency diagram and the

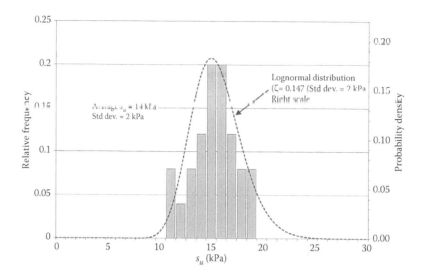

*Figure 3.5* Relative frequency diagram of measured undrained shear strength data with superimposed lognormal probability distribution.

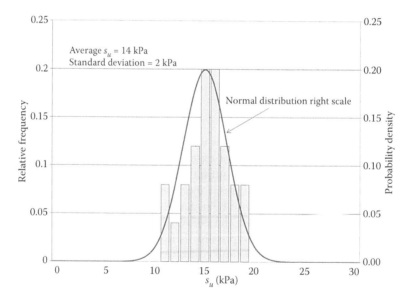

*Figure 3.6* Relative frequency diagram of measured undrained shear strength data with superimposed normal probability distribution.

theoretical distributions, it appears that the normal distribution in Figure 3.6 is a closer fit than the lognormal distribution in Figure 3.5, by a small margin.

## 3.3.10 Cumulative density function

CDF is the integral of the PDF, that is, it is the area under the PDF curve, computed from minus infinity to the value shown on the horizontal axis. The value of the CDF for any value

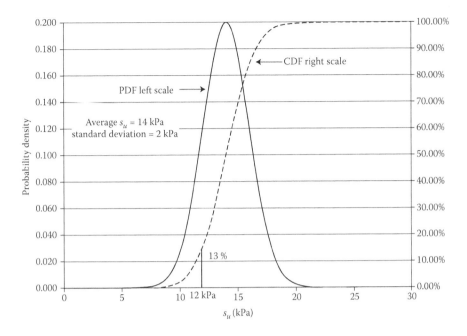

*Figure 3.7* PDF and CDF.

of x is the probability that x will be less than or equal to that value of x. A normal CDF for average $s_u$ = 14.36 kPa and standard deviation = 2.14 kPa is shown in Figure 3.7 together with the normal PDF. The probability that the value of $s_u$ may be less than or equal to a given value is indicated by the CDF. For example, based on the normal CDF, the probability that the value of $s_u$ may be less than 12 kPa is 13%.

Figure 3.8 shows the cumulative normal and lognormal CDFs for average $s_u$ = 14.36 kN/m² and standard deviation = 2.14 kPa. Although the two CDFs are similar, the tails of the distributions differ significantly, as can be seen from the low-value portions of the curves

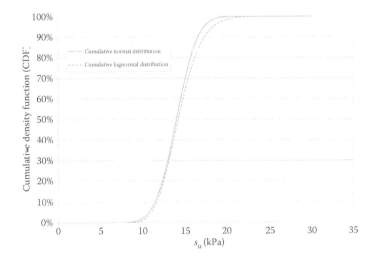

*Figure 3.8* Normal and lognormal CDFs.

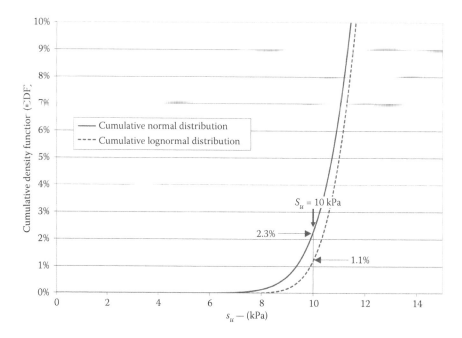

*Figure 3.9* Tails of normal and lognormal CDFs.

shown in Figure 3.9. According to the normal distribution, 2.3% of the values of $s_u$ would fall below 10 kPa; whereas according to the lognormal distribution, only 1.1% of the values of $s_u$ would fall below 10 kPa. These differences at the tails of the distributions can result in significant differences in computed values of probability of failure, as discussed below.

### 3.3.11 Probability of failure

The "probability of failure" is the probability that failure will occur. In terms of factor of safety, the probability of failure is defined as the probability that the factor of safety could be less than 1.0 given adverse values of the variables involved in its calculation.

Estimating the probability of failure is a major motive for applying probability theory in geotechnical engineering. Several methods for estimating probability of failure are discussed in a later section of this chapter.

### 3.3.12 Reliability

Reliability is the complement of the probability of failure. For example, if there is a 0.5% probability that the factor of safety against slope instability may be less than 1.0, the probability of failure is 0.5% and the reliability is 99.5%.

### 3.3.13 Reliability index

The reliability index ($\beta$) is the number of standard deviations between the most likely value of factor of safety and factor of safety = 1.0. The reliability index ($\beta$) is uniquely related to probability of failure, as shown in Table 3.2. The values of $P_f$ in Table 3.2 apply to both normally distributed and lognormally distributed factors of safety.

Table 3.2 Relationship between reliability index and
          probability of failure

| Reliability index $\beta_{Normal}$ or $\beta_{Lognormal}$ | Probability of failure, $P_f$ |
|---|---|
| 0.50 | 31% |
| 1.00 | 16% |
| 1.50 | 6.7% |
| 2.00 | 2.3% |
| 2.50 | 0.62% |
| 3.00 | 0.13% |
| 4.00 | 0.003% |
| 5.00 | 0.00003% |

The probability of failure for any value of $\beta$ can be evaluated using the NORMSDIST function in Excel, as shown by Equation 3.8

$$P_f = 1 - \text{NORMSDIST } (\beta) \tag{3.8}$$

Because $\beta$ is uniquely related to probability of failure, the value of $\beta$ has sometimes been used in lieu of the probability of failure as a measure of safety. However, in order to evaluate $\beta$, the distribution of the factor of safety must be assumed. Therefore, whether the value of $\beta$ or the value of $P_f$ is used as the safety criterion, it is necessary to assume a distribution for factor of safety. Most geotechnical engineers prefer to use $P_f$ rather than $\beta$, because $P_f$ has an easily understood relationship to likely performance and $\beta$ does not. Figure 3.10 illustrates

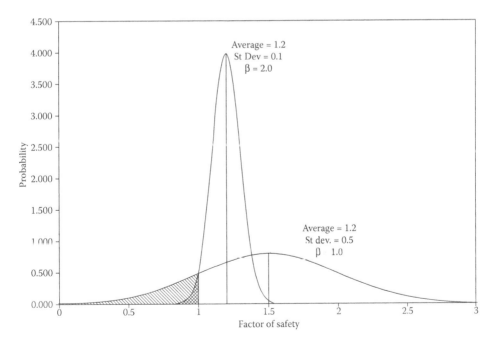

Figure 3.10 Factor of safety and reliability index. (After Christian, J.T., Ladd, C.C., and Baecher, G.B., 1994, *Journal of Geotechnical Engineering*, Vol. 120, No. 12, p. 2180–2207. Used with permission from ASCE.)

the relationships among most likely values of factor of safety ($F_{MLV}$), standard deviation, and β. In the figure the $F_{MLV}$ is the average (and in this case mode) of the PDF. The lower bell-shaped curve represents a case where $F_{MLV} = 1.5$ and standard deviation = 0.5. For this case $F_{MLV}$ is one standard deviation above the factor of safety, which is 1.0, which means β = 1. Table 3.2 shows that for β = 1, the probability of failure is 16%.

The narrow bell-shaped curve in Figure 3.10 represents a case where the most likely factor of safety is smaller ($F_{MLV} = 1.2$), but the standard deviation of the factor of safety too is much smaller ($\sigma_F = 0.1$). For this case $F_{MLV}$ is two standard deviations above the factor of safety = 1.0, which means β = 2. Table 3.2 shows that for β = 2, the probability of failure is 2.3%. Thus despite having a lower overall factor of safety, the probability of failure is less because the standard deviation is smaller.

The shaded areas beneath the PDF curves in Figure 3.10 can be shown to be equal to the probability of failure. The broader curve has a much greater area left of F = 1.0, corresponding to $P_f = 16\%$. The narrower curve has a smaller area left of F = 1.0, corresponding to $P_f = 2.5\%$.

### 3.3.14 Probability of failure on the CDF curve

The ordinate of the CDF is the area beneath the PDF curve, as discussed earlier. The probability of failure is the intercept of the CDF curve with the $F = 1.0$ line, as shown in Figure 3.11. This applies to both normal and lognormal distributions of factor of safety.

$P_f = 1.5\%$ based on normal distribution of F

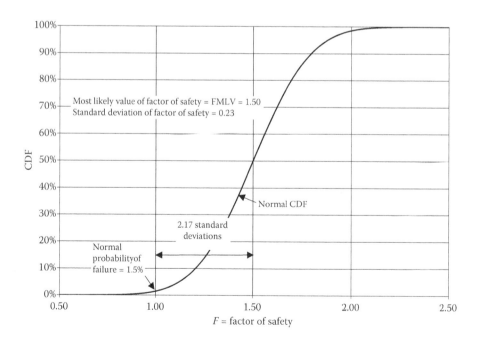

*Figure 3.11* Reliability index example for FMLV = 1.50, $\sigma_F = 0.23$, $\beta_{Normal} = 2.17$.

### 3.3.15 Reliability index for normally distributed factor of safety

If the factor of safety is assumed to be normally distributed, the reliability index is computed using Equation 3.9:

$$\beta_{\text{Normal}} = \frac{F_{\text{MLV}} - 1}{\sigma_F} \tag{3.9}$$

where

$\beta_{\text{Normal}}$ is the value of $\beta$ for the normal distribution assuming the factor of safety is normally distributed

$F_{\text{MLV}}$ is the most likely value of the factor of safety, calculated using the most likely values of all variables and $\sigma_F$ is the standard deviation of $F$.

This value of $\beta_{\text{Normal}}$ is shown graphically in Figure 3.11. For this example, with $F_{\text{MLV}} = 1.50$, and $\sigma_F = 0.23$, $\beta_{\text{Normal}}$ is equal to 2.17:

$$\beta_{\text{Normal}} = \frac{F_{\text{MLV}} - 1}{\sigma_F} = \frac{1.50 - 1}{0.23} = 2.17 \tag{3.10}$$

and the probability of failure calculated using the Excel function NORMSDIST (Equation 3.8) is 1.5%.

### 3.3.16 Reliability index for a lognormally distributed factor of safety

If the factor of safety is assumed to be lognormally distributed, the reliability index is computed using Equation 3.11:

$$\beta_{\text{Lognormal}} = \frac{\ln\left(\dfrac{F_{\text{MLV}}}{\sqrt{1 + (\text{COV}_F)^2}}\right)}{\sqrt{\ln(1 + (\text{COV}_F)^2)}} \tag{3.11}$$

For the same example ($F_{\text{MLV}} = 1.50$, and $\sigma_F = 0.23$), COV $= 0.23/1.50 = 15.3\%$, $\beta_{\text{Lognormal}}$ calculated using Equation 3.11 is equal to 2.58, and the probability of failure calculated using the Excel function NORMSDIST (Equation 3.8) is 0.5%, one-third of the value of $P_f$ for the normal distribution.

This example shows that the assumption regarding the distribution of factor of safety has an important effect on the computed probability of failure: $P_f = 1.5\%$ if the factor of safety is assumed to be normally distributed and $P_f = 0.5\%$ if the factor of safety is assumed to be lognormally distributed. This important factor is further discussed in the following sections.

### 3.3.17 Effect of standard deviation on estimated value of probability of failure

Figure 3.12 shows the CDF for two values of standard deviation of factor of safety, $\sigma_{F1} = 0.23$ and $\sigma_{F2} = 0.38$. As can be seen in Figure 3.12, an increase of 60% in standard deviation

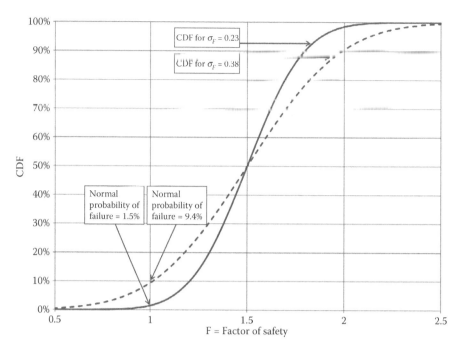

*Figure 3.12* Illustration of the effect of standard deviation on probability of failure for F = 1.50.

(1.6 * 0.23 = 0.38) increases the $P_f$ by more than 6 times. Thus, a 60% increase in uncertainty results in more than a six-fold increase in the probability of failure, all other things being equal.

## 3.4 PROBABILITY OF FAILURE AND FACTOR OF SAFETY

### 3.4.1 What is "failure?"

The phrase "probability of failure" comes from the theory of probability and reliability, which is neutral with regard to consequences of failure. The event described as "failure" need not necessarily be catastrophic. For example, if the factor of safety against a retaining wall sliding was less than 1.0, the wall might slide a short distance, the sliding movement would result in a decrease in the earth pressure, and sliding would stop. If the earth pressure increased at a later time due to creep in the backfill or due to some other cause, another increment of sliding might result. Eventually, if repeated episodes of sliding resulted in significant displacement of the wall, this behavior could be considered to be unsatisfactory performance of the wall, but probably not a catastrophic failure. However, if the factor of safety against instability of a soil slope was less than 1.0, large movements might occur, perhaps suddenly. Such movements could indeed be catastrophic, resulting in significant damage to structures on or below the slope, or even a threat to life. In recognition of the important difference between catastrophic failure and less significant performance problems, the Corps of Engineers use the term "probability of unsatisfactory performance" to describe more benign occurrences (U.S. Army Corps of Engineers, 1997).

Whatever terminology is used, it is important to keep in mind the real consequences of the event analyzed, and not to be blinded by the word "failure" when the term "probability of failure" is used. To avoid the possibility that the true significance of an event may be obscured by use of the term "probability of failure," it is preferable to be more specific, and to use descriptions such as "probability of retaining wall sliding," or "probability of slope instability," and to describe the likely consequences of the event together with the probability of its occurrence.

### 3.4.2 Assumed distribution of the factor of safety

The Taylor Series and Point Estimate methods (PEMs) for estimating the probability of failure require an assumption regarding the distribution of the factor of safety. Most often it is assumed that the distribution is either normal or lognormal. It is usual practice and usually conservative to assume that the factor of safety is normally distributed (Baecher and Christian 2003, Filz and Navin 2006).

Table 3.3 shows probabilities that the factor of safety may be smaller than 1.0 based on a normal distribution of the factor of safety. Table 3.4 shows the same information but assuming lognormal distribution of the factor of safety. Table 3.5 indicates which assumption results in a larger value of $P_f$; where the cells in Table 3.5 are shaded, assuming a normal distribution results in a greater value of $P_f$; where the table is not shaded, the normal $P_f$ is smaller than that the lognormal $P_f$. Table 3.5 shows that it is conservative in most cases to assume a normal distribution for the factor of safety. Only for low factors of safety and high coefficients of variation will the lognormal distribution of the factor of safety be higher.

## 3.5 METHODS OF ESTIMATING STANDARD DEVIATIONS

An essential element of the art of geotechnical engineering is the ability to estimate reasonable values of parameters, using meager data plus experience, or using correlations with results of *in situ* and index tests. In order to be able to estimate $P_f$, it is necessary first to estimate the standard deviations of the parameters involved in computing the factor of safety. Standard deviations can be estimated using the same types of judgment and experience used to estimate average values of parameters.

Depending on the amount of data available, various methods can be used to estimate standard deviations. Four methods that are applicable to various situations are described in the following sections.

### 3.5.1 Computation from data

When sufficient data are available, the formula definition of $\sigma$ can be used to calculate its value:

$$\sigma = \sqrt{\frac{\sum_{i=1}^{n} (x_i - \bar{x})^2}{n-1}} \tag{3.12}$$

in which $\sigma$ is the sample standard deviation, $x_i$ the $i^{th}$ value of the parameter $(x)$, $\bar{x}$ the average value of the parameter $x$, and $n$ the number of values of $x$, or the size of the sample.

Most scientific calculators, as well as spreadsheets, have routines for calculating standard deviation using Equation 3.12.

Table 3.3 Probabilities that the factor of safety is smaller than 1.0, based on a normal distribution of factor of safety

| $F_{MLV}$ | Coefficient of variation of factor of safety—$COV_F$ | | | | | | | | | | | | | | |
|---|---|---|---|---|---|---|---|---|---|---|---|---|---|---|---|
| | 2% | 4% | 6% | 8% | 10% | 12% | 14% | 16% | 20% | 25% | 30% | 40% | 50% | 60% | 80% |
| 1.05 | 0.9% | 11.7% | 21.4% | 27.6% | 31.7% | 34.6% | 36.7% | 38.3% | 40.6% | 42.4% | 43.7% | 45.3% | 46.2% | 46.8% | 47.6% |
| 1.10 | 0.0% | 1.2% | 6.5% | 12.8% | 18.2% | 22.4% | 25.8% | 28.5% | 32.5% | 35.8% | 38.1% | 41.0% | 42.8% | 44.0% | 45.5% |
| 1.15 | 0.0% | 0.1% | 1.5% | 5.2% | 9.6% | 13.9% | 17.6% | 20.7% | 25.7% | 30.1% | 33.2% | 37.2% | 39.7% | 41.4% | 43.5% |
| 1.16 | 0.0% | 0.0% | 1.1% | 4.2% | 8.4% | 12.5% | 16.2% | 19.4% | 24.5% | 29.1% | 32.3% | 36.5% | 39.1% | 40.9% | 43.2% |
| 1.18 | 0.0% | 0.0% | 0.6% | 2.8% | 6.4% | 10.2% | 13.8% | 17.0% | 22.3% | 27.1% | 30.6% | 35.1% | 38.0% | 40.0% | 42.4% |
| 1.20 | 0.0% | 0.0% | 0.3% | 1.9% | 4.8% | 8.2% | 11.7% | 14.9% | 20.2% | 25.2% | 28.9% | 33.8% | 36.9% | 39.1% | 41.7% |
| 1.25 | 0.0% | 0.0% | 0.0% | 0.6% | 2.3% | 4.8% | 7.7% | 10.6% | 15.9% | 21.2% | 25.2% | 30.9% | 34.5% | 36.9% | 40.1% |
| 1.30 | 0.0% | 0.0% | 0.0% | 0.2% | 1.1% | 2.7% | 5.0% | 7.5% | 12.4% | 17.8% | 22.1% | 28.2% | 32.2% | 35.0% | 38.6% |
| 1.35 | 0.0% | 0.0% | 0.0% | 0.1% | 0.5% | 1.5% | 3.2% | 5.3% | 9.7% | 15.0% | 19.4% | 25.8% | 30.2% | 33.3% | 37.3% |
| 1.40 | 0.0% | 0.0% | 0.0% | 0.0% | 0.2% | 0.9% | 2.1% | 3.7% | 7.7% | 12.7% | 17.0% | 23.8% | 28.4% | 31.7% | 36.0% |
| 1.50 | 0.0% | 0.0% | 0.0% | 0.0% | 0.0% | 0.3% | 0.9% | 1.9% | 4.8% | 9.1% | 13.3% | 20.2% | 25.2% | 28.9% | 33.8% |
| 1.60 | 0.0% | 0.0% | 0.0% | 0.0% | 0.0% | 0.1% | 0.4% | 1.0% | 3.0% | 6.7% | 10.6% | 17.4% | 22.7% | 26.6% | 32.0% |
| 1.70 | 0.0% | 0.0% | 0.0% | 0.0% | 0.0% | 0.0% | 0.2% | 0.5% | 2.0% | 5.0% | 8.5% | 15.2% | 20.5% | 24.6% | 30.3% |
| 1.80 | 0.0% | 0.0% | 0.0% | 0.0% | 0.0% | 0.0% | 0.1% | 0.3% | 1.3% | 3.8% | 6.9% | 13.3% | 18.7% | 22.9% | 28.9% |
| 1.90 | 0.0% | 0.0% | 0.0% | 0.0% | 0.0% | 0.0% | 0.0% | 0.2% | 0.9% | 2.9% | 5.7% | 11.8% | 17.2% | 21.5% | 27.7% |
| 2.00 | 0.0% | 0.0% | 0.0% | 0.0% | 0.0% | 0.0% | 0.0% | 0.1% | 0.6% | 2.3% | 4.8% | 10.6% | 15.9% | 20.2% | 26.6% |
| 2.20 | 0.0% | 0.0% | 0.0% | 0.0% | 0.0% | 0.0% | 0.0% | 0.0% | 0.3% | 1.5% | 3.5% | 8.6% | 13.8% | 18.2% | 24.8% |
| 2.40 | 0.0% | 0.0% | 0.0% | 0.0% | 0.0% | 0.0% | 0.0% | 0.0% | 0.2% | 1.0% | 2.6% | 7.2% | 12.2% | 16.5% | 23.3% |
| 2.60 | 0.0% | 0.0% | 0.0% | 0.0% | 0.0% | 0.0% | 0.0% | 0.0% | 0.1% | 0.7% | 2.0% | 6.2% | 10.9% | 15.3% | 22.1% |
| 2.80 | 0.0% | 0.0% | 0.0% | 0.0% | 0.0% | 0.0% | 0.0% | 0.0% | 0.1% | 0.5% | 1.6% | 5.4% | 9.9% | 14.2% | 21.1% |
| 3.00 | 0.0% | 0.0% | 0.0% | 0.0% | 0.0% | 0.0% | 0.0% | 0.0% | 0.0% | 0.4% | 1.3% | 4.8% | 9.1% | 13.3% | 20.2% |

Source:  Duncan, J.M., 2000, Factors of safety and reliability in geotechnical engineering, Journal of Geotechnical Engineering, Vol. 126, No. 4, p. 307–316. Used with permission from ASCE.

$F_{MLV}$ = factor of safety computed using most likely values of parameters.

Table 3.4 Probabilities that the factor of safety is smaller than 1.0, based on a lognormal distribution of factor of safety

| $F_{MLV}$ | Coefficient of variation of factor of safety—$COV_F$ | | | | | | | | | | | | | | |
| --- | --- | --- | --- | --- | --- | --- | --- | --- | --- | --- | --- | --- | --- | --- | --- |
| | 2% | 4% | 6% | 8% | 10% | 12% | 14% | 16% | 20% | 25% | 30% | 40% | 50% | 60% | 80% |
| 1.05 | 0.8% | 12% | 22% | 28% | 33% | 36% | 39% | 41% | 44% | 47% | 49% | 53% | 55% | 58% | 61% |
| 1.10 | 0.00% | 0.9% | 6% | 12% | 18% | 23% | 27% | 30% | 35% | 40% | 43% | 48% | 51% | 54% | 59% |
| 1.15 | 0.00% | 0.03% | 1.1% | 4% | 9% | 13% | 18% | 21% | 27% | 33% | 37% | 43% | 48% | 51% | 56% |
| 1.16 | 0.00% | 0.01% | 0.7% | 3% | 8% | 12% | 16% | 20% | 26% | 32% | 36% | 42% | 47% | 50% | 56% |
| 1.18 | 0.00% | 0.00% | 0.3% | 2% | 5% | 9% | 13% | 17% | 23% | 29% | 34% | 41% | 45% | 49% | 55% |
| 1.20 | 0.00% | 0.00% | 0.13% | 1.2% | 4% | 7% | 11% | 14% | 21% | 27% | 32% | 39% | 44% | 48% | 54% |
| 1.25 | 0.00% | 0.00% | 0.01% | 0.3% | 1.4% | 4% | 6% | 9% | 15% | 22% | 27% | 35% | 41% | 45% | 51% |
| 1.30 | 0.00% | 0.00% | 0.00% | 0.06% | 0.5% | 1.6% | 3% | 6% | 11% | 17% | 23% | 31% | 37% | 42% | 49% |
| 1.35 | 0.00% | 0.00% | 0.00% | 0.01% | 0.2% | 0.7% | 1.9% | 4% | 8% | 14% | 19% | 28% | 34% | 40% | 47% |
| 1.40 | 0.00% | 0.00% | 0.00% | 0.00% | 0.04% | 0.3% | 1.0% | 2% | 5% | 11% | 16% | 25% | 32% | 37% | 45% |
| 1.50 | 0.00% | 0.00% | 0.00% | 0.00% | 0.00% | 0.04% | 0.2% | 0.7% | 3% | 6% | 11% | 19% | 27% | 32% | 41% |
| 1.60 | 0.00% | 0.00% | 0.00% | 0.00% | 0.00% | 0.01% | 0.05% | 0.2% | 1.1% | 4% | 7% | 15% | 22% | 28% | 38% |
| 1.70 | 0.00% | 0.00% | 0.00% | 0.00% | 0.00% | 0.00% | 0.01% | 0.06% | 0.5% | 2% | 5% | 12% | 19% | 25% | 34% |
| 1.80 | 0.00% | 0.00% | 0.00% | 0.00% | 0.00% | 0.00% | 0.00% | 0.01% | 0.2% | 1.2% | 3% | 9% | 16% | 22% | 31% |
| 1.90 | 0.00% | 0.00% | 0.00% | 0.00% | 0.00% | 0.00% | 0.00% | 0.00% | 0.08% | 0.65% | 2% | 7% | 13% | 19% | 29% |
| 2.00 | 0.00% | 0.00% | 0.00% | 0.00% | 0.00% | 0.00% | 0.00% | 0.00% | 0.03% | 0.36% | 1.3% | 5% | 11% | 17% | 26% |
| 2.20 | 0.00% | 0.00% | 0.00% | 0.00% | 0.00% | 0.00% | 0.00% | 0.00% | 0.01% | 0.10% | 0.56% | 3% | 8% | 13% | 22% |
| 2.40 | 0.00% | 0.00% | 0.00% | 0.00% | 0.00% | 0.00% | 0.00% | 0.00% | 0.00% | 0.03% | 0.23% | 1.9% | 5% | 10% | 19% |
| 2.60 | 0.00% | 0.00% | 0.00% | 0.00% | 0.00% | 0.00% | 0.00% | 0.00% | 0.00% | 0.01% | 0.09% | 1.1% | 4% | 7% | 16% |
| 2.80 | 0.00% | 0.00% | 0.00% | 0.00% | 0.00% | 0.00% | 0.00% | 0.00% | 0.00% | 0.00% | 0.04% | 0.66% | 3% | 6% | 13% |
| 3.00 | 0.00% | 0.00% | 0.00% | 0.00% | 0.00% | 0.00% | 0.00% | 0.00% | 0.00% | 0.00% | 0.02% | 0.39% | 1.8% | 4% | 11% |

$F_{MLV}$ = factor of safety computed using most likely values of parameters

*Table 3.5* Probabilities that the factor of safety is smaller than 1.0 based on normal distribution of factor of safety

| $F_{MLV}$ | Coefficient of variation of factor of safety—$COV_F$ | | | | | | | | | | | | | | |
|---|---|---|---|---|---|---|---|---|---|---|---|---|---|---|---|
| | 2% | 4% | 6% | 8% | 10% | 12% | 14% | 16% | 20% | 25% | 30% | 40% | 50% | 60% | 80% |
| 1.05 | 0.9% | 11.7% | 21.4% | 27.6% | 31.7% | 34.6% | 36.7% | 38.3% | 40.6% | 42.4% | 43.7% | 45.3% | 46.2% | 46.8% | 47.6% |
| 1.10 | 0.0% | 1.2% | 6.5% | 12.8% | 18.2% | 22.4% | 25.8% | 28.5% | 32.5% | 35.8% | 38.1% | 41.0% | 42.8% | 44.0% | 45.5% |
| 1.15 | 0.0% | 0.1% | 1.5% | 5.2% | 9.6% | 13.9% | 17.6% | 20.7% | 25.7% | 30.1% | 33.2% | 37.2% | 39.7% | 41.4% | 43.5% |
| 1.16 | 0.0% | 0.0% | 1.1% | 4.2% | 8.4% | 12.5% | 16.2% | 19.4% | 24.5% | 29.1% | 32.3% | 36.5% | 39.1% | 40.9% | 43.2% |
| 1.18 | 0.0% | 0.0% | 0.6% | 2.8% | 6.4% | 10.2% | 13.8% | 17.0% | 22.3% | 27.1% | 30.6% | 35.1% | 38.0% | 40.0% | 42.4% |
| 1.20 | 0.0% | 0.0% | 0.3% | 1.9% | 4.8% | 8.2% | 11.7% | 14.9% | 20.2% | 25.2% | 28.9% | 33.8% | 36.9% | 39.1% | 41.7% |
| 1.25 | 0.0% | 0.0% | 0.0% | 0.6% | 2.3% | 4.8% | 7.7% | 10.6% | 15.9% | 21.2% | 25.2% | 30.9% | 34.5% | 35.9% | 40.1% |
| 1.30 | 0.0% | 0.0% | 0.0% | 0.2% | 1.1% | 2.7% | 5.0% | 7.5% | 12.4% | 17.8% | 22.1% | 28.2% | 32.2% | 35.0% | 38.6% |
| 1.35 | 0.0% | 0.0% | 0.0% | 0.1% | 0.5% | 1.5% | 3.2% | 5.3% | 9.7% | 15.0% | 19.4% | 25.8% | 30.2% | 33.3% | 37.3% |
| 1.40 | 0.0% | 0.0% | 0.0% | 0.0% | 0.2% | 0.9% | 2.1% | 3.7% | 7.7% | 12.7% | 17.0% | 23.8% | 28.4% | 31.7% | 36.0% |
| 1.50 | 0.0% | 0.0% | 0.0% | 0.0% | 0.0% | 0.3% | 0.9% | 1.9% | 4.8% | 9.1% | 13.3% | 20.2% | 25.2% | 28.9% | 33.8% |
| 1.60 | 0.0% | 0.0% | 0.0% | 0.0% | 0.0% | 0.1% | 0.4% | 1.0% | 3.0% | 6.7% | 10.6% | 17.4% | 22.7% | 26.6% | 32.0% |
| 1.70 | 0.0% | 0.0% | 0.0% | 0.0% | 0.0% | 0.0% | 0.2% | 0.5% | 2.0% | 5.0% | 8.5% | 15.2% | 20.5% | 24.6% | 30.3% |
| 1.80 | 0.0% | 0.0% | 0.0% | 0.0% | 0.0% | 0.0% | 0.1% | 0.3% | 1.3% | 3.8% | 6.9% | 13.3% | 18.7% | 22.9% | 28.9% |
| 1.90 | 0.0% | 0.0% | 0.0% | 0.0% | 0.0% | 0.0% | 0.0% | 0.2% | 0.9% | 2.9% | 5.7% | 11.8% | 17.2% | 21.5% | 27.7% |
| 2.00 | 0.0% | 0.0% | 0.0% | 0.0% | 0.0% | 0.0% | 0.0% | 0.1% | 0.6% | 2.3% | 4.8% | 10.6% | 15.9% | 20.2% | 26.6% |
| 2.20 | 0.0% | 0.0% | 0.0% | 0.0% | 0.0% | 0.0% | 0.0% | 0.0% | 0.3% | 1.5% | 3.5% | 8.6% | 13.8% | 18.2% | 24.8% |
| 2.40 | 0.0% | 0.0% | 0.0% | 0.0% | 0.0% | 0.0% | 0.0% | 0.0% | 0.2% | 1.0% | 2.6% | 7.2% | 12.2% | 16.5% | 23.3% |
| 2.60 | 0.0% | 0.0% | 0.0% | 0.0% | 0.0% | 0.0% | 0.0% | 0.0% | 0.1% | 0.7% | 2.0% | 6.2% | 10.9% | 15.3% | 22.1% |
| 2.80 | 0.0% | 0.0% | 0.0% | 0.0% | 0.0% | 0.0% | 0.0% | 0.0% | 0.1% | 0.5% | 1.6% | 5.4% | 9.9% | 14.2% | 21.1% |
| 3.00 | 0.0% | 0.0% | 0.0% | 0.0% | 0.0% | 0.0% | 0.0% | 0.0% | 0.0% | 0.4% | 1.3% | 4.8% | 9.1% | 13.3% | 20.2% |

$F_{MLV}$ = factor of safety computed using most likely values of parameters.

Note: The shaded areas show combinations of COV and $F_{MLV}$ for which the $P_f$ from the normal distribution is higher than the $P_f$ from the lognormal distribution.

If the only method of determining values of standard deviation was to use Equation 3.12, reliability analyses could not be used much in geotechnical engineering, because in many cases the values of some parameters are estimated using correlations or experience, and there is no real "data" that can be used with Equation 3.12. In order to be able to apply reliability analyses to these common situations, it is necessary to estimate values of standard deviation, rather than to calculate them. Three methods of estimating values of $\sigma$ are described in the following sections.

### 3.5.2 Published values

When correlations or experience is used to estimate parameter values, correlations or experience can also be used to estimate standard deviations. It is convenient in these cases to use the COV:

$$COV = \frac{\text{Standard deviation}}{\text{Average value}} = \frac{\sigma}{\bar{x}} \tag{3.13}$$

from which the standard deviation can be computed:

$$\sigma = (COV)(\bar{x}) \tag{3.14}$$

Values of COV for a number of geotechnical engineering parameters and *in situ* tests, compiled by Harr (1984), Kulhawy (1992), Lacasse and Nadim (1997), and the authors of this chapter are listed in Table 3.6. The best aspect of the values in Table 3.6 is that they are based on a large number of tests. However, the conditions of sampling and testing, which have a great influence on the variability of the test results, are not available for the data shown in Table 3.6, and the values therefore provide only a rough guide for estimating values of COV for any particular case. It is important to use judgment in applying values of COV from published sources, and to consider as well as possible the likely degree of uncertainty in the particular case at hand.

### 3.5.3 The "three-sigma rule"

Dai and Wang (1992) suggested that values of standard deviation could be estimated using what they called the "three-sigma rule." This rule uses the fact that 99.73% of all values of a normally distributed parameter fall within plus or minus three standard deviations (three sigma) from the average. Thus, an extreme low value would be three standard deviations below the average, and an extreme high value would be three standard deviations above the average. They suggested that, by estimating the extreme high and low values, and dividing the difference between them by six, the standard deviation could be estimated using Equation 3.15:

$$\sigma = \frac{HCV - LCV}{6} \tag{3.15}$$

where HCV is the highest conceivable value and LCV the lowest conceivable value.

The three-sigma rule has the advantage that it can be used to estimate values of standard deviation for parameters whose values are estimated based entirely on judgment, or on judgment plus meager data.

*Table 3.6* Values of coefficient of variation (COV) for geotechnical properties and *in situ* tests

| Property or in situ test result | Coefficient of variation— COV (%) | Source |
|---|---|---|
| Unit weight ($\gamma$) | 3%–7% | Harr (1984), Kulhawy (1992) |
| Buoyant unit weight ($\gamma_b$) | 0% 10% | This chapter, Lacasse and Nadim (1997) |
| Effective stress friction angle ($\varphi'$) | 2% 13% | Harr (1984), Kulhawy (1992) |
| Undrained shear strength ($S_u$) | 13%–40% | This chapter, Kulhawy (1992), Harr (1984), Lacasse and Nadim (1997) |
| Undrained shear strength ($S_u$) | Clay – UU Triaxial (10%–30%) | Phoon and Kulhawy (1999) |
| | Clay – UC Triaxial (20%–55%) | Phoon and Kulhawy (1999) |
| Undrained strength ratio ($S_u/\sigma_v'$) | 5%–15% | This chapter, Lacasse and Nadim (1997) |
| Compression index ($C_c$) | 10%–37% | This chapter, Kulhawy (1992), Harr (1984) |
| Preconsolidation pressure ($p_p$) | 10%–35% | This chapter, Harr (1984), Lacasse and Nadim (1997) |
| Coefficient of permeability of saturated clay (k) | 68%–90% | This chapter, Harr (1984) |
| Coefficient of permeability of partly saturated clay (k) | 130%–240% | Harr (1984), Benson et al. (1999) |
| Coefficient of consolidation ($c_v$) | 33%–68% | This chapter. |
| Standard Penetration Test blow count (N) | 15%–45% | Harr (1984), Kulhawy (1992) |
| Electric Cone Penetration Test ($q_c$) | 5%–15% | Kulhawy (1992) |
| Mechanical Cone Penetration Test ($q_c$) | 15%–37% | Harr (1984), Kulhawy (1992) |
| Dilatometer Test tip resistance ($q_{DMT}$) | 5%–15% | Kulhawy (1992) |
| Vane shear test undrained strength ($S_v$) | 10%–20% | Kulhawy (1992) |
| Plastic Limit | 6%–30% | Phoon and Kulhawy (1999) |
| Liquid Limit | 6%–30% | Phoon and Kulhawy (1999) |

Source: Duncan, J.M., 2000, Factors of safety and reliability in geotechnical engineering, *Journal of Geotechnical Engineering*, Vol. 126, No. 4, p. 307–316. Used with Permission from ASCE.

However, the accuracy of the three-sigma rule depends on the accuracy with which HCV and LCV can be estimated. Estimating these very extreme values has been found to be more difficult than it might at first appear (Christian and Baecher, 1999). Because the ±3σ range covers 99.73% of all values, only 0.27% of all values should lie outside the ±3σ range. This corresponds to only one instance in 370, a very rare occurrence.

Studies have shown that there is a strong tendency for engineers to estimate ranges of values between HCV and LCV that are too small. One such study, described by Folayan et al. (1970), involved asking a number of geotechnical engineers to estimate the possible range of values of $C_c/(1+e)$ for San Francisco Bay mud, with which they all had experience. The results of this exercise are summarized in Table 3.7. On average, these experienced engineers were able to estimate the average value of $C_c/(1+e)$ within about 15%, but they underestimated the COV by about 67% as compared with the results of 45 laboratory tests.

*Table 3.7* Estimated and measured values of Cc/(1 + e) and its coefficient of variation, for San Francisco Bay mud

| Estimated by | Estimated Cc/(1 + e) | Estimated COV% |
|---|---|---|
| Geotechnical Engineer #1 | 0.30 | 10 |
| Geotechnical Engineer #2 | 0.275 | 5 |
| Geotechnical Engineer #3 | 0.275 | 5.5 |
| Geotechnical Engineer #4 | 0.30 | 10 |
| Average – #1 through #4 | 0.29 | 8 |
| Measured values | Cc/(1 + e) = 0.34 | COV = 18 |

Source: Duncan, J.M., 2000, Factors of safety and reliability in geotechnical engineering, *Journal of Geotechnical Engineering*, Vol. 126, No. 4, p. 307–316. Used with permission from ASCE.

### 3.5.4 The "N-sigma rule"

Judgmental estimates of COV or standard deviation can be improved by recognizing that estimated values of LCV and HCV are unlikely to be sufficiently high and low to encompass $\pm 3\sigma$.

The N-sigma (Foye et al., 2006) rule provides a means of taking into account the fact that an engineer's experience and available information usually encompass considerably less than 99.73% of all possible values. The N-sigma rule is expressed as

$$\sigma = \frac{\text{HCV} - \text{LCV}}{\text{N}_\sigma} \tag{3.16}$$

where $\text{N}_\sigma$ is a number smaller than 6 that reflects the fact that estimates of LCV and HCV cannot be expected to span $\pm 3\sigma$. While there is no "one-size-fits-all" value of $\text{N}_\sigma$, a value of $\text{N}_\sigma = 4$ seems to be appropriate for many conditions. With $\text{N}_\sigma = 4$, the standard deviation would be calculated using Equation 3.17:

$$\sigma = \frac{\text{HCV} - \text{LCV}}{4} \tag{3.17}$$

*N-sigma rule example*: The concept of "equivalent fluid pressure" (Terzaghi et al., 1996; Clough and Duncan, 1991) is often used to estimate earth pressures on walls. The earth pressure distribution is approximated by a triangular pressure distribution that would be exerted by a fluid with a unit weight $\gamma_{eq}$. Values of $\gamma_{eq}$ are ascribed to various types of soils based on experience and judgment, as shown in Figure 3.13. Because values of $\gamma_{eq}$ are based on experience and judgment, the standard deviation of $\gamma_{eq}$ also has to be based on judgment. The N-sigma rule is convenient for this purpose.

Using Equation 3.17, the value of standard deviation of equivalent fluid unit weight ($\gamma_{eq}$) could be estimated as follows:

First, the most likely value of $\gamma_{eq}$ would be estimated using experience, tables, or charts of the type found in Terzaghi et al. (1996), Clough and Duncan (1991), or Figure 3.13. An example would be $\gamma_{eq} = 7.07$ kN/m³ for a compacted CL backfill, as shown in Figure 3.13.

Second, the highest and lowest conceivable values of $\gamma_{eq}$ would be estimated, considering the ways in which actual conditions might differ from the most likely conditions. These high and low values would depend on insights regarding the possibilities that the backfill might be of higher-or-lower quality than expected, that the backfill might contain materials other

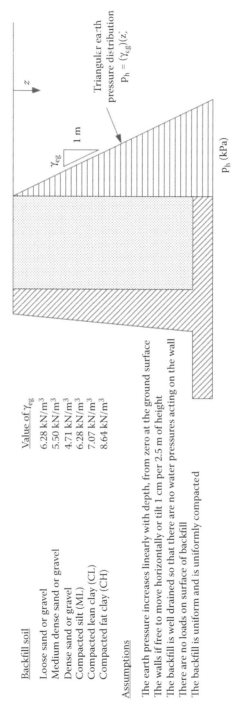

| Backfill soil | Value of $\gamma_{eg}$ |
|---|---|
| Loose sand or gravel | 6.28 kN/m³ |
| Medium dense sand or gravel | 5.50 kN/m³ |
| Dense sand or gravel | 4.71 kN/m³ |
| Compacted silt (ML) | 6.28 kN/m³ |
| Compacted lean clay (CL) | 7.07 kN/m³ |
| Compacted fat clay (CH) | 8.64 kN/m³ |

Assumptions

The earth pressure increases linearly with depth, from zero at the ground surface
The walls if free to move horizontally or tilt 1 cm per 2.5 m of height
The backfill is well drained so that there are no water pressures acting on the wall
There are no loads on surface of backfill
The backfill is uniform and is uniformly compacted

*Figure 3.13* Equivalent fluid pressures for design of retaining walls.

than CL clay, that the backfill might be less compacted or better compacted than expected, that the basis for estimating the average value of $\gamma_{eq}$ (experience, tables, or charts) might be biased, inappropriate or misinterpreted, and so forth.

If, in a particular instance, the value of LCV was judged to be 5.50 kN/m³, and the value of HCV was judged to be 8.64 kN/m³, the standard deviation of $\gamma_{eq}$ would be 0.76 kN/m³, calculated as follows:

$$\sigma = \frac{8.64 - 5.50}{4} = 0.76 \text{ kN/m}^3 \tag{3.18}$$

The COV of $\gamma_{eq}$ would be

$$COV = \frac{0.76}{7.07} = 11\% \tag{3.19}$$

### 3.5.6 Graphical N-sigma rule

The concept behind the N-sigma rule can be extended to situations where the parameter of interest, such as strength, varies with depth or pressure. Examples are shown in Figures 3.14 and 3.15.

The graphical N-sigma rule (with $N_\sigma = 2$) is applied as follows:

1. Draw a straight line or a curve through the data that represent the most likely average variation of the parameter with depth or pressure.
2. Draw straight lines or curves that represent the highest and lowest conceivable bounds on the data. These should be wide enough to include all valid data and an allowance for the fact that the natural tendency is to estimate such bounds too narrowly, as

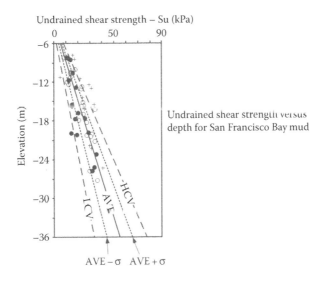

*Figure 3.14* Development of average plus one standard deviation and average minus one standard deviation variations of Su with depth using the 2σ rule. (After Duncan, J.M., and Buchignani, A.L. 1973, 'Failure of underwater slope in San Francisco Bay', *Journal of the Soil Mechanics and Foundations Division*, ASCE, Vol. 99, No. 9, p. 687–703. Used with permission from ASCE.)

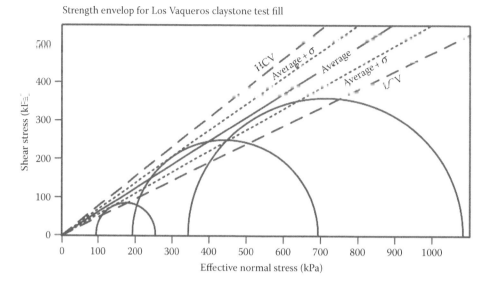

*Figure 3.15* Development of average plus one standard deviation and average minus one standard deviation strength envelopes using the 2σ rule. (After Duncan, J.M., 2000, Factors of safety and reliability in geotechnical engineering, *Journal of Geotechnical Engineering*, Vol. 126, No. 4, p. 307–316. Used with permission from ASCE.)

discussed previously. Note that some points in Figure 3.14 are outside the estimated highest and lowest values conceivable, indicating that these data points are believed to be erroneous.

3. Draw straight lines or curves that represent the average plus one standard deviation and the average minus one standard deviation halfway between the average and the lowest and highest conceivable lines. These lines are drawn halfway between the average and the LCV and HCV lines because $N_\sigma = 2$.

The use of this procedure to establish average $\pm\sigma$ variations of undrained strength with depth in San Francisco Bay mud is shown in Figure 3.14.

This same concept is useful in characterizing shear strength envelopes for soils. In this case the quantity (shear strength) varies with normal stress rather than depth, but the procedure is the same. Strength envelopes are drawn that represent the average and the highest and lowest conceivable bounds on the data, as shown in Figure 3.15. Then average plus one sigma and average minus one sigma envelopes are drawn halfway between the average envelope and the highest and lowest conceivable bounds.

These average plus one sigma and average minus one sigma envelopes are useful in calculating probability of failure. For example, the Taylor Series method involves calculating factors of safety with the values of each variable alternately increased and decreased by one standard deviation.

Using the graphical N-sigma rule to establish average-plus-one-sigma and average-minus-one-sigma strength envelopes is preferable to using separate standard deviations for the strength parameters c and φ. Strength parameters (c and φ) are useful empirical coefficients that characterize the variation of shear strength with normal stress, but they are not of fundamental significance or interest by themselves. The important variable is shear strength, and the graphical N-sigma rule provides a straightforward means for characterizing the uncertainty in the shear strength envelope.

## 3.6 COMPUTING PROBABILITY OF FAILURE

Four methods of computing probability of failure are described in this chapter. To illustrate these methods, they are used to calculate the probability of failure of the cantilever retaining wall shown in Figure 3.16. Three possible modes of failure will be considered for each method:

- sliding on the silty sand layer overlying the foundation clay,
- sliding beneath the silty sand layer, at the top of the clay foundation, and
- bearing capacity failure in the clay foundation.

Calculation of probabilities of failure for these modes of failure begins with conventional deterministic analyses to calculate factors of safety against sliding on the granular layer, sliding in the clay, and bearing capacity failure.

### 3.6.1 Deterministic analyses

The wall dimensions and properties for this example are shown in Figure 3.16. The reinforced concrete footing was cast on a 10 cm thick layer of silty sand to prevent softening of the clay foundation when the wet concrete was poured for the wall footing. As a result, sliding can occur on top of the silty sand layer or on top of the clay foundation, depending

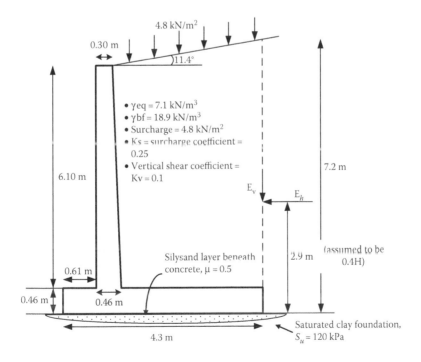

*Figure 3.16* Cantilever retaining wall with silty sand backfill.

on which offers less resistance to sliding. The bearing capacity failure mechanism would involve shearing within the clay foundation.

The deterministic calculations were performed using the CGPR Retaining Wall Stability Computation Sheet 2.03 (Tang and Duncan 2002, converted to the SI system for these examples). The factors of safety against sliding on the silty sand layer beneath the footing, sliding in the clay foundation, and bearing capacity in the clay foundation are summarized briefly in the following paragraphs.

### 3.6.2 Factor of safety against sliding on top of the silty sand layer

The factor of safety against sliding on the silty sand layer beneath the footing is given by the following formula:

$$F = \frac{N\mu}{E_h} \tag{3.20}$$

in which $N$ is the normal force on base (N/m), $\mu$ the base friction coefficient, which is the tangent of base friction angle, and $E_h$ the horizontal earth pressure force on vertical plane through heel of wall (N/m). The magnitudes of $N$, $\mu$, and $E_h$ are

$N = 536,769$ N/m,
$\mu = \tan \delta =$ tangent of the base friction angle $= 0.5$,
$E_h = 191,767$ N/m, and
$F = 1.40$ against sliding on the silty sand.

### 3.6.3 Factor of safety against sliding on the clay foundation

Meyerhoff's method (Meyerhoff 1953) is used to account for the fact that the footing is eccentrically loaded. A reduced effective footing width is used, such that the reduced width is centrally loaded. The reduced footing width is $2X = (2)(1.56) = 3.12$ m. The factor of safety against sliding on a clay layer can be calculated using this expression:

$$F = \frac{S_u(2x)}{E_h} \tag{3.21}$$

where $S_u$ is the undrained shear strength of the clay and $X$ is the distance from the front of the footing to the normal force on the base of the retaining wall. As shown in Figure 3.16:

$S_u = 120$ kPa
$X = 1.56$ m
$E_h = 191,767$ N/m
$F = 1.95$ against sliding at the top of the clay

### 3.6.4 Factor of safety against bearing capacity failure

The factor of safety against bearing capacity failure in the clay foundation is found from the following equation:

$$F = \frac{N_c S_u}{q} \tag{3.22}$$

where $N_c$ is the bearing capacity coefficient, $S_u$ is the undrained shear strength of the clay, and $q$ is the pressure at the base of the retaining wall. The pressure, $q$, is the normal force per meter divided by the reduced footing width.

$$q = \frac{N}{2X} \tag{3.23}$$

The bearing capacity factor, $N_c$ for a long footing, with reduction for load inclination, is found using Brinch Hansen's (Brinch Hansen, 1970) method.

As shown in Figure 3.16:

$N_c = 2.76$ for a long footing with load inclined at 19° from vertical,
$S_u = 120$ kPa,
$q = N/2X = 172$ kPa, and
$F = 1.93$ against bearing capacity failure in the clay foundation.

## 3.7 MONTE CARLO ANALYSIS USING @RISK™

The Monte Carlo method differs from the other methods discussed in this chapter in three ways:

1. It involves repeating the analysis many times (perhaps 5000 or 10,000 times) with randomly chosen values of the variables.
2. It uses assumptions regarding the distributions of the variables involved in the calculations rather than an assumption regarding the distribution of the factor of safety. For each calculation, randomly selected values of the variables are assigned based on the specified distributions and a random number selection procedure.
3. Because it involves a very large number of repeated calculations, it requires use of the computer program @Risk™, or another computer program that can automate the process. The Monte Caro analyses described here were performed using @Risk™.

With the Monte Carlo method, the probability of failure is determined by counting, among all the results, the number of times the computed factor of safety is less than or equal to 1.0. If 50 of 5000 analyses performed using randomly selected values of the variables result in $F \leq 1.0$, the estimate of the probability of failure is 50/5,000 or 1.0%.

While @Risk™ can perform Monte Carlo simulations for any spreadsheet-type analysis, Monte Carlo simulations methods are found with increasing frequency in geotechnical computer programs. For example, SLIDE by Rocscience and SLOPE/W by Geostudio both allow an engineer to perform Monte Carlo simulations for slope stability analyses within the respective program. While the following example is shown using @Risk™, the concepts shown are also applicable to any computer program using the Monte Carlo simulation method.

The computer program @Risk™ is designed to work with spreadsheet calculations. @Risk™ controls selection of the values of the variables for each of the many calculations, and keeps track of the results.

The steps involved in using @Risk™, and the retaining wall spreadsheet to compute the probability of failure for the retaining wall shown in Figure 3.16 are

Step 1: Estimate the COV (or, alternatively, the standard deviations) of the quantities involved in Equation 3.20: $N$, $\mu$, and $E_h$.

$N$, the normal force on the base of the footing, is proportional to the unit weight of backfill, and has the same COV. The COV of the backfill unit weight is small, estimated to be 3%, corresponding to standard deviation of $\gamma_{bf} = COV * \gamma_{bf} = (0.03)$ $(18.85 \text{ kN/m}^3) = 0.565 \text{ kN/m}^3$.

The base friction coefficient, $\mu = \tan \delta$, depends on the type of granular material on which the concrete footing is cast. According to the NAVFAC manual (NAVFAC 1986) the value of $\mu$ for silty sand ranges from 0.45 to 0.55. The $COV_{\tan\delta}$ is estimated to be 10%, corresponding to standard deviation = $(0.10)(0.50) = 0.05$.

$E_h$, the horizontal force on the wall, is proportional to the equivalent fluid pressure: coefficient $\gamma_{eq}$. The coefficient of variation $COV_{eq}$ is estimated to be 15%, corresponding to standard deviation $\sigma_{eq} = (0.15)(7.07 \text{ kN/m}^3) = 1.06 \text{ kN/m}^3$.

Step 2: Use a Microsoft Excel spreadsheet together with @Risk™ to compute the factor of safety against sliding on the silty sand layer a large number of times. For the example described here, Retaining Wall Stability Computation Sheet 2.05.1 (Yang and Duncan 2002) was used.

Step 3: One at a time, select the cells containing parameters whose values will be varied during the calculations. In this example these are the cells that contain $\gamma_{eq}$ (cell H19), $\gamma_{bf}$ (cell H20), and $\mu$ (cell H23). As each cell is selected the define distribution option in @Risk™ provides the opportunity to define the type of distribution to be used for the variable and its statistical parameters. Within @Risk™ the distributions of the variables are not limited to normal or lognormal. Thirty-seven different distribution types are available in @Risk™. Unless information is available that shows otherwise, however, a normal or lognormal distribution is probably the logical choice. The variables used in this retaining wall example are all assumed to be normally distributed.

Normal distribution has been selected, and the average value of $\gamma_{eq}$ (7.07 kN/m³), and the standard deviation (1.06 kN/m³) have been entered at the left of the screen. The corresponding PDF is displayed on the screen.

Step 4: The cell in the spreadsheet that calculates the factor of safety is defined as the output. Once the cell is selected, the "Define Output" button is pressed in @Risk™.

Step 5: When @Risk™ is run, the factor of safety is calculated for the number of times specified. As explained above, the probability of failure is equal to the number of times the calculated factor of safety is less than or equal to 1.0, divided by the total number of times the factor of safety is calculated. Examining the large number of calculated values and counting the number of times that the factor of safety is less than 1.0 would be very cumbersome. This can be avoided by adding a cell to the spreadsheet that keeps track of the number of safety factor values that are less than 1.0. This cell contains the following formula:

= if(referenced cell < 1,1,0)

This formula results in a value of either 0.0 (if the computed factor of safety is greater than 1.0) or 1.0 (if the computed factor of safety is less than or equal to 1.0). At the end of the analysis, @Risk™ computes and displays the average value of the cell for all the values computed. Thus, for example, if the spreadsheet is run 10,000 times and the calculated factor of safety is less than 1.0 in 245 of those calculations, the cell will contain the average of 9755 zeroes and 245 ones, or 0.0245. This is the probability of failure as computed by Monte Carlo simulation.

Because the parameter values are varied randomly, two @Risk™ runs with the same number of calculations will usually result in slightly different values of the factor of safety.

The retaining wall example was completed using @Risk™ for a total of 20,000 iterations. The calculated probability of failure is the mean value of the output cell formulated in step 5. The mean value is equal to 0.024 or 2.4% probability of failure.

The Monte Carlo analysis procedure was used to calculate probabilities of failure against sliding through the clay and bearing capacity failure in the clay beneath the retaining wall. The standard deviation of the undrained shear strength of the clay is estimated to be 24 kN/m². The calculated probability of failure for sliding in the clay is 2.5%. The calculated probability of failure for bearing capacity in the clay is 2.4%.

### 3.7.1  Accuracy of calculations

Because the accuracy with which values of $P_f$ can be computed is governed primarily by the accuracy with which the values of the parameters and their standard deviations can be estimated, it is clear that computed values of $P_f$ should not be considered to be highly precise. Variations in the values of $P_f$ by a factor of two or three due to reasonable changes in parameter values are to be expected. This degree of precision (or imprecision) should be kept in mind when considering the acceptability of the results of probability calculations.

The effect of the number of iterations used in @Risk™ is illustrated by Figure 3.17. To verify the accuracy of the calculated value of failure probability, it is useful to run several

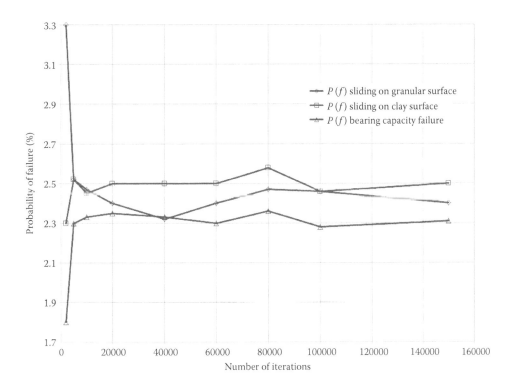

*Figure 3.17* Probability of failure for several Monte Carlo simulations of retaining wall stability.

Table 3.8 Monte Carlo simulation of sliding on a granular surface, sliding on a clay surface, and bearing capacity with changes in the standard deviation

| Normal Distribution 10,000 Iterations | |
| --- | --- |
| $\sigma_{su}$ | P(f) Bearing capacity (%) |
| 500 | 2.08 |
| 600 | 3.80 |
| 700 | 5.62 |
| 800 | 7.70 |
| $\sigma_{su}$ | P(f) Sliding on clay (%) |
| 500 | 2.33 |
| 600 | 4.18 |
| 700 | 6.13 |
| 800 | 8.34 |
| $\sigma_{\tan\delta}$ | P(f) Sliding on granular (%) |
| 0.05 | 2.58 |
| 0.06 | 3.65 |
| 0.07 | 4.93 |
| 0.08 | 6.59 |

simulations using different numbers of iterations. Considering that any reliability analysis is most highly dependent on the choice of the standard deviations of the variables; variations in the probability of failure like those shown in Figure 3.17 are not significant.

The influence of standard deviation is shown in Table 3.8. The probabilities of failure for all three failure modes were calculated with the values of standard deviations increased by 20%, 40%, and 60%. These increases in the standard deviation values increased the calculated probabilities of failure by as much as a factor of 3.7.

Although the Monte Carlo method can only be applied to computer analyses, it provides a useful standard of comparison for other methods because it does not involve assumptions that the factor of safety is normally or lognormally distributed.

## 3.8 HASOFER LIND METHOD

Filz and Navin (2006) found that the Hasofer Lind (1974) method is more accurate than the simpler Taylor Series or PEMs of reliability analyses, discussed subsequently. While the Taylor Series or the PEMs require an assumption on the distribution of the factor of safety, the Hasofer Lind method requires an assumption on the distributions of the variables involved in the analysis. This is considered more accurate because the distribution of the safety factor is difficult to predict and thus subject to wider variations (Baecher and Christian 2003). The Hasofer Lind method is used less than the Taylor Series or the PEMs because it requires more calculations, and in its original formulation required closed-form equations (Filz and Navin 2006). The Simplified Hasofer Lind method of reliability analysis developed by Filz and Navin (2006) is described here. The full method is presented well in Low (1996), Low and Tang (1997, 2004), and Low and Phoon (2002).

The Hasofer Lind method is a first-order reliability method (FORM). The application of the method to geotechnical engineering analysis has been described by Baecher and Christian (2003). The Simplified Hasofer Lind method presented here uses the factor of safety equal to 1.0 as the "performance function," and closed-form equations are not required. The method determines reliability geometrically as the shortest distance from the mean values of the variables to the performance function. This distance is essentially the reliability index, which can be used to find the probability of failure.

Three main stages are involved in the Simplified Hasofer Lind method. Stage one involves formulating trial values of the variables based on an assumed value of the reliability index. In most cases the initially assumed value of the reliability index is taken as 1.0 (Filz and Navin 2006). The factor of safety is calculated using reduced values of the variables corresponding to reliability index = 1.0. If the factor of safety calculated using these reduced values of the variables is not one, a different trial value of the reliability index is used to calculate new values of the variables. This iterative process is continued until the reliability index is found that will result in a factor of safety equal to 1.0.

Stage 2 is used to determine the "gradient of the failure function." This stage determines the change in the factor of safety with a change of the variables. This step requires N calculations of the factor of safety where N is the number of variables.

The final stage, stage 3, is similar to stage 1 where trial values of the reliability index are used to generate new values of the variables and the factor of safety is calculated. However, in this stage the values of the variables are also based on the gradients calculated in stage two. Like stage one, an iterative process is used until a factor of safety = 1.0 with reduced values of the variables is found.

This Simplified Hasofer Lind reliability method will be explained with the use of the same retaining wall sliding on a granular surface example used for the previous methods.

1. Estimate the standard deviations of the quantities involved in Equation 3.20.
   $\sigma_{efp}$ = standard deviation of the equivalent fluid pressure = 1.06 kN/m$^3$
   $\sigma_{\tan \delta}$ = standard deviation of tan $\delta$ = 0.05
   $\sigma_{\gamma bf}$ = standard deviation of the unit weight of backfill = 0.565 kN/m$^3$
2. Assume either normal or lognormal distributions of the variables, and indicate whether the variables are related to "capacity" or "demand," or in simple terms, load or resistance. If the variable is related to demand (load), a negative value of the standard deviation is used for the calculations. In this example problem the equivalent fluid pressure is related to demand, so the standard deviation used in the calculations is −1.06 kN/m$^3$.
3. Use an initial trial value of the reliability index to calculate factored values of the variables, and use these values to calculate the factor of safety. This is accomplished using Equations 3.24 or 3.25. Then calculate the factor of safety using these values of the variables. If the factor of safety is greater than one, a higher trial value of the reliability index should be used. A lower value of the reliability index should be used if the calculated factor of safety is less than one. This iterative process is continued until a factor of safety equal to one is calculated as shown in Table 3.9.

$$x_i = -\beta * \sigma_{xi} + \mu_{xi} \tag{3.24}$$

$$x_i = e^{(-\beta * \zeta_{xi} + \lambda_{xi})} \text{ where } \zeta_{xi} = \sqrt{\ln\left[1 + \frac{\sigma_{xi}^2}{\mu_{xi}^2}\right]} \text{ and } \lambda_{xi} = \ln(\mu_{xi}) - \frac{1}{2} * \zeta_{xi}^2 \tag{3.25}$$

Table 3.9 First stage of the Simplified Hasofer Lind method

| B | 1.0 | 1.5 | 1.28 | Assumed distribution (N or LN) | Capacity or demand? (C or D?) |
|---|---|---|---|---|---|
| $\gamma_{eq}$ | 8.1 | 8.7 | 8.4 | N | D |
| $\gamma_{bf}$ | 18.3 | 18.0 | 18.1 | N | C |
| tan δ | 0.45 | 0.43 | 0.44 | N | C |
| FS | 1.1 | 0.94 | 1.0 | | |

where Equation 3.13 is for a normal distribution of the factor of safety and Equation 3.25 is for a lognormal distribution of the factor of safety.

4. Stage two of the method is to calculate the "gradient" of the performance function. This is simply the change in the factor of safety for a given change in the variable. The value of each variable is changed by 10% (Filz and Navin 2006). Variables related to capacity are increased by 10% from the final values calculated in step 3, and variables related to demand are reduced by 10%. For example, $\gamma_{eq}$ at the end of step 3 has a value of 8.4 kN/m³. $\gamma_{eq}$ is related to demand, and therefore is decreased by 10%, to 7.6 kN/m³, while the other variables remain at the values shown in Table 3.10. Similarly, the values of $\gamma_{bf}$ and tanδ, both related to capacity are increased (one at a time) by 10%. The gradient is calculated as shown in Equations 3.26 and 3.27.

$$\frac{dg}{dx_i} = \frac{\Delta F}{\Delta x_i'} \tag{3.26}$$

$$x_i' = \frac{x_i - \mu_{xi}}{\sigma_{xi}} \tag{3.27}$$

5. Stage 3 is similar to stage one where trial values of the reliability index are used until the calculated factor of safety is equal to one. However, in this stage the variables are calculated using the gradients, the type of distribution of the variable, the standard deviation of the variable, and the trial value of the reliability index. The first value of the reliability index to use is the value found at the end of step 3 (Table 3.9). Then, the values of the variables are calculated using the value of $\alpha$ calculated from Equation 3.28 through 3.30. If the factor of safety is greater than one, a higher value of the reliability index should be used. A lower value of the reliability index should be used if the calculated factor of safety is less than one. This iterative process is continued until a factor of safety equal to one is calculated (Table 3.11).

Table 3.10 Stages 2 and 3 of the Simplified Hasofer Lind method

| | | | | Gradient = dg/dx' | α |
|---|---|---|---|---|---|
| $\gamma_{eq}$ | 7.6 | 8.4 | 8.4 | 0.13 | 0.78 |
| $\gamma_{bf}$ | 18.1 | 18.1 | 19.9 | 0.03 | 0.14 |
| Tan δ | 0.44 | 0.48 | 0.44 | 0.11 | 0.64 |
| FS | 1.1 | 1.1 | 1.1 | | |

Table 3.11 Final factor of safety calculations for the
Simplified Hasofer Lind method

| Assumed $\beta$ | 1.28 | 1.58 | 1.94 | 1.98 |
|---|---|---|---|---|
| $\gamma_{eq}$ | 8.11 | 8.34 | 8.62 | 8.66 |
| $\gamma_{bf}$ | 18.76 | 18.72 | 18.69 | 18.68 |
| $\mathrm{Tan}\,\delta$ | 0.46 | 0.45 | 0.44 | 0.44 |
| FS | 1.12 | 1.07 | 1.01 | 1.000 |

$$\alpha_i = \frac{\left(\dfrac{dg}{dx_i'}\right)}{\sqrt{\sum\left(\dfrac{dg}{dx_i'}\right)^2}} \tag{3.28}$$

$$x_i' = -\alpha_i\beta * \sigma_{xi} + \mu_{xi} \tag{3.29}$$

$$x_i = e^{(-\alpha_i\beta*\zeta_{xi}+\lambda_{xi})} \quad \text{where } \zeta_{xi} = \sqrt{\ln\left[1 + \frac{\sigma_{xi}^2}{\mu_{xi}^2}\right]} \quad \text{and} \quad \lambda_{xi} = \ln(\mu_{xi}) - \frac{1}{2} * \zeta_{xi}^2 \tag{3.30}$$

The $\alpha_1$ term is a unit vector and $g$ is the failure function, which in this case is the calculation of the factor of safety.

6. The final value of the reliability index, in this case 1.98, is the reliability index for the assumed failure mechanism. This can be converted into a probability of failure using Equation 3.8. For sliding on the silty sand layer, the Hasofer Lind probability of failure is 2.4%.

The Hasofer Lind method was also applied to the other failure mechanisms, sliding within the clay and bearing capacity failure in the clay. The standard deviation of the undrained shear strength of the clay was estimated to be 24 kN/m$^2$ in these analyses. The calculated probability of failure for sliding on the clay surface was 2.20%, and the calculated probability of bearing capacity failure was 1.80%.

The results of the Simplified Hasofer Lind method for sliding on the granular layer, sliding through the clay, and bearing capacity failure in the clay are compared with the results of the Monte Carlo simulation method in Table 3.12. Comparing the Simplified Hasofer Lind method with the Monte Carlo simulation method shows what effect the simplifying

Table 3.12 Comparison of the Simplified Hasofer Lind and Monte Carlo Simulation methods for the retaining wall example

| Failure mode | Deterministic factor of safety | Probability method | Probability of failure (%) |
|---|---|---|---|
| Sliding on sand | 1.40 | Monte Carlo | 2.4 |
| | | Simplified Hasofer Lind | 2.4 |
| Sliding in clay | 1.95 | Monte Carlo | 2.2 |
| | | Simplified Hasofer Lind | 2.5 |
| Bearing capacity | 1.97 | Monte Carlo | 1.8 |
| | | Simplified Hasofer Lind | 2.3 |

assumptions of the Simplified Hasofer Lind method has on the result. It can be seen that the results of the Simplified Hasofer Lind method are very similar to the results obtained with the Monte Carlo Simulation method.

### 3.8.1  Summary of the Hasofer Lind method

1. Estimate the standard deviations of the parameters that involve uncertainty, using the methods discussed in this chapter.
2. Compute the reliability index that gives a factor of safety equal to one. This is an iterative process where a trial value of the reliability index ($\beta$) is assumed, and values of the variables ($x_i$) are calculated using Equation 3.24 for normal distributions of variable or Equation 3.25 for lognormal distributions.

$$x_i = -\beta * \sigma_{xi} + \mu_{xi} \tag{3.24}$$

$$x_i = e^{(-\beta * \zeta_{xi} + \lambda_{xi})} \text{ where } \zeta_{xi} = \sqrt{\ln\left[1 + \frac{\sigma_{xi}^2}{\mu_{xi}^2}\right]} \text{ and } \lambda_{xi} = \ln(\mu_{xi}) - \frac{1}{2} * \zeta_{xi}^2 \tag{3.25}$$

3. Determine the change in the factor of safety for a 10% change in each of the variables. If the variables are related to capacity, increase the value by 10%, if the variables are related to demand, decrease the variable by 10%.
4. Calculate gradients using Equations 3.26 and 3.27

$$\frac{dg}{dx_i} = \frac{\Delta F}{\Delta x_i'} \tag{3.26}$$

$$x_i' = \frac{x_i - \mu_{xi}}{\sigma_{xi}} \tag{3.27}$$

5. Repeat step 2 to find a new value of $\beta$ by repeated trials, this time calculating the variables by using values of $\alpha_i$ calculated using Equation 3.28 and values of $x_i$ calculated using Equations 3.29 or 3.30

$$\alpha_i = \frac{\left(\frac{dg}{dx_i'}\right)}{\sqrt{\sum\left(\frac{dg}{dx_i'}\right)^2}} \tag{3.28}$$

$$x_i' = -\alpha_i \beta * \sigma_{xi} + \mu_{xi} \tag{3.29}$$

$$x_i = e^{(-\alpha_i \beta * \zeta_{xi} + \lambda_{xi})} \text{ where } \zeta_{xi} = \sqrt{\ln\left[1 + \frac{\sigma_{xi}^2}{\mu_{xi}^2}\right]} \text{ and } \lambda_{xi} = \ln(\mu_{xi}) - \frac{1}{2} * \zeta_{xi}^2 \tag{3.30}$$

   The value of $\beta$ is correct when the factor of safety calculated using the values of $x_i$ is equal to 1.00.
6. Calculate the probability of failure using Equation 3.8.

$$p_f = 1 - \text{NORMSDIST} (\beta) \tag{3.8}$$

## 3.9 TAYLOR SERIES METHOD WITH ASSUMED NORMAL DISTRIBUTION OF THE FACTOR OF SAFETY

The Taylor Series method has been described by Duncan (2000), Baecher and Christian (2003), Ang and Tang (2007), and Harr (1987). The method is simple, easy to understand, and easy to apply.

The Taylor Series method is a "first-order second moment" (FOSM) analysis. Only the first two "moments" (the mean and the standard deviation) are considered in the analysis. The application of the Taylor Series method in geotechnical engineering has been described by Wolff (1994), U.S. Army Corps of Engineers (1997), Duncan (2000), and by Baecher and Christian (2003). The method requires an assumption on the distribution of the factor of safety. When using the Taylor Series method, $2N + 1$ calculations of the factor of safety are required, where N is the number of variables.

The method consists of two main parts:

1. Use the Taylor Series method to compute the COV of the computed result (e.g., factor of safety or settlement), and
2. Assume a normal or lognormal distribution of the computed result, to determine the probability of failure.

The method is illustrated here by application to the cantilever retaining wall shown in Figure 3.18. This example was previously analyzed using the Monte Carlo and Hasofer Lind methods. Here, the probability of failure will be computed assuming first a normal distribution for the factor of safety and second a lognormal distribution of factor of safety. For the conditions shown in Figure 3.18, the value of the safety factor is 1.40, as shown on the top of Table 3.13. In the realm of probabilistic analyses, this is called the "most likely value" of factor of safety, $F_{MLV}$.

To calculate the probability of failure using the Taylor Series method, the following steps are used:

1. Estimate the standard deviations of the quantities involved in Equation 3.20. Simple methods for estimating standard deviations have been discussed previously in this chapter. Using those methods, the following values of standard deviation of the parameters involved in this example have been estimated:

   $\sigma\gamma_{eq}$ = standard deviation of the equivalent fluid pressure = 1.06 kN/m$^3$,
   $\sigma_{\tan\delta}$ = standard deviation of $\tan \delta$ = 0.05, and
   $\sigma\gamma_{bf}$ = standard deviation of the unit weight of backfill = 0.565 kN/m$^3$.
2. Use the Taylor Series technique (Wolff, 1994; U.S. Army Corps of Engineers 1997; Duncan 2000; and by Baecher and Christian, 2003) to estimate the standard deviation and the COV of the factor of safety using these formulas:

$$\sigma_F = \sqrt{\left(\frac{\Delta F_1}{2}\right)^2 + \left(\frac{\Delta F_2}{2}\right)^2 + \left(\frac{\Delta F_3}{2}\right)^2} \qquad (3.31)$$

$$V_F = \frac{\sigma_F}{F_{MLV}} \qquad (3.32)$$

in which $\Delta F_1 = (F_1^+ - F_1^-)$. $F_1^+$ is the factor of safety calculated with the value of the first parameter (in this case, the equivalent fluid pressure) increased by one standard

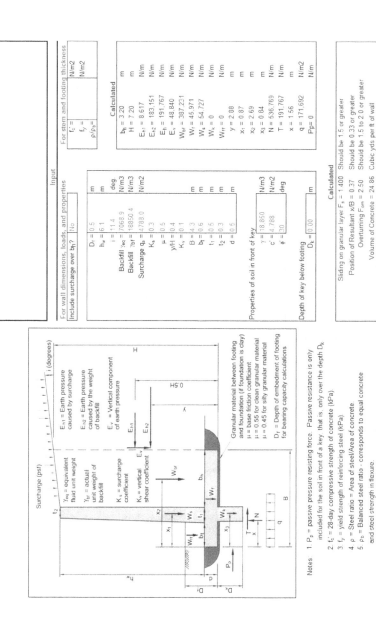

Figure 3.18 Retaining wall computation sheet used for the retaining wall example, used with permission from the Center for Geotechnical Practice and Research.

Table 3.13 Taylor Series reliability analysis of cantilever retaining wall

| With all variables assigned their most likely values | | F = 1.40 | |
| --- | --- | --- | --- |
| Variable | Values | Factors of safety | ΔF |
| Equivalent fluid unit weight, $\gamma_{eq}$ | | | |
| Most likely value plus σ | 8.13 kN/m$^3$ | F + =1.224 | |
| Most likely value minus σ | 6.01 kN/m$^3$ | F − =1.634 | −0.410 |
| Tangent of δ | | | |
| Most likely value plus σ | 0.55 | F + =1.539 | |
| Most likely value minus σ | 0.45 | F − =1.260 | 0.279 |
| Backfill unit weight, $\gamma_{bf}$ | | | |
| Most likely value plus σ | 19.42 kN/m$^3$ | F + =1.434 | |
| Most likely value minus σ | 18.28 kN/m$^3$ | F − =1.365 | 0.069 |

deviation from its best estimate value. $F_1^-$ is the factor of safety calculated with the value of the first parameter decreased by one standard deviation.

In calculating $F_1^+$ and $F_1^-$, the values of all of the other variables are kept at their most likely values.

The values of $\Delta F_2$, and $\Delta F_3$, are calculated by varying the values of the second and third variables (footing/sand friction angle, and backfill unit weight) by plus and minus one standard deviation from their most likely values. The results of these calculations are shown in Table 3.13.

3. Substituting the values of $\Delta F$ into Equation 3.31, the value of the standard deviation of the factor of safety ($\sigma_F$) is found to be 0.250, and the COV of the factor of safety ($V_F$), calculated using Equation 3.30, is found to be 18%.

4. With both $F_{MLV}$ and $V_F$ known, the probability of failure and the reliability of the factor of safety can be determined. First, the reliability index is calculated using Equation 3.33, assuming a normal distribution of the factor of safety.

$$\beta_{normal} = \frac{F_{MLV} - 1}{\sigma_{FS}} = \frac{1.400 - 1}{0.250} = 1.60 \tag{3.33}$$

The probability of failure can be calculated using the NORMSDIST function in Excel based on the reliability index using Equation 3.34.

$$P_{f(normal)} = 1 - NORMSDIST(\beta_{normal}) = 1 - NORMSDIST(1.60) = 5.50\% \tag{3.34}$$

The same analysis was used to compute the probability of failure by sliding on the clay, and by bearing capacity failure in the clay. The standard deviation of the undrained shear strength of the clay was estimated to be 24 kN/m$^2$. The calculated probability of failure for sliding in the clay was found to be 2.81%, and the calculated probability of failure for bearing capacity failure was found to be 1.47%.

The results of the Taylor Series method assuming a normal distribution for the factor of safety are compared to the results of the Monte Carlo method in Table 3.14. The results show that the probabilities of failure computed using the Taylor Series method, assuming a normal distribution for the factor of safety, differ from the values computed using the Monte Carlo method. For the mode of failure involving sliding on the sand layer at the base

*Table 3.14* Comparison of the Taylor Series method assuming normal and lognormal distribution of the factor of safety with the Monte Carlo methods for the retaining wall example

| Failure mode | Deterministic factor of safety | Probability method | Probability of failure (%) |
|---|---|---|---|
| Sliding on sand | 1.40 | Monte Carlo | 2.1 |
| | | Taylor Series, normal dist. of F | 5.5 |
| Sliding in clay | 1.95 | Monte Carlo | 2.2 |
| | | Taylor Series, normal dist. of F | 2.8 |
| Bearing capacity | 1.97 | Monte Carlo | 1.8 |
| | | Taylor Series, normal dist. of F | 1.5 |

of the footing, the difference is quite significant—more than a factor of two in the value of probability of failure. These results show that the assumptions involved in the Taylor Series method can result in considerable differences in results from the Monte Carlo method, which is considered the better method.

## 3.10  TAYLOR SERIES METHOD WITH A LOGNORMAL DISTRIBUTION OF THE FACTOR OF SAFETY

If a lognormal distribution for the factor of safety is assumed, the same procedure is followed until step 4. Then, equations are used to calculate the reliability index for a lognormal distribution of the factor of safety. The assumption of a lognormal distribution for factor of safety does not imply that the values of the individual variables ($\gamma_{eq}$, tan $\delta$, and $\gamma_{bf}$) must be distributed lognormally. The probability of failure using the Taylor Series method with a lognormal distribution for the factor of safety is calculated using the following steps:

1. Same as for a normal distribution for the factor of safety
2. Same as for a normal distribution for the factor of safety
3. Same as for a normal distribution for the factor of safety
4. With both $F_{MLV}$ and $V_F$ known, the probability of failure can be determined as follows. First, the reliability index is calculated using Equation 3.35.

$$\beta_{lognormal} = \frac{\ln \dfrac{F_{MLV}}{\sqrt{1 + (COV_F)^2}}}{\sqrt{\ln(1 + COV_F)^2}} = \frac{\ln \dfrac{1.400}{\sqrt{1 + (0.179)^2}}}{\sqrt{\ln(1 + 0.179)^2}} = 1.81 \tag{3.35}$$

The probability of failure can be calculated using the NORMSDIST function in Excel based on the reliability index using Equation 3.36.

$$P_{f(lognormal)} = 1 - NORMSDIST(\beta_{lognormal}) = 1 - NORMSDIST(1.81) = 3.53\% \tag{3.36}$$

The Taylor Series analysis was also used to compute the probability of failure by sliding on the clay, and by bearing capacity failure in the clay. The standard deviation of the undrained shear strength of the clay was estimated to be 24 kN/m². The calculated probability of failure for sliding in the clay was found to be 0.56%, and the calculated probability of failure for bearing capacity failure was found to be 0.17% when the factor of safety was assumed to be lognormally distributed.

*Table 3.15* Comparison of the Taylor Series method assuming normal and lognormal distribution of the factor of safety with the Monte Carlo methods for the retaining wall example

| Failure mode | Deterministic factor of safety | Probability method | Probability of failure (%) |
|---|---|---|---|
| Sliding on sand | 1.40 | Monte Carlo | 2.4 |
| | | Taylor Series, normal dist. of F | 5.5 |
| | | Taylor Series, lognormal F | 3.5 |
| Sliding in clay | 1.95 | Monte Carlo | 2.2 |
| | | Taylor Series, normal dist. of F | 2.8 |
| | | Taylor Series, lognormal F | 0.6 |
| Bearing capacity | 1.97 | Monte Carlo | 1.8 |
| | | Taylor Series, normal dist. of F | 1.5 |
| | | Taylor Series, lognormal F | 0.2 |

The results from the Taylor Series method assuming normal and lognormal distributions for the factor of safety are compared to the results of the Monte Carlo method in Table 3.15. None of the Taylor Series results agree very well with the Monte Carlo results, and in the case of the bearing capacity failure mode, they differ by nearly an order of magnitude.

Thus, while the Taylor Series method offers a relatively easy way to compute probabilities of failure, the results can differ significantly from the results of the Monte Carlo method, which is considered to be a better method. In addition, the Taylor Series method requires that the form of the safety factor distribution be assumed, and there is no logical way to determine the best form of this distribution. It seems reasonable to conclude that the strongest point in favor of using the Taylor Series method is its simplicity, but probabilities of failure computed this way should be viewed as rough estimates that may be higher or lower than values computed by better methods—the Monte Carlo method and the Hasofer Lind method.

### 3.10.1 Summary of the Taylor Series method

The steps involved in using the Taylor Series method are as follows:

1. Determine the most likely values of the parameters involved, and compute the factor of safety by the normal (deterministic) method. This is $F_{MLV}$.
2. Estimate the standard deviations of the parameters that involve uncertainty.
3. Compute the factor of safety with each parameter increased by one standard deviation and then decreased by one standard deviation from its most likely value, with the values of the other parameters equal to their most likely values. This involves 2N calculations, where N is the number of parameters whose values are being varied. These calculations result in N values of $F^+$ and N values of $F^-$. Using these values of $F^+$ and $F^-$, compute the values of $\Delta F$ for each parameter, and compute the standard deviation and COV of the factor of safety ($\sigma_F$ and $V_F$) using the following equations:

$$\sigma_F = \sqrt{\left(\frac{\Delta F_1}{2}\right)^2 + \left(\frac{\Delta F_2}{2}\right)^2 + \left(\frac{\Delta F_3}{2}\right)^2 + \left(\frac{\Delta F_4}{2}\right)^2} \qquad (3.31)$$

$$V_F = \frac{\sigma_F}{F_{MLV}} \qquad (3.32)$$

4. Use the value of $F_{MLV}$ from step 1 and the value of $V_F$ from step 3 to determine the value of $P_f$, by means of Table 3.4 or by determining the reliability index and the 'NORMSDIST' function in Excel.

$$\beta_{NORMAL} = \frac{F_{mean} - 1}{\sigma_{FS}} \qquad (3.37)$$

$$\beta_{Lognormal} = \frac{\ln \dfrac{F_{MLV}}{\sqrt{1 + (COV_F)^2}}}{\sqrt{\ln(1 + (COV_F)^2)}} \qquad (3.11)$$

$$P_f = 1 - NORMSDIST\,(\beta) \qquad (3.8)$$

## 3.11 PEM WITH A NORMAL DISTRIBUTION FOR THE FACTOR OF SAFETY

PEM is also a FOSM reliability method. The method was introduced by Rosenblueth (1975) and its use in geotechnical engineering is discussed by Baecher and Christian (2003), Harr (1987), and Wolff (1996). This method requires $2^N$ calculations of the factor of safety where N is the number of variables. The factor of safety is calculated using combinations of variables where each is a standard deviation above or below the mean. These combinations of variables are used to calculate the factor of safety. A table after Harr (1987) is shown below that aids in the factor of safety calculations. On the table the (−) and (+) signs indicate the mean value minus one standard deviation and the mean value plus one standard deviation. For example, if only one variable has uncertainty in the calculation for the factor of safety the number of cases to consider is $2^1 = 2$. The first case would be to calculate the factor of safety with the mean value of the variable minus the standard deviation. The second case would be to calculate the factor of safety with the mean value of the variable plus the standard deviation.

The same cantilever retaining wall example for the factor of safety against sliding used for the Taylor Series method will be presented here for the PEM assuming a normal distribution for the factor of safety.

The PEM follows these steps:

1. Estimate the standard deviations of the quantities involved in Equation 3.20. This is the same first step as the Taylor Series Method. The values of the standard deviations that will be used are
$\sigma\gamma_{eq}$ = standard deviation of the equivalent fluid pressure = 1.06 kN/m³,
$\sigma_{\tan}\delta$ = standard deviation of tan δ = 0.05, and
$\sigma\gamma_{bf}$ = standard deviation of the unit weight of backfill = 0.565 kN/m³.
2. Use the point estimate technique (Rosenblueth 1975) to estimate the standard deviation of the factor of safety. This is done by using Table 3.16 after Harr (1987) to generate variables that are the mean plus or minus the standard deviation. With three

Table 3.16 Values of variable used in the PEM, arranged using the procedure suggested by Harr (1987)

| Number of variables | Case | $\gamma_{eq}$ | $\gamma_{bf}$ | $\tan\delta$ | 4 |
|---|---|---|---|---|---|
| | | | Random variables | | |
| 2 | 1 | $(7.07 - 1.06) = 6.01$ | $(18.85 - 0.56) = 18.3$ | $(0.50 - 0.05) = 0.45$ | – |
| | 2 | $(7.07 + 1.06) = 8.13$ | $(18.85 - 0.56) = 18.3$ | $(0.50 - 0.05) = 0.45$ | – |
| $2^2 = 4$ | 3 | $(7.07 - 1.06) = 6.01$ | $(18.85 + 0.56) = 19.4$ | $(0.50 - 0.05) = 0.45$ | – |
| | 4 | $(7.07 + 1.06) = 8.13$ | $(18.85 + 0.56) = 19.4$ | $(0.50 - 0.05) = 0.45$ | – |
| $2^3 = 8$ | 5 | $(7.07-1.06) = 6.01$ | $(18.85-0.56) = 18.3$ | $(0.50 + 0.05) = 0.55$ | – |
| | 6 | $(7.07 + 1.06) = 8.13$ | $(18.85 - 0.56) = 18.3$ | $(0.50 + 0.05) = 0.55$ | – |
| | 7 | $(7.07 - 1.06) = 6.01$ | $(18.85 + 0.56) = 19.4$ | $(0.50 + 0.05) = 0.55$ | – |
| | 8 | $(7.07 + 1.06) = 8.13$ | $(18.85 + 0.56) = 19.4$ | $(0.50 + 0.05) = 0.55$ | – |
| $2^4 = 16$ | 9 | – | – | – | + |
| | 10 | + | – | – | + |
| | 11 | – | + | – | + |
| | 12 | + | + | – | + |
| | 13 | – | – | + | + |
| | 14 | + | – | + | + |
| | 15 | – | + | + | + |
| | 16 | + | + | + | + |

variables ($N = 3$), there are $2^3 = 8$ cases to calculate the factor of safety. For each case shown in Table 3.16, the factor of safety is calculated.

3. Each case is assigned a weighting probability. For uncorrelated variables, each case is assigned the same weighting probability from Equation 3.38.

$$P = \frac{1}{C} \tag{3.38}$$

where $C$ is the number of cases. With 8 cases to calculate the factor of safety, the weighting probability is $1/8 = 0.125$.

4. The mean factor of safety, $F_{mean}$ is then calculated. $F_{mean}$ can be calculated by either using the mean values of the variables, like the Taylor Series method, or by averaging the factors of safety calculated for each of the cases. These two methods do not give the same value because of the nonlinearity of the factor of safety. It is suggested to use the average of the factor of safety values calculated from the cases (Filz and Navin 2006).

$$F_{mean} = \frac{1.434 + 1.075 + 1.506 + 1.129 + 1.753 + 1.314 + 1.841 + 1.379}{8} = 1.43$$

5. With the value of $F_{mean}$ (Table 3.17), the standard deviation of the factor of safety can be calculated from the following equation:

$$\sigma_F = \sqrt{\sum P_i F_i^2 - F_{mean}^2} = \sqrt{2.11 - 1.43^2} = 0.253 \tag{3.39}$$

6. With the standard deviation of the factor of safety and the $F_{mean}$ value, the reliability index and probability of failure can be calculated. Assuming a normal distribution for the factor of safety, Equations 3.8 and 3.9 are used making use of the 'NORMSDIST' function in Microsoft's Excel to calculate these values.

Table 3.17 Factors of safety for each case using the PEM

| Case | $\gamma_{eq}$ | $\gamma_{bf}$ | $\tan\delta$ | $P_i$ | $F$ | $F_i^2 * P_i$ |
|---|---|---|---|---|---|---|
| 1 | 6.01 | 18.28 | 0.45 | 0.125 | 1.434 | 0.257 |
| 2 | 8.13 | 18.28 | 0.45 | 0.125 | 1.075 | 0.144 |
| 3 | 6.01 | 19.42 | 0.45 | 0.125 | 1.506 | 0.284 |
| 4 | 8.13 | 19.42 | 0.45 | 0.125 | 1.129 | 0.159 |
| 5 | 6.01 | 18.28 | 0.55 | 0.125 | 1.753 | 0.384 |
| 6 | 8.13 | 18.28 | 0.55 | 0.125 | 1.314 | 0.216 |
| 7 | 6.01 | 19.42 | 0.55 | 0.125 | 1.841 | 0.424 |
| 8 | 8.13 | 19.42 | 0.55 | 0.125 | 1.379 | 0.238 |
| | | | | Sum | 11.43 | 2.11 |
| | | | | $F_{mean}$ | 1.434 | |

Table 3.18 Comparison of the PEM assuming normal distribution of the factor of safety with the Monte Carlo methods for the retaining wall example

| Failure mode | Deterministic factor of safety | Probability method | Probability of failure (%) |
|---|---|---|---|
| Sliding on sand | 1.40 | Monte Carlo | 2.4 |
| | | Point Estimate, normal dist. of F | 4.5 |
| Sliding in clay | 1.95 | Monte Carlo | 2.2 |
| | | Point Estimate, normal dist. of F | 5.0 |
| Bearing capacity | 1.97 | Monte Carlo | 1.8 |
| | | Point Estimate, normal dist. of F | 4.3 |

Normal Distribution of the Factor of Safety:

$$\sigma_F = \sqrt{2.11 - 1.43^2} = 0.253 \quad \beta = \frac{1.43 - 1}{0.253} = 1.70 \quad P_f = 1 - \text{NORMSDIST}(1.70) = 4.5\%$$

The same analysis is applied to the factor of safety against sliding on a clay surface and bearing capacity failure. The standard deviation of the undrained shear strength of the clay is estimated to be 24 kN/m². The calculated probability of failure for sliding on a clay surface is 5.00%. The calculated probability of failure for bearing capacity is 4.34%.

The results of the PEM assuming a normal distribution for the factor of safety is compared to the Monte Carlo simulation method, showing how the simplifying assumptions of the PEM affect the result in Table 3.18.

## 3.12 PEM WITH A LOGNORMAL DISTRIBUTION FOR THE FACTOR OF SAFETY

If a lognormal distribution for the factor of safety is assumed instead of a normal distribution, these steps are followed:

1. Same as for a normal distribution for the factor of safety
2. Same as for a normal distribution for the factor of safety
3. Same as for a normal distribution for the factor of safety

Table 3.19 Comparison of the PEM assuming normal and lognormal distribution of the factor of safety with the Monte Carlo methods for the retaining wall example

| Failure mode | Deterministic factor of safety | Probability method | Probability of failure (%) |
|---|---|---|---|
| Sliding on sand | 1.40 | Monte Carlo | 2.4 |
| | | Point Estimate, normal dist. of F | 4.5 |
| | | Point Estimate, lognormal F | 2.6 |
| Sliding in clay | 1.95 | Monte Carlo | 2.2 |
| | | Point Estimate, normal dist. of F | 5.0 |
| | | Point Estimate, lognormal F | 1.6 |
| Bearing capacity | 1.97 | Monte Carlo | 1.8 |
| | | Point Estimate, normal dist. of F | 4.3 |
| | | Point Estimate, lognormal F | 1.1 |

4. Same as for a normal distribution for the factor of safety
5. Same as for a normal distribution for the factor of safety
6. If a lognormal distribution of the factor of safety is assumed, then the probability of failure can be calculated using Equation 3.40 and the 'NORMSDIST' function in Microsoft's Excel.

Lognormal distribution of the factor of safety:

$$\beta_{\text{Lognormal}} = \frac{\ln\left(F_{\text{MLV}}/\sqrt{1 + (\text{COV}_F)^2}\right)}{\sqrt{\ln(1 + (\text{COV}_F)^2)}} = \frac{\ln\left(1.43/\sqrt{1 + (0.253/1.43)^2}\right)}{\sqrt{\ln(1 + (0.253/1.43)^2)}} = 1.94 \qquad (3.40)$$

$$P_f = 1 - \text{NORMSDIST}(1.94) = 2.59\%$$

The same analysis is applied to the factor of safety against sliding on a clay surface and bearing capacity failure. The standard deviation of the undrained shear strength of the clay is estimated to be 24.0 kN/m². The calculated probability of failure for sliding on a clay surface is 1.58%. The calculated probability of failure for bearing capacity is 1.14%.

The results of the PEM assuming a lognormal distribution for the factor of safety is compared to the Monte Carlo simulation method to see how the simplifying assumptions of the PEM affect the result as shown in Table 3.19.

## 3.12.1 Summary of the PEM

The steps involved in using the PEM are

1. Estimate the standard deviations of the parameters that involve uncertainty, using the methods discussed in this chapter.
2. Compute the factor of safety with each parameter increased or decreased by one standard deviation. There are $2^N$ cases where $N$ is the number of parameters whose values are being varied.

3. Assign a weighting probability to each of the cases. When the variables are uncorrelated, the probability is the same for each case and calculated from

$$P = \frac{1}{C} \tag{3.38}$$

In this equation $C$ is the number of cases.

4. Calculate $F_{mean}$ by averaging the calculated values of the factor of safety from each of the cases considered.
5. Using the $F_{mean}$ value from step 4 and the results of the factor of safety calculations from each case, calculate the standard deviation from Equation 3.39:

$$\sigma_F = \sqrt{\sum P_i F_i^2 - F_{mean}^2} \tag{3.39}$$

6. Calculate the reliability index and probability of failure using Equations 3.37 or 3.12 assuming a normal or lognormal distribution of the factor of safety and the 'normsdist' function in excel.

$$\beta_{NORMAL} = \frac{F_{mean} - 1}{\sigma_{FS}} \tag{3.37}$$

$$\beta_{Lognormal} = \frac{\ln \dfrac{F_{MLV}}{\sqrt{1 + (COV_F)^2}}}{\sqrt{\ln(1 + (COV_F)^2)}} \tag{3.11}$$

$$P_f = 1 - \text{NORMSDIST}(\beta) \tag{3.8}$$

Today, when virtually all calculations of factor of safety are performed using spreadsheets or other computer programs, the additional calculations for reliability require little extra effort and little additional engineering time. These calculations can be done about as quickly as new parameter values can be entered into a spreadsheet or data file.

The bulk of the analysis effort is required to develop the data for the first calculation in all of the methods. Thus, although additional calculations must be performed, they involve little time and effort beyond estimating values of the standard deviations of the parameters. The use of prudent and informed judgment is as important in estimating values of standard deviations of parameters as it is in estimating most likely values of parameters.

The great advantage of computing $P_f$ (the probability that the factor of safety could be less than 1.0) is that it provides an overall measure of the uncertainty in the results of the analysis. Computing both $F$ and $P_f$ adds little to the time and effort required for the analysis, but adds greatly to the value of the result.

## 3.13 COMMENTS ON THE METHODS

The methods described in this chapter to compute the probability of failure have their own advantages and disadvantages. Because use of these methods are in addition to traditional geotechnical analyses, ease of use is important.

Table 3.20 details the number of calculations required for each method.

Table 3.20 Number of calculations required for the reliability
methods where N is the number of variables

| Reliability method | Number of calculations |
|---|---|
| Taylor Series | $2N + 1$ |
| Point Estimate | $2^N$ |
| Simplified Hasofer Lind | Depends on iterations, generally more than Taylor Series or Point Estimate |
| Monte Carlo Simulation | Typically >5000 |

## 3.13.1 Significance of the variables

Because of the few extra calculations required, the Taylor Series method is the simplest reliability method available to calculate probability of failure. Another advantage of the Taylor Series method is that it allows an engineer to see how significant each variable is to the overall factor of safety. As each variable is decreased or increased by the standard deviation, the factor of safety is calculated. It is therefore easy to see which variables have the largest effect on the factor of safety. Seeing these effects is difficult with the PEM because more than one variable is changed at a time. It is also difficult to see these effects with the Simplified Hasofer Lind method for the same reason. When performing Monte Carlo simulations using @Risk™ a sensitivity analysis is also performed by the program. This makes it very easy to see which variables influence the factor of safety calculation.

## 3.13.2 Accuracy

The precision with which $P_f$ can be computed depends primarily on the accuracy with which the most likely values and the standard deviations of the variables that define the problem can be estimated. In geotechnical engineering applications, where data are often sparse and estimated values of parameters and their standard deviations are themselves uncertain, computed values of $P_f$ cannot be expected to be highly precise. Here, accuracy of the methods is discussed independent of choosing the standard deviation. The following observations can be made when comparing the methods:

1. Values of $P_f$ computed assuming that the factors of safety are normally or lognormally distributed are both equally as accurate when compared to the Monte Carlo method results.
2. Values of $P_f$ computed by the PEM, assuming normal distribution of the factor of safety, are consistently higher than the Monte Carlo values.
3. The Simplified Hasofer Lind method more closely agrees with the results of the Monte Carlo simulation than the Taylor Series or the PEMs.

## 3.14 SUMMARY

It is important to remember that estimates of the probability of failure are just that—estimates. Their value lies in their order of magnitude—is it 0.1%, 1%, or 10%? These orders of magnitude, 0.1%, 1%, and 10% can be viewed as low, medium (or normal), and high. Additional digits, such as 1.76%, do not add more value to the estimate.

An additional benefit lies in the fact that the process of calculating probability of failure reveals which sources of uncertainty are most important, and which are unimportant. This understanding provides an effective guide to what improvements in knowledge will reduce overall uncertainty the most.

Although never highly precise, probabilities of failure provide a valuable supplement to other measures of safety, like factor of safety or margin of safety. For this reason, calculating probability of failure adds value to geotechnical engineering analyses.

## REFERENCES

Ang, A., and Tang, W.H., 2007, *Probability Concepts in Engineering*, 2nd edition, John Wiley & Sons, New York and London, pp. 403.

Baecher, G.B., and Christian, J.T., 2003, *Reliability and Statistics in Geotechnical Engineering*, John Wiley and Sons, London and New York, 605pp.

Benson, C., Daniel, D., and Boutwell, G., 1999, Field performance of compacted clay liners, *Journal of Geotechnical Engineering*, Vol. 125, No. 5, p. 390–403.

Brinch-Hansen, J., 1970, *A revised and extended formula for bearing capacity*, Bulletin No. 28, Danish Geotechnical Institute, Copenhagen, p. 5–11.

Christian, J.T., and Baecher, G.B., 1999, Point-estimate methods as numerical quadrature, *Journal of Geotechnical Engineering*, Vol. 125, No. 9, p. 779–786.

Christian, J.T., Ladd, C.C., and Baecher, G.B., 1994, Reliability applied to slope stability analysis, *Journal of Geotechnical Engineering*, Vol. 120, No. 12, p. 2180–2207.

Clough, G.W., and Duncan, J.M., 1991, Earth pressures, chapter in *Foundation Engineering Handbook*, 2nd edition, edited by Hsai-Yang Fand, van Nostrand Reinhold, New York, NY, p. 223–235.

Dai, S.-H., and Wang, M.O., 1992, *Reliability Analysis in Engineering Applications*. Van Nostrand Reinhold, New York.

Duncan, J.M., 2000, Factors of safety and reliability in geotechnical engineering, *Journal of Geotechnical Engineering*, Vol. 126, No. 4, p. 307–316.

Duncan, J.M., and Buchignani, A.L. 1973, 'Failure of underwater slope in San Francisco Bay', *Journal of the Soil Mechanics and Foundations Division, ASCE*, Vol. 99, No. 9, p. 687–703.

Filz, G.M., and Navin, M.P., 2006, *Stability of column supported embankments*, Virginia Transportation Research Council, 06-CR13, 73pp.

Folayan, J.I., Hoeg, K., and Benjamin, J.R., 1970, Decision theory applied to settlement predictions, *Journal of Soil Mechanics and Foundations*, ASCE, Vol. 94, No.4, p. 1127–1141.

Foye, K.C., Salgado, R., and Scott, B., 2006, Assessment of variable uncertainties for reliability-based design of foundations. *Journal of Geotechnical and Geoenvironmental Engineering*, Vol. 132, No. 9, p. 1197–1207.

Harr, M.E., 1984, *Reliability-based design in civil engineering*, Henry M. Shaw Lecture, Dept. of Civil Engineering, North Carolina State University, Raleigh, N.C.

Harr, M.E., 1987, *Reliability-Based Design in Civil Engineering*, Mc-Graw-Hill, New York.

Hasofer, A.M., and Lind, N.C., 1974, Exact and invariant second-moment code format, *Journal of Engineering Mechanics*, Vol. 100, p. 111–121.

Kulhawy, F.H., 1992, *On the evaluation of soil properties*, Geotechnical Special Publication, No. 31, p. 95–115.

Lacasse, S., and Nadim, F., 1997, *Uncertainties in Characterizing Soil Properties*, Publication No. 201, Norwegian Geotechnical Institute, Oslo, Norway, p. 49–75.

Low, B.K. 1996, Practical probabilistic approach using spreadsheet, Geotechnical Special Publication No. 58, *Proceedings, Uncertainty in the Geologic Environment: From Theory to Practice*, Madison, Wisconsin, ASCE, Vol. 2, p. 1284–1302.

Low, B.K., and Phoon, K.K., 2002, Practical first-order reliability computations using spreadsheet, *Proceedings of the International Conference on Probabilistic in Geotechnics: Technical and Economic Risk Estimation*, Graz, Austria, p. 39–46.

Low, B.K., and Tang, W.H., 1997, Efficient reliability evaluation using spreadsheet, *Journal of Engineering Mechanics*, Vol. 123, No. 7, p. 749–752.

Low, B.K., and Tang, W.H., 2004, Reliability analysis using object-oriented constrained optimization, *Structural Safety*, Vol. 26, No.1, p. 69–89.

Meyerhoff, G.G., 1953, The bearing capacity of foundation under eccentric and inclined loads, *Proceedings of the 3rd International Conference on Soil Mechanics and Foundation Engineering*, Zurich, Vol. 1, p. 440–445.

Naval Facilities Engineering Command (NAVFAC), 1986, *Soil mechanics design manual 7.01*, Alexandria, VA, pp. 389.

Phoon, K.K., and Kulhawy, F.H., 1999, Characterization of geotechnical variability, *Canadian Geotechnical Journal*, Vol. 36, No. 4, p. 612–624.

Rosenblueth, E., 1975, Point estimates for probability moments, *Proceedings of the National Academy of Science*, Vol. 72, No. 10, p. 3812–3814.

Tang, W.H., Stark, T.D., and Angulo, M. 1999, Reliability in back analysis of slope failures, *Soils and Foundations Journal*, Japanese Society of SMFE, Vol. 39, No. 5, p. 73–80.

Terzaghi, K., Peck, R.B., and Mesri, G., 1996, *Soil Mechanics in Engineering Practice*, John Wiley & Sons, London and New York, pp. 592.

U.S. Army Corps of Engineers. 1997, Introduction to probability and reliability methods for using in geotechnical engineering, ETL 1110-2-547.

Wolff, T.F., 1994, *Evaluating the reliability of existing levees*, Prepared for the US Army Engineers Waterways Experiment Station Geotechnical Lab, Vicksburg, MS.

Wolff, T.F., 1996, *Probabilistic slope stability in theory and practice*, Proceedings of Uncertainty '96, ASCE Geotechnical Special Publication No. 58, p. 419–433.

Yang, B., and Duncan, J.M., 2002, *Retaining wall stability 2.05—An Excel workbook and documentation*, Center for Geotechnical Practice and Research, Virginia Tech, Blacksburg, VA, pp. 20.

# Chapter 4

# Maximum likelihood principle and its application in soil liquefaction assessment

*Charng Hsein Juang, Sara Khoshnevisan, and Jie Zhang*

## 4.1 INTRODUCTION

The occurrence of soil liquefaction and ground failure during great earthquakes is one of the most crucial factors in the subsequent economic devastation and loss of lives that can result from such catastrophic events. Due to the difficulty and expense of securing and testing high-quality undisturbed samples of soils, empirical methods based on *in situ* tests such as the standard penetration test (SPT) or the cone penetration test (CPT) remain the dominant approaches in engineering practice for evaluating the liquefaction potential and its effect. Indeed, the simplified procedure pioneered by Seed and Idriss (1971) is perhaps the method most widely used over the last 40 years for evaluating liquefaction potential. In this procedure, developed from field observations and field and laboratory tests with a strong theoretical basis, the liquefaction potential of a soil is most often expressed as a factor of safety ($F_S$), which is defined as the ratio of cyclic resistance ratio ($CRR$) over cyclic stress ratio ($CSR$). In the context of liquefaction assessment, $CSR$ represents a dimensionless measure of the cyclic shear stress applied to a soil through seismic loading, and $CRR$ represents the corresponding measure of the cyclic shear resistance of the soil. In a deterministic assessment of the liquefaction potential, liquefaction occurs if $F_S \leq 1$ and does not occur if $F_S > 1$. In many situations, it is desirable to express the liquefaction potential in terms of probability of liquefaction ($P_L$) rather than with a factor of safety ($F_S$). Examples of how this expression of liquefaction potential may be used can involve: (1) mapping the liquefaction potential in a district where it is easier to interpret the liquefaction potential in terms of probability rather than factor of safety; (2) post-event investigations where the conservative bias that was typically built into existing deterministic models becomes undesirable, as it may mislead the assessment; and (3) performance-based earthquake engineering, where the unbiased probability at the component level is required. Thus, there is definitely a need for estimating the probability of liquefaction.

Many approaches have been taken to develop simplified probabilistic models for liquefaction potential evaluation; for example, discriminant analysis (Christian and Swiger 1975), logistic regression (Liao et al. 1988; Lai et al. 2006), Bayesian mapping (Juang et al. 1999, 2000, 2002), and the Bayesian regression approach (Cetin et al. 2002; Moss et al. 2006; Boulanger and Idriss 2012). Underlying many of these approaches is the maximum likelihood principle (Edwards 1974; Aldrich 1997; Stigler 2007), which is widely used in statistical estimation. The ability to consistently consider multiple types of data for model calibration makes the maximum likelihood method particularly suitable for developing liquefaction probability models. While the maximum likelihood method is increasingly used by researchers in liquefaction analysis and other fields in geotechnical engineering, there is no easy-to-access publication to elucidate how this method works and how it can be used efficiently. This missing element hinders the wider use of the maximum likelihood method in the geotechnical profession. Hence, the objectives of this chapter are to introduce the maximum likelihood principle with an emphasis on its application in geotechnical engineering, and to exemplify its use in the development of various probabilistic models for liquefaction probability prediction. This chapter thus consists of two parts. First, the principle of maximum likelihood is introduced with a focus on its applications in geotechnical engineering. Several examples are then presented to illustrate the development and application of the maximum likelihood-based models for the evaluation of the liquefaction potential and liquefaction-induced settlement.

## 4.2 PRINCIPLE OF MAXIMUM LIKELIHOOD

Let $\theta$ denote the parameters of an intended model to be estimated or calibrated, which is a vector, and let $\mathbf{D}$ denote the observed data, which is a vector or matrix. Let $f(\mathbf{D}|\theta)$ denote the joint probability density function of $\mathbf{D}$ given $\theta$, or equivalently, the chance to observe $\mathbf{D}$ given $\theta$. When viewed as a function of $\theta$, $f(\mathbf{D}|\theta)$ can be denoted as $l(\theta|\mathbf{D})$, which is known as the likelihood function. The maximum likelihood principle (e.g., Givens and Hoeting 2005) states that the optimal value of $\theta$ can be estimated by maximizing the likelihood function. In other words, a series of $\theta$ values are assumed and the corresponding values of the likelihood function $l(\theta|\mathbf{D})$ are computed. The $\theta$ value yielding the highest likelihood value has the greatest chance for observing the given (known) $\mathbf{D}$, and thus is the optimal value of $\theta$.

A maximization of the likelihood function $l(\theta|\mathbf{D})$ is equivalent to the maximization of the log-likelihood function $L(\theta|\mathbf{D})$, which is defined as $L(\theta|\mathbf{D}) = \ln l(\theta|\mathbf{D})$. In that the maximum of a log-likelihood function is more easily evaluated, the log-likelihood function is often maximized for estimating the parameters (e.g., Givens and Hoeting 2005). Under some general regular conditions and for most distributions used as models in practical applications, the maximum likelihood estimate has the following properties (e.g., Barnett 1999; Gentle 2002):

1. Consistency—As the number of observations increases, the maximum likelihood estimator will approach the true value.
2. Normality—As the number of observations increases, the estimate of $\theta$, denoted as $\theta^*$, tends toward a normal distribution with a mean of $\theta$ and a covariance matrix of $\mathbf{H}^{-1}$ with $\mathbf{H}_{ij} = \mathrm{E}(-\partial^2 L(\theta|\mathbf{D})/\partial\theta_i\partial\theta_j)$. A consistent estimator of $\mathbf{H}$ is the negative Hessian matrix of the log-likelihood function evaluated at $\theta^*$ (Givens and Hoeting 2005).
3. Invariance—If $\theta^*$ is the maximum likelihood estimate of $\theta$, then $\tau^* = g(\theta^*)$ is also the maximum likelihood estimate of $\tau = g(\theta)$.

Interested readers are referred to Lehmann and Casella (1998) for further discussions on the regular conditions where the maximum likelihood estimates exhibit good statistical properties, and to Cam (1990) and Cheng and Traylor (1995) for counterexamples in nonregular conditions. Edwards (1974), Aldrich (1997), and Stigler (2007) provide a review of the historical development of this maximum likelihood method as applicable to statistics.

Two major challenges are often inherent in the application of the maximum likelihood principle: (1) in constructing the likelihood function $l(\theta|\mathbf{d})$ and (2) in solving the maximization problem. In the following sections, we will illustrate how the likelihood function can be constructed for different types of data, and how the maximum likelihood point can be found for several typical geotechnical problems.

## 4.2.1 Independent observations

Let $d_1$, $d_2$, ..., and $d_n$ denote $n$ observed data. Assuming these observations are statistically independent, the likelihood of $\theta$, denoted as $l(\theta|D)$, which is the chance to observe the data $D = \{d_1, d_2, ..., d_n\}$, can be written as follows:

$$l(\theta|D) = \prod_{i=1}^{n} f(d_i|\theta) \qquad (4.1)$$

where $\prod_{i=1}^{n} f(d_i|\theta)$ is the joint probability of observing $D$ given $\theta$, which is the product of $n$ probability terms $[f(d_i|\theta), i = 1, n]$ since $d_1$, $d_2$, ..., and $d_n$ are statistically independent observations. Based on the maximum likelihood principle, the optimal values of $\theta$ can be obtained by maximizing the above likelihood function, or equivalently, the logarithm of the likelihood function is as follows:

$$L(\theta|D) = \sum_{i=1}^{n} \ln f(d_i|\theta) \qquad (4.2)$$

The same optimal values of $\theta$ will be obtained regardless of whether the likelihood function (Equation 4.1) or the logarithm of the likelihood function (Equation 4.2) is maximized. However, maximizing the latter is more efficient computationally.

### EXAMPLE 4.1

Table 4.1 lists the shear strength data of a clay core of a gravity dam. Suppose the friction angle ($\varphi$) follows the normal distribution and we are interested in its mean and standard deviation, which are denoted herein as $\mu$ and $\sigma$, respectively.

Let $\varphi_1$, $\varphi_2$, ..., $\varphi_n$ denote $n$ observed values of the friction angle $\varphi$, respectively. In this example, $d_i = \varphi_i$, and $D = \{d_1, d_2, ..., d_n\} = \{\varphi_1, \varphi_2, ..., \varphi_n\}$. Based on the probability density function of a normal distribution (e.g., Ang and Tang 2007), the chance to observe the data $D$ given $\theta$, which is the likelihood of interest, can be written as follows:

Table 4.1 Test data from the Ankang Hydropower site

| Test no. | c (kPa) | $\varphi$ (°) | Test no. | c (kPa) | $\varphi$ (°) |
|---|---|---|---|---|---|
| 1 | 165 | 11.86 | 14 | 85 | 20.81 |
| 2 | 127 | 14.04 | 15 | 18 | 25.64 |
| 3 | 253 | 13.50 | 16 | 15 | 22.29 |
| 4 | 427 | 10.20 | 17 | 78 | 24.70 |
| 5 | 106 | 11.31 | 18 | 12 | 26.10 |
| 6 | 242 | 12.95 | 19 | 34 | 22.78 |
| 7 | 209 | 12.41 | 20 | 70 | 19.80 |
| 8 | 328 | 13.50 | 21 | 20 | 17.74 |
| 9 | 98 | 12.95 | 22 | 20 | 20.81 |
| 10 | 10 | 15.64 | 23 | 217 | 20.30 |
| 11 | 213 | 16.17 | 24 | 221 | 20.30 |
| 12 | 365 | 17.22 | 25 | 254 | 27.47 |
| 13 | 324 | 20.81 | | | |

Source: Data from Tang, X.S. 2013. *Computers and Geotechnics*, 49, 264–278.

$$l(\theta|\mathbf{D}) = \prod_{i=1}^{n} f(\mathbf{d}_i|\theta) = \prod_{i=1}^{n} \frac{1}{\sqrt{2\pi}\sigma} \exp\left[-\frac{(\mathbf{d}_i - \mu)^2}{2\sigma^2}\right] \qquad (4.3)$$

Substituting Equation 4.3 into Equation 4.2, its log-likelihood can be written as follows:

$$L(\theta|\mathbf{D}) = \sum_{i=1}^{n} \left( \ln\left( \frac{1}{\sqrt{2\pi}\sigma} \right) - \frac{(\mathbf{d}_i - \mu)^2}{2\sigma^2} \right) \qquad (4.4)$$

For this particular problem, the optimal values of $\mu$ and $\sigma$ can be determined analytically by equating the gradients of Equation 4.4 with respect to $\mu$ and $\sigma$, respectively, equal to zero (e.g., Ang and Tang 2007), which yields the following analytical solutions:

$$\mu^* = \frac{1}{n}\sum_{i=1}^{n}\mathbf{d}_i \qquad (4.5)$$

$$\sigma^* = \sqrt{\frac{1}{n}\sum_{i=1}^{n}(\mathbf{d}_i - \mu^*)^2} \qquad (4.6)$$

where $\mu^*$ and $\sigma^*$ are the maximum likelihood estimates for $\mu$ and $\sigma$, respectively. Based on Equations 4.5 and 4.6, the analytical solution given data listed in Table 4.1 yields $\mu^* = 18.05°$ and $\sigma^* = 5.01°$.

In many cases, the analytical solution for maximizing the likelihood function is unattainable, which necessitates a numerical implementation of the maximum likelihood method. Many existing software packages such as MATLAB® and Microsoft Excel™ can be used to perform this numerical optimization. In the examples detailed in this chapter, Excel was used to maximize the likelihood function whenever possible. For example, Figure 4.1 shows an Excel spreadsheet that is configured to maximize the likelihood function using Solver, a built-in optimization tool in Excel. Here, the optimal values of $\mu$ and $\sigma$ are found to be 18.05° and 5.01°, respectively, which are exactly the same as that obtained from the analytical solution.

### EXAMPLE 4.2

Using the same data listed in Table 4.1, we now want to calibrate the parameters for the joint distribution of $c$ and $\varphi$, assuming it is a bivariate normal distribution.

Let $\mu_1$ and $\mu_2$ denote the mean values of $c$ and $\varphi$, respectively, and $\sigma_1$ and $\sigma_2$ denote the standard deviations of $c$ and $\varphi$, respectively. Let $\rho$ denote the correlation coefficient between $c$ and $\varphi$. In the bivariate normal distribution, the uncertain parameters to be calibrated can be denoted as $\theta = \{\mu_1, \sigma_1, \mu_2, \sigma_2, \rho\}$. Let $\mathbf{d}_i = \{d_{i1}, d_{i2}\}$ denote the $i$th observation of $\{c, \varphi\}$ and $\mathbf{D} = \{\mathbf{d}_1, \mathbf{d}_2, ..., \mathbf{d}_n\}$. Based on the probability density function of a bivariate normal distribution (e.g., Ang and Tang 2007), the chance to observe the $i$th data if the values of $\theta$ are known can be written as follows:

$$f(\mathbf{d}_i|\theta) = \frac{1}{2\pi\sigma_1\sigma_2\sqrt{1 - \rho^2}}$$

$$\times \exp\left\{ -\frac{1}{2(1-\rho^2)}\left[ \left( \frac{d_{i1} - \mu_1}{\sigma_1} \right)^2 - 2\rho\left( \frac{d_{i1} - \mu_1}{\sigma_1} \right)\left( \frac{d_{i2} - \mu_2}{\sigma_2} \right) + \left( \frac{d_{i2} - \mu_2}{\sigma_2} \right)^2 \right] \right\} \quad (4.7)$$

| A | B | C | D | E | F | G | H |
|---|---|---|---|---|---|---|---|
| 2 | | **Spreadsheet Template for Calibrating a Normal Distribution** | | | | | |
| 3 | | | | | | | |
| 4 | Parameters to | | | | | | |
| 5 | be estimated | | $\mu$ | $\sigma$ | | | |
| 6 | | | 18.052 | 5.014 | | | |
| 7 | | | | Oberved data | Eq. (4.3) | Log-likelihood | |
| 8 | | Test No. | $\phi\,(°)$ | $f(\mathbf{d}_i|\theta)$ | $\ln[f(\mathbf{d}_i|\theta)]$ | $L\,(\theta|\mathbf{D})$ | |
| 9 | | 1 | 11.86 | 0.037117612 | -3.293663704 | -75.78058016 | |
| 10 | | 2 | 14.04 | 0.057769078 | -2.851301635 | | |
| 11 | | 3 | 13.5 | 0.052693349 | -2.943266032 | | |
| 12 | | 4 | 10.2 | 0.023347321 | -3.757273035 | | |
| 13 | | 5 | 11.31 | 0.032221271 | -3.435128467 | | |
| 14 | | 6 | 12.95 | 0.04741316 | -3.04885546 | | |
| 15 | | 7 | 12.41 | 0.042246646 | -3.164230304 | | |
| 16 | | 8 | 13.5 | 0.052693349 | -2.943266032 | | |
| 17 | | 9 | 12.95 | 0.04741316 | -3.04885546 | | |
| 18 | | 10 | 15.64 | 0.070870141 | -2.646906075 | | |
| 19 | | 11 | 16.17 | 0.07415113 | -2.601649971 | | |
| 20 | | 12 | 17.22 | 0.078474352 | -2.544983438 | | |
| 21 | | 13 | 20.81 | 0.068391957 | -2.682500042 | | |
| 22 | | 14 | 20.81 | 0.068391957 | -2.682500042 | | |
| 23 | | 15 | 25.64 | 0.025317143 | -3.676273531 | | |
| 24 | | 16 | 22.29 | 0.055664491 | -2.888412834 | | |
| 25 | | 17 | 24.7 | 0.033036205 | -3.410151183 | | |
| 26 | | 18 | 26.1 | 0.021942924 | -3.819310554 | | |
| 27 | | 19 | 22.78 | 0.051007472 | -2.97578315 | | |
| 28 | | 20 | 19.8 | 0.074870812 | -2.591991157 | | |
| 29 | | 21 | 17.74 | 0.079408061 | -2.53315539 | | |
| 30 | | 22 | 20.81 | 0.068391957 | -2.682500042 | | |
| 31 | | 23 | 20.3 | 0.071954129 | -2.631726466 | | |
| 32 | | 24 | 20.3 | 0.071954129 | -2.631726466 | | |
| 33 | | 25 | 27.47 | 0.013634258 | -4.29516969 | | |
| 34 | | | | | | | |
| 35 | Notes: | | | | | | |
| 36 | (1) The setting in Solver is "Maximize Cell G9 by changing the values in Cells D6 and E6". | | | | | | |
| 37 | | | | | | | |

*Figure 4.1* Spreadsheet template for calibrating parameters of a normal distribution.

Assume that the observed data $\mathbf{d}_1, \mathbf{d}_2, \ldots,$ and $\mathbf{d}_n$ are statistically independent. The likelihood and log-likelihood functions in this problem can be formulated by substituting Equation 4.7 into Equations 4.1 and 4.2, respectively. By maximizing the likelihood, the optimal values of $\theta$ can then be determined. Figure 4.2 shows a spreadsheet template for estimating the values of $\theta$ based on the maximum likelihood principle. With Excel Solver, the maximum likelihood estimate of $\theta$ is found to be $\{\mu_1, \sigma_1, \mu_2, \sigma_2, \rho\} = \{156.44$ kPa, $122.23$ kPa, $18.05°, 5.01°, -0.41\}$.

| | | | | | | | | | | |
|---|---|---|---|---|---|---|---|---|---|---|

**Spreadsheet Template for Calibrating a Bivarate Normal Distribution**

Parameters to be estimated $\rightarrow$

| $\mu_1$ | $\sigma_1$ | $\mu_2$ | $\sigma_2$ | $\rho$ |
|---|---|---|---|---|
| 156.44 | 122.229 | 18.052 | 5.01425 | -0.4068 |

Log-likelihood
Eq. (4.7)

Observed data

| Test No. | $c$ (kPa) | $\phi$ (°) | A | B | C | D | $f(\mathbf{d}_i\mid\theta)$ | $\ln[f(\mathbf{d}_i\mid\theta)]$ | $L(\theta\mid\mathbf{D})$ |
|---|---|---|---|---|---|---|---|---|---|
| 1 | 165 | 11.86 | 0.0049 | 1.52492 | -0.0704 | 0.87448 | 0.000119 | -9.04006 | -229.14 |
| 2 | 127 | 14.04 | 0.05801 | 0.64019 | 0.15681 | 0.5123 | 0.00017 | -8.67789 | |
| 3 | 253 | 13.5 | 0.62409 | 0.82412 | -0.5836 | 0.51808 | 0.000169 | -8.68366 | |
| 4 | 427 | 10.2 | 4.89981 | 2.45212 | -2.8205 | 2.71516 | 1.88E-05 | -10.8807 | |
| 5 | 106 | 11.31 | 0.17029 | 1.80785 | 0.45149 | 1.45579 | 6.63E-05 | -9.62137 | |
| 6 | 242 | 12.95 | 0.49 | 1.0353 | -0.5796 | 0.56667 | 0.000161 | -8.73225 | |
| 7 | 209 | 12.41 | 0.18491 | 1.26605 | -0.3937 | 0.63349 | 0.000151 | -8.79907 | |
| 8 | 328 | 13.5 | 1.97008 | 0.82412 | -1.0368 | 1.05299 | 9.92E-05 | -9.21857 | |
| 9 | 98 | 12.95 | 0.2286 | 1.0353 | 0.39585 | 0.99449 | 0.000105 | -9.16007 | |
| 10 | 10 | 15.64 | 1.4354 | 0.23139 | 0.46894 | 1.27968 | 7.91E-05 | -9.44526 | |
| 11 | 213 | 16.17 | 0.21413 | 0.14087 | -0.1413 | 0.12803 | 0.00025 | -8.29361 | |
| 12 | 365 | 17.22 | 2.91148 | 0.02753 | -0.2304 | 1.62296 | 5.61E-05 | -9.78854 | |
| 13 | 324 | 20.81 | 1.87929 | 0.30254 | 0.61355 | 1.67493 | 5.33E-05 | -9.84051 | |
| 14 | 85 | 20.81 | 0.34161 | 0.30254 | -0.2616 | 0.22922 | 0.000226 | -8.3948 | |
| 15 | 18 | 25.64 | 1.28285 | 2.29004 | -1.3947 | 1.30515 | 7.71E-05 | -9.47073 | |
| 16 | 15 | 22.29 | 1.33905 | 0.71435 | -0.7958 | 0.75352 | 0.000134 | -8.9191 | |
| 17 | 78 | 24.7 | 0.41184 | 1.75781 | -0.6923 | 0.88518 | 0.000117 | -9.05076 | |
| 18 | 12 | 26.1 | 1.39646 | 2.57611 | -1.5433 | 1.45555 | 6.63E-05 | -9.62113 | |
| 19 | 34 | 22.78 | 1.00346 | 0.88909 | -0.7686 | 0.67346 | 0.000145 | -8.83904 | |
| 20 | 70 | 19.8 | 0.50013 | 0.12153 | -0.2006 | 0.25229 | 0.000221 | -8.41787 | |
| 21 | 20 | 17.74 | 1.24605 | 0.00387 | 0.05651 | 0.78279 | 0.00013 | -8.94837 | |
| 22 | 20 | 20.81 | 1.24605 | 0.30254 | -0.4996 | 0.62853 | 0.000152 | -8.79411 | |
| 23 | 217 | 20.3 | 0.24548 | 0.201 | 0.18075 | 0.37582 | 0.000195 | -8.5414 | |
| 24 | 221 | 20.3 | 0.27898 | 0.201 | 0.19268 | 0.40305 | 0.00019 | -8.56863 | |
| 25 | 254 | 27.47 | 0.63708 | 3.52782 | 1.21987 | 3.22644 | 1.13E-05 | -11.392 | |

Notes:

(1) To faciliate spreadsheet implementation, four intermedidate variables are defined as follows:

$$A = \left(\frac{d_{i1} - \mu_1}{\sigma_1}\right)^2, \ B = \left(\frac{d_{i2} - \mu_2}{\sigma_2}\right)^2, \ C = -2\rho\left(\frac{d_{i1} - \mu_1}{\sigma_1}\right)\left(\frac{d_{i2} - \mu_2}{\sigma_2}\right), \text{ and } D = \frac{A + B + C}{2(1 - \rho^2)}$$

(2) The setting in Solver is "Maximize Cell L9 by changing the values in Cells E6, F6, G6, H6, and I6."

Figure 4.2 Spreadsheet template for calibrating the bivariate normal distribution.

## 4.2.2 Correlated observations

The maximum likelihood method is also applicable to correlated observations, as illustrated in the following example.

### EXAMPLE 4.3

Due to the existence of spatial variability, the properties of soil samples taken from the same soil layer are not necessarily the same but are usually correlated. Consider a hypothetical saturated clay soil as shown in Figure 4.3. Suppose the undrained shear strength of the clay, $c_u$, is a realization of a stationary lognormal random field with a mean $\mu$, a standard deviation $\sigma$, and an exponential correlation function defined as follows (Vanmarcke 1983):

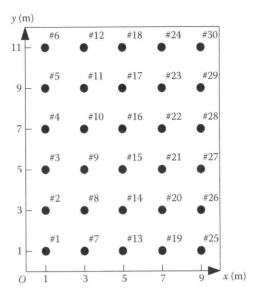

*Figure 4.3* Location of the soil samples.

$$\rho(\Delta) = \exp\left(-\frac{2\Delta}{\Omega}\right) \qquad (4.8)$$

where $\Omega$ is the correlation distance and $\Delta$ is the distance between two points in the random field.

In a stationary lognormal random field, the mean ($\mu$), standard deviation ($\sigma$), and the correlation distance ($\Omega$) are the three parameters to be calibrated, that is, $\theta = \{\mu, \sigma, \Omega\}$. Suppose 30 samples ($n = 30$) are taken from the ground with locations shown in Figure 4.3. The coordinates of the locations of the soil samples as well as the undrained shear strengths ($c_u$) measured at these locations are shown in Table 4.2. For convenience of presenting the likelihood function later, let $\lambda = \ln\{\mu/\sqrt{1 + (\sigma/\mu)^2}\}$ and $\xi = \sqrt{\ln[1 + (\sigma/\mu)^2]}$, where $\lambda$ and $\xi$ are the mean and standard deviation of $\ln(c_u)$, respectively.

Let $c_{ui}$ denote the $i$th measured value of $c_u$. In this example, $\mathbf{d}_i = \{c_{ui}\}$, and the observed data can be denoted as $\mathbf{D} = \{\mathbf{d}_1, \mathbf{d}_2, \dots, \mathbf{d}_n\} = \{c_{u1}, c_{u2}, \dots, c_{un}\}$. Let $\Lambda$ be an $n$-dimensional column vector with all elements being $\lambda$ and $\mathbf{T}$ be an $n \times n$ correlation matrix with $\mathbf{T}_{ij}$ being the correlation coefficient between $\ln(c_u)$ at location $i$ and location $j$. Based on the assumption of a lognormal random field, it can be shown that $\ln(\mathbf{D})$ follows the multivariate normal distribution with a mean of $\Lambda$ and a covariance matrix of $\xi^2\mathbf{T}$ (e.g., Fenton 1999). Based on the probability density function of a multivariate normal distribution, the chance to observe $\mathbf{D}$ given $\theta$, or the likelihood function of $\theta$, can be written as (e.g., Fenton 1999):

$$l(\theta|\mathbf{D}) = \frac{1}{(2\pi\xi^2)^{n/2}|\mathbf{T}|^{1/2}} \exp\left[-\frac{(\ln\mathbf{D} - \Lambda)^T \mathbf{T}^{-1}(\ln\mathbf{D} - \Lambda)}{2\xi^2}\right] \qquad (4.9)$$

where $\ln\mathbf{D}$ is a vector with the $i$th element being the logarithm of the $i$th element of $\mathbf{D}$.

*Table 4.2* Coordinates and measured values of $s_u$ of the 30 samples

| Sample no. | X (m) | Y (m) | $c_u$ (kPa) | Sample no. | X (m) | Y (m) | $c_u$ (kPa) |
|---|---|---|---|---|---|---|---|
| 1 | 1.0 | 1.0 | 24.6 | 16 | 5.0 | 7.0 | 19.8 |
| 2 | 1.0 | 3.0 | 23.4 | 17 | 5.0 | 9.0 | 22.7 |
| 3 | 1.0 | 5.0 | 31.0 | 18 | 5.0 | 11.0 | 32.3 |
| 4 | 1.0 | 7.0 | 25.9 | 19 | 7.0 | 1.0 | 31.4 |
| 5 | 1.0 | 9.0 | 22.9 | 20 | 7.0 | 3.0 | 36.5 |
| 6 | 1.0 | 11.0 | 27.1 | 21 | 7.0 | 5.0 | 32.4 |
| 7 | 3.0 | 1.0 | 24.1 | 22 | 7.0 | 7.0 | 35.8 |
| 8 | 3.0 | 3.0 | 32.8 | 23 | 7.0 | 9.0 | 19.7 |
| 9 | 3.0 | 5.0 | 46.4 | 24 | 7.0 | 11.0 | 26.5 |
| 10 | 3.0 | 7.0 | 21.8 | 25 | 9.0 | 1.0 | 27.9 |
| 11 | 3.0 | 9.0 | 32.1 | 26 | 9.0 | 3.0 | 29.8 |
| 12 | 3.0 | 11.0 | 31.8 | 27 | 9.0 | 5.0 | 37.2 |
| 13 | 5.0 | 1.0 | 24.5 | 28 | 9.0 | 7.0 | 26.8 |
| 14 | 5.0 | 3.0 | 35.6 | 29 | 9.0 | 9.0 | 29.2 |
| 15 | 5.0 | 5.0 | 26.6 | 30 | 9.0 | 11.0 | 19.8 |

The correlation coefficient between $\ln(c_{ui})$ and $\ln(c_{uj})$ can be calculated based on the correlation coefficient between $c_{ui}$ and $c_{uj}$ analytically (e.g., Law and Kelton 2000). In general, these two correlation coefficients are not exactly the same. For simplicity, however, it is assumed in this example that the two correlation coefficients are equal to each other.

The optimal values of $\theta$ are determined by maximizing the likelihood function in Equation 4.9, or equivalently, its log-likelihood function. When dealing with the spatial variability model as shown above, the log-likelihood function is often quite flat. This makes it difficult to maximize the likelihood with respect to $\mu$, $\sigma$, and $\Omega$ simultaneously. To this end, Fenton (1999) suggested that $\theta$ can be estimated using the following procedure:

1. Let $\Omega_i$ ($i = 1, 2, ..., m$) denote $m$ possible values of $\Omega$.
2. For $i = 1, 2, ..., m$ estimate the maximum likelihood of $\mu$ and $\sigma$ for the case of $\Omega = \Omega_i$. Denote the maximum likelihood estimates as $\mu_i$ and $\sigma_i$. Thus, $\{\mu_i, \sigma_i, \Omega_i\}$ is a local maximum likelihood point. Repeat this estimate for all $m$ possible values of $\Omega$.
3. Among the $m$ local maximum likelihood points, the one with the highest maximum likelihood value is the global maximum likelihood point.

The MATLAB code used to implement this concept is provided in Figure 4.4. As an example, the range of $\Omega$ is taken as [0.5 m, 10.0 m], and $\Omega$ values are assumed uniformly distributed within this interval with a spacing of 0.1 m. Based on the MATLAB code shown in Figure 4.4, the optimal values of $\mu$, $\sigma$, and $\Omega$ are 28.48 kPa, 5.93 kPa, and 1.60 m, respectively.

## 4.2.3 Censored observations

The observation occasionally presents in such a way that its quantity is larger or smaller than a threshold but the exact value is unknown. For instance, in a pile load test, a pile may survive the prescribed applied load. In such a case, the capacity of the pile is only known to be greater than the applied load. Such observations are known as censored data

```
function [mz,sz,Lz]=MLE(X,Y,D)
% X - vector of x coordinates of the locations of the test data
% Y - vector of y coordinates of the locations of the test data
% D - vector of measured data at {X, Y}
L1=0.5;
L2=10;
L=L1:0.1:L2;
NL=length(L); %search of grid of SOF
ND=length(D); %number of observations
lnD=log(D); %logarithm transformation of the data
U=ones(ND,1);
for i=1:NL
  for j=1:ND
    for k=1:ND
      dx=abs(X(j,1)-X(k,1));
      dy=abs(Y(j,1)-Y(k,1));
      djk=(dx^2+dy^2)^0.5;
      R(j,k)=exp(-2*djk/L(i)); %Eq. (4.8)
    end
  end
  r=inv(R)*lnD; % Eq.(33a) in Fenton (1999)
  s=inv(R)*U; % Eq.(33b) in Fenton (1999)
  ev=eig(R);
  mlnD=(transpose(U)*r)/(transpose(U)*s); %Eq.(34) in Fenton (1999)
  vlnD=transpose(lnD-mlnD*U)*r/ND; % Eq.(35) in Fenton (1999)
  like=-0.5*ND*log(vlnD)-0.5*sum(log(ev)); %Eq.(36) in Fenton (1999)
  MLERes(i,:)=[mlnD,vlnD,like,L(i)];
end
[MLEm,Ind]=max(MLERes(:,3));% get the global maximum likelihood
Lz=MLERes(Ind,4);
mlnz=MLERes(Ind,1);
slnz=sqrt(MLERes(Ind,2)); % std of lnZ
covz=sqrt(exp(slnz^2)-1); %COV of Z
mz=exp(mlnz+0.5*slnz^2); % Mean of Z
sz=mz*covz; %std of Z
end
```

*Figure 4.4* MATLAB® code for calibrating a stationary random field.

in the statistics. For example, a post-earthquake investigation at a given site often yields a simple observation of whether or not a soil has liquefied, while the actual factor of safety $F_S$ remains unknown. Hence, data censoring is commonly used in the development of liquefaction potential assessment models. One advantage of the maximum likelihood method over the method of moments in the model calibration is that it can deal with censored data effectively. In the following, we will use a slope example to illustrate model calibration with the censored data based on the maximum likelihood method.

## EXAMPLE 4.4

For simplification, assumptions are often made in a slope reliability analysis and thus, the calculated reliability index of a slope is not the actual reliability index (Zhang et al. 2007). When the performance of a large number of similar slopes in a region is known, it is possible to calibrate the error in the calculated reliability index. Let $\beta_c$ and $\beta_a$ denote the calculated and actual reliability indexes, respectively. Suppose the relationship between $\beta_c$ and $\beta_a$ can be linked with a correction factor $\varepsilon_\beta$ as follows:

$$\beta_a = \beta_c + \varepsilon_\beta$$

(4.10)

Although other forms of relationships between $\beta_a$ and $\beta_c$ may also be assumed, the simple additive relationship is considered here as an example for the illustration of model calibration with censored data. As will be shown later, effectiveness of different assumptions can be ranked using the Bayesian information criterion (Zhang et al. 2007) as needed. To calibrate the correction factor $\varepsilon_\beta$, suppose we have collected data of $n$ slopes (including $n_F$ failed slopes and $n_{NF}$ non-failed slopes; thus, $n = n_F + n_{NF}$) and calculated the reliability index for each of these $n$ slopes. As an example, Figure 4.5 shows the reliability indexes of 25 slopes. Let $I$ be an indicator variable with $I = 1$ denoting slope failure and $I = 0$ denoting no failure.

Let $I_i$ denote the indicator variable of the $i$th slope, that is, $\mathbf{d}_i = \{I_i\}$, and $\mathbf{D} = \{\mathbf{d}_1, \mathbf{d}_2, ..., \mathbf{d}_n\} = \{I_1, I_2, ..., I_n\}$ denote the calibration database. Assume the model parameter to be estimated is $\theta = \{\varepsilon_\beta\}$. For the $i$th slope, the probability to observe it as a failed slope if the value of $\varepsilon_\beta$ is known can be written as

$$P(\mathbf{d}_i = 1|\theta) = P(I_i = 1|\theta) = 1 - \Phi(\beta_{ci} + \varepsilon_\beta)$$

(4.11)

For the $j$th slope, the probability of observing it as a stable slope, if the value of $\varepsilon_\beta$ is known, can be written as

$$P(\mathbf{d}_j = 0|\theta) = P(I_j = 0|\theta) = \Phi(\beta_{cj} + \varepsilon_\beta)$$

(4.12)

Assuming the failure of slopes is statistically independent, the chance to observe the $n_F$ failed slopes and $n_{NF}$ stable slopes is then the product of the probability to observe the performance of each slope, which can be calculated based on Equations 4.11 and 4.12 for a failed and nonfailed slope, respectively. Hence, the chance to observe $\mathbf{D}$ given $\theta = \{\varepsilon_\beta\}$ or the likelihood function can be written as follows:

$$l(\theta|\mathbf{D}) = \prod_{i=1}^{n_F} \left[1 - \Phi(\beta_{ci} + \varepsilon_\beta)\right] \prod_{j=1}^{n_{NF}} \Phi(\beta_{cj} + \varepsilon_\beta)$$

(4.13)

where $\Phi$ is the cumulative density function of a standard normal variable. The log-likelihood function can then be written as

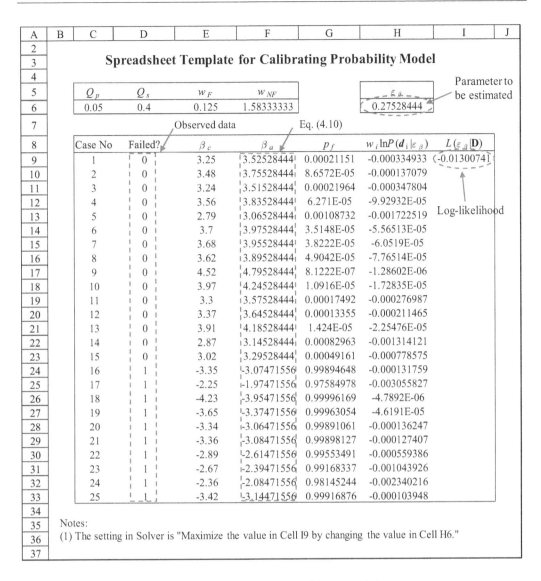

| A | B | C | D | E | F | G | H | I | J |
|---|---|---|---|---|---|---|---|---|---|
| 2 | | | | | | | | | |
| 3 | | **Spreadsheet Template for Calibrating Probability Model** | | | | | | | |
| 4 | | | | | | | | Parameter to | |
| 5 | | $Q_p$ | $Q_s$ | $w_F$ | $w_{NF}$ | | $\varepsilon_\beta$ | be estimated | |
| 6 | | 0.05 | 0.4 | 0.125 | 1.58333333 | | 0.27528444 | | |
| 7 | | | | Observed data | | Eq. (4.10) | | | |
| 8 | | Case No | Failed? | $\beta_c$ | $\beta_a$ | $p_f$ | $w_i \ln P(d_i \mid \varepsilon_\beta)$ | $L(\varepsilon_\beta \mid \mathbf{D})$ | |
| 9 | | 1 | 0 | 3.25 | 3.52528444 | 0.00021151 | -0.000334933 | -0.01300741 | |
| 10 | | 2 | 0 | 3.48 | 3.75528444 | 8.6572E-05 | -0.000137079 | | |
| 11 | | 3 | 0 | 3.24 | 3.51528444 | 0.00021964 | -0.000347804 | | |
| 12 | | 4 | 0 | 3.56 | 3.83528444 | 6.271E-05 | -9.92932E-05 | | |
| 13 | | 5 | 0 | 2.79 | 3.06528444 | 0.00108732 | -0.001722519 | Log-likelihood | |
| 14 | | 6 | 0 | 3.7 | 3.97528444 | 3.5148E-05 | -5.56513E-05 | | |
| 15 | | 7 | 0 | 3.68 | 3.95528444 | 3.8222E-05 | -6.0519E-05 | | |
| 16 | | 8 | 0 | 3.62 | 3.89528444 | 4.9042E-05 | -7.76514E-05 | | |
| 17 | | 9 | 0 | 4.52 | 4.79528444 | 8.1222E-07 | -1.28602E-06 | | |
| 18 | | 10 | 0 | 3.97 | 4.24528444 | 1.0916E-05 | -1.72835E-05 | | |
| 19 | | 11 | 0 | 3.3 | 3.57528444 | 0.00017492 | -0.000276987 | | |
| 20 | | 12 | 0 | 3.37 | 3.64528444 | 0.00013355 | -0.000211465 | | |
| 21 | | 13 | 0 | 3.91 | 4.18528444 | 1.424E-05 | -2.25476E-05 | | |
| 22 | | 14 | 0 | 2.87 | 3.14528444 | 0.00082963 | -0.001314121 | | |
| 23 | | 15 | 0 | 3.02 | 3.29528444 | 0.00049161 | -0.000778575 | | |
| 24 | | 16 | 1 | -3.35 | -3.07471556 | 0.99894648 | -0.000131759 | | |
| 25 | | 17 | 1 | -2.25 | -1.97471556 | 0.97584978 | -0.003055827 | | |
| 26 | | 18 | 1 | -4.23 | -3.95471556 | 0.99996169 | -4.7892E-06 | | |
| 27 | | 19 | 1 | -3.65 | -3.37471556 | 0.99963054 | -4.6191E-05 | | |
| 28 | | 20 | 1 | -3.34 | -3.06471556 | 0.99891061 | -0.000136247 | | |
| 29 | | 21 | 1 | -3.36 | -3.08471556 | 0.99898127 | -0.000127407 | | |
| 30 | | 22 | 1 | -2.89 | -2.61471556 | 0.99553491 | -0.000559386 | | |
| 31 | | 23 | 1 | -2.67 | -2.39471556 | 0.99168337 | -0.001043926 | | |
| 32 | | 24 | 1 | -2.36 | -2.08471556 | 0.98145244 | -0.002340216 | | |
| 33 | | 25 | 1 | -3.42 | -3.14471556 | 0.99916876 | -0.000103948 | | |
| 34 | | | | | | | | | |
| 35 | | Notes: | | | | | | | |
| 36 | | (1) The setting in Solver is "Maximize the value in Cell I9 by changing the value in Cell H6." | | | | | | | |
| 37 | | | | | | | | | |

*Figure 4.5* Spreadsheet template for model calibration using 25 censored data inputs.

$$L(\theta|\mathbf{D}) = \sum_{i=1}^{n_F} \ln\left[1 - \Phi(\beta_{ci} + \varepsilon_\beta)\right] + \sum_{j=1}^{n_{NF}} \ln \Phi(\beta_{cj} + \varepsilon_\beta) \qquad (4.14)$$

Maximizing Equation 4.14 with respect to $\varepsilon_\beta$ using data shown in Figure 4.5, the optimal value of $\varepsilon_\beta$ is −0.181. A negative value of $\varepsilon_\beta$ indicates that the slope reliability is overestimated. Thus, the procedure for calculating the slope reliability index is, on average, on the nonconservative side.

Adhering to the rule of random sampling (i.e., randomly selecting samples from the population) is important when applying statistical methods. For the slope example illustrated above, the random sampling criterion implies that the ratio of failed slopes to stable slopes in the calibration database should be as close to the true ratio in the region of interest as possible. The random sampling requirement can be compromised during the data collection process in practice, however. For example, in that it is possible to place more emphasis on the failed than stable slopes in a postevent investigation, the proportion of failed cases is larger than actual real-world failures in the calibration database. In such a case, the effect of the failed slopes on model calibration will be overrepresented in that database. Based on Equation 4.14, the average effect of a failed slope is $(1/n_F)\sum_{i=1}^{n_F}\ln[1 - \Phi(\beta_{ci} + \varepsilon_\beta)]$, and the average effect of a stable slope is $(1/n_{NF})\sum_{j=1}^{n_{NF}}\ln\Phi(\beta_{cj} + \varepsilon_\beta)$. If there was no sampling bias in the calibration database, the number of failed and stable slopes should have been $(n_F + n_{NF})Q_p$ and $(n_F + n_{NF})(1 - Q_p)$, respectively, where $Q_p$ is the true proportion of failed slopes. Thus, to remove this sampling bias, the contribution of likelihood regarding failed slopes and stable slopes should be adjusted to $[(n_F + n_{NF})Q_p/n_F]\sum_{i=1}^{n_F}\ln[1 - \Phi(\beta_{ci} + \varepsilon_\beta)]$ and $[(n_F + n_{NF})(1 - Q_p)/n_{NF}]\sum_{j=1}^{n_{NF}}\ln\Phi(\beta_{cj} + \varepsilon_\beta)$, respectively. Therefore, the likelihood function of an unbiased calibration database with a sample size of $n_F + n_{NF}$ is estimated as

$$L(\theta|D) = w_F \sum_{i=1}^{n_F} \ln\left[1 - \Phi(\beta_{ci} + \varepsilon_\beta)\right] + w_{NF} \sum_{j=1}^{n_{NF}} \ln\Phi(\beta_{cj} + \varepsilon_\beta) \tag{4.15}$$

$$w_F = \left[(n_F + n_{NF})Q_p/n_F\right] = \frac{Q_p}{Q_s} \tag{4.16}$$

$$w_{NF} = (n_F + n_{NF})(1 - Q_p)/n_{NF} = \frac{1 - Q_p}{1 - Q_s} \tag{4.17}$$

where $Q_s$ is the proportion of failed slopes in the database (i.e., $Q_s = (n_F/(n_F + n_{NF})) = (n_F/n)$). Equation 4.15 is often called the weighted likelihood function. The reasonableness of the above equations can be checked with the special case of $Q_s = Q_p$, where the problem of choice-based sampling bias vanishes, as $w_F = w_{NF} = 1$, and Equation 4.15 is reversed to the likelihood function of Equation 4.14.

Originally proposed by Manski and Lerman (1977) for use in social science research, this weighed likelihood method was later used in liquefaction probability analysis conducted by Cetin et al. (2002). Zhang et al. (2007) also used such a weighted likelihood method for calibrating several models for a reliability analysis of soil slopes in Hong Kong.

**EXAMPLE 4.5**

To illustrate this weighted likelihood method, assume that the problem described in Example 4.4 occurs in a region where the proportion of failed slopes is 5% ($Q_p = 0.05$). In the database shown in Figure 4.5, $Q_s = 0.4$, and based on Equations 4.16 and 4.17, $w_F = 0.125$, and $w_{NF} = 1.583$. The spreadsheet template for implementing the weighted likelihood method is shown in Figure 4.5. Maximizing the weighted log-likelihood function with respect to $\varepsilon_\beta$, the best estimate of $\varepsilon_\beta$ in this example is 0.275. A positive value of $\varepsilon_\beta$ indicates that the calculated reliability index underestimates the actual reliability index. As the conventional un-weighted likelihood function is only a special case of the weighted likelihood function with $Q_s = Q_p$, the spreadsheet template in Figure 4.5 can also be used to maximize the conventional un-weighted likelihood function.

## 4.2.4 Ranking of competing models

In reality, it is possible that several competing probabilistic models can be developed for the same problem. When it becomes difficult to justify which model is the best based on geotechnical knowledge, the Bayesian Information Criterion (BIC) (Schwarz 1978) is often used for comparing and ranking these models. Suppose there are $r$ models under investigation, which are denoted as $M_1$, $M_2$, ..., and $M_r$, respectively. The statistical BIC of the $i$th model $M_i$ can be calculated as follows (Schwarz 1978):

$$\text{BIC}_i = -2L(\theta^*|M_i, D) + k \ln n \tag{4.18}$$

where $\theta^*$ is the point where the likelihood function is maximized, $L(\theta^*|M_i, D)$ is the value of log-likelihood function of model $M_i$ at point $\theta^*$, $k$ is the number of parameters of the model, and $n$ is the number of measurements. In Equation 4.18, the first term accounts for the effect of fitting the model to the data and the second term for the complexity of the model. If the model fits the data better, the likelihood function will increase, and the first term on the right side of Equation 4.18 will decrease. Conversely, it is possible to raise the model fit by adding more parameters to the model, which will cause an increase in the second term of the BIC, however. The BIC value will decrease if the reduction of the value of the first term in Equation 4.18 overcomes the increase of the value of the second term in Equation 4.18, and vice versa. The smaller the BIC is, the stronger a model is supported by the data. The BIC can also be used to define a more intuitive quantity for model comparison, that is, the probability that a model is true if the true model is among the candidate models under consideration. The model probability of model $M_i$ given data $D$, $P(M_i|D)$, can be computed as follows (e.g., Burnham and Anderson 2004):

$$P(M_i|D) = \frac{\exp\left[-\Delta_i(\text{BIC})/2\right]}{\sum_{j=1}^{r} \exp\left[-\Delta_j(\text{BIC})/2\right]} \tag{4.19}$$

where

$$\Delta_i(\text{BIC}) = \text{BIC}_i - \min_{j=1,2,\ldots,r}\{\text{BIC}_j\} \tag{4.20}$$

If the true model is not among the candidate models, then $P(M_i|D)$ may be interpreted as the chance that model $M_i$ is the best among the candidate models considered.

### EXAMPLE 4.6

In Example 4.1, the friction angle is assumed to follow the normal distribution. Alternatively, the friction angle may be assumed to follow the lognormal distribution. As shown in Figure 4.1, the maximum value of the log-likelihood function for the normal distribution model is $-75.781$. In the normal distribution model, there are two uncertain parameters, that is, $k = 2$. The number of observations is $n = 25$. Hence, its BIC value is 158.000. Similarly, it can be shown that for the lognormal distribution model, the maximum value of the log-likelihood function is $-75.474$, the number of parameter is $k = 2$, and the number of observations is $n = 25$. Hence, its BIC value is 157.384. Based on Equations 4.19 and 4.20, the probability of the normal and lognormal distribution models as the best fit to the given data $D$ are 0.424 and 0.576, respectively. In this case, the lognormal assumption is more strongly supported by the data.

## 4.2.5 Limitations of the maximum likelihood method

Despite its versatility in a wide range of applications, it is important to recognize that the maximum likelihood method has several potential limitations. First, when the sample size is small, the estimate from the maximum likelihood method could be severely biased. Second, in some problems, the likelihood function may not have a maximum likelihood point. For example, in the back analysis of the shear strength parameters based on the slope failure information there are numerous combinations of cohesion and friction angles that yield the same maximum likelihood value and, thus, a distinct peak of the likelihood function cannot be found. In such a case, the Bayesian method with prior information may help regulate the problem to determine a unique peak for the posterior density function (e.g., Zhang et al. 2010). The likelihood function is an important element in such a Bayesian formulation.

Finally, it is interesting to recall some of the arguments presented by Cam (1990) regarding the principle of maximum likelihood. For example, it is not axiomatic that any principle, including the maximum likelihood principle, can always provide a sensible statistical estimation of model parameters. It is also important to be suspicious of a statistical estimation if the results change abruptly through the suppression of a single observation. Finally, it is often helpful to first try a crude but reliable procedure to locate the general ranges within which the parameters fall and then refine that estimate using a more elegant method such as the maximum likelihood principle.

In the remaining sections of this chapter, we will illustrate in detail three application examples of the maximum likelihood principle for assessing liquefaction hazards. The first example covers regression analysis of liquefaction cases for developing models for predicting liquefaction probability. The second example illustrates how to create a probabilistic version of a deterministic model for assessing liquefaction potential. The third example demonstrates how to create a probabilistic model for predicting liquefaction-induced settlement. The key in these applications lie in how the likelihood function is defined and developed appropriately. To ensure practicality, the maximization of the likelihood function in each of these examples is based on Excel Solver.

## 4.3 LIQUEFACTION PROBABILITY BASED ON GENERALIZED LINEAR REGRESSION

The logistic regression method is often used to construct probabilistic models for assessing liquefaction potential (e.g., Liao et al. 1988; Lai et al. 2006). The basic idea behind the logistic regression method is to view the soil subjected to seismic loading as a binary system: it will either liquefy or not liquefy. This idea can be extended to generalized linear models (e.g., Hoffmann 2004), which are a class of statistical theory models that are specifically used for the analysis of binary systems. In fact, the logistic regression is one of these generalized linear models (Table 4.3). While the logistic regression model has been frequently used in liquefaction analysis, the choice of this model instead of other generalized linear models seems quite arbitrary. To find the best regression model for liquefaction hazard assessment, these four generalized linear models in Table 4.3 will be evaluated and compared in the sections that follow.

### 4.3.1 Predicting liquefaction probability based on generalized linear models

Without loss of generality, let $x_1, x_2, ..., x_r$ denote explanatory variables in a regression model. In the logistic regression model, the liquefaction potential is measured by the probability of

*Table 4.3* Four commonly used generalized linear models

| Model | Link function | Equation for calculating $P_L$ |
|---|---|---|
| Logistic | $h(P_L) = \ln\left(\dfrac{P_L}{1 - P_L}\right)$ | $P_L = \dfrac{1}{1 + \exp\left\{-(b_0 + b_1 x_1 + b_2 x_2 + \cdots + b_r x_r)\right\}}$ |
| Probit | $h(P_L) = \Phi^{-1}(P_L)$ | $P_L = \Phi(b_0 + b_1 x_1 + b_2 x_2 + \cdots + b_r x_r)$ |
| Log-log | $h(P_L) = -\ln\left[-\ln(P_L)\right]$ | $P_L = \exp\left\{-\exp\left[-(b_0 + b_1 x_1 + b_2 x_2 + \cdots + b_r x_r)\right]\right\}$ |
| C-log-log | $h(P_L) = \ln\left[-\ln(1 - P_L)\right]$ | $P_L = 1 - \exp\left[-\exp(b_0 + b_1 x_1 + b_2 x_2 + \cdots + b_r x_r)\right]$ |

liquefaction, $P_L$, under the assumption that $\ln[P_L/(1 - P_L)]$ is a linear function of explanatory variables (e.g., Liao et al. 1988). Extending this idea, let $h(P_L)$ denote a function of $P_L$ and assume $h(P_L)$ is a linear function of the explanatorily variables:

$$h(P_L) = b_0 + b_1 x_1 + b_2 x_2 + \cdots + b_r x_r \qquad (4.21)$$

where $b_0$, $b_1$, ..., $b_r$ are the regression coefficients to be calibrated. Based on Equation 4.21, $P_L$ can be calculated as follows:

$$P_L = h^{-1}(b_0 + b_1 x_1 + b_2 x_2 + \cdots + b_r x_r) \qquad (4.22)$$

where $h^{-1}()$ is the inverse function of $h()$. The expressions for calculating $P_L$ based on various generalized linear models are also shown in Table 4.3.

As noted previously, whether a soil will liquefy or not is determined by the load on the soil and the resistance of the soil against liquefaction. The load on the soil that can cause liquefaction is represented by *CSR*. The resistance of the soil to liquefaction, expressed as *CRR*, may be assessed with such *in situ* tests as the SPT, CPT, or shear wave velocity ($V_s$) measurements (e.g., Youd et al. 2001; Juang et al. 2002).

The explanatory variables in Equation 4.21 depend on the type of *in situ* test data used to characterize the liquefaction resistance of a soil. As an illustration, the liquefaction probability is predicted herein based on CPT data. In the simplest form, *CRR* is a function of the clean-sand equivalence of normalized cone tip resistance, denoted as $q_{t1N,cs}$ (Robertson and Wride 1998; Robertson 2009). Note that historically, the normalized cone tip resistance ($q_{t1N}$), defined in Appendix I, was used for assessing liquefaction resistance *CRR* of clean sand (i.e., sands with less than 5% of fine-grained material). In other words, *CRR* model (see Appendix I) was first created for clean sands. To expand the use of this *CRR* model to soils with high fines content (FC), the normalized cone tip resistance of such soils is first converted into its clean-sand equivalence, denoted as $q_{t1N,cs}$. Thus, $q_{t1N,cs}$ and *CSR* can be used as explanatory variables in the analysis of liquefaction case histories. The equations used to calculate $q_{t1N,cs}$ and *CSR* that are recommended by Robertson and his colleagues are summarized in the Appendix I.

Generalized linear models are most appropriate when the explanatory variables follow the normal distribution (e.g., Hoffmann 2004). For an accurate evaluation of liquefaction probability, transformations can be applied to $q_{t1N,cs}$ and *CSR* such that the transformed variables are largely normal. Based on previous research (e.g., Liao et al. 1988; Toprak et al. 1999; Juang et al. 2002, 2003, 2006; Lai et al. 2006), four liquefaction models are investigated here using $q_{t1N,cs}$ and $\ln(CSR)$ as explanatory variables. Note that $\ln(CSR)$, in lieu of

*Table 4.4* Calibrated coefficients and model probabilities of the liquefaction models analyzed

| Model designation | Equation for calculating $P_L$ | Calibrated coefficients | | | Model probability posterior |
|---|---|---|---|---|---|
| | | $b_0$ | $b_1$ | $b_2$ | |
| $M_1$ (Logistic) | $P_L = \dfrac{1}{1+\exp\left\{-\left[b_0 + b_1 q_{t1N,cs} + b_2 \ln(CSR)\right]\right\}}$ | 21.971 | −0.123 | 6.181 | 0.17 |
| $M_2$ (Probit) | $P_L = \Phi\left[b_0 + b_1 q_{t1N,cs} + b_2 \ln(CSR)\right]$ | 12.939 | −0.072 | 3.635 | 0.33 |
| $M_3$ (Log-log) | $P_L = \exp\left\{-\exp\left[-\left(b_0 + b_1 q_{t1N,cs} + b_2 \ln(CSR)\right)\right]\right\}$ | 14.854 | −0.080 | 4.019 | 0.16 |
| $M_4$ (C-log-log) | $P_L = 1-\exp\left\{-\exp\left[b_0 + b_1 q_{t1N,cs} + b_2 \ln(CSR)\right]\right\}$ | 16.148 | −0.093 | 4.721 | 0.34 |

*CSR*, is used because it has been shown that the variable *CSR* is lognormally distributed (e.g., Hwang et al. 2005). Based on the generalized linear models in Table 4.3, the four liquefaction probability models shown in Table 4.4 (denoted as $M_1$, $M_2$, $M_3$, and $M_4$, respectively) are examined in the subsections that follow.

## 4.3.2 Calibration database

The database adopted for calibrating the generalized linear models consists of 152 cases taken from Robertson (2009) and an additional 13 cases taken from Moss et al. (2011). The 152 cases were derived by Robertson (2009) through a rescreening of the cases previously compiled and evaluated by Moss et al. (2006), which yielded 116 liquefied cases and 36 nonliquefied cases. The other 13 cases (9 liquefied cases and 4 nonliquefied cases) in the adopted database are taken from Moss et al. (2011). Thus, the new calibration database consists of a total of 165 cases (125 liquefied cases and 40 nonliquefied cases). The cases included in this database are listed in Table 4.5. Interested readers are referred to Ku et al. (2012) for further details.

## 4.3.3 Evaluation of sampling bias

As noticed in Cetin et al. (2002), sampling bias may exist in the database that is used for calibrating models that are in turn used to calculate liquefaction probabilities. Evaluating the effect of sampling bias in model calibration requires the knowledge of $Q_p$, which in practice is very difficult to obtain. Cetin et al. (2002) assessed the value of $Q_p$ through a survey of expert opinions. Although the exact value of $Q_p$ used in Cetin et al. (2002) was not directly reported, it can be back-calculated as follows: let $Q_{s0}$, $w_{L0}$, and $w_{NL0}$ denote $Q_s$, $w_L$, and $w_{NL}$ of the database used in Cetin et al. (2002). As there are 112 liquefied and 89 nonliquefied cases in the calibration database used in that database (Cetin et al. 2002), $Q_{s0} = 0.557$. The adjusting weights used in Cetin et al. (2002) satisfied the following relationship: $w_{NL0}/w_{L0} = 1.5$. Substituting Equations 4.16 and 4.17 and $Q_{s0} = 0.557$ into this relationship yields $Q_p = 0.456$. Although we adopt $Q_p = 0.456$ in this chapter, it is understood that the true value of $Q_p$ is very difficult to estimate correctly. Further discussions of this issue can be found in Juang et al. (2009) and Zhang et al. (2013).

The calibration database used in this example consists of 125 liquefied cases and 40 nonliquefied cases. Thus, $Q_s = 125/165 = 0.758$. Based on Equations 4.16 and 4.17, the adjusting weights are determined to be $w_L = 0.601$ and $w_{NL} = 2.247$.

Table 4.5 List of case histories

| Earthquake | Site | $M_w$ | Liq? | Critical layer (m) | | GWT (m) | $\sigma_{vo}$ (kPa) | $\sigma'_{vo}$ (kPa) | $a_{max}$ (g) | $q_{t1N}$ | $I_c$ | $q_{t1N,cs}$ |
|---|---|---|---|---|---|---|---|---|---|---|---|---|
| | | | | Upper | Lower | | | | | | | |
| 1999 Kocaeli | Hotel Sapanaca | 7.4 | Yes | 1.20 | 2.00 | 0.50 | 28.10 | 17.30 | 0.37 | 21.86 | 2.30 | 42.50 |
| | Soccer field | 7.4 | Yes | 1.20 | 2.40 | 1.00 | 30.30 | 22.45 | 0.37 | 26.56 | 2.42 | 62.98 |
| | Police station | 7.4 | Yes | 1.80 | 2.80 | 1.00 | 39.55 | 26.80 | 0.40 | 29.35 | 2.50 | 80.77 |
| | Yalova harbor | 7.4 | Yes | 3.00 | 4.50 | 1.00 | 63.60 | 39.40 | 0.37 | 73.84 | 1.81 | 82.09 |
| | Site B | 7.4 | Yes | 3.30 | 4.30 | 3.30 | 60.40 | 55.50 | 0.40 | 51.24 | 2.08 | 72.22 |
| | Site C2 | 7.4 | Yes | 3.30 | 4.80 | 0.44 | 73.61 | 38.19 | 0.40 | 26.55 | 2.39 | 59.82 |
| | Site D | 7.4 | Yes | 1.80 | 2.50 | 1.50 | 35.28 | 28.90 | 0.40 | 24.89 | 2.30 | 47.83 |
| | Site E | 7.4 | Yes | 1.50 | 3.00 | 0.50 | 40.13 | 22.96 | 0.40 | 40.92 | 2.04 | 54.90 |
| | Site F | 7.4 | Yes | 6.80 | 8.00 | 0.50 | 67.71 | 42.90 | 0.40 | 39.76 | 2.21 | 66.94 |
| | Site G | 7.4 | Yes | 1.50 | 2.70 | 0.45 | 37.50 | 21.31 | 0.40 | 28.42 | 2.14 | 43.42 |
| | Site H | 7.4 | Yes | 2.00 | 3.00 | 1.72 | 41.09 | 33.44 | 0.40 | 44.17 | 2.07 | 61.79 |
| | Site I | 7.4 | Yes | 3.00 | 3.50 | 0.71 | 58.00 | 33.08 | 0.40 | 28.71 | 2.23 | 49.95 |
| | Site J | 7.4 | Yes | 2.50 | 3.50 | 0.60 | 44.45 | 30.16 | 0.40 | 30.23 | 2.28 | 57.09 |
| | Site K | 7.4 | Yes | 2.00 | 3.00 | 0.80 | 43.85 | 27.17 | 0.40 | 34.48 | 2.26 | 62.96 |
| | Site L | 7.4 | Yes | 2.00 | 2.80 | 1.72 | 38.78 | 32.35 | 0.40 | 18.69 | 2.41 | 43.36 |
| 1999 Chi Chi | Nantou C | 7.6 | Yes | 2.00 | 4.50 | 1.00 | 58.75 | 36.68 | 0.38 | 40.46 | 2.25 | 72.53 |
| | WuFeng B | 7.6 | Yes | 2.50 | 5.00 | 1.12 | 77.39 | 46.68 | 0.60 | 50.00 | 2.33 | 102.55 |
| | WuFeng C | 7.6 | Yes | 2.50 | 5.50 | 1.20 | 72.40 | 44.93 | 0.60 | 51.00 | 2.10 | 73.72 |
| | WuFeng A | 7.6 | Yes | 5.50 | 8.50 | 0.80 | 130.60 | 69.78 | 0.60 | 88.00 | 1.95 | 108.46 |
| | WuFeng C-10 | 7.6 | Yes | 2.50 | 7.00 | 1.00 | 87.25 | 50.46 | 0.60 | 26.55 | 2.57 | 83.42 |
| | Yuanlin C-19 | 7.6 | Yes | 4.00 | 5.80 | 0.57 | 121.79 | 63.62 | 0.25 | 23.84 | 2.44 | 58.78 |
| | Yuanlin C-2 | 7.6 | Yes | 2.50 | 4.00 | 0.56 | 60.07 | 33.68 | 0.25 | 36.08 | 2.12 | 53.37 |
| | Yuanlin C-22 | 7.6 | Yes | 2.80 | 4.20 | 1.13 | 63.11 | 39.86 | 0.25 | 41.36 | 2.05 | 56.53 |
| | Yuanlin C-24 | 7.6 | Yes | 5.20 | 7.80 | 1.20 | 114.20 | 65.15 | 0.25 | 45.71 | 2.07 | 63.50 |
| | Yuanlin C-25 | 7.6 | Yes | 9.50 | 12.00 | 3.52 | 193.69 | 122.76 | 0.25 | 67.13 | 1.99 | 85.90 |
| | Yuanlin C-32 | 7.6 | Yes | 4.50 | 7.50 | 0.74 | 111.78 | 60.18 | 0.25 | 41.53 | 2.11 | 60.82 |
| | Yuanlin C-4 | 7.6 | Yes | 3.00 | 6.00 | 0.66 | 83.52 | 45.85 | 0.25 | 42.42 | 2.27 | 78.82 |

continued

Table 4.5 (Continued) List of case histories

| Earthquake | Site | $M_w$ | Liq? | Critical layer (m) | | GWT (m) | $\sigma_{vo}$ (kPa) | $\sigma'_{vo}$ (kPa) | $a_{max}$ (g) | $q_{t1N}$ | $I_c$ | $q_{t1Ncs}$ |
|---|---|---|---|---|---|---|---|---|---|---|---|---|
| | | | | Upper | Lower | | | | | | | |
| 1995 Hyogoken - Nanbu | Nantou C-8 | 7.6 | Yes | 5.00 | 9.00 | 1.00 | 130.00 | 71.14 | 0.38 | 30.57 | 2.51 | 42.03 |
| | Nantou C-7 | 7.6 | Yes | 2.50 | 4.50 | 1.00 | 63.50 | 38.98 | 0.38 | 16.78 | 2.45 | 42.03 |
| | Dust Management Center | 7.2 | Yes | 6.00 | 8.00 | 2.00 | 119.50 | 70.45 | 0.37 | 72.05 | 1.85 | 82.14 |
| | Koyo Junior High School | 7.2 | Yes | 6.50 | 7.50 | 4.00 | 124.50 | 95.07 | 0.45 | 78.00 | 2.05 | 106.63 |
| | Kobe Customs Maya Office A | 7.2 | Yes | 4.00 | 9.00 | 1.80 | 121.35 | 75.24 | 0.60 | 25.22 | 2.23 | 43.40 |
| | Kobe Customs Maya Office B | 7.2 | Yes | 2.00 | 6.00 | 1.80 | 82.35 | 55.86 | 0.60 | 66.19 | 2.02 | 86.93 |
| | Kobe Port Const. Office | 7.2 | Yes | 3.00 | 5.00 | 2.50 | 70.50 | 55.79 | 0.60 | 49.70 | 1.90 | 58.76 |
| | Koyo Pump Station | 7.2 | Yes | 5.00 | 6.00 | 2.60 | 99.45 | 71.00 | 0.45 | 21.10 | 2.60 | 69.44 |
| | Kobe Wharf Public Co. | 7.2 | Yes | 4.00 | 5.50 | 2.10 | 88.63 | 60.33 | 0.45 | 53.94 | 2.06 | 74.62 |
| | Koyo Elementary School | 7.2 | Yes | 6.50 | 7.00 | 4.20 | 119.03 | 94.01 | 0.45 | 27.62 | 2.56 | 84.43 |
| | Shiporex Kogyo Osaka Factory | 7.2 | Yes | 4.00 | 7.00 | 1.50 | 93.95 | 54.71 | 0.40 | 32.20 | 2.13 | 48.47 |
| | Hamakoshienn Housing Area | 7.2 | Yes | 2.50 | 5.00 | 2.00 | 67.13 | 49.96 | 0.50 | 63.90 | 1.96 | 79.39 |
| | Taito Kobe Factory | 7.2 | Yes | 3.20 | 4.20 | 1.60 | 62.73 | 42.13 | 0.45 | 37.49 | 2.06 | 51.79 |
| | Tokuyama Concrete Factory | 7.2 | Yes | 4.00 | 4.80 | 2.00 | 74.52 | 50.98 | 0.50 | 19.45 | 2.33 | 39.66 |
| | Nisseki Kobe Oil Tank A | 7.2 | Yes | 4.80 | 6.10 | 2.40 | 99.08 | 69.15 | 0.60 | 46.92 | 2.06 | 64.70 |
| | Nisseki Kobe Oil Tank B | 7.2 | Yes | 5.00 | 6.00 | 2.40 | 100.05 | 69.64 | 0.60 | 56.05 | 2.04 | 75.28 |
| | New Port No. 6 Pier | 7.2 | Yes | 3.50 | 5.50 | 2.50 | 70.50 | 55.79 | 0.60 | 82.03 | 1.77 | 88.85 |
| | Minatokima Junior High | 7.2 | Yes | 4.00 | 4.50 | 2.70 | 74.78 | 59.57 | 0.45 | 41.97 | 2.20 | 69.40 |

| Group | Site | | | | | | | | | | | |
|---|---|---|---|---|---|---|---|---|---|---|---|---|
| | New Wharf Const. Offices | 7.2 | Yes | 3.20 | 3.80 | 2.60 | 60.45 | 51.62 | 0.45 | 31.14 | 2.31 | 60.95 |
| | Honjyo Central Park | 7.2 | No | 4.00 | 6.00 | 2.50 | 95.00 | 70.48 | 0.70 | 165.98 | 1.609 | 159.58 |
| | Kobe Art Institute | 7.2 | No | 3.50 | 3.80 | 3.00 | 64.00 | 57.62 | 0.50 | 147.37 | 1.99 | 187.62 |
| | Yoshida Kogyo Factory | 7.2 | No | 3.00 | 5.00 | 3.00 | 69.00 | 59.19 | 0.50 | 100.56 | 2.21 | 169.13 |
| | Shimonakajima Park | 7.2 | No | 3.00 | 4.50 | 2.00 | 63.28 | 46.11 | 0.65 | 186.98 | 1.62 | 182.04 |
| | Sumiyoshi Elementary | 7.2 | No | 2.40 | 3.20 | 1.90 | 46.92 | 38.09 | 0.60 | 163.95 | 1.63 | 161.39 |
| 1994 Northridge | Nagashi Park | 7.2 | No | 1.10 | 1.80 | 1.00 | 26.00 | 21.59 | 0.65 | 140.48 | 1.81 | 156.04 |
| | Balboa Blvd | 6.7 | Yes | 8.30 | 9.80 | 7.19 | 162.74 | 144.99 | 0.69 | 58.38 | 2.30 | 113.05 |
| | Malden St | 6.7 | Yes | 9.20 | 10.70 | 3.90 | 169.80 | 110.45 | 0.51 | 27.69 | 2.58 | 88.07 |
| | Fotrero Canyon | 6.7 | Yes | 6.00 | 7.00 | 3.30 | 122.67 | 91.27 | 0.40 | 63.09 | 2.09 | 89.87 |
| | Wynne Ave | 6.7 | Yes | 5.80 | 6.50 | 4.30 | 112.76 | 94.85 | 0.54 | 87.77 | 1.99 | 112.04 |
| | Fory Lane | 6.7 | Yes | 3.00 | 5.00 | 2.70 | 66.60 | 53.85 | 0.77 | 47.80 | 2.32 | 95.71 |
| 1989 Loma Prieta | SFOBB-1 | 7 | Yes | 6.25 | 7.00 | 2.99 | 127.53 | 90.64 | 0.28 | 53.27 | 2.03 | 70.99 |
| | SFOBB-2 | 7 | Yes | 6.50 | 8.50 | 2.99 | 141.03 | 96.79 | 0.28 | 85.64 | 1.81 | 95.06 |
| | FOO7-2 | 7 | Yes | 5.50 | 6.80 | 2.30 | 111.18 | 73.41 | 0.28 | 64.34 | 1.88 | 74.72 |
| | FOO7-3 | 7 | Yes | 7.10 | 8.10 | 2.30 | 137.50 | 85.51 | 0.28 | 102.49 | 1.58 | 97.32 |
| | FOR-2 | 7 | Yes | 5.30 | 6.70 | 2.40 | 114.15 | 74.42 | 0.16 | 23.07 | 2.34 | 47.82 |
| | FOR-3 | 7 | Yes | 5.00 | 7.00 | 2.40 | 106.80 | 71.48 | 0.16 | 22.38 | 2.30 | 43.61 |
| | FOR-4 | 7 | Yes | 6.00 | 7.00 | 2.40 | 116.30 | 76.08 | 0.16 | 24.77 | 2.24 | 43.89 |
| | Marine Lab C4 | 7 | Yes | 5.20 | 5.80 | 2.50 | 95.75 | 66.32 | 0.25 | 24.43 | 2.28 | 45.74 |
| | Marine Lab UC-7 | 7 | Yes | 7.60 | 9.80 | 2.00 | 148.55 | 86.75 | 0.25 | 46.41 | 2.22 | 79.50 |
| | Sandholdt Rd. UC-4 | 7 | Yes | 2.40 | 4.60 | 2.70 | 56.40 | 48.55 | 0.25 | 106.14 | 1.68 | 108.39 |
| | MLSB UC-14 | 7 | Yes | 2.40 | 4.00 | 2.40 | 52.40 | 44.55 | 0.25 | 68.25 | 1.90 | 80.48 |
| | WM UC-11 | 7 | Yes | 2.50 | 3.40 | 2.50 | 46.65 | 43.22 | 0.25 | 81.66 | 1.80 | 89.88 |
| | HO UC 12,13 | 7 | Yes | 2.90 | 4.70 | 1.90 | 66.50 | 47.86 | 0.25 | 83.79 | 1.83 | 94.29 |
| | TI Naval Station | 7 | Yes | 3.50 | 7.00 | 1.50 | 97.43 | 60.64 | 0.16 | 46.12 | 2.14 | 70.28 |
| | Farris Farm | 7 | Yes | 6.00 | 7.00 | 4.50 | 106.75 | 87.13 | 0.31 | 41.59 | 2.14 | 63.19 |
| | Miller Farm CMF 8 | 7 | Yes | 6.80 | 8.00 | 4.91 | 123.42 | 98.99 | 0.30 | 46.11 | 1.91 | 54.89 |

continued

Table 4.5 (Continued) List of case histories

| Earthquake | Site | $M_w$ | Liq? | Critical layer (m) | | GWT (m) | $\sigma_{vo}$ (kPa) | $\sigma'_{vo}$ (kPa) | $a_{max}$ (g) | $q_{t1N}$ | $I_c$ | $q_{t1N,cs}$ |
|---|---|---|---|---|---|---|---|---|---|---|---|---|
| | | | | Upper | Lower | | | | | | | |
| | Miller Farm CMF 10 | 7 | Yes | 7.00 | 9.70 | 3.00 | 155.35 | 99.92 | 0.30 | 45.87 | 2.35 | 97.16 |
| | Miller Farm CMF 5 | 7 | Yes | 5.50 | 8.50 | 4.70 | 122.40 | 99.84 | 0.30 | 69.04 | 1.87 | 79.62 |
| | Miller Farm CMF 3 | 7 | Yes | 5.75 | 7.50 | 3.00 | 103.55 | 95.70 | 0.30 | 30.85 | 2.25 | 55.54 |
| | Model Airport 18 | 7 | Yes | 3.70 | 4.50 | 2.40 | 70.70 | 54.02 | 0.29 | 75.81 | 1.76 | 81.57 |
| | Model Airport 21 | 7 | Yes | 3.40 | 4.70 | 2.40 | 69.75 | 53.56 | 0.29 | 70.05 | 1.77 | 75.53 |
| | Farris 58 | 7 | Yes | 7.40 | 8.00 | 4.80 | 131.90 | 103.45 | 0.31 | 83.32 | 1.79 | 91.26 |
| | Farris 61 | 7 | Yes | 6.00 | 7.30 | 4.20 | 110.43 | 86.39 | 0.31 | 40.04 | 2.18 | 64.60 |
| | Granite 123 | 7 | Yes | 7.20 | 7.80 | 5.00 | 127.50 | 102.98 | 0.31 | 41.96 | 2.06 | 58.03 |
| | Jefferson 121 | 7 | Yes | 6.50 | 7.75 | 3.40 | 126.88 | 90.33 | 0.18 | 57.46 | 1.92 | 69.05 |
| | Jefferson 141 | 7 | Yes | 3.10 | 4.50 | 2.10 | 66.95 | 50.27 | 0.18 | 50.00 | 1.85 | 56.94 |
| | Jefferson 148 | 7 | Yes | 7.00 | 7.90 | 3.00 | 137.78 | 94.12 | 0.18 | 50.00 | 1.82 | 55.75 |
| | Jeff. Ranch 32 | 7 | Yes | 2.30 | 3.10 | 1.80 | 45.90 | 37.07 | 0.17 | 37.75 | 2.02 | 49.91 |
| | Kett 74 | 7 | Yes | 2.30 | 3.10 | 1.50 | 48.15 | 36.38 | 0.32 | 82.31 | 2.03 | 109.23 |
| | Leonardini 39 | 7 | Yes | 2.30 | 4.70 | 1.90 | 60.80 | 45.10 | 0.17 | 43.52 | 1.88 | 50.64 |
| | Leonardini 51 | 7 | Yes | 3.10 | 3.70 | 1.80 | 59.20 | 43.50 | 0.17 | 17.26 | 2.41 | 40.24 |
| | Leonardini 53 | 7 | Yes | 2.70 | 3.60 | 2.10 | 55.13 | 44.82 | 0.17 | 51.43 | 1.88 | 59.91 |
| | Marinovich 65 | 7 | Yes | 6.80 | 9.40 | 5.60 | 150.90 | 121.47 | 0.28 | 62.85 | 1.97 | 79.03 |
| | Radovich 99 | 7 | Yes | 4.75 | 6.90 | 4.10 | 79.38 | 72.26 | 0.28 | 59.46 | 2.02 | 78.04 |
| | Sea Mist 31 | 7 | Yes | 2.80 | 3.70 | 0.80 | 60.33 | 36.29 | 0.17 | 19.19 | 2.38 | 42.74 |
| | Silliman 68 | 7 | Yes | 4.70 | 7.10 | 3.50 | 103.37 | 79.83 | 0.28 | 51.94 | 2.05 | 70.70 |
| | SP Bridge 48 | 7 | Yes | 6.00 | 7.50 | 5.30 | 114.38 | 100.15 | 0.30 | 37.81 | 2.24 | 66.47 |
| | Alameda Bay | 7 | No | 5.00 | 6.00 | 2.50 | 103.75 | 74.32 | 0.24 | 80.23 | 2.20 | 133.83 |
| | MBAR RC-6 | 7 | No | 3.00 | 4.50 | 2.60 | 64.03 | 52.74 | 0.25 | 180.42 | 1.33 | 110.33 |
| | MBAR RC-7 | 7 | No | 4.00 | 5.00 | 3.70 | 74.80 | 66.95 | 0.25 | 111.32 | 1.58 | 105.72 |
| | Sandholdt UC-2 | 7 | No | 3.00 | 4.50 | 2.70 | 61.20 | 50.90 | 0.25 | 226.63 | 1.31 | 131.67 |
| | General Fish C6 | 7 | No | 2.20 | 3.20 | 1.70 | 48.90 | 39.09 | 0.25 | 146.59 | 1.49 | 126.21 |

| | Site | | | | | | | | | | | |
|---|---|---|---|---|---|---|---|---|---|---|---|---|
| | MBAR CPT-1 | 7 | No | 2.30 | 3.50 | 1.90 | 48.08 | 38.27 | 0.25 | 151.90 | 1.45 | 123.27 |
| | Sandholdt UC-6 | 7 | No | 6.20 | 7.00 | 2.70 | 123.90 | 85.64 | 0.25 | 199.03 | 1.36 | 134.53 |
| | MLSB 18 | 7 | No | 2.40 | 3.40 | 2.40 | 48.40 | 43.50 | 0.25 | 154.51 | 1.44 | 122.54 |
| | Leonardini 37 | 7 | No | 2.90 | 6.10 | 2.50 | 78.00 | 58.38 | 0.17 | 49.23 | 1.93 | 59.89 |
| | Leonardini 52a | 7 | No | 3.80 | 4.50 | 2.70 | 72.83 | 58.60 | 0.17 | 34.76 | 2.32 | 69.53 |
| | Matella 111 | 7 | No | 1.70 | 5.10 | 1.70 | 60.18 | 43.50 | 0.15 | 41.75 | 2.05 | 57.09 |
| | McGowan 136 | 7 | No | 2.40 | 3.10 | 2.40 | 46.36 | 42.92 | 0.26 | 55.07 | 2.13 | 83.08 |
| | Marinovich 67 | 7 | No | 6.20 | 7.00 | 6.20 | 113.40 | 109.48 | 0.28 | 139.75 | 1.70 | 144.33 |
| | Radovich 98 | 7 | No | 5.10 | 8.75 | 3.50 | 124.54 | 90.94 | 0.28 | 80.05 | 1.89 | 93.70 |
| | Salinas River 117 | 7 | No | 6.40 | 7.40 | 6.40 | 113.97 | 109.97 | 0.12 | 51.17 | 2.27 | 94.61 |
| | Tanimura 105 | 7 | No | 4.20 | 6.80 | 4.20 | 92.29 | 79.54 | 0.15 | 41.35 | 2.03 | 55.23 |
| 1987 Edgecumbe | Robinson Farm E. | 6.6 | Yes | 2.00 | 5.50 | 0.76 | 57.67 | 28.03 | 0.44 | 90.42 | 1.71 | 93.80 |
| | Cordon Farm 1 | 6.6 | Yes | 1.20 | 2.40 | 0.47 | 41.38 | 19.50 | 0.43 | 74.70 | 1.90 | 88.41 |
| | Morris Farm 1 | 6.6 | Yes | 7.00 | 8.50 | 1.63 | 118.50 | 58.46 | 0.42 | 96.61 | 1.68 | 98.58 |
| | Awaroa Farm | 6.6 | Yes | 2.30 | 3.30 | 1.15 | 42.25 | 26.06 | 0.37 | 120.00 | 1.70 | 123.95 |
| | Keir Farm | 6.6 | Yes | 6.50 | 9.50 | 2.54 | 121.46 | 67.90 | 0.31 | 85.94 | 1.69 | 88.35 |
| | James St. Loop | 6.6 | Yes | 3.40 | 6.80 | 1.15 | 77.90 | 39.15 | 0.28 | 85.88 | 1.82 | 95.52 |
| | Landing Rd. Bridge | 6.6 | Yes | 4.80 | 6.20 | 1.15 | 84.10 | 41.43 | 0.27 | 91.60 | 1.67 | 92.97 |
| | Whakatane Pony Club | 6.6 | Yes | 3.60 | 4.60 | 2.35 | 61.20 | 44.03 | 0.27 | 60.94 | 1.70 | 62.92 |
| | Sewage Pumping Station | 6.6 | Yes | 2.00 | 8.00 | 1.29 | 76.21 | 39.81 | 0.26 | 61.72 | 1.82 | 68.77 |
| | Edgecumbe Pipe Breaks | 6.6 | Yes | 5.00 | 5.90 | 2.50 | 81.98 | 53.04 | 0.39 | 80.71 | 1.76 | 86.66 |
| | Cordon farm 2 | 6.6 | No | 1.70 | 1.90 | 0.90 | 27.00 | 18.17 | 0.37 | 212.30 | 1.47 | 176.50 |
| | Brady farm 4 | 6.6 | No | 3.40 | 5.00 | 1.53 | 63.57 | 37.38 | 0.40 | 122.08 | 1.61 | 118.80 |

continued

Table 4.5 (Continued) List of case histories

| Earthquake | Site | $M_w$ | Liq? | Critical layer (m) | | GWT (m) | $\sigma_{vo}$ (kPa) | $\sigma'_{vo}$ (kPa) | $a_{max}$ (g) | $q_{t1N}$ | $I_c$ | $q_{t1N,cs}$ |
|---|---|---|---|---|---|---|---|---|---|---|---|---|
| | | | | Upper | Lower | | | | | | | |
| | Morris farm 3 | 6.6 | No | 5.20 | 6.60 | 2.10 | 89.35 | 52.07 | 0.41 | 108.03 | 1.60 | 104.15 |
| | Whakatane hospital | 6.6 | No | 4.40 | 5.00 | 4.40 | 68.45 | 65.51 | 0.26 | 167.47 | 1.54 | 152.93 |
| | Whakatane board mill | 6.6 | No | 7.00 | 8.00 | 1.44 | 114.81 | 55.36 | 0.27 | 96.39 | 1.71 | 100.44 |
| 1987 Elmore Ranch | Wildlife B | 6.2 | No | 3.70 | 6.70 | 0.90 | 98.70 | 56.52 | 0.17 | 60.00 | 2.00 | 77.52 |
| 1987 Superstition Hills | Wildlife B | 6.6 | Yes | 3.70 | 6.70 | 0.90 | 98.70 | 56.52 | 0.21 | 60.00 | 2.00 | 77.52 |
| 1983 Nihonkai | Akita B | 7.7 | Yes | 3.30 | 6.70 | 1.03 | 91.91 | 52.96 | 0.17 | 37.04 | 2.27 | 68.38 |
| | Akita C | 7.7 | No | 2.00 | 4.00 | 2.40 | 49.80 | 43.91 | 0.17 | 39.81 | 2.38 | 87.93 |
| 1983 Borah Peak | Pence ranch | 6.9 | Yes | 1.50 | 4.00 | 1.55 | 49.75 | 37.98 | 0.30 | 78.92 | 2.08 | 111.22 |
| | Whiskey springs site | 6.9 | Yes | 1.60 | 3.20 | 0.80 | 44.80 | 29.10 | 0.50 | 46.00 | 2.07 | 64.40 |
| | Whiskey springs wite 2 | 6.9 | Yes | 2.40 | 4.30 | 2.40 | 59.33 | 50.01 | 0.50 | 73.15 | 2.11 | 107.07 |
| | Whiskey springs site 3 | 6.9 | Yes | 6.80 | 7.80 | 6.80 | 125.45 | 120.45 | 0.50 | 73.63 | 2.11 | 107.78 |
| 1981 Westmorland | Wildlife B | 5.9 | Yes | 2.70 | 6.70 | 0.91 | 89.31 | 51.93 | 0.23 | 60.00 | 2.00 | 77.52 |
| | Kornbloom B | 5.9 | Yes | 2.80 | 5.80 | 2.74 | 73.48 | 58.18 | 0.19 | 20.00 | 2.40 | 45.97 |
| | Radio Tower B1 | 5.9 | Yes | 2.00 | 5.50 | 2.00 | 72.50 | 50.43 | 0.17 | 40.00 | 2.10 | 57.82 |
| | McKim Ranch A | 5.9 | No | 1.50 | 5.20 | 1.50 | 57.30 | 39.15 | 0.09 | 51.31 | 2.17 | 81.62 |
| | Radio Tower B2 | 5.9 | No | 2.00 | 3.00 | 2.01 | 40.98 | 36.17 | 0.16 | 103.50 | 1.99 | 132.05 |
| 1980 Mexicali | Delta Site 2 | 6.2 | Yes | 2.20 | 3.20 | 2.20 | 44.20 | 39.30 | 0.19 | 48.51 | 1.79 | 53.16 |
| | Delta Site 3 | 6.2 | Yes | 2.00 | 3.80 | 2.00 | 48.20 | 39.37 | 0.19 | 26.14 | 2.33 | 53.64 |
| | Delta Site 3p | 6.2 | Yes | 2.20 | 3.80 | 2.20 | 49.60 | 41.75 | 0.19 | 28.57 | 2.34 | 59.03 |
| | Delta Site 4 | 6.2 | Yes | 2.00 | 2.60 | 2.00 | 37.40 | 34.46 | 0.19 | 49.83 | 2.00 | 64.38 |
| | Delta Site 1 | 6.2 | No | 4.80 | 5.30 | 2.30 | 86.30 | 59.32 | 0.19 | 46.85 | 2.35 | 98.76 |

| | | | | | | | | | | | | |
|---|---|---|---|---|---|---|---|---|---|---|---|---|
| 1979 Imperial Valley | Radio Tower B1 | 6.5 | Yes | 3.00 | 5.50 | 2.01 | 74.72 | 52.75 | 0.18 | 40.00 | 2.10 | 57.82 |
| | McKim Ranch A | 6.5 | Yes | 1.50 | 4.00 | 1.50 | 47.75 | 35.49 | 0.51 | 43.77 | 2.10 | 63.27 |
| | Kornbloom B | 6.5 | No | 2.60 | 5.20 | 2.74 | 65.88 | 54.50 | 0.13 | 36.51 | 2.50 | 99.71 |
| | Wildlife B | 6.5 | No | 3.70 | 6.70 | 0.90 | 98.70 | 56.52 | 0.17 | 60.00 | 2.00 | 77.52 |
| | Radio Tower B2 | 6.5 | No | 2.00 | 3.00 | 2.01 | 41.47 | 36.66 | 0.16 | 93.19 | 2.03 | 124.40 |
| 1976 Tangshan | T1 | 7.8 | Yes | 4.75[a] | 4.75[a] | 3.70 | 83.30 | 73.00 | 0.64 | 68.5 | 2.27 | 127.23 |
| | T2 | 7.8 | Yes | 7.40[a] | 7.40[a] | 1.25 | 141.00 | 81.00 | 0.53 | 45.5 | 2.54 | 135.95 |
| | T6 | 7.8 | Yes | 5.10[a] | 5.10[a] | 1.50 | 95.70 | 60.00 | 0.64 | 123.7 | 1.80 | 136.66 |
| | T7 | 7.8 | Yes | 6.40[a] | 6.40[a] | 3.00 | 117.30 | 84.00 | 0.64 | 56.8 | 2.22 | 97.93 |
| | T8 | 7.8 | Yes | 5.25[a] | 5.25[a] | 2.20 | 96.88 | 67.00 | 0.64 | 103.7 | 1.85 | 118.78 |
| | T10 | 7.8 | Yes | 8.00[a] | 8.00[a] | 1.45 | 152.38 | 88.00 | 0.64 | 58.6 | 2.26 | 107.88 |
| | T11 | 7.8 | Yes | 2.10[a] | 2.10[a] | 0.85 | 38.83 | 27.00 | 0.61 | 66.5 | 2.13 | 100.79 |
| | T12 | 7.8 | Yes | 3.10[a] | 3.10[a] | 1.55 | 56.58 | 41.00 | 0.58 | 32 | 2.38 | 71.50 |
| | T13 | 7.8 | Yes | 7.00[a] | 7.00[a] | 1.05 | 133.88 | 76.00 | 0.58 | 141.2 | 1.79 | 154.80 |
| | T4 | 7.8 | No | 3.40[a] | 3.40[a] | 1.10 | 63.55 | 41.00 | 0.64 | 162.6 | 1.77 | 176.89 |
| | T5 | 7.8 | No | 4.50[a] | 4.50[a] | 3.00 | 80.25 | 66.00 | 0.64 | 125.8 | 1.85 | 144.22 |
| | T9 | 7.8 | No | 4.00[a] | 4.00[a] | 1.10 | 75.25 | 47.00 | 0.64 | 171.6 | 1.68 | 175.82 |
| | T16 | 7.8 | No | 7.50[a] | 7.50[a] | 3.50 | 137.50 | 98.00 | 0.26 | 108.8 | 1.86 | 125.94 |
| 1968 Inaguaha | Reedy's Farm | 7.4 | Yes | 1.80 | 1.00 | 0.10 | 26.66 | 14.10 | 0.20 | 18.54 | 2.47 | 48.25 |
| 1964 Niigata | Site F | 7.5 | No | 2.20 | 1.70 | 1.70 | 31.95 | 29.50 | 0.16 | 106.90 | 1.99 | 136.18 |
| | Site D | 7.5 | Yes | 6.00 | 2.70 | 1.12 | 47.94 | 32.44 | 0.16 | 64.37 | 2.10 | 92.51 |
| | Site E | 7.5 | Yes | 4.80 | 1.80 | 0.67 | 68.00 | 44.46 | 0.16 | 45.13 | 2.24 | 78.99 |

Source: Adapted from Ku, C.S. et al. 2012. *Canadian Geotechnical Journal*, 49(1), 27–44. With permission from NRC Research Press.

[a] Median depth to critical layer (m).

### 4.3.4 Calibration of liquefaction models

In this calibration study (involving database shown in Table 4.5, and models shown in Table 4.4), $\theta = \{b_0, b_1, b_2, ..., b_r\}$ are the uncertain parameters to be calibrated (in this case, $r = 2$). To begin with, let $I$ denote the indicator variable, with $I = 1$ denoting liquefaction and $I = 0$ denoting nonliquefaction, respectively. Let $I_i$ denote the indicator variable of the $i$th case. The observed data can be denoted as $d_i = \{I_i\}$, and the calibration database is $\mathbf{D} = \{d_1, d_2, ..., d_n\}$. Suppose in a calibration database, there are $n_L$ liquefied cases and $n_{NL}$ nonliquefaction cases. Based on a generalized linear model, the probability of observing a liquefied case is $P_L$, and the probability of observing a nonliquefied case is $1 - P_L$. The likelihood function, or the probability to observe $n_L$ liquefied cases and $n_{NL}$ nonliquefied cases, can be calculated by multiplying the probability to observe each case in the database. Using model $M_1$ (logistic regression model; see Table 4.4) as an example, the weighted likelihood function can be written as follows:

$$
\begin{aligned}
L(\theta|\mathbf{D}) = w_L \sum_{i=1}^{n_L} \ln \frac{1}{1 + \exp\left\{-[b_0 + b_1 q_{t1N,cs} + b_2 \ln(CSR)]\right\}} \\
+ w_{NL} \sum_{j=1}^{n_{NL}} \ln \left( 1 - \frac{1}{1 + \exp\left\{-[b_0 + b_1 q_{t1N,cs} + b_2 \ln(CSR)]\right\}} \right)
\end{aligned}
\tag{4.23}
$$

The spreadsheet template for maximizing Equation 4.23 is shown in Figure 4.6. With Excel Solver, the optimal values of $\theta$ are obtained as $\theta = \{b_0, b_1, b_2\} = \{21.932, -0.123, 6.182\}$.

### 4.3.5 Ranking of liquefaction models

Using the spreadsheet template shown in Figure 4.6, all four models are calibrated and the optimal regression coefficients for these models are summarized in Table 4.4 along with the model probabilities that are computed with Equation 4.19. Note that since the weighted likelihood function is used for model calibration to remove the effect of sampling bias, BIC, and model probability should be calculated accordingly (i.e., based on the weighted likelihood function). Model ranking based on BIC is likely supported by Laplace's approximation of the weighted log-likelihood function, although further research is needed. The results in Table 4.4 show that $M_4$ (c-log-log regression) has the highest model probability, indicating that it is the most suitable for the database shown in Table 4.5. The commonly used logistic regression model, however, has the least model probability with the database shown in Table 4.5.

It should be noted that the results obtained in the above study are not exactly the same as those obtained by Zhang et al. (2013) based on an earlier database. In that similar study, they determined that while the c-log-log model was indeed most suitable for constructing liquefaction probability models, the logistic regression model was the next best for such constructions. The model ranking results are thus shown as dependent on the adopted calibration database, and the commonly used logistic regression model is far from being the best option.

## 4.4 CONVERTING A DETERMINISTIC LIQUEFACTION MODEL INTO A PROBABILISTIC MODEL

The Robertson and Wride method (Robertson and Wride 1998), which is later updated in Robertson (2009), is one of the most widely used CPT-based simplified models for evaluating the potential of soil liquefaction. The Robertson and Wride method is a deterministic model, which yields a factor of safety ($F_S$) as the outcome of its evaluation of liquefaction

| 1 | B | C | D | E | F | G | H | I | J | K |
|---|---|---|---|---|---|---|---|---|---|---|
| 2 | | | | | | | | | | |
| 3 | | **Spreadsheet Template for Calibrating Generalized Linear Models** | | | | | | | | |
| 4 | | | | | | | | Parameters to be estimated | | |
| 5 | | $Q_p$ | $Q_s$ | $w_L$ | $w_{NL}$ | | | $b_0$ | $b_1$ | $b_2$ |
| 6 | | 0.456 | 0.758 | 0.601583 | 2.247934 | | | 21.932 | -0.123 | 6.182 |
| 7 | | | | Observed data | | | | $M_1$ in Table 4.4 | | |
| 8 | | Case No. | I | $CSR$ | $q_{t1n,cs}$ | Y | $P_{Li}$ | $w_i \ln P(\mathbf{d}_i|\boldsymbol{\theta})$ | $L(\boldsymbol{\theta}|\mathbf{D})$ | |
| 9 | | 1 | 1 | 0.40 | 42.50 | 11.00 | 0.999983 | -1.008E-05 | -41.11 | |
| 10 | | 2 | 1 | 0.33 | 62.98 | 7.33 | 0.999343 | -0.0003954 | | |
| 11 | | 3 | 1 | 0.35 | 80.77 | 5.50 | 0.995923 | -0.0024576 | | |
| 12 | | 4 | 1 | 0.38 | 82.09 | 5.83 | 0.997074 | -0.0017629 | | |
| 13 | | 5 | 1 | 0.24 | 72.22 | 4.29 | 0.986523 | -0.0081624 | Log-likelihood | |
| 14 | | 6 | 1 | 0.43 | 59.82 | 9.31 | 0.999909 | -5.45E-05 | | |
| 15 | | 7 | 1 | 0.29 | 47.83 | 8.41 | 0.999778 | -0.0001337 | | |
| 16 | | 8 | 1 | 0.42 | 54.90 | 9.77 | 0.999943 | -3.438E-05 | | |
| 17 | | 9 | 1 | 0.37 | 66.94 | 7.53 | 0.999463 | -0.0003234 | | |
| 18 | | 10 | 1 | 0.42 | 43.42 | 11.18 | 0.999986 | -8.409E-06 | | |
| 169 | | 161 | 0 | 0.29 | 187.62 | -8.77 | 0.000155 | -0.0003485 | | |
| 170 | | 162 | 0 | 0.30 | 169.13 | -6.31 | 0.001808 | -0.004067 | | |
| 171 | | 163 | 0 | 0.45 | 182.04 | -5.33 | 0.004824 | -0.0108703 | | |
| 172 | | 164 | 0 | 0.39 | 161.39 | -3.73 | 0.023489 | -0.0534314 | | |
| 173 | | 165 | 0 | 0.44 | 156.04 | -2.26 | 0.094122 | -0.2222098 | | |

| | |
|---|---|
| 174 | Notes: |
| 175 | (1) To facilitate the spreadsheet implementation, an intemideiate variable $Y$ is defined as follows: |
| 176 | $$Y = b_0 + b_1 q_{t1N,cs} + b_2 \ln(CSR_{7.5,\sigma})$$ |
| 177 | (2) Rows 19 through 168 are skipped to save space. |
| 178 | (3) The setting in Solver is "Maximize the value in Cell J9 by changing the values in Cells H6, I6, and |
| 179 | J6, respectively." |

*Figure 4.6* Spreadsheet template for calibrating the generalized linear models.

potential. In this section, we want to show how a probabilistic version of this popular deterministic model can be created. In the following example, we will illustrate how to develop a $P_L$–$F_S$ mapping function that relates the probability of liquefaction ($P_L$) to the factor of safety ($F_S$) computed with the deterministic Robertson and Wride method. The equations that are required for computing $CSR$ and $CRR$, and thus $F_S$, using the deterministic Robertson and Wride method are listed in Appendix I.

## 4.4.1 Probabilistic model

The semiempirical nature of the Robertson and Wride method ensures an inherent uncertainty within this model. The calculated factor of safety $F_S$ can be related to the actual factor of safety $F_{Sa}$ as follows:

$$F_{Sa} = \frac{F_S}{z} \tag{4.24}$$

where $z$ is a random variable characterizing the modeling error of the factor of safety. With Equation 4.24, the liquefaction probability can be expressed as follows:

$$P_L = P(F_{Sa} < 1) = P\left(\frac{F_S}{z} < 1\right) = P(z > F_S) = 1 - F(F_S) \tag{4.25}$$

where $F()$ is the cumulative density function of $z$. Let $\mu$ and $\delta$ denote the mean and coefficient of variation (COV) of $z$, respectively. The task of calibrating the $P_L$–$F_S$ relationship, which is in the form of $P_L = f(F_S)$, is then reduced to the task of calibrating $\mu$ and $\delta$, or in the notation in this chapter, $\theta = \{\mu, \delta\}$.

The shape of the assumed probability distribution of $z$ may affect the results of the maximum-likelihood analysis. To derive the best $P_L$ – $F_S$ relationship, the following cumulative distribution functions (denoted as $Q_1$, $Q_2$, $Q_3$, and $Q_4$) are examined and compared:

$Q_1$—Gaussian (normal):

$$F(z) = \Phi\big[(z - \mu)/(\mu \cdot \delta)\big] \tag{4.26}$$

$Q_2$—Lognormal:

$$F(z) = \Phi\left(\left[\ln(z) - \ln\left(\mu/\sqrt{1 + \delta^2}\right)\right]\Big/\sqrt{\ln(1 + \delta^2)}\right) \tag{4.27}$$

$Q_3$—Minimum Gumbel:

$$F(z) = 1 - \exp\left\{-\exp\left[\frac{(z - \mu)(\pi - 0.5772\sqrt{6} \cdot \delta)}{\sqrt{6} \cdot \mu \cdot \delta}\right]\right\} \tag{4.28}$$

$Q_4$—Maximum Gumbel:

$$F(z) = \exp\left\{-\exp\left[\frac{-(z - \mu)(\pi - 0.5772\sqrt{6} \cdot \delta)}{\sqrt{6} \cdot \mu \cdot \delta}\right]\right\} \tag{4.29}$$

Based on the above assumptions about the cumulative density function of $z$, four $P_L$ – $F_S$ relationships for $z$ are generated based on Equation 4.25, as summarized in Table 4.6. Note that in these derivations, COV of $z$ is treated as a constant.

### 4.4.2 Calibration and ranking of $P_L$–$F_s$ relationships

The four probability models as shown in Table 4.6 can be calibrated using the database as shown in Table 4.5 (Ku et al. 2012). As an example, the analysis is conducted with the assumption that the variable $z$ follows the lognormal distribution which is designated as $Q_2$. The weighted likelihood function can be expressed as follows:

$$L(\theta|\mathbf{D}) = w_L \sum_{i=1}^{n_L} \ln\left\{1 - \Phi\left(\left[\ln(F_{si}) - \ln\left(\mu/\sqrt{1 + \delta^2}\right)\right]\Big/\sqrt{\ln(1 + \delta^2)}\right)\right\}$$
$$+ w_{NL} \sum_{j=1}^{n_{NL}} \ln \Phi\left(\left[\ln(F_{sj}) - \ln\left(\mu/\sqrt{1 + \delta^2}\right)\right]\Big/\sqrt{\ln(1 + \delta^2)}\right) \tag{4.30}$$

*Table 4.6* Results of the maximum likelihood analyses with various assumptions

| Distribution type for z | Model form | $\mu$ | $\delta$ | BIC | Model probability |
|---|---|---|---|---|---|
| Normal($Q_1$) | $P_L = 1 - \Phi\left(\dfrac{F_S - \mu}{\mu \cdot \delta}\right)$ | 0.905 | 0.273 | 107.16 | 0.13 |
| Lognormal($Q_2$) | $P_L = 1 - \Phi\left(\dfrac{\left[\ln(F_S) - \ln\left(\mu/\sqrt{1+\delta^2}\right)\right]}{\sqrt{\ln(1+\delta^2)}}\right)$ | 0.914 | 0.281 | 104.94 | 0.40 |
| Minimum Gumbel($Q_3$) | $P_L = \exp\left\{-\exp\left[\dfrac{(F_S - \mu)(\pi - 0.5772\sqrt{6}\cdot\delta)}{\sqrt{6}\cdot\mu\cdot\delta}\right]\right\}$ | 1.028 | 0.337 | 112.38 | 0.01 |
| Maximum Gumbel($Q_4$) | $P_L = 1 - \exp\left\{-\exp\left[\dfrac{-(F_S - \mu)(\pi - 0.5772\sqrt{6}\cdot\delta)}{\sqrt{6}\cdot\mu\cdot\delta}\right]\right\}$ | 0.800 | 0.282 | 104.71 | 0.45 |

As analyzed in the previous section, the adjusting weights for the calibration database are $w_L = 0.601$ and $w_{NL} = 2.247$, respectively. Figure 4.7 shows the spreadsheet template to calibrate model $Q_2$ by applying the principle of maximum likelihood based on Equation 4.30. With Solver, the optimal values of $\theta$ are $\theta^* = \{\mu^*, \delta^*\} = \{0.914, 0.281\}$.

Based on Equation 4.27 and the knowledge of model error (bias), which is characterized with $\theta^* = \{\mu^*, \delta^*\} = \{0.914, 0.281\}$, the following $P_L - F_S$ mapping function can be established:

$$P_L = 1 - \Phi\left[\frac{0.128 + \ln(F_S)}{0.276}\right] \tag{4.31}$$

Thus, the probability of liquefaction can be computed once the nominal factor of safety is calculated using the Robertson and Wride method.

The calibrated results and model probabilities of the four models are also shown in Table 4.6. Among the four models considered, the lognormal model and maximum Gumbel model are the most supported by the adopted database, whereas the normal model and minimum Gumbel model are much less supported by the database. Thus, either the lognormal model or the maximum Gumbel model may be used for future applications. It should be noted that this particular conclusion differs somewhat from that obtained by Ku et al. (2012), as they employed a different set of adjusting weights ($w_L = 0.660$ and $w_{NL} = 2.063$). The recommended $P_L - F_S$ mapping function (Equation 4.31) obtained through the calibration in this study is practically identical, however, to that obtained and recommended by Ku et al. (2012).

## 4.5 ESTIMATION OF LIQUEFACTION-INDUCED SETTLEMENT

### 4.5.1 Probabilistic model for predicting liquefaction-induced settlement

Liquefaction-induced settlement has been the subject of multiple investigations, from which many semiempirical methods have been developed based on results of both laboratory testing

| A | B | C | D | E | F | G | H | I | J | K |
|---|---|---|---|---|---|---|---|---|---|---|
| 2 | | spreadsheet | | | | | | | | |
| 3 | | | | | | | | | | |
| 4 | | | | | | | | | | |
| 5 | | $Q_p$ | $Q_s$ | $w_L$ | $w_{NL}$ | | | $\mu$ | $\delta$ | |
| 6 | | 0.456 | 0.758 | 0.601583 | 2.247934 | | | 0.914337 | 0.28116825 | |
| 7 | | | | Observed data | | | | | $Q_2$ in Table 4.6 | |
| 8 | | Case No. | I | $CSR_{7.5}$ | $CRR$ | $F_S$ | $P_L$ | $w_i \ln P(d_i|\theta)$ | $L(\theta|D)$ | |
| 9 | | 1 | 1 | 0.396 | 0.085 | 0.215 | 1 | -1.013E-07 | -47.3646 | |
| 10 | | 2 | 1 | 0.329 | 0.103 | 0.314 | 0.999906 | -5.637E-05 | | |
| 11 | | 3 | 1 | 0.348 | 0.129 | 0.371 | 0.99914 | -0.0005174 | | |
| 12 | | 4 | 1 | 0.377 | 0.131 | 0.349 | 0.999606 | -0.000237 | | |
| 13 | | 5 | 1 | 0.242 | 0.115 | 0.476 | 0.987077 | -0.0078249 | | |
| 14 | | 6 | 1 | 0.425 | 0.100 | 0.235 | 0.999999 | -5.043E-07 | | |
| 15 | | 7 | 1 | 0.290 | 0.090 | 0.310 | 0.999923 | -4.618E-05 | | |
| 16 | | 8 | 1 | 0.416 | 0.095 | 0.230 | 0.999999 | -3.303E-07 | | |
| 17 | | 9 | 1 | 0.367 | 0.108 | 0.294 | 0.999965 | -2.088E-05 | | |
| 18 | | 10 | 1 | 0.416 | 0.086 | 0.207 | 1 | -4.786E-08 | | |
| 168 | | 160 | 0 | 0.433 | 0.458 | 1.059 | 0.25156 | -0.65137 | | |
| 169 | | 161 | 0 | 0.288 | 0.694 | 2.407 | 0.000132 | -0.0002974 | | |
| 170 | | 162 | 0 | 0.297 | 0.530 | 1.782 | 0.005272 | -0.0118826 | | |
| 171 | | 163 | 0 | 0.451 | 0.641 | 1.423 | 0.040832 | -0.0937148 | | |
| 172 | | 164 | 0 | 0.387 | 0.471 | 1.215 | 0.121059 | -0.290067 | | |
| 173 | | 165 | 0 | 0.442 | 0.433 | 0.981 | 0.346507 | -0.9563221 | | |
| 174 | | Notes: | | | | | | | | |
| 175 | | (1) Rows 19 through 167 are skipped to save space. | | | | | | | | |
| 176 | | (2) The setting in Solver is "Maximize the value in Cell J9 by changing the values in Cells H6 and I6, | | | | | | | | |
| 177 | | respectively." | | | | | | | | |

Parameters to be estimated

Log-likelihood

*Figure 4.7* Spreadsheet template for calibrating the probabilistic model that is based on the Robertson and Wride method.

and field case histories (e.g., Lee and Albaisa 1974; Tokimatsu and Seed 1984; Ishihara and Yoshimine 1992; Zhang et al. 2002; Dashti et al. 2010). For a site with level ground, far from any free water surface, it is reasonable to assume little or no lateral displacement occurs after an earthquake. Thus, the volumetric strain is approximately equal to the vertical strain. The liquefaction-induced settlement caused by a given earthquake can then be determined as a summation of the product of the volumetric strain in each liquefied soil layer and the corresponding depth, symbolically expressed as follows (Juang et al. 2013):

$$s_p = \sum_{i=1}^{N} \varepsilon_{vi} \cdot \Delta z_i \cdot I_i \tag{4.32}$$

where $s_p$ = predicted settlement at the ground surface, $N$ = total number of soil layers, $\Delta z_i$ = thickness of the $i$th layer, $\varepsilon_{vi}$ = volumetric strain of the $i$th layer, and $I_i$ = an indicator of liquefaction occurrence in the $i$th layer, which is equal to 0 if the $i$th layer does not liquefy and equal to 1 if the $i$th layer liquefies. Equation 4.32 is an extension of the method suggested by Zhang et al. (2002) by including a likelihood of soil liquefaction (Juang et al. 2013). Zhang et al. (2002) coupled the CPT-based method by Robertson and Wride (1998) with the volumetric strain relationship defined by Ishihara and Yoshimine (1992) to provide a design chart for estimating the volumetric strain $\varepsilon_v$. Through curve fitting, the chart by Zhang et al. (2002) is approximated with the following equation (Juang et al. 2013):

$$\varepsilon_v(\%) = \begin{cases} 0 & \text{if} \quad F_S \geq 2 \\ \min\left(\dfrac{a_0 + a_1 \ln(q)}{1/(2 - F_S) - (a_2 + a_3 \ln(q))}, b_0 + b_1 \ln(q) + b_2 \ln(q)^2\right) & \text{if} \quad 2 - \dfrac{1}{a_2 + a_3 \ln(q)} < F_S < 2 \\ b_0 + b_1 \ln(q) + b_2 \ln(q)^2 & \text{if} \quad F_S \leq 2 - \dfrac{1}{a_2 + a_3 \ln(q)} \end{cases}$$

(4.33)

where

$$a_0 = 0.3773, \quad a_1 = -0.0337, \quad a_2 = 1.5672, \quad a_3 = -0.1833,$$
$$b_0 = 28.45, \quad b_1 = -9.3372, \quad b_2 = 0.7975,$$
$$q = q_{t1N,cs}, \text{ in kg/cm}^2 (\approx 100 \text{kPa}).$$

where $F_S$ = factor of safety calculated using the Robertson and Wride method.

As it is difficult to predict with certainty whether the soil will liquefy, the liquefaction indicator $I_i$ in Equation 4.32 is modeled as a binomial random variable. Let $P_{Li}$ denote the probability of liquefaction of layer $i$. Based on the property of a binomial distribution (e.g., Ang and Tang 2007), the first two moments of $I_i$ can be calculated as follows: $E[I_i] = P_{Li}$ and $Var[I_i] = P_{Li}(1 - P_{Li})$. According to Equation 4.31, $P_L$ can be calculated based on the factor of safety $F_S$ computed with the Robertson and Wride method.

In Equation 4.33, deterministic nominal values are adopted for $q_{t1N,cs}$ and $F_S$. Hence, $\varepsilon_{vi}$ in Equation 4.32 is also a deterministic value. Thus, $I$ is the only random variable in Equation 4.32. Based on Equation 4.32, the mean of $s_p$ can be determined as follows:

$$\mu_p = E\left[\sum_{i=1}^{N} \varepsilon_{vi} \cdot \Delta z_i \cdot I_i\right] = \sum_{i=1}^{N} \varepsilon_{vi} \cdot \Delta z_i \cdot E[I_i] = \sum_{i=1}^{N} \varepsilon_{vi} \cdot \Delta z_i \cdot P_{Li}$$

(4.34)

Further, if the indicator functions ($I_i$, $i = 1, \ldots, N$) are assumed independent from each other, then the variance of the predicted settlement can be determined as follows:

$$\sigma_p^2 = Var\left[\sum_{i=1}^{N} \varepsilon_{vi} \cdot \Delta z_i \cdot I_i\right] = \sum_{i=1}^{N} \varepsilon_{vi}^2 \cdot \Delta z_i^2 \cdot Var[I_i] = \sum_{i=1}^{N} \varepsilon_{vi}^2 \cdot \Delta z_i^2 \cdot P_{Li}(1 - P_{Li})$$

(4.35)

As the simplified modeling assumptions are involved, it is reasonable to expect that Equation 4.32 is not perfect. To consider the model error in settlement prediction, a model bias factor $\alpha$ can be applied to Equation 4.32 as follows:

$$s_a = \alpha s_p$$

(4.36)

where $s_a$ is the actual settlement for a future case. To avoid negative values in settlement, it is reasonable to assume that $\alpha$ is lognormally distributed.

Let $\mu_\alpha$ and $\sigma_\alpha$, respectively, denote the mean and standard deviation of $\alpha$. In Equation 4.36, $s_p$ can be calculated with Equation 4.32. Hence, the uncertain parameters to be calibrated can be denoted as $\theta = \{\mu_\alpha, \sigma_\alpha\}$. Assuming that $\alpha$ is independent from $I_i$, and thus is independent from $s_p$, then the mean and variance of $s_a$ can be determined, respectively, as follows (Ang and Tang 2007):

$$\mu_a = E[s_a] = E[\alpha s_p] = \mu_\alpha \cdot \mu_p \tag{4.37}$$

and

$$\sigma_a^2 = \text{Var}[s_a] = \text{Var}[\alpha s_p] = \mu_\alpha^2 \cdot \sigma_p^2 + \sigma_\alpha^2 \cdot \mu_p^2 + \sigma_\alpha^2 \cdot \sigma_p^2 \tag{4.38}$$

The COV of $s_a$, denoted as $\delta_a$, can then be computed as

$$\delta_a = \sigma_a / \mu_a = \left(\mu_\alpha^2 \cdot \sigma_p^2 + \sigma_\alpha^2 \cdot \mu_p^2 + \sigma_\alpha^2 \cdot \sigma_p^2\right)^{0.5} / (\mu_\alpha \cdot \mu_p) \tag{4.39}$$

### 4.5.2 Calibration database

To calibrate the model bias factor $\alpha$, a database consisting of 64 case histories was compiled and summarized in Juang et al. (2013). Among the 64 cases, 21 were obtained from the 1989 Loma Prieta, California, earthquake; 19 case histories were obtained from the 1999 Kocaeli, Turkey, earthquake; and 24 case histories were from the 1999 Chi-Chi, Taiwan, earthquake. The database consists of 32 liquefaction-induced free-field settlement observations and 32 liquefaction-induced building settlement observations. The settlement behaviors of the two groups of observations are different because the building settlement observations are complicated by the soil–structure interaction. Interested readers are referred to Juang et al. (2013) for further discussions. For our purposes here, the 32 observations on liquefaction-induced free-field settlement, the details of which are summarized in Table 4.7, were used for our model calibration. In this database, observations of the post-liquefaction ground settlement may be categorized as either a fixed value or a range. How these two types of data can be used together for model calibration will be explained in the following section.

### 4.5.3 Maximum likelihood estimation of statistics of model bias factor

Consider a general situation where the database consists of $m + n$ case histories of liquefaction-induced settlement, where $m$ is the number of cases with a fixed-value settlement observation and $n$ is the number of cases in which the settlement encompasses a range of values. Thus, in a special situation when $n = 0$, the database will consist of only cases with fixed-value observations; and when $m = 0$, the database will consist of only cases with range observations. In this section, we deal with a database of both fixed-value observations and range observations. In the following, we will use $i$ and $j$ as indexes for a fixed-value observation and a range observation, respectively. For a fixed-value observation, $\mathbf{d}_i = \{s_i\}$. Assuming $s_i$ follows lognormal distribution, the chance to observe a fixed value observation $s_i$ is

$$f(\mathbf{d}_i | \theta) = f(s_i | \theta) = \frac{1}{\sqrt{2\pi} s_i \xi_i} \exp\left[-\frac{1}{2}\left(\frac{\ln s_i - \lambda_i}{\xi_i}\right)^2\right] \tag{4.40}$$

Table 4.7 Summary of case histories for post-liquefaction settlement

| CPT designation | GW (m) | $a_{max}$ (g) | $M_w$ | Observed settlement (cm) | $\mu_p$ by Equation 4.34 (cm) | $\sigma_p^2$ by Equation 4.35 (cm²) | $\mu_a$ by Equation 4.37 (cm) | $\delta_a$ by Equation 4.39 | Earthquake and site |
|---|---|---|---|---|---|---|---|---|---|
| C2 | 2.3 | 0.24 | 7.0 | 9.6–10.7 | 10.82 | 0.64 | 11.31 | 0.33 | 1989 Loma Prieta Marina District |
| C8 | 2.7 | 0.16 | 7.0 | 1.9 | 1.31 | 0.29 | 1.37 | 0.54 | 1989 Loma Prieta Marina District |
| C9 | 2.6 | 0.16 | 7.0 | 0–3.4 | 0.02 | 0 | 0.02 | 0.32 | 1989 Loma Prieta Marina District |
| C12 | 2.3 | 0.24 | 7.0 | 7.0–10.7 | 6.83 | 0.73 | 7.14 | 0.34 | 1989 Loma Prieta Marina District |
| M1 | 2.3 | 0.16 | 7.0 | 0–3.4 | 4.19 | 0.29 | 4.38 | 0.34 | 1989 Loma Prieta Marina District |
| M2 | 2.7 | 0.16 | 7.0 | 0–3.4 | 0.94 | 0.09 | 0.98 | 0.46 | 1989 Loma Prieta Marina District |
| M3 | 2.7 | 0.16 | 7.0 | 1.1 | 0.6 | 0.06 | 0.63 | 0.53 | 1989 Loma Prieta Marina District |
| M4 | 2.4 | 0.24 | 7.0 | 9.6 | 10.75 | 0.35 | 11.23 | 0.32 | 1989 Loma Prieta Marina District |
| M6 | 5.5 | 0.12 | 7.0 | 0–1.6 | 0.04 | 0 | 0.04 | 0.32 | 1989 Loma Prieta Marina District |
| C28 | 2 | 0.16 | 7.0 | 5–10 | 10.42 | 0.5 | 10.89 | 0.33 | 1989 Loma Prieta Treasure Island Zone one |
| C32 | 2 | 0.16 | 7.0 | 5–10 | 9.6 | 0.4 | 10.03 | 0.32 | 1989 Loma Prieta Treasure Island Zone one |
| C33 | 2 | 0.16 | 7.0 | 5–10 | 6.82 | 0.38 | 7.13 | 0.33 | 1989 Loma Prieta Treasure Island Zone one |
| C34 | 2 | 0.16 | 7.0 | 5–10 | 5.48 | 0.32 | 5.73 | 0.34 | 1989 Loma Prieta Treasure Island Zone one |
| C35 | 2 | 0.16 | 7.0 | 5–10 | 5.66 | 0.28 | 5.92 | 0.33 | 1989 Loma Prieta Treasure Island Zone one |
| C37 | 2 | 0.16 | 7.0 | 5–10 | 5.78 | 0.3 | 6.04 | 0.33 | 1989 Loma Prieta Treasure Island Zone one |
| C42 | 2 | 0.16 | 7.0 | 5–10 | 10.28 | 0.53 | 10.74 | 0.33 | 1989 Loma Prieta Treasure Island Zone one |

*continued*

Table 4.7 (Continued) Summary of case histories for post-liquefaction settlement

| CPT designation | GW (m) | $a_{max}$ (g) | $M_w$ | Observed settlement (cm) | $\mu_p$ by Equation 4.34 (cm) | $\sigma_p^2$ by Equation 4.35 (cm²) | $\mu_a$ by Equation 4.37 (cm) | $\delta_a$ by Equation 4.39 | Earthquake and site |
|---|---|---|---|---|---|---|---|---|---|
| C29A | 2 | 0.16 | 7.0 | 10–15 | 9.26 | 0.48 | 9.68 | 0.33 | 1989 Loma Prieta Treasure Island Zone two |
| C30 | 2 | 0.16 | 7.0 | 10–15 | 14.59 | 0.72 | 15.25 | 0.32 | 1989 Loma Prieta Treasure Island Zone two |
| C31 | 2 | 0.16 | 7.0 | 10–15 | 21.82 | 0.92 | 22.80 | 0.32 | 1989 Loma Prieta Treasure Island Zone two |
| C39 | 2 | 0.16 | 7.0 | 10–15 | 15.56 | 0.63 | 16.26 | 0.32 | 1989 Loma Prieta Treasure Island Zone two |
| UM10 | 2 | 0.16 | 7.0 | 10–15 | 18.35 | 0.64 | 19.18 | 0.32 | 1989 Loma Prieta Treasure Island Zone two |
| TW-WF-C7 | 3.2 | 0.79 | 7.6 | 20 | 12.51 | 0.01 | 13.07 | 0.32 | 1999 Chi-Chi Wufeng |
| TW-WF-C9 | 1.38 | 0.79 | 7.6 | 14–28 | 15.03 | 0 | 15.71 | 0.32 | 1999 Chi-Chi Wufeng |
| TW-WF-C10 | 1.4 | 0.79 | 7.6 | 15 | 14.13 | 0 | 14.77 | 0.32 | 1999 Chi-Chi Wufeng |
| TW-WF-C15 | 2 | 0.79 | 7.6 | 15–16 | 11.16 | 0.01 | 11.66 | 0.32 | 1999 Chi-Chi Wufeng |
| TW-YL-C12 | 1.58 | 0.19 | 7.6 | 8–13 | 13.72 | 0.23 | 14.34 | 0.32 | 1999 Chi-Chi Yuanlin |
| TW-YL-C20 | 1.4 | 0.19 | 7.6 | 20 | 21.76 | 0.15 | 22.74 | 0.32 | 1999 Chi-Chi Yuanlin |
| TW-YL-C32 | 0.7 | 0.19 | 7.6 | 10–20 | 22.56 | 0.16 | 23.58 | 0.32 | 1999 Chi-Chi Yuanlin |
| TW-YL-C45 | 2.3 | 0.19 | 7.6 | 15–20 | 20.34 | 0.47 | 21.26 | 0.32 | 1999 Chi-Chi Yuanlin |
| TW-LW-C1 | 2 | 0.12 | 7.6 | 36.8 | 21.37 | 1.2 | 22.33 | 0.32 | 1999 Chi-Chi Lunwei |
| TW-LK-EQ3 | 1.5 | 0.12 | 7.6 | 43 | 24.59 | 1.33 | 25.70 | 0.32 | 1999 Chi-Chi Lukang |
| TW-DN-D1 | 1.5 | 0.18 | 7.6 | 10 | 13.94 | 0.35 | 14.57 | 0.32 | 1999 Chi-Chi Dounan |

Source: Adapted from Juang, C.H. et al. 2013. *Canadian Geotechnical Journal*, 50(10), 1055–1066. With permission from NRC Research Press.

where $\lambda_i = \ln\left\{\mu_{ai}/\sqrt{1 + \delta_{ai}^2}\right\}$ is the mean value of the variable $\ln(s_i)$, and $\xi_i = \sqrt{\ln\left(1 + \delta_{ai}^2\right)}$ is the standard deviation of the variable $\ln(s_i)$.

Let $s_{jl}$ and $s_{ju}$ denote the lower and upper bound values of observed settlement for the $j$th range observation, respectively. For a range observation, the observed data can be denoted as $\mathbf{d}_j = \{s_{jl}, s_{ju}\}$. Based on the lognormal distribution assumption, the chance to observe such a range observation can be written as

$$P(\mathbf{d}_j|\theta) = P(s_{jl} \leq s_j \leq s_{ju}|\theta) = \Phi\left(\frac{\ln s_{ju} - \lambda_j}{\xi_j}\right) - \Phi\left(\frac{\ln s_{jl} - \lambda_j}{\xi_j}\right) \tag{4.41}$$

Note that the database with range observations analyzed herein may be considered as a generalization of the database with only censored data analyzed previously. For censored data, only the lower bound or the upper bound is known; for a range observation, both the lower and the upper bounds are known. For the database with $m$ fixed-value observations and $n$ range observations, the likelihood function can be obtained by multiplying the chance for each observation together (regardless of whether an observation is a fixed-value observation or a range observation), as long as these observations are statistically independent. Note the chance for a fixed-value observation and a range observation can be calculated with Equations 4.40 and 4.41, respectively. Therefore, the log-likelihood function of $\theta$ can then be written as follows:

$$L(\theta|\mathbf{D}) = \sum_{i=1}^{m}\left\{-\ln\left(\sqrt{2\pi}s_i\xi_i\right) - \frac{1}{2}\left(\frac{\ln s_i - \lambda_i}{\xi_i}\right)^2\right\} + \sum_{j=1}^{n}\ln\left[\Phi\left(\frac{\ln s_{ju} - \lambda_j}{\xi_j}\right) - \Phi\left(\frac{\ln s_{jl} - \lambda_j}{\xi_j}\right)\right]$$
$$\tag{4.42}$$

The spreadsheet template for maximizing likelihood in Equation 4.42 is shown in Figure 4.8. With Solver, the optimal values of $\theta$ are $\theta^* = \{\mu_\alpha^*, \sigma_\alpha^*\} = \{1.045, 0.318\}$.

To determine the liquefaction-induced settlement in a future case, Equations 4.34 and 4.35 are first used to compute $\mu_p$ and $\sigma_p$, respectively. With the knowledge of bias factor, $\{\mu_\alpha^*, \sigma_\alpha^*\} = \{1.045, 0.318\}$, we can compute the mean and standard deviation of the actual settlement using Equations 4.37 and 4.38.

## 4.6 SUMMARY AND CONCLUSIONS

We illustrated the effectiveness of the maximum likelihood principle for calibrating probabilistic models for liquefaction hazard evaluation using observed field performance data. In this chapter, we show how the likelihood function can be constructed for different types of data with an emphasis on applications in liquefaction evaluation. For most of the examples explained in this chapter, the likelihood functions are well behaved and the maximum likelihood point can be found through a convenient use of optimization tools such as Excel Solver. A detailed formulation of the model calibration with illustrated examples presented in this chapter clearly show the development of these probabilistic models for evaluating liquefaction hazards, and how they may be used in future events.

When developing models for liquefaction probability, it is generally understood that existing censored databases used for model calibration are subjected to sampling bias. Although the weighted likelihood method can be used to adjust the effect of sampling bias, the exact value of sampling bias is usually hard to assess due to the difficulties encountered when

| A | B | C | D | E | F | G | H | I | J | K | L | M | N | O | P | Q |
|---|---|---|---|---|---|---|---|---|---|---|---|---|---|---|---|---|

**Spreadsheet Template for Calibrating a Liquefaction Induced Settlement Prediction Model**

Parameters to be estimated

| $\mu_o$ | $\delta_o$ |
|---|---|
| 1.0451 | 0.3175 |

Observed data    Eqs. (4.34) and (4.35)    Eqs. (4.37) and (4.39)    Eq. (4.41)

| | Case No. | Range ? | $s_{il}$ (cm) | $s_{iu}$ (cm) | $\mu_p$ (cm) | $\sigma_p^2$ (cm²) | $\delta_p$ | $\mu_a$ (cm) | $\delta_a$ | $\lambda_a$ | $\xi_a$ | $P(d_i|\theta)$ | $\ln P(d_i|\theta)$ | $L(\theta|D)$ |
|---|---|---|---|---|---|---|---|---|---|---|---|---|---|---|
| 9 | 1 | 1 | 9.60 | 10.70 | 10.82 | 0.64 | 0.07 | 11.308 | 0.3268 | 2.375 | 0.319 | 0.1329 | 2.017874 | -26.758609 |
| 10 | 2 | 0 | 1.90 | | 1.31 | 0.29 | 0.41 | 1.3691 | 0.5356 | 0.188 | 0.502 | 0.5281 | 0.638443 | |
| 11 | 3 | 1 | 0.00 | 3.40 | 0.02 | 0 | 0.00 | 0.0209 | 0.3175 | -3.92 | 0.31 | 1 | 0 | Log-likelihood |
| 12 | 4 | 1 | 7.00 | 10.70 | 6.83 | 0.73 | 0.13 | 7.138 | 0.3435 | 1.91 | 0.334 | 0.3728 | 0.986635 | |
| 13 | 5 | 1 | 0.00 | 3.40 | 4.19 | 0.29 | 0.13 | 4.379 | 0.3449 | 1.421 | 0.335 | 0.2786 | 1.278018 | |
| 14 | 6 | 1 | 0.00 | 3.40 | 0.94 | 0.09 | 0.32 | 0.9824 | 0.4614 | -0.11 | 0.439 | 0.9988 | 0.001162 | |
| 15 | 7 | 0 | 1.10 | | 0.6 | 0.06 | 0.41 | 0.6271 | 0.5332 | -0.59 | 0.5 | 0.3105 | 1.169726 | |
| 16 | 8 | 0 | 9.60 | | 10.75 | 0.35 | 0.06 | 11.235 | 0.3227 | 2.369 | 0.315 | 1.1954 | 0.178512 | |
| 17 | 9 | 1 | 0.00 | 1.60 | 0.04 | 0 | 0.00 | 0.0418 | 0.3175 | -3.22 | 0.31 | 1 | 0 | |
| 18 | 10 | 1 | 5.00 | 10.00 | 10.42 | 0.5 | 0.07 | 10.89 | 0.3254 | 2.338 | 0.317 | 0.4453 | -0.80903 | |
| 36 | 28 | 1 | 10.00 | 20.00 | 22.56 | 0.16 | 0.02 | 23.577 | 0.318 | 3.112 | 0.31 | 0.3493 | 1.051828 | |
| 37 | 29 | 1 | 15.00 | 20.00 | 20.34 | 0.47 | 0.03 | 21.257 | 0.3194 | 3.008 | 0.312 | 0.3163 | 1.151112 | |
| 38 | 30 | 0 | 36.80 | | 21.37 | 1.2 | 0.05 | 22.334 | 0.322 | 3.057 | 0.314 | 0.2761 | 1.286883 | |
| 39 | 31 | 0 | 43.00 | | 24.59 | 1.33 | 0.05 | 25.699 | 0.3213 | 3.197 | 0.313 | 0.2523 | 1.377071 | |
| 40 | 32 | 0 | 10.00 | | 13.94 | 0.35 | 0.04 | 14.569 | 0.3206 | 2.63 | 0.313 | 0.7376 | 0.304418 | |

Notes:
(1) Rows 19 through 35 are skipped to save space.
(2) The setting in Solver is "Maximize the value in Cell P9 by changing the values in Cells I6 and J6".

*Figure 4.8* Spreadsheet template for calibrating the model correction factor for predicting liquefaction-induced settlement.

evaluating the proportion of liquefied soils ($Q_p$) in the real world (population). Nevertheless, using the same value of $Q_p$ consistently when developing liquefaction prediction models can mitigate the effect of sampling bias and make more comparable those models derived based on different databases.

It is worth repeating the limitations of the principle of maximum likelihood that were articulated by Cam (1990). For example, it is not axiomatic that any principle, including the maximum likelihood principle, can always provide a sensible statistical estimation of model parameters. It is also important to be suspicious of a statistical estimation if the results change abruptly through the suppression of a single observation. Finally, it is often helpful to first try a crude but reliable procedure to locate the general area in which the parameters lie and then refine that estimate using a more elegant method such as the maximum likelihood principle.

## ACKNOWLEDGMENTS

The first author acknowledges the National Science Foundation and U.S. Geological Survey for their multiple grants that have supported his ongoing studies on soil liquefaction since 1997. The third author thanks the National 973 Basic Research Program of China (2011CB013800, 2014CB049100) for the support of his research detailed in this chapter. Much material presented in this chapter was the results of re-analysis of the first author's prior work and the collaborative work by the first author and his collaborators, Dr. Jianye Ching and Dr. Chieh-Sheng Ku. The authors thank Dr. Ching and Dr. Ku for their contributions, Mr. Wenping Gong, Dr. Zhe Luo and Dr. Lei Wang for reviewing the manuscript, and Mr. Godfrey Kimball for providing the requisite editorial assistance.

## APPENDIX 4A: MODEL OF ROBERTSON AND WRIDE (1998) AND ROBERTSON (2009)

$CSR$ is the adjusted cyclic stress ratio defined as

$$CSR = 0.65 \left( \frac{\sigma_{vo}}{\sigma'_{vo}} \right) \left( \frac{a_{max}}{g} \right) (r_d) \left( \frac{1}{\text{MSF}} \right) \left( \frac{1}{K_\sigma} \right)$$

where

$\sigma_{vo}$, $\sigma'_{vo}$ are the vertical total and effective overburden stresses, respectively
$g$ the acceleration of gravity
$a_{max}$ the peak ground surface acceleration (unit: g)
$r_d$ the depth-dependent shear stress reduction factor
MSF the magnitude scaling factor
$K_\sigma$ the overburden correction factor for $CSR$ ($K_\sigma = 1$ for $\sigma'_{vo} < 1$ atm)

The terms $r_d$, MSF, and $K_\sigma$ follow the definitions provided in Youd et al. (2001). For MSF, the lower-bound equation is used

$$\text{MSF} = 10^{2.24}/M_w^{2.56}$$

where $M_w$ = moment magnitude.

Though $CSR$ as defined above is often denoted as $CSR_{7.5}$, for simplicity, $CSR$ is used in this study. In a deterministic model, as in the Robertson and Wride method, $CRR$ is evaluated and compared with $CSR$ to obtain a measure of the liquefaction potential.

Let $q_c$ denote the raw cone resistance, which is generally in MPa, but may require conversion into kPa when used with other parameters to derive the dimensionless parameters such as $Q_t$ and $q_{t1N}$. Let $u_2$ denote the penetration pore pressure (kPa). Let $q_t = q_c + (1 - a)u_2$ denote the corrected cone resistance, where $a$ is the area ratio of the cone used (in this study, $a = 0.85$). In the Robertson and Wride model, $CRR$ is computed as

$$CRR = \begin{cases} 0.833[(q_{t1N,cs})/1000] + 0.05 & q_{t1N,cs} \le 50 \\ 93[(q_{t1N,cs})/1000]^3 + 0.08 & 50 \le q_{t1N,cs} < 160 \end{cases}$$

where

$$q_{t1N,cs} = K_c \, q_{t1N}$$

$q_{t1N}$ = normalized cone tip resistance = $[(q_t - \sigma_{vo})/p_a](p_a/\sigma'_{vo})^n$

$$K_c = \begin{cases} 1.0, & \text{for } I_c \leq 1.64 \\ -0.403I_c^4 + 5.581I_c^3 - 21.63I_c^2 + 33.75I_c - 17.88, & \text{for } I_c > 1.64 \end{cases}$$

$$I_c = \sqrt{(3.47 - \log_{10} Q_t)^2 + (\log_{10} F_r + 1.22)^2}$$

$Q_t$ = normalized cone resistance = $(q_t - \sigma_{vo})/\sigma'_{vo}$
$F_r$ = normalized friction ratio = $f_s/(q_t - \sigma_{vo}) \times 100\%$

## APPENDIX 4B: NOTATION

$a_{max}$ = peak ground surface acceleration
$b_0, b_1, \ldots b_r$ = regression coefficients to be calibrated in a generalized linear model
BIC = Bayesian information criterion
CPT = cone penetration test
$CRR$ = cyclic resistance ratio
$CSR$ = cyclic stress ratio
$c_u$ = undrained shear strength
$c_{ui}$ = $i$th measured value of $c_u$
$\mathbf{D}$ = observed database
$F_S$ = factor of safety
$F_{Sa}$ = actual factor of safety
$F(z)$ = cumulative distribution function of random variable $z$
$g$ = acceleration of gravity
$I$ = indicator factor
$k$ = number of parameters of the model to be estimated
$K_\sigma$ = overburden correction factor for $CSR$
$l(\theta|\mathbf{D})$ = likelihood function of $\theta$ given data $\mathbf{D}$
$MSF$ = magnitude scaling factor
$n_F$ = number of failed slopes
$n_L$ = number of liquefied cases
$n_{NF}$ = number of nonfailed slopes
$n_{NL}$ = number of nonliquefied cases
$P_L$ = probability of liquefaction
$q_{t1N}$ = normalized cone tip resistance
$q_{t1N,cs}$ = clean sand equivalence of normalized cone tip resistance
$Q_p$ = proportion of failed slopes or liquefied cases in real world
$Q_s$ = proportion of failed slopes or liquefaction sites in a calibration database
$r_d$ = depth-dependent shear stress reduction factor
$s_a$ = actual total settlement at the ground surface
$s_i$ = the $i$th fixed observation of liquefied-induced settlement
$s_j$ = the $j$th range observation of liquefied-induced settlement
$s_{jl}$ = lower bound of $s_j$
$s_{ju}$ = upper bound of $s_j$
$s_p$ = predicted total settlement at the ground surface

$SPT$ = standard penetration test

$T = n \times n$ correlation matrix with element in the $i$th row and $j$th column denoted by $\mathbf{T}_{ij}$

$\mathbf{T}_{ij}$ = correlation coefficient between $\ln(c_u)$ at location $i$ and location $j$

$w_L$ = weighting factor to apply to liquefied cases in the weighted likelihood function

$w_{NL}$ = weighting factor to apply to non-liquefied cases in the weighted likelihood function

$\alpha$ = model correction factor

$\beta_a$ = actual reliability index

$\beta_c$ = calculated reliability index

$\delta_a$ = COV of $S_a$

$\delta_\alpha$ = COV of $\alpha$

$\Delta$ = distance between two points in a random field

$\Delta z_i$ = thickness of the $i$th layer

$\varepsilon_\beta$ = correction factor for reliability index

$\varepsilon_{vi}$ = volumetric strain of the $i$th layer

$\theta$ = vector denoting uncertain parameters to be estimated

$\theta^*$ = value of $\theta$ where the likelihood function is maximized

$\lambda$ = mean of the logarithm of a lognormal random variable

$\Lambda$ = n-dimensional column vector with all elements being $\lambda$

$\mu$ = mean of a random variable

$\mu_a$ = mean of $S_a$

$\mu_p$ = mean of $S_p$

$\mu_\alpha$ = mean of $\alpha$

$\xi$ = standard deviation of the logarithm of a lognormal random variable

$\rho$ = correlation coefficient

$\rho(\Delta)$ = correlation function

$\sigma$ = standard deviation of a random variable

$\sigma_a$ = standard deviation of $S_a$

$\sigma_p$ = standard deviation of $S_p$

$\sigma_\alpha$ = standard deviation of $\alpha$

$\sigma_{vo}, \sigma'_{vo}$ = vertical total and effective overburden stresses, respectively

$\varphi$ = friction angle

$\Omega$ = correlation distance

## REFERENCES

Aldrich, J. 1997. R. A. Fisher and the making of maximum likelihood 1912–1922. *Statistical Science*, 12(3), 162–176.

Ang, A.H.S., and Tang, W.H., 2007. *Probability Concepts in Engineering: Emphasis on Applications to Civil and Environmental Engineering*, 2nd edition, Wiley, New York.

Barnett, V. 1999. *Comparative Statistical Enference*, 3rd edition, Wiley, New York.

Boulanger, R.W., and Idriss, I.M. 2012. Probabilistic SPT-based liquefaction triggering procedure. *Journal of Geotechnical and Geoenvironmental Engineering*, 138(10), 1185–1195.

Burnham, K.P., and Anderson, D.R. 2004. Multimodel inference: Understanding AIC and BIC in model selection. *Sociological Methods & Research*, 33(2), 261–304.

Cam, L.L. 1990. Maximum Likelihood: An Introduction. *International Statistical Review*, 58(2), 153–171.

Cetin, K.O., Der Kiureghian, A., and Seed, R.B. 2002. Probabilistic models for the initiation of seismic soil liquefaction. *Structural Safety*, 24(1), 67–82.

Cheng, R.C.H., and Traylor, L. 1995. Non-regular maximum likelihood problems. *Journal of the Royal Statistical Society*, Series B, 57(1), 3–44.

Christian, J.T., and Swiger, W.F. 1975. Statistics of liquefaction and SPT results. *Journal of Geotechnical Engineering*, 101(11), 1135–1150.

Dashti, S., Bray, J., Pestana, J., Riemer, M., and Wilson, D. 2010. Mechanisms of seismically induced settlement of buildings with shallow foundations on liquefiable soil. *Journal of Geotechnical and Geoenvironmental Engineering*, 136(1), 151–164.

Edwards, A.W.F. 1974. The history of likelihood. *International Statistical Review*, 42(1), 9–15.

Fenton, G. 1999. Estimation for stochastic soil models. *Journal of Geotechnical and Geoenvironmental Engineering*, 125(6), 470–485.

Gentle, J.E. 2002. *Elements of Computational Statistics*, Springer-Verlag, New York.

Givens, G., and Hoeting, J. 2005. *Computational Statistics*, Wiley, New York.

Hoffmann, J.P. 2004. *Generalized Linear Models: An Applied Approach*, Pearson, Boston.

Hwang, J.H., Chen, C.H., and Juang, C.H. 2005. *Liquefaction Hazard Analysis: A Fully Probabilistic Method*, Geotechnical Special Publication No. 133, ASCE, Reston, VA.

Ishihara, K., and Yoshimine, M. 1992. Evaluation of settlements in sand deposits following liquefaction during earthquakes. *Soils and Foundations*, 32(1), 173–188.

Juang, C.H., Rosowsky, D.V., and Tang, W.H. 1999. Reliability-based method for assessing liquefaction potential of sandy soils. *Journal of Geotechnical and Geoenvironmental Engineering*, 125(8), 684–689.

Juang, C.H., Chen, C.J., Rosowsky, D.V., and Tang, W.H. 2000. CPT-based liquefaction analysis, Part 2: Reliability for design, *Geotechnique*, 50(5), 593–599.

Juang, C., Jiang, T., and Andrus, R. 2002. Assessing probability-based methods for liquefaction potential evaluation. *Journal of Geotechnical and Geoenvironmental Engineering*, 128(7), 580–589.

Juang, C.H., Yuan, H., Lee, D.H., and Lin, P.S. 2003. Simplified cone penetration test-based method for evaluating liquefaction resistance of soils. *Journal of Geotechnical and Geoenvironmental Engineering*, 129(1), 66–80.

Juang, C.H., Fang, S.Y., and Khor, E.H. 2006. First order reliability method for probabilistic liquefaction triggering analysis using CPT, *Journal of Geotechnical and Geoenvironmental Engineering*, 132(3), 337–350.

Juang, C.H., Fang, S.Y., Tang, W.H., Khor, E.H., Kung, G.T.C., and Zhang, J. 2009. Evaluating model uncertainty of an SPT-based simplified method for reliability analysis for probability of liquefaction, *Soils and Foundations*, 49(1), 147–152.

Juang, C.H., Ching, J., Wang, L., Khoshnevisan, S., and Ku, C.S. 2013. Simplified procedure for estimation of liquefaction-induced settlement and site-specific probabilistic settlement exceedance curve using cone penetration test (CPT). *Canadian Geotechnical Journal*, 50(10), 1055–1066.

Ku, C.S., Juang, C.H., Chang, C.W., and Ching, J. 2012. Probabilistic version of the Robertson and Wride method for liquefaction evaluation: Development and application. *Canadian Geotechnical Journal*, 49(1), 27–44.

Lai, S.Y., Chang, W.J., and Lin, P.S. 2006. Logistic regression model for evaluating soil liquefaction probability using CPT data. *Journal of Geotechnical and Geoenvironmental Engineering*, 132(6), 694–704.

Law, A.M., and Kelton, W.D. 2000. *Simulation Modeling and Analysis*, 3rd edition, McGraw-Hill, New York.

Lee, K.L., and Albaisa, A. 1974. Earthquake-induced settlement in saturated sand. *Journal of Geotechnical Engineering Division*, 100(4), 387–406.

Lehmann, E., and Casella, G. 1998. *Theory of Point Estimation*, 2nd edition, Springer-Verlag, New York.

Liao, S.S.C., Veneziano, D., and Whitman, R.V. 1988. Regression model for evaluating liquefaction probability. *Journal of Geotechnical Engineering*, 114(4), 389–410.

Manski, C.F., and Lerman, S.R. 1977. The estimation of choice probabilities from choice-based samples. *Econometrica*, 45(8), 1977–1988.

Moss, R.E.S., Seed, R.B., Kayen, R.E., Stewart, J.P., Der Kiureghian, A., Cetin, K.O. 2006. CPT-based probabilistic and deterministic assessment of *in situ* seismic soil liquefaction potential. *Journal of Geotechnical and Geoenvironmental Engineering*, 132(8), 1032–1051.

Moss, R.E.S., Kayen, R.E., Tong, L.Y., Liu, S.Y., Cai, G.J., and Wu, J. 2011. Retesting of liquefaction and nonliquefaction case histories from the 1976 Tangshan Earthquake. *Journal of Geotechnical and Geoenvironmental Engineering*, 137(4), 334–343.

Robertson, P.K. 2009. Performance based earthquake design using the CPT. *Proc., IS Tokyo Conf.*, CRC Press/Balkema, Taylor & Francis Group, Tokyo, pp. 3–20.

Robertson, P.K., and Wride, C.E. 1998. Evaluating cyclic liquefaction potential using the cone penetration test. *Canadian Geotechnical Journal*, 35(3), 442–459.

Schwarz, G. 1978. Estimating the dimension of a model. *Annals of Statistics*, 6(2), 461–464.

Seed, H.B., Idriss, I.M. 1971. Simplified procedure for evaluating soil liquefaction potential. *Journal of the Soil Mechanics and Foundation Division*, 97(9), 1249–1273.

Stigler, S.M. 2007. The epic story of maximum likelihood. *Statistical Science*, 22(4), 598–620.

Tang, X.S., Li, D.Q., Rong, G., Phoon, K.K., and Zhou, C.B. 2013. Impact of copula selection on geotechnical reliability under incomplete probability information. *Computers and Geotechnics*, 49, 264–278.

Tokimatsu, K., and Seed, H.B. 1984. Simplified procedures of the evaluation of settlements in clean sands. Earthquake Engineering Research Center Report CB/EERC-84/16, Univ. of California, Berkeley, Calif.

Toprak, S., Holzer, T.L., Bennett, M.J., and Tinsley, J.C.I. 1999. CPT- and SPT-based probabilistic assessment of liquefaction potential. *Proceedings of the 7th US - Japan Workshop on Earthquake Resistant Design of Lifeline Facilities and Countermeasures against Soil Liquefaction* (ed. by T.D. O'Rourke, J.P. Bardet, and M. Hamada), Seattle, August 1999, State University of New York at Buffalo, Buffalo, New York, 69–86.

Vanmarcke, E.H. 1983. *Random Fields—Analysis and Synthesis*, MIT-Press, Cambridge, Massachusetts.

Youd, T.L., Idriss, I.M., Andrus, R.D., Arango, I., Castro, G., Christian, J.T., Dobry, R., Liam Finn, W.D., Harder, L.F., Jr., Hynes, M.E. et al. 2001. Liquefaction resistance of soils: summary report from the 1996 NCEER and 1998 NCEER/NSF workshops on evaluation of liquefaction resistance of soils. *Journal of Geotechnical and Geoenvironmental Engineering*, 127(10), 817–833.

Zhang, G., Robertson, P.K., and Brachman, R.W.I. 2002. Estimating Liquefaction-induced ground settlements from CPT for level ground. *Canadian Geotechnical Journal*, 39(5), 1168–1180.

Zhang, J., Tang, W.H., and Zhang, L.M. 2010. Efficient probabilistic back-analysis of slope stability model parameters. *Journal of Geotechnical and Geoenvironmental Engineering*, 136(1), 99–109.

Zhang, J., Tang, W.H., Zhang, L.M., and Zheng, Y.R. 2007. Calibrating and comparing reliability analysis procedures for slope stability problems. *Proc. 1st Int. Symp. Geotech. Safety and Risk*, Tongji University, Shanghai, China, 205–216.

Zhang, J., Zhang, L.M., Huang, H.W. 2013. Evaluation of generalized linear models for soil liquefaction probability prediction. *Environmental Earth Sciences*, 68, 1925–1933.

# Chapter 5

# Bayesian analysis for learning and updating geotechnical parameters and models with measurements

*Daniel Straub and Iason Papaioannou*

## 5.1 INTRODUCTION

Geotechnical planning and construction is typically associated with large uncertainties and limited data on site conditions. To describe the geotechnical performance as accurately as possible, it is thus necessary to combine information from different sources (site measurements, expert knowledge, and data from literature). The engineers collect a few hypotheses about site conditions and then gather field observations (e.g., measurements of deformations, stresses, or other relevant data) to identify the correct hypothesis. As we show in this chapter, this process can be formalized through Bayesian updating as part of a probabilistic reliability and risk assessment. Thereby, a prior probabilistic model is updated with the new data to a posterior probabilistic model, which is then the basis for further reliability and risk assessments. Bayesian updating has significant advantages over other methods for learning geotechnical models, due to its flexibility and the possibility to consistently combine data and observations from various sources with mechanical models and expert estimates.

In this chapter, we present the basic concepts and theory of Bayesian updating, together with simple and advanced computational methods and algorithms. Following an introduction to the theory in Section 5.2, a hands-on presentation of the method is provided in Section 5.3. By means of didactical examples, it will be demonstrated how the procedure is implemented through the following steps:

1. Establishing an initial prior model
2. Computing the reliability and risk based on the prior model
3. Describing new observations and data
4. Updating the model
5. Updating the reliability and risk
6. Communicating the results

In the general case, Bayesian updating is performed numerically on computationally demanding models, and efficient algorithms are thus required. Section 5.4 provides an introduction to the state-of-the-art algorithms, starting out with conceptually easy-to-understand algorithms and closing with an overview on computationally more efficient algorithms. In Sections 5.5 and 5.6, we present two applications of Bayesian updating, which highlight different aspects of the theory. In Section 5.5, we consider a transmission tower foundation, where the probabilistic model and the reliability estimate are updated with site measurements and with observed performance under past loadings. In Section 5.6, we demonstrate Bayesian updating in the context of finite-element models; deformation measurements are used to update a spatial random field model of soil parameters as well as the reliability. We conclude with some final remarks on the potential and challenges faced by Bayesian updating in the context of geotechnical risk and reliability applications.

## 5.2 BAYESIAN ANALYSIS

Bayesian analysis allows one to consistently and effectively combine new information (measurement, data) with existing models. As illustrated in Figure 5.1, Bayesian updating can be applied to learn model parameters (a) from direct sampling and measurements of soil and loading parameters, and (b) from measurements and observations of system parameters. Task (a) corresponds to classical statistical inference, where the probability distribution of the model parameters is learned based on samples; task (b) corresponds to a probabilistic solution of an inverse analysis or parameter identification. In this chapter, we will consider both applications of Bayesian updating.

For illustration purposes, consider the following situation: A construction is planned with a geotechnical model, using initial estimates of soil and hydraulic parameters. The model predicts stability, deformations, and groundwater flow at the site. Such a model and its parameters are subject to uncertainty, which in the classical geotechnical design are addressed through the use of safety factors, characteristic values, and conservative assumptions. Alternatively, it is possible to quantify the uncertainty explicitly through a probabilistic analysis. Thereby, the main uncertain or random parameters of the model are represented by random variables with corresponding probability distributions. We denote the set of these random variables with $\mathbf{X}$. Through the geotechnical model, all events of interest can be expressed as a function of $\mathbf{X}$. As an example, the event $F =$ "loss of stability" is expressed through a limit state function (LSF) $g(\mathbf{X})$, so that $F = \{g(\mathbf{X}) \leq 0\}$, that is, failure corresponds to the LSF taking negative values.

During the construction, additional measurements become available, for example, additional soil probes are taken and tested, deformations are measured, or water ingress is monitored. These measurements provide information on $\mathbf{X}$, either directly or indirectly. The information is direct if one or more of the parameters in $\mathbf{X}$ are measured, for example, measurements of soil parameters. The information is indirect if the measurement outcomes are related to the parameters $\mathbf{X}$ through some model, as is, for example, the case if deformations are measured at the construction site. Mathematically, these measurements are events, which we here denote with $Z$. In the above example, let the measurement outcome be = "deformation between 20 and 30 mm."

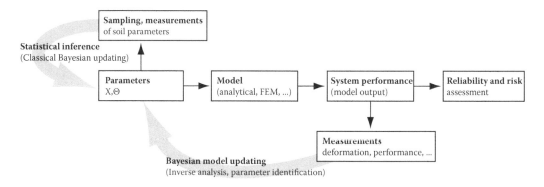

*Figure 5.1* Overview on Bayesian updating in the context of geotechnical models.

The goal is now to quantify the impact of the measurement outcome $Z$ on the parameters $\mathbf{X}$ and, ultimately, on the event of interest $F$. In Bayesian analysis, this is carried out by computing the conditional probability of $F$ given the information $Z$: $\Pr(F|Z)$. For the above example, we assume that the measured deformation is more than initially predicted. In this case, the new information will increase the estimated probability of $F$, that is, $\Pr(F|Z) > \Pr(F)$.

The conditional probability is defined as

$$\Pr(F|Z) = \frac{\Pr(F \cap Z)}{\Pr(Z)}. \tag{5.1}$$

Following Bayes' rule, this conditional probability can be calculated by

$$\Pr(F|Z) = \frac{\Pr(Z|F)\Pr(F)}{\Pr(Z)}. \tag{5.2}$$

In Bayes' rule (and in Bayesian analysis, in general), the terms have the following meaning:

$\Pr(F)$: prior probability;

$\Pr(F|Z)$: posterior probability;

$\Pr(Z|F)$: likelihood;

$\Pr(Z)$: probability of making the observation.

Bayes' rule tells us how a prior probability $\Pr(F)$ is updated to a posterior probability $\Pr(F|Z)$ when making the observation $Z$. The observation is described by the likelihood term $\Pr(Z|F)$, which is introduced in more detail in Section 5.3.3.

In many instances, one is not only interested in a single event $F$, but in the distribution of all parameters $\mathbf{X}$. Bayes' rule can be extended to random variables $\mathbf{X}$:

$$f_{\mathbf{X}}''(\mathbf{x}) = aL(\mathbf{x})f_{\mathbf{X}}'(\mathbf{x}). \tag{5.3}$$

where $f_{\mathbf{X}}'$ is the prior probability density function (PDF) and $f_{\mathbf{X}}''$ is the posterior PDF, that is, $f_{\mathbf{X}}'' = f_{\mathbf{X}|Z}$ is the conditional PDF of $\mathbf{X}$ given the observation $Z$. $L(\mathbf{x})$ is the likelihood function, which, in analogy to Equation 5.2, is defined as $L(\mathbf{x}) \propto \Pr(Z|\mathbf{X} = \mathbf{x})$; here, the proportionality is with respect to $\mathbf{x}$ (see Section 5.3.3). The constant $a$ is

$$a = \frac{1}{\displaystyle\int_{-\infty}^{\infty} L(\mathbf{x})f_{\mathbf{X}}'(\mathbf{x})d\mathbf{x}}. \tag{5.4}$$

Here, we use the convention $\int_{\mathbf{X}} d\mathbf{x} = \int_{-\infty}^{\infty}...\int_{-\infty}^{\infty} dx_1...dx_n$. $a$ is often called the normalizing constant, since it ensures that $f_{\mathbf{X}}''(\mathbf{x})$ integrates to 1. By comparing with Equation 5.2, it can be observed that $a$ corresponds to $1/\Pr(Z)$, that is, it is an indication for how likely the observation is.

The application of Bayes' rule for updating a single random variable is presented in the following illustration.

## Illustration 1: Bayesian updating of a soil parameter

A geotechnical site consists of a silty soil, whose mean friction angle $\mu_\varphi$ is to be estimated. It has been found that the cohesion is close to zero and can be neglected. It is assumed that the normal distribution describes the variability of the friction angle well (this choice is for illustration purposes). On the basis of the previous measurements on similar soils in the vicinity, it has been found that the mean value is commonly between $25°$ and $31°$. We assume that these values correspond to the 10 and 90% quantiles and we fit a normal distribution to these values. It follows that the prior distribution of $\mu_\varphi$ is the normal distribution with mean $\mu'_\mu = 28°$ and standard deviation $\sigma'_\mu = 2.34°$:

$$f'_{\mu_\varphi}(\mu_\varphi) = \frac{1}{2.34°\sqrt{2\pi}} \exp\left[-\frac{1}{2}\left(\frac{\mu_\varphi - 28°}{2.34°}\right)^2\right]. \tag{5.5}$$

Additionally, the engineer decides to take three samples from the site and carry out direct shear tests. These result in the following observed values: $\varphi_1 = 25.6°$, $\varphi_2 = 25.5°$, $\varphi_3 = 24°$. These values are taken from Oberguggenberger and Fellin (2002). The likelihood function describing these three measurements is[*]

$$L(\mu_\varphi) = \frac{1}{\left(3°\sqrt{2\pi}\right)^3} \exp\left[-\frac{1}{2}\sum_{i=1}^{3}\left(\frac{\varphi_i - \mu_\varphi}{3°}\right)^2\right]. \tag{5.6}$$

The constant $a$ can now be computed numerically following Equation 5.4 to $a = 1313.4$. Following Equation 5.3, the resulting posterior distribution is

$$f''_{\mu_\varphi}(\mu_\varphi) = aL(\mu_\varphi)f'_{\mu_\varphi}(\mu_\varphi)$$

$$= a\frac{1}{\left(3°\sqrt{2\pi}\right)^3}\frac{1}{2.34°\sqrt{2\pi}}\exp\left[-\frac{1}{2}\left(\frac{\mu_\varphi - 28°}{2.34°}\right)^2 - \frac{1}{2}\sum_{i=1}^{3}\left(\frac{\varphi_i - \mu_\varphi}{3°}\right)^2\right]$$

$$= \frac{1}{1.39°\sqrt{2\pi}}\exp\left[-\frac{1}{2}\left(\frac{\mu_\varphi - 26.08°}{1.39°}\right)^2\right]. \tag{5.7}$$

As seen from the last line, the resulting posterior distribution of $\mu_\varphi$ is again a normal distribution, with posterior mean $\mu''_\mu = 26.08°$ and standard deviation $\sigma''_\mu = 1.39°$. Such special cases, where the prior and the posterior distributions are of the same type, are known as conjugate priors and are discussed in Section 5.3.4.

The prior, the likelihood, and the posterior are shown in Figure 5.2. It can be observed that the posterior PDF is in-between the prior PDF and the likelihood. In this example, the prior and the measurements have approximately equal influence on the posterior PDF. In many instances, the posterior is mainly determined by the measurements, that is, the posterior PDF is very similar to the likelihood, as discussed later.   ∎

---

[*] This likelihood is based on assuming that the variability of $\varphi$ at the site is described by a normal distribution with standard deviation $\sigma_\varphi = 3°$. The derivation of this likelihood function is explained in Section 5.3.3.

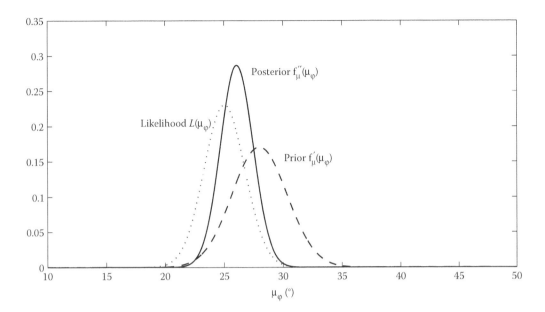

*Figure 5.2* Prior PDF of the mean friction angle, likelihood function describing the joint samples, and posterior PDF of the mean friction angle. The likelihood function is scaled by $[\int_{-\infty}^{\infty} L(\mu_\varphi)d\mu_\varphi]^{-1}$.

The difficulty in Bayesian updating stems from the need to compute the integral $\Pr(Z) = \int_{-\infty}^{\infty} L(\mathbf{x})f_X'(\mathbf{x})d\mathbf{x}$. Analytical solutions are available only for special cases, namely for the so-called conjugate priors as discussed in Section 5.3.4. Efficient numerical integration using quadrature rules, as implemented in standard mathematical software, is applicable only when the number of random variables in $\mathbf{X}$ is small, in the order of 3. For most realistic problems, tailor-made numerical methods are required to solve the Bayesian updating problem. The most popular among these is the family of Markov chain Monte Carlo (MCMC) methods, which allow sampling directly from $f_X''$ without determining the constant $a$, but many alternative methods have been proposed. Methods for the numerical solution are presented in Section 5.4.

Ultimately, the goal of the analysis is to make predictions on the performance of the geotechnical system and, sometimes, on individual parameters of the model. In case the interest is in the performance expressed through one or more failure events $F$, the prediction is in terms of the updated $\Pr(F|Z)$. This can be obtained through direct application of Equations 5.1 or 5.2, which is the preferred approach for many problems and which is presented later in Section 5.3.5. Alternatively, it is possible to first update the probability distribution of the model parameters $\mathbf{X}$ to $f_X''$ following Equation 5.3 and then perform a probabilistic reliability analysis with $f_X''$. This approach is illustrated in the following.

**Illustration 2: Bayesian updating of the geotechnical reliability**

We compute the reliability of a centrically loaded square footing with dimensions $l = 3$ m and $t = 1$ m, illustrated in Figure 5.3. This example is modified from Oberguggenberger and Fellin (2002). The footing will be in the silty soil described in Illustration 1, with a mean friction angle $\mu_\varphi$ and cohesion $c = 0$.

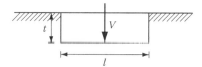

Figure 5.3 Footing design.

The LSF describing failure of this foundation is

$$g(\mathbf{X}) = Q_u - \frac{V}{l^2}. \tag{5.8}$$

where $Q_u$ is the ultimate bearing capacity and $V$ is the applied vertical loading, which includes the weight of the foundation. For simplicity, we assume a deterministic $V = 3000$ kN. With cohesion $c = 0$, the ultimate bearing capacity is

$$Q_u = \gamma t N_q s_q + \gamma l N_\gamma s_\gamma, \tag{5.9}$$

where $\gamma = 19.8$ kN/m³ is the unit weight of the soil, and the coefficients are $N_q = \exp(\pi + \tan\varphi)$ $\tan^2(45° + (\varphi/2))$, $N_\gamma = (N_q + 1)\tan\varphi$, and shape factors $s_q = 1 + \sin\varphi$, $s_\gamma = 0.7$.

The friction angle $\varphi$ is a spatially variable parameter. Exact computations should explicitly address this spatial variability through random fields, as discussed in Section 5.3.1. However, if we assume highly fluctuating soil properties, then as a first approximation, one can take the mean value of the friction angle as representative (the approximation is exact for an uncorrelated soil, i.e., for a correlation length zero; (Griffiths and Fenton 2001)). We make this assumption here, that is, we replace $\varphi$ in the above expressions by $\mu_\varphi$. However, it must be reminded that this is an unconservative approximation, and one should make adjustments when using this approximation in real problems (Griffiths and Fenton 2001).

Owing to the fact that the only random variable in this problem is $\mu_\varphi$, a simple solution for the probability of failure $\Pr(F) = \Pr(g(\mathbf{X}) \le 0)$ is available: First, we find the value of the friction angle for which $g = 0$ as $\varphi_F = 21.36°$. Failure occurs if the mean friction angle $\mu_\varphi$ takes a value smaller than $\varphi_F$, that is, an equivalent expression for the failure event is $F = \{\mu_\varphi \le \varphi_F\}$. Hence, the probability of failure a priori is

$$\Pr(F) = \Pr(g(\mathbf{X}) \le 0) = \Pr(\mu_\varphi \le \varphi_F) = F'_{\mu_\varphi}(\varphi_F) = 2.3 \times 10^{-3}. \tag{5.10}$$

where $F'_{\mu_\varphi}$ is the prior cumulative distribution function (CDF); here, the normal CDF with mean $\mu'_\mu = 28°$ and standard deviation $\sigma'_\mu = 2.34°$. The corresponding reliability index is $\beta' = -\Phi^{-1}(2.3 \times 10^{-3}) = 2.84$, with $\Phi^{-1}$ being the inverse normal CDF. Note that, because $F'_{\mu_\varphi}$ is a normal CDF, the reliability index can also be computed directly as $\beta' = (28° - \varphi_F)/2.34° = 2.84$.

Since the posterior distribution of $\mu_\varphi$ conditional on the measurements is the normal distribution with $\mu''_\mu = 26.08°$ and $\sigma''_\mu = 1.94°$, the posterior reliability index is $\beta'' = (26.08° - \varphi_F)/1.39° = 3.40$ and the corresponding posterior probability of failure given the measurements is

$$\Pr(F|Z) = F''_{\mu_\varphi}(\varphi_F) = 3.4 \times 10^{-4}. \tag{5.11}$$

Note that the measurements of Illustration 1 resulted in friction angles that are lower than the prior mean and hence the posterior mean (26.08°) is below the prior mean (28°).

Nevertheless, the reliability is significantly higher following the measurements. The reason is the reduced uncertainty, which can be observed in Figure 5.2. The probability of values $\mu_\varphi < 21.36°$ is reduced from the prior to the posterior distribution.  ∎

As in the example above, the quantities of interest are generally integral functions of $f_X''$; more precisely, they are expected values. This includes the mean value and covariance of the uncertain parameters $X$ as well as the probability of failure. It follows that prediction requires methods that can perform such integrations efficiently, similar to the computation of the constant $a$. This will be discussed in Section 5.3.5.

## 5.3 GEOTECHNICAL RELIABILITY BASED ON MEASUREMENTS: STEP-BY-STEP PROCEDURE FOR BAYESIAN ANALYSIS

### 5.3.1 Initial probabilistic model: Prior distribution

The first step in the analysis is to establish the geotechnical model, that is, the ultimate and serviceability LSFs, and to propose an initial probabilistic model of its parameters. This is the prior distribution $f_X'$. Once the relevant uncertain parameters $X$ are identified, their probability distributions are established as in classical geotechnical reliability analysis (e.g., Rackwitz 2000). Thereby, all relevant information prior to making the measurements is collected and assessed and appropriate models are found.

In many instances, information on the parameters is available from knowledge on the geology or geotechnical conditions at nearby sites. Information may also be available from the previous geotechnical assessments, as well as from earlier measurements at the site. Additionally, literature sources can provide probabilistic models of soil conditions. Finally, expert estimations can and should also be included, but thereby, it has to be ensured that information is not used twice. (The expert's estimate may be founded in past measurements. If these measurements are considered as additional separate independent information, then the information content is overestimated, resulting in overconfidence.) When combining different sources of information, it is also possible to use Bayesian analysis, whereby one source of information (e.g., literature) is taken as a prior and the remaining information (e.g., previous measurements) are modeled by the likelihood function following Section 5.3.3.

If the available information is vague, then the prior distribution should reflect this. In Bayesian analysis, one often uses the so-called "non-informative priors," that can, for example, be a uniform distribution from 0 to ∞ for a non-negative real-valued parameter $Y$. However, the term "non-informative prior" is misleading, since it is not possible to put no information into a prior distribution. For example, the prior distribution for the non-negative real-valued parameter $Y$ might also be chosen as a uniform distribution on log $Y$, defined from −∞ to ∞; clearly, this would differ from the choice above, but both choices seem to reflect ignorance on the value of this parameter. For this reason, a number of different definitions of noninformative priors have been proposed in the mathematical literature, for example, using the maximum entropy principle (Jaynes 1968) or the so-called reference prior (Bernardo 1979). The latter is defined as the prior that maximizes the expected Kullback–Leibler (KL) divergence of the posterior distribution with respect to the prior distribution. Since the KL divergence is a measure of information gain, this will lead to the choice of the prior distribution with the least information relative to the posterior. This definition of the noninformative prior leads to different prior distributions for different likelihood functions.

We note that for engineering applications, the discussion on what constitutes a noninformative prior is of little practical relevance. First, one almost always has some, if only

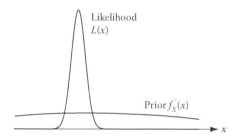

*Figure 5.4* Illustration of a weakly informative prior.

vague, information on the parameters, in particular, for parameters with physical meaning. Second, one can typically choose a prior distribution that has no or only very limited effect on the posterior distribution and predictions, as illustrated in Figure 5.4. We use the more loosely defined term *weakly informative prior* to describe such probability distributions. Note that whether or not a probability distribution falls in this category is dependent on the likelihood function, that is, on the measurements, in analogy to the non-informative prior. If the measurements are very informative, the likelihood function will be highly peaked and even a prior distribution with a moderate variance may have no effect on the posterior. In practice, whether or not a prior is weakly informative can be found from a simple sensitivity analysis: Does the posterior distribution change notably when the mean or standard deviation of the prior distribution is changed?

It is possible to choose so-called *improper* prior distributions. These are probability distributions that do not integrate to 1. The uniform distribution from 0 to $\infty$ is an example of such a probability distribution. It is $f'_X(x) = b$, $y \geq 0$. If the constant $b$ is zero, the integral $\int_0^\infty f'_X(x)dx$ results in zero; if $b > 0$ it is $\int_0^\infty f'_X(x)dx = \infty$. However, such a prior distribution is admissible in Bayesian analysis, as long as the resulting posterior distribution is proper. This is due to the normalization constant $a$ in Equation 5.3: As long as the integral $a = \int_0^\infty L(x)f'_X(x)dx$ results in a positive finite number, the posterior distribution will be proper.

### Illustration 3: Weakly informative prior

We repeat the analysis of Illustrations 1 and 2 with a weakly informative prior. We now use a normal distribution with mean $\mu'_\mu = 30°$ and standard deviation $\sigma'_\mu = 100°$ as a prior distribution. Such a distribution obviously defies the physical boundaries at 0° and 90°, but this is not critical as long as we ensure that the prior PDF has no influence on the final result. When repeating the Bayesian updating of Illustration 1, the posterior PDF is obtained as a normal distribution with mean $\mu''_\mu = 25.035°$ and standard deviation $\sigma''_\mu = 1.732°$ You can easily verify that the prior has no influence on the posterior in this case. In the extreme, one can select a normal distribution with standard deviation $\sigma'_\mu = \infty$ as a prior. (The prior mean is irrelevant in this case.) This results in a normal posterior PDF with mean $\mu''_\mu = 25.033°$ and standard deviation $\sigma''_\mu = 1.732°$. This clearly demonstrates that the weakly informative prior has no influence on the final result in this case, and the posterior is determined by the likelihood.

We also recompute the posterior reliability following Illustration 2. The posterior reliability index calculated with the assumption of the weakly informative prior is $\beta' = 2.12$ and the corresponding probability of failure is $\Pr(F|Z) = 1.7 \times 10^{-2}$. This probability is significantly higher than was obtained in Illustration 2 with the informative prior. Therefore, the assumption of a weakly informative prior is conservative here. In fact, it is conservative in most applications. However, if there are good reasons to assume an informative prior distribution, then one should do so, to obtain a geotechnical design that is both safe and economical.   ■

Many parameters have physical boundaries. Most geotechnical parameters are bounded at zero, as they are not defined for negative values. Others also have upper boundaries, such as the friction angle, which cannot exceed 90°. When using informative priors, these boundaries should ideally be included in the prior distribution to ensure physical consistency. For example, a parameter bounded at zero should not have a normal distribution as a prior, as this choice would imply a nonzero probability of values smaller than zero. (It is pointed out that this is in contradiction to Illustration 1, where a normal prior was chosen for the friction angle. For practical purposes, however, the choice is not critical, since with the given mean and variance, the probability of a negative value is $10^{-33}$.) For parameters with an upper and lower bound, the beta-distribution, which includes the uniform distribution as a special case, is often a good choice.

Note that in many instances, random variables in $\mathbf{X}$ will be correlated a priori, such as the soil parameters friction angle $\varphi$ and cohesion $c$. Correlation among random variables a priori can be represented by the Nataf model, also known as the Gaussian copula (Der Kiureghian and Liu 1986). It is based on marginally transforming all random variables into standard normal random variables $\mathbf{Y}$. The Nataf model prescribes that these underlying variables $\mathbf{Y}$ have the standard multinormal distribution, that is, they are normal random variables with zero mean, unit standard deviation, and correlation coefficient $\mathbf{R}_{YY}$. The Nataf model is described in detail in Ditlevsen and Madsen (1996). Alternative copula models exist; for their application in geotechnical reliability, see Tang et al. (2013) or Chapter 2 of this book. Finally, an alternative, more flexible model for dependence among the random variables is given by the Rosenblatt transformation (Hohenbichler and Rackwitz 1981).

### 5.3.1.1 Modeling spatially variable parameters

Soil properties are varying in space, even within one soil type (Baecher and Christian 2008). As an example, the shear strength and cohesion parameters within a silty soil layer will be different at different locations. Such variability can be modeled in two fundamentally different ways.

In the first—and in practice more common—modeling approach, the soil property within an area is modeled through a single random variable $X$. This random variable represents the inherent variability of the soil property within the area. With this modeling approach, the property at a specific location is not explicitly modeled; hence, it will always be subject to the inherent variability as described by the probability distribution of $X$, $f_X$. This corresponds to the situation in classical statistics, where a random variable $X$ describes the randomness within a population. This randomness cannot be further reduced, but the parameters $\theta$ of the distribution $f_X$ can be learned. In a Bayesian analysis, this is reflected by defining a prior distribution $f'_\Theta$ on $\theta$ and then updating this distribution when samples of $X$ are obtained. This prior distribution $f'_\Theta$ can be selected following the principles outlined earlier. Since the distribution of $X$ is now defined conditional on its parameters $\theta$, which are themselves uncertain, the unconditional distribution of $X$ must be calculated by integration over $\theta$. This unconditional distribution is termed *predictive distribution* and is explained in more detail in Section 5.3.4.

Note that this random variable modeling of the spatial variability was underlying Illustrations 1 and 2. The spatial variability of the friction angle was not explicitly considered, but was modeled through a normal distribution with parameters $\theta = [\mu_\varphi, \sigma_\varphi]$. The standard deviation was assumed as fixed $\sigma_\varphi = 3°$ and only the mean friction angle $\mu_\varphi$ was learned. For the reliability analysis, that is, for prediction, $\mu_\varphi$ was taken as the representative characteristic. As was discussed, this approach is a nonconservative approximation, which is exact for a soil with correlation length zero. In real applications, this approximation must be addressed by correcting for the effect of a larger correlation length (Griffiths and Fenton 2001). Optimally, however, an explicit random field modeling approach is selected.

The second modeling approach for spatially variable soil properties is to model them at each location explicitly. This is the *random field approach*, where a property $X$ is modeled through a random variable $X(z)$ at each location $z$ (Rackwitz 2000; Baecher and Christian 2008). To numerically represent such random fields, it is necessary to discretize them, for example, by the Karhunen–Loève expansion (e.g., Betz et al. 2014b). The random field is then described by a discrete set of random variables, which are part of $X$; the number of random variables in this set can be considerable.

If the soil is modeled through random fields, that is, by explicitly modeling the probabilistic spatial variation of the soil, it is necessary to define the correlation structure. This a priori correlation model should be selected carefully, as it can have a significant influence on the results of the Bayesian analysis: Shorter correlation lengths (i.e., larger spatial fluctuations) signify that the effect of a measurement beyond its immediate vicinity is limited, and longer correlation lengths (smaller spatial fluctuations) lead to measurements having a more global effect, that is, the impact of the measurement becomes larger. Note that a model without random field implicitly assumes full correlation among the soil properties in the areas represented by the same random variable. The reader is referred to Rackwitz (2000) for more information on modeling soil properties with random fields. Once a prior random field is established, the Bayesian analysis follows the general procedure presented in this chapter, with the parameters describing the random field being part of $X$. This is independent of the way the random field is discretized.

### 5.3.2 Computing the reliability and risk based on the prior model

Once the prior probabilistic model of $X$ is defined, it is possible to compute the a priori reliability of the geotechnical construction. In the general case, the failure event (both ultimate and serviceability) is described through LSFs $g(X)$. The probability of failure is

$$\Pr(F) = \Pr(g(X) \leq 0), \tag{5.12}$$

which is computed from

$$\Pr(F) = \int_{g(x)\leq 0} f_X(x)dx. \tag{5.13}$$

The corresponding reliability index is

$$\beta = -\Phi^{-1}[\Pr(F)]. \tag{5.14}$$

$\Phi^{-1}$ is the inverse of the standard normal CDF.

$\Pr(F)$ and $\beta$ can be computed by approximating the integral in Equation 5.13 using the classical methods of structural reliability, such as first- and second-order reliability methods (FORM/SORMs) or sampling-based methods including crude Monte Carlo, subset simulation, and importance sampling. The application of these methods to geotechnical problems is presented in detail in Phoon (2008).

It is noted that when weakly informative priors are selected for some of the random variables in $X$, it is to be expected that the a priori reliability is low. In these situations, it is necessary to collect further information and thus to reduce the uncertainty before the final design and construction can be decided. Exceptions from this low reliability can occur only when the random variables with weakly informative priors have no or only limited influence on the limit state.

The a priori reliability can also provide an indication on how much measurements are necessary and optimal. This is formalized in the value-of-information concept of the Bayesian decision analysis (Straub 2014).

### 5.3.3 Describing observations and data: The likelihood

Measurements (as well as other observations) $Z$ are described by the likelihood. It is defined as the conditional probability of making the measurement given a particular system state, $Pr(Z|system state)$. The system state can be, for example, the failure event $F$. In case the system state is defined by the continuous random variables $\mathbf{X}$, the likelihood function is defined as being proportional to the probability of making the measurement when the uncertain parameters $\mathbf{X}$ take a value $\mathbf{x}$:

$$L(\mathbf{x}) \propto Pr(Z|\mathbf{X} = \mathbf{x}). \tag{5.15}$$

Note that the likelihood is a function of $\mathbf{x}$, even though it describes the probability of the measurement outcome. Sometimes the likelihood is also denoted as $L(\mathbf{x}|\mathbf{d})$, where $\mathbf{d}$ is the measured data (it is $Z = \{\mathbf{D} = \mathbf{d}\}$).

The likelihood is not only used for Bayesian analysis, but is also a cornerstone of classical statistics, where the maximum likelihood estimator (MLE) is the most common approach to statistical inference (Fisher 1922); see also Chapter 4 of this book.

If multiple measurements or observations $Z_1, Z_2, \ldots, Z_m$ are available, likelihood functions $L_i, i = 1, \ldots, m$, can be established for all of them individually. If these measurements are statistically independent for the given model parameters $\mathbf{X}$, then it follows that the joint probability of all measurements is $Pr(Z | \mathbf{X} = \mathbf{x}) = Pr(Z_1 \cap Z_2 \cap \ldots \cap Z_m|\mathbf{X} = \mathbf{x}) = \prod_{i=1}^{m} Pr(Z_i|\mathbf{X} = \mathbf{x})$ and hence the joint likelihood is

$$L(\mathbf{x}) = \prod_{i=1}^{m} L_i(\mathbf{x}). \tag{5.16}$$

The assumption of independence is commonly made in practice. Situations where this assumption does not hold are discussed later.

In the following, the derivation of the likelihood function for single measurements $Z_i$ is described for different classes of measurements.

#### 5.3.3.1 Measurement $x_i$ of a parameter X

In some applications, it is possible to directly measure a model parameter $X$. If the measurement was perfect, then there would be no more uncertainty on $X$. In this case, the random variable would become a deterministic parameter with value* $X = x_i$. However, in almost all real applications, measurements are subject to measurement error or uncertainty, often because the measurement is only indirect. The measurement error $\epsilon$ is modeled probabilistically by

---

* The corresponding likelihood function would be the Dirac delta function with argument $x_i - x : \delta(x_i - x)$. However, there is no need to perform Bayesian updating, and it is sufficient and straightforward to replace $X$ with $x_i$ in this case.

its PDF $f_\epsilon$. Let us first consider the case that the measurement error is additive, that is, the measured value is equal to the true value plus the error, $x_i = X + \epsilon$. It follows that $\epsilon = x_i - X$. The probability of observing a measurement $x_i$ given that the true value is $X = x$ is equal to $\epsilon$ taking the value $x_i - X$. Therefore, the likelihood is

$$L_i(x) = f_\epsilon(x_i - x). \tag{5.17}$$

Accordingly, if the measurement error is multiplicative, it is $x_i = X \times \epsilon$, so that $\epsilon = x_i/X$ and the likelihood function is

$$L_i(x) = f_\epsilon\left(\frac{x_i}{x}\right). \tag{5.18}$$

### 5.3.3.2 Samples of a spatially variable parameter

If samples of a spatially variable parameter $X$ are taken, it must be carefully considered how this parameter is modeled probabilistically, as discussed in Section 5.3.1. If the parameter is modeled spatially explicit by means of a random field $\{X\}$, then the sample taken at a location $z$ is actually a measurement of the random variable $X(z)$. In this case, the likelihood describing the sample is as presented in the previous paragraph, where $X$ is replaced by $X(z)$.

More commonly, the spatially variable parameter is not modeled explicitly, but through a single random variable $X$ with corresponding PDF $f_X$ describing the population. In this case, the samples are measurements of realizations of $X$, which are used to learn the parameters $\theta$ of the distribution of $f_X$. Hence, the likelihood function would be defined on $\theta$. To make the dependence of $f_X$ on its parameters explicit, we write it as $f_X(x|\theta)$. The likelihood describing a sample $x_i$ of $X$ is then

$$L_i(\theta) = f_X(x_i|\theta). \tag{5.19}$$

As an example, if the variability of $X$ is modeled through a normal distribution, then the parameters are $\theta = [\mu_X; \sigma_X]$, the mean and standard deviation of $X$. In this case, the likelihood function describing a sample $x_i$ is $L_i(\mu_X, \sigma_X) = (\sqrt{2\pi}\sigma_X)^{-1} \exp[-(1/2)((x_i - \mu_X/\sigma_X))^2]$. In Illustration 1, we considered this case, but with only the mean $\mu_X$ being uncertain and $\sigma_X$ being known.

It is common to assume independence among multiple samples; thus, the joint likelihood describing samples $x_1, \ldots, x_m$ is given according to Equation 5.16 as

$$L(\theta) = \prod_{i=1}^{m} L_i(\theta). \tag{5.20}$$

### 5.3.3.3 Measurement of site performance parameters

In many instances, measurements of performances of the geotechnical construction are made, such as deformation measurements. These measurements are related to the model parameters $\mathbf{X}$ by means of the geotechnical model. Therefore, the likelihood function must include this model. As an example, if deformations at a site are measured, the model predictions of these deformations for given values of $\mathbf{X}$ are required. Let $h_i(\mathbf{X})$ denote such a model prediction. Furthermore, let $y_i$ denote the corresponding observed deformation and let $\epsilon_i$ denote the deviation of the model prediction from the observation. This deviation is typically due to measurement error, but sometimes also includes model errors;[*] it is modeled through the PDF $f_{\epsilon_i}(\cdot)$. Assuming an additive error, the following relationship holds: $y_i - h_i(\mathbf{X}) = \epsilon_i$. The likelihood function $L_i(\mathbf{x})$ describing this observation is therefore

$$L_i(\mathbf{x}) = f_{\epsilon_i}(y_i - h_i(\mathbf{x})). \tag{5.21}$$

More generally, the likelihood function for measurements $y_i$ of continuous quantities is

$$L_i(\mathbf{x}) = f_{Y_i|\mathbf{X}}(y_i|\mathbf{x}), \tag{5.22}$$

where $f_{Y_i|\mathbf{X}}$ is the conditional PDF of the measured quantity given $\mathbf{X} = \mathbf{x}$, and, as in Equation 5.21, involves the outcome $h_i(\mathbf{x})$ of the geotechnical model.

In case the measurements or observations $Z$ are on discrete quantities or events, they are characterized by a finite probability of occurrence, for example, observations of categorical values or observations of system performances such as failure/survival, and also censored data. One then refers to *inequality information* (Madsen et al. 1985; Straub 2011). Such information can be described through

$$Z_i = \{\mathbf{x} \in \mathbb{R}^n : h_i(\mathbf{x}) \leq 0\}, \tag{5.23}$$

where $h_i$ is a function that describes the relation between the observed event and the model parameters $\mathbf{X}$. In structural reliability, $h_i$ is known as an LSF. Inequality information can be interpreted as an observation that $\mathbf{X}$ must be in the domain $\{h_i(\mathbf{x}) \leq 0\}$. The corresponding likelihood function is then

$$L_i(\mathbf{x}) = \Pr(Z_i|\mathbf{X} = \mathbf{x}) = I(h_i(\mathbf{x}) \leq 0). \tag{5.24}$$

where $I$ is the indicator function. This likelihood thus takes on values 0 or 1, because for given $\mathbf{X} = \mathbf{x}$, the event $Z_i$ either occurs, $L_i(\mathbf{x}) = 1$, or does not occur, $L_i(\mathbf{x}) = 0$.

### Illustration 4: Likelihood of a spatially distributed soil parameter

Reconsider the example given in Illustration 1, where the likelihood function was introduced without a detailed explanation. In Illustration 1, the measurements are samples of the spatially variable friction angle $\varphi$. In constructing the likelihood function, it was assumed that the variability of $\varphi$ within a site could be described by a normal distribution with fixed standard deviation $\sigma_\varphi = 3°$ and mean value $\mu_\varphi$. Therefore, following Equation 5.19, the likelihood function for learning $\mu_\varphi$ is the conditional PDF of $\varphi$ given $\mu_\varphi$, that is, $L_i(\mu_\varphi) = f_\varphi(\varphi|\mu_\varphi)$, which is the normal distribution with parameters $\mu_\varphi$ and $\sigma_\varphi = 3°$: $L_i(\mu_\varphi) = (1/3°\sqrt{2\pi})\exp[-(1/2)(\varphi_i - \mu_\varphi/3°)^2]$. The combined likelihood of $m$ samples is given by Equation 5.20:

$$L(\mu_\varphi) = \prod_{i=1}^{m} L_i(\mu_\varphi) = \frac{1}{\left(3°\sqrt{2\pi}\right)^m} \exp\left[-\frac{1}{2}\sum_{i=1}^{m}\left(\frac{\varphi_i - \mu_\varphi}{3°}\right)^2\right]. \tag{5.25}$$

The full measurements reported in Oberguggenberger and Fellin (2002) consist of 20 samples, of which only three were considered in Illustration 1. The full samples are reported in Table 5.1. To illustrate the effect of the numbers of measurements on the likelihood function,

---

[*] The representation of model errors is further discussed at the end of this section.

*Table 5.1* Measured friction angles (°)

| 25.6 | 25.5 | 24 | 26 | 24.1 | 24 | 28.5 | 25.3 | 23.4 | 26.5 |
|------|------|-----|-----|------|-----|------|------|------|------|
| 23.2 | 25 | 22 | 24 | 24.2 | 30 | 27 | 21.1 | 21.3 | 29.5 |

Source: Oberguggenberger, M. and W. Fellin. 2002. From probability to fuzzy sets: The struggle for meaning in geotechnical risk assessment. *Probabilistics in Geotechnics: Technical and Economic Risk Estimation.* R. Pötter, H. Klapperich and H. F. Schweiger. Essen, Germany, Verlag Glückauf GmbH: 29–38.

Figure 5.5 shows the likelihood functions describing the first 1, 3, 5, 10, and 20 samples from this complete set. It can be clearly observed how the increased number of samples reduces the width of the likelihood function. It is pointed out that in real situations, the number of measurements is limited and it is unlikely that more than five samples are available.   ■

### Illustration 5: Likelihood describing deformation measurements at the site

Consider deformation measurements made at locations $i = 1, \ldots, m$ on a site. Let $y_i$ denote these measurements, which are associated with additive measurement errors $\epsilon_i$, all of which have the same PDF $f_\epsilon$. Furthermore, let $h_i(\mathbf{x})$ be the results of a numerical model, predicting deformation at the locations for given model parameters $\mathbf{x}$. In this case, the likelihood function describing these deformation measurements is

$$L(\mathbf{x}) = \prod_{i=1}^{m} f_\epsilon(y_i - h_i(\mathbf{x})). \tag{5.26}$$

This likelihood is applied in the application presented in Section 5.6.   ■

### Illustration 6: Censored measurement

A measurement of a strength parameter $X$ results in the outcome "$X$ is larger than $x_m$". Such an outcome may occur, for example, when the material strength exceeds the capacity

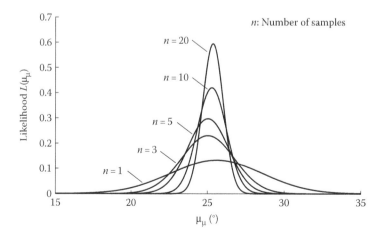

*Figure 5.5* Likelihood function of the mean friction angle, for different number of samples. The likelihood functions are scaled by $[\int_{-\infty}^{\infty} L(\mu_\varphi)d\mu_\varphi]^{-1}$.

of the testing equipment. This is inequality information, which can be written according to Equation 5.23 with LSF $h_i(x) = x_m - x$. The corresponding likelihood function is

$$L_i(x) = I(x_m - x \leq 0) \tag{5.27}$$

with $I$ being the indicator function.  ∎

It is commonly assumed, and often reasonably so, that the measurement outcomes are independent for given model parameters $\mathbf{X} = \mathbf{x}$ or $\Theta = \theta$. However, situations may arise where this does not hold. The accuracy of measurements is frequently influenced by factors that are common to multiple measurements, for example, temperature. In such cases, the measurement errors will be correlated, and Equation 5.16 is no longer valid. If this correlation is known, it is possible to include it in the formulation of the joint likelihood function. For example, if the individual measurement errors are represented by a normal distribution, the joint probability distribution of the errors, and therefore the likelihood function, can be expressed by the multinormal PDF. An example of such a correlated error model is given in Straub and Papaioannou (2014).

In many instances, the likelihood includes model errors, that is, it reflects deviations of the model predictions from the true value. As an example, reconsider Illustration 5. There, it was implicitly assumed that the model prediction of the deformations $h_i(\mathbf{x})$ is exact for correct values of $\mathbf{x}$. As every engineer knows, no deterministic model is 100% correct; nevertheless, the assumption of a correct model can be reasonable in a probabilistic context if one or more random variables are included in $\mathbf{X}$ to represent the model errors (Ditlevsen 1982; Zhang et al. 2009). If the model does not include error terms, then this model error must be included in the likelihood function. The same holds if the model does contain terms that describe the errors in the prediction of the quantity of interest (e.g., of the ultimate limit state describing failure), but which do not adequately describe the error in the prediction of the measured quantity, such as a deformation. If model errors must be included in the likelihood, the deviation $\epsilon_i$ of the measurement $y_i$ from the model prediction $h_i(\mathbf{x})$ is due to a combination of the measurement error and the model error. The problem with this representation is that the model errors related to different predictions $h_i(\mathbf{x})$, $i = 1, \ldots, m$, will generally be dependent. Unfortunately, this dependence is difficult or even impossible to estimate. By neglecting it, however, one generally overestimates the effect of the measurements, which can lead to strong overconfidence in the model and its parameters. The appropriate representation of the model error is probably the most critical challenge for Bayesian updating in practice and is the subject of ongoing research (Beven 2010; Goulet and Smith 2013; Simoen et al. 2013).

### 5.3.4  Updating the model

Once the prior distribution and the likelihood are established, Bayesian updating of the model parameters $\mathbf{X}$ or their distribution parameters $\Theta$ is performed following the basic Equations 5.2 or 5.3. However, in practice, the application of these equations is not generally straightforward and is often computationally demanding. For this reason, it is necessary to choose an effective computational strategy. This choice will be a function of the probabilistic model, in particular the number of random variables and the type of their distribution, and the computational cost of the geotechnical model. In the following section, different computational strategies for implementing Bayesian updating are presented.

#### 5.3.4.1  Conjugate priors

When performing Bayesian analysis of parameters $\Theta$ based on soil samples, it is often possible to select the so-called conjugate priors for the parameters, which lead to analytical solutions of the posterior distribution. For a given likelihood function, the conjugate prior distribution will lead to a posterior distribution of the same analytical form (Raiffa and Schlaifer 1961). One example of such conjugate priors was already seen in Illustration 1: If the likelihood of a parameter $\theta$ is the normal distribution with mean $\theta$ and fixed standard deviation $\sigma_X$, the conjugate prior is the normal distribution with mean $\mu_0'$ and standard deviation $\sigma_\theta'$. If $n$ samples are taken with sample mean $\bar{x}$, the posterior distribution of $\theta$ is the normal distribution with parameters:

$$\mu_\theta'' = \frac{(\mu_\theta'/\sigma_\theta'^2) + (n\bar{x}/\sigma_X^2)}{(1/\sigma_\theta'^2) + (n/\sigma_X^2)}, \tag{5.28}$$

$$\sigma_\theta'' = \left[\frac{1}{\sigma_\theta'^2} + \frac{n}{\sigma_X^2}\right]^{-(1/2)}. \tag{5.29}$$

The reader is asked to apply these results to the data in Illustration 1. It can be observed that the posterior mean depends only on the number and mean of the samples (in statistical jargon: the sample number and sample mean are a sufficient statistic of the samples). The posterior standard deviation depends only on the number of samples, but not on the actual outcome of the samples. This shows that—for these distributions—the uncertainty in the posterior will always be less than the uncertainty of the prior, independent of the measurement outcome.

Conjugate priors facilitate the choice of noninformative priors. For the above example, a noninformative prior is obtained by selecting $\sigma_\theta'^2 = \infty$. In this case, the posterior parameters are $\mu_\theta'' = \bar{x}$ and $\sigma_\theta'' = \sigma_X/\sqrt{n}$. This was utilized in Illustration 3.

Additional examples of conjugate priors include the $\gamma$-distribution as the conjugate prior of the parameter $\lambda$ of the exponential distribution; and the $\beta$-distribution as the conjugate prior of the parameter $p$ of the binomial distribution. These and additional conjugate priors are described in detail in Raiffa and Schlaifer (1961) and Fink (1997).

### Illustration 7: Conjugate prior of the normal distribution when both mean and standard deviation are uncertain

We extend Illustration 1, considering now both the mean $\mu_\varphi$ and the standard deviation $\sigma_\varphi$ of $\varphi$ to be uncertain. $\varphi$ is still modeled as normal distributed. For mathematical convenience, it is beneficial to replace $\sigma_\varphi$ by the precision $\tau_\varphi = (1/\sigma_\varphi^2)$, the reciprocal of the variance. Therefore, the parameters to estimate are $\theta = [\mu_\varphi, \tau_\varphi]$. The conjugate prior for a normal distribution with unknown mean and precision is the normal-gamma-distribution, whose PDF is (Raiffa and Schlaifer 1961; DeGroot 1969)

$$f_\theta(\mu_\varphi, \tau_\varphi) = \frac{\beta^\alpha \sqrt{\lambda}}{\Gamma(\alpha)\sqrt{2\pi}} \tau_\varphi^{\alpha-0.5} \exp(-\beta\tau_\varphi) \exp\left[-\frac{1}{2}\lambda\tau_\varphi(\mu_\varphi - \nu)^2\right] \tag{5.30}$$

The normal-gamma-distribution has parameters $\nu, \lambda, \alpha, \beta$. It is obtained by modeling $\tau_\varphi$ as a gamma-distributed random variable with parameters $\alpha$ and $\beta$, and $\mu_\varphi$ by a conditional normal distribution given $\tau_\varphi$, with mean $\nu$ and precision $\tau_\varphi\lambda$.

The posterior marginal distribution of the mean $\mu_\varphi$ is Student's $t$ distribution with $2\alpha$ degrees of freedom:

$$f_{\mu_\varphi}''(\mu_\varphi) = \frac{\sqrt{(\alpha''\lambda''/\beta'')}\Gamma(\alpha'' + (1/2))}{\sqrt{2\alpha''\pi}\,\Gamma(\alpha'')}\left[1 + \frac{1}{2}\frac{\lambda''(\mu_\varphi - \nu'')^2}{\beta''}\right]^{-(\alpha''+(1/2))}. \tag{5.31}$$

The posterior parameters are calculated as

$$v'' = \frac{\lambda'v' + n\bar{\varphi}}{\lambda' + n}, \lambda'' = \lambda' + n, \alpha'' = \alpha' + \frac{n}{2}, \beta'' = \beta' + \sum_{i=1}^{n}\frac{(\varphi_i - \bar{\varphi})^2}{2} + \frac{n\lambda'}{\lambda' + n}\frac{(\bar{\varphi} - v')^2}{2}.$$

A noninformative prior is obtained by choosing $\lambda' = 0$, $\alpha' = -(1/2)$, and $\beta' = 0$.

We consider three direct shear tests (as in Illustration 1), resulting in friction angles $\varphi_1 = 25.6°$, $\varphi_2 = 25.5°$, $\varphi_3 = 24°$. Inserting the non-informative prior parameters and these samples into the above expressions, we obtain the posterior parameters as

$$v'' = 25.03°, \quad \lambda'' = 3, \alpha'' = 1, \quad \beta'' = 0.803$$

The resulting marginal distribution of the mean, $f''_{\mu_\varphi}(\mu_\varphi)$, is shown in Figure 5.6, together with the posterior distribution obtained when assuming that the standard deviation is fixed at $\sigma_\varphi = 3°$ (also with a noninformative prior, as calculated in Illustration 3). Note that in this case, the assumption of an uncertain standard deviation of $\varphi$ leads to a lower standard deviation in the estimate of $\mu_\varphi$. However, the tail of $f''_{\mu_\varphi}(\mu_\varphi)$ is heavier in this case, as is evident when plotting the PDF in log scale (right-hand side of Figure 5.6). As an example, the probability $\Pr(\mu_\varphi \leq 20°)$ is $2 \times 10^{-3}$ in the case of $\sigma_\varphi = 3°$ and $4 \times 10^{-3}$ in the case of an uncertain $\sigma_\varphi$.  ■

### 5.3.4.2 Numerical integration to determine the proportionality constant

A possible solution to the Bayesian updating problem is the numerical evaluation of the proportionality constant $a = [\int_{-\infty}^{\infty} L(\mathbf{x})f'_X(\mathbf{x})d\mathbf{x}]^{-1}$. If the number of random variables in $\mathbf{X}$ is small, classical integration schemes based on quadrature rules are applicable. The full posterior PDF is then available through Equation 5.3.

For larger numbers of random variables, it is still possible to evaluate $a$ approximately using sampling schemes (see Section 5.4). However, making predictions with the posterior distribution also requires integration over $\mathbf{X}$. If a sampling method is used to determine $a$, it is therefore more convenient to directly work with these samples when doing predictions

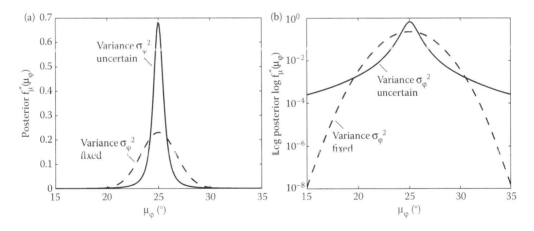

*Figure 5.6* Posterior distributions of the mean friction angle, once with uncertain mean and standard deviation, and once with uncertain mean and fixed standard deviation $\sigma_\varphi = 3°$. (a) Normal scale; (b) log scale.

with the posterior distributions. That is, the posterior distribution is represented through these samples, and $a$ is not needed in these cases. This is described in the next section.

### 5.3.4.3 Advanced sampling methods

When Bayesian updating is to be performed for multiple random variables jointly, advanced sampling methods are often the only computationally feasible approach. Arguably, the most popular approach is the MCMC method, which allows to sample directly from the posterior distribution without the need to compute the proportionally constant $a$, thus avoiding the integration (Gilks et al. 1998; Gelman 2004). Many authors have applied and adopted MCMC to Bayesian updating of mechanical models in general and geotechnical models in particular, including Beck and Au (2002), Cheung and Beck (2009), Zhang et al. (2010), and Sundar and Manohar (2013). The main problem of MCMC methods is that it cannot generally be ensured that the samples have reached the stationary distribution of the Markov chain, that is, the posterior distribution (Plummer et al. 2006). Various alternatives to MCMC exist, which are mostly based on rejection sampling and importance sampling, for example, the adaptive rejection sampling from a log-concave envelope distribution, which is effective for updating single random variables (Gilks and Wild 1992), or generalized sequential particle filter methods for updating arbitrary static or dynamic systems (Chopin 2002; Ching and Chen 2007). These methods are often combined with MCMC. The authors have recently proposed a novel approach termed Bayesian updating with structural reliability methods (BUS), which is based on principles from structural reliability (Straub and Papaioannou 2014), but may also be considered as an extension of the rejection-sampling approach, and which is particularly efficient for high-dimensional applications.

In Section 5.4, the three main strategies are presented in more detail: the MCMC approach, the particle filter or sequential Monte Carlo approach, and the BUS approach.

### 5.3.4.4 Multinormal approximation of the posterior

For cases where large numbers of measurements are available, which contain strong information relative to the prior distribution, the posterior can be approximated in terms of an asymptotic expression (Beck and Katafygiotis 1998; Papadimitriou et al. 2001). Thereby, the posterior PDF is approximated around its mode

$$\mathbf{x}_{PM} = \arg\max f_X''(\mathbf{x}), \tag{5.32}$$

where the index PM stands for posterior maximum. Let $\mathbf{A}$ be the negative Hessian of the logarithm of the posterior PDF evaluated at $\mathbf{x}_{PM}$, that is, $\mathbf{A} = -\nabla\nabla \ln f_X''(\mathbf{x})$. Noting that the first partial derivatives of $f_X''(\mathbf{x})$ at $\mathbf{x}_{PM}$ are zero, the second-order Taylor expansion of the log posterior around $\mathbf{x}_{PM}$ is obtained as

$$\ln f_X''(\mathbf{x}) \approx \ln f_X''(\mathbf{x}_{PM}) - \frac{1}{2}(\mathbf{x} - \mathbf{x}_{PM})^T \mathbf{A}(\mathbf{x} - \mathbf{x}_{PM}). \tag{5.33}$$

Therefore,

$$f_X''(\mathbf{x}) \approx f_X''(\mathbf{x}_{PM}) \exp\left[-\frac{1}{2}(\mathbf{x} - \mathbf{x}_{PM})^T \mathbf{A}(\mathbf{x} - \mathbf{x}_{PM})\right] \tag{5.34}$$

This is the multinormal distribution with mean $\mathbf{x}_{PM}$ and covariance matrix $\mathbf{A}^{-1}$. Locally, the posterior distribution can thus be approximated by this distribution. It requires that the maximization problem of Equation 5.32 is solved and it necessitates the Hessian, that is, the second-order derivatives of $\ln f_X''(\mathbf{x})$ around the mode. For a limited number of random variables, this solution can be computationally efficient. The main advantage of this approach is the fact that the multinormal approximation strongly facilitates posterior predictions.

Note that the above result only holds if the posterior distribution has only one mode (sometimes this is referred to as globally identifiable problems). For posterior distributions with multiple modes, see Papadimitriou et al. (2001). Since the above expression is an asymptotic result, the approximation is improving with an increasing number of measurements. It should be pointed out that for most geotechnical applications, the validity of the above approximation is not given because not sufficient information is available. We therefore do not consider this approach further.

### 5.3.4.5 Direct updating of the reliability

It is possible to directly update the reliability without the need to first update the random variables $\mathbf{X}$. First, approaches based on first-order approximations were presented in Madsen (1987) and Schall et al. (1989). A completely general formulation, which is compatible with modern sampling-based structural reliability methods, was introduced in Straub (2011). This approach is presented in Section 5.3.5.

### 5.3.4.6 Predictive distributions

When Bayesian updating is applied to learn the parameters $\Theta$ of the probability distribution of $\mathbf{X}$, one is ultimately interested in predicting $\mathbf{X}$ probabilistically. This is achieved through the posterior predictive distribution of $\mathbf{X}$, which is defined as

$$\tilde{f}_X(\mathbf{x}) = \int_{\Theta} f_X(\mathbf{x}|\theta) f_{\Theta}''(\theta) d\theta. \tag{5.35}$$

The predictive $\tilde{f}_X(\mathbf{x})$ is the expected value of $f_X(\mathbf{x}|\theta)$ with respect to the posterior distribution of $\theta$. The necessary integration over $\theta$ can be performed following the same approach as the integration required for Bayesian updating. That is, if samples of the posterior $f_{\Theta}''(\theta)$ are available, these will be used to compute the integral. If the posterior $f_{\Theta}''(\theta)$ is obtained analytically through the use of a conjugate prior, then an analytical expression is often available for the predictive PDF as well. This is shown in Illustration 8.

### Illustration 8: Predictive distribution of friction angle

Reconsider the probabilistic analysis of the friction angle $\varphi$ from Illustration 1. The friction angle at a specific location is described by the normal distribution with mean $\mu_{\varphi}$ and standard deviation $\sigma_{\varphi} = 3°$. The posterior predictive distribution of $\varphi$ is given through Equation 5.35 as $\tilde{f}_{\varphi}(\mathbf{x}) = \int_{-\infty}^{\infty} f_{\varphi}(\varphi|\mu_{\varphi}) f_{\mu_{\varphi}}''(\mu_{\varphi}) d\mu_{\varphi}$. In the general case, this integral cannot be solved analytically. Here, because $\mu_{\varphi}$ has the normal distribution, which is the conjugate prior, the predictive distribution is also normal with mean value $\mu_{\mu}'' = 26.08°$ and standard deviation $\sqrt{\sigma_{\mu}''^2 + \sigma_{\varphi}^2} = 3.31°$. ■

## Illustration 9: Random field modeling and updating of friction angle

We show the random field modeling approach to the friction angle presented earlier, and its updating. We recall that the variability of the friction angle $\varphi$ was assumed to follow a normal distribution with uncertain mean $\mu_\varphi$ and fixed standard deviation $\sigma_\varphi = 3°$. A priori, the mean $\mu_\varphi$ was modeled through a normal distribution with parameters $\mu'_\mu - 28°$ and $\sigma'_\mu = 2.34°$ (see Illustration 1). This can be translated into an equivalent random field model for $\varphi$. Such a random field model additionally requires that a correlation function representing the spatial fluctuation of $\varphi$ is prescribed. We assume a homogeneous random field with an exponential correlation model with correlation length 2 m:

$$r_\varphi(\Delta z) = \exp\left(-\frac{\Delta z}{2\,\mathrm{m}}\right), \tag{5.36}$$

where $\Delta z = |z_1 - z_2|$ is the distance between two points and $r_\varphi$ is the correlation coefficient between $\varphi$ at these points. The marginal distribution must be equal to the prior predictive distribution (analog Equation 5.35), which is the normal distribution with mean $28°$ and standard deviation $\sqrt{(2.34°)^2 + (3°)^2} = 3.8°$. The covariance function of the random field is

$$\mathrm{Cov}\big[\varphi(z_1),\ \varphi(z_2)\big] = (2.34°)^2 + \exp\left(-\frac{|z_1 - z_2|}{2\,\mathrm{m}}\right)(3°)^2. \tag{5.37}$$

We note that, due the uncertainty in the mean, the covariance becomes $\sigma'^2_\mu = (2.34°)^2$ as $\Delta z \to \infty$. Since the marginal is the normal distribution, it is consistent (and convenient) to model the random field $\{\varphi\}$ as a normal (Gaussian) random field.

For illustration purposes, we model only a one-dimensional random field $\{\varphi\}$ in the domain $z \in [-10,10\mathrm{m}]$. We assume that the three measurements considered in Illustration 1 ($\varphi_1 = 25.6°$, $\varphi_2 = 25.5°$, $\varphi_3 = 24°$) are taken at locations $z_1 = -2m$, $z_2 = 0m$, and $z_3 = 2m$. The likelihood function describing these measurements is given according to Section 5.3.3 as

$$L(\{\varphi\}) = \prod_{i=1}^{3} f_\varepsilon(\varphi_i - \varphi(z_i)). \tag{5.38}$$

For simplicity, we assume that the measurements are perfect (this was also implicitly assumed in the previous illustrations). In this case, $f_\varepsilon$ is the Dirac delta function $\delta$.

Since the random field is Gaussian, its posterior is also Gaussian and an analytical solution is available for updating it with the observations (e.g., Straub 2012). The resulting posterior mean and standard deviation is shown in Figure 5.7.

The posterior random field is no longer homogeneous. The posterior uncertainty is smallest at the locations where the samples are taken and largest at the points furthest away from the sample locations. Here, because of the assumption of no measurement error, the posterior standard deviation at the location of the samples is zero. At a distance $\Delta z \to \infty$ from the measurements, the posterior mean is $26.36°$ and the posterior standard deviation is $3.39°$. These values are slightly higher than those obtained earlier for the predictive posterior distribution calculated in Illustration 8. The reason for this difference is that the prior random field model accounts for the correlation among the friction angle values at the measured locations, $\varphi(z_1)$, $\phi(z_2)$, and $\varphi(z_3)$ due to their proximity. This correlation reduces the information content of the samples, to a correlation among measurement errors. If we considered

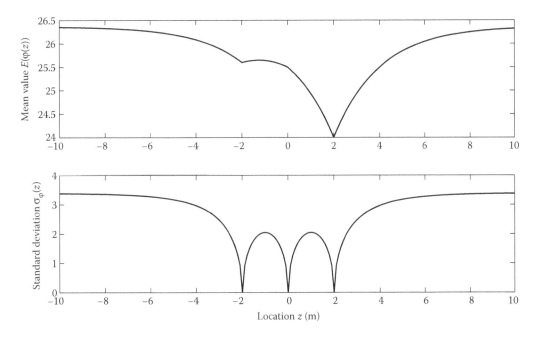

*Figure 5.7* Posterior mean and standard deviation of the friction angle $\{\varphi\}$.

a correlation length of zero instead of 2 m in Equations 5.36 and 5.37, the results would be identical to the random variable case. ∎

As a last note, it is pointed out that Bayesian analysis can be carried out sequentially, as long as measurements are independent for given $\mathbf{X} = \mathbf{x}$. The model can first be updated with one measurement alone, the resulting posterior can then be updated with the second measurement, and so on. This becomes obvious from noting that, for independent measurements, the likelihood $L(\mathbf{x})$ is defined as the product of the likelihoods describing the individual measurements, Equation 5.16, $L(\mathbf{x}) = \Pi_{i=1}^{m} L_i(\mathbf{x})$. Considering the case of two measurements, $m = 2$, and inserting this expression in the general Bayesian updating formulation, Equation 5.3, it follows

$$f_X''(\mathbf{x}) = aL(\mathbf{x})f_X'(\mathbf{x}) = a\prod_{i=1}^{2}L_i(\mathbf{x})f_X'(\mathbf{x}) = a_2L_2(\mathbf{x})[\underbrace{a_1L_1(\mathbf{x})f_X'(\mathbf{x})}_{\text{First updating}}].$$

$$\underbrace{\phantom{f_X''(\mathbf{x}) = a_2L_2(\mathbf{x})[a_1L_1(\mathbf{x})f_X'(\mathbf{x})]}}_{\text{Second updating}}$$

(5.39)

In other words, it is possible to initially compute a first posterior conditional on the first measurement, $f_{X|Z_1}(\mathbf{x}) = a_1L_1(\mathbf{x})f_X'(\mathbf{x})$. This posterior then becomes the new prior and is again updated with the second measurement to $f_X''(\mathbf{x}) = a_2L_2(\mathbf{x})f_{X|Z_1}(\mathbf{x})$. This result can be extended to $m$ measurements. Therefore, it is always possible to consider measurements sequentially, as long as they are independent for given $\mathbf{X} = \mathbf{x}$. The ordering of the measurements is irrelevant. This possibility to do sequential updating is of particular use for monitoring geotechnical structures in-service. The posterior distribution at time $t$ is a sufficient description of all measurements made up to time $t$ and it is not necessary to store all measurement results explicitly.

### 5.3.5 Updating reliability and risk estimates

Bayesian analysis can be applied to compute characteristic values of soil parameters. These are defined as lower or upper quantiles of the posterior distribution, depending on whether the parameters act as a load or resistance variable in the relevant limit state. These characteristic values are then used as an input to a geotechnical assessment based on codes and standards, typically following the partial safety factor format (LRFD). This approach is straightforward for individual parameters, but may be less trivial if the parameters are statistically dependent after the measurements Z.

The fact that the parameters are already available in a probabilistic format facilitates the application of a reliability-based design or assessment. This approach should be selected when the relevant code does not cover the specific design situation, but can also be preferable in other situations; for example, when uncertain parameters are statistically dependent as discussed above. As noted by Kulhawy and Phoon (2002, p. 32), "reliability-based design (RBD) is the only methodology available to date that can ensure self-consistency from both physical and probabilistic requirements."

The computation of the reliability a priori was outlined in Section 5.3.2. We now consider the computation of the posterior reliability conditional on the measurement outcomes Z, expressed in terms of the conditional probability of failure $\Pr(F|Z)$ or the posterior reliability index $\beta''$. Following Section 5.3.2, failure is expressed through the LSF $g(\mathbf{x})$ as $F = \{g(\mathbf{X}) \leq 0\}$. If the posterior PDF of $\mathbf{X}$ is available in analytical form, then the computation of $\Pr(F|Z)$ can be carried out with the classical structural reliability methods. It is only necessary to replace the prior joint PDF of $\mathbf{X}$ with the posterior $f_{\mathbf{X}}''(\mathbf{x})$ in Equation 5.13: $\Pr(F|Z) = \int_{g(\mathbf{x}) \leq 0} f_{\mathbf{X}}''(\mathbf{x}) d\mathbf{x}$.

For the case that the posterior PDF $f_{\mathbf{X}}''$ is not available in explicit form and can only be computed numerically, a number of approaches have been proposed that are based on first computing samples of $f_{\mathbf{X}}''$ and then applying these samples for determining the reliability using a Monte Carlo or other simulation approaches (e.g., Jensen et al. 2013; Sundar and Manohar 2013). These methods can lack efficiency or require problem-specific adjustments that can make their application cumbersome. An efficient and simple alternative has been proposed in Straub (2011), which is based on translating the likelihood function into an LSF, and then performing two reliability computations.

Following Straub (2011), an LSF $h(\mathbf{x}, p)$ can be defined as

$$h(\mathbf{x}, p) = p - cL(\mathbf{x}). \tag{5.40}$$

This LSF is defined in the augmented outcome space of $\mathbf{X}$ and $P$. The latter is a standard uniform random variable $f_P(p) = 1$, $0 \leq p \leq 1$. The constant $c$ in Equation 5.40 is added to ensure that the product $cL(\mathbf{x})$ is not larger than 1; it can be chosen freely as long as it is ensured that $cL(\mathbf{x}) \leq 1$ for any $\mathbf{x}$. This LSF provides an alternative definition of the measurements Z described by the likelihood function $L(\mathbf{x})$. It can be shown that the conditional probability of failure given Z is

$$\Pr(F|Z) = \frac{\int_{g(\mathbf{x}) \leq 0 \cap h(\mathbf{x}, p) \leq 0} f_{\mathbf{X}}(\mathbf{x}) d\mathbf{x} dp}{\int_{h(\mathbf{x}, p) \leq 0} f_{\mathbf{X}}(\mathbf{x}) d\mathbf{x} dp}. \tag{5.41}$$

Both the numerator and the denominator in Equation 5.41 are structural reliability problems and can be solved with any of the available structural reliability methods. Note that

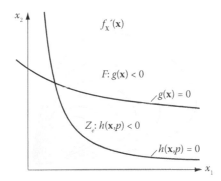

*Figure 5.8* Illustration of the Bayesian updating of the reliability. The observation $Z_e$ reduces the sample space to the domain $\{h(x,p) \leq 0\}$. The posterior probability of failure $\Pr(F|Z) = \Pr(F|Z_e)$ is computed as $\Pr(F|Z_e) = \Pr(F \cap Z_e) / \Pr(Z_e)$ by solving the integrals in Equation 5.41. (After Straub, D. 2011. *Probabilistic Engineering Mechanics* **26**(2): 254–258.)

this approach does not necessitate to first compute the posterior PDF of **X**. Therefore, the complete Bayesian updating and reliability computation is performed by solving two standard structural reliability problems.

An intuitive interpretation of the solution provided in Equation 5.41 is as follows. The LSF $h(\mathbf{x}, p)$ defines an inequality observation $Z_e$ in the outcome space of **x** and $p$ (see Equation 5.23). This inequality observation $Z_e$ is equivalent to the original measurement $Z$ in the sense that $\Pr(F|Z) = \Pr(F|Z_e)$. According to the definition of the conditional probability, it is $\Pr(F|Z_e) = \Pr(F \cap Z_e)/\Pr(Z_e)$. The numerator in Equation 5.41 is $\Pr(F \cap Z_e)$ and the denominator is $\Pr(Z_e)$. The approach is illustrated in Figure 5.8.

### 5.3.6 Communicating the results

One of the difficulties in applying Bayesian analysis—as with any probabilistic analysis—is the communication of the results and their interpretation. The client or other stakeholders are not generally adept at probabilistic analysis. It is therefore important that significant efforts are made to communicate the results and the assumptions underlying these results in a clear manner. Graphical representations of inputs and results can help in this task, as can sensitivity studies.

A problem that is specific to Bayesian analysis is that the theory is often criticized for being "subjective," due to the need for selecting a prior distribution.* This criticism is an artifact from the time when the frequentist interpretation of probability was the leading school of thought. In the mathematical community, Bayesian analysis has long since become an accepted method for probabilistic inference. It has been recognized that, when solving practical problems, subjective model choices cannot be avoided. The advantage of Bayesian analysis is that these choices are made more explicit. It should also be noted that the Bayesian interpretation of probability is much closer to the intuitive understanding of engineers than the frequentist interpretation. Nevertheless, as with any method, it is important to not only communicate the results, but also the limitations of the modeling and the assumptions underlying the obtained results.

---

* Interestingly, those who make this criticism often have no difficulty in making deterministic model choices based on their subjective experience.

The key assumptions that are made in a Bayesian analysis and which should be addressed are

- *The selected deterministic geotechnical model used in the analysis.* A Bayesian analysis must always be seen in the context of the underlying deterministic geotechnical model used for the prediction.
  *The selected probabilistic model framework.* This concerns mainly the representation of the spatial distribution of parameters, through a random field model or a population approach (see Section 5.3.1).
- *The selected prior probability distributions of parameters.* In the case of a probabilistic reliability analysis, this also includes the choice of those parameters that are not updated.
- *The representation of the model error.* As discussed in Section 5.3.3, the representation of the model error can be crucial, in particular, when updating models with observed performances, for example, measured deformations.

A common approach to investigating and documenting the effect of assumptions on the results are sensitivity analyses or parameter studies. These are particularly helpful in communicating and also facilitate checking the plausibility of the results. Some care is needed in selecting the parameter cases to be analyzed and the way they are presented. When presented with parameter studies, clients often tend to select the most conservative results. In this way, when performing an extensive parameter study, one runs some risk of ending up with a conservative suboptimal solution. In this way, advanced probabilistic approaches are punished compared to simplistic code-based approaches, where even crude assumptions are often not investigated further.

### Illustration 10: Effect of the probability distribution model on the reliability

We reconsider the reliability problem described in Illustrations 1 and 2. There it was originally assumed that the friction angle follows a normal distribution with standard deviation $\sigma_\varphi = 3°$. We now consider additionally the lognormal and the $\beta$-distribution, and we test the effect of changing $\sigma_\varphi = 3°$ to $\sigma_\varphi = 2.5°$ and $\sigma_\varphi = 3.5°$. We assume a weakly informative prior in all cases, a normal distribution with $\mu'_\mu = 30°$, and standard deviation $\sigma'_\mu = 100°$ (see Illustration 3). The resulting posterior reliability indexes $\beta''$ are presented in Table 5.2; they are computed following Illustration 2 as $\beta'' = -\Phi^{-1}[F''_{\mu_\varphi}(21.36°)]$.

Evidently, the choice of the distribution type for the friction angle $\varphi$ has a minor effect on the reliability. This is due to the fact that the capacity is (assumed to be) determined by the mean friction angle. On the other hand, the standard deviation $\sigma_\varphi$ has a significant effect on the posterior reliability, since it determines the sampling uncertainty (the width of the likelihood function). ∎

It is possible to formalize the analysis of uncertainties in model choices by adding an additional layer of random variables. These additional random variables represent the

*Table 5.2* Resulting posterior reliability indexes for different model assumptions on the distribution of $\varphi$

|  | Normal | Lognormal | Beta |
|---|---|---|---|
| $\sigma_\varphi = 2.5°$ | 2.54 | 2.56 | 2.53 |
| $\sigma_\varphi = 3°$ | 2.12 | 2.22 | 2.15 |
| $\sigma_\varphi = 3.5°$ | 1.82 | 1.99 | 1.87 |

uncertainty in model choices. These random variables are not included in the reliability analysis; instead, the reliability is computed conditional on these variables. These random variables then lead to the reliability index being a random variable itself (Kiureghian and Ditlevsen 2009).

One example of this approach is found in Bayesian model class selection (Yuen 2010). Thereby, different models are assigned prior probabilities, which can be updated if measurements of system performances are available. The resulting prediction is made by combining predictions from different models through the total probability theorem. However, it is also possible to explicitly communicate the results obtained with the different models.

It has been proposed to model the uncertainty on model choices by nonprobabilistic approaches, including, for example, interval theory or fuzzy set theory (Rubio et al. 2004). This is not advocated here, since we believe that probabilistic methods are entirely capable of dealing with these uncertainties without the need to resort to another technique. Also, while the researchers developing such alternative methods are mostly aware of the underlying limitations these model, other engineers are not. Hence, we believe that these methods just add to the confusion. However, independent of which method one uses to represent the uncertainty on model choices, the main difficulty in explicitly showing the uncertainty associated with the model lies in selecting which parameters are model choices and which are uncertainties inherent to the problem. As an example, reconsider the simple example of Illustrations 1, 2, and 10. The parameter $\sigma_\varphi$, the standard deviation of $\varphi$, can be selected by the modeler, as we have done here; in this case, it is a model choice. Alternatively, the parameter $\sigma_\varphi$ can also be considered as a random variable, as we have done in Illustration 7; in this case, it is no longer a model choice. Therefore, in the second case, the model uncertainty would be reduced, without really changing the computations. For this reason, for practical applications, it is advocated that the uncertainty on model choices is not further formalized. Instead, the crucial model selections should be made explicit by parameter studies.

## 5.4 ADVANCED ALGORITHMS FOR EFFICIENT AND EFFECTIVE BAYESIAN UPDATING OF GEOTECHNICAL MODELS

In this section, three main strategies for the numerical solution of the Bayesian updating problem are outlined. These are the MCMC approach (in Section 5.4.1), the particle filter or sequential Monte Carlo approach (in Section 5.4.2), and the BUS approach, including the basic rejection sampling (in Section 5.4.3). The principles of the methods are introduced through basic algorithms, and more advanced and efficient algorithms are briefly described. A detailed presentation of these methods is well beyond the scope of this chapter, but the relevant references are given for readers who want to implement them.

### 5.4.1 Markov chain Monte Carlo

MCMC is a powerful approach for generating samples from distributions that are difficult to sample from directly. The main advantage of MCMC methods is that they do not require complete specification of the distribution from which one wants to sample. This is particularly useful for Bayesian updating, whereby the posterior distribution is known only up to a normalizing constant.

The basic idea of MCMC is to construct a stationary Markov chain with invariant distribution equal to the target distribution (Tierney 1994; Gelman 2004). MCMC methods have their roots in the Metropolis algorithm (Metropolis et al. 1953) that was developed for computing complex integrals with application to statistical physics. Hastings (1970) presented a generalization of the original Metropolis algorithm that encompasses several

MCMC algorithms as special cases. The Metropolis–Hastings algorithm proceeds by generating each new state of the Markov chain from a proposal distribution $q(\cdot|\mathbf{x}^{(k)})$ conditional on the current state $\mathbf{x}^{(k)}$ and then accepts or rejects the sample with a certain acceptance probability that depends on the current and proposed state. The Metropolis–Hastings algorithm for generating $K$ states of a Markov chain with stationary distribution equal to the posterior distribution $f_X''(\mathbf{x})$ can be summarized as follows.

## Metropolis–Hastings algorithm

1. Choose an arbitrary point $\mathbf{x}^{(0)}$.
2. $k = 1$.
3. Generate a candidate $\mathbf{y}$ from $q(\mathbf{y}|\mathbf{x}^{(k-1)})$.
4. Compute the acceptance probability

$$a(\mathbf{x}^{(k-1)}, \mathbf{y}) = \min\left\{\frac{f_X''(\mathbf{y})q(\mathbf{x}^{(k-1)}|\mathbf{y})}{f_X''(\mathbf{x}^{(k-1)})q(\mathbf{y}|\mathbf{x}^{(k-1)})}, 1\right\}.$$

5. Generate a sample $p$ from the standard uniform distribution in $[0,1]$.
6. If $p < a(\mathbf{x}^{(k-1)}, \mathbf{y})$, set $\mathbf{x}^{(k)} = \mathbf{y}$; otherwise, set $\mathbf{x}^{(k)} = \mathbf{x}^{(k-1)}$.
7. $k = k + 1$.
8. Stop if $k = K$, else go to 2.

It is noted that evaluation of the acceptance probability does not require the knowledge of the proportionality constant. The main computational effort consists of the evaluation of the likelihood function for each candidate sample. Starting from a state that may or may not be distributed according to the target distribution $f_X''(\mathbf{x})$, it can be shown that, under certain restrictions on the choice of the proposal distribution, the algorithm will asymptotically converge to the target distribution (Hastings 1970; Tierney 1994). However, the initial samples may follow a significantly different distribution. Therefore, an initial number of samples (typically in the order of 1000), constituting the so-called burn-in period, need to be discarded. It is often nontrivial to assess whether the chain has converged after the initial burn-in phase (Plummer et al. 2006).

Typical choices of the proposal distribution $q(\cdot|\mathbf{x}^{(k)})$ are symmetric distributions centered on the current state $\mathbf{x}^{(k)}$, such as the normal or uniform distribution, leading to the so-called random-walk samplers. The generated samples are correlated; however, their correlation can be controlled by the choice of the dispersion parameter of the proposal distribution (Gelman et al. 1996). However, the sample correlation increases significantly with the increase of the dimension of the target distribution (Gelman et al. 1996). As a result, the classical Metropolis–Hastings algorithm cannot be applied efficiently to high-dimensional problems. Some MCMC algorithms (Duane et al. 1987; Roberts and Tweedie 1996; Haario et al. 2005; Cheung and Beck 2009) can reduce the correlation of the Markov samples in high dimensions; however, they require additional evaluations of the likelihood function or its gradient at each sample.

High-dimensional problems can be tackled efficiently in cases where a conjugate prior describes each component of the prior distribution and hence its conditional distribution given all other components can be derived analytically (see, e.g., Ching et al. 2006). In such cases, it is advantageous to apply the so-called Gibbs sampling (Geman and Geman 1984). Gibbs sampling, which can be understood as a special case of the Metropolis–Hastings algorithm with acceptance probability of one, does not require parameter tuning and is shown to accelerate the convergence of the chain to its stationary distribution as compared to the classical Metropolis–Hastings algorithm (Gelman 2004).

## Illustration 11: Bayesian updating of friction angle with MCMC

We consider again the Bayesian updating problem of Illustration 1, where measurements of the spatially variable friction angle of a silty soil are used to update the distribution of its mean $\mu_\varphi$. We apply here the random-walk Metropolis–Hastings algorithm to obtain samples of the posterior distribution of $\mu_\varphi$. As the proposal distribution, we choose the normal distribution centered at the current sample. Figure 5.9 shows the obtained samples from three different choices of the standard deviation of the proposal PDF $\sigma_q$, namely $\sigma_q = 0.5°$, $2°$, and $8°$. We choose the same starting point $\mu_\varphi^{(0)} = 0°$ for all three cases. It is shown that all three cases include an initial burn-in period, whose length increases with decreasing $\sigma_q$. Moreover, if the variance of the proposal distribution is chosen too large (Case $\sigma_q = 8°$ in Figure 5.9), then the generated Markov chain is characterized by long flat periods that correspond to low acceptance probabilities of the candidate states. Conversely, if the variance is chosen too small (Case $\sigma_q = 0.5°$ in Figure 5.9), then the acceptance probability increases

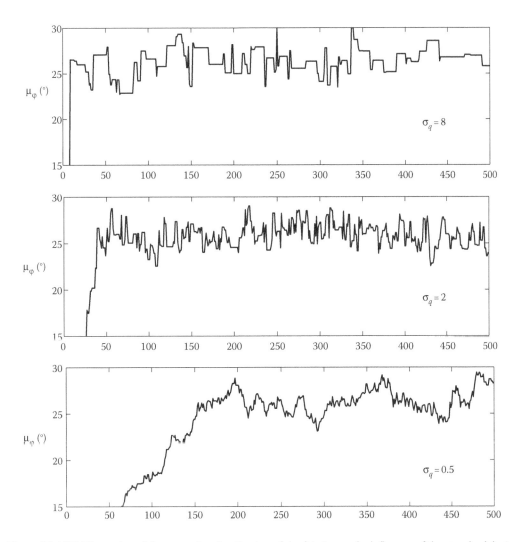

*Figure 5.9* MCMC samples of the posterior distribution of the friction angle: Influence of the standard deviation of the proposal PDF $\sigma_q$.

and the resulting Markov chain is highly correlated. Both of these cases result in the so-called poorly mixing chains. Well-mixing chains have low correlation and hence resemble white noise (Case $\sigma_\eta = 2°$ in Figure 5.9). The sample statistics obtained from the latter case after discarding the initial 50 samples are $\mu_\mu'' = 26.09°$ and $\sigma_\mu'' = 1.44°$. These values are close to the true statistics $\mu_\mu'' = 26.08°$ and $\sigma_\mu'' = 1.39°$. It is noted that if the Markov chain is started close to the center of the target distribution, then burn-in is typically not required.   ∎

## 5.4.2 Sequential Monte Carlo

During the last decade, a number of methods have been developed in the statistical community for exploring posterior distributions that can be classified under the umbrella term sequential Monte Carlo (Del Moral et al. 2006). These methods include annealed importance sampling (Neal 2001) and sequential particle filtering (Chopin 2002). In the engineering community, this approach is known as transitional MCMC (Ching and Chen 2007). The basic idea behind all these approaches is to gradually translate samples from the prior distribution to samples from the posterior distribution through a sequential reweighting operation. Reweighting is based on importance sampling from a sequence of distributions $\{f_i(\mathbf{x}), i = 1, \ldots, M\}$ constructed such that they gradually approach the target posterior distribution. The intermediate distributions are chosen as

$$f_i(\mathbf{x}) \propto L(\mathbf{x})^{q_i} f_X'(\mathbf{x}), \tag{5.42}$$

wherein $0 = q_0 < \cdots < q_M = 1$. For $i = 0$, $f_0(\mathbf{x})$ equals the prior distribution and for $i = M$, $f_M(\mathbf{x})$ is proportional to the posterior distribution. Initially, sequential Monte Carlo generates samples from the prior distribution $f_0(\mathbf{x}) = f_X'(\mathbf{x})$. At each subsequent step $i$, available samples from the distribution $f_{i-1}(\mathbf{x})$ are transformed into weighted samples from the distribution $f_i(\mathbf{x})$ by application of importance sampling with importance-sampling function equal to $f_{i-1}(\mathbf{x})$. To obtain unweighted samples, a resampling scheme must be applied (Doucet et al. 2001). The derived unweighted samples are then moved by application of MCMC with invariant distribution $f_i(\mathbf{x})$, to decrease the sample correlation. The sequential Monte Carlo algorithm for generating $K$ samples from the posterior distribution $f_X''(\mathbf{x})$ can be summarized as follows.

### Sequential Monte Carlo algorithm

1. Generate $K$ samples $\mathbf{x}_0^{(k)}$, $k = 1, \ldots, K$, from the prior distribution $f'(\mathbf{x})$.
2. $i = 1$.
3. Compute the weights $w_i^{(k)}$, $k = 1, \ldots, K$, by applying $w_i^{(k)} = L\left(\mathbf{x}_{i-1}^{(k)}\right)^{q_i - q_{i-1}}$.
4. Resample: Generate samples $\tilde{\mathbf{x}}_i^{(k)}$, $k = 1, \ldots, K$, by sampling from the discrete distribution $\{\mathbf{x}_{i-1}^{(k)}, w_i^{(k)}\}$.
5. Move: From each sample $\tilde{\mathbf{x}}_i^{(k)}$, $k = 1, \ldots, K$, perform an MCMC transition with invariant distribution $f_i(\mathbf{x})$ to obtain samples $\mathbf{x}_i^{(k)}$, $k = 1, \ldots, K$.
6. Stop if $i = M$, else go to 3.

The parameters $\{q_i, i = 1, \ldots, M\}$ should be chosen such that each pair of consecutive distributions does not vary considerably. One approach is to compute the $q_i$s adaptively such that the coefficient of variation (COV) of the weights $\{w_i^{(k)}, k = 1, \ldots, K\}$ equals a prescribed value, for example, 100% (Ching and Chen 2007). The performance of the algorithm is sensitive to the choice of the proposal distribution of the MCMC transition step. Typically, statistics of the weighted samples can be used to obtain a distribution that approximates the target distribution (Chopin 2002; Ching and Chen 2007).

### 5.4.3 Bayesian updating with structural reliability methods

The BUS method proposed in Straub and Papaioannou (2014) is based on ideas already presented in Section 5.3.5. The likelihood function $L(\mathbf{x})$ can be represented in the form of an LSF $h(\mathbf{x}, p) = p - cL(\mathbf{x})$ in the augmented outcome space of $\mathbf{X}$ and $P$, where $P$ is a standard uniform random variable and c is a constant that ensures $cL(\mathbf{x}) \leq 1$ for any $\mathbf{x}$. It can be shown that the posterior distribution of $\mathbf{X}$ corresponds to the prior distribution of $\mathbf{X}$ censored in the domain $\{h(\mathbf{x}, p) \leq 0\}$. The BUS approach uses structural reliability methods to efficiently evaluate the posterior distribution defined in this way.

In its simplest form, the BUS approach reduces to the classical rejection-sampling algorithm where the prior distribution is applied as an envelope distribution and the likelihood is applied as a filter (Smith and Gelfand 1992). It can be summarized as follows (where $K$ = total number of samples from the posterior).

Simple rejection-sampling algorithm

1. $k = 1$.
2. Generate a sample $\mathbf{x}^{(k)}$ from $f'_X(\mathbf{x})$.
3. Generate a sample $p^{(k)}$ from the standard uniform distribution in [0,1].
4. If $h(\mathbf{x}^{(k)}, p^{(k)}) \leq 0$
   a. Accept $\mathbf{x}^{(k)}$
   b. $k = k + 1$.
5. Stop if $k = K$, else go to 2.

The rejection-sampling approach is illustrated in Figure 5.10 for the example of Illustration 1. Here, the constant c is chosen as $[\max L(\mu_\varphi)]^{-1} = 0.0022$. The observation domain is shown by the shaded area on the left side. All samples that fall into this domain are accepted; these accepted samples follow the posterior distribution. This is verified on the right side of Figure 5.10, where the empirical CDF of the accepted samples is compared with the analytical solution of Illustration 1.

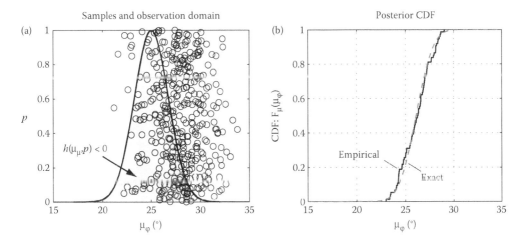

*Figure 5.10* Simple rejection sampling for the example of Illustration 1. (a) Samples from the prior distribution of $\mu_\mu$ and P. The shaded area is the domain $\{h(\mu_\varphi, p) \leq 0\}$; all samples within this domain are accepted. (b) Empirical CDF obtained from the accepted samples, together with the exact solution (normal CDF with mean 26.08° and standard deviation 1.39°).

While this rejection-sampling algorithm is easy to implement, it is generally inefficient, since the acceptance rate is small. In the example shown in Figure 5.10, the acceptance rate is 35%, but this rate decreases with increasing dimension (number of random variables) and with an increasing number of measurements. The BUS approach overcomes this problem by resorting to structural reliability methods for representing the observation domain $h(\mathbf{x}, p) \leq 0$. In Straub (2011), importance sampling is applied for this purpose. Straub and Papaioannou (2014) demonstrated how subset simulation is effective for solving this problem. As also demonstrated in Betz et al. (2014a), the algorithm is particularly efficient in high dimensions, making it suitable for Bayesian updating of models involving random fields. This is shown in the application presented in Section 5.6.

## 5.5 APPLICATION: FOUNDATION OF TRANSMISSION TOWERS UNDER TENSILE LOADING

In this example, we examine a shallow square foundation of a transmission tower embedded in a sandy soil. We consider two different information sets: cone penetration tests (CPTs) at the site of the foundation, and the observation that the system survived an extreme loading condition. We utilize the information to update the distribution of the material parameters of the soil as well as to update the reliability of the foundation subjected to tensile loading.

The base width of the footing and the depth of the foundation are taken as $B = D = 2$ m (see Figure 5.11); both the base thickness and pedestal width equal 0.6 m. Transmission towers apply tensile loads to their foundations, mainly due to transient wind loads. The LSF expressing the performance of the foundation under tensile force can be expressed as

$$g = Q_u - F, \tag{5.43}$$

where $F$ is the applied tensile loading and $Q_u$ is the uplift capacity of the foundation. It is noted that for the design of transmission tower foundations, other limit states must also be checked, but Equation 5.43 is often the critical one (Pacheco et al. 2008). The tensile loading is modeled as $F = kV^2$, where $V$ is a random variable modeling the maximum annual wind speed and $k$ is a deterministic coefficient incorporating several factors, such as geometry of the structure and geographical terrain. For simplicity here, we neglect the uncertainty in this coefficient, it is $k = 0.2$ tn/m. Considering drained conditions, the uplift capacity of a spread foundation can be expressed as (Kulhawy et al. 1983)

$$Q_u = W + Q_{su}, \tag{5.44}$$

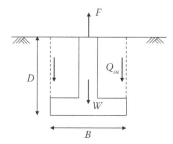

*Figure 5.11* Spread foundation of a transmission tower subjected to the uplift load.

where $W$ is the weight of the foundation and enclosed soil, and $Q_{su}$ is the side resistance (see Figure 5.7). $Q_{su}$ is evaluated applying the effective stress model (Kulhawy et al. 1983) by

$$Q_{su} = p_b \int_0^D K(z)\sigma'(z)\tan \varphi(z)dz, \tag{5.45}$$

where $p_b = 4B$ is the perimeter of the foundation, $\varphi(z)$ is the drained friction angle, $\sigma'(z) = z\gamma$ is the effective stress, $\gamma$ is the unit weight of the soil, and $K(z)$ is the operative horizontal stress coefficient. The friction angle and horizontal stress coefficient are spatially variable quantities. We adopt the common (but possibly unconservative) approach discussed in Section 5.3.1 and instead of explicitly modeling the quantities $\varphi$ and $K$ as random fields, we model their mean values $\mu_\varphi$ and $\mu_K$ as random variables. The spatial variability of the unit weight $\gamma$ is neglected, that is, $\gamma$ is assumed to vary uniformly within the soil profile. Equation 5.45 can then be written as

$$Q_{su} = 2B\gamma D^2 \mu_K \tan \mu_\varphi. \tag{5.46}$$

## 5.5.1 Prior probabilistic model

The wind speed $V$ is modeled by the Gumbel distribution with a COV of 30% and a mean value of 16.88 m/s corresponding to a 50-year return period wind speed of 30 m/s. In sandy soils, the mean friction angle $\mu_\varphi$ is expected to vary between 30° and 45°. These values are taken as the 10 and 90% quantiles of $\mu_\varphi$. We model $\mu_\varphi$ with the β-distribution with bounds 10° and 80°. The parameters of the beta-distribution are evaluated by matching the 10 and 90% quantiles to the aforementioned values, for which we obtain a mean value of 37.40° and a standard deviation of 5.80°. The mean horizontal stress coefficient $\mu_K$ typically depends on the mean friction angle and can vary between $K_0 = 1 - \sin\mu_\varphi$ and $(2/3)K_p = (2/3)\tan^2(45 + \mu_\varphi/2)$, where $K_0$ is the in situ coefficient of horizontal stress and $K_p$ is the coefficient of passive soil stress (Kulhawy et al. 1991). $K_0$ and $(2/3)K_p$ are taken as the 10 and 90% quantiles of the distribution of $\mu_K$ conditional on $\mu_\varphi$ and are used to fit the parameters of the lognormal distribution. Hence, the conditional distribution of $\mu_K$ given $\mu_\varphi$ is obtained as

$$f_{\mu_K}(\mu_K|\mu_\varphi) = \frac{1}{\mu_K \zeta \sqrt{2\pi}} \exp\left[-\frac{1}{2}\left(\frac{\ln\mu_K - \lambda}{\zeta}\right)^2\right], \tag{5.47}$$

wherein

$$\lambda = 0.5\ln\left(\frac{2}{3}\tan^2\left(45 + \frac{\mu_\varphi}{2}\right)(1 - \sin\mu_\varphi)\right), \tag{5.48}$$

$$\zeta = 0.39\ln\left(\frac{2}{3}\frac{\tan^2(45 + (\mu_\varphi/2))}{1 - \sin\mu_\varphi}\right). \tag{5.49}$$

The prior joint distribution of $\mu_K$ and $\mu_\varphi$ can then be obtained as

$$f'_{\mu_K \mu_\varphi}(\mu_K, \mu_\varphi) = f_{\mu_K}(\mu_K | \mu_\varphi) f'_{\mu_\varphi}(\mu_\varphi). \tag{5.50}$$

The unit weight of the soil $\gamma$ follows the normal distribution with mean value 19 kN/m³ and a COV of 5%. The unit weight of concrete that enters the evaluation of the self-weight of the foundation in Equation 5.46 is assumed deterministic and taken as $\gamma_c = 23$ kN/m³.

## 5.5.2 Reliability analysis based on the prior model

The reliability analysis is performed by application of the subset simulation (Au and Beck 2001). The method is implemented in the standard normal space. The Rosenblatt transformation (Hohenbichler and Rackwitz 1981) is applied to transform the samples of the independent standard normal distribution into samples of the joint distribution of $V$, $\gamma$, $\mu_K$ and $\mu_\varphi$. The computed prior failure probability is $Pr(F) = 1.70 \times 10^{-3}$ and the corresponding reliability index is $\beta' = 2.93$.

## 5.5.3 Updating with CPT test outcomes

We assume that a CPT is performed at the site of the foundation. Typically, data from electronic CPT tests are recorded at 5-cm depth intervals (Mayne 2007). Here, we assume for simplicity that five recordings of the tip resistance $q_t$ are obtained along the depth of the foundation. The recordings after normalization read: $Q_{tn1} = 38.2$, $Q_{tn2} = 43.5$, $Q_{tn3} = 45.8$, $Q_{tn4} = 35.4$, $Q_{tn5} = 41.1$, with $Q_{tn} = (q_t/p_a)(p_a/\sigma')^{0.5}$; $p_a$ is the atmospheric pressure and $\sigma'$ is the effective stress. Each measurement outcome is associated with an additive measurement error $\epsilon_m$ modeled by a zero mean normal random variable with a COV of 5%, which lies within the typical range of the COV of the measurement error of CPT tests (Orchant et al. 1988; Phoon and Kulhawy 1999). The recordings of the CPT test can be used to update the joint distribution of $\mu_K$ and $\mu_\varphi$ employing the following correlation between the normalized tip resistance and the friction angle (Kulhawy and Mayne 1990):

$$\varphi = 17.6 + 11.0 \log_{10} Q_{tn} + \epsilon_t. \tag{5.51}$$

The zero mean random variable $\epsilon_t$ expresses the transformation uncertainty of the correlation model and has a standard deviation of 2.8° (Kulhawy and Mayne 1990). We further assume that $\epsilon_t$ follows the normal distribution. Equation 5.51 describes a point-wise correlation between the friction angle and the normalized tip resistance $Q_{tn}$. To construct the likelihood of the measurement of $Q_{tn}$ for learning $\mu_\varphi$, we model the conditional distribution of $\varphi$ given $\mu_\varphi$ at each point by a beta-distribution with mean $\mu_\varphi$, COV 10%, and bounds 0° and 90°. The chosen COV agrees with the typical COV of inherent variability of the friction angle of sandy soils (Phoon and Kulhawy 1999). The likelihood of each measurement $Q_{tni}$ for learning $\mu_\varphi$ can then be expressed as

$$
\begin{aligned}
L_i(\mu_\varphi) &= f_{Q_{tni}}(Q_{tni} | \mu_\varphi) \\
&= \int_0^\infty \int_{-\infty}^\infty f_{Q_{tni}}(Q_{tni} | Q_{tn}) f_{Q_{tn}}(Q_{tn} | \varphi) f_\varphi(\varphi | \mu_\varphi) d\varphi dQ_{tn} \\
&= \int_0^\infty \int_{-\infty}^\infty f_{\epsilon_m}(Q_{tni} - Q_{tn}) f_{\epsilon_t}(\varphi - 17.6 - 11.0 \log_{10} Q_{tn}) f_\varphi(\varphi | \mu_\varphi) d\varphi dQ_{tn}.
\end{aligned}
\tag{5.52}
$$

Assuming that the measurement errors $\epsilon_m$ of the CPT recordings, the transformation uncertainties $\epsilon_t$, and the friction angle $\varphi$ at different locations are pairwise-independent random variables, we can obtain the likelihood describing the combined information from all five recordings, as follows:

$$L(\mu_\varphi) = \prod_{i=1}^{5} L_i(\mu_\varphi).$$ (5.53)

Note that $\mu_K$ and the CPT recordings are conditionally independent given $\mu_\varphi$, which implies that $f_{Q_{tni}}(Q_{tni}|\mu_\varphi,\mu_K) = f_{Q_{tni}}(Q_{tni}|\mu_\varphi)$ for all $Q_{tni}$. Therefore, we can update the joint distribution of $\mu_K$ and $\mu_\varphi$ with the likelihood of Equation 5.53, which gives

$$f''_{\mu_K\mu_\varphi}(\mu_K,\mu_\varphi) = aL(\mu_\varphi)f'_{\mu_K\mu_\varphi}(\mu_K,\mu_\varphi),$$ (5.54)

where $a$ is the proportionality constant.

We perform the updating by application of the BUS approach with subset simulation (see Section 5.4.3). We also update the reliability of the foundation by application of Equation 5.41. It is noted that the denominator in Equation 5.41 is computed as a by-product of the Bayesian updating algorithm. Therefore, for updating the reliability, only the numerator has to be computed additionally. The latter is computed by application of the subset simulation.

The updated mean and COV of $\mu_\varphi$ and $\mu_K$ as well as the updated probability of failure are displayed in Table 5.3 together with the corresponding priors. Figure 5.12 shows the prior joint PDF of $\mu_\varphi$ and $\mu_K$ and their posterior joint PDF conditional on the CPT measurements. It is shown that the means of the two parameters decrease compared to the priors. Moreover, their COVs also decrease, which demonstrate the effect of the CPT test on the reduction of the uncertainty of both $\mu_\varphi$ and $\mu_K$. The reduction is larger for $\mu_\varphi$ than for $\mu_K$. This is due to the fact that the tip resistance of the CPT test correlates with the friction angle, while $\mu_K$ is influenced indirectly due to its dependence with $\mu_\varphi$. The posterior probability of failure decreases slightly compared to the prior due to the combined effect of the reduction of the mean and of the uncertainty of the random variables. The posterior reliability index is $\beta'' = 3.01$.

## 5.5.4 Updating with survived loading conditions

We now assume a different scenario: the foundation survived an extreme loading condition caused by a wind speed of $V_m = 40$ m/s, which is significantly higher than the design 50-year return period wind speed. We can use this information to update the distribution of the

Table 5.3 Prior and posterior of the mean and COV of the soil properties and probability of failure given CPT measurement outcomes

| Parameter | Prior | | Posterior | |
|---|---|---|---|---|
| | Mean | COV | Mean | COV |
| Mean friction angle $\mu_\varphi$ (°) | 37.40 | 15.5% | 35.41 | 5.4% |
| Mean uplift coefficient $\mu_K$ | 1.43 | 114.6% | 1.32 | 78.6% |
| Probability of failure | $1.70 \times 10^{-3}$ | – | $1.31 \times 10^{-3}$ | – |

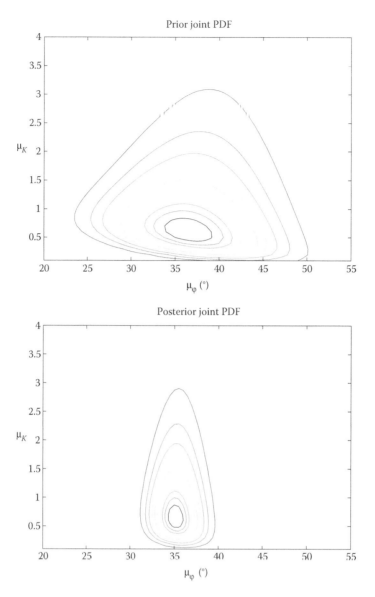

*Figure 5.12* Contour plots of the prior and posterior joint PDF of $\mu_\varphi$ and $\mu_K$ given CPT measurement outcomes.

random variables $\gamma$, $\mu_K$, and $\mu_\varphi$ that characterize the soil material. The likelihood function of this information can be expressed as follows:

$$L(\gamma, \mu_K, \mu_\varphi) = I(h(\gamma, \mu_K, \mu_\varphi) \leq 0), \tag{5.55}$$

where $I$ is the indicator function and $h(\gamma, \mu_K, \mu_\varphi)$ is an LSF describing the information and is given by

$$h(\gamma, \mu_K, \mu_\varphi) = k V_m^2 - Q_u(\gamma, \mu_K, \mu_\varphi). \tag{5.56}$$

*Table 5.4* Prior and posterior of the mean and COV of the soil properties and probability of failure given information on a survived loading condition

| Parameter | Prior | | Posterior | |
|---|---|---|---|---|
| | Mean | COV | Mean | COV |
| Unit weight $\gamma$ (kN/m³) | 19.0 | 5.0% | 19.04 | 4.8% |
| Mean friction angle $\mu_\varphi$ (°) | 37.40 | 15.5% | 38.22 | 14.4% |
| Mean uplift coefficient $\mu_K$ | 1.43 | 114.6% | 1.85 | 97.5% |
| Probability of failure | $1.70 \times 10^{-3}$ | – | $3.58 \times 10^{-4}$ | – |

To obtain samples from the posterior distribution of $\gamma$, $\mu_K$, and $\mu_\varphi$, it suffices to sample their prior distribution conditional on $\{h(\gamma, \mu_K, \mu_\varphi) \leq 0\}$. The samples are obtained through subset simulation. We also compute the updated failure probability by application of Equation 5.41. The numerator in Equation 5.41 is again computed with subset simulation. The prior and posterior statistics of the random variables and probability of failure are given in Table 5.4.

It is shown that the means of $\mu_\varphi$ and $\mu_K$ are increased compared to their priors, while their COVs are decreased implying reduction of uncertainty. This is reflected in the decrease of the probability of failure. The corresponding posterior reliability index is $\beta'' = 3.38$.

## 5.6 APPLICATION: FINITE-ELEMENT-BASED UPDATING OF SOIL PARAMETERS AND RELIABILITY

This example is based on the previous stochastic model and reliability updating presented in (Papaioannou and Straub 2012; Straub and Papaioannou 2014). We update the material properties of the soil surrounding a geotechnical site based on a deformation measurement performed in situ. Also, deformation measurements made at an intermediate excavation depth are utilized to update the reliability of the construction site at the stage of full excavation.

The site consists of a 5.0 m deep trench with cantilever sheet piles in a homogeneous soil layer of dense cohesionless sand with uncertain spatially varying mechanical properties (see Figure 5.13). The soil is modeled in two dimensions (2D) with plane-strain finite elements. For simplicity, neither groundwater nor external loading is considered. Additionally, we

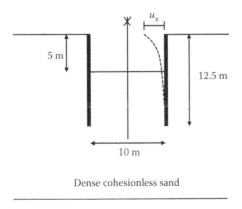

*Figure 5.13* Sheet pile wall in sand.

take advantage of the symmetry of the trench and model just one-half of the soil profile, although this implies an approximation when randomness in the soil material is taken into account. The material model used is an elasto-plastic model with a prismatic yield surface according to the Mohr–Coulomb criterion and a nonassociated plastic flow. The sheet pile is modeled using beam elements and the interaction between the retaining structure and the surrounding soil is modeled using nonlinear interface elements. The corresponding finite element (FE) model is implemented in the SOFiSTiK program (SOFiSTiK 2012). The finite-element mesh used in the analysis is shown in Figure 5.14b.

The sheet pile dimension and profile is determined analytically using the conventional method for the cantilever sheet pile design in granular soils, which requires equilibrium of the active and passive lateral pressures (see, e.g., Tschebotarioff 1951). Applying a global safety factor of 1.5, the design results in sheet piles of depth of 7.5 m and profile PZC 13. Young's modulus of steel is 210 GPa. The pile is modeled using beam elements with an equivalent rectangular cross section that behaves as the sheet pile under bending and axial loading. The interaction between the retaining structure and the surrounding soil is modeled using nonlinear interface elements. An elasto-plastic model with a yield surface defined by the Mohr–Coulomb criterion is used to describe the interface behavior. The elastic properties of the interface elements are taken from the mean values of the adjacent soil, while the strength properties are reduced by a factor 2/3 and a zero dilatancy is chosen.

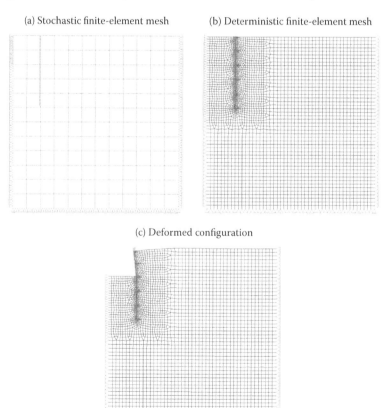

*Figure 5.14* (a) Stochastic and (b) deterministic finite-element mesh of the geotechnical site, shown for the situation prior to the excavation; (c) deformed configuration at full excavation for the mean values of the material properties. (From Straub, D. and I. Papaioannou. 2014. Doi: 10.1061/(asce) em.1943-7889.0000839.)

The finite-element analysis is performed stepwise, following the construction process. First, the modeling of the in situ stress state is carried out by means of the $K_0$-procedure, where $K_0 = 1 - \sin\varphi$ is the lateral earth pressure coefficient at rest for normally consolidated soils. Next, the sheet pile is installed by activating the corresponding beam and interface elements. Finally, the excavation is modeled by removing the plane-strain elements corresponding to the trench and applying the necessary loading to establish equilibrium.

### 5.6.1  Prior probabilistic model

Homogeneous non-Gaussian random fields describe the prior distributions of the uncertain material properties: Young's modulus $E$, friction angle $\varphi$, and unit weight $\gamma$. The joint distribution at each pair of locations is modeled by the Nataf distribution (Der Kiureghian and Liu 1986) with marginal distributions according to Table 5.5. The autocorrelation coefficient function is given by a separable exponential model $\rho_{xz}(\tau_x, \tau_z) = \exp(-(\tau_x/\lambda_x) - (\tau_z/\lambda_z))$, where $\tau_x$, $\tau_z$ are the absolute distances in the x (horizontal) and z (vertical) directions. The correlation lengths are $\lambda_x = 20$ m and $\lambda_z = 5$ m for all uncertain soil material properties. Cross-correlation between the different material properties is not included. The random fields are discretized by the midpoint method (Der Kiureghian and Ke 1988) using a stochastic mesh, consisting of 144 deterministic FE patches. The stochastic discretization resulted in a total of $3 \times 144 = 432$ basic random variables gathered in a vector $\mathbf{X}$. In Figure 5.14, the stochastic and deterministic FE meshes are shown. Figure 5.14c shows the deformed configuration at the final excavation stage computed with the mean values of the random fields.

### 5.6.2  Updating the soil parameters with deformation measurements

We assume that a measurement of the horizontal displacement at the top of the trench $u_{x,m} = 60$ mm is made at full excavation. The measurement is subjected to an additive error $\epsilon_m$, which is described by a normal PDF $f_{\epsilon_m}$ with zero mean and standard deviation $\sigma_{\epsilon_m} = 5$ mm. The likelihood function describing the measurement is

$$L(\mathbf{x}) = f_{\epsilon_m}[u_{x,m} - u_x(\mathbf{x})], \tag{5.57}$$

where $\mathbf{x}$ describes the material properties at the midpoints of the stochastic elements and $u_x(\mathbf{x})$ is the displacement evaluated by the FE program. Bayesian updating of the vector $\mathbf{X}$ is performed with BUS in conjunction with subset simulation. The constant $c$ is selected as $c = \sigma_{\epsilon_m}$, which satisfies the condition $cL(\mathbf{x}) \le 1$.

Table 5.5  Prior marginal distributions of the material properties of the soil

| Parameter | Distribution | Mean | COV |
|---|---|---|---|
| Unit weight $\gamma$ (kN/m³) | Normal | 19.0 | 5% |
| Young's modulus $E$ (MPa) | Lognormal | 125.0 | 25% |
| Poisson's ratio $\nu$ | – | 0.35 | – |
| Friction angle $\varphi$ (°) | Beta(0.0, 45.0) | 35.0 | 10% |
| Cohesion $c$ (MPa) | – | 0.0 | – |
| Dilatancy angle $\psi$ (°) | – | 5.0 | – |

The prior mean of $u_x(\mathbf{X})$ is 50.2 mm, which indicates that the prior model underestimates the measured tip displacement. Figures 5.15 and 5.16 show the posterior mean of Young's modulus $E$ and friction angle $\varphi$, respectively. The posterior means of the elements in the vicinity of the trench are smaller than the prior mean, which reflects the effect of the measured displacement. The effect is local, since the values of the stiffness and strength of the soil farther away from the trench have limited influence on the deformation at the location

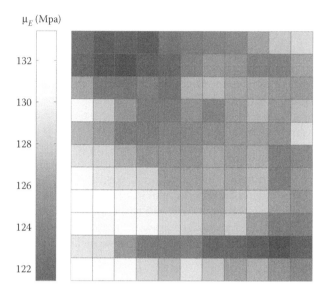

*Figure 5.15* Posterior mean of Young's modulus $E$ of the soil. (From Straub, D. and I. Papaioannou. 2014. Doi: 10.1061/(asce)em.1943-7889.0000839.)

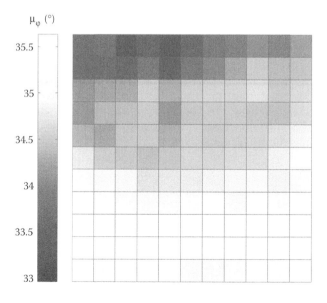

*Figure 5.16* Posterior mean of the friction angle $\varphi$ of the soil. (From Straub, D. and I. Papaioannou. 2014. Doi: 10.1061/(asce)em.1943-7889.0000839.)

of the measurement. Moreover, the results show the influence of the autocorrelation of the prior distribution. The change of the posterior means is steeper in the vertical than in the horizontal direction, which is due to the fact that the prior correlation length in the horizontal direction is larger than in the vertical direction. The low values of Young's modulus observed in the bottom right of Figure 5.15 cannot be explained by the measurement and hence are attributed to sampling error.

### 5.6.3 Updating the reliability with deformation measurements

We now consider that deformation measurements are taken at an intermediate excavation step and utilize the measurements to update the reliability at the final excavation state. The failure event $F$ is defined as the event of the horizontal displacement at the top of the trench $u_x(\mathbf{X})$ exceeding a threshold of $u_{x,t} = 100$ mm at full excavation. Mathematically, this is expressed through the following LSF:

$$g(\mathbf{x}) = u_{x,t} - u_x(\mathbf{x}),\tag{5.58}$$

This is a serviceability limit state, reflecting the assumed serviceability design requirements. The reliability analysis is performed by means of subset simulation. Without measurements, the computed failure probability is $\Pr(F) = 1.36 \times 10^{-2}$ with a corresponding reliability index $\beta' = 2.21$.

We assume that a measurement of the displacement $u_{x,2.5m}$ is made at an intermediate excavation step of 2.5 m depth. The measurement is subjected to an additive error $\epsilon_m$, which is described by a normal PDF $f_{\epsilon_m}$ with zero mean and standard deviation $\sigma_{\epsilon_m}$. This information is expressed by an event $Z$, described by the following likelihood function:

$$L(\mathbf{x}) = f_{\epsilon_m}[u_{x,m} - u_{x,2.5m}(\mathbf{x})].\tag{5.59}$$

For the estimation of the updated failure probability conditional on the measurement event $Z$, we apply the BUS approach combined with subset simulation. The constant $c$ is again selected as $c = \sigma_{\epsilon_m}$, which satisfies the condition $cL(\mathbf{x}) \leq 1$. The reliability updating was performed for different measurement outcomes $u_{x,m}$, and different values of the standard deviation $\sigma_{\epsilon_m}$ of the measurement error. The results are summarized in Table 5.6 and the computed reliability indices are plotted in Figure 5.17. For comparison, the (a priori) expected value of the measurement outcome $u_{x,m}$ is computed as 2.6 mm.

Not surprisingly, for measurements significantly higher than the expected value, the updated failure probability is higher than the prior probability. This difference is more pronounced when the measurement device is more accurate, that is, when $\sigma_{\epsilon_m}$ is smaller. For measurements lower than the expected value, the updated failure probability is lower than

Table 5.6 Updated probability of failure (prior probability of failure $\Pr(F) = 1.36 \times 10^{-2}$)

| Measurement | $\sigma_{\epsilon_m} = 2\,mm$ | | $\sigma_{\epsilon_m} = 1\,mm$ | |
|---|---|---|---|---|
| | $Pr(F|Z)$ | $\beta''$ | $Pr(F|Z)$ | $\beta''$ |
| $u_{x,m} = 10$ mm | $2.18 \times 10^{-1}$ | 0.78 | $3.31 \times 10^{-1}$ | 0.44 |
| $u_{x,m} = 5$ mm | $2.09 \times 10^{-2}$ | 2.04 | $3.59 \times 10^{-2}$ | 1.80 |
| $u_{x,m} = 2$ mm | $6.74 \times 10^{-3}$ | 2.47 | $1.84 \times 10^{-3}$ | 2.90 |

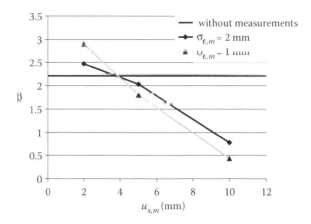

*Figure 5.17* Reliability index against measured displacement. Effect of measurement accuracy. (From Papaioannou, I. and D. Straub. 2012. *Computers and Geotechnics* **42**: 44–51.)

the prior probability. Again, the difference increases with the decreasing value of $\sigma_{\epsilon_m}$, because this implies a higher information content of the measurement. A measurement that exactly corresponds to the expected value of the deformation would lead to a posterior failure probability that is lower than the prior probability, due to a reduction of uncertainty.

For practical implementation, the reliability can be computed conditional on different hypothetical measurement outcomes, prior to the in situ measurement. Then a threshold value for the actual measurement may be obtained as a function of the target reliability index $\beta^T$ as illustrated in Figure 5.18. Assuming that the target reliability is $\beta^T = 2.5$ and the measurement accuracy is $\sigma_{\epsilon_m} = 1\,\mathrm{mm}$, the threshold value is 3.1 mm. Any measurement larger than this value corresponds to a reliability index less than the acceptable one. This would indicate that the retaining wall would not satisfy the reliability requirements at the final excavation stage and additional measures (e.g., anchors) would be necessary.

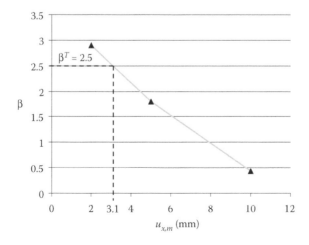

*Figure 5.18* Reliability index against measured displacement. Measurement to comply with target reliability. (From Papaioannou, I. and D. Straub. 2012. *Computers and Geotechnics* **42**: 44–51.)

## 5.7 CONCLUDING REMARKS

Bayesian analysis and updating of geotechnical models is a flexible and consistent framework for combining information from different sources in a single model and prediction tool. Its application in research is quickly growing, and geotechnical engineers in practice start to understand its relevance and potential, as they realize that purely deterministic models do not facilitate including measurements and observations in a consistent manner. There is also a growing acceptance that increasing the accuracy of deterministic models is not sufficient to explain the behavior of an inherently uncertain and random material such as soil.

As demonstrated in this chapter, the application of Bayesian updating is not very difficult from a mathematical or even computational point of view. Nevertheless, Bayesian updating is not easy and requires training and experience. The difficulties in Bayesian updating lie in the probabilistic modeling of (a) the randomness and uncertainty in the material and loading parameters, (b) the measurements, and (c) the model errors. A particular difficulty in geotechnical engineering, upon which we have touched occasionally, is the need to represent the spatially variable soil properties. Ideally, a random field modeling is applied, but for practical purposes (to reduce the computational effort), it is often necessary to represent the soil by representative values, possibly in combination with correction factors. While this reduces the computational efforts, it does not simplify the probabilistic modeling, even though this is— unfortunately—believed by many. If the variability of the soil is not represented consistently, results are obtained that are misleading and wrong. Multiple criticisms that were previously made of probabilistic analysis, in general, and Bayesian analysis, in particular, stem from problems that originate in incorrect probabilistic modeling. It has to be understood by geotechnical engineers that a basic introduction to statistics is not sufficient for carrying out probabilistic and reliability analysis of geotechnical systems.

The computational aspects of Bayesian updating are more straightforward and hence less critical than the modeling aspects. As we have shown in this chapter, once the deterministic and probabilistic models and likelihood functions are defined, computations can be performed relatively easily, both conceptually and implementation wise. For updating individual parameters, it is sufficient to numerically solve integrals or use analytical solutions through conjugate priors. For updating of the models or their parameters by means of a probabilistic inverse analysis, the algorithms presented in Section 5.4 are readily available. For example, the simple rejection-sampling approach can be implemented with just a few lines of code. If more efficient algorithms are needed, the BUS approach only necessitates structural reliability algorithms, which are available in many free and commercial codes.

While we have focused on Bayesian updating of mechanical (geotechnical) models, Bayesian updating is also a viable tool for other probabilistic models in the context of geotechnical construction. In particular, probabilistic models of construction performance and risk can be updated with observations made during the construction progress. As an example, Špačková et al. (2013) and Špačková and Straub (2013) show how observed progress in tunnel construction performance can be used to update the prediction of future performance and the total construction time through Bayesian analysis.

## ACKNOWLEDGMENT

We acknowledge insightful discussions we had with Dr. Olga Špačková on the application and the interpretation of the Bayesian updating results as well as her input on the illustrations.

## REFERENCES

Au, S.-K. and J. L. Beck. 2001. Estimation of small failure probabilities in high dimensions by subset simulation. *Probabilistic Engineering Mechanics* 16(4). 263–277.

Baecher, G. B. and J. T. Christian. 2008. Spatial variability and geotechnical reliability. *Reliability-Based Design in Geotechnical Engineering*. Taylor & Francis, London and New York, 76–133.

Beck, J. and L. Katafygiotis. 1998. Updating models and their uncertainties. I. Bayesian statistical framework. *Journal of Engineering Mechanics* 124(4): 455–461.

Beck, J. L. and S. K. Au. 2002. Bayesian updating of structural models and reliability using Markov chain Monte Carlo simulation. *Journal of Engineering Mechanics-ASCE* 128(4): 380–391.

Bernardo, J. M. 1979. Reference posterior distributions for Bayesian inference. *Journal of the Royal Statistical Society. Series B (Methodological)* 41(2): 113–147.

Betz, W., I. Papaioannou and D. Straub. 2014a. Adaptive variant of the BUS approach to Bayesian updating. *Eurodyn 2014*. Porto, Portugal.

Betz, W., I. Papaioannou and D. Straub. 2014b. Numerical methods for the discretization of random fields by means of the Karhunen–Loève expansion. *Computer Methods in Applied Mechanics and Engineering* 271(0): 109–129.

Beven, K. 2010. *Environmental Modelling: An Uncertain Future?* CRC Press, London.

Cheung, S. H. and J. L. Beck. 2009. Bayesian model updating using hybrid Monte Carlo simulation with application to structural dynamic models with many uncertain parameters. *Journal of Engineering Mechanics-ASCE* 135(4): 243–255.

Ching, J. and Y. Chen. 2007. Transitional Markov chain Monte Carlo method for Bayesian model updating, model class selection, and model averaging. *Journal of Engineering Mechanics* 133(7): 816–832.

Ching, J., M. Muto and J. L. Beck. 2006. Structural model updating and health monitoring with incomplete modal data using Gibbs sampler. *Computer-Aided Civil and Infrastructure Engineering* 21(4): 242–257.

Chopin, N. 2002. A sequential particle filter method for static models. *Biometrika* 89(3): 539–552.

DeGroot, M. H. 1969. *Optimal Statistical Decisions*. New York, McGraw-Hill.

Del Moral, P., A. Doucet and A. Jasra. 2006. Sequential Monte Carlo samplers. *Journal of the Royal Statistical Society: Series B (Statistical Methodology)* 68(3): 411–436.

Der Kiureghian, A. and J.-B. Ke. 1988. The stochastic finite element method in structural reliability. *Probabilistic Engineering Mechanics* 3(2): 83–91.

Der Kiureghian, A. and P.-L. Liu. 1986. Structural reliability under incomplete probability information. *Journal of Engineering Mechanics* 112(1): 85–104.

Ditlevsen, O. 1982. Model uncertainty in structural reliability. *Structural Safety* 1(1): 73–86.

Ditlevsen, O. and H. O. Madsen. 1996. *Structural Reliability Methods*. Chichester [u.a.], Wiley.

Doucet, A., N. De Freitas and N. Gordon. 2001. *Sequential Monte Carlo Methods in Practice*. New York, Springer.

Duane, S., A. D. Kennedy, B. J. Pendleton and D. Roweth. 1987. Hybrid Monte Carlo. *Physics Letters B* 195(2): 216–222.

Fink, D. 1997. A compendium of conjugate priors. *Technical Report*. Montana State University.

Fisher, R. A. 1922. On the mathematical foundations of theoretical statistics. *Philosophical Transactions of the Royal Society of London. Series A* 222(594–604): 309–368.

Gelman, A. 2004. *Bayesian Data Analysis*. Boca Raton, FL, Chapman & Hall/CRC.

Gelman, A., G. Roberts and W. Gilks. 1996. Efficient Metropolis jumping rules. *Bayesian Statistics* 5: 599–608.

Geman, S. and D. Geman. 1984. Stochastic relaxation, Gibbs distributions, and the Bayesian restoration of images. *Pattern Analysis and Machine Intelligence, IEEE Transactions on* 6(6): 721–741.

Gilks, W. R., S. Richardson and D. J. Spiegelhalter. 1998. *Markov Chain Monte Carlo in Practice*. Boca Raton, FL, Chapman & Hall.

Gilks, W. R. and P. Wild. 1992. Adaptive rejection sampling for Gibbs sampling. *Applied Statistics* 41(2): 337–348.

Goulet, J.-A. and I. F. C. Smith. 2013. Structural identification with systematic errors and unknown uncertainty dependencies. *Computers and Structures* **128**(0): 251–258.

Griffiths, D. and G. A. Fenton. 2001. Bearing capacity of spatially random soil: The undrained clay Prandtl problem revisited. *Geotechnique* **51**(4): 351–359.

Haario, H., E. Saksman and J. Tamminen. 2005. Componentwise adaptation for high dimensional MCMC. *Computational Statistics* **20**(2): 265–273.

Hastings, W. K. 1970. Monte Carlo sampling methods using Markov chains and their applications. *Biometrika* **57**(1): 97–109.

Hohenbichler, M. and R. Rackwitz. 1981. Non-normal dependent vectors in structural safety. *Journal of the Engineering Mechanics Division-ASCE* **107**(6): 1227–1238.

Jaynes, E. T. 1968. Prior probabilities. *Systems Science and Cybernetics, IEEE Transactions on* **4**(3): 227–241.

Jensen, H., C. Vergara, C. Papadimitriou and E. Millas. 2013. The use of updated robust reliability measures in stochastic dynamical systems. *Computer Methods in Applied Mechanics and Engineering* **267**: 293–317.

Der Kiureghian, A. and O. Ditlevsen. 2009. Aleatory or epistemic? Does it matter? *Structural Safety* **31**(2): 105–112.

Kulhawy, F. H. and P. W. Mayne. 1990. *Manual on Estimating Soil Properties for Foundation Design*, Electric Power Research Institute, Palo Alto, CA (USA); Cornell University, Ithaca, NY (USA). Geotechnical Engineering Group.

Kulhawy, F. H., C. N. Nikolaides and C. H. Trautmann. 1991. Experimental investigation of the uplift behavior of spread foundations in cohesionless soil, Report Number TR-100220, Electric Power Research Institute, Palo Alto, CA.

Kulhawy, F. H. and K.-K. Phoon. 2002. Observations on geotechnical reliability-based design development in North America. Foundation design codes and soil investigation in view of International Harmonization and Performance Based Design. *Proceedings of IWS Kamakura*, Japan, 31–48.

Kulhawy, F. H., C. H. Trautmann, J. F. Beech, T. D. O'Rourke, W. McGuire, W. A. Wood and C. Capano. 1983. Transmission line structure foundations for uplift-compression loading, Report Number EL-2870, Electric Power Research Institute, Palo Alto, CA.

Madsen, H. O. 1987. Model updating in reliability theory. *ICASP 5, International Conference on Application of Statistics and Probability in Soil and Structures*. Vancouver: 565–577.

Madsen, H. O., S. Krenk and N. C. Lind. 1985. *Methods of Structural Safety*. Englewood Cliffs, NJ, Prentice-Hall.

Mayne, P. W. 2007. *Cone Penetration Testing*. Washington, DC, NCHRP Synthesis 368, Transportation Research Board.

Metropolis, N., A. W. Rosenbluth, M. N. Rosenbluth, A. H. Teller and E. Teller. 1953. Equation of state calculations by fast computing machines. *Journal of Chemical Physics* **21**(6): 1087–1092.

Neal, R. M. 2001. Annealed importance sampling. *Statistics and Computing* **11**(2): 125–139.

Oberguggenberger, M. and W. Fellin. 2002. From probability to fuzzy sets: The struggle for meaning in geotechnical risk assessment. *Probabilistics in Geotechnics: Technical and Economic Risk Estimation*. R. Pötter, H. Klapperich and H. F. Schweiger. Essen, Germany, Verlag Glückauf GmbH: 29–38.

Orchant, C. J., F. H. Kulhawy and C. H. Trautmann. 1988. Reliability-based foundation design for transmission line structures: Critical evaluation of *in-situ* test methods. Report Number EL-5507(2), Electric Power Research Institute, Palo Alto, CA.

Pacheco, M. P., F. A. B. Danziger and C. Pereira Pinto. 2008. Design of shallow foundations under tensile loading for transmission line towers: An overview. *Engineering Geology* **101**(3): 226–235.

Papadimitriou, C., J. L. Beck and L. S. Katafygiotis. 2001. Updating robust reliability using structural test data. *Probabilistic Engineering Mechanics* **16**(2): 103–113.

Papaioannou, I. and D. Straub. 2012. Reliability updating in geotechnical engineering including spatial variability of soil. *Computers and Geotechnics* **42**: 44–51.

Phoon, K.-K. 2008. Numerical recipes for reliablity analysis—A primer. *Reliability-Based Design in Geotechnical Engineering: Computations and Applications*. K.-K. Phoon. CRC Press.

Phoon, K.-K. and F. H. Kulhawy. 1999. Characterization of geotechnical variability. *Canadian Geotechnical Journal* 36(4): 612–624.

Plummer, M., N. Best, K. Cowles and K. Vines. 2006. CODA: Convergence diagnosis and output analysis for MCMC. *R News* 6(1): 7–11.

Rackwitz, R. 2000. Reviewing probabilistic soils modelling. *Computers and Geotechnics* 26(3–4): 199–223.

Raiffa, H. and R. Schlaifer. 1961. *Applied Statistical Decision Theory*. Boston, Division of Research, Graduate School of Business Administration, Harvard University.

Roberts, G. O. and R. L. Tweedie. 1996. Exponential convergence of Langevin distributions and their discrete approximations. *Bernoulli* 2: 341–363.

Rubio, E., J. W. Hall and M. G. Anderson. 2004. Uncertainty analysis in a slope hydrology and stability model using probabilistic and imprecise information. *Computers and Geotechnics* 31(7): 529–536.

Schall, G., S. Gollwitzer and R. Rackwitz. 1989. Integration of multinormal densities on surfaces. *Reliability and Optimization of Structural Systems' 88*, Springer, Berlin, 235–248.

Simoen, E., C. Papadimitriou and G. Lombaert. 2013. On prediction error correlation in Bayesian model updating. *Journal of Sound and Vibration* 332(18): 4136–4152.

Smith, A. F. M. and A. E. Gelfand. 1992. Bayesian statistics without tears: A sampling–resampling perspective. *The American Statistician* 46(2): 84–88.

SOFiSTiK. 2012. *SOFiSTiK Analysis Programs. Version 2012*. Oberschleissheim, SOFiSTiK AG.

Špačková, O., J. Šejnoha and D. Straub. 2013. Probabilistic assessment of tunnel construction performance based on data. *Tunnelling and Underground Space Technology* 37(0): 62–78.

Špačková, O. and D. Straub. 2013. Dynamic Bayesian network for probabilistic modeling of tunnel excavation processes. *Computer-Aided Civil and Infrastructure Engineering* 28(1): 1–21.

Straub, D. 2011. Reliability updating with equality information. *Probabilistic Engineering Mechanics* 26(2): 254–258.

Straub, D. 2012. *Lecture Notes in Engineering Risk Analysis*. München, Engineering Risk Analysis Group, Technische Universität.

Straub, D. 2014. Value of information analysis with structural reliability methods. *Structural Safety* 49: 75–86.

Straub, D. and I. Papaioannou. 2014. Bayesian updating with structural reliability methods. *Journal of Engineering Mechanics*, Trans ASCE. Doi: 10.1061/(asce)em.1943-7889.0000839.

Sundar, V. S. and C. S. Manohar. 2013. Updating reliability models of statically loaded instrumented structures. *Structural Safety* 40: 21–30.

Tang, X.-S., D.-Q. Li, G. Rong, K.-K. Phoon and C.-B. Zhou. 2013. Impact of copula selection on geotechnical reliability under incomplete probability information. *Computers and Geotechnics* 49(0): 264–278.

Tierney, L. 1994. Markov chains for exploring posterior distributions. *The Annals of Statistics* 22(4): 1701–1728.

Tschebotarioff, G. P. 1951. *Soil Mechanics, Foundations, and Earth Structures*. New York, McGraw-Hill.

Yuen, K.-V. 2010. Recent developments of Bayesian model class selection and applications in civil engineering. *Structural Safety* 32(5): 338–346.

Zhang, J., L. Zhang and W. H. Tang. 2009. Bayesian framework for characterizing geotechnical model uncertainty. *Journal of Geotechnical and Geoenvironmental Engineering* 135(7): 932–940.

Zhang, L. L., J. Zhang, L. M. Zhang and W. H. Tang. 2010. Back analysis of slope failure with Markov chain Monte Carlo simulation. *Computers and Geotechnics* 37(7–8): 905–912.

Chapter 6

# Polynomial chaos expansions and stochastic finite-element methods

*Bruno Sudret*

## 6.1 INTRODUCTION

Soil and rock masses naturally present heterogeneity at various scales of description. This heterogeneity may be of two types. On a large scale, soil properties may be considered piece-wise homogeneous once regions (e.g., layers) have been identified. On a lower scale, the local spatial variability of the properties shall be accounted for. In any case, the use of deterministic values for representing the soil characteristics is poor, since it ignores the natural randomness of the medium. Alternatively, this randomness may be properly modeled using probabilistic models.

In the first of the two cases identified above, the material properties (e.g., Young's modulus, cohesion, friction angle, etc.) may be modeled in each region as *random variables* whose distributions (and possibly mutual correlation) have to be specified. In the second case, the introduction of random fields is necessary. In this respect, probabilistic soil modeling is a long-term story, see, for example, Vanmarcke (1977); DeGroot and Baecher (1993); Fenton (1999a,b); Rackwitz (2000); and Popescu et al. (2005).

Usually, soil characteristics are investigated to feed models of geotechnical structures in the context of the engineering design. Examples of such structures are dams, embankments, pile or raft foundations, tunnels, and so on. The design then consists of choosing characteristics of the structure (dimensions, material properties) so that the latter fulfills some requirements (e.g., retain water, support a building, etc.) under a given set of environmental actions that we will call "loading." The design is practically carried out by satisfying some *design criteria* that usually apply onto model response quantities (e.g., global equilibrium equation, settlements, bearing capacity, etc.). The conservatism of the design according to codes of practice is ensured first by introducing safety coefficients, and second by using penalized values of the model parameters. In this approach, the natural spatial variability of the soil is completely hidden.

From another point of view, when the uncertainties and variability of the soil properties have been identified, methods that allow propagating these uncertainties throughout the model have to be used. Perturbation methods used in the 1980s and 1990s (Baecher and Ingra, 1981; Phoon et al., 1990) allow estimating the mean value and standard deviation of the system response. First-/second-order reliability methods (FORM/SORMs) are used for assessing the probability of failure of the system with respect to performance criteria (Ditlevsen and Madsen, 1996). Numerous applications of the latter can be found, for example, in Phoon (2003); Low (2005); Low and Tang (2007); and Li and Low (2010) among others.

In the early 1990s, a new approach called *stochastic finite-element method* (SFEM) has emerged, which allows one to solve boundary value problems with uncertain coefficients and is especially suited to spatially variable inputs (Ghanem and Spanos, 1991). The key ingredient in this approach is the so-called *polynomial chaos expansions* (PCEs), which allow one to represent a random output (e.g., the nodal displacement vector resulting from a finite-element analysis) as a polynomial series in the input variables. Early applications of such an SFEM to geotechnics can be found in Ghanem and Brzkala (1996); Sudret and Der Kiureghian (2000); Ghiocel and Ghanem (2002); Clouteau and Lafargue (2003); Sudret et al. (2004, 2006); and Berveiller et al. (2006).

During the past 10 years, PCEs have become a cross-field key approach to uncertainty quantification in engineering problems ranging from computational fluid dynamics and heat-transfer problems to electromagnetism. The associated computational methods have also been somewhat simplified due to the emergence of *nonintrusive* spectral approaches, as shown later.

The goal of this chapter is to give an overview on stochastic finite-element analysis using PCEs, focusing more specifically on nonintrusive computation schemes. The chapter is organized as follows. Section 6.2 presents a versatile uncertainty quantification framework that is now widely used by both researchers and practitioners (Sudret, 2007; De Rocquigny, 2012). Section 6.3 presents the machinery of PCEs in a step-by-step approach: how to construct the polynomial chaos basis, how to compute the coefficients, how to estimate the quality of the obtained series, and how to address large dimensional problems using *sparse* expansions. Section 6.4 shows how to postprocess a PC expansion for different applications, that is, compute statistical moments of the response quantities, estimate the model output distribution, or carry out sensitivity analysis. Finally, Section 6.5 presents different application examples in the field of geotechnics.

## 6.2 UNCERTAINTY PROPAGATION FRAMEWORK

### 6.2.1 Introduction

Let us consider a physical system (e.g., a foundation on a soft soil layer, a retaining wall, etc.) whose mechanical behavior is represented by a computational model $\mathcal{M}$:

$$x \in \mathcal{D}_X \subset \mathbb{R}^M \mapsto y = \mathcal{M}(x) \in \mathbb{R}. \tag{6.1}$$

In this equation, $x = \{x_1, ..., x_M\}^T$ gathers the $M$ input parameters of the model while $y$ is the *quantity of interest* (QoI) in the analysis, for example, a load-carrying capacity, a limit state equation for stability, and so on. In the sequel, only models having a single (scalar) QoI are presented, although the derivations hold component-wise in case of vector-valued models $y = \mathcal{M}(x) \in \mathbb{R}^q$.

As shown in Figure 6.1, once a computational model $\mathcal{M}$ is chosen to evaluate the performance of the system of interest (Step A), the sources of uncertainty are to be quantified (Step B): in this step, the available information (expert judgment on the problem, databases and literature, and existing measurements) is used to build a proper probabilistic model of the input parameters, which is eventually cast as a random vector $X$ described by a joint probability density function (PDF) $f_X$. When the parameters are assumed statistically independent, this joint distribution is equivalently defined by the set of marginal distribution of all input parameters, say $\{f_{X_i}, i = 1, ..., M\}$. If dependence exists, the copula formalism may be used, see Nelsen (1999); Caniou (2012). As a consequence, the QoI becomes a random variable

$$Y = \mathcal{M}(X), \tag{6.2}$$

Step B                          Step A                          Step C

Quantification of              Model(s) of the                 Uncertainty propagation

sources of uncertainty         assessment criteria

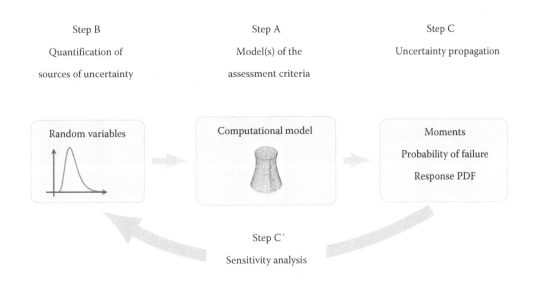

Step C´

Sensitivity analysis

*Figure 6.1* Uncertainty quantification framework.

whose properties are implicitly defined by the *propagation* of the uncertainties described by the joint distribution $f_X$ through the computational model (Step C). This step consists of characterizing the probabilistic content of $Y$, that is, its statistical moments, quantiles, or full distribution, to derive confidence intervals (CIs) around the mean QoI for robust predictions, or to carry out reliability assessment.

When the spatial variability of soil properties is to be modeled, *random fields* have to be used. The mathematical description of random fields and their discretization is beyond the scope of this chapter. For an overview, the interested reader is referred to Vanmarcke (1983); Sudret and Der Kiureghian (2000); and Sudret and Berveiller (2007). In any case, and whatever the random field discretization technique (e.g., Karhunen–Love [KL] expansion, expansion optimal linear estimation [EOLE], etc.), the problem eventually reduces to an input random vector (which is usually Gaussian) appearing in the discretization. Then the uncertainty propagation issue identically suits the framework described above. For the sake of illustration, an application example involving spatial variability will be addressed in Section 6.6.3.

## 6.2.2 Monte Carlo simulation

The Monte Carlo simulation (MCS) is a well-known technique for estimating statistical properties of the random response $Y = \mathcal{M}(X)$: realizations of the input vector $X$ are sampled according to the input distribution $f_X$, then the computational model $\mathcal{M}$ is run for each sample, and the resulting set of QoI is post-processed (Rubinstein and Kroese, 2008). Although rather universal, MCS suffers from low efficiency. Typically, $10^{3–4}$ samples are required to reach an acceptable accuracy. The cost even blows up when probabilities of failure are to be computed for the sake of reliability assessment, since $10^{k+2}$ samples are required when estimating a probability of $10^{-k}$. Thus, alternative methods have to be devised for addressing uncertainty quantification problems that involve computationally demanding models such as finite-element models. In the past decade, PCEs have become a popular approach in this respect.

## 6.3 POLYNOMIAL CHAOS EXPANSIONS

### 6.3.1 Mathematical setting

Consider a computational model $\mathcal{M}$ whose input parameters are represented by a random vector $X$, and the associated (random) QoI $Y = \mathcal{M}(X)$. Assuming that $Y$ has a finite variance (which is a physically meaningful assumption when dealing with geotechnical systems), it belongs to the so-called Hilbert space of second-order random variables, which allows for the following representation (Soize and Ghanem, 2004):

$$Y = \sum_{j=0}^{\infty} y_j Z_j. \tag{6.3}$$

In Equation 6.3 the random variable $Y$ is cast as an infinite series, in which $\{Z_j\}_{j=0}^{\infty}$ is a numerable set of random variables (which form a basis of the Hilbert space), and $\{y_j\}_{j=0}^{\infty}$ are coefficients. The latter may be interpreted as the *coordinates* of $Y$ in this basis. Hilbertian analysis guarantees the existence of such bases and representation; however, many choices are possible. In the sequel, we focus on PCEs, in which the basis terms $\{Z_j\}_{j=0}^{\infty}$ are multivariate orthonormal polynomials in the input vector $X$, that is, $Z_j = \Psi_j(X)$.

### 6.3.2 Construction of the basis

#### 6.3.2.1 Univariate orthonormal polynomials

For the sake of simplicity, we assume that the input random vector has *independent* components denoted by $\{X_i, i = 1, ..., M\}$, meaning that the joint distribution is simply the product of the $M$ marginal distributions $\{f_{X_i}\}_{i=1}^{M}$:

$$f_X(x) = \prod_{i=1}^{M} f_{X_i}(x_i), \quad x_i \in \mathcal{D}_{X_i}, \tag{6.4}$$

where $\mathcal{D}_{X_i}$ is the support of $X_i$. For each single variable $X_i$ and any two functions $\phi_1, \phi_2 : x \in \mathcal{D}_{X_i} \mapsto \mathbb{R}$, we define a functional inner product by the following integral (provided it exists):

$$\langle \phi_1, \phi_2 \rangle_i = \int_{\mathcal{D}_{X_i}} \phi_1(x)\phi_2(x)f_{X_i}(x)dx. \tag{6.5}$$

Equation 6.5 is nothing but the expectation $\mathbb{E}[\phi_1(X_i)\,\phi_2(X_i)]$ with respect to the marginal distribution $f_{X_i}$. Two such functions are said to be *orthogonal* with respect to the probability measure $\mathbb{P}(dx) = f_{X_i}(x)dx$ if $\mathbb{E}[\phi_1(X_i)\,\phi_2(X_i)] = 0$. Using the above notation, classical algebra allows one to build a family of *orthogonal polynomials* $\{\pi_k^{(i)}, k \in \mathbb{N}\}$ satisfying

$$\langle \pi_j^{(i)}, \pi_k^{(i)} \rangle_i \stackrel{\text{def}}{=} \mathbb{E}\left[\pi_j^{(i)}(X_i)\pi_k^{(i)}(X_i)\right] = \int_{\mathcal{D}_{X_i}} \pi_j^{(i)}(x)\pi_k^{(i)}(x)f_{X_i}(x)\,dx = a_j^i \delta_{jk}, \tag{6.6}$$

where subscript $k$ denotes the degree of the polynomial $\pi_k^{(i)}$, $\delta_{jk}$ is the Kronecker symbol equal to 1 when $j = k$ and 0 otherwise, and $a_j^i$ corresponds to the squared norm of $\pi_j^{(i)}$:

$$a_j^i \overset{\text{def}}{=} \left\| \pi_j^{(i)} \right\|_i^2 \overset{\text{def}}{=} \left\langle \pi_j^{(i)}, \pi_j^{(i)} \right\rangle_i. \tag{6.7}$$

This family can be obtained by applying the Gram–Schmidt orthogonalization procedure to the canonical family of monomials $\{1, x, x^2, \ldots\}$. For standard distributions, the associated family of orthogonal polynomials is well known. For instance, if $X_i \sim \mathcal{U}(-1, 1)$ has a uniform distribution over $[-1, 1]$, the resulting family is that of the so-called *Legendre polynomials*. If $X_i \sim \mathcal{N}(0, 1)$ has a standard normal distribution with a zero mean value and unit standard deviation, the resulting family is that of *Hermite polynomials*. The families associated to standard distributions are summarized in Table 6.1 (Xiu and Karniadakis, 2002).

Note that the obtained family is usually not orthonormal. By enforcing the normalization, an *orthonormal family* $\{\psi_j^{(i)}\}_{j=0}^\infty$ is obtained from Equations 6.6 and 6.7 by

$$\psi_j^{(i)} = \pi_j^{(i)} \Big/ \sqrt{a_j^i} \quad i = 1, \ldots, M, \ j \in \mathbb{N}. \tag{6.8}$$

The normalizing coefficients are listed in Table 6.1 for the standard families. For the sake of illustration, Hermite polynomials up to degree 4 are plotted in Figure 6.2.

### 6.3.2.2 Multivariate polynomials

To build up a basis such as in Equation 6.3, *tensor products* of univariate orthonormal polynomials are built up. For this purpose, let us define multi-indices (also called tuples) $\alpha \in \mathbb{N}^M$ that are ordered lists of integers

$$\alpha = (\alpha_1, \ldots, \alpha_M), \quad \alpha_i \in \mathbb{N}. \tag{6.9}$$

One can associate a multivariate polynomial $\Psi_\alpha$ to any multi-index $\alpha$ by

$$\Psi_\alpha(x) \overset{\text{def}}{=} \prod_{i=1}^M \psi_{\alpha_i}^{(i)}(x_i), \tag{6.10}$$

Table 6.1 Classical families of orthogonal polynomials

| Type of variable | Distribution | Orthogonal polynomials | Hilbertian basis $\psi_k(x)$ |
|---|---|---|---|
| Uniform $\mathcal{U}(-1, 1)$ | $\mathbf{I}_{[-1,1]}(x)/2$ | Legendre $P_k(x)$ | $P_k(x) \Big/ \sqrt{\dfrac{1}{2k+1}}$ |
| Gaussian $\mathcal{N}(0, 1)$ | $\dfrac{1}{\sqrt{2\pi}} e^{-x^2/2}$ | Hermite $H_{e_k}(x)$ | $H_{e_k}(x)/\sqrt{k!}$ |
| Gamma $\Gamma(a, \lambda = 1)$ | $x^a e^{-x} \mathbf{I}_{\mathbb{R}^+}(x)$ | Laguerre $L_k^a(x)$ | $L_k^a(x) \Big/ \sqrt{\dfrac{\Gamma(k+a+1)}{k!}}$ |
| Beta $\mathcal{B}(a, b)$ | $\mathbf{I}_{[-1,1]}(x) \dfrac{(1-x)^a (1+x)^b}{B(a)B(b)}$ | Jacobi $J_k^{a,b}(x)$ | $J_k^{a,b}(x)/\mathfrak{I}_{a,b,k}$ |

$$\mathfrak{I}_{a,b,k}^2 = \frac{2^{a+b+1}}{2k+a+b+1} \frac{\Gamma(k+a+1)\Gamma(k+b+1)}{\Gamma(k+a+b+1)\Gamma(k+1)}$$

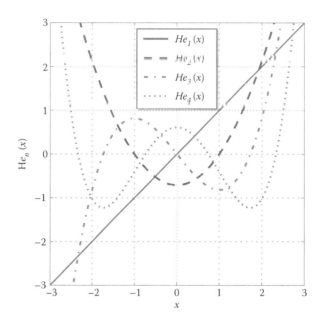

*Figure 6.2* Univariate Hermite polynomials.

where the univariate polynomials $\{\psi_k^{(i)}, k \in \mathbb{N}\}$ are defined according to the $i$th marginal distribution, see Equations 6.6 through 6.8. By virtue of Equation 6.6 and the above tensor product construction, the multivariate polynomials in the input vector $X$ are also orthonormal, that is,

$$\mathbb{E}\left[\Psi_\alpha(X)\,\Psi_\beta(X)\right] \overset{\text{def}}{=} \int_{\mathcal{D}_X} \Psi_\alpha(x)\Psi_\beta(x)\,f_X(x)dx = \delta_{\alpha\beta} \quad \forall\alpha, \beta \in \mathbb{N}^M, \tag{6.11}$$

where $\delta_{\alpha\beta}$ is the Kronecker symbol that is equal to 1 if $\alpha = \beta$ and zero otherwise. With this notation, it can be proven that the set of all multivariate polynomials in the input random vector $X$ forms a basis of the Hilbert space in which $Y = \mathcal{M}(X)$ is to be represented (Soize and Ghanem, 2004):

$$Y = \sum_{\alpha \in \mathbb{N}^M} y_\alpha \Psi_\alpha(X). \tag{6.12}$$

This equation may be interpreted as an intrinsic representation of the random response $Y$ in an abstract space through an orthonormal basis and coefficients that are the *coordinates* of $Y$ in this basis.

### 6.3.3  Practical Implementation

#### 6.3.3.1  Isoprobabilistic transform

In practical uncertainty quantification problems, the random variables that model the input parameters (e.g., material properties, loads, etc.) are usually *not* standardized as those shown in Table 6.1. Thus, it is necessary to first transform the random vector $X$ into a set of *reduced variables* $U$ through an isoprobabilistic transform

$$X = \mathcal{T}(U). \tag{6.13}$$

Depending on the marginal distribution of each input variable $X_i$, the associated reduced variable $U_i$ may be standard normal $\mathcal{N}(0, 1)$, standard uniform $\mathcal{U}(-1, 1)$, and so on. Then the random model response $Y$ is cast as a function of the reduced variables by composing the computational model $\mathcal{M}$ and the transform $\mathcal{T}$:

$$Y = \mathcal{M}(X) = \mathcal{M} \circ \mathcal{T}(U) = \sum_{\alpha \in \mathbb{N}^M} y_\alpha \Psi_\alpha(U). \tag{6.14}$$

Note that the isoprobabilistic transform also allows one to address the case of correlated input variables through, for example, Nataf transform (Ditlevsen and Madsen, 1996).

### EXAMPLE 6.1

Suppose $X = \{X_1, \ldots, X_M\}^{\mathrm{T}}$ is a vector of independent Gaussian variables $X_i \sim \mathcal{N}(\mu_i, \sigma_i)$ with the respective mean value $\mu_i$ and standard deviation $\sigma_i$. Then a one-to-one mapping $X = \mathcal{T}(U)$ is obtained by

$$X_i = \mu_i + \sigma_i U_i, \quad i = 1, \ldots, M. \tag{6.15}$$

where $U = \{U_1, \ldots, U_M\}^{\mathrm{T}}$ is a standard normal vector.

### EXAMPLE 6.2

Suppose $X = \{X_1, X_2\}^{\mathrm{T}}$ where $X_1 \sim \mathcal{LN}(\lambda, \varsigma)$ is a lognormal variable and $X_2 \sim \mathcal{U}(a, b)$ is a uniform variable. It is natural to transform $X_1$ into a standard normal variable and $X_2$ into a standard uniform variable. The isoprobabilistic transform $X = \mathcal{T}(U)$ then reads:

$$\begin{cases} X_1 = e^{\lambda + \varsigma U_1} \\ X_2 = \dfrac{b-a}{2} U_2 + \dfrac{b+a}{2} \end{cases} \tag{6.16}$$

#### 6.3.3.2 Truncation scheme

The representation of the random response in Equation 6.12 is exact when the infinite series is considered. However, in practice, only a finite number of terms may be computed. For this purpose, a *truncation scheme* has to be adopted. Since the polynomial chaos basis is made of polynomials, it is natural to consider as a truncated series all polynomials up to a certain degree. Let us define the *total degree* of a multivariate polynomial $\Psi_\alpha$ by

$$|\alpha| \overset{\mathrm{def}}{=} \sum_{i=1}^{M} \alpha_i. \tag{6.17}$$

The standard truncation scheme consists of selecting all polynomials such that $|\alpha|$ is smaller than a given $p$, that is,

$$\mathcal{A}^{M,p} = \{\alpha \in \mathbb{N}^M : |\alpha| \le p\}. \tag{6.18}$$

The number of terms in the truncated series is

$$\text{card } \mathcal{A}^{M,p} = \begin{pmatrix} M + p \\ p \end{pmatrix} = \frac{(M + p)!}{M!p!}. \tag{6.19}$$

The maximal polynomial degree $p$ may typically be equal to 3–5 in practical applications. The question on how to define the suitable $p$ for obtaining a given accuracy in the truncated series will be addressed later in Section 6.3.6. Note that the cardinality of $\mathcal{A}^{M,p}$ increases polynomially with $M$ and $p$. Thus, the number of terms in the series, that is, the number of coefficients to be computed, increases dramatically, say when $M > 10$. This complexity is referred to as the *curse of dimensionality*. Other advanced truncation schemes that allow one to bypass this problem will be considered later on in Section 6.3.6. As a conclusion, the construction of a truncated PC expansion requires to

- Transform the input random vector $\boldsymbol{X}$ into reduced variables;
- Compute the associated families of univariate orthonormal polynomials;
- Compute the set of multi-indices corresponding to the truncation set (Equation 6.18). For this purpose, two different algorithms may be found in Sudret et al. (2006, Appendix I) and Blatman (2009, Appendix C).

### 6.3.3.3 Application example

Let us consider a computational model $y = \mathcal{M}(x_1, x_2)$ involving two random parameters $\{X_1, X_2\}$ that are modeled by *lognormal* distributions, for example, the load-carrying capacity of a foundation in which the soil cohesion and friction angle are considered uncertain. Denoting by $(\lambda_i, \zeta_i)$ the parameters of each distribution (i.e., the mean and standard deviation of the logarithm of $X_i$, $i = 1, 2$), the input variables may be transformed into reduced standard normal variables $\boldsymbol{X} = \mathcal{T}(\boldsymbol{U})$ as follows:

$$\begin{array}{ll} X_1 \sim \mathcal{LN}(\lambda_1, \zeta_1) & X_1 = e^{\lambda_1 + \zeta_1 U_1} \\ X_2 \sim \mathcal{LN}(\lambda_2, \zeta_2) & X_2 = e^{\lambda_2 + \zeta_2 U_2} \end{array}. \tag{6.20}$$

The problem reduces to representing a function of two standard normal variables onto a PCE:

$$Y = \mathcal{M}\big(\mathcal{T}(U_1, U_2)\big) = \sum_{\alpha \in \mathbb{N}^2} y_\alpha \, \Psi_\alpha(U_1, U_2). \tag{6.21}$$

Since the reduced variables are standard normal, Hermite polynomials are used. Their derivation is presented in detail in Appendix A. For the sake of illustration, the orthonormal Hermite polynomials up to degree $p = 3$ read (see Equation 6.8):

$$\psi_0(x) = 1 \quad \psi_1(x) = x \quad \psi_2(x) = (x^2 - 1)/\sqrt{2} \quad \psi_3(x) = (x^3 - 3x)/\sqrt{6}. \tag{6.22}$$

Suppose a standard truncation scheme of maximal degree $p = 3$ is selected. This leads to a truncation set $\mathcal{A}^{2,3}$ of size $P = \binom{2+3}{3} = 10$. The set of multi-indices $(\alpha_1, \alpha_2)$ such that $\{\alpha_i \geq 0, \alpha_1 + \alpha_2 \leq 3\}$ is given in Table 6.2 together with the expression of the resulting multivariate polynomials.

Table 6.2 Hermite polynomial chaos basis – $M = 2$
standard normal variables, $p = 3$

| $j$ | $\alpha$ | $\Psi_\alpha \equiv \Psi_j$ |
|---|---|---|
| 0 | $(0,0)$ | $\Psi_0 = 1$ |
| 1 | $(1,0)$ | $\Psi_1 = U_1$ |
| 2 | $(0,1)$ | $\Psi_2 = U_2$ |
| 3 | $(2,0)$ | $\Psi_3 = (U_1^2 - 1)/\sqrt{2}$ |
| 4 | $(1,1)$ | $\Psi_4 = U_1 U_2$ |
| 5 | $(0,2)$ | $\Psi_5 = (U_2^2 - 1)/\sqrt{2}$ |
| 6 | $(3,0)$ | $\Psi_6 = (U_1^3 - 3U_1)/\sqrt{6}$ |
| 7 | $(2,1)$ | $\Psi_7 = (U_1^2 - 1)U_2/\sqrt{2}$ |
| 8 | $(1,2)$ | $\Psi_8 = (U_2^2 - 1)U_1/\sqrt{2}$ |
| 9 | $(0,3)$ | $\Psi_9 = (U_2^3 - 3U_2)/\sqrt{6}$ |

As a conclusion, the random response of our computational model $Y = \mathcal{M}(\mathcal{T}(U_1, U_2))$ will be approximated by a 10-term polynomial series expansion in $(U_1, U_2)$:

$$\tilde{Y} \overset{\text{def}}{=} \mathcal{M}^{\text{PC}}(U_1, U_2) = y_0 + y_1 U_1 + y_2 U_2 + y_3 (U_1^2 - 1)/\sqrt{2} + y_4 U_1 U_2$$
$$+ y_5 (U_2^2 - 1)/\sqrt{2} + y_6 (U_1^3 - 3U_1)/\sqrt{6} + y_7 (U_1^2 - 1)U_2/\sqrt{2}$$
$$+ y_8 (U_2^2 - 1)U_1/\sqrt{2} + y_9 (U_2^3 - 3U_2)/\sqrt{6}. \tag{6.23}$$

## 6.3.4 Computation of the coefficients

### 6.3.4.1 Introduction

Once the truncated basis has been selected, the coefficients $\{y_\alpha\}_{\alpha \in A^{M,p}}$ shall be computed. Historically, the so-called *intrusive* computation schemes have been developed in the context of stochastic finite-element analysis (Ghanem and Spanos, 1991). In this setup, the constitutive equations of the physical problem (e.g., linear elasticity for estimating the settlement of foundations) are discretized both in the physical space (using standard finite-element techniques) and in the random space using the PCE. This results in coupled systems of equations that require ad-hoc solvers, thus the term "intrusive." The application of such approaches to geotechnical problems may be found in Ghiocel and Ghanem (2002); Berveiller et al. (2004b); Sudret et al. (2004, 2006); and Sudret and Berveiller (2007, Chapter 7).

In the past decade, alternative approaches termed *nonintrusive* have been developed for computing the expansion coefficients. The common point of these techniques is that they rely upon the *repeated run* of the computational model for selected realizations of random vector $X$, exactly as in MCS. Thus, the computational model may be used without modification. The main techniques are now reviewed with an emphasis on *least-square minimization*.

### 6.3.4.2 Projection

Owing to the orthogonality of the PC basis (Equation 6.11), one can compute each expansion coefficient as follows:

$$\mathbb{E}\left[Y\,\Psi_\alpha(X)\right] = \mathbb{E}\left[\Psi_\alpha(X)\cdot\sum_{\beta\in\mathbb{N}^M}y_\beta\Psi_\beta(X)\right] = \sum_{\beta\in\mathbb{N}^M}y_\beta\,\overbrace{\mathbb{E}\left[\Psi_\alpha(X)\Psi_\beta(X)\right]}^{\delta_{\alpha\beta}} = y_\alpha. \tag{6.24}$$

Thus, each coefficient $y_\alpha$ is nothing but the *orthogonal projection* of the random response $Y$ onto the corresponding basis function $\Psi_\alpha(X)$. The latter may be further elaborated as

$$y_\alpha = \mathbb{E}\left[Y\,\Psi_\alpha(X)\right] = \int_{D_X}\mathcal{M}(x)\Psi_\alpha(x)f_X(x)\,dx. \tag{6.25}$$

The numerical estimation of $y_\alpha$ may be carried out with either one of the two expressions, namely:

- By MCS allowing one to estimate the expectation in Equation 6.25 (Ghiocel and Ghanem, 2002). This technique however shows low efficiency, as does MCS in general;
- By the numerical integration of the right-hand side of Equation 6.25 using Gaussian quadrature (Le Maître et al., 2002; Berveiller et al., 2004a; Matthies and Keese, 2005).

The quadrature approach has been extended using *sparse grids* for a more efficient integration, especially in large dimensions. The so-called *stochastic collocation methods* have also been developed. The reader is referred to the review paper by Xiu (2009) for more details.

### 6.3.4.3 Least-square minimization

Instead of devising numerical methods that directly estimate each coefficient from the expression $y_\alpha = \mathbb{E}[Y\Psi_\alpha(X)]$, an alternative approach based on least-square minimization and originally termed "regression approach" has been proposed in Berveiller et al. (2004b, 2006). The problem is set up as follows. Once a truncation scheme $\mathcal{A}\subset\mathbb{N}^M$ is chosen (for instance, $\mathcal{A}=\mathcal{A}^{M,p}$ as in Equation 6.18), the infinite series is recast as the sum of the truncated series and a residual:

$$Y = \mathcal{M}(X) = \sum_{\alpha\in\mathcal{A}}y_\alpha\,\Psi_\alpha(X) + \varepsilon, \tag{6.26}$$

in which $\varepsilon$ corresponds to all those PC polynomials whose index $\alpha$ is *not* in the truncation set $\mathcal{A}$. The least-square minimization approach consists of finding the set of coefficients $y = \{y_\alpha, \alpha\in\mathcal{A}\}$ that minimizes the mean-square error

$$\mathbb{E}[\varepsilon^2] \stackrel{\text{def}}{=} \mathbb{E}\left[\left(Y - \sum_{\alpha\in A}y_\alpha\,\Psi_\alpha(X)\right)^2\right], \tag{6.27}$$

that is

$$
y = \arg \min_{y \in \mathbb{R}^{\text{card} \mathcal{A}}} \mathbb{E}\left[ \left( \mathcal{M}(X) - \sum_{\alpha \in \mathcal{A}} y_\alpha \Psi_\alpha(X) \right)^2 \right]. \tag{6.28}
$$

The residual in Equation 6.27 is nothing but a quadratic function of the (still unknown) coefficients $\{y_\alpha\}_{\alpha \in \mathcal{A}}$. By simple algebra, it can be proven that the solution is identical to that obtained by projection in Equation 6.25. However, the setup in Equation 6.28 that is similar to regression opens to new computational schemes.

For this purpose, the *discretized version* of the problem is obtained by replacing the expectation operator in Equation 6.28 by the empirical mean over a sample set:

$$
\hat{y} = \arg \min_{y \in \mathbb{R}^{\text{card} \mathcal{A}}} \frac{1}{n} \sum_{i=1}^{n} \left( \mathcal{M}\left(x^{(i)}\right) - \sum_{\alpha \in \mathcal{A}} y_\alpha \Psi_\alpha\left(x^{(i)}\right) \right)^2. \tag{6.29}
$$

In this expression, $\chi = \{x^{(i)}, i = 1, \ldots, n\}$ is a sample set of points (also called *experimental design* [ED]), that is typically obtained by MCS of the input random vector $X$. The least-square minimization problem in Equation 6.29 is solved as follows:

- The computational model $\mathcal{M}$ is run for each point in the ED, and the results are stored in a vector

$$
\mathcal{Y} = \left\{ y^{(1)} = \mathcal{M}(x^{(1)}), \ldots, y^{(n)} = \mathcal{M}(x^{(n)}) \right\}^{\top}. \tag{6.30}
$$

- The *information matrix* is calculated from the evaluation of the basis polynomials onto each point in the ED:

$$
\mathbf{A} = \left\{ \mathbf{A}_{ij} \overset{\text{def}}{=} \Psi_j(x^{(i)}), \ i = 1, \ldots, n, \ j = 1, \ldots, \text{card } \mathcal{A} \right\}. \tag{6.31}
$$

- The solution of the least-square minimization problem reads

$$
\hat{y} = (\mathbf{A}^{\top}\mathbf{A})^{-1}\mathbf{A}^{\top}\mathcal{Y}. \tag{6.32}
$$

To be well posed, the least-square minimization requires that the number of unknown $P = \text{card } \mathcal{A}$ is smaller than the size of the ED $n = \text{card } \mathcal{X}$. The empirical thumb rule $n \approx 2 P - 3 P$ is often mentioned (Sudret, 2007; Blatman, 2009). To overcome the potential ill-conditioning of the information matrix, a singular-value decomposition shall be used (Press et al., 2001).

The points used in the ED may be obtained from crude MCS. However, other types of designs are of common use, especially Latin Hypercube sampling (LHS), see McKay et al. (1979), or quasi-random sequences such as the Sobol' or Halton sequence (Niederreiter, 1992). From the author's experience, the latter types of design provide rather equivalent accuracy in terms of the resulting mean-square error, for the same sample size $n$. Note that deterministic designs based on the roots of the orthogonal polynomials have also been proposed earlier in Berveiller et al. (2006) based on Isukapalli (1999).

Once the coefficients have been evaluated (Equation 6.32), the approximation of the random response is the random variable

$$\hat{Y} = \mathcal{M}^{PC}(X) = \sum_{\alpha \in \mathcal{A}} \hat{y}_{\alpha} \Psi_{\alpha}(X). \tag{6.33}$$

The above equation may also be interpreted as a response surface, that is, a function $x \mapsto \mathcal{M}^{PC}(x) = \sum_{\alpha \in \mathcal{A}} \hat{y}_{\alpha} \Psi_{\alpha}(x)$ that allows one to surrogate (fast, although approximately) the original model $y = \mathcal{M}(x)$.

### 6.3.5 Validation

#### 6.3.5.1 Error estimators

As mentioned already, it is not possible to know in advance how to choose the maximal polynomial degree in the standard truncation scheme (Equation 6.18). A crude approach would consist of testing several truncation schemes of increasing degree (e.g., $p = 2, 3, 4$) and observe if there is some convergence for the quantities of interest. Recently, *a posteriori* error estimates have been proposed by Blatman and Sudret (2010) that allow for an objective evaluation of the accuracy of any truncated PCE.

First of all, it is reminded that a good measure of the error committed by using a truncated series expansion is the mean-square error of the residual (which is also called *generalization error* in statistical learning theory):

$$Err_{G} \overset{\text{def}}{=} \mathbb{E}\left[\varepsilon^{2}\right] = \mathbb{E}\left[\left(Y - \sum_{\alpha \in \mathcal{A}} \hat{y}_{\alpha} \Psi_{\alpha}(X)\right)^{2}\right]. \tag{6.34}$$

In practice, the latter is not known analytically; yet, it may be estimated by an MCS using a large sample set, say $\mathcal{X}_{\text{val}} = \{x_{1}, \ldots, x_{n_{\text{val}}}\}$:

$$\widehat{Err_{G}} \overset{\text{def}}{=} \frac{1}{n_{\text{val}}} \sum_{i=1}^{n_{\text{val}}} \left(\mathcal{M}(x_{i}) - \sum_{\alpha \in \mathcal{A}} \hat{y}_{\alpha} \Psi_{\alpha}(x_{i})\right)^{2}. \tag{6.35}$$

The so-called *validation set* $\mathcal{X}_{\text{val}}$ shall be large enough to get an accurate estimation, for example, $n_{\text{val}} = 10^{3-5}$. However, as the computation of $\widehat{Err_{G}}$ requires evaluating $\mathcal{M}$ for each point in $\mathcal{X}_{\text{val}}$, this is not affordable in real applications and would ruin the efficiency of the approach. Indeed, the purpose of using PCEs is to avoid MCS, that is, to limit the number of runs of the computational model $\mathcal{M}$ in Equation 6.30 to the smallest possible number, typically $n = 50$ to a few hundreds.

As a consequence, to get an estimation of the generalization error (Equation 6.34) at an affordable computational cost, the points in the ED $\mathcal{X}$ could be used in Equation 6.35 instead of the validation set, leading to the so-called *empirical error* $\widehat{Err_{E}}$ defined by

$$\widehat{Err_{E}} \overset{\text{def}}{=} \frac{1}{n} \sum_{i=1}^{n} \left(\mathcal{M}(x^{(i)}) - \sum_{\alpha \in \mathcal{A}} \hat{y}_{\alpha} \Psi_{\alpha}(x^{(i)})\right)^{2}, \quad x^{(i)} \in \mathcal{X} \tag{6.36}$$

This empirical error now only uses the values $\mathcal{M}(\boldsymbol{x}^{(i)})$ that are already available from Equation 6.30 and is thus readily computable. Note that the normalized quantity

$$R^2 = 1 - \frac{\widehat{Err}_E}{\operatorname{Var}[\mathcal{Y}]}, \tag{6.37}$$

is the well-known *coefficient of determination* in regression analysis, where Var $[\mathcal{Y}]$ is the empirical variance of the set of response quantities in Equation 6.30.

However, $\widehat{Err}_E$ usually underestimates (sometimes severely) the real generalization error $Err_G$. As an example, in the limit case, when an interpolating polynomial would be fitted to the ED, $\widehat{Err}_E$ would be exactly zero while $Err_G$ in Equation 6.34 would probably not: this phenomenon is known as *overfitting*.

### 6.3.5.2 Leave-one-out cross-validation

A compromise between fair error estimation and affordable computational cost may be obtained by *leave-one-out* (LOO) cross-validation, which was originally proposed by Allen (1971); Geisser (1975). The idea is to use different sets of points to (i) build a PC expansion and (ii) compute the error with the original computational model. Starting from the full ED $\mathcal{X}$, LOO cross-validation sets one point apart, say $\boldsymbol{x}^{(i)}$ and builds a PC expansion denoted by $\mathcal{M}^{\text{PC}\backslash i}(.)$ from the $n-1$ remaining points, that is, from the ED $\mathcal{X}\backslash\boldsymbol{x}^{(i)} \stackrel{\text{def}}{=} \{\boldsymbol{x}^{(1)}, ..., \boldsymbol{x}^{(i-1)}, \boldsymbol{x}^{(i+1)}, ..., \boldsymbol{x}^{(n)}\}$. The predicted residual error at that point reads:

$$\Delta_i \stackrel{\text{def}}{=} \mathcal{M}(\boldsymbol{x}^{(i)}) - \mathcal{M}^{\text{PC}\backslash i}(\boldsymbol{x}^{(i)}). \tag{6.38}$$

The PRESS coefficient (predicted *residual sum of squares*) and the LOO *error* respectively read:

$$PRESS = \sum_{i=1}^{n} \Delta_i^2, \tag{6.39}$$

$$\widehat{Err}_{LOO} = \frac{1}{n}\sum_{i=1}^{n} \Delta_i^2. \tag{6.40}$$

Similar to the determination coefficient in Equation 6.37, the $Q^2$ indicator defined by

$$Q^2 = 1 - \frac{\widehat{Err}_{LOO}}{\operatorname{Var}[\mathcal{Y}]}, \tag{6.41}$$

is a normalized measure of the accuracy of the metamodel. From the above equations, one could think that evaluating $\widehat{Err}_{LOO}$ is computationally demanding since it is based on the sum of $n$ different predicted residuals, each of them obtained from a *different* PC expansion. However, algebraic derivations may be carried out to compute $\widehat{Err}_{LOO}$ from *a single* PC expansion analysis using the full original ED $\mathcal{X}$ (details may be found in

Blatman (2009, Appendix D)) as follows. The predicted residual in Equation 6.38 eventually reads:

$$\Delta_i = \mathcal{M}(\boldsymbol{x}^{(i)}) - \mathcal{M}^{\mathrm{PC}\backslash i}(\boldsymbol{x}^{(i)}) = \frac{\mathcal{M}(\boldsymbol{x}^{(i)}) - \mathcal{M}^{\mathrm{PC}}(\boldsymbol{x}^{(i)})}{1 - h_i}, \tag{6.42}$$

where $h_i$ is the $i$th diagonal term of matrix $\mathbf{A}(\mathbf{A}^{\mathrm{T}}\mathbf{A})^{-1}\mathbf{A}^{\mathrm{T}}$. The LOO error estimate eventually reads

$$\widehat{Err}_{LOO} = \frac{1}{n}\sum_{i=1}^{n}\left(\frac{\mathcal{M}(\boldsymbol{x}^{(i)}) - \mathcal{M}^{\mathrm{PC}}(\boldsymbol{x}^{(i)})}{1 - h_i}\right)^2, \tag{6.43}$$

where $\mathcal{M}^{\mathrm{PC}}$ has been built up from the *full* ED. As a conclusion, from a single resolution of a least-square problem using the ED $\mathcal{X}$, a fair error estimate of the mean-square error is available *a posteriori* using Equation 6.43. Note that in practice, a normalized version of the LOO error is obtained by dividing $\widehat{Err}_{LOO}$ by the sample variance Var $[\mathcal{Y}]$. A correction factor that accounts for the limit size of the ED is also added (Chapelle et al., 2002) that eventually leads to

$$\hat{\epsilon}_{LOO} = \frac{1}{n - P}\left(\frac{1 + \frac{1}{n}\,\mathrm{tr}\,\mathrm{C}_{emp}^{-1}}{\mathrm{Var}[\mathcal{Y}]}\right)\sum_{i=1}^{n}\left(\frac{\mathcal{M}(\boldsymbol{x}^{(i)}) - \mathcal{M}^{\mathrm{PC}}(\boldsymbol{x}^{(i)})}{1 - h_i}\right)^2, \tag{6.44}$$

where $\mathrm{tr}(\cdot)$ is the trace, $P = \mathrm{card}\,\mathcal{A}$, and $\mathrm{C}_{emp} = (1/n)\mathbf{\Psi}^{\mathrm{T}}\,\mathbf{\Psi}$.

### 6.3.6 Curse of dimensionality

Common engineering problems are solved with computational models having typically $M = 10\text{--}50$ input parameters. Even when using a low-order PC expansion, this leads to a large number of unknown coefficients (size of the truncation set $\mathcal{A}^{M,p}$), which is equal to, for example, $286 - 23{,}426$ terms when choosing $p = 3$. As explained already, the suitable size of the ED shall be 2–3 times those numbers, which may reveal as unaffordable when the computational model $\mathcal{M}$ is, for example, a finite-element model.

On the other hand, most of the terms in this large truncation set $\mathcal{A}^{M,p}$ correspond to polynomials representing *interactions* between input variables. Yet, it has been observed in many practical applications that only the low-interaction terms have coefficients that are significantly nonzero. Taking again the example $\{M = 50, p = 3\}$, the number of *univariate polynomials* in the input variables (i.e., depending *only* on $U_1$, on $U_2$, etc.) is equal to $p \cdot M = 150$, that is, <1% of the total number of terms in $\mathcal{A}^{50,3}$. As a conclusion, the common truncation scheme in Equation 6.18 leads to compute a large number of coefficients, whereas most of them may reveal negligible once the computation has been carried out.

Owing to this *sparsity-of-effect principle*, Blatman (2009) has proposed to use truncation schemes that favor the low-interaction (also called low-rank) polynomials. Let us define the *rank r* of a multi-index $\alpha$ by the number of nonzero integers in $\alpha$, that is,

$$r \overset{\mathrm{def}}{=} \|\alpha\|_0 = \sum_{i=1}^{M} 1_{\{\alpha_i > 0\}}. \tag{6.45}$$

This rank corresponds to the number of input variables a polynomial $\Psi_\alpha$ depends on. For instance, $\Psi_1$, $\Psi_3$, and $\Psi_9$ in Table 6.2 are of rank 1 (they only depend on one single variable) whereas $\Psi_4$, $\Psi_7$, and $\Psi_8$ are of rank 2. Blatman and Sudret (2010) propose to fix *a priori* the maximum rank $r_{max}$ and define truncation sets as follows:

$$\mathcal{A}^{M,p,r_{max}} = \{\alpha \in \mathbb{N}^M : |\alpha| \leq p, \|\alpha\|_0 \leq r_{max}\}. \tag{6.46}$$

Another approach that has proven to be more relevant in the context of adaptive algorithms has been introduced in Blatman and Sudret (2011a) and is referred to as the *hyperbolic truncation scheme*. Let us define for any multi-index $\alpha$ and $0 < q \leq 1$ the $q$-norm:

$$\|\alpha\|_q \overset{\text{def}}{=} \left(\sum_{i=1}^M \alpha_i^q\right)^{1/q}. \tag{6.47}$$

The hyperbolic truncation scheme corresponds to selecting all multi-indices of $q$-norm less than or equal to $p$:

$$\mathcal{A}^{M,p,q} = \{\alpha \in \mathbb{N}^M : \|\alpha\|_q \leq p\}. \tag{6.48}$$

As shown in Figure 6.3, for a two-dimensional problem (2D) ($M = 2$), such a truncation set contains all univariate polynomials up to degree $p$ (since tuples of the form $(0, ..., \alpha_i \neq 0, ..., 0)$ belong to it as long as $\alpha_i \leq p$). The case $q = 1$ corresponds to the standard truncation set $\mathcal{A}^{M,p}$ defined in Equation 6.18. When $q < 1$ though, the polynomials of rank $r > 1$ (corresponding to the blue points that are not on the axes in Figure 6.3) are less numerous than in $\mathcal{A}^{M,p}$. The gain in the basis size is all the more important when $q$ is small and $M$ is large. In the limit when $q \to 0^+$ only univariate polynomials (rank 1, no interaction terms) are retained leading to an additive surrogate model, that is, a sum of univariate functions. As shown in Blatman and Sudret (2011a), the size of the truncation set in Equation 6.48 may be smaller by 2–3 orders of magnitude than that of the standard truncation scheme for large $M$ and $p \geq 5$.

## 6.3.7 Adaptive algorithms

The use of hyperbolic truncation schemes $\mathcal{A}^{M,p,q}$ as described above allows one to *a priori* decrease the number of coefficients to be computed in a truncated series expansion. This automatically reduces the computational cost since the minimal size of the ED $\mathcal{X}$ shall be equal to $k \cdot \text{card } \mathcal{A}^{M,p,q}$, where $k = 2$–3. However, this may remain too costly when largely dimensional, highly nonlinear problems are to be addressed.

Moreover, it is often observed *a posteriori* that the nonzero coefficients in the expansion form a *sparse* subset of $\mathcal{A}^{M,p,q}$. Thus came the idea to build, on-the-fly, the suitable sparse basis instead of computing useless terms in the expansions that are eventually negligible. For this purpose, adaptive algorithms have been introduced in Blatman and Sudret (2008) and further improved in Blatman and Sudret (2010, 2011a). In the latter publication, the question of finding a suitable truncated basis is interpreted as a *variable selection* problem (Figure 6.4). The so-called *least-angle regression* (LAR) algorithm (Efron et al., 2004) has proven to be remarkably efficient in this respect (see also Hastie et al. (2007)).

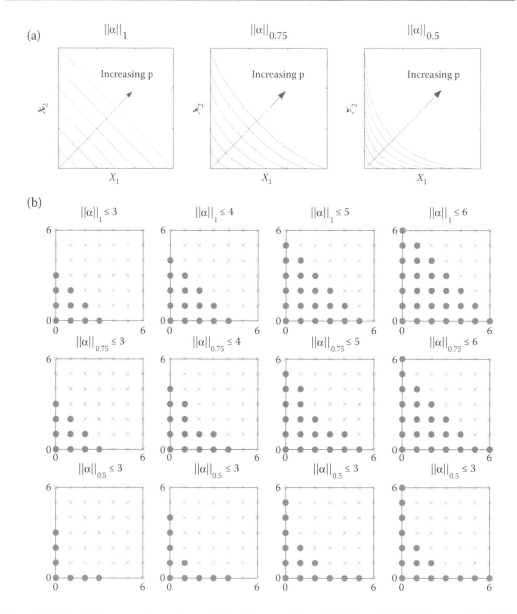

*Figure 6.3* Hyperbolic truncation scheme. (After Blatman, G. and B. Sudret. 2011a. *J. Comput. Phys. 230*, 2345–2367.)

The principle of LAR is to (i) select a *candidate set* of polynomials $\mathcal{A}$, for example, a given hyperbolic truncation set as in Equation 6.48, and (ii) build up from scratch a sequence of sparse bases having 1, 2, ..., card $\mathcal{A}$ terms. The algorithm is initialized by looking for the basis term that is the most correlated with the response vector Y. The correlation is practically computed from the realizations of Y (i.e., the set $\mathcal{Y}$ of QoI in Equation 6.30) and the realizations of the $\Psi_\alpha$'s, namely the information matrix in Equation 6.31. This is carried out by normalizing each column vector into a zero-mean, unit variance vector, such that the correlation is then obtained by a mere scalar product of the normalized vector. Once the first basis term $\Psi_{\alpha_1}$ is identified, the associated coefficient is computed such that the residual $Y - y_{\alpha_1}^{(1)} \Psi_{\alpha_1}(X)$ becomes equicorrelated with two basis terms $(\Psi_{\alpha_1}, \Psi_{\alpha_2})$. This will define the

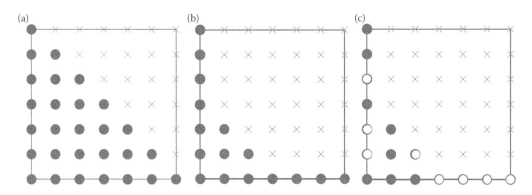

*Figure 6.4* Sketch of the different truncation sets (black full circles). (a) Standard truncation $\mathcal{A}^{M,p}$, (b) hyperbolic truncation $\mathcal{A}^{M,p,q}$, (c) sparse truncation $\mathcal{A}$.

best one-term expansion. Then the current approximation is improved by moving along the direction $(\Psi_{\alpha_1} + \Psi_{\alpha_2})$ up to a point where the residual becomes equicorrelated with a third polynomial $\Psi_{\alpha_3}$, and so on.

In the end, the LAR algorithm has produced a sequence of less and less sparse expansions. The LOO error of each expansion can be evaluated by Equation 6.43. The sparse model providing the smallest error is retained. The great advantage of LAR is that it can also be applied in the case when the size of the candidate basis $\mathcal{A}$ is larger than the cardinality of the experimental design, card $\mathcal{X}$. Usually, the size of the optimal sparse truncation is smaller than card $\mathcal{X}$ in the end. Thus, the coefficients of the associated PC expansion may be recomputed by least-square minimization for a better accuracy (Efron et al., 2004).

Note that all the above calculations are conducted from a prescribed initial ED $\mathcal{X}$. It may be that this size is too small to address the complexity of the problem, meaning that there is not enough information to find a sparse expansion with a sufficiently small LOO error. In this case, overfitting appears, which can be detected automatically as shown in Blatman and Sudret (2011a). At that point, the ED shall be enriched by adding new points (Monte Carlo samples or nested LHS).

All in all, a fully automatic "basis-and-ED" adaptive algorithm may be devised that solely requires to prescribe the target accuracy of the analysis, that is, the maximal tolerated LOO error, and an initial ED. The algorithm then automatically runs LAR analysis with increasingly larger candidate sets $\mathcal{A}$, and possibly by increasing large EDs so as to reach the prescribed accuracy (see Figure 6.5). Note that extensions to vector-valued models have been recently proposed in Blatman and Sudret (2011b, 2013).

## 6.4 POST-PROCESSING FOR ENGINEERING APPLICATIONS

The PCE technique presented in the previous sections leads to cast the QoI of a computational model $Y = \mathcal{M}(X)$ through a somewhat abstract representation by a polynomial series $\hat{Y} = \Sigma_{\alpha \in \mathcal{A}} \hat{y}_{\alpha} \Psi_{\alpha}(X)$ (Equation 6.33). Once the PC basis has been set up (*a priori* or using an adaptive algorithm such as LAR) and once the coefficients have been calculated, the series expansion shall be post-processed so as to provide engineering-wise meaningful numbers and statements: what is the mean behavior of the system (mean QoI), scattering (variance of the QoI), confidence intervals, or probability of failure (i.e., the probability that the QoI exceeds an admissible threshold)? In this section, the various ways of post-processing a PC expansion are reviewed.

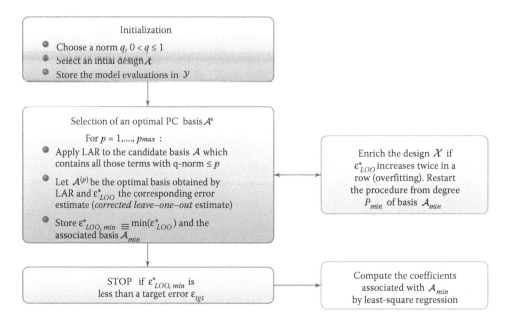

*Figure 6.5* Basis-and-ED adaptive algorithm for sparse PC expansions. (After Blatman, G. and B. Sudret. 2011a. *J. Comput. Phys. 230*, 2345–2367.)

### 6.4.1 Moment analysis

From the orthonormality of the PC basis shown in Equation 6.11, one can easily compute the mean and standard deviation of a truncated series $\hat{Y} = \sum_{\alpha \in \mathcal{A}} \hat{y}_\alpha \Psi_\alpha(X)$. Indeed, each polynomial shall be orthogonal to $\Psi_0 = 1$, meaning that $\mathbb{E}[\Psi_\alpha(X)] = 0;\ \forall\ \alpha \neq 0$. Thus, the mean value of $\hat{Y}$ is the first term of the series:

$$\mathbb{E}\left[\hat{Y}\right] = \mathbb{E}\left[\sum_{\alpha \in \mathcal{A}} \hat{y}_\alpha \Psi_\alpha(X)\right] = y_0. \tag{6.49}$$

Similarly, due to Equation 6.11, the variance reads

$$\sigma_{\hat{Y}}^2 \overset{\text{def}}{=} \text{Var}\left[\hat{Y}\right] = \mathbb{E}\left[(\hat{Y} - y_0)^2\right] = \sum_{\substack{\alpha \in \mathcal{A} \\ \alpha \neq 0}} \hat{y}_\alpha^2. \tag{6.50}$$

Higher-order moments such as the skewness and kurtosis coefficients $\delta_{\hat{Y}}$ and $\kappa_{\hat{Y}}$ may also be computed, which however, requires the expectation of products of three (resp. four) multivariate polynomials:

$$\delta_{\hat{Y}} \overset{\text{def}}{=} \frac{1}{\sigma_{\hat{Y}}^3} \mathbb{E}\left[(\hat{Y} - y_0)^3\right] = \frac{1}{\sigma_{\hat{Y}}^3} \sum_{\alpha \in \mathcal{A}} \sum_{\beta \in \mathcal{A}} \sum_{\gamma \in \mathcal{A}} \mathbb{E}\left[\Psi_\alpha(X)\Psi_\beta(X)\Psi_\gamma(X)\right] \hat{y}_\alpha \hat{y}_\beta \hat{y}_\gamma.$$

$$\kappa_{\hat{Y}} \overset{\text{def}}{=} \frac{1}{\sigma_{\hat{Y}}^4} \mathbb{E}\left[(\hat{Y} - y_0)^4\right] = \frac{1}{\sigma_{\hat{Y}}^4} \sum_{\alpha \in \mathcal{A}} \sum_{\beta \in \mathcal{A}} \sum_{\gamma \in \mathcal{A}} \sum_{\delta \in \mathcal{A}} \mathbb{E}\left[\Psi_\alpha(X)\Psi_\beta(X)\Psi_\gamma(X)\Psi_\delta(X)\right] \hat{y}_\alpha \hat{y}_\beta \hat{y}_\gamma \hat{y}_\delta. \tag{6.51}$$

The above expectations of products can be given analytical expressions only when Hermite polynomials are used (see Sudret et al. (2006, Appendix I)). Otherwise, they may be computed numerically by quadrature.

## 6.4.2 Distribution analysis and confidence intervals

As shown from Equation 6.33, the PC expansion $\mathcal{M}^{\text{PC}}$ can be used as a polynomial response surface. Thus, the output PDF of $\hat{Y} = \sum_{\alpha \in \mathcal{A}} \hat{y}_\alpha \Psi_\alpha(X)$ can be obtained by merely sampling the input random vector $X$, say $\mathcal{X}_{\text{MCS}} = \{x_1, \ldots, x_{n_{\text{MCS}}}\}$ and evaluating the PC expansion onto this sample, that is,

$$\mathcal{Y}_{\text{MCS}}^{\text{PC}} = \{\mathcal{M}^{\text{PC}}(x_1), \ldots, \mathcal{M}^{\text{PC}}(x_{n_{\text{MCS}}})\}. \tag{6.52}$$

Using a sufficiently large sample set (e.g., $n_{\text{MCS}} = 10^{5-6}$), one can then compute and plot the almost-exact PDF of $\hat{Y}$ by using a kernel density estimator (Wand and Jones, 1995):

$$\hat{f}_{\hat{Y}}(y) = \frac{1}{n_{\text{MCS}} h} \sum_{i=1}^{n_{\text{MCS}}} K\left(\frac{y - \mathcal{M}^{\text{PC}}(x_i)}{h}\right). \tag{6.53}$$

In this equation, the *kernel function K* is a positive definite function integrating to one (e.g., the standard normal PDF $\phi(y) = e^{-y^2/2}/\sqrt{2\pi}$) and $h$ is the *bandwidth*. The latter can be taken, for instance, from Silverman's equation:

$$h = 0.9 n_{\text{MCS}}^{-1/5} \min(\hat{\sigma}_{\hat{y}}, (Q_{0.75} - Q_{0.25})/1.34), \tag{6.54}$$

where $\sigma_{\hat{y}}$ (resp. $Q_{0.25}$, $Q_{0.75}$) is the empirical standard deviation of $\mathcal{Y}_{\text{MCS}}^{\text{PC}}$ (resp. the first and third quartile of $\mathcal{Y}_{\text{MCS}}^{\text{PC}}$). Note that these quantiles as well as any other can be obtained from the large sample set in Equation 6.52. Having first reordered it in ascending order, say $\{\hat{y}_{(1)}, \ldots, \hat{y}_{(n_{\text{MCS}})}\}$, the empirical $p$-quantile $Q_{p\%}$, $0 < p < 1$ is the $\lfloor p\% \cdot n_{\text{MCS}} \rfloor$-th point in the ordered sample set, that is,

$$\tilde{Q}_{p\%} = \hat{y}_{(\lfloor p\% \cdot n_{\text{MCS}} \rfloor)}, \tag{6.55}$$

where $\lfloor u \rfloor$ is the largest integer that is smaller than $u$. This allows one to compute confidence intervals (CIs) on the QoI $\hat{Y}$. For instance, the 95% centered CI, whose bounds are defined by the 2.5% and 97.5% quantile, is

$$CI_{\hat{Y}}^{95\%} = \left[\hat{y}_{(\lfloor 2.5\% \cdot n_{\text{MCS}} \rfloor)}, \hat{y}_{(\lfloor 97.5\% \cdot n_{\text{MCS}} \rfloor)}\right]. \tag{6.56}$$

Note that all the above post-processing may be carried out on large Monte Carlo samples since the function to evaluate in Equation 6.52 is the polynomial surrogate model and not the original model $\mathcal{M}$. Such an evaluation is nowadays a matter of seconds on standard computers, even with $n_{\text{MCS}} = 10^{5-6}$.

### 6.4.3 Reliability analysis

Reliability analysis aims at computing the probability of failure associated to a performance criterion related to the QoI $Y = \mathcal{M}(X)$. In general, the failure criterion under consideration is represented by a *limit state function* $g(X)$ defined in the space of parameters as follows (Ditlevsen and Madsen, 1996):

- $\mathcal{D}_s = \{x \in \mathcal{D}_X : g(x) > 0\}$ is the *safe domain* of the structure;
- $\mathcal{D}_f = \{x \in \mathcal{D}_X : g(x) < 0\}$ is the *failure domain*;
- The set of realizations $\{x \in \mathcal{D}_X : g(x) = 0\}$ is the so-called *limit state surface*.

Typical performance criteria are defined by the fact that the QoI shall be smaller than an admissible threshold $y_{\mathrm{adm}}$. According to the above definition, the limit state function then reads

$$g(x) = y_{\mathrm{adm}} - \mathcal{M}(x). \tag{6.57}$$

Then the probability of failure of the system is defined as the probability that $X$ belongs to the failure domain:

$$P_f = \int_{\{x: y_{\mathrm{adm}} - \mathcal{M}(x) \leq 0\}} f_X(x) dx = \mathbb{E}\left[ 1_{\{x: y_{\mathrm{adm}} - \mathcal{M}(x) \leq 0\}}(X) \right], \tag{6.58}$$

where $f_X$ is the joint PDF of $X$ and $1_{\{x: y_{\mathrm{adm}} - \mathcal{M}(x) \leq 0\}}$ is the indicator function of the failure domain. In all but academic cases, this integral cannot be computed analytically, since the failure domain is defined from a QoI $Y = \mathcal{M}(X)$ (e.g., displacements, strains, stresses, etc.), which is obtained by means of a computer code (e.g., finite-element code) in industrial applications.

Once a PC expansion of the QoI is available though, the probability of failure may be obtained by substituting $\mathcal{M}$ by $\mathcal{M}^{\mathrm{PC}}$ in Equation 6.57:

$$P_f^{\mathrm{PC}} = \int_{\{x: y_{\mathrm{adm}} - \mathcal{M}^{\mathrm{PC}}(x) \leq 0\}} f_X(x) dx = \mathbb{E}\left[ 1_{\{x: \mathcal{M}^{\mathrm{PC}}(x) \geq y_{\mathrm{adm}}\}}(X) \right]. \tag{6.59}$$

The latter can be estimated by crude MCS. Using the sample set in Equation 6.52, one computes the number $n_f$ of samples such that $\mathcal{M}^{\mathrm{PC}}(x_i) \geq y_{\mathrm{adm}}$. Then the estimate of the probability of failure reads:

$$\widehat{P_f^{\mathrm{PC}}} = \frac{n_f}{n_{\mathrm{MCS}}}. \tag{6.60}$$

This crude Monte Carlo approach will typically work efficiently if $P_f \leq 10^{-4}$, that is, if at most $10^6$ runs of PC expansion are required. Note that any standard reliability method such as importance sampling (IS) or subset simulation could also be used.

### 6.4.4 Sensitivity analysis

#### 6.4.4.1 Sobol decomposition

Global sensitivity analysis (GSA) aims at quantifying which input parameters $\{X_i, i = 1, \ldots, M\}$ or combinations thereof best explain the variability of the QoI $Y = \mathcal{M}(X)$ (Saltelli et al., 2000, 2008). This variability being well described by the variance of $Y$, the question reduces to apportioning Var $[Y]$ to each input parameter $\{X_1, ..., X_M\}$, second-order interactions $X_i X_j$, and so on. For this purpose, variance decomposition techniques have gained interest since the mid-1990s. The Sobol' decomposition (Sobol', 1993) states that any square-integrable function $\mathcal{M}$ with respect to a probability measure associated with a PDF $f_X(x) = \prod_{i=1}^{M} f_{X_i}(x_i)$ (independent components) may be cast as

$$\mathcal{M}(x) = \mathcal{M}_0 + \sum_{i=1}^{M} \mathcal{M}_i(x_i) + \sum_{1 \le i < j \le M} \mathcal{M}_{ij}(x_i, x_j) + \cdots + \mathcal{M}_{12,\dots,M}(x), \tag{6.61}$$

that is, as a sum of a constant, univariate functions $\{\mathcal{M}_i(x_i), 1 \le i \le M\}$, bivariate functions $\{\mathcal{M}_{ij}(x_i, x_j), 1 \le i < j \le M\}$, and so on. Using the *set notation* for indices

$$\mathbf{u} \stackrel{\text{def}}{=} \{i_1, ..., i_s\} \subset \{1, ..., M\}, \tag{6.62}$$

the Sobol' decomposition in Equation 6.61 reads:

$$\mathcal{M}(x) = \mathcal{M}_0 + \sum_{\substack{\mathbf{u} \subset \{1,\dots,M\} \\ \mathbf{u} \ne \emptyset}} \mathcal{M}_\mathbf{u}(x_\mathbf{u}), \tag{6.63}$$

where $x_\mathbf{u}$ is a subvector of $x$ that only contains the components that belong to the index set $\mathbf{u}$. It can be proven that the Sobol' decomposition is unique when the orthogonality between summands is required, namely:

$$\mathbb{E}[\mathcal{M}_\mathbf{u}(x_\mathbf{u})\mathcal{M}_\mathbf{v}(x_\mathbf{v})] = 0 \ \forall \mathbf{u}, \mathbf{v} \subset \{1, ..., M\}, \mathbf{u} \ne \mathbf{v}. \tag{6.64}$$

A recursive construction is obtained by the following recurrence relationship:

$$\begin{aligned}
\mathcal{M}_0 &= \mathbb{E}[\mathcal{M}(X)]. \\
\mathcal{M}_i(x_i) &= \mathbb{E}[\mathcal{M}(X)|X_i = x_i] - \mathcal{M}_0. \\
\mathcal{M}_{ij}(x_i, x_j) &= \mathbb{E}[\mathcal{M}(X)|X_i = x_i, X_j = x_j] - \mathcal{M}_i(x_i) - \mathcal{M}_j(x_j) - \mathcal{M}_0.
\end{aligned} \tag{6.65}$$

The latter equation is of little interest in practice since the integrals required to compute the various conditional expectations are cumbersome. Nevertheless, the existence and unicity of Equation 6.61 together with the orthogonality property in Equation 6.64 now allow one to decompose the variance of $Y$ as follows:

$$D \stackrel{\text{def}}{=} \mathrm{Var}[Y] = \mathrm{Var}\left[ \sum_{\substack{\mathbf{u} \subset \{1,\dots,M\} \\ \mathbf{u} \ne \emptyset}} \mathcal{M}_\mathbf{u}(x_\mathbf{u}) \right] = \sum_{\substack{\mathbf{u} \subset \{1,\dots,M\} \\ \mathbf{u} \ne \emptyset}} \mathrm{Var}[\mathcal{M}_\mathbf{u}(X_\mathbf{u})], \tag{6.66}$$

where the *partial variances* read:

$$D_\mathbf{u} \stackrel{\text{def}}{=} \mathrm{Var}[\mathcal{M}_\mathbf{u}(X_\mathbf{u})] = \mathbb{E}[\mathcal{M}_\mathbf{u}^2(X_\mathbf{u})]. \tag{6.67}$$

### 6.4.4.2 Sobol indices

The so-called *Sobol' indices* are defined as the ratio of the partial variances $D_u$ to the total variance $D$. The so-called *first order indices* correspond to single-input variables, that is, $\mathbf{u} = \{i\}$:

$$S_i = \frac{D_i}{D} = \frac{\text{Var}[\mathcal{M}_i(X_i)]}{\text{Var}[Y]}. \tag{6.68}$$

The second-order indices ($\mathbf{u} = \{i, j\}$) read:

$$S_{ij} = \frac{D_{ij}}{D} = \frac{\text{Var}[\mathcal{M}_{ij}(X_i, X_j)]}{\text{Var}[Y]}, \tag{6.69}$$

and so on. Note that the *total Sobol' index* $S_i^T$, which quantifies the total impact of a given parameter $X_i$ including all interactions, may be computed by the sum of the Sobol' indices of any order that contain $X_i$:

$$S_i^T = \sum_{i \in \mathbf{u}} S_{\mathbf{u}}. \tag{6.70}$$

### 6.4.4.3 Sobol indices from PC expansions

Sobol' indices have proven to be the most efficient sensitivity measures for general computational models (Saltelli et al., 2008). However, they are traditionally evaluated by MCS (possibly using quasi-random sequences) (Sobol' and Kucherenko, 2005), which make them difficult to use when costly computational models $\mathcal{M}$ are used. To bypass the problem, Sudret (2006, 2008) has proposed an original post-processing of PCEs for sensitivity analysis. Indeed, the Sobol' decomposition of a truncated PC expansion $\hat{Y} = \mathcal{M}^{\text{PC}}(\mathbf{X}) = \sum_{\alpha \in \mathcal{A}} \hat{y}_\alpha \Psi_\alpha(\mathbf{X})$ can be established *analytically*, as shown below.

For any subset of variables $\mathbf{u} = \{i_1, \ldots, i_s\} \subset \{1, \ldots, M\}$, let us define the set of multivariate polynomials $\Psi_\alpha$ that depend *only* on $\mathbf{u}$:

$$\mathcal{A}_u = \{\alpha \in \mathcal{A} : \alpha_k \neq 0 \quad \text{if and only if } k \in \mathbf{u}\}. \tag{6.71}$$

It is clear that the $A_u$'s form a partition of $\mathcal{A}$ since

$$\bigcup_{\mathbf{u} \subset \{1, \ldots, M\}} \mathcal{A}_{\mathbf{u}} = \mathcal{A}. \tag{6.72}$$

Thus, a truncated PC expansion such as in Equation 6.33 may be rewritten as follows by simple reordering of the terms:

$$\mathcal{M}^{\text{PC}}(\mathbf{x}) = y_0 + \sum_{\substack{\mathbf{u} \subset \{1, \ldots, M\} \\ \mathbf{u} \neq \emptyset}} \mathcal{M}_{\mathbf{u}}^{\text{PC}}(\mathbf{x}_{\mathbf{u}}), \tag{6.73}$$

where

$$\mathcal{M}_{\mathbf{u}}^{\text{PC}}(\mathbf{x}_{\mathbf{u}}) \overset{\text{def}}{=} \sum_{\alpha \in \mathcal{A}_{\mathbf{u}}} y_\alpha \Psi_\alpha(\mathbf{x}). \tag{6.74}$$

Consequently, due to the orthogonality of the PC basis, the partial variance $D_u$ reduces to

$$D_u = \text{Var}\left[\mathcal{M}_u^{\text{PC}}(\boldsymbol{X}_u)\right] = \sum_{\alpha \in \mathcal{A}_u} y_\alpha^2. \tag{6.75}$$

In other words, from a given PC expansion, the Sobol' indices *at any order* may be obtained by a mere combination of the squares of the coefficients. As an illustration, the first-order PC-based Sobol' indices read:

$$S_i^{\text{PC}} = \sum_{\alpha \in \mathcal{A}_i} y_\alpha^2 / D \quad \text{where} \quad \mathcal{A}_i = \{\alpha \in \mathcal{A} : \alpha_i > 0,\ \alpha_{j \neq i} = 0\}, \tag{6.76}$$

whereas the total PC-based Sobol' indices are

$$S_i^{T,\text{PC}} = \sum_{\alpha \in \mathcal{A}_i^T} y_\alpha^2 / D \quad \text{where} \quad \mathcal{A}_i^T = \{\alpha \in \mathcal{A} : \alpha_i > 0\}. \tag{6.77}$$

PCEs and the various types of post-processing presented above are now applied to different classical geotechnical problems.

## 6.5 APPLICATION EXAMPLES

### 6.5.1 Load-carrying capacity of a strip footing

#### 6.5.1.1 Independent input variables

Let us consider the strip footing of width $B = 10$ m sketched in Figure 6.6 that is embedded at depth $D$. We assume that the ground water table is far below the surface. The soil layer is assumed homogeneous with cohesion $c$, friction angle $\phi$, and unit weight $\gamma$.

The ultimate bearing capacity reads (Lang et al., 2007):

$$q_u = cN_c + \gamma D N_q + \frac{1}{2} B\gamma N_\gamma, \tag{6.78}$$

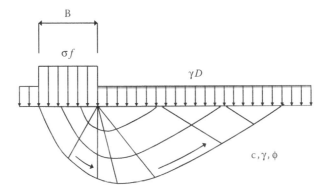

*Figure 6.6* Example #1: Strip footing.

*Table 6.3* Ultimate bearing capacity of a strip foundation—probabilistic model

| Parameter | Notation | Type of PDF | Mean value | Coefficient of variation |
|---|---|---|---|---|
| Foundation width | $B$ | Deterministic | 10 m | – |
| Foundation depth | $D$ | Gaussian | 1 m | 15% |
| Unit soil weight | $\gamma$ | Lognormal | 20 kN/m³ | 10% |
| Cohesion | $c$ | Lognormal | 20 kPa | 25% |
| Friction angle | $\phi$ | Beta | Range: $[0, 45°]$, $\mu = 30°$ | 10% |

where the bearing capacity factors read:

$$N_q = e^{\pi \tan\phi} \tan^2 (\pi/4 + \phi/2).$$

$$N_c = (N_q - 1) \cot \phi.$$ (6.79)

$$N_\gamma = 2(N_q - 1) \tan \phi.$$

The soil parameters and the foundation depth are considered as independent random variables, whose properties are listed in Table 6.3. Let us denote the model input vector by $X = \{D, \gamma, c, \phi\}^T$. The associated random-bearing capacity is $q_u(X)$.

Using the mean values of the parameters in Table 6.3, the ultimate bearing capacity is equal to $\overline{q_u}$ = 2.78 MPa. We now consider several design situations with applied loads $q_{des} = \overline{q_u}/SF$ where $SF = 1.5, 2, 2.5$, and 3 would be the global safety factor obtained from a deterministic design. Then we consider the reliability of the foundation with respect to the ultimate bearing capacity. The limit state function reads:

$$g(X) = q_u(X) - q_{des} = q_u(X) - \overline{q_u}/SF.$$ (6.80)

Classical reliability methods are used, namely FORM, SORM, and crude MCS with $10^7$ samples to get a reference solution. The uncertainty quantification software UQLab is used (Marelli and Sudret, 2014). Alternatively, a PC expansion $q_u^{PC}(X)$ of the ultimate bearing capacity is first computed using an LHS ED of size $n = 500$. Then the PC expansion is substituted for in Equation 6.80 and the associated probability of failure is computed by MCS ($10^7$ samples), now using only the PC expansion (and for the different values of SF). The results are reported in Table 6.4.

From Table 6.4, it is clear that the results obtained by PC expansion are almost equal to those obtained by the reference MCS. The relative error in terms of the probability of failure is all in all <1% (the corresponding error on the generalized reliability index $\beta_{gen} = -\Phi^{-1}(P_f)$ is negligible).

*Table 6.4* Ultimate bearing capacity of a strip foundation—probability of failure (resp. generalized reliability index $\beta_{gen} = -\Phi^{-1}(P_f)$ between parentheses)—case of independent variables

| SF | FORM | SORM | MCS ($10^7$ Runs) | PCE + MCS[a] |
|---|---|---|---|---|
| 1.5 | $1.73 \cdot 10^{-1}$ (0.94) | $1.70 \cdot 10^{-1}$ (0.96) | $1.69 \cdot 10^{-1}$ (0.96) | $1.70 \cdot 10^{-1}$ (0.96) |
| 2.0 | $5.49 \cdot 10^{-2}$ (1.60) | $5.30 \cdot 10^{-2}$ (1.62) | $5.30 \cdot 10^{-2}$ (1.62) | $5.29 \cdot 10^{-2}$ (1.62) |
| 2.5 | $1.72 \cdot 10^{-2}$ (2.11) | $1.65 \cdot 10^{-2}$ (2.13) | $1.63 \cdot 10^{-2}$ (2.14) | $1.65 \cdot 10^{-2}$ (2.13) |
| 3.0 | $5.54 \cdot 10^{-3}$ (2.54) | $5.23 \cdot 10^{-3}$ (2.56) | $5.24 \cdot 10^{-3}$ (2.56) | $5.20 \cdot 10^{-3}$ (2.56) |

[a] $n = 500, n_{MCS} = 10^7$.

*Table 6.5* Ultimate bearing capacity of a strip foundation—probability of failure (resp. generalized reliability index $\beta_{gen} = \Phi^{-1}(P_f)$ between parentheses)—case of dependent $(c, \phi)$

| SF | FORM | SORM | MCS ($10^7$ Runs) | PCE + MCS[a] |
|---|---|---|---|---|
| 1.5 | $1.55 \cdot 10^{-1}$ (1.01) | $1.52 \cdot 10^{-1}$ (1.03) | $1.51 \cdot 10^{-1}$ (1.03) | $1.52 \cdot 10^{-1}$ (1.03) |
| 2.0 | $4.02 \cdot 10^{-2}$ (1.75) | $3.85 \cdot 10^{-2}$ (1.77) | $3.85 \cdot 10^{-2}$ (1.77) | $3.85 \cdot 10^{-2}$ (1.77) |
| 2.5 | $9.60 \cdot 10^{-3}$ (2.34) | $8.98 \cdot 10^{-3}$ (2.37) | $8.98 \cdot 10^{-3}$ (2.37) | $8.99 \cdot 10^{-3}$ (2.37) |
| 3.0 | $2.20 \cdot 10^{-3}$ (2.85) | $2.00 \cdot 10^{-3}$ (2.88) | $2.01 \cdot 10^{-3}$ (2.88) | $2.00 \cdot 10^{-3}$ (2.88) |

[a] $n = 500, n_{MCS} = 10^7$.

The number of runs associated to FORM (resp. SORM) are 31 for $SF = 1.5$ and 2, and 35 for $SF = 2.5$ and 3 (resp. 65 for $SF = 1.5$ and 2, and 69 for $SF = 2.5$ and 3). Note that for each value of $SF$, a new analysis FORM/SORM analysis shall be run. In contrast to FORM/SORM, a *single* PC expansion has been used to obtain the reliability associated to all safety factors. Using 500 points in the ED, that is, 500 evaluations of $q_u(X)$, the obtained PC expansion provides a normalized LOO error equal to $1.7 \cdot 10^{-7}$ (the maximal PC degree is 6 and the number of terms in the sparse expansion is 140).

### 6.5.1.2 Correlated input variables

For a more realistic modeling of the soil properties, one now considers the statistical dependence between the cohesion $c$ and the friction angle $\phi$. From the literature (see a review in Al Bittar and Soubra (2012)), the correlation between these parameters is negative with a value around $-0.5$. In this section, we model the dependence between $c$ and $\phi$ by a Gaussian copula that is parameterized by the *rank correlation coefficient* $\rho_R = -0.5$. Owing to the choice of marginal distributions in Table 6.3, this corresponds to a linear correlation of $-0.512$. Using 500 points in the ED, that is, 500 evaluations of $q_u(X)$, the obtained PC expansion provides an LOO error equal to $6.8 \cdot 10^{-7}$ (the maximal PC degree is 6 and the number of terms in the sparse expansion is 172). The reliability results accounting for correlation are reported in Table 6.5.

These results show that PCEs may be applied to also solve reliability problems when the variables in the limit state function are correlated. In terms of accuracy, the PCE results compare very well with the reference results obtained by MCS, the error on the probability of failure being, again, <1%. SORM provides accurate results as well, at a cost of 65, 65, 72, and 83 runs when SF = 1.5, 2, 2.5, and 3 (the associated FORM analysis required are 31, 31, 38, and 49 runs). Moreover, it clearly appears that neglecting the correlation between $c$ and $\phi$ leads to a conservative estimation of the probability of failure, for example, by a factor 2.5 for $SF = 3$ (10% underestimation of the generalized reliability index).

## 6.5.2 Settlement of a foundation on an elastic two-layer soil mass

Let us consider an elastic soil mass made of two layers of different isotropic linear elastic materials lying on a rigid substratum. A foundation on this soil mass is modeled by a uniform pressure $P_1$ applied over a length $2 B_1 = 10$ m of the free surface. An additional load $P_2$ is applied over a length $2 B_2 = 5$ m (Figure 6.7a).

Owing to the symmetry, half of the structure is modeled by finite elements (Figure 6.7b). The mesh comprises 500 QUAD4 isoparametric elements. A plane strain analysis is carried out. The geometry is considered as deterministic. The elastic material properties of both layers and the applied loads are modeled by random variables, whose PDF is specified in Table 6.6. All six random variables are supposed to be independent.

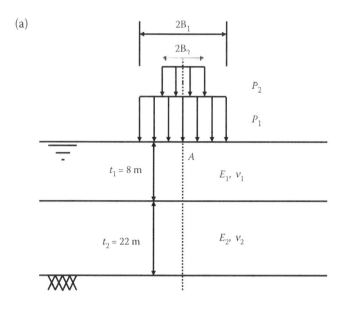

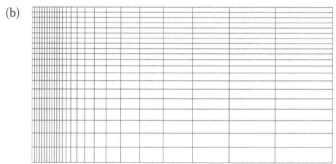

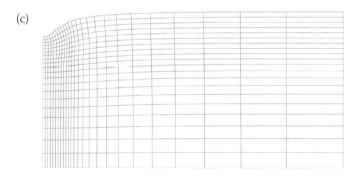

*Figure 6.7* Example #2: Foundation on a two-layer soil mass. (a) Foundation on a two-layer soil mass, (b) finite element mesh, (c) deformed mesh under loading.

The QoI is the maximum vertical displacement $u_A$ at point A, that is, on the symmetry axis of the problem. The finite-element model is thus considered a black-box $\mathcal{M}^{FE}$ that computes $u_A$ as a function of the six input parameters $X = \{E_1, E_2, \nu_1, \nu_2, P_1, P_2\}^T$:

$$u_A = \mathcal{M}^{FE}(E_1, E_2, \nu_1, \nu_2, P_1, P_2).$$ \hfill (6.81)

*Table 6.6* Example #2: Two-layer soil layer mass—parameters of the model

| Parameter | Notation | Type of PDF | Mean Value | Coefficient of Variation |
|---|---|---|---|---|
| Upper layer soil thickness | $t_1$ | Deterministic | 8 m | – |
| Lower layer soil thickness | $t_2$ | Deterministic | 22 m | – |
| Upper layer Young's modulus | $E_1$ | Lognormal | 50 MPa | 20% |
| Lower layer Young's modulus | $E_2$ | Lognormal | 100 MPa | 20% |
| Upper layer Poisson ratio | $v_1$ | Uniform | 0.3 | 15% |
| Lower layer Poisson ratio | $v_2$ | Uniform | 0.3 | 15% |
| Load #1 | $P_1$ | Gamma | 0.2 MPa | 20% |
| Load #2 | $P_2$ | Weibull | 0.4 MPa | 20% |

*Table 6.7* Settlement of a foundation on a two-layer soil mass—reliability results

| Threshold | FORM | | SORM | | Importance Sampling | |
|---|---|---|---|---|---|---|
| $u_{adm}$ (cm) | $P_{f,FORM}$ | $\beta$ | $P_{f,SORM}$ | $\beta$ | $P_{f,IS}$ | $\beta$ |
| 12 | $1.54 \cdot 10^{-1}$ | 1.02 | $1.38 \cdot 10^{-1}$ | 1.09 | $1.42 \cdot 10^{-1}$ [CoV = 1.27%] | 1.07 |
| 15 | $1.64 \cdot 10^{-2}$ | 2.13 | $1.36 \cdot 10^{-2}$ | 2.21 | $1.43 \cdot 10^{-2}$ [CoV = 1.67%] | 2.19 |
| 20 | $1.54 \cdot 10^{-4}$ | 3.61 | $1.17 \cdot 10^{-4}$ | 3.68 | $1.23 \cdot 10^{-4}$ [CoV = 2.23%] | 3.67 |
| 21 | $5.57 \cdot 10^{-5}$ | 3.86 | $4.14 \cdot 10^{-5}$ | 3.94 | $4.53 \cdot 10^{-5}$ [CoV = 2.27%] | 3.91 |

The serviceability of this foundation on a layered soil mass vis-à-vis an admissible settlement is studied. The limit state function is defined by

$$g(\boldsymbol{X}) = u_{adm} - u_A = u_{adm} - M^{FE}(E_1, E_2, v_1, v_2, P_1, P_2), \qquad (6.82)$$

in which the admissible settlement is chosen to be 12, 15, 20, and 21 cm. First, FORM and SORM are applied, together with IS at the design point using $n_{MCS} = 10^4$ samples. The latter is considered as the reference solution. Results are reported in Table 6.7.

Then PC expansions of the maximal settlement $\mathcal{M}^{PC}$ are computed using different EDs of increasing size, namely $n = 100, 200, 500,$ and $1000$ using the UQLab platform (Marelli and Sudret, 2014). The obtained expansion is then substituted for in the limit state function (Equation 6.82) and the associated reliability problem is solved using crude MCS ($n_{MCS} = 10^6$ samples). The results are reported in Table 6.8.

*Table 6.8* Settlement of a foundation on a two-layer soil mass—reliability results from PC expansion (MCS with $n_{MCS} = 10^7$)

| Threshold | $n = 100$ | | $n = 200$ | | $n = 500$ | | $n = 1000$ | |
|---|---|---|---|---|---|---|---|---|
| $u_{adm}$ (cm) | $P_f$ | $\beta$ | $P_f$ | $\beta$ | $P_f$ | $\beta$ | $P_f$ | $\beta$ |
| 12 | $1.44 \cdot 10^{-1}$ [CoV = 0.08%] | 1.06 | $1.45 \cdot 10^{-1}$ [CoV = 0.08%] | 1.06 | $1.44 \cdot 10^{-01}$ [CoV = 0.08%] | 1.06 | $1.44 \cdot 10^{-1}$ [CoV = 0.08%] | 1.06 |
| 15 | $1.33 \cdot 10^{-2}$ [CoV = 0.27%] | 2.22 | $1.44 \cdot 10^{-2}$ [CoV = 0.26%] | 2.19 | $1.45 \cdot 10^{-02}$ [CoV = 0.26%] | 2.18 | $1.45 \cdot 10^{-2}$ [CoV = 0.26%] | 2.18 |
| 20 | $5.84 \cdot 10^{-5}$ [CoV = 4.14%] | 3.85 | $1.13 \cdot 10^{-4}$ [CoV = 2.97%] | 3.69 | $1.26 \cdot 10^{-04}$ [CoV = 2.82%] | 3.66 | $1.24 \cdot 10^{-4}$ [CoV = 2.84%] | 3.66 |
| 21 | $1.66 \cdot 10^{-5}$ [CoV = 7.76%] | 4.15 | $3.64 \cdot 10^{-5}$ [CoV = 5.24%] | 3.97 | $4.72 \cdot 10^{-05}$ [CoV = 4.60%] | 3.90 | $4.56 \cdot 10^{-5}$ [CoV = 4.68%] | 3.91 |

*Table 6.9* Settlement of a foundation on a two-layer soil mass—PC expansion features ($p$ stands for the maximal degree of polynomials and $P$ is the number of nonzero polynomials in the sparse expansion)

|  | $n = 100$ | $n = 200$ | $n = 500$ | $n = 1000$ |
|---|---|---|---|---|
| $1 - R^2$ (Equation 6.37) | $8.45 \cdot 10^{-5}$ | $6.76 \cdot 10^{-6}$ | $1.25 \cdot 10^{-6}$ | $1.14 \cdot 10^{-7}$ |
| $\hat{\varepsilon}_{LOO}$ (Equation 6.44) | $1.20 \cdot 10^{-3}$ | $2.39 \cdot 10^{-4}$ | $1.33 \cdot 10^{-5}$ | $2.08 \cdot 10^{-6}$ |
| $p$ | 4 | 4 | 4 | 5 |
| $P$ | 59 | 126 | 152 | 225 |

It can be observed that the results obtained from the PC expansion compare very well to the reference as soon as $n = 200$ points are used in the ED. The error is <1% in the generalized reliability index, for values as large as $\beta = 4$, that is, for probabilities of failure in the order of $10^{-5}$. The detailed features of the PC expansions built for each ED of size $n = 100$, 200, 500, and 1000 are reported in Table 6.9.

Again, the sparsity of the expansions is clear: a full expansion with all polynomials up to degree $p = 4$ in $M = 6$ variables has $P = \binom{6+4}{4} = 210$ terms, which would typically require an ED of size $2 \times 210 = 440$. Using only $n = 100$ points, a sparse PC expansion having 59 terms could be built up. It is also observed that the classical (normalized) empirical error $1 - R^2$ is typically one order of magnitude smaller than the LOO normalized error, the latter being a closer estimate of the real generalization error.

### 6.5.3 Settlement of a foundation on soil mass with spatially varying Young's modulus

Let us now consider a foundation on an elastic soil layer showing spatial variability in its material properties (after Blatman, 2009). A structure to be founded on this soil mass is idealized as a uniform pressure $P$ applied over a length $2B = 20$ m of the free surface (Figure 6.8).

The soil layer thickness is equal to 30 m. The soil mesh width is equal to 120 m. The soil layer is modeled as an elastic linear isotropic material with Poisson's ratio equal to 0.3. A plane strain analysis is carried out. The finite-element mesh is made of 448 QUAD4 elements. Young's modulus is modeled by a 2D homogeneous *lognormal* random field with mean value $\mu_E = 50$ MPa and a coefficient of variation of 30%. The underlying Gaussian random field $\log E(x, \omega)$ has a square-exponential autocorrelation function:

$$\rho_{\log E}(x, x') = \exp\left(-\frac{\|x - x'\|^2}{\ell^2}\right), \tag{6.83}$$

where $\ell = 15$ m. The Gaussian random field $\log E(x, \omega)$ is discretized using the KL expansion (Loève, 1978; Sudret and Der Kiureghian, 2000; Sudret and Berveiller, 2007, Chapter 7):

$$\log E(x, \omega) = \mu_{\log E} + \sigma_{\log E} \sum_{i=1}^{\infty} \sqrt{\lambda_i} \xi_i(\omega) \phi_i(x), \tag{6.84}$$

where $\{\xi_i(\omega)\}_{i=1}^{\infty}$ are independent standard normal variables and the pairs $\{(\lambda_i, \phi_i)\}_{i=1}^{\infty}$ are the solution of the following eigenvalue problem (Fredholm integral of the second kind):

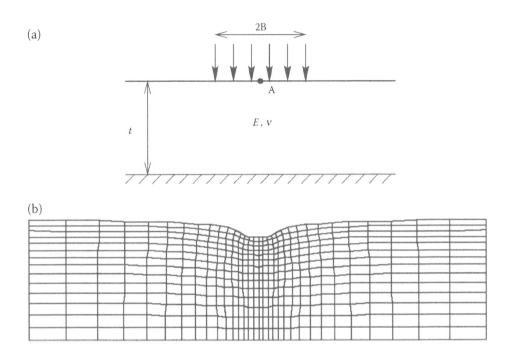

*Figure 6.8* Example #3: Foundation on a soil mass with spatially varying Young's modulus. (a) Scheme of the foundation, (b) finite element mesh.

$$\int \rho_{\log E}(\boldsymbol{x}, \boldsymbol{x}')\phi_i(\boldsymbol{x}')\,d\boldsymbol{x}' = \lambda_i\phi_i(\boldsymbol{x}). \tag{6.85}$$

As no analytical solution to the Fredholm equation exists for this type of autocorrelation function, the latter is solved by expanding the eigenmodes onto an orthogonal polynomial basis, see details in Blatman (2009, Appendix B). Note that other numerical methods have been proposed in the literature (Phoon et al., 2002a,b, 2005; Li et al., 2007). Eventually, 38 modes are retained in the truncated expansion:

$$\log E(\boldsymbol{x}, \omega) \approx \mu_{\log E} + \sigma_{\log E} \sum_{i=1}^{38} \sqrt{\lambda_i}\,\xi_i(\omega)\phi_i(\boldsymbol{x}), \tag{6.86}$$

where $\mu_{\log E} = 3.8689$ and $\sigma_{\log E} = 0.2936$ in this application. The first nine eigenmodes are plotted in Figure 6.9 for the sake of illustration. Note that these 38 modes allow one to account for 99% of the variance of the Gaussian field.

The *average settlement under the foundation* is computed by finite-element analysis. It may be considered as a random variable $Y = \mathcal{M}(\xi)$, where $\xi$ is the standard normal vector of dimension 38 that enters the truncated KL expansion. Of interest is the sensitivity of this average settlement to the various modes appearing in the KL expansion. To address this problem, a sparse PC expansion is built using an LHS ED of size $n = 200$. It allows one to get a LOO error <5%. From the obtained expansion, the total Sobol' indices related to each input variable $\xi_i$ (i.e., each mode in the KL expansion) are computed and plotted in Figure 6.10.

It appears that only seven modes contribute to the variability of the settlement. This may be explained by the fact that the model response is an averaged quantity over the domain of the application of the load, which is therefore rather insensitive to small-scale fluctuations

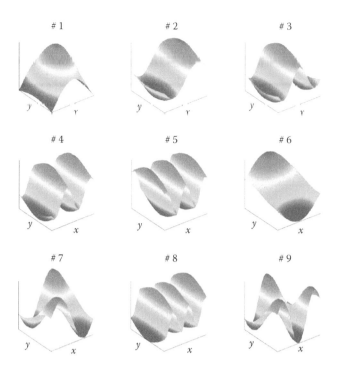

*Figure 6.9* Example #3: First modes of the KL expansion of Young's modulus. (After Blatman, G. and B. Sudret. 2011a. *J. Comput. Phys. 230*, 2345–2367.)

of the spatially variable random Young's modulus. Note that some modes have a zero total Sobol' index, namely modes #2, 4, 7, and 8. From Figure 6.10, it appears that they correspond to *antisymmetric modes* with respect to the vertical axis (see Figure 6.9). This means that the symmetry of the problem is accounted for in the analysis. It is now clear why the PC expansion is sparse since roughly half of the modes (i.e., half of the input variables of the uncertainty quantification problem) do not play any role in the analysis.

## 6.5.4 Conclusions

In this section, three different problems of interest in geotechnical engineering have been addressed, namely, the bearing capacity of a strip footing, the maximal settlement of a foundation on a two-layer soil mass, and the settlement in case of a single layer with spatially varying Young's modulus. In the first two cases, reliability analysis is carried out as a post-processing of a PC expansion. The results in terms of probability of failure compare very well with those obtained by reference methods such as IS at the design point. From a broader experience in structural reliability analysis, it appears that PC expansions are suitable for reliability analysis as long as the probability of failure to be computed is larger than $10^{-5}$. For very small probabilities, suitable methods such as adaptive Kriging would rather be used (Dubourg et al., 2011; Sudret, 2012).

In the last example, the spatial variability of the soil properties is introduced. The purpose is to show that problems involving a large number of random input variables (here, 38) may be solved at an affordable cost using sparse PC expansions (here, 200 samples in the ED). Recent applications of this approach to the bearing capacity of 2D and three-dimensional (3D) foundations can be found in Al Bittar and Soubra (2012); Mao et al. (2012).

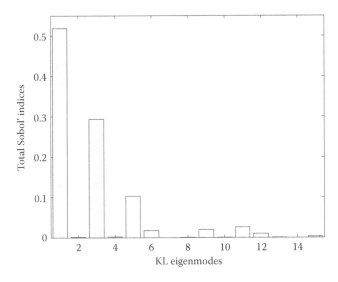

*Figure 6.10* Example #3: Foundation on a soil mass with spatially varying Young's modulus—total Sobol' indices. (After Blatman, G. and B. Sudret. 2011a. *J. Comput. Phys.* 230, 2345–2367.)

## 6.6 SUMMARY AND OUTLOOK

Accounting for uncertainties has become a crucial issue in modern geotechnical engineering due to the large variability of soil properties as well as the associated limited information. The uncertainty analysis framework that is nowadays widely used in many fields applies equally to geotechnics. Starting from a computational model used for assessing the system performance, the input parameters are represented by random variables or fields. The effect of the input uncertainty onto the model response (i.e., the system performance) can be assessed by a number of numerical methods.

MCS offers a sound framework for uncertainty propagation; however, its low efficiency precludes its use for analyses involving finite-element models. Classical structural reliability methods such as FORM and SORM may be used, at the price of some linearizations and approximations. In contrast, the so-called PCEs allow for an accurate, intrinsic representation of the model output. The series expansion coefficients may be computed using nonintrusive schemes that are similar in essence to MCS: a sample set of input vectors is simulated and the corresponding model output is evaluated. From these data, algorithms such as least-square minimization or LAR may be used.

The resulting series expansion can be post-processed to compute the statistical moments of the model response (mean value, standard deviation, etc.), quantiles, and confidence intervals, or even the PDF. In the latter case, the PC expansion is used as a surrogate to the original computational model. Owing to its polynomial expression, it can be used together with MCS and kernel-smoothing techniques.

PC expansions also appear extremely efficient in the context of sensitivity analysis. The Sobol' indices, which are considered as the most accurate sensitivity measures, may be analytically computed from PC expansions coefficients by simple algebra. The PC expansion itself may be reordered so as to exhibit the so-called Sobol' decomposition, which allows for detecting the linear and nonlinear effects of the input variables as well as their interaction.

Problems involving spatial variability of the material properties may be efficiently addressed by introducing random fields and using discretization techniques. The discretized field is represented by a vector of (usually) standard normal variables, whose size may be large (e.g., 20–100), especially when the correlation length of the field is small. However, the so-called sparse PC expansion technique may be used so that the full analysis can be carried out using a few hundred to a thousand model runs.

## ACKNOWLEDGMENTS

The author thanks Dr. Géraud Blatman (EDF R&D, France) for the original joint work on sparse PC expansions and ongoing discussions. The support of Dr. Stefano Marelli and Roland Schöbi (ETH Zürich, Chair of Risk, Safety, and Uncertainty Quantification) for preparing the numerical examples is also gratefully acknowledged.

## APPENDIX 6A: HERMITE POLYNOMIALS

The Hermite polynomials $He_n(x)$ are the solution of the following differential equation:

$$y'' - xy' + ny' = 0. \quad n \in \mathcal{N} \tag{A.1}$$

They may be generated in practice by the following recurrence relationship:

$$He_0(x) = 1. \tag{A.2}$$

$$He_{n+1}(x) = x\, He_n(x) - n\, He_{n-1}(x). \tag{A.3}$$

They are orthogonal with respect to the Gaussian probability measure:

$$\int_{-\infty}^{\infty} He_m(x)\, He_n(x)\, \phi(x)\, dx = n!\, \delta_{mn}, \tag{A.4}$$

where $\phi(x) = 1/\sqrt{2\pi}\, e^{-x^2/2}$ is the standard normal PDF. If $U$ is a standard normal random variable, the following relationship holds:

$$\mathbb{E}\left[He_m(U)\, He_n(U)\right] = n!\, \delta_{mn}. \tag{A.5}$$

The first four Hermite polynomials are

$$He_0(x) = 1 \quad He_1(x) = x \quad He_2(x) = x^2 - 1 \quad He_3(x) = x^3 - 3x. \tag{A.6}$$

Owing to Equation A.5, the orthonormal polynomials read:

$$\psi_0(x) = 1 \quad \psi_1(x) = x \quad \psi_2(x) = (x^2 - 1)/\sqrt{2} \quad \psi_3(x) = (x^3 - 3x)/\sqrt{6}. \tag{A.7}$$

## LIST OF SYMBOLS

| | |
|---|---|
| $\alpha$ | Multi-index of $\Psi_\alpha$ |
| $\delta_{\hat{Y}}^2$ | PC-based response skewness |
| $D_i, D_u$ | Partial variance |
| $\widehat{Err}_E$ | Empirical error |
| $\widehat{Err}_{LOO}$ | Leave-one-out error |
| $f_X$ | Probability density function of the input vector |
| $f_{X_i}$ | Marginal distribution of $X_i$ |
| $g(.)$ | Limit state function |
| $h$ | Kernel bandwidth |
| $H_{e_i}$ | Hermite polynomial |
| $\kappa_{\hat{Y}}^2$ | PC-based response kurtosis |
| $K(.)$ | Kernel function |
| $M$ | Dimension of the input random vector $X$ |
| MCS | Abbreviation of "Monte Carlo simulation" |
| $P_j$ | Legendre polynomials |
| $P_f$ | Probability of failure |
| PC | Abbreviation of "polynomial chaos" |
| $\pi_j^{(i)}, \psi_j^{(i)}$ | Univariate orthogonal (resp. orthonormal) polynomials |
| $Q_{p\%}$ | $p\%$-quantile |
| $\sigma_{\hat{Y}}^2$ | PC-based response variance |
| $S_i, S_{ij}$ | Sobol' index |
| $S_i^T$ | Total Sobol' index |
| $X$ | Input random vector |
| $y = \mathcal{M}(x)$ | Model response (quantity of interest) |
| $y_\alpha$ | Coefficient of the polynomial chaos expansion |
| $\Psi_\alpha(x)$ | Multivariate orthonormal polynomials |
| $\mathcal{A}, \mathcal{A}^{M,p}$ | Truncation set |
| $\mathcal{A}^{M,p,q}$ | Hyperbolic truncation set |
| $\mathcal{D}_s, \mathcal{D}_f$ | Safe (resp. failure) domain |
| $\mathcal{LN}(\lambda, \zeta)$ | Lognormal distribution |
| $\mathcal{M}$ | Computational model |
| $\mathcal{M}_i(x_i), \mathcal{M}_{ij}(x_i, x_j), \mathcal{M}_u(x_u)$ | Terms in the Sobol'–Hoeffding decomposition |
| $\mathcal{T}$ | Isoprobabilistic transform |
| $\mathcal{U}(a, b)$ | Uniform distribution on $[a, b]$ |
| $\mathcal{X}$ | Experimental design of computer simulations |

## REFERENCES

Al Bittar, T. and A. Soubra. 2012. Bearing capacity of strip footings on spatially random soils using sparse polynomial chaos expansion. *Int. J. Num. Anal. Meth. Geomech.* 37(13), 2039–2060.

Allen, D. 1971. The prediction sum of squares as a criterion for selecting prediction variables. Technical Report 23, Department of Statistics, University of Kentucky.

Baecher, G. and T. Ingra. 1981. Stochastic finite element method in settlement predictions. *J. Geotech. Eng. Div.* 107(4), 449–463.

Berveiller, M., B. Sudret, and M. Lemaire. 2004a. Comparison of methods for computing the response coefficients in stochastic finite element analysis. In *Proceedings of the 2nd International ASRANet Colloquium*, Barcelona, Spain.

Berveiller, M., B. Sudret, and M. Lemaire. 2004b. Presentation of two methods for computing the response coefficients in stochastic finite element analysis. In *Proceedings of the 9th ASCE Specialty Conference on Probabilistic Mechanics and Structural Reliability*, Albuquerque, USA.

Berveiller, M., B. Sudret, and M. Lemaire. 2006. Stochastic finite elements. A nonintrusive approach by regression. *Eur. J. Comput. Mech.* 15(1–3), 81–92.

Blatman, G. 2009. *Adaptive Sparse Polynomial Chaos Expansions for Uncertainty Propagation and Sensitivity Analysis*. PhD thesis, Université Blaise Pascal, Clermont-Ferrand.

Blatman, G. and B. Sudret. 2008. Sparse polynomial chaos expansions and adaptive stochastic finite elements using a regression approach. *Comptes Rendus Mécanique* 336(6), 518–523.

Blatman, G. and B. Sudret. 2010. An adaptive algorithm to build up sparse polynomial chaos expansions for stochastic finite element analysis. *Prob. Eng. Mech.* 25(2), 183–197.

Blatman, G. and B. Sudret. 2011a. Adaptive sparse polynomial chaos expansion based on least angle regression. *J. Comput. Phys.* 230, 2345–2367.

Blatman, G. and B. Sudret. 2011b. Principal component analysis and least angle regression in spectral stochastic finite element analysis. In M. Faber (ed.), *Proceedings of the 11th International Conference on Applications of Statistics and Probability in Civil Engineering (ICASP11)*, Zurich, Switzerland.

Blatman, G. and B. Sudret. 2013. Sparse polynomial chaos expansions of vector-valued response quantities. In G. Deodatis (ed.), *Proceedings of the 11th International Conference on Structural Safety and Reliability (ICOSSAR'2013)*, New York, USA.

Caniou, Y. 2012. *Global Sensitivity Analysis for Nested and Multiscale Models*. PhD thesis, Université Blaise Pascal, Clermont-Ferrand.

Chapelle, O., V. Vapnik, and Y. Bengio. 2002. Model selection for small sample regression. *Mach. Learn.* 48(1), 9–23.

Clouteau, D. and R. Lafargue. 2003. An iterative solver for stochastic soil-structure interaction. In P. Spanos and G. Deodatis (eds.), *Proceedings of the 4th International Conference on Computer Stochastic Mechanics (CSM4)*, pp. 119–124. Corfu.

De Rocquigny, E. 2012. *Modelling Under Risk and Uncertainty: An Introduction to Statistical, Phenomenological and Computational Methods*. Wiley series in probability and statistics. John Wiley & Sons, Chichester.

DeGroot, D. and G. Baecher. 1993. Estimating autocovariance of *in-situ* soil properties. *J. Geo. Eng.* 119(1), 147–166.

Ditlevsen, O. and H. Madsen. 1996. *Structural Reliability Methods*. John Wiley & Sons, Chichester.

Dubourg, V., B. Sudret, and J.-M. Bourinet. 2011. Reliability-based design optimization using Kriging and subset simulation. *Struct. Multidisc. Optim.* 44(5), 673–690.

Efron, B., T. Hastie, I. Johnstone, and R. Tibshirani. 2004. Least angle regression. *Ann. Stat.* 32, 407–499.

Fenton, G.-A. 1999a. Estimation for stochastic soil models. *J. Geo. Eng.* 125(6), 470–485.

Fenton, G.-A. 1999b. Random field modeling of CPT data. *J. Geo. Eng.* 125(6), 486–498.

Geisser, S. 1975. The predictive sample reuse method with applications. *J. Am. Stat. Assoc.* 70, 320–328.

Ghanem, R. and V. Brzkala. 1996. Stochastic finite element analysis of randomly layered media. *J. Eng. Mech.* 122(4), 361–369.

Ghanem, R. and P. Spanos. 1991. *Stochastic Finite Elements: A Spectral Approach*. Springer Verlag, New York. (Reedited by Dover Publications, 2003).

Ghiocel, D. and R. Ghanem. 2002. Stochastic finite element analysis of seismic soil-structure interaction. *J. Eng. Mech.* 128, 66–77.

Hastie, T., J. Taylor, R. Tibshirani, and G. Walther. 2007. Forward stagewise regression and the monotone lasso. *Electron. J. Stat.* 1, 1–29.

Isukapalli, S. S. 1999. *Uncertainty Analysis of Transport-Transformation Models*. PhD thesis, The State University of New Jersey.

Lang, H.-J., J. Huder, P. Amann, and A.-M. Puzrin. 2007. *Bodenmechanik und Grundbau*. Springer, Berlin, Heidelberg

Le Maître, O., M. Reagan, H. Najm, R. Ghanem, and O. Knio. 2002. A stochastic projection method for fluid flow—II. Random process. *J. Comput. Phys.* 181, 9–44.

Li, H. and B. Low. 2010. Reliability analysis of circular tunnel under hydrostatic stress field. *Comput. Geotech. 37*, 50–58.

Li, L., K. Phoon, and S. Quek. 2007. Comparison between Karhunen–Loève expansion and translation-based simulation of non-Gaussian processes. *Comput. Struct. 85*(5–6), 263–276.

Loève, M. 1978. *Probability Theory: Graduate Texts in Mathematics* (4th ed.), Volume 2. Springer Verlag, New York.

Low, B. 2005. Reliability-based design applied to retaining walls. *Géotechnique 55*(1), 63–75.

Low, B. and W. Tang. 2007. Efficient spreadsheet algorithm for first-order reliability. *J. Eng. Mech. (ASCE) 133*(12), 1378–1387.

Mao, N., T. Al-Bittar, and A.-H. Soubra. 2012. Probabilistic analysis and design of strip foundations resting on rocks obeying Hoek–Brown failure criterion. *Int. J. Rock Mech. Min. Sci. 49*, 45–58.

Marelli, S. and Sudret, B. 2014. UQLab: A framework for uncertainty quantification in MATLAB, *Proceedings of the 2nd International Conference on Vulnerability, Risk Analysis and Management (ICVRAM2014)*, Liverpool, United Kingdom.

Matthies, H. and A. Keese. 2005. Galerkin methods for linear and nonlinear elliptic stochastic partial differential equations. *Comput. Meth. Appl. Mech. Eng. 194*, 1295–1331.

McKay, M. D., R. J. Beckman, and W. J. Conover. 1979. A comparison of three methods for selecting values of input variables in the analysis of output from a computer code. *Technometrics 2*, 239–245.

Nelsen, R. 1999. *An Introduction to Copulas. Volume 139 of Lecture Notes in Statistics*. Springer-Verlag, New York.

Niederreiter, H. 1992. *Random Number Generation and Quasi-Monte Carlo Methods*. Society for Industrial and Applied Mathematics, Philadelphia, PA, USA.

Phoon, K. 2003. Representation of random variables using orthogonal polynomials. In A. Der Kiureghian, S. Madanat, and J. Pestana (eds.), *Applications of Statistics and Probability in Civil Engineering*, Millpress Science Publishers, pp. 97–104.

Phoon, K., S. Huang, and S. Quek. 2002a. Implementation of Karhunen–Loève expansion for simulation using a wavelet-Galerkin scheme. *Prob. Eng. Mech. 17*(3), 293–303.

Phoon, K., S. Huang, and S. Quek. 2002b. Simulation of second-order processes using Karhunen–Loève expansion. *Comput. Struct. 80*(12), 1049–1060.

Phoon, K., H. Huang, and S. Quek. 2005. Simulation of strongly non Gaussian processes using Karhunen–Loève expansion. *Prob. Eng. Mech. 20*(2), 188–198.

Phoon, K., S. Quek, Y. Chow, and S. Lee. 1990. Reliability analysis of pile settlements. *J. Geotech. Eng. 116*(11), 1717–1735.

Popescu, R., G. Deodatis, and A. Nobahar. 2005. Effects of random heterogeneity of soil properties on bearing capacity. *Prob. Eng. Mech. 20*, 324–341.

Press, W., W. Vetterling, S.-A. Teukolsky, and B.-P. Flannery. 2001. *Numerical Recipes*. Cambridge University Press, New York.

Rackwitz, R. 2000. Reviewing probabilistic soil modeling. *Comput. Geotech. 26*, 199–223.

Rubinstein, R. and D. Kroese. 2008. *Simulation and the Monte Carlo Method*. Wiley Series in Probability and Statistics. Wiley, New York.

Saltelli, A., K. Chan, and E. Scott. (Eds.) 2000. *Sensitivity Analysis*. John Wiley & Sons, New York.

Saltelli, A., M. Ratto, T. Andres, F. Campolongo, J. Cariboni, D. Gatelli, M. Saisana, and S. Tarantola. 2008. *Global Sensitivity Analysis—The Primer*. Wiley, New York.

Sobol', I. 1993. Sensitivity estimates for nonlinear mathematical models. *Math. Modeling Comput. Exp. 1*, 407–414.

Sobol', I. and S. Kucherenko. 2005. Global sensitivity indices for nonlinear mathematical models. Review. *Wilmott Mag. 1*, 56–61.

Soize, C. and R. Ghanem. 2004. Physical systems with random uncertainties: Chaos representations with arbitrary probability measure. *SIAM J. Sci. Comput. 26*(2), 395–410.

Sudret, B. 2006. Global sensitivity analysis using polynomial chaos expansions. In P. Spanos and G. Deodatis (eds.), *Proceedings of the 5th International Conference on Computer Stochastic Mechanics (CSM5)*, Rhodes, Greece.

Sudret, B. 2007. *Uncertainty Propagation and Sensitivity Analysis in Mechanical Models—Contributions to Structural Reliability and Stochastic Spectral Methods*. Université Blaise Pascal, Clermont-Ferrand, France. Habilitation à diriger des recherches, 173p.

Sudret, B. 2008. Global sensitivity analysis using polynomial chaos expansions. *Reliab. Eng. Sys. Safety* 93, 964–979.

Sudret, B. 2012. Meta-models for structural reliability and uncertainty quantification. In K. Phoon, M. Beer, S. Quek, and S. Pang (eds.), *Proceedings of the 5th Asian Pacific Symposium of Structural Reliability (APSSRA'2012)*, Singapore, pp. 53–76. Keynote lecture.

Sudret, B. and M. Berveiller. 2007. Stochastic finite element methods in geotechnical engineering. In K. Phoon (ed.), *Reliability-Based Design in Geotechnical Engineering: Computations and Applications*, Chapter 7. Taylor & Francis, London.

Sudret, B., M. Berveiller, and M. Lemaire. 2004. A stochastic finite element method in linear mechanics. *Comptes Rendus Mécanique 332*, 531–537.

Sudret, B., M. Berveiller, and M. Lemaire. 2006. A stochastic finite element procedure for moment and reliability analysis. *Eur. J. Comput. Mech. 15*(7–8), 825–866.

Sudret, B. and A. Der Kiureghian. 2000. Stochastic finite elements and reliability: A state-of-the-art report. Technical Report UCB/SEMM-2000/08, University of California, Berkeley. 173 p.

Vanmarcke, E. 1977. Probabilistic modeling of soil profiles. *J. Geo. Eng. Div. 103*(GT11), 1227–1246.

Vanmarcke, E. 1983. *Random Fields: Analysis and Synthesis*. The MIT Press, Cambridge, Massachussets.

Wand, M. and M. Jones. 1995. *Kernel Smoothing*. Chapman and Hall, Boca Raton.

Xiu, D. 2009. Fast numerical methods for stochastic computations: A review. *Comm. Comput. Phys. 5*(2–4), 242–272.

Xiu, D. and G. Karniadakis. 2002. The Wiener–Askey polynomial chaos for stochastic differential equations. *SIAM J. Sci. Comput. 24* (2), 619–644.

# Chapter 7

# Practical reliability analysis and design by Monte Carlo Simulation in spreadsheet

*Yu Wang and Zijun Cao*

## 7.1 INTRODUCTION

Uncertainties are unavoidable in geotechnical engineering, and they arise from loads, geotechnical properties, calculation models, and so on (e.g., Baecher and Christian, 2003; Ang and Tang, 2007). To deal rationally with these uncertainties in geotechnical analysis and design, several reliability (probability)-based analysis and design approaches have been developed for geotechnical structures (e.g., Tang et al., 1976; Christian et al., 1994; Phoon et al., 1995; Low and Tang, 1997; El-Ramly et al., 2005; Wang, 2011; Wang et al., 2011a,b). Although these efforts significantly facilitate the understanding and application of geotechnical reliability-based approaches, practicing engineers are reluctant to adopt them in geotechnical practice, at least, due to two reasons: (1) the training of geotechnical practitioners in probability and statistics is often limited and, hence, they feel less comfortable dealing with probabilistic modeling than working with deterministic modeling (El-Ramly et al., 2002); and (2) the reliability algorithms are often mathematically and computationally sophisticated and become a major hurdle for geotechnical practitioners when using geotechnical reliability-based approaches. It is, therefore, worthwhile for geotechnical practitioners to have a practical and conceptually simple framework that is directly extended from conventional deterministic modeling and removes the hurdle of reliability algorithms.

This chapter presents a Monte Carlo Simulation (MCS)-based practical framework for reliability analysis and design of geotechnical structures in a commonly available spreadsheet platform, such as Microsoft Excel (Microsoft Corporation, 2012). The MCS-based practical framework deliberately decouples the conventional deterministic modeling from the probabilistic modeling and effectively removes the computational hurdle of reliability algorithms. MCS is a numerical process of repeatedly calculating a mathematical or empirical operator, in which the variables within the operator are random or contain uncertainty with prescribed probability distributions (e.g., Ang and Tang, 2007). The repeated calculations lead to a large number of sets of operator outputs that can be used in statistical analysis for directly estimating the failure probability $P(F)$ and probabilistic properties (e.g., mean, standard deviation, probability density function (PDF), cumulative distribution function (CDF)) of the operator outputs. MCS is conceptually simple and it can be treated by geotechnical practitioners as repetitive computer executions of the conventional deterministic modeling in a systematic manner. When compared with analytical reliability methods (e.g., first- or second-order reliability method, FORM, or SORM), MCS has wide applicability to complex engineering problems and systems that defy analytical solutions of the probabilities associated with their responses to random inputs (Baecher and Christian, 2003; Fenton and Griffiths, 2008). It has been widely used in probabilistic analysis of geotechnical engineering problems, such as slope stability analysis (e.g., El-Ramly et al.,

2002, 2005; Wang, 2012), retaining structures (e.g., Chalermyanont and Benson, 2004, 2005; Fenton et al., 2005; Goh et al., 2009; Wang, 2013), and foundations (e.g., Fenton and Griffiths, 2002; Phoon et al., 2006; Wang and Kulhawy, 2008; Wang, 2011; Wang et al., 2011a).

Although MCS provides a simple and robust way to assess failure probability or obtain other reliability analysis results, it is well recognized that it suffers from two obvious and significant drawbacks: (1) lack of resolution and efficiency, particularly at small probability levels, and (2) no insight into the relative contributions of various uncertainties to the reliability analysis results. This chapter addresses the first drawback of MCS by introducing an advanced MCS method called "Subset Simulation" (Au and Beck, 2001, 2003; Au and Wang, 2014). The Subset Simulation is integrated with MCS-based reliability analysis and design approaches to improve the efficiency and resolution of estimating failure probability at small probability levels. This chapter also deals with the second drawback of MCS using probabilistic failure analysis. The probabilistic failure analysis approach makes use of the failure samples generated in MCS and analyzes these failure samples to assess the effects of various uncertainties on failure probability. It may be further integrated with Subset Simulation to improve the efficiency of generating failure samples.

This chapter starts with a brief review of Subset Simulation, followed by integration of Subset Simulation with an MCS-based reliability approach, called expanded reliability-based design (expanded RBD) approach, and probabilistic failure analysis approach. Then, these approaches are implemented in a Microsoft Excel spreadsheet environment. A drilled shaft design example and a slope stability analysis example are used to illustrate such implementation.

## 7.2 SUBSET SIMULATION

### 7.2.1 Algorithm

Subset Simulation is an advanced MCS method that makes use of conditional probability and Markov chain Monte Carlo (MCMC) method to efficiently compute a small tail probability (Au and Beck, 2001, 2003; Au and Wang, 2014). It expresses a rare event $E$ with a small probability as a sequence of intermediate events $\{E_1, E_2, ..., E_m\}$ with larger conditional probabilities and employs specially designed Markov chains to generate conditional samples of these intermediate events until the target sample domain is achieved. Let $Y$ be the output parameter that is of interest and increases monotonically and defines the rare event $E$ as $E = Y > y$, in which $y$ is a given threshold value for determining whether $E$ occurs. The choice of $Y$ is pivotal to the efficient generation of conditional samples of interest in Subset Simulation. As $Y$ increases, Subset Simulation gradually drives the sampling space to the target sample domain (e.g., failure domain in reliability analysis) of interest. Hence, $Y$ is referred to as "driving variable" in this chapter. Since the conditional samples of interest are different for different problems (e.g., expanded RBD and probabilistic failure analysis in this chapter), different driving variables shall be adopted for different applications.

Let $y = y_m > y_{m-1} > \cdots > y_2 > y_1$ is a decreasing sequence of intermediate threshold values. Then, the intermediate events $\{E_i, i = 1, 2, ..., m\}$ are defined as $E_i = \{Y > y_i, i = 1, 2, ..., m\}$. They are a sequence of nested events, that is, $E_1 \supset E_2 \supset \cdots \supset E_{m-1} \supset E_m$. Hence, the probability of event $E$, that is, $P(E) = P(Y > y)$, can be written as

$$P(E) = P(E_m) = P(E_m E_{m-1} ... E_1) \tag{7.1}$$

Using the product rule of probability (e.g., Ang and Tang, 2007), Equation 7.1 is rewritten as

$$P(E) = P(E_m) = P(E_1)\prod_{i=2}^{m} P(E_i | E_{i-1} E_{i-2} \dots E_1) = P(E_1)\prod_{i=2}^{m} P(E_i | E_{i-1}) \tag{7.2}$$

where $P(E_1) = P(Y > y_1)$ and $P(E_i | E_{i-1}) = \{P(Y > y_i | Y > y_{i-1}), i = 2, \dots, m\}$ is equal to $P(E_i | E_{i-1} E_{i-2} \dots E_1)$ in Equation 7.2 because $E_{i-1}, E_{i-2}, \dots, E_1$ are a sequence of nested events, that is, $E_1 \supset E_2 \supset \dots \supset E_{i-1}$. In implementations, $y_1, y_2, \dots, y_m$ are generated adaptively using information from simulated samples so that the sample estimate of $P(E_1)$ and $\{P(E_i | E_{i-1}), i = 2, \dots, m\}$ always corresponds to a common specified value of conditional probability $p_0$ (Au and Beck, 2001, 2003; Au et al., 2010). The efficient generation of conditional samples is pivotal to the success of Subset Simulation, and it is made possible through the machinery of MCMC. In MCMC, Metropolis algorithm (Metropolis et al., 1953) is used here, and successive samples are generated from a specially designed Markov chain whose limiting stationary distribution tends to the target PDF as the length of Markov chain increases.

### 7.2.2 Simulation procedures

Subset Simulation starts with Direct MCS, in which N direct MCS samples are generated. The Y values of the N samples are calculated and ranked in an ascending order. The $(1 - p_0)$ Nth value in the ascending list of Y values is chosen as $y_1$, and hence, the sample estimate for $P(E_1) = P(Y > y_1)$ is $p_0$. In other words, there are $p_0 N$ samples with $E_1 = Y > y_1$ among the N samples generated from Direct MCS. Then, the $p_0 N$ samples with $E_1 = Y > y_1$ are used as "seeds" for the application of MCMC to simulate N additional conditional samples given $E_1 = Y > y_1$. The $p_0 N$ seed samples are then discarded so that there are a total of N samples with $E_1 = Y > y_1$. The Y values of the N samples with $E_1 = Y > y_1$ are ranked again in an ascending order, and the $(1 - p_0)$Nth value in the ascending list of Y values is chosen as $y_2$, which defines the $E_2 = Y > y_2$. Note that the sample estimate for $P(E_2 | E_1) = P(Y > y_2 | Y > y_1)$ is also equal to $p_0$. Similarly, there are $p_0 N$ samples with $E_2 = Y > y_2$. These samples provide "seeds" in MCMC to simulate additional N conditional samples with $E_2 = Y > y_2$. Then, the $p_0 N$ seed samples are discarded, so that there are N conditional samples with $E_2 = Y > y_2$. The procedure is repeated m times until the probability space of interest (i.e., the sample domain with $Y > y_m$) is achieved. Note that the Subset Simulation procedures contain $m + 1$ steps, including one Direct MCS to generate unconditional samples and m steps of MCMC to simulate conditional samples. The $m + 1$ steps of simulations are referred to as "$m + 1$ levels" in Subset Simulation (Au and Beck, 2001, 2003; Au et al., 2010). Finally, $N + m(1 - p_0)N$ samples are obtained from the $m + 1$ levels of simulations.

A polynomial function example is used here to demonstrate the Subset Simulation procedures described above. Consider a reliability analysis with a performance function $G = 3X_1 + 9X_2^2$, where $X_1 = $ a Gaussian random variable with a mean of 1 and standard deviation of 0.05 and $X_2 = $ a Gaussian random variable with a mean of 1 and standard deviation of 0.1. Failure occurs when $G \geq 15$. A Subset Simulation run with $N = 50$, $m = 1$, $p_0 = 0.1$, and $Y = G$ is performed to evaluate the failure probability in the example. Detailed steps of Subset Simulation are described below:

1. Generate 50 MCS samples of $(X_1, X_2)$, as shown in Figure 7.1a by circles;
2. Calculate the Y (i.e., $Y = G = 3X_1 + 9X_2^2$) values of the 50 MCS samples;

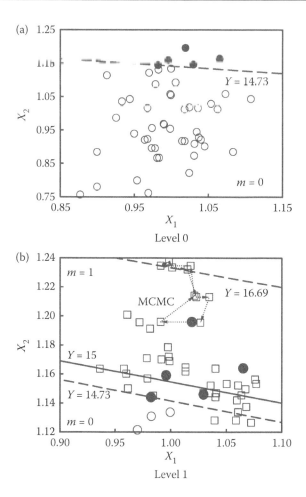

*Figure 7.1* Illustration of Subset Simulation procedure.

3. Sort 50 $Y$ values in an ascending order and determine the 45th largest value (i.e., 14.73, as shown in Figure 7.1a by the dashed line) among the 50 $Y$ values, which is taken as $y_1$, that is, $y_1 = 14.73$;
4. Determine five seed samples of $(X_1, X_2)$ with $Y$ values greater than $y_1$ (i.e., 14.73), which are highlighted in Figure 7.1a by solid circles;
5. Start with each seed sample to generate a Markov chain consisting of 10 conditional samples for $Y > y_1$ using MCMC. The seed samples are then discarded. For example, Figure 7.1b shows the transition between the current sample to the next sample on a Markov chain by arrows. A Markov chain is generated for each seed sample. As a result, 50 conditional samples with $Y > y_1$ are obtained, as shown in Figure 7.1b by squares;
6. Sort the 50 $Y$ values for the 50 conditional samples with $Y > y_1$ in an ascending order and determine the 45th largest value among the 50 $Y$ values, that is, 16.69 in this example (see Figure 7.1b), which is greater than 15. The target failure domain is hence achieved, and the simulation is stopped.

Since five seed samples used to generate conditional samples in the next simulation level (i.e., $m = 1$) are discarded, 45 MCS samples are obtained for $m = 0$. In addition, 50 conditional samples are generated for $m = 1$. Therefore, a total of 95 samples are obtained from these two levels of simulation. Note that, for illustration purposes, a relatively small number (i.e., $N = 50$) of samples are used in each simulation level. Such a small sample number in each simulation level may not be sufficient to ensure the accuracy of results (i.e., $P(F)$) estimated from the simulation.

Subset Simulation has been applied to various geotechnical engineering problems, such as slope stability and foundations (Santoso et al., 2009; Wang et al., 2011b; Ahmed and Soubra, 2011; Wang and Cao, 2013). The next section integrates Subset Simulation with the expanded RBD approach (Wang, 2011, 2013; Wang et al., 2011a) to improve the efficiency and resolution of MCS at small probability levels.

## 7.3 EXPANDED RBD WITH SUBSET SIMULATION

### 7.3.1 Expanded RBD approach

The expanded RBD approach formulates the design process of geotechnical structures as an augmented reliability problem (Au, 2005) or expanded reliability problem (Wang et al., 2011a). The expanded RBD problem refers to a reliability analysis of a system, in which a set of system design parameters are artificially considered as uncertain with probability distributions specified by the user for design exploration purposes (Wang, 2011, 2013; Wang and Cao, 2013; Wang et al., 2011a). For example, consider designing a drilled shaft with a diameter $B$ and depth $D$. The design process is one of finding a set of $B$ and $D$ values that satisfy both the ultimate limit state (ULS) and serviceability limit state (SLS) requirements and achieve the design target failure probability $p_T$ or target reliability index $\beta_T$. In the context of expanded RBD, the design parameters $B$ and $D$ are considered to be independent discrete random variables with uniformly distributed probability mass function $P(B,D)$, which is given by

$$P(B,D) = \frac{1}{n_B n_D} \tag{7.3}$$

in which $n_B$ and $n_D$ are the number of possible discrete values for $B$ and $D$, respectively. Note that $P(B,D)$ does not reflect the uncertainty in $B$ and $D$, because $B$ and $D$ represent design decisions and no uncertainty is to be associated with them. Instead, it is used to yield desired design information. The drilled shaft design process is then considered as a process of finding failure probabilities corresponding to designs with various combinations of $B$ and $D$ (i.e., conditional probability $P(F|B,D)$) and comparing them with $p_T$. Feasible designs are those with $P(F|B,D) \leq p_T$. Failure herein refers to events in which the load exceeds capacity. In other words, failure occurs when the factor of safety (FS) (i.e., $FS_{uls}$ or $FS_{sls}$ for ULS or SLS requirements, respectively) is less than 1. Using Bayes' Theorem (e.g., Ang and Tang, 2007), the conditional probability $P(F|B,D)$ is given by

$$P(F|B,D) = \frac{P(B,D|F)P(F)}{P(B,D)} \tag{7.4}$$

in which $P(B,D|F)$ is the conditional joint probability of $B$ and $D$ given failure and $P(F)$ is the probability of failure. The expanded RBD approach employs a single run of MCS to estimate $P(B,D|F)$ and $P(F)$, which are further used in Equation 7.4 to obtain $P(F|B,D)$ (Wang, 2011, 2013; Wang et al., 2011a). Since MCS is simply a repetitive computer execution of traditional deterministic design calculation with different combinations of input parameters, it is equivalent to a sensitivity study. From this perspective, the MCS-based expanded RBD approach can be considered as a process of performing systematic design sensitivity studies (i.e., varying design parameters $B$ and $D$) and determining the final design by comparing the sensitivity study results with a predefined target failure probability $p_T$.

## 7.3.2 Desired sample number in direct MCS

The resolution and accuracy of $P(B,D|F)$ and $P(F)$ from MCS are pivotal for the $P(F|B,D)$ obtained (see Equation 7.4) and depend on the number of failure samples generated in MCS. As the number of failure samples increases, the resolution and accuracy improve. Since the target failure probability $p_T$ is predefined and generally small (e.g., $p_T = 0.001$) in RBD, a large number of failure samples necessitate an increase in the total sample number in MCS. For example, the minimum MCS sample number ($n_{min}$) in the drilled shaft design example is estimated as (Ang and Tang, 2007; Wang et al., 2011a)

$$n_{min} = \frac{((1/p_T) - 1)n_B n_D}{(COV_T)^2} \tag{7.5}$$

in which $COV_T$ is the target coefficient of variation for the failure probability estimated from Direct MCS. Equation 7.5 shows that the value of $n_{min}$ increases dramatically as the probability level of interest (i.e., $p_T$) decreases, the desired level of accuracy (i.e., $COV_T$) improves, or the number ($n_B n_D$) of $B$ and $D$ combinations increases. For $p_T = 0.001$, $COV_T = 30\%$, and $n_B n_D = 1000$, Equation 7.5 leads to a relatively large $n_{min}$ value of 11,100,000. If $p_T$ and $COV_T$ are further decreased to 0.0001 and 10%, respectively, the $n_{min}$ value increases rapidly to 999,900,000. Such a large number of MCS samples require extensive computational efforts and might lead to computational difficulties (e.g., long computational time and insufficient computer memory), particularly for design situations with a small target failure probability, a high-accuracy requirement, and a large number of design parameter combinations. In this chapter, Subset Simulation is employed to calculate $P(B,D|F)$, $P(F)$, and $P(F|B,D)$ for the expanded RBD and to improve efficiency and resolution at small failure probability levels, as discussed in the next section.

## 7.3.3 Integration of expanded RBD approach with Subset Simulation

Consider designing a drilled shaft by the expanded RBD approach and performing a Subset Simulation with $m + 1$ level of simulations to estimate $P(B,D|F)$, $P(F)$, and $P(F|B,D)$ in Equation 7.4. Since the design parameters $B$ and $D$ are artificially treated as uncertain parameters in the expanded RBD approach, their random samples are generated during Subset Simulation. The sample space $\Omega$ of uncertain parameters (including $B$ and $D$ in the expanded RBD of drilled shafts) is divided into $m + 1$ individual subsets $\{\Omega_i, i = 0, 1, 2, \ldots, m\}$ by the intermediate threshold values $\{y_i, i = 1, 2, \ldots, m\}$ of the driving variable $Y$ (Au and Wang, 2014). As mentioned before, the driving variable $Y$ is a key factor that affects the generation of conditional samples of interest in Subset Simulation. For the drilled shaft design problem herein, $Y$ is defined as $BD/FS_{min}$, in which $FS_{min}$ is the minimum FS among the $FS_{uls}$ and $FS_{sls}$ for the respective ULS and SLS requirements. The effects of the driving variable on the expanded RBD will be further discussed in Section 7.6.6.

In Subset Simulation, the intermediate threshold values $\{y_i, i = 1, 2, \ldots, m\}$ of $Y = BD/FS_{min}$ are adaptively determined to generate $m + 1$ individual subsets $\{\Omega_i, i = 0, 1, 2, \ldots, m\}$ of $B$ and $D$, and samples in different subsets are generated level by level and correspond to different conditional probabilities. According to the Theorem of Total Probability (e.g., Ang and Tang, 2007), the failure probability $P(F)$ is, therefore, expressed as

$$P(F) = \sum_{i=0}^{m} P(F|\Omega_i)P(\Omega_i) \tag{7.6}$$

where $P(F|\Omega_i)$ is the conditional failure probability given sampling in $\Omega_i$ and $P(\Omega_i)$ is the probability of the event $\Omega_i$. $P(F|\Omega_i)$ is estimated as the ratio of the failure sample number in $\Omega_i$ over the total sample number in $\Omega_i$. $P(\Omega_i)$ is calculated as

$$\begin{aligned}
P(\Omega_0) &= 1 - p_0 \\
P(\Omega_i) &= p_0^i - p_0^{i+1}, \quad i = 1, \dots m - 1 \\
P(\Omega_m) &= p_0^m
\end{aligned} \tag{7.7}$$

Note that $\Omega_i$, $i = 0, 1, 2, \dots, m$, are mutually exclusive and collectively exhaustive events, that is, $P(\Omega_k \cap \Omega_j) = 0$ for $k \neq j$ and $\sum_{i=0}^{m} P(\Omega_i) = 1$. When $P(F|\Omega_i)$, $P(\Omega_i)$, and $P(F)$ are obtained, the conditional probability $P(\Omega_i|F)$ is calculated using the Bayes' Theorem:

$$P(\Omega_i|F) = \frac{P(F|\Omega_i)P(\Omega_i)}{P(F)} \tag{7.8}$$

Then, the conditional probability $P(B,D|F)$ of a specific combination of $B$ and $D$ is given using the Theorem of Total Probability as

$$P(B,D|F) = \sum_{i=0}^{m} P(B,D|F \cap \Omega_i)P(\Omega_i|F) \tag{7.9}$$

where $P(\Omega_i|F)$ is estimated from Equation 7.8; and $P(B,D|F \cap \Omega_i)$ is the conditional failure probability of a combination of $B$ and $D$ given sampling in $\Omega_i$, and it is expressed as the ratio of the number ($n_{\Omega_i,BD}$) of failure samples in $\Omega_i$ with a combination of $B$ and $D$ over the total failure sample number ($n_{\Omega_i}$) in $\Omega_i$, that is, $P(B,D|F \cap \Omega_i) = n_{\Omega_i,BD} / n_{\Omega_i}$.

Using Equations 7.6 through 7.9, $P(F)$ and $P(B,D|F)$ are calculated using simulation samples from Subset Simulation. Subsequently, $P(F|B,D)$ is obtained in accordance with Equation 7.4. Since the failure is defined as $FS_{uls} < 1$ or $FS_{sls} < 1$ for ULS and SLS requirements, respectively, two sets of conditional probabilities $P(F|B,D)$ (i.e., the respective failure probabilities $p_f^{ULS}$ and $p_f^{SLS}$ of ULS and SLS failures for given $B$ and $D$ combinations) are calculated for the drilled shaft design. Finally, the feasible design values of $B$ and $D$ are determined by comparing the $P(F|B,D)$ with the target failure probability $p_T$. This section only provides the conceptual framework. More details of the approach are illustrated through a drilled shaft design example in Section 7.6.

## 7.4 PROBABILISTIC FAILURE ANALYSIS USING SUBSET SIMULATION

MCS can be treated as a "black box" that takes samples of the uncertain parameters as input and returns failure probability or other reliability analysis results (e.g., complementary cumulative distribution function (CCDF) and PDF) as output. It does not provide information on uncertainty propagation through the analysis. How the uncertainty propagates from the input parameters, through the deterministic analysis models, to the output of reliability analysis and affects reliability analysis results (e.g., failure probability) is unclear. To address this drawback of MCS, this section presents a probabilistic failure analysis approach, in which the failure samples generated in MCS are collected and reanalyzed to assess the effects of various uncertainties on failure probability. The probabilistic failure analysis approach contains two major components: hypothesis tests for prioritizing effects of various uncertain parameters and Bayesian analysis for further quantifying their effects. The hypothesis tests and Bayesian analysis are described in the following two sections, respectively.

## 7.4.1 Hypothesis testing

The effects of various uncertain parameters $\mathbf{X} = [X_1, X_2, \ldots, X_n]$ on failure probability $P(F)$ are prioritized by comparing, statistically, failure samples with their respective nominal (unconditional) samples. When the distribution of failure samples of an input parameter significantly deviates from that of unconditional samples, the uncertainty of the input parameter has a significant effect on $P(F)$, as further discussed in the next section. The deviation between the distribution of failure samples and that of unconditional samples can be quantified by the difference between the mean $\mu_f$ of failure samples of the parameter and the mean $\mu$ of its unconditional samples. When $\mu_f$ deviates significantly from $\mu$, the uncertainty of the input parameter has a significant effect on $P(F)$. The statistical difference between $\mu_f$ and $\mu$ is evaluated by hypothesis tests. A null hypothesis $H_0$ and alternative hypothesis $H_A$ are defined as (e.g., Walpole et al., 1998)

$$
\begin{aligned}
H_0 &: \mu = \mu_f \\
H_A &: \mu \neq \mu_f
\end{aligned}
\tag{7.10}
$$

A hypothesis test statistic $Z_H$ of the parameter is then formulated as

$$
Z_H = \frac{\mu - \mu_f}{\sigma / \sqrt{n_f}}
\tag{7.11}
$$

where $\sigma$ is the unconditional standard deviation of the uncertain parameter concerned and $n_f$ is the number of failure samples. On the basis of the Central Limit Theorem, $Z_H$ follows the standard normal distribution when $n_f$ is large (e.g., $n_f \geq 30$) (e.g., Walpole et al., 1998). When the failure sample mean $\mu_f$ deviates statistically from the unconditional mean $\mu$ of the parameter, the absolute value of $Z_H$ is relatively large. As the absolute value of $Z_H$ increases, the statistical difference between $\mu_f$ and $\mu$ becomes growingly significant. The effect of the uncertain parameter on failure probability also becomes growingly significant. The absolute value of $Z_H$ can therefore be used as an index to measure the effects of the uncertain parameters on failure probability and to prioritize their relative effects on failure probability. By comparing the absolute values of $Z_H$ for various uncertain parameters, the important uncertain parameters that have significant effects on failure probability are identified.

Consider the polynomial function example described in Section 7.2.2 and a direct Monte Carlo run with 100,000 samples to evaluate the $P(F)$ in the example. 6213 samples are identified as failure samples (i.e., $n_f = 6213$), leading to $P(F) = 6213/100{,}000 = 0.06213$. Then, these 6213 failure samples are collected to calculate their respective $\mu_f$ values for $X_1$ and $X_2$, as summarized in Table 7.1. Equation 7.11 is subsequently used to calculate the absolute values of $Z_H$ for $X_1$ and $X_2$, respectively. As shown in the fifth row of Table 7.1, the absolute

Table 7.1  Summary of hypothesis test results for the polynomial function example

| Uncertain parameter | $X_1$ | $X_2$ |
|---|---|---|
| Unconditional mean $\mu$ | 1.00 | 1.00 |
| Standard deviation $\sigma$ | 0.05 | 0.10 |
| Failure sample mean $\mu_f$ | 1.01 | 1.20 |
| Absolute value of $Z_H$ | 15.76 | 157.65 |
| Ranking | 2 | 1 |

values of $Z_H$ are 15.76 and 157.65 for $X_1$ and $X_2$, respectively. Based on the absolute values of $Z_H$, the uncertain parameter $X_2$ has more significant effects on the failure probability than $X_1$. The effect of the important uncertain parameters (e.g., $X_2$ in the polynomial function example) on failure probability can be further quantified using a Bayesian analysis described in the next section.

## 7.4.2  Bayesian analysis

For the important uncertain parameter identified from hypothesis tests, Bayesian analysis can be performed to explore how the failure probability varies as the important parameter changes. Let $X_k$ be an important uncertain parameter identified from the hypothesis test. In the context of the Bayes' theorem (e.g., Ang and Tang, 2007):

$$P(F|x_k) = \frac{p(x_k|F)P(F)}{p(x_k)} \tag{7.12}$$

where $P(F|x_k)$ is the conditional failure probability at $X_k = x_k$, $p(x_k|F)$ is the conditional PDF of $x_k$ given that failure occurs, $P(F)$ is the failure probability, and $p(x_k)$ is the unconditional PDF of $x_k$ that is given before simulation and can be determined analytically. Both $P(F)$ and $p(x_k|F)$ are estimated from failure samples of MCS. Consider an MCS run with a total sample number of $n_t$. $P(F)$ is estimated using the following equation:

$$P(F) = \frac{n_f}{n_t} \tag{7.13}$$

in which $n_f$ is the number of failure samples in the simulation. $p(x_k|F)$ is estimated from an $x_k$ histogram in which $x_k$ is divided into a number of bins (e.g., $n_b$ bins) and $p(x_k|F)$ for $x_k$ within a bin $j$ (i.e., $P(x_k \in bin_j|F)$, for $j = 1, 2, ..., n_b$) is estimated using

$$p(x_k|F) = P(x_k \in bin_j|F) = \frac{n_j}{n_f} \tag{7.14}$$

in which $n_j$ is the number of simulation samples where failure occurs and the $x_k$ value falls into bin $j$. In this way, the "interval" probability $P(x_k \in bin_j|F)$ (instead of $p(x_k|F)$) for $x_k \in bin_j$ is obtained, yielding an interval analog of Equation 7.12:

$$P(F | x_k \in bin_j) = \frac{P(x_k \in bin_j | F)P(F)}{P(x_k \in bin_j)}, \quad \text{for } j = 1, 2, ..., n_b \tag{7.15}$$

where

$$P(x_k \in bin_j) = \int\limits_{bin_j} p(x_k) dx_k \qquad (7.16)$$

Since $p(x_k)$ is known before the simulation, $P(x_k \in bin_j)$ can be determined analytically without any information from MCS.

Combining Equations 7.13 and 7.14 leads to

$$P(x_k \in bin_j | F) P(F) = \frac{n_j}{n_t}, \quad \text{for } j = 1, 2, \ldots, n_b \qquad (7.17)$$

Since both $n_f$ and $n_j$, for $j = 1, 2, \ldots, n_b$, are obtained directly from a single simulation run by simply counting the numbers of failure samples, $P(F | x_k \in bin_j)$, for $j = 1, 2, \ldots, n_b$, from Equations 7.15 and 7.17 can be estimated directly from a single simulation run.

Note that counting the sample numbers for Equations 7.15 and 7.17 can be considered to contain two steps: first, grouping the simulation samples in accordance with the bins of $x_k$ and second, counting the sample number for each bin of $x_k$. From this perspective, the Bayesian analysis herein is equivalent to grouping all simulation samples into $n_b$ subsets (i.e., $n_b$ bins) in accordance with the $x_k$ value and to determining the $P(F)$ in each subset (i.e., $P(F | x_k \in bin_j)$). When the bin interval is small, the $x_k$ value is virtually deterministic and constant. Note that $P(F | x_k \in bin_j)$, for $j = 1, 2, \ldots, n_b$, is a variation of the failure probability $P(F)$ as a function of the value of the $x_k$ bin. It can be considered as results of the $n_b$ repeated simulation runs in which the $x_k$ value adopted in each simulation run is deterministic but different (i.e., the value of $bin_j$, for $j = 1, 2, \ldots, n_b$), and the other uncertain parameters (i.e., $X_1, X_2, \ldots, X_{k-1}, X_{k+1}, \ldots, X_n$) remain random. In other words, the Bayesian analysis approach, which makes use of the failure samples generated in a single simulation run for assessment of failure probability, provides results that are equivalent to those from a sensitivity study, which frequently includes many repeated simulation runs with different given values of $x_k$ in each run. Additional computational times and efforts for repeated simulation runs in the sensitivity study can be avoided when using the Bayesian analysis described herein. In addition, it is also worthwhile to point out that the Bayesian analysis results can be used to further provide the sensitivity on failure probability or PDF of the operator output to $X_k$ through sample reassembling when $X_k$ is considered as random, but with statistical parameters or the distribution type which are different from those adopted in the nominal case. Interested readers are referred to Wang (2012) for details on sampling reassembling in MCS.

Equation 7.12 implies that the comparison between the conditional PDF $p(x_k | F)$ and its unconditional one $p(x_k)$ provides an indication of the effect of the uncertain parameter $X_k$ on failure probability. Generally speaking, $P(F | x_k)$ changes as the values of the uncertain parameter $x_k$ change. However, when $p(x_k | F)$ is similar to $p(x_k)$, $P(F | x_k)$ remains more or less constant regardless of the values of $x_k$. This implies that the effect of $X_k$ on the failure probability is minimal. Such implication can be used to validate the results obtained from hypothesis tests.

Consider again the polynomial function example and the direct Monte Carlo run with 100,000 samples in Sections 7.2.2 and 7.4.1. The hypothesis tests in the previous section have shown that the uncertain parameter $X_2$ has the most significant effect on the failure probability. Bayesian analysis is therefore performed to further quantify the variation of $P(F)$ as the $X_2$ value changes. The 6213 failure samples are collected to analyze their

*Table 7.2* Summary of Bayesian analysis for the polynomial function example

| $X_2$ value | 1.14 | 1.16 | 1.18 |
|---|---|---|---|
| $X_2$ $bin_j$ | (1.13, 1.15] | (1.15, 1.17] | (1.17, 1.19] |
| Failure sample number $n_j$ within the jth bin (i.e., $x_2 \in bin_j$) | 166 | 1574 | 1562 |
| $P(x_2 \in bin_j)$ using Equation 7.16 | 0.02999 | 0.02224 | 0.01585 |
| $P(x_2 \in bin_j|F)P(F)$ using Equation 7.17 | 166/100,000 | 1574/100,000 | 1562/100,000 |
| $P(F|x_2 \in bin_j)$ using Equation 7.15 | 0.055 | 0.708 | 0.985 |

*Note:*   Total sample number $n_t = 100,000$.

corresponding $X_2$ values. The minimum and maximum $X_2$ values are found to be 1.13 and 1.46, respectively. Consider estimating the $P(F)$ value when the $X_2$ value is taken as a deterministic value (e.g., 1.14, 1.16, or 1.18 in Table 7.2) and $X_1$ remains random. A relatively small bin size (e.g., 0.02) is first selected, such as (1.13, 1.15] for 1.14 in the second column of Table 7.2. The failure samples that fall within the corresponding bins are then identified from the 6213 failure samples. The number of failure samples within each bin is subsequently counted as 166, 1574, and 1562 for the $X_2$ value of 1.14, 1.16, or 1.18, respectively (see the third row in Table 7.2). Equations 7.16 and 7.17 are used to calculate $P(x_2 \in bin_j)$ and $P(x_2 \in bin_j|F)P(F) = (n_j/n_t)$ in the fourth and fifth rows, respectively. Note that $P(x_2 \in bin_j)$ is calculated analytically without any information from MCS. For instance, $P(x_2 \in (1.13, 1.15]) = CDF(1.15) - CDF(1.13)$, where CDF in this example is a Normal distribution CDF with a mean and standard deviation of 1 and 0.1, respectively. Finally, Equation 7.15 is used to estimate the $P(F)$ value at a given deterministic $X_2$ value, such as $P(F \mid x_2 = 1.14) \approx P(F \mid x_2 \in (1.13, 1.15]) = 0.055$ in the sixth row of Table 7.2.

Figure 7.2 shows a $X_2$ histogram from the 6213 failure samples, together with the nominal (unconditional) probability distribution of $X_2$ (i.e., a normal probability distribution with a mean and standard deviation of 1 and 0.1, respectively). The majority of the 6213 failure samples has $X_2$ values larger than 1.13, leading to a histogram peaking at the upper tail of the nominal distribution. The failure sample distribution of $X_2$ and its nominal probability distribution are quite different, indicating that the effect of $X_2$ on failure probability is significant. This is consistent with the ranking obtained from the hypothesis tests (see the last row in Table 7.1). In contrast, Figure 7.3 shows a $X_1$ histogram from the 6213 failure

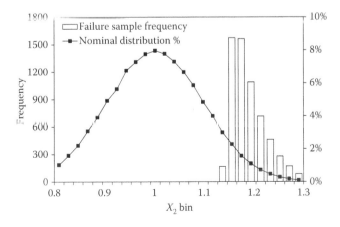

*Figure 7.2* Histogram of the $X_2$ failure samples from direct MCS.

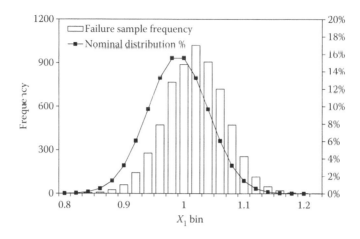

*Figure 7.3* Histogram of the $X_1$ failure samples from direct MCS.

samples, together with the nominal (unconditional) probability distribution of $X_1$ (i.e., a normal probability distribution with a mean and standard deviation of 1 and 0.05, respectively). The $X_1$ histogram from the 6213 failure samples deviates slightly from its nominal (unconditional) probability distribution. This suggests that the effect of $X_1$ on failure probability is rather minimal. Such results agree very well with the results from the hypothesis tests (see the last row in Table 7.1).

The resolution of $P(F)$ and $p(x_k|F)$ is pivotal to obtain $P(F|x_k)$ and it depends on the number of failure samples generated in MCS (see Equations 7.13 and 7.14). As the number of failure samples increases, the resolution improves. Note that the value of $P(F)$ is constant, although unknown before MCS. Therefore, increasing the number $n_f$ of failure samples necessitates an increase in the total number $n_t$ of samples in MCS based on Equation 7.13. One possible way to improve the resolution is, therefore, to increase the total number of samples in MCS at the expense of computational time. Alternatively, advanced MCS methods (e.g., Subset Simulation) can be employed to improve efficiency and resolution at small failure probability levels. The next section integrates Subset Simulation and probabilistic failure analysis for generating failure samples efficiently.

### 7.4.3 Integration of probabilistic failure analysis with Subset Simulation

Consider a Subset Simulation that performs $m + 1$ levels of simulations with a driving variable $Y$. The first level of Subset Simulation is Direct MCS. Samples of the next level are then generated conditional on the samples collected from the previous level. As mentioned above, the conditional samples generated in Subset Simulation rely on $Y$. In probabilistic failure analysis, the conditional samples of interest are failure samples. For most geotechnical engineering problems (e.g., slope stability analysis), the failure samples have relatively small $FS$ (e.g., $FS < 1$). On the other hand, the convention of Subset Simulation is to define $Y$ as a variable that monotonically increases as the simulation level increases. Following the convention of Subset Simulation, $Y$ in probabilistic failure analysis is defined as $1/FS$. As $1/FS$ increases, $FS$ decreases and the simulation gradually approaches the failure domain. The intermediate threshold values $\{y_i, i = 1, 2, ..., m\}$ of $Y$ are adaptively determined to divide the sample space $L$ of an uncertain parameter $X_k$ into $m + 1$ individual sets $\{L_i, i = 0, 1, 2, ..., m\}$. According to the Total Probability Theorem (e.g., Ang and Tang, 2007), the failure probability can be written as

$$P(F) = \sum_{i=0}^{m} P(F|L_i)P(L_i) \tag{7.18}$$

where $L_0 = \{Y \leq y_1\}$, $L_i = \{y_i \leq Y \leq y_{i+1}\}$ for $i = 1, \ldots, m-1$, $L_m = \{Y \geq y_m\}$, $P(F|L_i)$ is the conditional failure probability given sampling in $L_i$, and $P(L_i)$ is the probability of the event $L_i$. $P(F|L_i)$ is estimated as the fraction of the failure samples in $L_i$. The failure samples are collected from samples generated by Subset Simulation and are based on the performance failure criteria (e.g., $FS < 1$). $P(L_i)$ is calculated as

$$\begin{aligned} P(L_0) &= 1 - p_0 \\ P(L_i) &= p_0^i - p_0^{i+1}, \quad i = 1, \ldots, m-1 \\ P(L_m) &= p_0^m \end{aligned} \tag{7.19}$$

Note that $P(L_i \cap L_j) = 0$ for $i \neq j$ and $\sum_{i=0}^{m} P(L_i) = 1$. When $P(F)$, $P(F|L_i)$, and $P(L_i)$ are obtained, the conditional probability $P(L_i|F)$ is calculated using the Bayes' theorem as

$$P(L_i|F) = \frac{P(F \mid L_i)P(L_i)}{P(F)} \tag{7.20}$$

The conditional PDF $p(x_k|F)$ of an uncertain parameter $X_k$ is then given by the Total Probability Theorem as

$$p(x_k|F) = \sum_{i=0}^{m} p(x_k|L_i \cap F)P(L_i|F) \tag{7.21}$$

where $p(x_k|L_i \cap F)$ is the conditional PDF of $x_k$ estimated from failure samples that lie in $L_i$ through the histogram. $x_k$ is divided into a number of bins (e.g., $n_b$ bins), and $p(x_k|L_i \cap F)$ for $x_k$ within a bin $j$ (i.e., $P(x_k \in bin_j \mid L_i \cap F)$, for $j = 1, 2, \ldots, n_b$) is estimated as

$$p(x_k \mid L_i \cap F) \sim P(x_k \in bin_j \mid L_i \cap F) - \frac{n_{ji}}{n_{fi}} \tag{7.22}$$

in which $n_{ji}$ is the number of failure samples at the simulation level 'i' with the $x_k$ value falling into bin $j$ and $n_{fi}$ is the total failure sample number at the simulation level 'i'. Using Equations 7.18, 7.21, and 7.22, $P(F)$ and $p(x_k|F)$ can be calculated from Subset Simulation. $P(F|x_k)$ is then estimated using Equation 7.12 accordingly. This is further illustrated using a slope stability analysis example later in Section 7.7.

## 7.5 SPREADSHEET IMPLEMENTATION OF MCS-BASED RELIABILITY ANALYSIS AND DESIGN

This section implements the MCS-based reliability analysis and design procedures (e.g., expanded RBD and probabilistic failure analysis described above) in a spreadsheet environment, such as Microsoft Excel (Microsoft Corporation, 2012), by a package of worksheets and Visual Basic for Applications (VBA) functions/Add-Ins. The implementation is deliberately divided into three uncoupled modules, namely deterministic modeling, uncertainty modeling, and uncertainty propagation. The reliability analysis (including uncertainty modeling and propagation) is decoupled from the conventional deterministic geotechnical analysis so that the reliability analysis can proceed as an extension of the deterministic analysis. This allows the deterministic geotechnical analysis and the reliability analysis to be performed in a nonintrusive manner through computer simulation and avoids additional conceptual and mathematical complexity when the reliability algorithms are coupled with geotechnical deterministic models (e.g., FORM or SORM). This permits geotechnical practitioners, who are not necessarily probabilistic experts, to instruct the spreadsheet to repeatedly calculate the geotechnical deterministic models in a systematic manner for assessing the effects of uncertainties and making risk-informed decisions conveniently. The three modules of the spreadsheet implementation are further discussed in the following three sections, respectively.

## 7.5.1 Deterministic modeling

Deterministic modeling is the process of calculating system responses (e.g., $FS$, $1/FS$, or $BD/FS_{min}$) of interest for a given nominal set of values of system parameters. The system parameters include, but are not limited to, design parameters (e.g., $B$ and $D$ for drilled shafts), design loads, soil properties, and profile of soil layers. The calculation process of the deterministic model is implemented in a series of worksheets assisted by some VBA functions/Add-In (Au et al., 2010; Wang and Cao, 2013; Wang et al., 2011b). From an input–output perspective, the deterministic analysis worksheets take a given set of values as input, calculate the system responses, and return system responses as an output. No probability concept is involved in the deterministic model worksheet, and it can be developed by practitioners without reliability analysis background.

## 7.5.2 Uncertainty modeling

An uncertainty model worksheet is developed to define the uncertain system parameters that are treated as random variables in the reliability-based analysis and design. Based on the distribution type and statistics defined in the worksheet for each random variable, random samples of the random variables are generated in the worksheet. In Microsoft Excel, the generation of random samples starts with a built-in function "RAND()" for generating uniform random samples, which are then transformed into random samples of the target distribution type (e.g., normal distribution or discrete uniform distribution). Detailed examples of the random sample generation process are further illustrated in the next section. From the input–output perspective, the uncertainty model worksheet takes no input but returns a set of random samples of the uncertain system parameters as its output.

When deterministic model worksheet and uncertainty model worksheet are developed, they are linked together through their input/output cells to perform the analysis and design. The connection is carried out by simply setting the cell references for nominal values of uncertain parameters in the deterministic model worksheet to be the cell references for the random samples in the uncertainty model worksheet in Excel. After this task, the values of uncertain system parameters shown in the deterministic model worksheet are equal to those generated in the uncertainty model worksheet, and the values of the system response (e.g., *FS*, 1/*FS*, or *BD*/*FS*$_{min}$) calculated in the deterministic modeling worksheet are random.

### 7.5.3 Uncertainty propagation

When the deterministic analysis and uncertainty model worksheets are completed and linked together, MCS or Subset Simulation procedure is invoked for uncertainty propagation. An Excel Add-In called UPSS (Uncertainty Propagation using Subset Simulation) has been developed for implementing Direct MCS and Subset Simulation (Au and Wang, 2014; Au et al., 2010). UPSS can be obtained through the following webpage: https://sites.google.com/site/upssvba/. Figure 7.4 shows a variant of UPSS that is tailored to the integration of Subset Simulation with the expanded RBD approach (Wang and Cao, 2013). The Subset Simulation userform is shown in Figure 7.4a. The upper four input fields of the userform (i.e., number of Subset Simulation runs, number of samples per level *N*, conditional probability $p_0$ from one level to the next level, and the highest Subset Simulation level *m*) control the number of samples generated by Subset Simulation. The lower four input fields of the userform record the cell references of the random variables, their PDF values, and the cell references of the system response (e.g., driving variable *Y*), and other variables *V* that will be recorded during the simulation, respectively. After setting up the userform, Subset Simulation can be performed by clicking the "Run" button.

After each simulation run, the Add-In provides the CCDF of the driving variable versus the threshold level, that is, the estimate for $P(Y > y)$ versus *y*, into a new spreadsheet, and a plot is produced. Then, based on the output information from Subset Simulation, failure probability $P(F)$ or its conditional counterparts (e.g., $P(F|B,D)$ for the expanded RBD of drilled shafts and $P(F|x_i)$ for probabilistic failure analysis) are calculated using the procedures and equations described in Sections 7.3 and 7.4.

To facilitate the integration of Subset Simulation with the expanded RBD approach, Equations 7.4 and 7.6 through 7.9 are implemented in Excel spreadsheet as a VBA Add-In for calculating the conditional failure probability (e.g., $P(F|B,D)$) for an expanded RBD. Figure 7.4b shows the RBD userform of the Add-In. The checkbox "RBD" at the top of the userform is used to enable/disenable the input fields of "Failure Modes," "Values of Design Parameters," and "Samples of Design Parameters" for RBD. The Add-In can calculate conditional failure probability for up to five failure modes simultaneously, for example, ULS and SLS failures in the RBD of drilled shafts. Each failure mode is defined by three input fields, including the system response (e.g., values of 1/*FS*$_{uls}$ or 1/*FS*$_{sls}$ of random samples generated during Subset Simulation) of interest, the type of the failure criterion (i.e., >, =, or <), and a critical value (e.g., one for 1/*FS*$_{uls}$ and 1/*FS*$_{sls}$) for judging the occurrence of failure. The two input fields (i.e., "Values of Design Parameters" and "Samples of Design Parameters") at the bottom require cell references of possible values of design parameters (e.g., *B* and *D* in the RBD of drilled shafts) and their random samples generated during Subset Simulation, respectively. After setting up the userform, the conditional failure probability is calculated by clicking the "Run" button. After the calculation, the RBD Add-In generates a worksheet for each failure mode, which contains the conditional failure probability of designs with various combinations of design parameters for the failure mode and a plot of the conditional failure probability versus design parameters.

The next two sections illustrate how the spreadsheet Add-In is used in expanded RBD with Subset Simulation and probabilistic failure analysis with Subset Simulation, respectively. The three modules of the spreadsheet implementation are illustrated in detail through a drilled shaft design example in the next section.

(a)

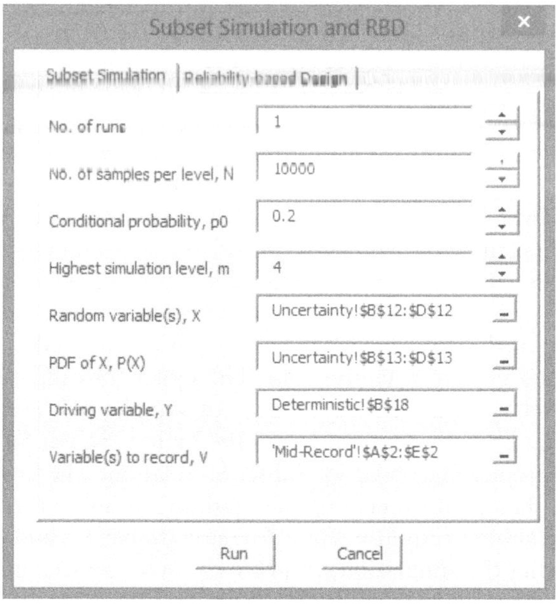

Subset Simulation userform

(b)

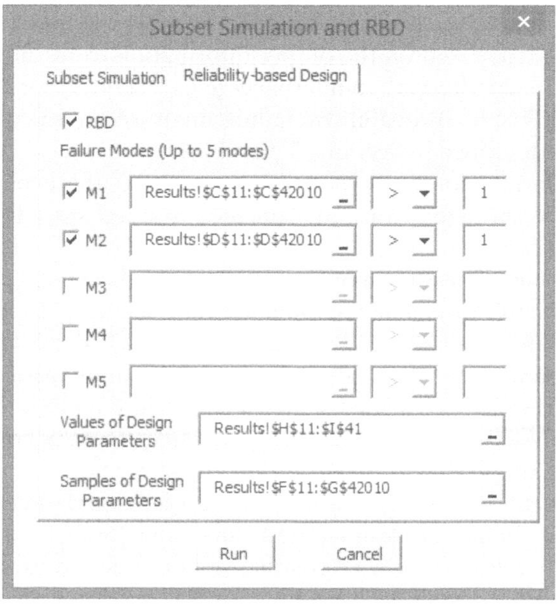

Reliability-based design userform

*Figure 7.4* Subset Simulation and RBD Add-In.

## 7.6 ILLUSTRATIVE EXAMPLE I: DRILLED SHAFT DESIGN

Phoon et al. (1995) and Wang et al. (2011a) used a drilled shaft design example shown in Figure 7.5 to illustrate the multiple resistance factor design (MRFD) approach and the expanded RBD approach with Direct MCS. This design example is redesigned in this section to illustrate the integration of the expanded RBD approach with Subset Simulation

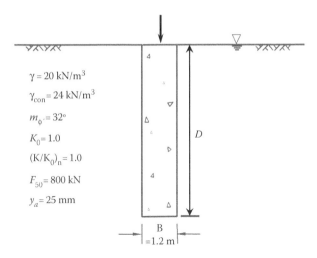

*Figure 7.5* Design example of a drilled shaft. (Adapted from Phoon, K. K., Kulhawy, F. H. and Grigoriu, M. D. 1995. *Reliability-Based Design of Foundations for Transmission Line Structures, Report TR-105000,* Electric Power Research Institute, Palo Alto.)

in a spreadsheet environment. The drilled shaft is installed in loose sand with a total unit weight $\gamma = 20.0$ kN/m³ and mean effective friction angle $m_{\phi'} = 32°$. The shaft diameter $B$, concrete unit weight $\gamma_{con}$, nominal at-rest horizontal soil stress coefficient $K_0$, and nominal operative *in situ* horizontal stress coefficient ratio $(K/K_0)_n$ are 1.2 m, 24.0 kN/m³, 1.0, and 1.0, respectively. The water table is at the ground surface. The shaft is assumed to fail in drained general shear under a design compression load $F_{50} = 800$ kN with an allowable displacement $y_a = 25$ mm. For the ULS and SLS target reliability indices $\beta_T{}^{ULS} = 3.2$ (i.e., $p_T{}^{ULS} = 0.00069$) and $\beta_T{}^{SLS} = 2.6$ (i.e., $p_T{}^{SLS} = 0.0047$) adopted for transmission line structure foundations in North America, Phoon et al. (1995) showed that $D = 4.2$ m is a feasible design when the coefficient of variation (COV) of $\phi'$, $COV_{\phi'} = 5$–10%, and $B = 1.2$ m. In addition, Wang et al. (2011a) showed that the feasible designs of the example are drilled shafts with $D \geq 4.4$ m when $COV_{\phi'} = 7\%$ and $B = 1.2$ m. This drilled shaft example is redesigned using the above-mentioned Excel spreadsheet Add-In, as discussed in the following three sections.

## 7.6.1 Deterministic model worksheet

The deterministic model is largely identical to the traditional drilled shaft design calculation. It takes a given set of design parameters $B$ and $D$, design loads, and soil properties as input, and produces an output of the ULS and SLS factors of safety (i.e., $FS_{uls}$ and $FS_{sls}$) and the driving variable (i.e., $BD/FS_{min}$) for Subset Simulation. To enable a consistent comparison with the previous results by Phoon et al. (1995) and Wang et al. (2011a), $FS_{uls}$ and $FS_{sls}$ of the drilled shaft are calculated from

$$FS_{uls} = Q_{uls}/F_{50} = (Q_{side} + Q_{tip} - W)/F_{50} \tag{7.23}$$

$$FS_{sls} = Q_{sls}/F_{50} = 0.625a\left(\frac{y_a}{B}\right)^b Q_{uls}/F_{50} \tag{7.24}$$

in which $Q_{uls}$ and $Q_{sls}$ = ULS and SLS capacity, respectively; $Q_{side}$, $Q_{tip}$, and $W$ = side resistance, tip resistance, and effective shaft weight, respectively; and $a = 4.0$ and $b = 0.4$ = curve-fitted parameters for the load–displacement model. The $Q_{side}$, $Q_{tip}$, and $W$ are calculated from

$$Q_{side} = \pi BD(K/K_0)_n K_0 \sigma'_{vm} \tan \delta \tag{7.25}$$

$$Q_{tip} = 0.25\pi B^2 [0.5B(\gamma - \gamma_w)N_\gamma \zeta_{\gamma s}\zeta_{\gamma d}\zeta_{\gamma r} + (\gamma - \gamma_w)DN_q \, \zeta_{qs} \, \zeta_{qd} \, \zeta_{qr}] \tag{7.26}$$

$$W = 0.25\pi B^2 D(\gamma_{con} - \gamma_w) \tag{7.27}$$

in which $\sigma'_{vm}$ is the mean vertical effective stress along the shaft depth, $\delta$ is the friction angle at the soil–shaft interface = soil effective stress friction angle $\phi'$ (for a rough interface), $\gamma_w$ is the unit weight of water, $N_\gamma$ and $N_q$ are the bearing capacity factors, and $\zeta_{\gamma s}$, $\zeta_{\gamma d}$, $\zeta_{\gamma r}$, $\zeta_{qs}$, $\zeta_{qd}$, and $\zeta_{qr}$ are the correction factors for the respective bearing capacity factors. The bearing capacity and correction factors are calculated using the basic Vesić (1975) model with minor updates (Kulhawy, 1991). Failure occurs when $F_{50}$ exceeds $Q_{uls}$ or $Q_{sls}$ (i.e., $FS_{uls} < 1$ or $FS_{sls} < 1$). The driving variable $BD/FS_{min}$ is also calculated in the deterministic model worksheet for Subset Simulation.

Figure 7.6 shows an example of the deterministic model worksheet in an Excel spreadsheet. The worksheet is divided into three parts: an input data zone from Rows 2 to 6, a capacity calculation zone from Rows 7 to 11, and an output zone starting from Row 12. The input data consist of soil properties (e.g., $\phi'$ in Cell B4 in degrees or Cell C4 in radians), design parameters, design requirements, unit weight of water and concrete, mean vertical effective stress along the shaft depth, bearing capacity factors, correction factors for the respective bearing capacity factors, and curve-fitted parameters for the load–displacement model. In the capacity calculation zone, Equations 7.25 through 7.27 are implemented in Cells B9–D9 to calculate the side resistance, tip resistance, and effective shaft weight, respectively. Then, the ULS and SLS capacities are calculated in Cells B11 and C11, respectively. In the output

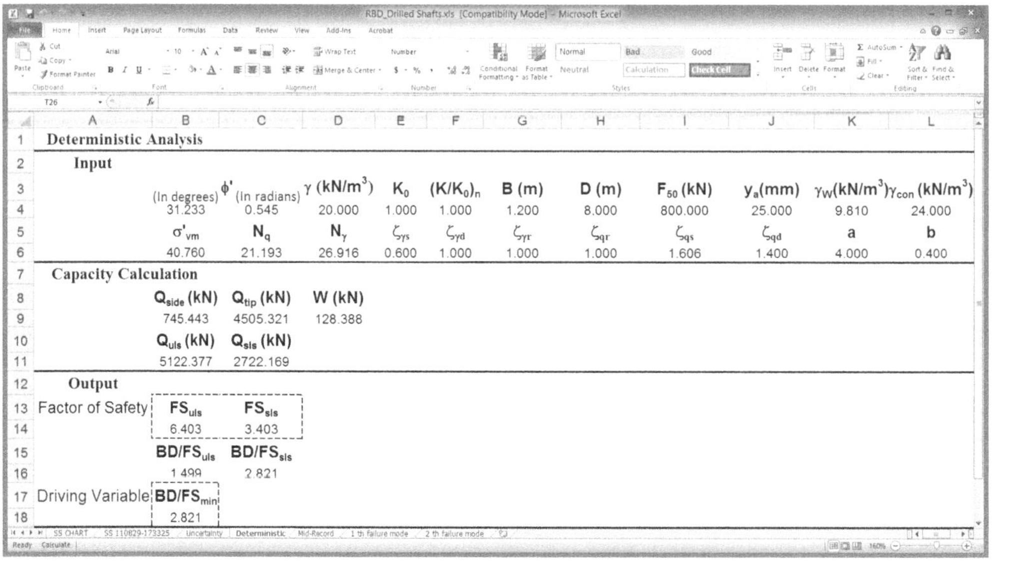

*Figure 7.6* Deterministic model worksheet for the drilled shaft design example.

zone, $FS_{uls}$ and $FS_{sls}$ are calculated using Equation 7.23 in Cell B14 and Equation 7.24 in Cell C14, respectively. The smaller one among $FS_{uls}$ and $FS_{sls}$ is used to calculate the driving variable $BD/FS_{min}$ in Cell B18.

### 7.6.2 Uncertainty model worksheet

To enable a consistent comparison with the designs by Phoon et al. (1995) and Wang et al. (2011a), the uncertainty modeling in this section follows those adopted in their studies. Since only the uncertainty on $\phi'$ is considered explicitly in their study, this section only models the $\phi'$ as a random variable that is lognormally distributed with a mean = 32° and COV = 7% (between 5 and 10%). In addition, since the design parameters (i.e., $B$ and $D$) of the drilled shaft are artificially treated as uniformly distributed variables in the expanded RBD approach, their random samples are generated in the uncertainty model worksheet. In this study, three possible $B$ values of 0.9, 1.2, and 1.5 m (i.e., $n_B = 3$) are considered, and the possible $D$ values vary from 2.0 to 8.0 m with an increment of 0.2 m (i.e., $n_D = 31$), which are consistent with those adopted by Wang et al. (2011a).

Figure 7.7 shows the uncertainty model worksheet, which consists of two parts: a variable description zone from Rows 2 to 8 and a random sample generation zone starting from Row 9. The variable description zone is used to define the variables $\phi'$, $B$, and $D$. Their distribution types (Cells B4, E4, and F4) and statistics (i.e., mean (Cell B5) and COV (Cell B6)) of $\phi'$, and the minimum values (Cells E5 and F5), maximum values (Cells E6 and F6), and numbers (Cells E8 and F8) of possible values of $B$ and $D$ are defined in the description zone. Using the information of variables $\phi'$, $B$, and $D$, their random samples are generated in the second zone (Rows 9–14) of the uncertainty model worksheet. The generation of random samples starts with generating uniform random samples using an Excel built-in function "RAND()." In this study, "RAND()" is implemented in Cells B11–D11 to generate three random samples

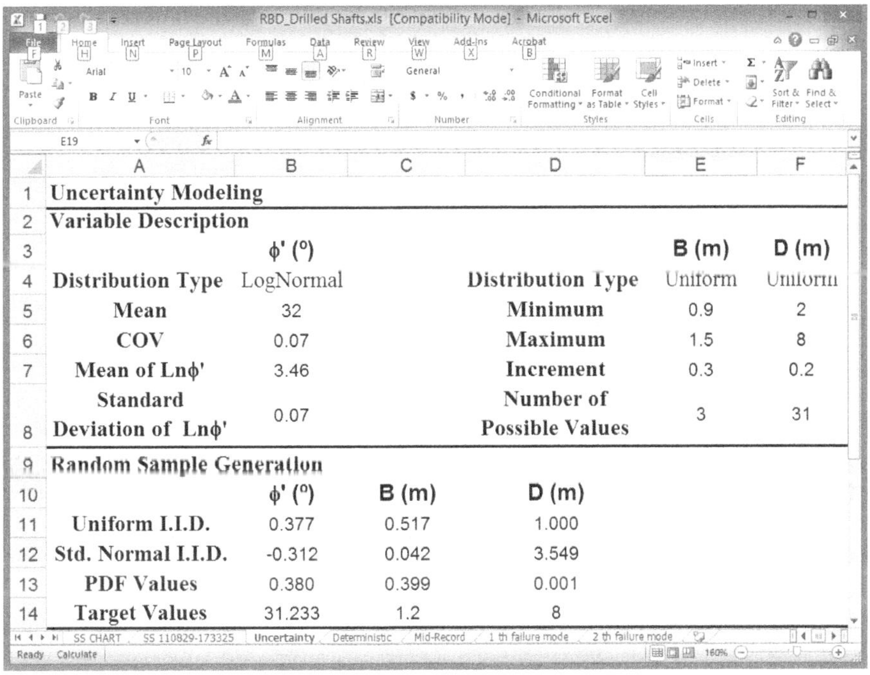

| | A | B | C | D | E | F |
|---|---|---|---|---|---|---|
| 1 | Uncertainty Modeling | | | | | |
| 2 | Variable Description | | | | | |
| 3 | | ϕ' (°) | | | B (m) | D (m) |
| 4 | Distribution Type | LogNormal | | Distribution Type | Uniform | Uniform |
| 5 | Mean | 32 | | Minimum | 0.9 | 2 |
| 6 | COV | 0.07 | | Maximum | 1.5 | 8 |
| 7 | Mean of Lnϕ' | 3.46 | | Increment | 0.3 | 0.2 |
| 8 | Standard Deviation of Lnϕ' | 0.07 | | Number of Possible Values | 3 | 31 |
| 9 | Random Sample Generation | | | | | |
| 10 | | ϕ' (°) | B (m) | D (m) | | |
| 11 | Uniform I.I.D. | 0.377 | 0.517 | 1.000 | | |
| 12 | Std. Normal I.I.D. | -0.312 | 0.042 | 3.549 | | |
| 13 | PDF Values | 0.380 | 0.399 | 0.001 | | |
| 14 | Target Values | 31.233 | 1.2 | 8 | | |

*Figure 7.7* Uncertainty model worksheet for the drilled shaft design example.

uniformly varying from 0 to 1 for $\phi'$, $B$, and $D$, respectively. These three uniform random samples are then transformed into three standard Gaussian random samples using an Excel built-in function "NORMSINV()," which is the inverse function of the standard Gaussian CDF. These three standard Gaussian random samples are generated in Cells B12–D12, and their PDF values are calculated by invoking a built-in function "NORMSDIST()" in Cells B13–D13. Then, these three standard Gaussian random samples are transferred into random samples of the lognormal variable $\psi'$ and discrete uniform random variables $B$ and $D$. The lognormal variable $\phi'$ is expressed as (e.g., Ang and Tang, 2007; Au et al., 2010)

$$\phi' = \exp(\mu_N + \sigma_N w_1) \tag{7.28}$$

in which $\mu_N$ and $\sigma_N$ are the mean and standard deviation of $\ln(\phi')$ (i.e., Cells B7 and B8), respectively, and $w_1$ is the a standard Gaussian variable (i.e., Cell B12). For discrete uniform random variables $B$ and $D$, they are expressed as

$$B = B_{min} + INT[\Phi(w_2)N_B]d_B \tag{7.29}$$

$$D = D_{min} + INT[\Phi(w_3)N_D]d_D \tag{7.30}$$

in which $B_{min}$ and $D_{min}$ are the respective minimum possible values for $B$ and $D$ (i.e., Cells E5 and F5 for $B$ and $D$, respectively), $w_2$ and $w_3$ are the independent standard Gaussian random variables; $d_B$ and $d_D$ are the respective increments of $B$ and $D$ (i.e., Cells E7 and F7 for $B$ and $D$, respectively), $N_B$ and $N_D$ are the respective numbers of possible values of $B$ and $D$ (i.e., Cells E8 and F8 for $B$ and $D$, respectively), $\Phi()$ is the CDF of the standard Gaussian variable, which is implemented by a built-in function "NORMSDIST()" in Excel, and $INT()$ is an Excel built-in function that rounds the number in the bracket down to the nearest integer. In the uncertainty model worksheet, Equations 7.28 through 7.30 are used to generate, respectively, random samples of $\phi'$, $B$, and $D$ in Cells B14–D14 from random samples of standard Gaussian variables in Cells B12–D12. Note that $\Phi(w_2)$ in Equation 7.29 and $\Phi(w_3)$ in Equation 7.30 vary uniformly from 0 to 1. They are first transformed into discrete integer random variables $INT(\Phi(x_2)N_B)$ and $INT(\Phi(x_3)N_D)$ ranging from 0 to $N_B$ and from 0 to $N_D$ in Equations 7.29 and 7.30, respectively, with an increment of 1. Then, $INT(\Phi(x_2)N_B)$ and $INT(\Phi(x_3)N_D)$ are further transformed into the discrete random variables $B$ and $D$ in Equations 7.29 and 7.30 using their respective minimum values (i.e., $B_{min}$ and $D_{min}$) and increments (i.e., $d_B$ and $d_D$).

### 7.6.3 Subset Simulation and RBD Add-In

After the deterministic and uncertainty model worksheet are developed and linked together, the Subset Simulation Add-In shown in Figure 7.4a is invoked for uncertainty propagation. Using the Add-In, a Subset Simulation run with $N = 10,000$, $p_0 = 0.2$, and $m = 4$ is executed, and a total of $N + mN(1 - p_0) = 42,000$ samples are obtained from the simulation. Note that the choice of $p_0$ and $m$ affects the efficiency of Subset Simulation. A reasonable range of $p_0$ suggested in literature (e.g., Zuev et al., 2012) is from 0.1 to 0.3. For a given $p_0$ value, the number (i.e., $m$) of Subset Simulation levels is selected to ensure that the target failure probability level is reached, that is, $p_0^{m+1} < p_T$.

In addition, the "Random variable(s), $X$" input field in Figure 7.4a records the cell references of random variables $\phi'$, $B$, and $D$ or their respective equivalents in standard normal space, that is, $w_1$, $w_2$, and $w_3$ in Cells B12–D12 of the uncertainty model worksheet shown in Figure 7.7. The "PDF of $X$, $P(X)$" input field records the respective cell references of the PDF values of the random variables (e.g., Cells B13–D13 in the uncertainty model worksheet for $w_1$, $w_2$, and $w_3$). The "Driving variable, $Y$" input field contains the cell reference (i.e., Cell B18 in the deterministic model worksheet shown in Figure 7.6) of the driving variable $BD/FS$. The "Variable(s) to record, $Y$" input field contains the cell references of Cells A2–E2 in a worksheet named "Mid-Record," which are used to record $1/FS_{uls}$, $1/FS_{sls}$, $BD/FS_{min}$, $\phi'$, $B$, and $D$ during the simulation. After the input fields are set up, Subset Simulation is performed by clicking the "Run" button.

After the simulation, an output worksheet is created by the Subset Simulation Add-In to record the results, including the random samples generated in simulation and their corresponding values of $1/FS_{uls}$, $1/FS_{sls}$, and $BD/FS_{min}$. The simulation results are used as input of the RBD Add-In (see Figure 7.4b) to calculate the respective failure probability $p_f^{ULS}$ and $p_f^{SLS}$ of ULS and SLS failures for given $B$ and $D$ combinations, as shown in Figure 7.8a and b.

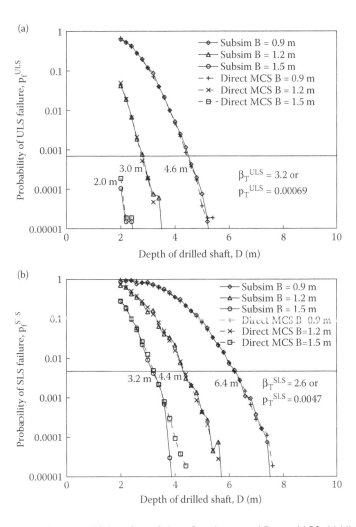

*Figure 7.8* Conditional probability of failure from Subset Simulation and Direct MCS. (a) Ultimate limit state (ULS) failure. (b) Serviceability limit state (SLS) failure.

### 7.6.4 Determination of feasible designs

Figure 7.8a and b shows, by solid lines, the conditional probability $P(F|B,D)$ obtained from a single run of Subset Simulation for ULS and SLS failures, respectively. Note that $P(F|B,D)$ is a variation of failure probability as a function of $(B,D)$. The horizontal axis in Figure 7.8 represents the variation of $D$, and the values of $p_f^{ULS}$ or $p_f^{SLS}$ for three different values of $B$ are included in the figure. For a given value of $B$, $p_f^{ULS}$ and $p_f^{SLS}$ for the respective ULS and SLS failures decrease as $D$ increases. Similarly, for a given value of $D$, $p_f^{ULS}$ and $p_f^{SLS}$ decrease as $B$ increases. Figure 7.8 also includes $p_T^{ULS} = 0.00069$ and $p_T^{SLS} = 0.0047$ adopted for transmission line foundations in North America (Phoon et al., 2003a,b), and feasible designs are those that fall below the $p_T^{ULS}$ and $p_T^{SLS}$ shown in the figure.

For the ULS requirement (Figure 7.8a), the feasible designs include the drilled shafts with $B = 1.5$ m and $D \geq 2.0$ m, $B = 1.2$ m and $D \geq 3.0$ m, or $B = 0.9$ m and $D \geq 4.6$ m. For the SLS requirement (Figure 7.8b), the feasible designs include those with $B = 1.5$ m and $D \geq 3.2$ m, $B = 1.2$ m and $D \geq 4.4$ m, or $B = 0.9$ m and $D \geq 6.4$ m. For a given value of $B$, the minimum feasible values (i.e., 3.2, 4.4, and 6.4 m for $B = 1.5$, 1.2, and 0.9 m, respectively) of $D$ for the SLS requirement are larger than those (i.e., 2.0, 3.0, and 4.6 m for $B = 1.5$, 1.2, and 0.9 m, respectively) for the ULS requirement. The SLS requirement is therefore the critical one that controls the design, and the feasible designs are the same as those for the SLS requirement. Phoon et al. (1995) and Wang et al. (2011a) also found in their design calculations for this example that the SLS requirement is the critical one, which is consistent with the observation herein.

### 7.6.5 Results comparison

Figure 7.8 also shows, by dashed lines, the values of $p_f^{ULS}$ or $p_f^{SLS}$ estimated from Direct MCS with 10,000,000 (Wang et al., 2011a). It is evident that the dashed lines almost overlap the solid lines, and the results from Subset Simulation are in good agreement with those from Direct MCS. Tables 7.3 and 7.4 summarize the respective feasible designs for ULS and SLS requirements from Subset Simulation and Direct MCS. For ULS requirement, the feasible designs obtained from Direct MCS are those with $B = 1.5$ m and $D \geq 2.0$ m, $B = 1.2$ m and $D \geq 2.8$ m, or $B = 0.9$ m and $D \geq 4.6$ m, which agree well with those (i.e., $B = 1.5$ m and $D \geq 2.0$ m, $B = 1.2$ m and $D \geq 3.0$ m, or $B = 0.9$ m and $D \geq 4.6$ m) obtained from Subset Simulation (see Table 7.3). Similar observations are also found from the SLS feasible designs obtained from Direct MCS and Subset Simulation (see Table 7.4). Such good agreement suggests that the integration of the expanded RBD approach with Subset Simulation works well. When compared with the expanded RBD approach with Direct MCS, only 42,000 random samples are, however, needed in Subset Simulation, which are much less than the 10,000,000 samples used in Direct MCS. Subset Simulation substantially improves the computational efficiency at small probability levels, and it significantly enhances the expanded RBD approach in design situations with a small target failure probability (e.g., $p_T^{ULS} = 0.00069$ and $p_T^{SLS} = 0.0047$ in this example).

*Table 7.3* Feasible designs for ULS requirement (i.e., $\beta_T^{ULS} = 3.2$ and $p_T^{ULS} = 0.00069$)

| Design approach | B = 0.9 m | B = 1.2 m | B = 1.5 m |
|---|---|---|---|
| Expanded RBD with Subset Simulation | D ≥ 4.6 m | D ≥ 3.0 m | D ≥ 2.0 m |
| Expanded RBD with Direct MCS | D ≥ 4.6 m | D ≥ 2.8 m | D ≥ 2.0 m |

*Table 7.4* Feasible designs for SLS requirement (i.e., $\beta_T^{SLS} = 2.6$ and $p_T^{SLS} = 0.0047$ )

| Design approach | B = 0.9 m | B = 1.2 m | B = 1.5 m |
|---|---|---|---|
| Expanded RBD with Subset Simulation | D ≥ 6.4 m | D ≥ 4.4 m | D ≥ 3.2 m |
| Expanded RBD with Direct MCS | D ≥ 6.2 m | D ≥ 4.4 m | D ≥ 3.4 m |

## 7.6.6 Effects of the driving variable

Since the driving variable $Y$ is a key factor that affects the generation of conditional samples of interest in Subset Simulation, proper selection of $Y$ plays a pivotal role in the integration of the expanded RBD approach with Subset Simulation. Note that the convention of Subset Simulation is to define $Y$ as a variable that monotonically increases as the simulation level $m$ increases. In the expanded RBD approach, the conditional samples of interest are failure samples conditional on design parameters (e.g., $B$ and $D$ for drilled shaft). Since the failure is defined as $FS_{uls} < 1$ or $FS_{sls} < 1$, failure samples have relative small values of $FS$. To assure that $Y$ is a monotonic variable that increases as $FS$ decreases, $Y$ is defined to be proportional to the reciprocal of $FS$ (i.e., $1/FS$). As two $FS$ are calculated in the deterministic model, the minimum (i.e., $FS_{min}$) of these two $FS$ is used to define $Y$.

In addition, the deterministic calculation model (i.e., Equations 7.23 through 7.27) for drilled shaft design shows that the values of $FS_{min}$ decrease as the design parameters $B$ and $D$ decrease. Using $1/FS_{min}$ as a driving variable tends to drive the sampling space to samples with low $FS_{min}$ values, the overwhelming majority of which correspond to relatively small values of $B$ and $D$. On the other hand, failure samples with relatively large values of $B$ and $D$ are also of interest in the expanded RBD approach. If there is an insufficient number of failure samples with relatively large $B$ or $D$ values, the resolution and accuracy of the estimates of their conditional probabilities (e.g., $P(B,D|F)$ and $P(F|B,D)$) would be poor, and hence, some feasible designs with relatively large $B$ or $D$ values would not be identified properly. To assure that feasible designs with a wide range of $B$ and $D$ values are all covered properly in the expanded RBD approach, it is necessary to define the driving variable as a combination of failure criterion (e.g., $FS_{min}$) and design parameters of interest (e.g., $B$ and $D$ in drilled shaft design).

This study defines the driving variable as $Y = BD/FS_{min}$. Subset Simulation generates samples with increasing values of $BD/FS_{min}$ as the level $m$ increases. The increase of $BD/FS_{min}$ is attributed to two factors: decrease in the denominator $FS_{min}$ and increase in the numerator $BD$. Thus, the effect of driving variable $Y = BD/FS_{min}$ on the sampling process is two-fold. On the one hand, due to the effect of denominator $FS_{min}$ (i.e., the effect of $1/FS_{min}$), Subset Simulation drives the sampling space to the failure domain with relatively small $FS_{min}$ values that usually correspond to relatively small $B$ and $D$ values. On the other hand, because of the effect of the numerator $BD$, Subset Simulation generates samples with relatively large $B$ and $D$ values. The combined effects of the denominator $FS_{min}$ and numerator $BD$ in the driving variable $Y = BD/FS_{min}$ improve the efficiency of generating failure samples that cover a wide range of $B$ and $D$ values, particularly those with relatively large values of $B$ and $D$.

To illustrate the effect of the driving variable, two Subset Simulations are performed with two different driving variables: one with $Y = BD/FS_{min}$ and the other with $Y = 1/FS_{min}$. Defining the driving variable as a function of the failure criterion (e.g., $Y = 1/FS_{min}$) is a common practice in Subset Simulation (e.g., Au et al., 2010). Two Subset Simulations both have $N = 10,000$ samples per level, $p_0 = 0.2$, and the highest simulation level $m = 4$, resulting in 42,000 samples per simulation.

Table 7.5 summarizes the failure samples generated in these two Subset Simulation runs. The Subset Simulation with $Y = 1/FS_{min}$ generates 22,253 ULS failure samples, among which,

*Table 7.5* Summary of failure samples generated by different driving variables

| Limit state | Driving variable Y | Number of failure samples | Percentage of failure samples (%) | | | $D_{max}$ (m) | | |
|---|---|---|---|---|---|---|---|---|
| | | | B = 0.9 m | B = 1.2 m | B = 1.5 m | B = 0.9 m | B = 1.2 m | B = 1.5 m |
| ULS | $1/FS_{min}$ | 22,253 | 98.3 | 1.7 | 0.0 | 4.4 | 2.8 | N/A |
| | $BD/FS_{min}$ | 12,844 | 93.4 | 6.5 | 0.1 | 5.2 | 3.4 | 2.4 |
| SLS | $1/FS_{min}$ | 31,042 | 90.4 | 8.3 | 1.3 | 6.6 | 4.6 | 3.4 |
| | $BD/FS_{min}$ | 24,708 | 82.7 | 15.0 | 2.3 | 7.4 | 5.6 | 3.8 |

98.3% have $B = 0.9$ m, 1.7% have $B = 1.2$ m, and none of them have $B = 1.5$ m. Since there is no failure sample with $B = 1.5$ m, it is difficult to estimate failure probability for designs with $B = 1.5$ m. In contrast, the Subset Simulation with $Y = BD/FS_{min}$ generates 12,844 ULS failure samples. The percentage of ULS failure samples with $B = 0.9$ m decreases to 93.4%, while the percentages of ULS failure samples with $B = 1.2$ m and $B = 1.5$ m increase to 6.5 and 0.1%, respectively. The simulation with $Y = BD/FS_{min}$ generates more ULS failure samples with relatively large $B$ values (e.g., $B = 1.2$ m and $B = 1.5$ m) than the simulation with $Y = 1/FS_{min}$. In addition, Table 7.5 shows that the maximum values ($D_{max}$) of $D$ among the ULS failure samples at $Y = BD/FS_{min}$ are also larger than those at $Y = 1/FS_{min}$. Similar observations are also found for the SLS failure samples summarized in the fourth and fifth rows of Table 7.5. It is evident that using $Y = BD/FS_{min}$ leads to failure samples with relatively large $B$ and $D$ values. It is more appropriate to use $Y = BD/FS_{min}$ than $Y = 1/FS_{min}$ in the expanded RBD of drilled shafts.

## 7.7 ILLUSTRATIVE EXAMPLE II: JAMES BAY DIKE DESIGN SCENARIO

In this section, a design scenario of the James Bay Dike in Canada is reanalyzed using the probabilistic failure analysis approach for identifying the uncertain parameter that is the most influential to the dike performance and for understanding how the performance (e.g., failure probability $P(F)$) of the designed dike changes as this important uncertain parameter changes. The James Bay Dike is a 50-km-long earth dike of the James Bay hydroelectric project in Canada. Soil properties and various design scenarios of the dike were studied by Ladd et al. (1983), Soulié et al. (1990), Christian et al. (1994), El-Ramly et al. (2002), Xu and Low (2006), and Wang et al. (2010). As shown in Figure 7.9, the embankment is 12 m

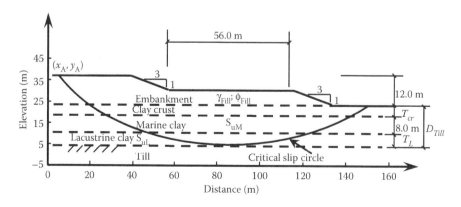

*Figure 7.9* Illustration of a James Bay Dike design scenario.

high with a 56-m-wide berm at mid-height. The slope angle of the embankment is about 18.4° (3H:1 V). The embankment is overlying on a clay crust with a thickness $T_{cr}$. The clay crust is underlain by a layer of 8.0-m-thick sensitive marine clay and a layer of lacustrine clay with a thickness $T_L$. The undrained shear strength (i.e., $S_{uM}$ and $S_{uL}$) of the marine clay and the lacustrine clay was measured by field vane tests (Ladd et al., 1983; Soulié et al., 1990; Christian et al., 1994). The lacustrine clay is overlying on a stiff till layer, the depth to the top of which is $D_{Till}$.

Six system parameters have been identified as uncertain parameters in the previous studies (e.g., El-Ramly et al., 2002; Xu and Low, 2006; Wang et al., 2010), including the friction angle $\phi_{Fill}$ and unit weight $\gamma_{Fill}$ of embankment material, the thickness $T_{cr}$ of clay crust, the undrained shear strength $S_{uM}$ of the marine clay, the undrained shear strength $S_{uL}$ of the lacustrine clay, and the depth of the till layer $D_{Till}$. In the reliability analysis of the design scenario, these six uncertain parameters are represented by six independent Gaussian random variables (El-Ramly, 2001), respectively. Table 7.6 summarizes the statistics (i.e., mean, standard deviation, and COV) of these six random variables. These statistics are used to generate random samples for each random variable in an uncertain model worksheet, as shown in Figure 7.10. Note that the thickness of the lacustrine clay $T_L$ is an uncertain variable that depends on $T_{cr}$ and $D_{Till}$ and has a mean of about 6.5 m (see Figure 7.9). In addition to these uncertain parameters, other system parameters are taken as deterministic, including an undrained shear strength of 41 kPa for the clay crust and unit weights of 19, 19, and 20.5 kN/m³ for the clay crust, marine clay, and lacustrine clay (El-Ramly, 2001), respectively. With these soil properties, a circular critical slip surface is used together with a simplified Bishop method to calculate FS along the critical slip surface in the deterministic slope stability analysis (El-Ramly, 2001; Wang et al., 2010). The critical slip surface is always tangential to the top of the till layer and passes through the point A ($x_A = 4.9$ m, $y_A = 36.0$ m), as shown in Figure 7.9. The x-coordinate of the center is fixed at 85.9 m. For each set of simulation samples, the critical slip surface is uniquely specified by the value of $D_{Till}$, and its corresponding FS is calculated in a deterministic model worksheet, as shown in Figure 7.11.

After the uncertain model worksheet (see Figure 7.10) and deterministic model worksheet (see Figure 7.11) are set up, the Subset Simulation Add-In (see Figure 7.12) is invoked for uncertainty propagation. A Subset Simulation run is performed with the highest simulation level $m = 3$, $p_0 = 0.1$, and $N = 1000$ samples per level. The driving variable Y in Subset Simulation is defined as 1/FS to drive the sampling space to gradually approach the failure domain, which is of particular interest in probabilistic failure analysis. After the simulation, a total of 3700 samples are generated from the Subset Simulation, including 900 samples for simulation levels '0', '1', and '2', respectively, and 1000 samples for simulation level '3'.

*Table 7.6* Uncertainty characterization of the James Bay Dike design example

| Soil layers | Uncertain parameters[a] | Mean | Standard deviation | COV (%) |
|---|---|---|---|---|
| Embankment | $\phi_{Fill}$ (°) | 30.0 | 1.79 | 6.0 |
| Embankment | $\gamma_{Fill}$ (kN/m³) | 20.0 | 1.10 | 5.5 |
| Clay crust | $T_{cr}$ (m) | 4.0 | 0.48 | 12.0 |
| Marine clay | $S_{uM}$ (kN/m²) | 34.5 | 3.95 | 11.5 |
| Lacustrine clay | $S_{uL}$ (kN/m²) | 31.2 | 6.31 | 20.2 |
| Till | $D_{Till}$ (m) | 18.5 | 1.00 | 5.4 |

[a] All uncertain parameters follow normal distributions. The thickness of the lacustrine clay layer $T_L$ is an uncertain parameter that depends on $T_{cr}$ and $D_{Till}$ and has a mean of about 6.5 m.

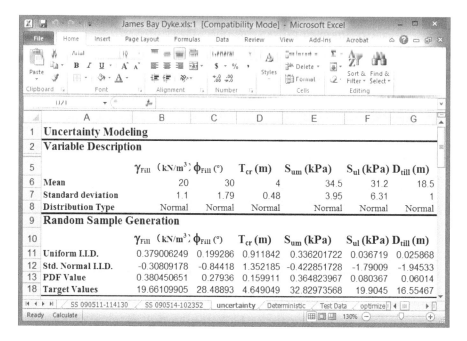

*Figure 7.10* Uncertainty model worksheet for the James Bay Dike example.

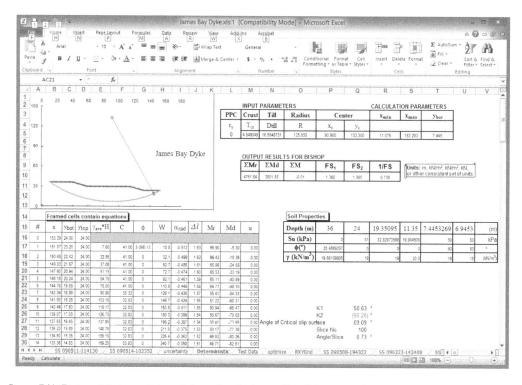

*Figure 7.11* Deterministic model worksheet for the James Bay Dike example.

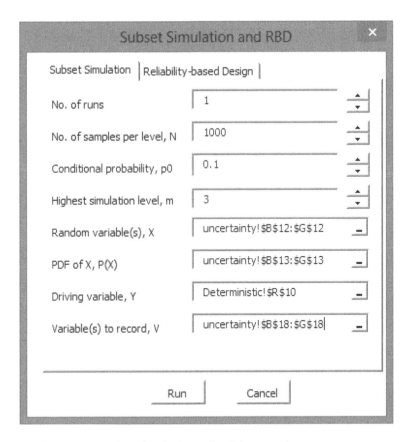

*Figure 7.12* Subset Simulation userform for the James Bay Dike example.

### 7.7.1 Subset Simulation results

Among the 3700 samples from Subset Simulation, 1134 samples are identified as failure samples (i.e., $FS < 1$ or $Y > 1$). These 1134 failure samples include 1000 samples in simulation level '3' (i.e., all samples in simulation level '3' fail) and 134 samples in simulation level '2'. As summarized in Table 7.7, Equation 7.18 is used to estimate the failure probability. Since $p_0$ is taken as 0.1 in Subset Simulation, $P(L_i)$ is 0.9, 0.09, 0.009, and 0.001 for simulation levels '0', '1', '2', and '3', respectively. Since no failure sample occurs in simulation levels '0' and '1', the $P(F|L_i)$ values at these two simulation levels are 0. On the other hand, all samples fail in simulation level '3', leading to $P(F|L_i) = 1000/1000 = 1$ at the simulation level '3'. The $P(F|L_i)$ at simulation level '2' is 134/900. As shown in the last column of Table 7.7, the failure probability is calculated as 0.234%. Figure 7.13 shows a CCDF for

*Table 7.7* Failure probability of the James Bay Dike design example

| Simulation level i | $P(L_i)$ | $P(F|L_i)$ | $P(F)$ |
|---|---|---|---|
| 0 | 0.9 | 0/900 | $0.9 \times 0/900 + 0.09 \times 0/900 + 0.009 \times 134/900 + 0.001 \times 1000/1000 = 0.234\%$ |
| 1 | 0.09 | 0/900 | |
| 2 | 0.009 | 134/900 | |
| 3 | 0.001 | 1000/1000 | |

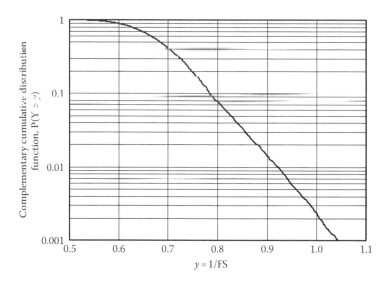

*Figure 7.13* CCDF plot from Subset Simulation.

$Y = 1/FS$ (i.e., $P(Y > y)$ vs. $y$) from Subset Simulation. The failure probability can also be directly obtained from the plot.

## 7.7.2 Hypothesis test results

With a large number of failure samples generated from Subset Simulation, hypothesis testing is performed for the James Bay Dike design scenario for identifying the important uncertain parameters. Using the 1134 failure samples generated from Subset Simulation, the $\mu_f$ values for $\phi_{Fill}$, $\gamma_{Fill}$, $T_{cr}$, $S_{uM}$, $S_{uL}$, and $D_{Till}$ are calculated. This can be achieved using an Excel built-in function "AVERAGE ()," which uses 1134 failure samples as input. Then, hypothesis tests are performed accordingly, and the results are summarized in Table 7.8. As shown in the fifth row of Table 7.8, the absolute values of $Z_H$ vary from about 7 for the thickness of clay crust $T_{cr}$ to about 93 for undrained shear strength of the lacustrine clay $S_{uL}$. The decreasing order of the $Z_H$ absolute values is $S_{uL}$, $D_{Till}$, $\gamma_{Fill}$, $S_{uM}$, $\phi_{Fill}$, and $T_{cr}$. This implies that the uncertain parameter $S_{uL}$ has the most significant effect on the slope failure probability, while the uncertainty on $T_{cr}$ only has minimal influence on the failure probability.

## 7.7.3 Bayesian analysis results

Since $S_{uL}$ is identified as the most important uncertain parameter from hypothesis tests, Bayesian analysis is performed to quantify the variation of $P(F)$ as the $S_{uL}$ value changes.

*Table 7.8* Summary of hypothesis test results for the James Bay Dike design example

| Uncertain parameter | $\phi_{Fill}$ | $\gamma_{Fill}$ | $T_{cr}$ | $S_{uM}$ | $S_{uL}$ | $D_{Till}$ |
|---|---|---|---|---|---|---|
| Unconditional mean $\mu$ | 30.00 | 20.00 | 4.00 | 34.50 | 31.20 | 18.50 |
| Standard deviation $\sigma$ | 1.79 | 1.10 | 0.48 | 3.95 | 6.31 | 1.00 |
| Failure sample mean $\mu_f$ | 29.12 | 20.85 | 3.89 | 31.87 | 13.76 | 19.84 |
| Absolute value of $Z_H$ | 16.52 | 26.05 | 7.49 | 22.43 | 93.08 | 45.28 |
| Ranking | 5 | 3 | 6 | 4 | 1 | 2 |

*Table 7.9* Calculation of $p(S_{uL} = 13|F)$ in the James Bay Dike design example

| Simulation level $i$ | $P(L_i)$ | $P(F|L_i)$ | $P(L_i|F)$ | $n_{ji}$ | $n_{fi}$ | $p(S_{uL} = 13|F)$ |
|---|---|---|---|---|---|---|
| 0 | 0.9 | 0 | 0 | 0 | 0 | 0.269 |
| 1 | 0.09 | 0 | 0 | 0 | 0 | |
| 2 | 0.009 | 0.1489 | 0.5727 | 27 | 134 | |
| 3 | 0.001 | 1 | 0.4274 | 359 | 1000 | |

The $S_{uL}$ values of the 1134 failure samples are collected and analyzed. The maximum and minimum $S_{uL}$ values are found to be 21.7 and 8.6 kPa, respectively. Consider, for example, estimating the failure probability $P(F|S_{uL} = 13)$ when the $S_{uL}$ value is taken as a deterministic value of 13 kPa and the other uncertain parameters (i.e., $D_{Till}$, $\gamma_{Fill}$, $S_{uM}$, $\phi_{Fill}$, and $T_{cr}$) remain as random. An $S_{uL}$ bin of (12, 14] is adopted in the Bayesian analysis. $P(L_i|F)$ is estimated using Equation 7.20, as summarized in the fourth column of Table 7.9. Among all 134 failure samples that occur at the simulation level '2' (i.e., $n_{f2} = 134$), 27 samples are found to have $S_{uL}$ values that lie within the bin (12, 14] (i.e., $n_{j2} = 27$). Similarly, among all 1000 failure samples that occur at the simulation level '3' (i.e., $n_{f3} = 1000$), 359 samples are found to have $S_{uL}$ values that lie within the bin (12, 14] (i.e., $n_{j3} = 359$). As shown in the last column of Table 7.9, Equation 7.21 is used to calculate $p(S_{uL} = 13|F) = 0.269$. $P(S_{uL} \in (12,14])$ is then calculated analytically as $P(S_{uL} \in (12,14]) = CDF_{S_u}(14) - CDF_{S_u}(12)$, where $CDF_{S_u}$ in this example is a normal CDF with a mean and standard deviation of 31.2 and 6.31 kPa, respectively. Finally, Equations 7.12 and 7.15 are used to estimate the failure probability at $S_{uL} = 13$ kPa:

$$P(F | S_{uL} = 13) \approx P(F | S_{uL} \in (12,14]) = \frac{0.269 \times 0.234\%}{0.00204} = 0.3091 \qquad (7.31)$$

The Bayesian analysis is repeated for various $S_{uL}$ values. Figure 7.14 shows the Bayesian analysis results by open squares for various $S_{uL}$ values. Note that the conditional probability

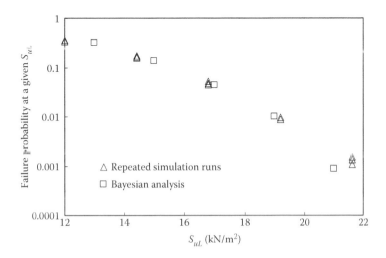

*Figure 7.14* Bayesian analysis results for $S_{uL}$.

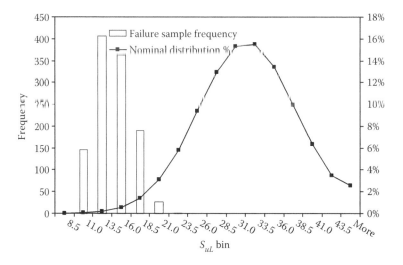

*Figure 7.15* Histogram of the $S_{uL}$ failure samples from Subset Simulation.

(i.e., $P(F|S_{uL})$ in Figure 7.14) obtained from Bayesian analysis is a variation of failure probability as a function of $S_{uL}$. Figure 7.14 shows that, as $S_{uL}$ increases from 13 to 21 kPa, the slope failure probability decreases from more than 10% to about 0.1%. It is obvious that the values of $S_{uL}$ have significant effects on slope failure probability. Such effects are explicitly quantified from the Bayesian analysis of failure samples. Variations of failure probability as a function of $S_{uL}$ shown in Figure 7.14 can also be obtained from repeated simulation runs with different deterministic $S_{uL}$ value in each run. Figure 7.14 also includes results from such repeated simulation runs by open triangles. The open triangles follow a trend similar to the open squares (i.e., the Bayesian analysis results). This validates the Bayesian analysis results.

Figure 7.15 shows an $S_{uL}$ histogram of the 1134 failure samples from Subset Simulation and a nominal (unconditional) probability distribution of $S_{uL}$ (i.e., a normal probability

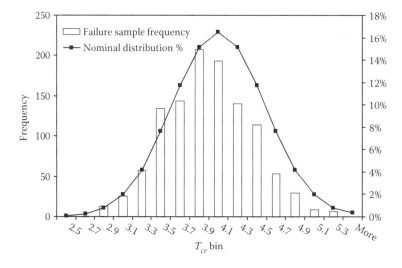

*Figure 7.16* Histogram of the $T_{cr}$ failure samples from Subset Simulation.

distribution with a mean and standard deviation of 31.2 and 6.31 kPa, respectively). Although the failure samples still more or less follow a normal distribution, its mean, standard deviation, and range are obviously different from those of the nominal distribution. All 1134 failure samples have $S_{uL}$ values smaller than 21.7 kPa, leading to a histogram peaking in the lower tail of the nominal distribution. The $S_{uL}$ distribution from failure samples and the $S_{uL}$ nominal probability distribution are quite different. This suggests that the effect of $S_{uL}$ on failure probability is significant. This result agrees well with the hypothesis test results (see the ranking in Table 7.8). In contrast, Figure 7.16 shows a $T_{cr}$ histogram of the 1134 failure samples and a nominal probability distribution of $T_{cr}$ (i.e., a Normal probability distribution with a mean and standard deviation of 4.0 and 0.48 m, respectively). The $T_{cr}$ histogram of the 1134 failure samples has a pattern similar to its nominal probability distribution. This suggests that the effect of $T_{cr}$ on failure probability is rather minimal. The results shown in Figure 7.16 are consistent with the hypothesis test results.

## 7.8 SUMMARY AND CONCLUDING REMARKS

This chapter presented an MCS-based practical framework for reliability analysis and design of geotechnical structures in a commonly available spreadsheet environment. Two drawbacks of MCS were highlighted: (1) lack of resolution and efficiency, particularly at small probability levels; and (2) no insight into the relative contributions of various uncertainties to the reliability analysis results. These two drawbacks of MCS were addressed by introducing an advanced MCS called "Subset Simulation" and probabilistic failure analysis, respectively. Subset Simulation significantly improves the computational efficiency of MCS at small probability levels. The probabilistic failure analysis approach makes use of the failure samples generated in MCS and analyzes these failure samples to assess the effects of various uncertainties on failure probability. Subset Simulation can also be used together with the probabilistic failure analysis to improve the efficiency of generating failure samples.

To further remove the hurdle of reliability algorithms for geotechnical practitioners, the MCS-based reliability analysis and design approaches were implemented in an Excel spreadsheet platform. The implementation is deliberately divided into three uncoupled modules: uncertainty modeling, deterministic modeling, and uncertainty propagation. The reliability analysis (including uncertainty modeling and propagation) is decoupled from the conventional deterministic geotechnical analysis so that the reliability analysis can proceed as an extension of the deterministic analysis in a nonintrusive manner. This allows the deterministic and reliability analysis to be performed separately by personnel with different expertise and in a parallel manner.

The MCS-based approaches and their implementations in the spreadsheet were illustrated through a drilled shaft design example and reliability analysis of a James Bay Dike scenario. The results obtained from the drilled shaft example showed that Subset Simulation significantly improves the computational efficiency at small probability levels and substantially reduces the sample numbers or computational efforts required in the expanded RBD approach. The reliability analysis results of the James Bay Dike scenario showed that the effects of various uncertainties on slope failure probability are properly prioritized and quantified by the probabilistic failure analysis approach. In addition, the probabilistic failure analysis approach gives results equivalent to those from sensitivity studies using repeated MCS runs, and hence, saves additional computational time and efforts for sensitivity studies.

## ACKNOWLEDGMENT

The work described in this chapter was supported by a grant from National Natural Science Foundation of China (Project Number 51208446). The financial supports are gratefully acknowledged.

## LIST OF SYMBOLS

| | |
|---|---|
| $\beta_T$ | Target reliability index |
| $\beta_T{}^{ULS}, \beta_T{}^{SLS}$ | Target reliability indices for ULS and SLS |
| $\delta$ | Friction angle at the soil–shaft interface |
| $\phi_{Fill}, \gamma_{Fill}$ | Friction angle and unit weight of embankment material |
| $\phi'$ | Effective friction angle of soil |
| $\Phi()$ | CDF of the standard Gaussian variable |
| $\gamma$ | Total unit weight of soil |
| $\gamma_{con}$ | Concrete unit weight |
| $\gamma_w$ | Unit weight of water |
| $\mu$ | Unconditional mean |
| $\mu_f$ | Mean of failure samples |
| $\mu_N, \sigma_N$ | Mean and standard deviation of $\ln(\phi')$ |
| $\sigma$ | Unconditional standard deviation |
| $\sigma'_{vm}$ | Mean vertical effective stress along the shaft depth |
| $\Omega$ | Sample space of $B$ and $D$ |
| $\Omega_i$ | $i$th subset of sample space of $B$ and $D$ |
| $\zeta_{\gamma s}, \zeta_{\gamma d}, \zeta_{\gamma r}, \zeta_{qs}, \zeta_{qd}, \zeta_{qr}$ | Correction factors for bearing capacity factors |
| $a, b$ | Curve-fitted parameters for the load–displacement model |
| $B, D$ | Diameter and depth of the drilled shaft |
| $B_{min}, D_{min}$ | Respective minimum possible values for B and D |
| $bin_j$ | $j$th bin |
| CCDF | Complementary cumulative distribution function |
| CDF | Cumulative distribution function |
| $CDF_{Su}$ | Cumulative distribution function of undrained shear strength |
| $COV_T$ | Target coefficient of variation of failure probability |
| $COV_{\phi'}$ | Coefficient of variation of $\phi'$ |
| $d_B, d_D$ | Increment of B and D |
| $D_{Till}$ | Depth to the top of the stiff till layer |
| E | Event |
| $E_m$ | $m$th intermediate event |
| $F_{50}$ | Design compression load |
| FORM | First-order reliability method |
| FS | Factor of safety |
| $FS_{min}$ | Minimum factor of safety |
| $FS_{uls}, FS_{sls}$ | Factor of safety for ULS and SLS |
| G | Performance function |
| $H_0$ | Null hypothesis |
| $H_A$ | Alternative hypothesis |
| $K_0$ | Nominal at-rest horizontal soil stress coefficient |
| $(K/K_0)_n$ | Nominal operative *in situ* horizontal stress coefficient ratio |
| L | Sample space of $x_k$ |

| | |
|---|---|
| $L_i$ | $i$th subset of sample space of $x_k$ |
| $m$ | Highest Subset Simulation level |
| $m_{\phi'}$ | Mean effective friction angle of soil |
| MCMC | Markov chain Monte Carlo |
| MCS | Monte Carlo Simulation |
| MRFD | Multiple resistance factor design |
| $n_{\Omega i}$ | Number of failure samples in $\Omega_i$ |
| $n_{\Omega i, BD}$ | Number of failure samples in $\Omega_i$ with a combination of $B$ and $D$ |
| $n_B, n_D$ | Number of possible discrete values for $B$ and $D$ |
| $n_f$ | Number of failure samples in the simulation |
| $n_{fi}$ | Number of failure samples at the simulation level '$i$' |
| $n_j$ | Number of simulation samples where failure occurs and the $x_k$ value falls into bin $j$ |
| $n_{ji}$ | Number of failure samples at the simulation level '$i$' with the $x_k$ value falling into bin $j$ |
| $n_{min}$ | Minimum MCS sample number |
| $n_t$ | Total number of samples in the MCS |
| $N$ | Number of samples per Subset Simulation level |
| $N_\gamma, N_q$ | Bearing capacity factors |
| $N_B, N_D$ | Number of possible values for B and D |
| $p()$ | Probability density function |
| $p(\|)$ | Conditional probability density function |
| $p_0$ | Conditional probability adopted in Subset Simulation |
| $p_f^{ULS}, p_f^{SLS}$ | Failure probability for ULS and SLS |
| $p_T$ | Target failure probability |
| $P()$ | Probability |
| $P(\|)$ | Conditional probability |
| PDF | Probability density function |
| $Q_{side}, Q_{tip}$ | Side resistance and tip resistance |
| $Q_{uls}, Q_{sls}$ | ULS and SLS capacity |
| RBD | Reliability-based design |
| SLS | Serviceability limit state |
| SORM | Second-order reliability method |
| $S_{uM}, S_{uL}$ | Undrained shear strength of the marine clay and the lacustrine clay |
| $T_{cr}, T_L$ | Thickness of the clay crust and the lacustrine clay |
| ULS | Ultimate limit state |
| UPSS | Uncertainty propagation using Subset Simulation |
| $V$ | Variable needed to be recorded during Subset Simulation |
| VBA | Visual Basic for Applications |
| $w$ | Standard Gaussian variable |
| $W$ | Effective shaft weight |
| $x_A, y_A$ | x and y coordinates of point A |
| $x_k$ | Value of random variable $X_k$ |
| $\mathbf{X}$ | Uncertain parameter vector |
| $X_k$ | Random variable |
| $Y$ | Driving variable |
| $y$ | Value of the driving variable |
| $y_m$ | $m$th intermediate threshold value during Subset Simulation |
| $y_a$ | Allowable displacement |
| $Z_H$ | Hypothesis testing statistic |

# REFERENCES

Ahmed, A. and Soubra, A.-H. 2011. Subset simulation and its application to a spatially random soil. *Geotechnical Risk Assessment and Management, GeoRisk 2011*. Atlanta, Georgia, GSP No. 224, 209–216.

Ang, H. S. and Tang, W. H. 2007. *Probability Concepts in Engineering: Emphasis on Applications to Civil and Environmental Engineering*, 2nd ed., New York, John Wiley and Sons.

Au, S. K. 2005. Reliability-based design sensitivity by efficient simulation. *Computers and Structures*, 83(14), 1048–1061.

Au, S. K. and Beck, J. L. 2001. Estimation of small failure probabilities in high dimensions by subset simulation. *Probabilistic Engineering Mechanics*, 16(4), 263–277.

Au, S. K. and Beck, J. L. 2003. Subset simulation and its application to probabilistic seismic performance assessment. *Journal of Engineering Mechanics*, 129(8), 1–17.

Au, S. K., Cao, Z. and Wang, Y. 2010. Implementing advanced Monte Carlo simulation under spreadsheet environment. *Structural Safety*, 32, 281–292.

Au, S. K. and Wang, Y. 2014. *Engineering Risk Assessment with Subset Simulation*, Singapore, John Wiley and Sons.

Baecher, G. B. and Christian, J. T. 2003. *Reliability and Statistics in Geotechnical Engineering*, New Jersey, John Wiley and Sons.

Chalermyanont, T. and Benson, C. H. 2004. Reliability-based design for internal stability of mechanically stabilized earth walls. *Journal of Geotechnical and Geoenvironmental Engineering*, 130(2), 163–173.

Chalermyanont, T. and Benson, C. H. 2005. Reliability-based design for external stability of mechanically stabilized earth walls. *International Journal of Geomechanics*, 5(3),196–205.

Christian, J. T., Ladd, C. C. and Baecher, G. B. 1994. Reliability applied to slope stability analysis. *Journal of Geotechnical Engineering*, 120(12), 2180–2207.

El-Ramly, H. 2001. Probabilistic analysis of landslide hazards and risks bridging theory and practice. PhD thesis of University of Alberta.

El-Ramly, H., Morgenstern, N. R. and Cruden, D. M. 2002. Probabilistic slope stability analysis for practice. *Canadian Geotechnical Journal*, 39, 665–683.

El-Ramly, H., Morgenstern, N. R. and Cruden, D. M. 2005. Probabilistic assessment of stability of a cut slope in residual soil. *Geotechnique*, 55(1), 77–84.

Fenton, G. A. and Griffiths, D. V. 2002. Probabilistic foundation settlement on spatially random soil. *Journal of Geotechnical and Geoenvironmental Engineering*, 128(5), 381–390.

Fenton, G. A. and Griffiths, D. V. 2008. *Risk Assessment in Geotechnical Engineering*, Hoboken, New Jersey, John Wiley and Sons.

Fenton, G. A., Griffiths, D. V. and Williams, M. B. 2005. Reliability of traditional retaining wall design. *Geotechnique*, 55(1), 55–62.

Goh, A. T. C., Phoon, K. K. and Kulhawy, F. H. 2009. Reliability analysis of partial safety factor design method for cantilever retaining walls in granular soils. *Journal of Geotechnical and Geoenvironmental Engineering*, 135(5), 616–622.

Kulhawy, F. H. 1991. Drilled shaft foundations. *Chapter 14 in Foundation Engineering Handbook*, 2nd ed. (ed. by H. Y. Fang), New York, Van Nostrand Reinhold, 537–552.

Ladd, C. C., Dascal, O., Law, K. T., Lefebrve, G., Lessard, G., Mesri, G. and Tavenas, F. 1983. Report of the subcommittee on embankment stability—Annex II. Committee of Specialists on Sensitive Clays on the NBR Complex. Société d'Energie de la BaieJames, Montréal, Que.

Low, B. K. and Tang, W. H. 1997. Reliability analysis of reinforced embankments on soft ground. *Canadian Geotechnical Journal*, 34(5), 672–685.

Metropolis, N., Rosenbluth, A., Rosenbluth, M. and Teller, A. 1953. Equations of state calculations by fast computing machines. *Journal of Chemical Physics*, 21(6), 1087–1092.

Microsoft Corporation. 2012. Microsoft Office EXCEL 2010, http://www.microsoft.com/en-us/default.aspx.

Phoon, K. K., Chen, J. R. and Kulhawy, F. H. 2006. Characterization of model uncertainties for augered cast-in-place (ACIP) piles under axial compression. *Foundation Analysis and Design: Innovative*

*Methods* (ed. by R. L. Parsons, L. Zhang, W. D. Guo, K. K. Phoon and M. Yang) (GSP 153) [*Proceedings of GeoShanghai*]. Reston: ASCE, 82–89.

Phoon, K. K., Kulhawy, F. H. and Grigoriu, M. D. 1995. *Reliability-Based Design of Foundations for Transmission Line Structures, Report TR-105000*, Electric Power Research Institute, Palo Alto.

Phoon, K. K., Kulhawy, F. H. and Grigoriu, M. D. 2003a. Development of a reliability-based design framework for transmission line structure foundations. *Journal of Geotechnical and Geoenvironmental Engineering*, 129(9), 798–806.

Phoon, K. K., Kulhawy, F. H. and Grigoriu, M. D. 2003b. Multiple resistance factor design for shallow transmission line structure foundations. *Journal of Geotechnical and Geoenvironmental Engineering*, 129(9), 807–818.

Santoso, A., Phoon, K. K. and Quek, S.-T. 2009. Reliability analysis of infinite slope using subset simulation. Contemporary topics in *in situ* testing, analysis, and reliability of foundations (GSP 186), *2009 International Foundation Congress and Equipment Expo*; March 15–19, 2009, Orlando, FL; ASCE, 278–285.

Soulié, M., Montes, P. and Silvestri, V. 1990. Modeling spatial variability of soil parameters. *Canadian Geotechnical Journal*, 27, 617–630.

Tang, W. H., Yucemen, M. S. and Ang, A. H. S. 1976. Probability based short-term design of slope. *Canadian Geotechnical Journal*, 13, 201–215.

Vesić, A. S. 1975. Bearing capacity of shallow foundations, *Chapter 3 in Foundation Engineering Handbook* (ed. by H. F. Winterkorn and H. Y. Fang), New York, Van Nostrand Reinhold, 121–147.

Walpole, R. E., Myers, R. H. and Myers, S. L. 1998. *Probability and Statistics for Engineers and Scientists*, Upper Saddle River, New Jersey, Prentice-Hall.

Wang, Y. 2011. Reliability-based design of spread foundations by Monte Carlo simulations. *Geotechnique*, 61(8), 677–685.

Wang, Y. 2012. Uncertain parameter sensitivity in Monte Carlo simulation by sample reassembling. *Computers and Geotechnics*, 46, 39–47.

Wang, Y. 2013. MCS-based probabilistic design of embedded sheet pile walls. *Georisk: Assessment and Management of Risk for Engineered Systems and Geohazards*, 7(3), 151–162.

Wang, Y. and Cao, Z. 2013. Expanded reliability-based design of piles in spatially variable soil using efficient Monte Carlo simulations. *Soils and Foundations*, 53(6), 820–834.

Wang, Y., Au, S. K. and Kulhawy, F. H. 2011a. Expanded reliability-based design approach for drilled shafts. *Journal of Geotechnical and Geoenvironmental Engineering*, 137(2), 140–149.

Wang, Y., Cao, Z. and Au, S. K. 2010. Efficient Monte Carlo simulation of parameter sensitivity in probabilistic slope stability analysis. *Computers and Geotechnics*, 37(7–8), 1015–1022.

Wang, Y., Cao, Z. and Au, S. K. 2011b. Practical analysis of slope stability by advanced Monte Carlo simulation in spreadsheet. *Canadian Geotechnical Journal*, 48(1), 162–172.

Wang, Y. and Kulhawy, F. H. 2008. Reliability index for serviceability limit state of building foundations. *Journal of Geotechnical and Geoenvironmental Engineering*, 134(11), 1587–1594.

Xu, B. and Low, B. K. 2006. Probabilistic stability analyses of embankments based on finite element method. *Journal of Geotechnical and Geoenvironmental Engineering*, 132(11), 1444–1454.

Zuev, K. M., Beck, J. L., Au, S. K. and Katafygiotis, L. S. 2012. Bayesian post-processor and other enhancements of subset simulation for estimating failure probabilities in high dimensions. *Computers and Structures*, 92, 283–296.

# Part III

# Design

# Chapter 8

# LRFD calibration of simple limit state functions in geotechnical soil-structure design

*Richard J. Bathurst*

## 8.1 INTRODUCTION

The North American practice for design of soil structures in geotechnical applications is based on a load and resistance factor design (LRFD) approach. Limit state design equations for failure or serviceability modes can be expressed as follows (AASHTO 2012; CSA 2006):

$$\varphi R_n \geq \sum \gamma_{Qi} Q_{ni} \tag{8.1}$$

where $R_n$ is the nominal resistance for a particular limit state, $\varphi$ is the resistance factor, $Q_{ni}$ is the nominal load contribution, and $\gamma_{Qi}$ is the (corresponding) load factor. The load terms are due to permanent and live load contributions. For typical limit state design equations found in North American design codes, $\varphi \leq 1$ and $\gamma_{Qi} \geq 1$. The expectation is that by satisfying limit state design equations expressed as Equation 8.1, design outcomes will have a probability of failure that is acceptable (i.e., small). For the case of a single load term, the limit state design equation can be expressed as

$$\varphi R_n \geq \gamma_Q Q_n \tag{8.2}$$

The nominal values in the above equations are typically computed using closed-form equations that are deterministic and/or prescribed values (e.g., yield strength of a steel element). In LRFD *design*, the load terms are typically determined first and then the resistance value ($R_n$) is adjusted so that the limit state design equation is satisfied. LRFD *calibration* involves selecting values $\varphi \leq 1$ and $\gamma_{Qi} \geq 1$ so that the probability of failure for multiple nominally identical structures does not exceed an accepted value.

This chapter describes the computational details of rigorous reliability theory-based calibration to select load and resistance factors for simple limit state design equations (i.e., Equation 8.2). Computational details are presented to facilitate implementation within Excel spreadsheets. An example of the general approach is demonstrated for the steel strip reinforcement pullout failure limit state used in the internal stability design of mechanically stabilized earth (MSE) walls.

## 8.2 PRELIMINARIES

As a starting point, consider the following simple limit state equation (performance function) with one load term:

$$g = R_m - Q_m \tag{8.3}$$

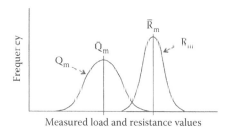

*Figure 8.1* Frequency distributions for measured load and resistance values.

Here g is a random variable representing the margin of safety and $R_m$ and $Q_m$ are random uncorrelated measured (actual) resistance and load values, respectively. Idealized distributions for $R_m$ and $Q_m$ are shown in Figure 8.1. An estimate of the factor of safety used in allowable stress design (ASD) could be computed from these two data sets as $F = \bar{R}_m / \bar{Q}_m$, where $\bar{R}_m$ and $\bar{Q}_m$ are the mean values of the two distributions. The limitations of the factor of safety approach are well known and are not repeated here.

Most often, nominal (predicted) values of resistance and load used in limit state design equations can be expected to vary from measured values for soil-structure limit states. The ratio of measured to nominal (predicted) value is called "bias." In this chapter, bias values are denoted as X. In the context of geotechnical soil-structure design, the magnitude of bias values will depend on model accuracy (the intrinsic accuracy of the deterministic theoretical, semi-empirical, or empirical model representing the mechanics of the limit state under investigation), random variation in input parameter values, spatial variation in input values, quality of data and, consistency in interpretation of data when data are gathered from multiple sources, which is the typical case (Allen et al. 2005). Bias values and bias statistics can be used to transform nominal (predicted) values to measured values so that estimates of actual probability of failure are realistic. Resistance bias $(X_R)$ and load bias $(X_Q)$ are random variables and are computed as

$$X_R = R_m/R_n \tag{8.4a}$$

$$X_Q = Q_m/Q_n \tag{8.4b}$$

Here, $R_n$ and $Q_n$ are nominal (predicted) resistance and load values, respectively. These transformations are only valid if the bias values are uncorrelated with predicted values. Substitution into Equation 8.3 gives

$$g = X_R R_n - X_Q Q_n \tag{8.5}$$

Equations 8.3 and 8.5 with $g = 0$ are of particular interest since these equations delineate between safe and unsafe outcomes (i.e., the failure line). The minimum value of $R_n$ to satisfy Equation 8.2 is

$$R_n = \frac{\gamma_Q Q_n}{\varphi} \tag{8.6}$$

Substituting into Equation 8.5 gives

$$g = X_R \frac{\gamma_Q Q_n}{\varphi} - X_Q Q_n \tag{8.7}$$

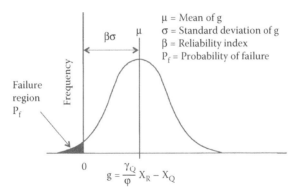

*Figure 8.2* Distribution of factored limit state function g.

Or, equivalently by redefining $g/Q_n$ as g

$$g = \frac{\gamma_Q}{\varphi} X_R - X_Q \tag{8.8}$$

The mean values of resistance and load bias values are denoted as $\mu_R$ and $\mu_Q$, respectively, and the corresponding coefficient of variation (COV) values as $COV_R$ and $COV_Q$.

LRFD calibration can now be understood to be the selection of resistance and load factors such that the probability that g is less than zero ($P_f (g < 0)$) in Equation 8.8 does not exceed a prescribed value. The distribution of g values for candidate values of $\gamma_Q$ and $\varphi$ and random variables $X_R$ and $X_Q$ is illustrated in Figure 8.2. $P_f (g < 0)$ is related to the highlighted area of the frequency distribution curve with g < 0.

Figure 8.2 also shows the definition of reliability index ($\beta$) as the number of standard deviations ($\sigma$) between the mean ($\mu$) of the distribution and g = 0. The relationship between probability of failure and reliability index is

$$P_f = 1 - \Phi(\beta) \tag{8.9}$$

Here, $\Phi$ is the standard normal cumulative distribution function (CDF) (NORM.DIST in Excel). Common practice is to use the reliability index $\beta$ to quantify margins of safety. Equation 8.6 shows that the conventional meaning of factor of safety appears as $F = \gamma_Q/\varphi$. This equivalency provides a method to relate past practice using ASD to LRFD calibration outcomes as discussed later in the chapter.

## 8.3 BIAS VALUE DISTRIBUTIONS

The distributions for load and resistance bias values are best described using standard normal CDF plots in which the horizontal axis is bias X and the vertical axis is the standard normal variable (z) computed as

$$z_i = \Phi^{-1}(p_i) \tag{8.10}$$

Here, $p_i = i/(n + 1)$, n is the number of bias values and $i = rank(X_i)$. Normal distributions of random variable X will plot as a straight line with $\mu_X$ (mean of bias values) intersecting the horizontal axis at z = 0 and the slope of the line will be the inverse of the standard deviation ($1/\sigma_X$) of the X distribution.

Lognormal distributions of X will plot as a curve. However, these X values can be transformed to log values (LNX) and the transformed values will then plot as a straight line on a CDF plot of $z_i$ versus $\ln(X_i)$. The mean of log values of X, denoted as $\mu_{LNX}$, will intersect the horizontal axis at z = 0 and the slope of the line will be the inverse of the standard deviation ($1/\sigma_{LNX}$) of the $\ln(X)$ distribution.

## 8.4 CALCULATION OF $\beta$, $\Upsilon_Q$, AND $\varphi$

Monte Carlo (MC) simulation can be used to compute an array of g values using random values of $X_R$ and $X_Q$ taken from frequency distribution functions with mean and standard deviation matching CDF plots of z versus bias values. Closed-form solutions can be used to compute $\beta$ and $\varphi$ if $X_R$ and $X_Q$ are both normal distributed or both are lognormal distributed. If this is not the case, then MC simulation must be used. The following sections describe the steps to process data and to carry out calibration.

### 8.4.1 Generation of bias values

The first step in LRFD calibration is to gather measured load ($Q_m$) data and measured resistance ($R_m$) data for the limit state under investigation. For each measured value, compute the corresponding nominal (predicted) value ($Q_n$ or $R_n$). For example, for each load measurement in a reinforcement layer in a reinforced soil wall structure, compute the matching predicted load value using a closed-form solution that corresponds to that limit state. An example for the resistance side is the measured pullout capacity ($R_m$) of the same type of reinforcement material from a pullout box test (or in situ pullout test) and the predicted pullout capacity ($R_n$) of the same reinforcement specimen computed using a closed-form solution for this ultimate limit state. The ratios of matching pairs of measured and predicted values (Equation 8.4) are the load and resistance bias values.

The following steps are used to create normal and lognormal CDF plots (z versus bias) for each set bias of values in an Excel spreadsheet (Nowak and Collins 2000):

- Place bias values (X) in column 1 of the spreadsheet.
- Sort the values from lowest to highest ($X_1$ to $X_n$) where n is the number of values.
- Calculate lognormal bias values $LNX_i$ using the Excel LN() function; column 2.
- Create (rank) column of integer values from i = 1 to n; column 3.
- Compute cumulative probability of bias values $p_i = i/(n + 1)$; column 4.
- Calculate the standard normal variable using the inverse of the standard normal cumulative distribution for each bias value as $z_i = \Phi^{-1}(p_i)$; column 5 = NORM.S.INV (column 4).
- Plot $z_i$ versus $X_i$.
- Plot $z_i$ versus $LNX_i$.

The two plots can be visually examined to decide whether the data are generally normal or lognormal distributed. While it is tempting to use quantitative statistical tools and forgo generation of a CDF plot, it is better to plot and visually examine the data because there are often deviations from an idealized normal or lognormal plot, particularly at the tails. It is the lower tail of the resistance bias distribution and the upper tail of the load bias distribution that contribute largely to the estimate of $P_f$ (g < 0). Hence, an accurate fit to these locations may be required to better estimate the true probability of failure rather than simply fitting a normal or lognormal distribution to the entire data set. Most often it is a lognormal CDF plot that best fits bias values using the entire data or fitting to the tails in most soil-structure cases.

The next step is to fit a curve (or curves) to the CDF plot that matches the entire distribution of data points and/or is a good fit to the bottom of the resistance bias plot or the top of the load bias plot based on a visual inspection.

As a starting point, the mean ($\mu_X$) and standard deviation ($\sigma_X$) of the bias distribution X can be computed for all data (AVERAGE and STDEVP functions in Excel). The coefficient of variation of bias values $COV_X$ is computed as

$$COV_X = \sigma_X / \mu_X \tag{8.11}$$

Approximations to the measured data assuming a normal distribution can be determined by computing the predicted bias value as

$$X_i = \mu_X(1 + z_i \times COV_X) \tag{8.12}$$

For a lognormal approximation:

$$LNX_i - EXP(\mu_{LNX} + z_i \times \sigma_{LNX}) \tag{8.13}$$

Values of $X_i$ and $LNX_i$ can be placed in columns 6 and 7, respectively, to facilitate plotting.

A useful alternative method is to use the following approximations to estimate $\sigma_{LNX}$ and $\mu_{LNX}$ using normal bias statistics:

$$\sigma_{LNX} = \sqrt{LN(1 + COV_X^2)} \tag{8.14}$$

$$\mu_{LNX} = LN(\mu_X) - \frac{1}{2}\sigma_{LNX}^2 \tag{8.15}$$

These substitutions can be used in Equation 8.13 to estimate $LNX_i$ and these values placed in column 8.

Values of $COV_X$ and $\mu_X$ can be changed by trial and error to explore the visual fit to the bias data over the entire range of data points or at the tails as demonstrated in the examples given later in this chapter.

## 8.4.2 Selection of load factor

For a prescribed target probability of failure (or target $\beta$ value), either $\varphi$ or $\gamma_Q$ must be assumed to give a unique solution. Typically, $\gamma_Q$ is selected first based on a load exceedance criterion. For example, during the development of the AASHTO (2012) and Canadian (CSA 2006) LRFD highway bridge design codes, the load factor for vehicle loads on bridges was selected so that the factored nominal load would not exceed measured load data in 97.7% of cases (i.e., two standard deviations below the mean of load bias values) (Nowak 1999; Nowak and Collins 2000). This value corresponds to a load factor computed using the following equation with $n_Q = 2$ and a distribution of vehicle loads that is normal distributed:

$$\gamma_Q = \mu_Q(1 + n_Q \times COV_Q) \tag{8.16}$$

Alternatively, the load factor corresponding to a specified load exceedance criterion can be computed as the bias value for which the fraction of bias values greater than that value equals the exceedance value. This is the same as plotting a cumulative probability plot (e.g., $p_i$ versus $X_{Qi}$) and finding the value of $X_{Qi}$ that corresponds to the target exceedance value on the vertical axis. The selection of the load factor could also be based on past practice. Bathurst et al. (2013) studied the case of steel reinforced soil walls where, consciously or not, a past practice led to measured reinforcement loads that exceeded predicted loads in 37% of cases as opposed to 3% of the cases in the vehicle traffic load example described at the beginning of this section. Clearly, these two exceedance values will give different load factors for the same load bias distributions.

### 8.4.3 Selection of target reliability index

Past geotechnical design practice has led to a target probability of failure for foundations, in general, of approximately 1 in 1000, or equivalently $\beta = 3.09$. For highly strength-redundant systems, such as pile groups and reinforced soil walls that have multiple layers of reinforcement, a target reliability index value of $\beta = 2.0$–$2.5$ is often used (Barker et al. 1991; Paikowsky et al. 2004; Allen 2005). The justification for the lower $\beta$ value is that if one load-carrying element in these systems fails, the load can be shed to other elements in the system. The final choice of $\beta$ is often decided by the regulators who are responsible for LRFD design guidance documents.

### 8.4.4 Calculation of $\varphi$

There are two general approaches to calculate $\varphi$ for a given load factor $\gamma_Q$ and a target $\beta$ value. The most robust method is MC simulation since this method can be used for any distributions of load and resistance bias values. For the simple linear limit state equation introduced earlier, there are closed-form solutions for $\varphi$ if both bias value distributions are normal distributed or both distributions are lognormal distributed.

#### 8.4.4.1 MC simulation

MC simulation in the context of this chapter involves randomly sampling distributions of load and resistance bias values using bias statistics that match the range on the CDF plots that are of interest (i.e., the entire range of measured data or the tail). There are software programs that can perform MC simulation very easily. However, in the text to follow, a methodology that can be implemented within an Excel spreadsheet is described because this approach highlights details that are valuable to novices and numerical outcomes can be easily plotted.

- Select the number of simulations n and place $i = 1$ to n in column 1.
- Compute a column of random values of load bias $X_{Qi}$ and place in column 2. If the distribution is normal, use NORM.INV(RAND(), $\mu_{XQ}$, $\sigma_{XQ}$). If the distribution is lognormal, use LOGINV(RAND(), $\mu_{LNXQ}$, $\sigma_{LNXQ}$).
- Compute a column of random values of resistance bias $X_{Ri}$ and place in column 3. If the distribution is normal, use NORM.INV(RAND(), $\mu_{XR}$, $\sigma_{XR}$). If the distribution is lognormal, use LOGINV(RAND(), $\mu_{LNXR}$, $\sigma_{LNXR}$).
- Select values of load factor and resistance factor ($\gamma_Q$ and $\varphi$) and compute $g_i = (\gamma_Q/\varphi)X_{Ri} - X_{Qi}$ in column 4 for each pair of random numbers.
- Copy column of numbers in column 4 to column 5 and sort from minimum to maximum.

- Compute the cumulative probability of each $g_i$ value as $p_i = i/(n+1)$ and place in column 6.
- Compute the standard normal variable for each value of $g_i$ as $z_i = \Phi^{-1}(p_i)$ using NORM.S.INV $(p_i)$ and place in column 7.
- Plot column 6 data versus column 5 data as a cumulative probability distribution ($p_i$ versus $g_i$) plot.
- Where the plot crosses the vertical axis at $g = 0$ corresponds to the probability of failure $P_f$ ($g < 0$).
- Plot column 7 data versus column 5 data as a standard normal CDF plot ($z_i$ versus $g_i$).
- Where the plot crosses the vertical axis at $g = 0$ corresponds to $-\beta$ for $P_f$ ($g < 0$).

Typically, the calculations described above are repeated with different trial resistance factors until a target $\beta$ value is achieved. Small differences between simulation outcomes will occur using the same mean and COV input parameters, but with different arrays of randomly generated variables $X_Q$ and $X_R$. Similarly, the number of simulation runs will influence computed values for $P_f$ and $\beta$. As a guide, the number of simulations N to reach a target probability of failure $P_f$ (target) can be estimated as (Nowak and Collins 2000)

$$N = \frac{1 - P_f(\text{target})}{COV_f^2 \times P_f(\text{target})} \tag{8.17}$$

Here, $COV_f$ is the desired target coefficient of variation in numerical outcomes.

### 8.4.4.2 Closed-form solutions

The closed-form solution for reliability index $\beta$ for normal distributions of load and resistance bias values can be expressed as

$$\beta = \frac{(\gamma_Q/\varphi)\mu_R - \mu_Q}{\sqrt{(COV_R \times (\gamma_Q/\varphi)\mu_R)^2 + (COV_Q \times \mu_Q)^2}} \tag{8.18}$$

Substituting Equations 8.14 and 8.15 for lognormal distributions of bias values into Equation 8.18, the expression for the reliability index is

$$\beta = \frac{LN\left((\gamma_Q/\varphi)(\mu_R/\mu_Q)\sqrt{(1 + COV_Q^2)/(1 + COV_R^2)}\right)}{\sqrt{LN[(1 + COV_Q^2)(1 + COV_R^2)]}} \tag{8.19}$$

Rearrangement leads to

$$\varphi = \frac{\gamma_Q(\mu_R/\mu_Q)\sqrt{(1 + COV_Q^2)/(1 + COV_R^2)}}{\exp\left\{\beta\sqrt{LN[(1 + COV_Q^2)(1 + COV_R^2)]}\right\}} \tag{8.20}$$

An advantage of these equations is that they can be used to quickly generate smooth distributions of $\beta$ and $\varphi$ that are useful for sensitivity analysis and presentation of results.

## 8.5 EXAMPLE

### 8.5.1 General

LRFD concepts and procedures for the case of a simple linear limit state function with one load term are demonstrated in this section using the example of one internal limit state for one class of mechanically stabilized earth (MSE) wall structures. MSE walls are particularly well suited for reliability-theory-based calibration because there are many instrumented walls reported in the literature. These data can be used to compare measured loads to nominal (predicted) loads in the soil-reinforcing layers under operational conditions and thus generate load bias data. Similarly, there are substantial collections of pullout (resistance) data available from published and unpublished sources for different soil-reinforcing types (e.g., steel strips, steel grids, steel anchors, and polymeric geogrids). Predicted pullout capacities at failure can be generated using any number of pullout models available in the literature. Predicted values can then be used with measured data from laboratory or in situ pullout tests to generate resistance bias values.

In the example to follow, the focus is on the treatment of bias data and LRFD calibration rather than the details of the actual load and resistance models. Load and resistance bias values used in this example are given in rank order in Table 8.1.

### 8.5.2 Load data

Measured reinforcement loads from steel strip reinforced soil walls have been collected by Miyata and Bathurst (2012a), Allen et al. (2001, 2004), and Bathurst et al. (2008b, 2009). Miyata and Bathurst (2012a) fitted an exponential equation to the measured data to improve the accuracy of the model compared to existing bi-linear load models. Fitting was carried out by selecting empirical coefficients such that the mean of the load bias values was 1 and the COV of load bias values was as small as possible. This explains why the mean of the load bias $\mu_Q = 1$ using all the load bias data in this example. In most LRFD calibration exercises, the mean of load bias values is unlikely to be unity and is most often less than 1. This is because most load models were developed for ASD with factors of safety and thus, they cautiously overestimate expected actual load values.

Load bias data are presented as normal and lognormal CDF plots in Figure 8.3a and 8.3b, respectively. Approximations are superimposed on the plots using the computed values of the mean and COV for all the load bias values ($\mu_Q$ and $COV_Q$). Visually, the normal distribution does well except at the top end of the data. The approximation to the data using a lognormal CDF plot does better at the top end of the measured data, but poorly at the bottom. Four data points are identified as possible outliers. Careful attention must be paid to any potential outliers at the end of CDF plots since it is the tails of these plots that can have a large influence on LRFD calibration outcomes. If these data are not representative of the sample population used to plot the entire data set, then they should be discarded. As an example, the outliers are removed in Figure 8.3c and approximations are made to the entire filtered data set using the mean and COV of the filtered data set and just the upper tail using $\mu_Q = 1.20$ and $COV_Q = 0.108$. The tail statistics were selected by manual adjustment until the approximation through the upper tail of the filtered data set was visually judged to be good. Load bias statistics are summarized in Table 8.1.

Filtered load bias data are plotted with cumulative and exceedance fraction axes in Figure 8.3d. Trial load factor values ($\gamma_Q$) can be selected for different exceedance values as illustrated in the figure. In the calculations to follow, a range of load factors from 1.00 to 1.50 is considered.

*Table 8.1* Bias data

| | Resistance bias | | Load bias (unfiltered) | | | | | | | | |
|---|---|---|---|---|---|---|---|---|---|---|---|
| Data points | 71 | | Data points | 93 | | | | | | | |
| Mean | 1.002 | | Mean | 1.002 | | | | | | | |
| S. dev. | 0.357 | | S. dev. | 0.316 | | | | | | | |
| COV | 0.356 | | COV | 0.316 | | | | | | | |
| Rank | $X_R$ | $X_Q$ | Rank | $X_R$ | $X_Q$ | Rank | $X_R$ | $X_Q$ | | | |
| 1 | 0.457 | 0.369 | 36 | 0.947 | 0.881 | 71 | 2.173 | 1.275 | | | |
| 2 | 0.495 | 0.373 | 37 | 0.949 | 0.887 | 72 | | 1.281 | | | |
| 3 | 0.545 | 0.475 | 38 | 0.952 | 0.890 | 73 | | 1.284 | | | |
| 4 | 0.604 | 0.519 | 39 | 0.955 | 0.897 | 74 | | 1.288 | | | |
| 5 | 0.632 | 0.548 | 40 | 0.960 | 0.907 | 75 | | 1.306 | | | |
| 6 | 0.644 | 0.558 | 41 | 0.986 | 0.911 | 76 | | 1.310 | | | |
| 7 | 0.656 | 0.560 | 42 | 0.988 | 0.918 | 77 | | 1.313 | | | |
| 8 | 0.673 | 0.562 | 43 | 1.001 | 0.919 | 78 | | 1.313 | | | |
| 9 | 0.689 | 0.589 | 44 | 1.017 | 0.924 | 79 | | 1.343 | | | |
| 10 | 0.691 | 0.596 | 45 | 1.029 | 0.947 | 80 | | 1.358 | | | |
| 11 | 0.692 | 0.599 | 46 | 1.046 | 0.957 | 81 | | 1.365 | | | |
| 12 | 0.698 | 0.601 | 47 | 1.050 | 0.963 | 82 | | 1.371 | | | |
| 13 | 0.701 | 0.603 | 48 | 1.057 | 0.970 | 83 | | 1.373 | | | |
| 14 | 0.721 | 0.607 | 49 | 1.065 | 0.985 | 84 | | 1.384 | | | |
| 15 | 0.728 | 0.636 | 50 | 1.078 | 0.993 | 85 | | 1.396 | | | |
| 16 | 0.735 | 0.662 | 51 | 1.086 | 1.041 | 86 | | 1.404 | | | |
| 17 | 0.739 | 0.664 | 52 | 1.088 | 1.052 | 87 | | 1.412 | | | |
| 18 | 0.744 | 0.680 | 53 | 1.091 | 1.066 | 88 | | 1.427 | | | |
| 19 | 0.761 | 0.683 | 54 | 1.097 | 1.083 | 89 | | 1.471 | | | |
| 20 | 0.768 | 0.705 | 55 | 1.115 | 1.098 | 90 | | 1.485 | | | |
| 21 | 0.776 | 0.711 | 56 | 1.153 | 1.122 | 91 | | 1.530 | | | |
| 22 | 0.779 | 0.753 | 57 | 1.160 | 1.125 | 92 | | 1.839 | | | |
| 23 | 0.791 | 0.759 | 58 | 1.179 | 1.128 | 93 | | 1.851 | | | |
| 24 | 0.795 | 0.789 | 59 | 1.207 | 1.130 | | | | | | |
| 25 | 0.807 | 0.808 | 60 | 1.345 | 1.133 | | | | | | |
| 26 | 0.816 | 0.814 | 61 | 1.388 | 1.134 | | | | | | |
| 27 | 0.829 | 0.814 | 62 | 1.406 | 1.141 | | | | | | |
| 28 | 0.844 | 0.820 | 63 | 1.420 | 1.159 | | | | | | |
| 29 | 0.850 | 0.844 | 64 | 1.429 | 1.201 | | | | | | |
| 30 | 0.851 | 0.850 | 65 | 1.514 | 1.217 | | | | | | |
| 31 | 0.875 | 0.855 | 66 | 1.595 | 1.223 | | | | | | |
| 32 | 0.889 | 0.857 | 67 | 1.736 | 1.231 | | | | | | |
| 33 | 0.914 | 0.860 | 68 | 1.799 | 1.245 | | | | | | |
| 34 | 0.933 | 0.865 | 69 | 1.919 | 1.248 | | | | | | |
| 35 | 0.937 | 0.869 | 70 | 2.077 | 1.262 | | | | | | |

Figure 8.4 shows a plot of load bias values versus nominal (predicted) load $Q_n$. The visual impression is that load bias values are uncorrelated with predicted load values. This is a necessary condition to allow the limit state function to be expressed in terms of bias values (Equation 8.8). If this condition is not satisfied, then the load model can be improved to remove this dependency or different load factors can be assigned to different ranges of predicted load. Statistical tests such as Spearman's rank correlation test can be used to examine the hypothesis that $X_Q$ and $Q_n$ are uncorrelated at (say) a level

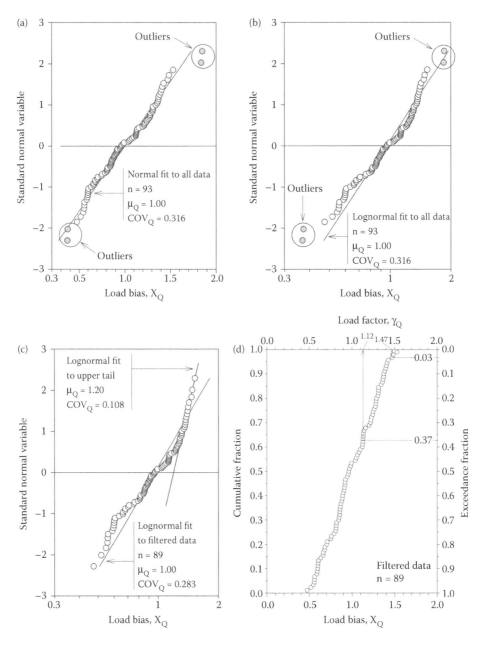

Figure 8.3 Load bias data: (a) Unfiltered data with normal fit, (b) unfiltered data with lognormal fit, (c) filtered data with lognormal fit, (d) cumulative and exceedance fractions.

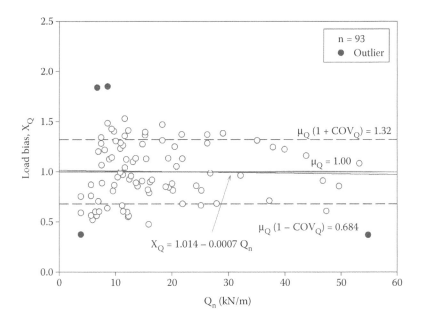

*Figure 8.4* Load bias versus nominal (predicted) load, $Q_n$.

of significance of 5% (Bathurst et al. 2008a). Another simple test is to fit a first-order polynomial to the data as shown in the figure and test the hypothesis that a zero slope cannot be excluded at a level of significance of 5%. In this example, regardless of whether or not the outlier data are considered, both quantitative tests lead to the conclusion that $X_Q$ and $Q_n$ are uncorrelated.

### 8.5.3 Pullout (resistance) data

Miyata and Bathurst (2012b) collected data from laboratory pullout tests and fitted an exponential model to the data to give improved predictions of pullout failure loads. As in the previous section, the new model was also developed to give a mean resistance bias value of 1 which explains why $\mu_R = 1$ in Figure 8.5. Figure 8.5a shows that the data are not normal distributed. A lognormal fit to the data using mean and COV statistics for the entire data set is visually judged to be a good fit over the entire range (Figure 8.5b). In particular, the fit is good through the bottom tail of the resistance bias data. Recall that it is the lower tail of the $X_R$ distribution that is most important in LRFD calibration. Resistance (pullout) bias statistics are summarized in Table 8.2.

*Table 8.2* Bias statistics

| Parameter | Load bias, $X_Q$ | | | | Resistance bias, $X_R$ | |
| --- | --- | --- | --- | --- | --- | --- |
| | Fit to all data | Fit to filtered data | Fit to upper tail | | Fit to all data | |
| n | 93 | 89 | 89 | n | | 71 |
| $\mu_Q$ | 1.00 | 1.00 | 1.20 | $\mu_R$ | | 1.00 |
| $COV_Q$ | 0.316 | 0.283 | 0.108 | $COV_R$ | | 0.356 |

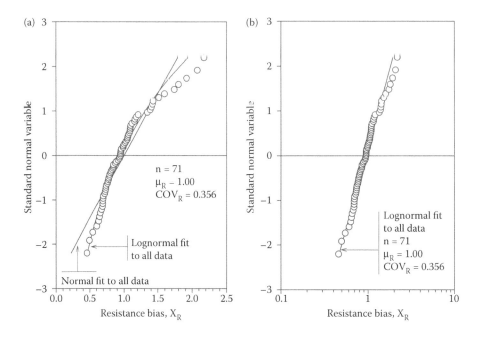

*Figure 8.5* Resistance bias data: (a) CDF plot with normal $X_R$ axis and (b) CDF plot with lognormal $X_R$ axis.

Figure 8.6 shows resistance bias data plotted against nominal (predicted) resistance values. In this particular example, only 50 of 71 tests in the source documents reported the actual pullout test capacity. Nevertheless, quantitative tests of the type previously described for the load data confirmed that there are no hidden dependencies between $X_R$ and $R_n$, which is also consistent with the visual impression in this figure.

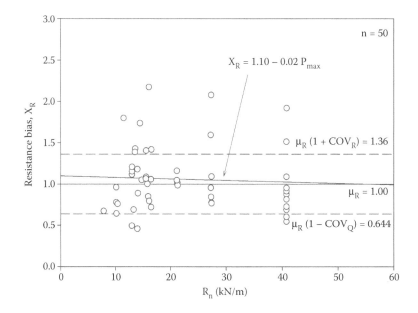

*Figure 8.6* Resistance bias versus (nominal) predicted resistance (pullout capacity), $R_n$.

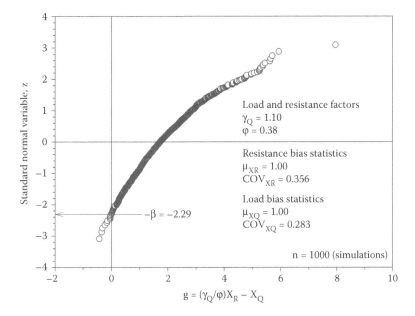

*Figure 8.7* Example plot of MC simulation results showing location of $\beta$ for $g = 0$.

## 8.5.4 Calibration

### 8.5.4.1 Resistance factor using MC simulation

Figure 8.7 shows the results of MC simulation using one set of load and resistance factors, and random values taken from lognormal distributions of $X_R$ and $X_Q$ with the mean and COV values shown in the figure. In this example, only 1000 simulations were carried out. Nevertheless, the $\beta$ value corresponding to $g = 0$ is easily detectable and corresponds to $\beta = 2.29$ which is equivalent to $P_f = 0.011$. Additional simulations with 3000 and 5000 runs gave essentially the same value.

### 8.5.4.2 Resistance factor using closed-form solution

Target reliability index values of $\beta = 2.33$ ($P_f = 0.01$) and $\beta = 3.09$ ($P_f = 0.001$) were selected and values of resistance factor computed for a range of load factors using Equation 8.20. The numerical results are presented in Figure 8.8. The value of $\varphi = 0.384$ shown in the figure is in close agreement with the result from MC simulation for the same input parameters. In practice, load and resistance factors are typically computed to two decimal places. In LRFD design codes for geotechnical structures, the values are reported to the nearest 0.05 (AASHTO 2012; CSA 2006).

## 8.6 ADDITIONAL CONSIDERATIONS

This chapter has highlighted the fundamental steps to carry out LRFD calibration for a simple linear limit state design function. The example is the pullout limit state for the internal stability design of MSE walls constructed with steel strip soil reinforcement.

Figure 8.8 shows that there are detectable differences in numerical outcomes using bias statistics for fit-to-all load bias data and fit-to-upper tail only. The differences here can be

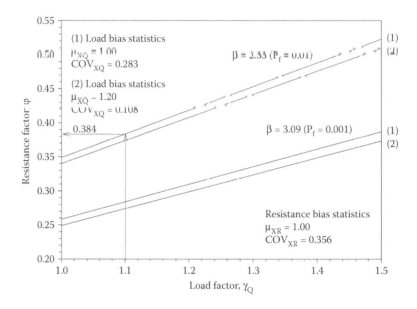

*Figure 8.8* Example outcomes for resistance factor φ using closed-form solution.

argued to be small, but recall that the load and resistance models used here were judged to be good models because they were originally calibrated to: (a) give mean bias values close to 1; (b) minimize the spread in bias values; and (c) remove bias dependencies with predicted values. Poorer models with greater differences between fit-to-tail and fit-to-all data can be expected to give larger differences in calculated resistance factor. Furthermore, many load and resistance models used in ASD past practice are poor candidates for LRFD calibration because the mean load bias values are so low that in order to satisfy a reasonable target β value during calibration, the resistance factor value must be greater than 1. This is contrary to expectations of design engineers using current LRFD design codes in North America.

The inverse of the slopes of the lines plotted in Figure 8.8 are equivalent to the factor of safety $F = \gamma_Q/\varphi$ in ASD past practice. For β = 2.33, F = 2.87–2.94 and for β = 3.09, F = 3.87–4.02. For steel strip walls, AASHTO (2012) recommends $\gamma_Q = 1.35$ and φ = 0.90. These values were selected to give F = 1.5 in order to be consistent with ASD past practice. This value is very much lower than the values computed using $F = \gamma_Q/\varphi$. The difficulty of relating LRFD calibration outcomes that are based on reliability-based theory to factor of safety values used in ASD past practice is typical. Nevertheless, in this particular example, it can be argued that the new load and pullout models will result in safer structures based on the conventional notion of factor of safety in ASD past practice.

This chapter has demonstrated that judgment plays an important role in the selection of data used to perform LRFD calibration. One example is the treatment of possible outliers. Outliers located at the lower tail of a resistance bias CDF plot and the upper tail of a load bias CDF plot should receive special attention because it is the distribution of bias values at these locations that largely influence calibration outcomes. Removing bias values at these tails when they have been taken from different case studies with no documented evidence for rejection could lead to non-conservative LRFD calibration outcomes. Proper documentation of measurement data cannot be over-stated in order that proper assessment of the validity of all data points in measured load and resistance sample populations can be made (Allen et al. 2005).

The results of LRFD calibration are sensitive to the database of measured values used to perform the calibration. Hence, calibrated LRFD limit state design equations should be restricted to project-specific designs that fall within the database envelope that was used to perform calibration. For the example used in this chapter, this means projects that fall within the same range of problem geometry, reinforcement type, and soil properties. As more data become available, it may be necessary to periodically recalibrate load and resistance factors.

## 8.7 CONCLUSIONS

The focus of this chapter has been on a methodology to carry out LRFD calibration of simple linear limit state design functions for which quality load and resistance bias data are available and the underlying deterministic models to compute load and resistance values are reasonably accurate. In order to demonstrate concepts and to facilitate plotting, the calculations have been described with implementation within Excel spreadsheets in mind. Once the reader is comfortable with the calibration methodology, the calculation steps can be easily implemented within simple numerical codes (e.g., Visual Basic, MATLAB®).

While the example used in this chapter to illustrate fundamental steps is for the pullout limit state in an MSE wall structure, the general approach is valid for other types of soil structures for which databases of physical load and resistance measurements are available. Examples can be found in publications by Allen (2005), Bathurst et al. (2011a, b, c, 2012, 2013) and Huang et al. (2012).

## REFERENCES

AASHTO 2012. *AASHTO LRFD Bridge Design Specifications*. American Association of State Highway and Transportation Officials, 6th Edition, Washington, DC, USA.

Allen, T., Christopher, B., Elias, V. and DeMaggio, J. 2001. *Development of the Simplified Method for Internal Stability*. Report WA-RD 513.1 July 2001, Washington State Department of Transportation, Olympia, WA, USA.

Allen, T.M. 2005. *Development of Geotechnical Resistance Factors and Downdrag Load Factors for LRFD Foundation Strength Limit State Design*. Publication No. FHWA-NHI-05-052, Federal Highway Administration, Washington, DC, USA.

Allen, T.M., Bathurst, R.J., Holtz, R.D., Lee, W.F. and Walters, D.L. 2004. A new working stress method for prediction of loads in steel reinforced soil walls. *ASCE Journal of Geotechnical and Geoenvironmental Engineering* 130(11): 1109–1120.

Allen, T.M., Nowak, A.S. and Bathurst, R.J. 2005. *Calibration to Determine Load and Resistance Factors for Geotechnical and Structural Design*. Transportation Research Board Circular E-C079, Washington, DC, USA.

Barker, R.M., Duncan, J.M., Rojiani, K.B., Ooi, P.S.K., Tan, C.K. and Kim. S.G. 1991. *Manuals for the Design of Bridge Foundations*. NCHRP Report 343, National Cooperative Highway Research Program, Transportation Research Board, Washington, DC, USA.

Bathurst, R.J., Allen, T.M., Miyata, Y. and Huang, B. 2013. LRFD calibration of metallic reinforced soil walls. *ASCE Geotechnical Special Publication No. 229, "Foundation Engineering in the Face of Uncertainty" Honoring Fred H. Kulhawy*, 585–601.

Bathurst, R.J., Allen, T.M. and Nowak, A.S. 2008a. Calibration concepts for load and resistance factor design (LRFD) of reinforced soil walls. *Canadian Geotechnical Journal* 45(10): 1377–1392.

Bathurst, R.J., Huang, B. and Allen, T.M. 2011a. Load and resistance factor design (LRFD) calibration for steel grid reinforced soil walls. *Georisk* 5(3–4): 218–228.

Bathurst, R.J., Huang, B. and Allen, T.M. 2011b. Analysis of installation damage tests for LRFD calibration of reinforced soil structures. *Geotextiles and Geomembranes* 29(3): 323–334.

Bathurst, R.J., Huang, B. and Allen, T.M. 2012. Interpretation of laboratory creep testing for reliability-based analysis and load and resistance factor design (LRFD) calibration. *Geosynthetics International* 19(1): 39–53.

Bathurst, R.J., Miyata, Y. and Konami, T. 2011c. Limit states design calibration for internal stability of multi-anchor walls. *Soils and Foundations* 51(6): 1051–1064.

Bathurst, R.J., Nernheim, A. and Allen, T.M. 2008b. Comparison of measured and predicted loads using the coherent gravity method for steel soil walls. *Ground Improvement* 161(3): 113–120.

Bathurst, R.J., Nernheim, A. and Allen, T.M. 2009. Predicted loads in steel reinforced soil walls using the AASHTO simplified method. *ASCE Journal of Geotechnical and Geoenvironmental Engineering* 135(2): 177–184.

CSA. 2006. *Canadian Highway Bridge Design Code (CHBDC)*. CSA Standard S6-06. Canadian Standards Association (CSA), Toronto, Ontario.

Huang, B., Bathurst, R.J. and Allen, T.M. 2012. Load and resistance factor design (LRFD) calibration for steel strip reinforced soil walls. *ASCE Journal of Geotechnical and Geoenvironmental Engineering* 138(8): 922–933.

Miyata, Y. and Bathurst, R.J. 2012a. Measured and predicted loads in steel strip reinforced c-φ soil walls in Japan. *Soils and Foundations* 52(1): 1–17.

Miyata, Y. and Bathurst, R.J. 2012b. Analysis and calibration of default steel strip pullout models used in Japan. *Soils and Foundations* 52(3): 481–497.

Nowak, A.S. 1999. *Calibration of LRFD Bridge Design Code*. NCHRP Report 368, National Cooperative Highway Research Program, Transportation Research Board, Washington, DC, USA.

Nowak, A.S. and Collins, K.R. 2000. *Reliability of Structures*. McGraw-Hill, New York, NY, USA.

Paikowsky, S.G., Birgisson, B., McVay, M., Nguyen, T., Kuo, C., Baecher, G., Ayyub, B. et al. 2004. *Load and Resistance Factor Design (LRFD) for Deep Foundations*. NCHRP Report 507, National Cooperative Highway Research Program, Transportation Research Board, Washington, DC, USA.

## Chapter 9

# Reliability-based design

## Practical procedures, geotechnical examples, and insights

*Bak-Kong Low*

### 9.1 INTRODUCTION

The design approach based on the overall factor of safety has long been used by geotechnical engineers. More recent alternatives are the load and resistance factor design (LRFD) approach in North America, and the characteristic values and partial factors used in the limit state design approach in Eurocode 7 (EC7). Yet another approach can play at least a useful complementary role to LRFD and EC7, namely the design based on a target reliability index that explicitly reflects the uncertainty of the parameters and their correlation structure. Among the various versions of reliability indices, that based on the first-order reliability method (FORM) for correlated non-normals is most consistent. A special case of FORM is the earlier Hasofer and Lind index (1974) for correlated normal random variables. These reliability methods are described in Ditlevsen (1981), Ang and Tang (1984), Madsen et al. (1986), Haldar and Mahadevan (1999), Melchers (1999), and Baecher and Christian (2003), for example. In addition, Low and Tang (1997a, 2004, 2007) presented spreadsheet-based practical procedures for FORM reliability-based analysis and design, with the 2007 approach being an equally efficient alternative as the 2004 approach, and the 2004 approach being much more versatile and efficient over the 1997 prototype.

In many geotechnical problems, the limit state surface (LSS, which separates safe combinations of parametric values from unsafe combinations) is practically plane so that the probability of failure inferred from FORM based on the hyperplane assumption is sufficiently accurate. Nevertheless, the FORM results can be easily extended to the second-order reliability method (SORM) that accounts for the curvature of the LSS near the design point.

This chapter illustrates geotechnical reliability-based design (RBD)/analysis for a spread footing, a rock slope, three soil slopes, a laterally loaded single pile, an anchored sheet pile wall, a rockbolt-reinforced tunnel, and soft clay consolidation accelerated by prefabricated vertical drains, all adapted from papers authored/coauthored by the writer. The focus is on practical procedures in the Excel spreadsheet platform for FORM, SORM, system FORM and importance sampling (IS), subtleties that require attention, and some interesting insights. In FORM, the design point is a point on the boundary (the LSS) that separates safe combinations of parametric values (e.g., the mean-value point) from the unsafe combinations of parametric values. The design point is the most probable failure combination of parametric values. The similarities and differences between the ratios of mean values to design point values and the partial factors of the limit state design in EC7 will be discussed. Guidance on EC7 is provided in Simpson and Driscoll (1998) and Frank et al. (2005), for example.

This chapter is at a deeper level than the author's earlier chapter (Low 2008a), which can be read in conjunction.

This chapter only deals with certain aspects of reliability, namely methodology and concepts and insights, and not reliability in its broadest sense. Some of the RBD examples in Sections 9.2 through 9.10 have explicit closed-form performance functions, while other examples have implicit performance functions based on numerical methods. The spreadsheet based reliability computational procedures presented herein can be coupled with stand-alone numerical geotechnical packages (e.g., FEM, FDM) via bridging techniques such as response surface or artificial neural network methods.

A brief summary is given next for two relatively intuitive and transparent FORM computational procedures of Low and Tang (2004, 2007) and its relationship to the **u**-space computational approach, together with the dispersion ellipsoid perspective in the original space of random variables, aiming at overcoming the language and conceptual barriers (aptly noted by Whitman 1984) surrounding reliability analysis.

### 9.1.1 Three spreadsheet **FORM** procedures and intuitive dispersion ellipsoid perspective

The matrix formulation (Veneziano 1974, Ditlevsen 1981) of the Hasofer and Lind (1974) index $\beta$ is

$$\beta = \min_{\mathbf{x} \in F} \sqrt{(\mathbf{x} - \mu)^T \mathbf{C}^{-1} (\mathbf{x} - \mu)} \tag{9.1a}$$

where $\mathbf{x}$ is a vector representing the set of random variables $x_i$, $\mu$ is the vector of mean values $\mu_i$, $\mathbf{C}$ is the covariance matrix, and $F$ is the failure domain. The notations "$T$" and "$-1$" denote *transpose* and *inverse*, respectively.

The following alternative formulation, which is mathematically equivalent to Equation 9.1a, was used in Low and Tang (1997b), because the correlation matrix $\mathbf{R}$ is easier to set up, and conveys the correlation structure more explicitly than the covariance matrix $\mathbf{C}$:

$$\beta = \min_{\mathbf{x} \in F} \sqrt{\left[ \frac{x_i - \mu_i}{\sigma_i} \right]^T \mathbf{R}^{-1} \left[ \frac{x_i - \mu_i}{\sigma_i} \right]} \tag{9.1b}$$

where $\mathbf{R}$ is the correlation matrix, $\sigma_i$ are the standard deviations, and other symbols as defined for Equation 9.1a. The point denoted by the $x_i$ values, which minimize Equation 9.1 and satisfy $\mathbf{x} \in F$, is the design point. This is the point of tangency of an expanding dispersion ellipsoid with the LSS, which separates safe combinations of parametric values from unsafe combinations (Figure 9.1). The one-standard-deviation (1-$\sigma$) dispersion ellipse and the $\beta$-ellipse in Figure 9.1 are tilted by virtue of cohesion $c$ and friction angle $\phi$ being negatively correlated. The quadratic form in Equation 9.1 also appears in the negative exponent of the established probability density function (PDF) of the multivariate normal distribution. As a multivariate normal dispersion ellipsoid expands from the mean-value point, its expanding surfaces are contours of decreasing probability values. Hence, to obtain $\beta$ by Equation 9.1 means maximizing the value of the multivariate normal PDF (at the most probable failure combination of parametric values) and is graphically equivalent to finding the smallest ellipsoid tangent to the LSS at the most probable failure point (the *design point*). This intuitive and visual understanding of the *design point* is consistent with the more mathematical approach in Shinozuka (1983), in which all variables were standardized and the limit state equation was written in terms of standardized variables.

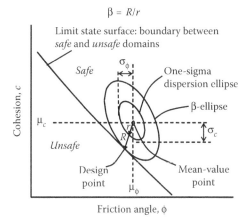

$$\beta = R/r$$

Limit state surface: boundary between *safe* and *unsafe* domains

*Figure 9.1* Illustration of the reliability index $\beta$ in the plane when $c$ and $\phi$ are negatively correlated.

In FORM, one can rewrite Equation 9.1b as follows (Low and Tang 2004) and regard the computation of $\beta$ as that of finding the smallest equivalent hyperellipsoid (centered at the equivalent normal mean-value point $\mu^N$ and with equivalent normal standard deviations $\sigma^N$) that is tangent to the LSS:

$$\beta = \min_{\mathbf{x} \in F} \sqrt{\left[ \frac{x_i - \mu_i^N}{\sigma_i^N} \right]^T \mathbf{R}^{-1} \left[ \frac{x_i - \mu_i^N}{\sigma_i^N} \right]} \tag{9.2}$$

where $\mu_i^N$ and $\sigma_i^N$ can be calculated by the Rackwitz and Fiessler (1978) transformation. Hence, for correlated non-normals, the ellipsoid perspective still applies in the original coordinate system, except that the non-normal distributions are replaced by an equivalent normal ellipsoid, centered not at the original mean values of the non-normal distributions, but at the equivalent normal mean $\mu^N$.

Equation 9.2 and the Rackwitz–Fiessler equations (for $\mu_i^N$ and $\sigma_i^N$) were used in the spreadsheet-automated constrained optimization FORM computational approach in Low and Tang (2004). An alternative to the 2004 FORM procedure is given in Low and Tang (2007), which uses the following equation for the reliability index $\beta$:

$$\beta = \min_{\mathbf{x} \in F} \sqrt{\mathbf{n}^T \mathbf{R}^{-1} \mathbf{n}} \tag{9.3}$$

The computational approaches of Equations 9.1b through 9.3 and associated ellipsoidal perspectives are complementary to the classical $\mathbf{u}$-space computational approach and may help to reduce the conceptual and language barriers of FORM.

The two spreadsheet-based computational approaches of FORM are compared in Figure 9.2. Either method can be used as an alternative to the classical $\mathbf{u}$-space FORM procedure. A third alternative (illustrated in Lü and Low 2011) is also shown in Figure 9.2, for which the Microsoft Excel's built-in constrained optimization routine (*Solver*) is invoked to automatically vary the $\mathbf{u}$ vector so that $\beta$ and the design point are obtained. This requires only adding one $\mathbf{u}$ column to the 2007 procedure, and expressing the unrotated $\mathbf{n}$ vector in terms of $\mathbf{u}$, where $\mathbf{u}$ is the uncorrelated standard equivalent normal vector in the rotated space of

Low and Tang (2004) FORM procedure:
minimize β by directly varying **x**

$$\beta = \min_{\mathbf{x} \in F} \sqrt{\left[ \frac{x_i - \mu_i^N}{\sigma_i^N} \right]^T \mathbf{R}^{-1} \left[ \frac{x_i - \mu_i^N}{\sigma_i^N} \right]}$$

$$\sigma^N = \frac{\phi \left\{ \Phi^{-1} [\Gamma(x)] \right\}}{f(x)}$$

$$\mu^N = x - \sigma^N \times \Phi^{-1} [F(x)]$$

Use Excel's *Solver* to change the **x** vector.
Subject to $g(\mathbf{x}) = 0$

Low and Tang (2007) FORM procedure:
minimize β by varying **n**, on which **x** depends

$$\beta = \min_{\mathbf{x} \in F} \sqrt{\mathbf{n}^T \mathbf{R}^{-1} \mathbf{n}}$$

Use Excel's *Solver* to change the **n** vector.

Subject to $g(\mathbf{x}) = 0$

For each trial **n**, get $x_i = F^{-1} [\Phi(n_i)]$

Third spreadsheet-based FORM procedure:
minimize β by varying **u**, from which **n** and **x** are readily obtainable

$$\beta = \min_{\mathbf{x} \in F} \sqrt{\mathbf{u}^T \mathbf{u}} \qquad \text{Use Excel's } \textit{Solver} \text{ to change the } \mathbf{u} \text{ vector, subject to } g(\mathbf{x}) = 0$$

For each automated trial **u**, get $\mathbf{n} = \mathbf{L}\mathbf{u}$, and $x_i = F^{-1} [\Phi(n_i)]$

*Figure 9.2* Comparison of the two FORM computational approaches of Low and Tang (2004, 2007), and the additional **u**-to-**n**-to-**x** approach. All three procedures use the optimization routine Solver resident in the Microsoft Excel spreadsheet.

the classical mathematical approach of FORM. The vectors **n** and **u** can be obtained from one another, $\mathbf{n} = \mathbf{L}\mathbf{u}$ and $\mathbf{u} = \mathbf{L}^{-1}\mathbf{n}$, as follows (e.g., Low et al. 2011):

$$\beta = \min_{\mathbf{x} \in F} \sqrt{\mathbf{n}^T \mathbf{R}^{-1} \mathbf{n}} = \min_{\mathbf{x} \in F} \sqrt{\mathbf{n}^T (\mathbf{L}\mathbf{U})^{-1} \mathbf{n}} = \min_{\mathbf{x} \in F} \sqrt{(\mathbf{L}^{-1}\mathbf{n})^T (\mathbf{L}^{-1}\mathbf{n})} \qquad (9.4a)$$

$$\text{that is, } \beta = \min_{\mathbf{x} \in F} \sqrt{\mathbf{u}^T \mathbf{u}}, \quad \text{where } \mathbf{u} = \mathbf{L}^{-1}\mathbf{n}, \quad \text{and} \quad \mathbf{n} = \mathbf{L}\mathbf{u}, \qquad (9.4b)$$

in which **L** is the lower triangular matrix of **R**. When the random variables are uncorrelated, $\mathbf{u} = \mathbf{n}$ by Equation 9.4, because then $\mathbf{L}^{-1} = \mathbf{L} = \mathbf{I}$ (the identity matrix).

The probability of failure can be approximately estimated as follows:

$$P_f \approx 1 - \Phi(\beta) = \Phi(-\beta) \qquad (9.5)$$

where $\Phi$ is the cumulative distribution function (CDF) of the standard normal random variable.

Equation 9.5 is exact when the LSS is planar and the parameters follow normal distributions. Inaccuracies in $P_f$ estimation may arise when the LSS is significantly nonlinear. More refined alternatives have been proposed, for example, the SORM, by Fiessler et al. (1979), Tvedt (1983, 1988, 1990), Breitung (1984), Hohenbichler and Rackwitz (1988), Koyluoglu and Nielsen (1994), Cai and Elishakoff (1994), Hong (1999), and Zhao and Ono (1999).

SORM analysis requires the FORM β value and design point values as inputs, and therefore is an extension dependent on FORM results. Hence, the SORM results are displayed alongside the FORM results in some of the examples to follow. In general, the SORM attempts to assess the curvatures of the LSS near the FORM design point in the dimensionless and rotated **u**-space. The failure probability is calculated from the FORM reliability index β and estimated principal curvatures of the LSS using established SORM equations.

Some files illustrating the Low and Tang (2004, 2007) approaches are available at http://alum.mit.edu/www/bklow. Step-by-step guidance that enables hands-on appreciations of the Low and Tang (2004, 2007) spreadsheet procedures are available in the Low (2008a) chapter, which will not be repeated here. Instead, the following sections discuss insights, advantages, and subtleties associated with FORM RBD in various geotechnical engineering examples, and the complementary roles that FORM RBD can play to LRFD or the design based on EC7.

## 9.2 EXAMPLE OF RELIABILITY-BASED SHALLOW FOUNDATION DESIGN

Tomlinson (1995)'s Example 2.2 determines the factor of safety against bearing capacity failure of a retaining wall that carries a horizontal load ($Q_h$) of 300 kN/m applied at a point 2.5 m above the base and a centrally applied vertical load ($Q_v$) of 1100 kN/m. The base (5 × 25 m) of the retaining wall is founded at a depth of 1.8 m in a silty sand with friction angle $\phi = 25°$, cohesion $c = 15$ kN/m², and unit weight $\gamma = 21$ kN/m³.

With respect to bearing capacity failure, the performance function (*PerFunc*) is

$$PerFunc = q_u - q \tag{9.6a}$$

where

$$q_u = cN_c s_c d_c i_c + p_o N_q s_q d_q i_q + \frac{B'}{2}\gamma N_\gamma s_\gamma d_\gamma i_\gamma \tag{9.6b}$$

$$q = Q_v/B' \tag{9.6c}$$

in which $q_u$ is the ultimate bearing capacity, $q$ is the applied bearing pressure, $c$ is the cohesion of soil, $p_o$ is the effective overburden pressure at the foundation level, $B'$ is the effective width of the foundation, $\gamma$ is the unit weight of soil below the base of the foundation, and $N_c$, $N_q$, and $N_\gamma$ are bearing capacity factors, which are functions of the friction angle ($\phi$) of soil:

$$N_c = (N_q - 1)\cot(\phi) \tag{9.7a}$$

$$N_q = e^{\pi\tan\phi}\tan^2\left(45 + \frac{\phi}{2}\right) \tag{9.7b}$$

$$N_\gamma = 2(N_q + 1)\tan\phi \tag{9.7c}$$

Several expressions for $N_\gamma$ exist. The above $N_\gamma$ is attributed to Vesic in Bowles (1996). The nine factors $s_i$, $d_i$, and $i_i$ in Equation 9.6b account for the shape and depth effects of the foundation and the inclination effect of the applied load. The formulas for these factors are based on Tables 4.5a and b of Bowles (1996).

To illustrate RBD, Tomlinson's deterministic example is extended probabilistically here (Figure 9.3). The width B of the foundation is to be determined based on a reliability index β = 3.0 against bearing capacity failure. The parameters $c$, $\phi$, $Q_h$, and $Q_v$ are assumed to be lognormal random variables with mean values equal to the values in Tomlinson's deterministic example, and with coefficient of variation equal to 0.20, 0.1, 0.15, and 0.10, respectively. The mean and standard deviation of these four variables are shown in Figure 9.3. The random variables are partially correlated, with the correlation matrix as shown in the figure. The foundation width required to achieve a reliability index β of 3.0 is B = 4.51 m. The similarities and differences between this RBD and the design based on EC7 or LRFD are discussed next.

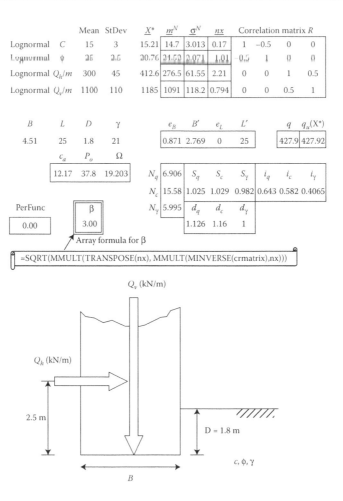

| | Mean | StDev | $X^*$ | $m^N$ | $\sigma^N$ | $nx$ | Correlation matrix $R$ | | | |
|---|---|---|---|---|---|---|---|---|---|---|
| Lognormal $C$ | 15 | 3 | 15.21 | 14.7 | 3.013 | 0.17 | 1 | −0.5 | 0 | 0 |
| Lognormal $\phi$ | 26 | 2.6 | 30.76 | 24.59 | 2.971 | 1.91 | 0.5 | 1 | 0 | 0 |
| Lognormal $Q_h/m$ | 300 | 45 | 412.6 | 276.5 | 61.55 | 2.21 | 0 | 0 | 1 | 0.5 |
| Lognormal $Q_v/m$ | 1100 | 110 | 1185 | 1091 | 118.2 | 0.794 | 0 | 0 | 0.5 | 1 |

| $B$ | $L$ | $D$ | $\gamma$ | | $e_B$ | $B'$ | $e_L$ | $L'$ | | $q$ | $q_u(X^*)$ |
|---|---|---|---|---|---|---|---|---|---|---|---|
| 4.51 | 25 | 1.8 | 21 | | 0.871 | 2.769 | 0 | 25 | | 427.9 | 427.92 |

| | $c_a$ | $P_o$ | $\Omega$ |
|---|---|---|---|
| | 12.17 | 37.8 | 19.203 |

| $N_q$ | 6.906 | $S_q$ | $S_c$ | $S_\gamma$ | $i_q$ | $i_c$ | $i_\gamma$ |
|---|---|---|---|---|---|---|---|
| $N_c$ | 15.58 | 1.025 | 1.029 | 0.982 | 0.643 | 0.582 | 0.4065 |
| $N_\gamma$ | 5.995 | $d_q$ | $d_c$ | $d_\gamma$ | | | |
| | | 1.126 | 1.16 | 1 | | | |

| PerFunc | $\beta$ |
|---|---|
| 0.00 | 3.00 |

Array formula for $\beta$

=SQRT(MMULT(TRANSPOSE(nx), MMULT(MINVERSE(crmatrix),nx)))

*Figure 9.3* Determining foundation width B, for a reliability index $\beta$ of 3.0.

## 9.2.1 RBD compared with EC7 or LRFD design, and complementary roles of RBD to EC7 and LRFD design

In the geotechnical design based on EC7, the characteristic values of resistance parameters are divided by partial factors to obtain the design values, while the characteristic values of action (load) parameters are multiplied by partial factors to obtain the design values. The resistance thus diminished is required to be greater or equal to the amplified actions (loadings), Figure 9.4, which also summarizes the three design approaches DA1, DA2, and DA3. The DA2 is akin in principle to the LRFD of North America. With respect to the characteristic value, Clause 2.4.5.2(2)P EN 1997-1 defines it as being "selected as a cautious estimate of the value affecting the occurrence of the limit state," and Clause 2.4.5.2(10) of EC7 states that statistical methods may be used when selecting characteristic values of geotechnical parameters, but they are not mandatory.

One may note the following similarities and differences between an RBD and that based on EC7 and LRFD, in the context of the RBD of Figure 9.3:

i. The $x^*$ column denotes the point where the four-dimensional equivalent dispersion ellipsoid touches the LSS. It is the most probable failure combination of the parameters.

General concepts of ultimate limit state design in Eurocode 7:

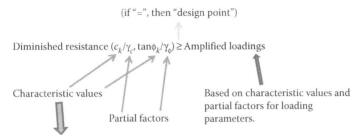

(if "=", then "design point")

Diminished resistance ($c_k/\gamma_c$, $\tan\phi_k/\gamma_\phi$) ≥ Amplified loadings

Characteristic values

Partial factors

Based on characteristic values and partial factors for loading parameters.

"Conservative", for example, 10 percentile for strength parameters, 90 percentile for loading parameters

The three sets of partial factors (on resistance, actions, and material properties) are not necessarily all applied at the same time.

In EC7, there are three possible design approaches:

- Design Approach 1 (DA1): (a) factoring actions only; (b) factoring materials only.
- Design Approach 2 (DA2): factoring actions and resistances (but not materials).
- Design Approach 3 (DA3): factoring structural actions only (geotechnical actions from the soil are unfactored) and materials.

*Figure 9.4* Characteristic values, partial factors, design point, and design approaches (DA) in Eurocode 7.

ii. The mean value, standard deviation, and design point value of a parameter can be related to the characteristic value and partial factor of the same parameter in the EC7 design approach. For example, for the given mean and standard deviations of the angle of friction $\phi$, the characteristic value $\phi_k$ (assuming at 10 percentile of the lognormal distribution) is 21.9°. The partial factor of $\phi$ (denoted by the symbol $\gamma_\phi$) implied in the design point (the $x^*$ values) is $\gamma_\phi = \phi_k/\phi^* = 21.9°/20.76° = 1.055$. (Note that in EC7, the partial factor $\gamma_\phi$ is applied to $\tan\phi_k$, and not to $\phi_k$. Had the mean and standard deviation been given for $\tan\phi$ rather than $\phi$, the characteristic value $\tan\phi_k$ and its corresponding partial factor can be back calculated from the FORM results by the same principle.)

iii. On the other hand, for the load parameter $Q_h$, with mean 300 and standard deviation 45, the characteristic value $Q_{hk}$ (assuming at 90 percentile of the lognormal distribution) is 359. The partial factor of $Q_h$ (denoted by the symbol $\gamma_{Qh}$) implied in the design point (the $x^*$ values) is $\gamma_{Qh} = Q_h^*/Q_{hk} = 412.6/359 = 1.15$.

iv. For lower mean value and standard deviation of $Q_h$, or for problems in different realms (e.g., slope stability or earth-retaining walls), the back-calculated partial factors could be different. To apply the same rigid partial factors across different realms or different levels of parametric uncertainty may not imply the same target failure probability. In an RBD, the same target reliability index (with its implied probability of failure) can be used across different realms and different levels of parametric uncertainty and correlations, including asymmetric probability distribution if appropriate. Also, if higher reliability is required where the consequence of failure is severe, a safer design can be obtained by raising the target reliability index. Such flexibility and automatic reflections of parametric sensitivities and correlations are not found in a Eurocode design based on code-recommended partial factors.

v. Note that the back-calculated partial factors are corollary by-products of an RBD, which requires statistical inputs (mean values, standard deviations, parametric correlations, and probability distributions) but not partial factors or characteristic values.

The design point is obtained as the most probable failure combinations of parametric values, and the reliability index $\beta$, being the distance (in units of directional standard deviations) from the safe mean-value point to the nearest failure boundary (the LSS, Figure 9.1), conveys information on the probability of failure. In LRFD or Eurocode design, there is no explicit information on the probability of failure.

vi. It is clear from the above discussions that in an RBD, one does not use or specify the partial factors. The design point values $x^*$ (and implied characteristic values and partial factors) are determined automatically and reflect sensitivities, standard deviations, correlation structure, and probability distributions in a way that prescribed partial factors and "conservative" characteristic values cannot.

vii. In Figure 9.3, by comparing the values under the $nx$ column, it is evident that bearing capacity is more sensitive to $Q_h$ than to $Q_v$, for the case in hand with its statistical inputs.

viii. The $x^*$ value for $c$ is slightly higher than the original mean value of $c$, due to the negative correlation between $c$ and $\phi$. In this case, the response is far more sensitive to $\phi$ than to $c$. Negatively correlated $c$ and $\phi$ means low values of $\phi$ tend to occur with high values of c, and vice versa. Under such circumstances, an RBD will automatically reflect sensitivities in obtaining design point values of $c$ and $\phi$ that can be both lower than their respective mean values, or one above average and the other below average as in this case. For the case in hand, to have both design values of $c$ and $\phi$ below their mean values using partial factors are unrealistic.

The above discussions suggest that RBD can be a complementary alternative to EC7 design or LRFD design *when the statistical information (mean values, standard deviations, correlations, and probability distributions) of the key parameters affecting the design are known* and one or more of the following circumstances apply:

- When partial factors have yet to be proposed by EC7 to cover uncertainties of the less common parameters, for example, *in situ* stress coefficient K in underground excavations in rocks, dip direction and dip angles of rock discontinuity planes, and smear effect of vertical drains installed in soft clay.
- When output has different sensitivity to an input parameter depending on the engineering problems (e.g., shear strength parameters in bearing capacity, retaining walls, and slope stability).
- When input parameters are by their physical nature either positively or negatively correlated. One may note that EC7 does not provide for the use of different characteristic values and partial factors when some parameters are correlated. The same design is obtained in EC7 with or without modeling of correlations among parameters.
- When spatially autocorrelated soil properties need to be modeled. This will be illustrated in Section 9.5 on a Norwegian slope with spatially autocorrelated soil unit weight and undrained shear strength.
- When there is a target reliability index or probability of failure for the design in hand. In this regard, one may note that a design by EC7 or LRFD provides no explicit information on the probability of failure.
- When uncertainty in unit weight $\gamma$ of soil needs to be modeled. This will be illustrated in the RBD of the anchored sheet pile wall in Section 9.8.

The following sections revisit—with more elaborations and discussions—some of the geotechnical examples of RBD and analysis from papers that the writer authored or coauthored, focusing on subtleties and insights. The cases involve both ultimate limit states (rock slope, a slope failure in San Francisco Bay mud, a Norwegian slope, a two-layered clay slope, an anchored sheet pile wall, and a rock tunnel) and serviceability limit states (a laterally loaded pile, and consolidation settlement of soft clay accelerated by vertical drains).

Figure 9.3 uses the Low and Tang (2004) spreadsheet-based FORM procedure by automatically changing the **x** vector using the constrained optimization routine *Solver* that resides in Microsoft Excel. The other two methods, namely the Low and Tang (2007) **x**-via-**n** method and the **x**-via-**n**-via-**u** method, as summarized in Figure 9.2, will obtain the same reliability index and design point for this example and for the examples of the following sections. Each of these three spreadsheet-based FORM computational procedures will be illustrated in different sections.

## 9.3 SORM ANALYSIS ON THE FOUNDATION OF FORM RESULTS FOR A ROCK SLOPE

This case is adapted from Low (2008b) and Low (2013).

### 9.3.1 Constrained optimizational FORM spreadsheet approach with respect to the u vector

Figure 9.5 shows the FORM and SORM analyses of a two-dimensional rock slope with five correlated random variables, two of which obey the highly asymmetric truncated

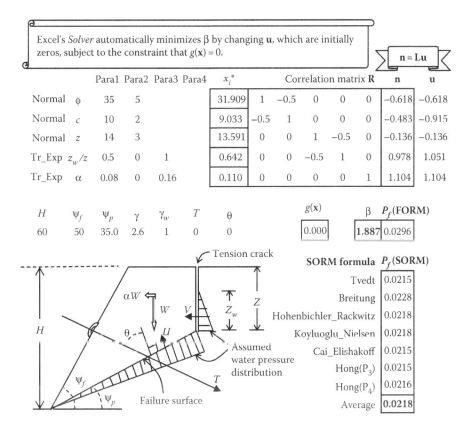

*Figure 9.5* FORM based on automatically changing the **u** vector in the Microsoft Excel platform, and SORM estimation of curvatures and $P_f$ on the basis of FORM $\beta$ and design point for a rock slope with five correlated random variables, two of which ($z_w/z$ and $\alpha$) obey the truncated exponential distribution.

exponentials. The five random variables are the shear strength parameters $c$ and $\phi$ of the failure surface (joint plane), tension crack depth $z$, height of water $z_w$ in the tension crack, and the coefficient of horizontal earthquake acceleration $\alpha$. The deterministic formulations are in Hoek (2007). The FORM analysis invokes Excel Solver to automatically change the $\mathbf{u}$ vector (initially zeros) so as to obtain the reliability index $\beta$ and the design point $x^*$ via the $\mathbf{n}$ vector and the $\mathbf{u}$ vector. The SORM analysis uses the Chan and Low (2012a) Excel spreadsheet approach. Five Monte Carlo simulations (MCS), each with 500,000 trials yielded $P_f$ values within the range 2.24–2.28%. The FORM $P_f$ is 2.96%, higher than the Monte Carlo average $P_f$ of about 2.25%. In contrast, the average SORM $P_f$ on the basis of FORM $\beta$ and four estimated components of curvature at the FORM design point is 2.18% (as shown) if the eight discrete points selected for curvature estimation correspond to $k = 2$ (coarser grid), and 2.05% if $k = 1$. The Breitung result of 2.28% is the closest to the Monte Carlo $P_f$ for this case where the value of $\beta$ is nearer to the practical higher design range of $\beta$.

There is no unique SORM $P_f$ value. It depends on the method used for estimating the curvatures at the design point and on the formula used to compute $P_f$ based on FORM $\beta$ and the curvatures at the FORM design point. Nevertheless, in the practical high reliability range ($P_f < 2\%$), the seven SORM formulas give consistent $P_f$ values for the case in Figure 9.5 and are more accurate than FORM $P_f$. One can as a rule extend the FORM analysis into the SORM analysis. Should the curvatures of the LSS turn out to be negligible, all the SORM formulas will approach Equation 9.5, with the result that the computed SORM probability of failure will be the same as FORM probability of failure.

An alternative solution of the slope of Figure 9.5 was given in Li et al. (2011), using 378 collocation sampling points, obtaining a $P_f$ of 2.30%.

It may be noted that $\mathbf{n} = \mathbf{Lu}$ by Equation 9.4b, and hence, $\mathbf{n} \neq \mathbf{u}$ as shown in the top two right-most columns of Figure 9.5 when some of the random variables are correlated.

The Low and Tang (2004) $\mathbf{x}$ space approach and the Low and Tang (2007) $\mathbf{n}$ space approach obtain the same reliability index $\beta$ and the design point $x^*$. Hence, all three methods of Figure 9.2 are efficient for the case in hand.

### 9.3.2 Positive reliability index only if the mean-value point is in the safe domain

In reliability analysis and RBD, one needs to distinguish negative from positive reliability index. The computed $\beta$ index can be regarded as positive only if the performance function value is positive at the mean-value point. Although the discussions in the next paragraph assume normally distributed random variables, they are equally valid for the equivalent normals of non-normal random variables in FORM.

The five random variables of the rock slope in Figure 9.5 include the shear strength parameters $c$ and $\phi$ of the discontinuity plane that is inclined at $\psi_p$. In the two-dimensional schematic illustration of Figure 9.6, the LSS is defined by performance function $g(\mathbf{x}) = 0$. The safe domain is where $g(\mathbf{x}) > 0$, and the unsafe domain is where $g(\mathbf{x}) < 0$. The mean-value point of Case I is in the safe domain and at the center of a one-standard-deviation dispersion ellipse, or ellipsoid in higher dimensions. As the dispersion ellipsoid expands, the probability density on its surface diminishes. The first point of contact with the LSS is the most probable failure point, also called the design point. The reliability index $\beta$ of Case I is therefore positive and represents the distance (in units of directional standard deviations) *from the safe mean-value point to the unsafe boundary* (the LSS) in the space of the random variables. The corresponding probability of failure $\Phi(-\beta)$ is <0.5. If the mean-value point sits right on the LSS, the probability of failure is 0.5 (if the LSS is planar) because $\beta = 0$ when the $\beta$-ellipsoid reduces to a point on the LSS. In contrast, Case II's mean-value point

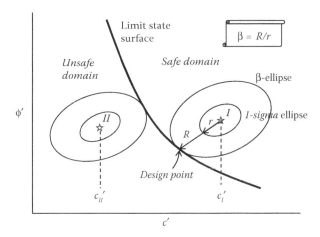

*Figure 9.6* Distinguishing negative from positive reliability index. Increasing the magnitude of reinforcing force T will increase the β value of Case II from *negative* to *zero* to *target* β.

is already in the unsafe domain, and the computed β must be given a negative sign because it is the distance *from the unsafe mean-value point to the safe boundary* (the LSS). Case II's probability of failure is >0.5.

The reinforcing force T required to achieve a target reliability index β (e.g., 2.5) for the slope of Figure 9.5 can be obtained as follows:

  i. If the performance function $g(\mathbf{x})$ is positive at the mean-value point (as in Case I), β is positive. Perform FORM reliability analysis with increasing $T$ until $\beta = \beta_{\text{target}}$.
  ii. If the performance function $g(\mathbf{x})$ is negative at the mean-value point (as in Case II), choose a value of $T$ so that $g(\mathbf{x})$ is > 0 at the mean-value point, then perform FORM reliability analysis by increasing $T$ until $\beta = \beta_{\text{target}}$.

The RBD of the embedment depth of an anchored sheet pile wall in Low (2005) provides another example of the need to distinguish negative β from positive β values.

## 9.4 PROBABILISTIC ANALYSES OF A SLOPE FAILURE IN SAN FRANCISCO BAY MUD

The failure of a slope excavated underwater in San Francisco Bay has been described in Duncan and Buchignani (1973), Duncan (2000, 2001), and Duncan and Wright (2005). The slope was part of a temporary excavation and was designed with an unusually low factor of safety to minimize construction costs. During construction, a portion of the excavated slope failed.

Low and Duncan (2013) analyzed the same underwater slope that failed, first deterministically using data from field vane shear and laboratory triaxial tests, Figure 9.7, then probabilistically, accounting for parametric uncertainty and positive correlation of the undrained shear strength and soil unit weight.

In the deterministic analysis, the factors of safety were computed with search for *critical noncircular slip surface* based on a reformulated Spencer method. The $F_s$ values obtained were 1.20, 1.16, and 1.00, for three types of undrained shear strength data, namely field vane tests, unconsolidated-undrained (UU) triaxial tests on trimmed 35 mm specimens, and

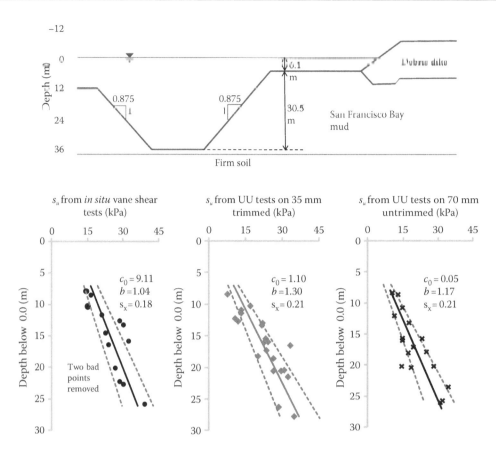

*Figure 9.7* Underwater excavated slope in San Francisco Bay mud, and the average $s_u$ profile, described by $s_u = c_0 + by$, and the ± 1 Std. Dev. lines.

UU triaxial tests on untrimmed 70 mm specimens, respectively. For comparison, the corresponding $F_s$ values for *critical circular slip surfaces* were 1.23, 1.19, and 1.03, respectively, slightly higher than those of the critical noncircular slip surfaces. It was noted that measured strength values were affected by the disturbance and rate of loading effects. Subtle errors were also caused by extrapolation of the undrained shear strength (*in situ* and lab tests data, available only for the upper 21 m of the Bay mud, from depth 6 m to depth 27 m; Figure 9.7) to the full depth of underwater excavation. Since the midpoint of a slip circular arc is at about the two-thirds depth, this means that in limit equilibrium slope stability analysis, half the slip surface was based on extrapolated strength.

In the probabilistic analyses by Low and Duncan (2013), FORM analyses based on lognormal distributions and *critical circular slip surfaces* produced failure probability values (by Equation 9.5) of 9.7, 19.4, and 45.6%, for undrained shear strength from field vane tests, UU triaxial tests on trimmed 35 mm specimens, and UU triaxial tests on untrimmed 70 mm specimens, respectively. These $P_f$ values are much higher than the $P_f$ of about 0.6% implied by the commonly required β of 2.5, or $P_f$ of 0.14% for a target β of 3.0. Hence, a failure was not unlikely, and did happen.

FORM analyses can be performed easily with search for the reliability-based *critical noncircular slip surfaces*, producing probabilities of failure (by Equation 9.5) a few percentages higher than the above-mentioned 9.7–45.6% range (circular slip surfaces) for the three types of undrained shear strength data. However, obtaining probability of failure from MCS for critical noncircular slip surfaces (for comparison with FORM $P_f$) would be very time-consuming because a search for critical noncircular slip surface is required for each set of random numbers generated in MCS. In contrast, it is much simpler to obtain Monte Carlo $P_f$ values with search for the *critical circular slip surface*, and these (Table 5 in Low and Duncan 2013) are practically identical to the FORM $P_f$ values of 9.7, 19.4, and 45.6% mentioned above.

## 9.5 RELIABILITY ANALYSIS OF A NORWEGIAN SLOPE ACCOUNTING FOR SPATIAL AUTOCORRELATION

Spatial autocorrelation (also termed spatial variability) arises in the geological material by virtue of its formation by natural processes acting over unimaginably long time (millions of years). This endows the geomaterial with some unique statistical features (e.g., spatial autocorrelation) not commonly found in the structural material manufactured under strict quality control. For example, by the nature of the slow precipitation (over many seasons) of fine-grained soil particles in water in nearly horizontal layers, two points in close horizontal or vertical proximity to one another are likely to be more positively correlated (e.g., likely to have similar undrained shear strength $c_u$ values) than two points further apart in the vertical direction.

A clay slope in southern Norway was analyzed deterministically and probabilistically by Low et al. (2007) using the Low (2003a) spreadsheet-based reformulations of the Spencer method and the intuitive FORM of Low and Tang (2004). The reformulation allows switching among the Spencer, Bishop simplified, and wedge methods on the same template, by specifying different side-force inclination options and different constraints of optimization. Search for the critical circular or noncircular slip surface is possible. The deterministic procedure was extended probabilistically by implementing the FORM via constrained optimization of the equivalent dispersion ellipsoid in the original space of the random variables. The procedure was illustrated for an embankment on soft ground and for a clay slope in southern Norway, both involving spatially correlated soil properties. The effects of autocorrelation distance on the results of reliability analysis were studied. Shear strength anisotropy was modeled via user-created simple function codes in the programming environment of the spreadsheet.

Figure 9.8 shows the results of reliability analysis involving 24 spatially correlated $c_u$ values and 24 spatially correlated unit weight values. The size of the correlation matrix is $48 \times 48$. The design point obtained by Excel Solver represents the most probable combination of the 24 values of $c_u$ and the 24 values of $\gamma$ that would cause failure. As expected for resistance parameters, the 24 values of undrained shear strength $c_u$ at the design point are all lower than their respective mean values. On the other hand, when the autocorrelation distance $\delta$ is 10 m or lower, as shown in Figure 9.8, the design point index of $\gamma$—defined as $(\gamma_i^* - \mu_\gamma)/\sigma_\gamma$, where $\gamma_i^*$ is the design point value of unit weight $\gamma$ for $i$ from 1 to 24—shows most values of unit weight $\gamma^*$ above their mean value $\mu_\gamma$, as expected for loading parameters, but, somewhat paradoxically, there are some design point values of $\gamma$ near the toe that are below their mean values. The implication is that the slope is less safe when the unit weights near the toes are lower. This implication can be verified by deterministic runs using higher $\gamma$ values near the toe, resulting in higher factors of safety. It would be difficult for the design code committee to recommend partial factors such that the design values of $\gamma$ are above the mean along some portions of the slip surface and below their mean along other portions. In contrast, the design point is automatically located in FORM analysis by the spreadsheet's

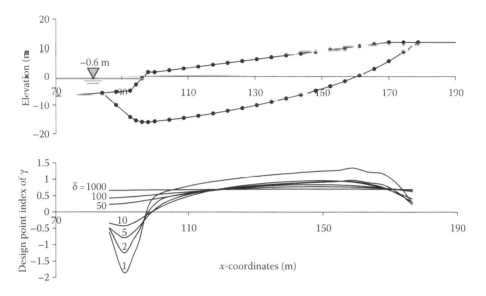

*Figure 9.8* Design point index of clay unit weight as a function of horizontal autocorrelation distance δ in meter.

built-in constrained optimization program. It reflects sensitivity and the underlying statistical assumptions from case to case in a way that the specified partial factors cannot.

At higher values of autocorrelation distance δ, the correlation coefficients approach 1.0; the design point indices of γ of the 24 slices approach a common value, as shown by the nearly horizontal line in Figure 9.8 for δ = 1000 m. The design point indices of $c_u$—defined as $(c_{ui}{}^* - \mu_{cui})/\sigma_{cui}$—of the 24 slices also approach a uniform common value when δ = 1000 m; however, the individual design point values of $c_u$ differ from slice to slice because the mean, $\mu_{cui}$, and standard deviation, $\sigma_{cui}$, vary from slice to slice.

The implications of not considering seabed erosion versus treating seabed as random (to account for uncertain depth of erosion) were discussed in Low et al. (2007).

The results of reliability analysis are only as good as the statistical input and reliability method used (e.g., FORM or SORM), in the same way that the results of deterministic analysis are only as good as the deterministic input and method used (e.g., Spencer method or other methods). A reliability analysis requires additional statistical input information that is not required in a deterministic factor-of-safety approach, but results in richer information pertaining to the performance function and the design point that is missed in a deterministic analysis.

The FORM reliability approach reflects the underlying analytical formulations and statistical assumptions and is able to locate the most probable combination of parametric values that would cause failure and the corresponding reliability index, without relying on rigid partial factors.

## 9.6 SYSTEM FORM RELIABILITY ANALYSIS OF A SOIL SLOPE WITH TWO EQUALLY LIKELY FAILURE MODES

A slope in two clayey soil layers was analyzed in Ching et al. (2009) using MCS and IS methods. The same two-layered slope was analyzed in Low et al. (2011), using system reliability bounds for multiple failure modes. It was shown that when two (rather than one) reliability-based critical slip surfaces are considered, the system failure probability bounds obtained in Low et al. (2011) agreed well with the MCS and IS results of Ching et al. (2009).

As shown in Figure 9.9, the upper clay layer is 18 m thick, with undrained shear strength $c_{u1}$; the lower clay layer is 10 m thick, with undrained shear strength $c_{u2.}$ The undrained shear strengths are normally distributed and independent. A hard layer exists below the second clay layer.

Since the shear strengths are characterized by $c_{u1}$ and $c_{u2}$, with $\phi_u = 0$, Bishop's simplified method and the ordinary method of slices will yield the same factor of safety, and either method can be used. Also, in this case where the upper clay layer is weaker than the lower clay layer, it is logical to locate two reliability-based critical slip circles, as shown in Figure 9.9, one entirely in the upper clay layer and the other passing through both layers. The FORM reliability indices for the two modes are 2.795 and 2.893, respectively. It is interesting to note that although $c_{u1}$ and $c_{u2}$ are uncorrelated, there is correlation between the two failure modes ($\rho_{12} = 0.4535$), because $c_{u1}$ affects both slip circles. The bounds on system failure probability, computed in two cells in Figure 9.9 by efficient implementation (in a ubiquitous spreadsheet platform) of the Kounias–Ditlevsen bimodal bounds for systems with multiple failure modes, are 0.432–0.441%, compared with the MCS estimated range of 0.37–0.506% (from Ching et al.'s reported MCS mean of 0.44% and COV of 15.04%).

The two reliability-based critical slip circles in Figure 9.9 have the smallest $\beta$ values among all possible slip circles tangent to the bottoms of the upper and lower clay layers, respectively. One can search for more reliability-based critical slip circles corresponding to different trial-tangent depths. Alternatively, a series of $\beta$ values can be obtained as a function of the x-coordinate values of the lower exit end of critical slip circles, as shown in Figure 9.10, where the existence of two stationary values ("troughs") of $\beta$ is obvious. It would be interesting to investigate the effect on the bounds of system failure probability when more reliability-based modes are considered. This is done in Figure 9.11, which, in contrast to Figure 9.9, has three additional modes ($\beta_3$, $\beta_6$, $\beta_7$) adjacent to the mode corresponding to the local minimum $\beta_1$, and three additional modes ($\beta_4$, $\beta_5$, $\beta_8$) adjacent to the mode corresponding to the local minimum $\beta_2$.

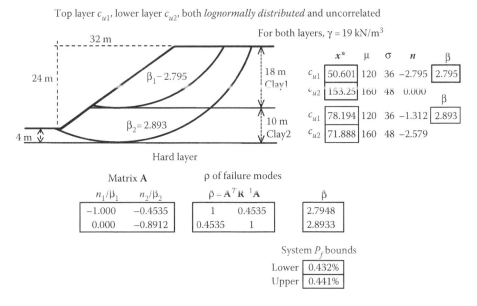

*Figure 9.9* FORM results for two reliability-based critical slip circles followed by system reliability analysis.

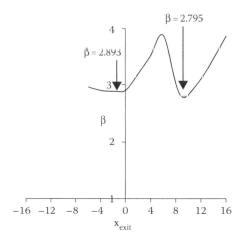

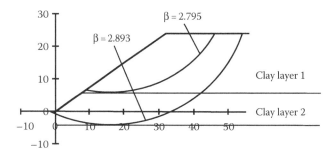

*Figure 9.10* Variation of reliability indices with the x-coordinate of the lower exit point of slip circles, and the two reliability-based critical slip circles with lower exit points at $x_{exit}$ of −1.5 m and +9 m, respectively.

It was noted, for example, in Ang and Tang (1984) that the bimodal bounds on failure probability of systems with multiple failure modes will depend on the ordering of the individual failure modes. It was suggested, for example, in Madsen et al. (1986), Melchers (1999), and Haldar and Mahadevan (1999) that ordering the failure modes in decreasing probabilities of failure will lead to closer bounds. This has been done in Figure 9.11, yielding $0.416\% \leq P_{F,sys} \leq 0.441\%$, practically the same range as that in Figure 9.9 when only the two local minimum modes were considered. A simple Visual Basic for Applications (VBA) code was also created to investigate the effects of all possible permutations (8!) of the failure modes on the system failure probability bounds: the same bounds as in Figure 9.11 were obtained.

That the system reliability bounds of the eight modes in Figure 9.11 differ little from the bounds of the two local minimum modes of Figure 9.9 can be attributed to the strong correlations among modes 1, 3, 6, and 7, and among modes 2, 4, 5, and 8, as seen from the very high (≈1.0) intermodal correlation coefficients of $\rho_{13}$, $\rho_{16}$, $\rho_{17}$, and of $\rho_{24}$, $\rho_{25}$, $\rho_{28}$. Physically, this means that direction vectors (linking the mean-value point and design points of the failure modes) are nearly parallel for modes 1, 3, 6, and 7, and for modes 2, 4, 5, and 8. The implied overlapping of the failure probability contents for modes 1, 3, 6, and 7, and also for modes 2, 4, 5, and 8 means that it is sufficiently accurate to calculate the bounds for the system failure probability by considering only the two stationary values of reliability index, namely $\beta_1$ and $\beta_2$ in Figure 9.9.

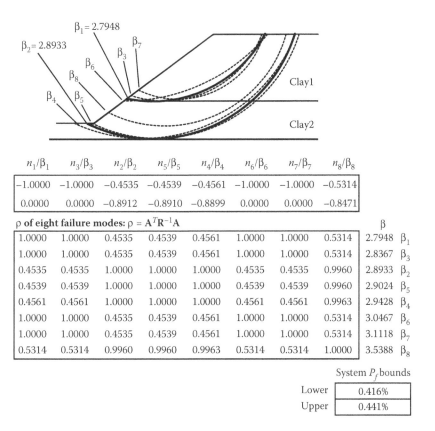

| $n_1/\beta_1$ | $n_3/\beta_3$ | $n_2/\beta_2$ | $n_5/\beta_5$ | $n_4/\beta_4$ | $n_6/\beta_6$ | $n_7/\beta_7$ | $n_8/\beta_8$ |
|---|---|---|---|---|---|---|---|
| −1.0000 | −1.0000 | −0.4535 | −0.4539 | −0.4561 | −1.0000 | −1.0000 | −0.5314 |
| 0.0000 | 0.0000 | −0.8912 | −0.8910 | −0.8899 | 0.0000 | 0.0000 | −0.8471 |

$\rho$ of eight failure modes: $\rho = \mathbf{A}^T\mathbf{R}^{-1}\mathbf{A}$  $\quad\quad\quad\quad\quad\quad\quad\quad\quad\quad\quad\quad\beta$

| | | | | | | | | $\beta$ | |
|---|---|---|---|---|---|---|---|---|---|
| 1.0000 | 1.0000 | 0.4535 | 0.4539 | 0.4561 | 1.0000 | 1.0000 | 0.5314 | 2.7948 | $\beta_1$ |
| 1.0000 | 1.0000 | 0.4535 | 0.4539 | 0.4561 | 1.0000 | 1.0000 | 0.5314 | 2.8367 | $\beta_3$ |
| 0.4535 | 0.4535 | 1.0000 | 1.0000 | 1.0000 | 0.4535 | 0.4535 | 0.9960 | 2.8933 | $\beta_2$ |
| 0.4539 | 0.4539 | 1.0000 | 1.0000 | 1.0000 | 0.4539 | 0.4539 | 0.9960 | 2.9024 | $\beta_5$ |
| 0.4561 | 0.4561 | 1.0000 | 1.0000 | 1.0000 | 0.4561 | 0.4561 | 0.9963 | 2.9428 | $\beta_4$ |
| 1.0000 | 1.0000 | 0.4535 | 0.4539 | 0.4561 | 1.0000 | 1.0000 | 0.5314 | 3.0467 | $\beta_6$ |
| 1.0000 | 1.0000 | 0.4535 | 0.4539 | 0.4561 | 1.0000 | 1.0000 | 0.5314 | 3.1118 | $\beta_7$ |
| 0.5314 | 0.5314 | 0.9960 | 0.9960 | 0.9963 | 0.5314 | 0.5314 | 1.0000 | 3.5388 | $\beta_8$ |

| | System $P_f$ bounds |
|---|---|
| Lower | 0.416% |
| Upper | 0.441% |

*Figure 9.11* System reliability analysis considering eight failure modes, including the two reliability-based critical modes of Figure 9.10.

## 9.7 MULTICRITERIA RBD OF A LATERALLY LOADED PILE IN SPATIALLY AUTOCORRELATED CLAY

Low et al. (2001) studied the deflection and bending moment of laterally loaded single piles in which soil–pile interaction was based on the nonlinear and strain-softening Matlock (1970) $p$–$y$ curves (Figure 9.12) The soil–pile interaction problem was solved using a rigorous numerical procedure based on constrained optimization in the spreadsheet platform. The numerical procedure was then extended to reliability analysis in which the Hasofer and Lind index was computed. The soil resistance was modeled stochastically to reflect spatial variation. Multicriteria RBD of a laterally loaded pile was also illustrated. The work was further extended in Chan and Low (2012b) to a probabilistic analysis of laterally loaded piles using response surface and neural network approaches. A case of deterministic analysis followed by multicriteria RBD from Low et al. (2001) is presented in this section.

This deterministic problem in Figure 9.12a is described in Tomlinson (1994, Example 8.2). A steel tubular pile having an outside diameter $d$ of 1.3 m and a wall thickness of 0.03 m forms part of a pile group in a breasting dolphin. The flexural rigidity $E_pI_p$ of the pile is 4,829,082 kNm². The pile is embedded 23 m in the stiff overconsolidated clay with undrained shear strength $c_u = 150$ kN/m² and protrudes 26 m above the seabed. For the case where a cyclic force of 421 kN is applied at 26 m above the seabed, only the embedded portion of the pile requires soil–structure interaction analysis. The deflection of the pile head

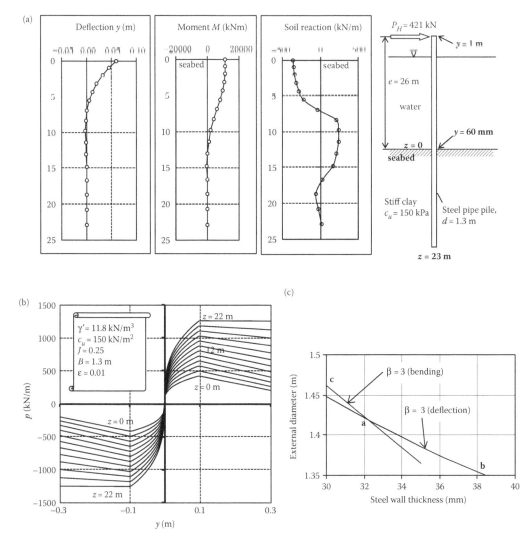

*Figure 9.12* Multi-criteria reliability-based design of a laterally loaded pile in spatially autocorrelated clay: (a) Numerical procedure for nonlinear $p$-$y$ analysis of a steel tubular pile in a breasting dolphin; (b) Matlock's nonlinear $p$-$y$ curves; (c) combinations of pile external diameter and wall thickness for reliability index $\beta$ of 3.

at 26 m above the seabed can be inferred from statics once the deflection and rotation of the pile at seabed level are known. The Matlock $p$-$y$ curves for clays, as shown in Figure 9.12b, have been used. The $p$-$y$ curves are nonlinear, exhibit strain softening, and vary with depth. The calculated pile deflection ($y_i$) at seabed level (where $z = 0$) is 0.0602 m. The pile head deflection (at 26 m above the seabed) is, by integrating the moment–curvature equation, 0.994 m or about 1 m. Separate analysis using a specially written Fortran program to perform the finite-element analysis using 60 equally spaced elements yielded a pile deflection of 0.0596 m at seabed level, compared with 0.0602 m in Figure 9.12, and practically identical shear and moment distribution along the pile length.

For reliability analysis, the 23-m embedded length of the pile was discretized into 30 segments of progressively greater length. The random variables are the lateral load $P_H$ at pile head and the undrained shear strength $c_u$ at 31 nodal points along the embedded pile length below the seabed. The $P_H$ was assumed to be normally distributed, with a mean value of 421 kN and a coefficient of variation of 25%. Since the mean undrained shear strength $c_u$ typically exhibits an increasing trend with depth, it was assumed that $\mu_{cu} = 150 + 2z$ kPa. The standard deviations of the 31 $c_u$ random variables were equal to 30% of their respective mean values. The following established negative exponential model was adopted to model the spatial variation of the $c_u$ values:

$$\rho_{ij} = e^{-(|Depth(i)-Depth(j)|)/\delta} \tag{9.8}$$

The $\beta$ index obtained was 1.514 with respect to yielding at the outer edge of the annular steel cross-section.

At the most probable failure point (where the ellipsoid touches the LSS), the value of the lateral load at pile head is $P_H = 580.3$ kN, that is, at $1.513\sigma_{PH}$ from the mean value of $P_H$, while the 31 autocorrelated $c_u$ values deviate only very slightly from their mean values. This is not surprising given the $e = 26$ m cantilever length above the seabed; in fact, the maximum bending moment occurs at a depth of only 1.36 m below the seabed, or 27.36 m from the pile head. Hence, for the case in hand, pile yielding caused by bending moment is sensitive to the applied load at pile head and not sensitive to the uncertainty of the shear strength below the seabed. However, separate reliability analysis for cases where the lateral load acts on the pile head near the ground surface (with zero cantilever length) indicates that the response is sensitive to both the lateral load at pile head and the soil-shearing resistance within the first few meters of the ground surface. The different sensitivities from case to case are automatically reflected in reliability analysis aiming at a target index value (presented next), but will be difficult to consider in codes based on partial factors.

## 9.7.1 Illustrative example of multicriteria RBD of a laterally loaded pile

The steel tubular pile that forms part of a pile group in a breasting dolphin (Figure 9.12) has been examined probabilistically above based on the single performance function of the bending failure mode. The reliability index is only 1.514. In design, a reliability index of 2.5 or 3.0 is often stipulated. Further, deflection criterion also needs to be considered. In the following illustrative design example, it is assumed that 1.4 m is the maximum tolerable pile head deflection, which means a secant-tilt angle of about 1.4 in 26, or about 3°, with respect to the seabed.

Analysis using mean parametric values (including $\mu_{cu} = 150 + 2z$ kPa) results in a pile head deflection of 0.986 m (slightly smaller than the 0.994 m of the above-mentioned deterministic case for which a $c_u$ value of 150 kPa was used as in Tomlinson (1994)). This average deflection of 0.986 m does not provide information on the reliability of not exceeding the 1.4 m tolerable limit, because the uncertainties of the random variables have not been reflected in estimates based on mean values.

To illustrate multicriteria RBD, suppose it is desired to select the external diameter $d$ and steel wall thickness $t$ of the tubular pile so as to achieve a reliability index of 3.0 with respect to both the pile head deflection limit state and pile-bending moment limit state. The mean and covariance structure of $P_H$ and $c_u$ profile are as in the previous section. Pile embedment length is 23 m. Note that the external diameter $d$ and wall thickness $t$ affect both ultimate limit state and serviceability limit state functions, by affecting the moment of inertia $I$, the Matlock $p$-$y$ curves (Figure 9.12b, in which $B = d = 1.3$ m), and the yield moment $M_y = 2\sigma_y I/d$, where $\sigma_y$ is the yield stress (417 MPa) of high-tensile alloy steel.

Figure 9.12c shows the curves of reliability index $\beta = 3$ as a function of the thickness $t$ and external diameter $d$ of the tubular pile, for the deflection performance criterion and pile-bending failure mode. The configuration at location $a$ (i.e., 32 mm annular wall thickness and external diameter 1.42 m) yields a reliability index of 3 with respect to both the deflection and the bending moment modes. Configurations along line $ac$ will have a reliability index of 3 for the bending moment mode, and >3 for the deflection mode, whereas configurations along line $ab$ mean a reliability index of 3 for the deflection mode, and >3 for the bending moment mode.

## 9.8 FORM DESIGN OF AN ANCHORED SHEET PILE WALL

The Microsoft Excel constrained optimization approach for RBD of the anchored sheet pile wall in Figure 9.13 was presented in Low (2006). Another example was presented in Low (2005). Given the uncertainties and correlation structure in Figure 9.13, one wishes to find the required total wall height $H$ so as to achieve a reliability index of 3.0 against rotational failure about point "$A$." The solution (Figure 9.13) indicates that a total height of 12.15 m would give a reliability index $\beta$ of 3.0 against rotational failure. With this wall height, the mean-value point is safe against rotational failure, but failure occurs when the mean values descend/ascend to the values indicated under the $x^*$ column. These $x^*$ values

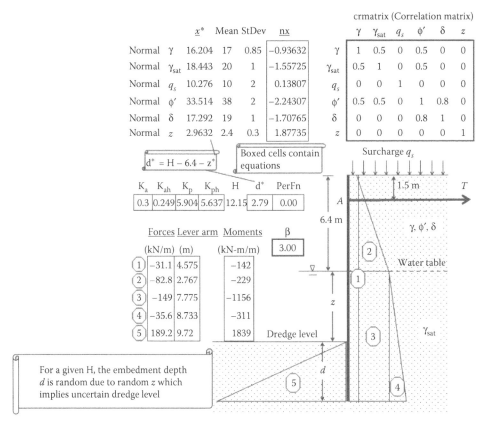

Figure 9.13 Design total wall height for a reliability index of 3.0 against rotational failure. Dredge level and hence $z$ and $d$ are random variables.

denote the *design point* on the LSS and represent the most likely combination of parametric values that will cause failure. The expected embedment depth is $d = 12.15–6.4–\mu_z = 3.35$ m. At the failure combination of parametric values, the design value of $z$ is $z^* = 2.9632$, and $d^* = 12.15–6.4–z^* = 2.79$ m. This corresponds to an "overdig" allowance of 0.56 m. Unlike EC7, this "overdig" is determined automatically and reflects uncertainties and sensitivities from case to case in a way that specified "overdig" cannot.

The *nx* column indicates that, for the given mean values and uncertainties, rotational stability is, not surprisingly, most sensitive to $\phi'$ and the dredge level (which affects $z$ and $d$ and hence the passive resistance). It is interesting to note that at the design point where the six-dimensional dispersion ellipsoid touches the LSS, both unit weights $\gamma$ and $\gamma_{sat}$ (16.20 and 18.44) are lower than their corresponding mean values, contrary to the expectation that higher unit weights will increase active pressure and hence greater instability. This apparent paradox is resolved if one notes that smaller $\gamma_{sat}$ will (via smaller $\gamma'$) reduce passive resistance, smaller $\phi'$ will cause greater active pressure and smaller passive pressure, and that $\gamma$, $\gamma_{sat}$, and $\phi'$ are logically positively correlated.

In an RBD (such as the case in Figure 9.13), one does not prescribe the ratios *mean/x\**— such ratios, or ratios of (*characteristic values*)/*x\**, are prescribed in EC7 limit state design— but leave it to the expanding dispersion ellipsoid to seek the most probable failure point on the LSS, a process that automatically reflects the sensitivities of the parameters. Besides, one can associate a probability of failure for each target reliability index value. The ability to seek the most probable design point without presuming any partial factors and to automatically reflect sensitivities from case to case is a desirable feature of the RBD approach.

## 9.9 RELIABILITY ANALYSIS OF ROOF WEDGES AND ROCKBOLT FORCES IN TUNNELS

A symmetric roof wedge of central height $h$ and apical angle $2\alpha$ in a circular tunnel of radius $R$ is shown in Figure 9.14. An analytical approach for assessing the stability can be based on Bray's 1977 two-stage relaxation procedure, as described, for example, in Sofianos et al. (1999) and Brady and Brown (2006). The first stage computes the confining lateral force $H_0$ on the wedge from the stress field and the geometries of the wedge and tunnel, for an assumed homogeneous, isotropic, linearly elastic, and weightless medium. The second stage then assumes deformable joints and a rigid rock mass, to arrive at the normal force N acting on each joint surface.

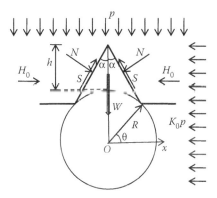

*Figure 9.14* Notation for symmetric roof wedge in a circular tunnel.

Two different definitions of the factor of safety against wedge falling have been reported in the literature, each with its own rationale. The first appeared in Sofianos et al. (1999) and Brady and Brown (2006), for example. It is the ratio of the pull-out resistance of the wedge to the weight of the wedge and was expressed as follows:

$$FS_1 = \frac{2MH_0}{W}, \quad \text{where } M = f(\phi, i, \alpha, k_s/k_n) \quad \text{and} \quad H_0 = f(p, R, K_0, h/R) \tag{9.9}$$

In the above equations, $W$ is the weight of the wedge, $\alpha$ is the semiapical angle of the wedge, $\phi$ and $i$ are the effective friction and dilation angles of the joints, $k_s$ and $k_n$ are the shear stiffness and normal stiffness of the joints, $R$ is the radius of the tunnel, $p$ and $K_0$ are the vertical *in situ* stress and the coefficient of the horizontal *in situ* stress, $h$ is the clear height of the wedge (measured from the tunnel crown), and $\theta$ is the angle denoted in Figure 9.14.

The second definition is similar in principle to that which has been long and widely used in soil and rock slope stability analysis, in the *Unwedge* program of *Rocscience.com*, and in Asadollahi and Tonon (2010), for example. It is the ratio of the available shear strength to the shear strength required for equilibrium. In this context of tunnel roof wedge, this definition was given in Asadollahi and Tonon (2010) as follows (assuming the dilation angle of the joints $i = 0$):

$$FS_2 = \frac{2S\cos\alpha}{2N\sin\alpha + W}, \quad \text{where } N = f(H_0, \phi, \alpha, k_s, k_n) \quad \text{and} \quad S = f(H_0, \phi, \alpha, k_s, k_n) \tag{9.10}$$

where $N$ and $S$ are normal and shear forces, and other symbols as defined for Equation 9.9.

Low and Einstein (2013) showed that the two definitions, Equations 9.9 and 9.10, can be recast in terms of $N$, $W$, $\alpha$, and $\phi$, as follows:

$$FS_1 = \frac{\text{Limiting wedge weight}}{\text{Actual wedge weight}} = \frac{2N\tan\phi\cos\alpha - 2N\sin\alpha}{W} = \frac{\tan\phi/\tan\alpha - 1}{W/(2N\sin\alpha)} \tag{9.11}$$

$$FS_2 = \frac{\text{Maximum available resisting forces}}{\text{Downward driving forces}} = \frac{2N\tan\phi\cos\alpha}{2N\sin\alpha + W} = \frac{\tan\phi/\tan\alpha}{1 + W/(2N\sin\alpha)} \tag{9.12}$$

The "Limiting wedge weight" in Equation 9.11 means the wedge weight at limiting equilibrium, that is, the wedge weight that just causes failure. It is negative if $\phi < \alpha$.

The same $FS_1$ is obtained whether computed from Equation 9.9 or 9.11, and the same $FS_2$ is obtained whether computed from Equation 9.10 or 9.12. Nevertheless, the rationales, similarities, and differences between $FS_1$ and $FS_2$ are rendered much more transparent in Equations 9.11 and 9.12 than in Equations 9.9 and 9.10. That $FS_1$ can be negative when $\phi < \alpha$ is also readily appreciated from Equation 9.11. One may note that $FS_1$ by Equation 9.11—which is mathematically equivalent to Equation 9.9—can be very large and positive if $W/N$ is small and $\phi > \alpha$, and negative if $\phi < \alpha$.

The two definitions as given by Equations 9.11 and 9.12, and hence Equations 9.9 and 9.10, are mathematically equivalent when $FS_1 = FS_2 = 1$. This is shown in Figure 9.15a, where the $FS_1$ contours (solid lines) of 1, 10, 20, and 30 are shown together with the $FS_2$ contours (dashed lines) of 1.0, 1.15, 1.30, 1.48, and 1.60. The $FS_1 = 1.0$ and $FS_2 = 1.0$ contours coincide perfectly. For the input values given in the figure caption, the factors of safety at the mean-value point $\alpha = 25°$ and $\phi = 35°$ are $FS_1 = 30.1$ and $FS_2 = 1.48$, but Figure 9.15a

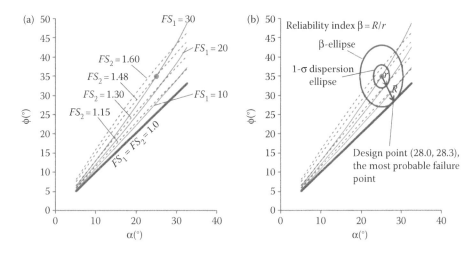

*Figure 9.15* (a) Contours of $FS_1$ and $FS_2$ and the mean-value point where $FS_1 = 30.1$ and $FS_2 = 1.48$; (b) dispersion ellipses in the space of $\alpha$ and $\phi$. The plots are for i =0, $k_s/k_n = 0.1$, p = 1 MPa, $K_0 = 1.5$, R = 2 m, $h/R = 0.85$, and $\gamma = 0.027$ MN/m³.

clearly shows that the mean $FS_1$ is only as safe as the mean $FS_2$, despite its being 20 times higher than $FS_2$, because the same ($\alpha$, $\phi$) point for both is at the same distance from the LSS (where $FS_1 = FS_2 = 1$), which separates the safe combinations (FS > 1.0) of values of ($\alpha$, $\phi$) from the unsafe combinations ($FS < 1.0$) of ($\alpha$, $\phi$).

This problematic aspect of using safety factors can be avoided if one uses the reliability index, as shown in Figure 9.15b. The FORM computes a reliability index, $\beta$, which is the distance from the safe mean-value point ($\alpha$, $\phi$) = ($25°$, $35°$) to the most probable failure point ($\alpha = 28.01°$, $\phi = 28.33°$), also called the design point, in units of directional standard deviations. The same $\beta$ value is computed regardless of whether $FS_1$ or $FS_2$ is used to define the LSS FS = 1.0.

Figure 9.16a shows the results of SORM reliability analysis on the foundation of FORM results. The FORM results were obtained automatically using the practical constrained optimization approach of Low and Tang (2007). The SORM procedure in the spreadsheet was described in Chan and Low (2012a). The parameters $\alpha$, $\phi$, $k_s/k_n$, $p$, and $K_0$ (defined in connection with Equation 9.9) are assumed to be normally distributed random variables, with mean values μ and standard deviations σ as shown in the labeled columns. Other input values (with zero standard deviations) are $i = 0$, $R = 2.0$ m, $h/R = 0.85$, and $\gamma = 0.027$ MN/ m³. The mean ratio of normal force N to wedge weight W is 71 for this case. The mean $FS_1 = 30.1$, and mean $FS_2 = 1.48$, with $\alpha = 25°$, $\phi = 35°$, $k_s/k_n = 0.1$, $p = 1$ MPa, and $K_0 = 1.5$. Hence, at these mean input values, the wedge is safe. However, wedge failure will occur when the mean values descend/ascend to the design point values shown in the $x^*$ column in Figure 9.16a, analogous to Figure 9.1 but in five-dimensional (5D) space. The design point values are $\alpha^* = 28.01°$, $\phi^* = 28.34°$, $(k_s/k_n)^* = 0.1$, $p^* = 0.9975$ MPa, and $K_0^* = 1.474$. The wedge is on the verge of failure at these design values, mainly due to a greater semi-apical angle ($\alpha^* = 28.01°$ > mean $\alpha$ value of $25°$) and a lower friction angle of the joints ($\phi^* = 28.34°$ is smaller than the mean $\phi$ value of $35°$). The distance from the safe mean-value point to this most probable failure combination of parametric values, in units of directional standard deviations, is the reliability index $\beta$, equal to 2.684 in this case.

The **n** in Equation 9.3 are the dimensionless equivalent standard normal random variables under the square root sign of Equation 9.2, and their values in Figure 9.16a suggest that $\phi$ and

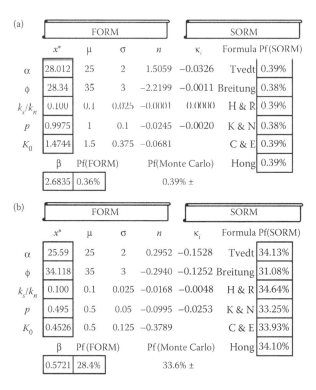

Figure 9.16 FORM and SORM reliability analysis of tunnel roof wedges: (a) R = 2 m, high p and $K_0$; (b) R = 6 m, low p and $K_0$.

$\alpha$ are the most sensitive of the five random variables. The insensitivity of wedge stability to the *in situ* stress parameters p and $K_0$ in this case is due to the large mean N/W value of 2.779 MN/0.0391MN = 71. Given the large gripping force arising from *in situ* stresses, and in the presence of the assumed parametric uncertainties for the five random variables ($\alpha$, $\phi$, $k_s/k_n$, p, and $K_0$), the most probable failure point (such as Figure 9.15, but in the 5D space) is with increasing $\alpha$ and decreasing $\phi$. In contrast, Figure 9.16b has mean values of p = 0.5 MPa and $K_0 = 0.5$, instead of the mean values of p = 1.0 MPa and $K_0 = 1.5$ used in Figure 9.16a, but the same coefficients of variations ($\sigma/\mu$) as Figure 9.16a. Also, R = 6.0 m instead of 2.0 m (but h/R remains at 0.85). This reduces the value of mean N/W to 1.131MN/0.352MN = 3.214 (vs. 71 of Figure 9.16a). The mean factors of safety are $FS_1 = 1.36$ and $FS_2 = 1.10$. The reliability index is 0.572 (vs. 2.684 of Figure 9.16a), and $K_0$, with the largest numerical value of n, is the most sensitive random variable of the five. The ability of the FORM reliability index to reflect different parametric sensitivities and uncertainties from case to case without relying on rigid partial factors is an important advantage of reliability analysis and RBD.

For the case in Figure 9.16a, the probability of roof wedge failure based on Equation 9.5 is $P_f = 0.36\%$, compared with an SORM $P_f$ of about 0.39% using the Chan and Low (2012a) spreadsheet codes. The SORM $P_f$ agrees very well with the $P_f$ values (0.39, 0.39, and 0.40%) from three MCS of the wedge stability problem, each with 100,000 realizations using the software @Risk (http://www.palisade.com).

For the case in Figure 9.16b, the probability of roof wedge failure based on Equation 9.5 is $P_f = 28.4\%$, compared with an SORM $P_f$ of about 33.5%. The SORM $P_f$ agrees very well with the $P_f$ values (33.84, 33.73, and 33.37%) from three MCS, each with 50,000 realizations.

For the multidimensional LSS of the roof wedge stability model above and the rock-bolt-reinforced tunnel below, one may extend the FORM analysis into the SORM analysis. Should the curvatures of the LSS turn out to be negligible, the $\kappa_i$ values in Figure 9.16a and b will be practically zeros, and all the SORM formulas will reduce to Equation 9.5, with the result that the computed SORM probability of failure will be the same as FORM probability of failure. As a practical alternative, one may also note that RBD typically requires a target $\beta$ index of 2.5 or 3.0, corresponding to FORM $P_f$ of 0.62 and 0.14%, respectively, and hence, some inaccuracy in the FORM $P_f$ (from Equation 9.5, $P_f \approx 1 - \Phi(\beta)$) is of no practical concern because the correct $P_f$ will still be below 1%. For the case in Figure 9.16a, the $P_f$ of 0.36%, estimated based on FORM $\beta$ of 2.68, is lower than the more accurate 0.39%, but this discrepancy may not justify much practical concern.

Low and Einstein (2013) also investigated FORM design of the length and spacings $S_z S_\theta$ of rockbolts for a reliability index $\beta$ of 2.5 against bolt tension exceeding the bolt rupture strength. The implicit mechanical model for the mobilized bolt force as formulated in Bobet and Einstein (2011) was used. Four lognormally distributed random variables were modeled, namely the three-dimensional (3D) effect parameter $\beta_\sigma$, and the shear strength parameters $\phi$ and c and modulus E of the rock mass, with a negative correlation coefficient $\rho$ of $-0.5$ between $\phi$ and c. Figure 9.17 shows the FORM probability of failure for different

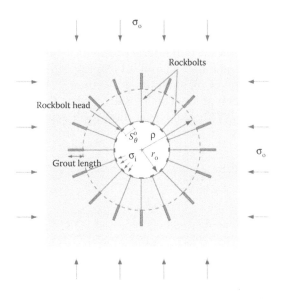

| Rock-bolt length $(\rho - r_0)$ | Required $S_z S_\theta$ for $\beta = 2.5$ | Probability of failure | | | |
|---|---|---|---|---|---|
| | | FORM $P_f$ $= \Phi(-\beta)$ | SORM $P_f$ | Six simulations with importance sampling each 3000 random sets | Six Monte Carlo simulations each 3000 random sets |
| 2 m | 0.59 m² | 0.62% | 0.44% | 0.43%, 0.47%, 0.43%, 0.46%, 0.45%, 0.41% | 0.50%, 0.33%, 0.40%, 0.40%, 0.57%, 0.47% |
| 3 m | 0.70 m² | 0.62% | 0.45% | 0.46%, 0.48%, 0.44%, 0.48%, 0.47%, 0.45% | 0.57%, 0.53%, 0.60%, 0.40%, 0.43%, 0.50% |
| 4 m | 0.81 m² | 0.62% | 0.47% | 0.47%, 0.47%, 0.48%, 0.47%, 0.46%, 0.46% | 0.37%, 0.60%, 0.73%, 0.53%, 0.30%, 0.47% |

*Figure 9.17* Required length and spacing of rockbolts for a target reliability index $\beta$ of 2.5, and comparisons of probability of failure based on FORM, SORM, importance sampling, and ordinary Monte Carlo simulations.

combinations of bolt length and spacing. For comparison, probabilities of failure are also obtained from SORM, IS, and ordinary MCS.

The results of SORM and the averages of IS simulations in Figure 9.17 are in very good agreement and indicate a probability of failure of about 0.4–0.5% depending on rock-bolt length, compared with a FORM $P_f$ of about 0.6% based on β – 2.5. Engineers have to decide whether for practical purposes (i) it is adequate (e.g., using a target FORM β of 2.5) that the implied *small* probability of failure (0.6% by Equation 9.5) is approximate but similarly small as the more accurate probability of failure (about 0.45% for the cases in Figure 9.17), or (ii) engage in more precise RBD via SORM or IS for a target probability of failure. If the latter is the desired action, a practical approach is as follows. In the light of the comparison of FORM $P_f$ of 0.62% versus actual $P_f$ of 0.45% in Figure 9.17, suppose a probability of failure of 0.2% is the target $P_f$ for the case in hand, one may do a FORM design of the rock-bolt length and spacing for a reliability index corresponding to a probability of failure (by Equation 9.5) of $(0.62\%/0.45\%)*0.2\% = 0.28\%$, that is, $β = \Phi^{-1}(1 - 0.0028) =$ Excel function NormSInv(0.9972) = 2.77. For the case with rock-bolt length of 3 m, the required $S_zS_\theta$ is found (via Excel Solver runs) to be about 0.67 m² for a target β = 2.77. This solution of 0.67 m² takes about a minute. To verify, with the input $S_zS_\theta = 0.67$ m² and rock-bolt length of 3 m, three MCS incorporating IS each of 10,000 random sets yielded (in 25 min) probabilities of failure of 0.206, 0.199, and 0.203%, respectively; the average is practically the target $P_f$ of 0.2%.

The above RBD procedure may be summarized as follows:

i. Conduct RBD based on the initial target reliability index $β_i$ of 2.5 or 3.0. The implied FORM $P_f$ is $\Phi(-β_i)$, equal to 0.62 and 0.14%, respectively.
ii. For the design solution of (i), obtain more accurate $P_f$ using SORM or IS on the foundation of FORM design point. This $P_f$ is denoted as $P_{f,SORM}$.
iii. Decide on a target $P_f$ (<1%), and repeat the RBD based on an updated target reliability index ($β_{rev}$) as follows:

$$\beta_{rev} = \Phi^{-1}\left(1 - \frac{\Phi(-\beta_i)}{P_{f,SORM}} \times \text{target } P_f\right) \tag{9.13}$$

The final design solution is obtained via repeated FORM analyses using different designs until the reliability index is equal to $β_{rev}$. Linear interpolation between two designs and their respective reliability indices can also be done. The final design based on $β_{rev}$ will have failure probability close to the target $P_f$.

If the design $S_zS_\theta$ is to be obtained by direct trial and error of different designs, each checked against MCS (with or without IS), the computation time will be much longer than the FORM-and-SORM approach involving Equation 9.13.

## 9.10 PROBABILISTIC SETTLEMENT ANALYSIS OF A HONG KONG TRIAL EMBANKMENT ON SOFT CLAY

The Chek Lap Kok test reclamation fill on marine clay was part of a larger reclamation project for the construction of a Hong Kong replacement airport (which started operation in 1997). The objective of the test fill was to investigate the feasibility of reclamation over soft marine clay and the effectiveness of vertical drains in accelerating consolidation. The main test area, located about 200 m offshore and 100 m square in plan, was divided into quadrants: one was a control area, with no treatment of the marine clay, and the remaining three quadrants were installed with vertical drains at different spacing through about 7-m-thick marine clay. The offshore geotechnical investigations, the test fill, and the instrumentation program were described in substantial detail by Foott et al. (1987) and Koutsoftas et al. (1987), and further studied deterministically in Choa et al. (1990).

Low (2003b) presented a deterministic numerical method written in the VBA programming environment of Microsoft Excel spreadsheet for consolidation analysis involving vertical drains. The program uses Barron's solution for equal vertical strain of consolidation due to radial drainage and Carillo's equation for combined radial and vertical drainage. The program accounts—in an approximate manner—for stage loading, load reduction due to fill submergence, delayed vertical drain installation, changes in length of vertical drainage path with time, and variation of soil stress history with depth. A practical algorithm for prediction of the rate of settlement was adopted because, even in the relatively simple approach adopted, 15 or more values of individual input parameters were required. The uncertainties associated with some of these parameters will limit the accuracy of prediction even if sophisticated models are used. The computational algorithm in the program is not fully rigorous, because idealizations and approximations have been made. Nevertheless, limited comparisons made by the author suggest that the degree of accuracy achieved is adequate for the purpose in hand. Its relative simplicity also gave rise to some insights on parametric relationships and sensitivities. The results of deterministic analysis using the program compared well with the instrumented settlement records of the soft clay beneath the Chek Lap Kok test fills.

Figure 9.18 shows the results of a deterministic analysis using the program, for one of the four test fills of the Chek Lap Kok land reclamation project where the prefabricated vertical band drains were installed at a spacing of 1.5 m. There was staged loading due to the increase in fill thickness with time as shown. The clay was overconsolidated since the initial effective vertical stress profile was less than the profile of maximum past pressures (preconsolidation pressures). The discrepancies between the computed curve and the measured range are partly attributable to the underestimation of the final consolidation settlement by the program. The final consolidation settlements are functions of compression ratio $C_R$, recompression ratio $C_{RR}$, stress history, and applied loadings, but are not affected by the values of the rate parameters $c_v$ and $c_h$, which are the coefficients of consolidation for vertical and horizontal flow, respectively, nor by the idealizations and approximations in the modeling of excess pore pressure dissipation in the program. The program-computed results will be even closer to the observed settlement curves if different values of $c_h$ and compression ratios are used, but such speculations belong to the realm of back analysis and hindsight and will not be pursued here. Instead, the Low and Tang (2007) efficient spreadsheet FORM procedure will be illustrated for the case in hand to highlight some subtleties and insights of RBD involving serviceability limit states. The material in Sections 9.10.1 through 9.10.5 is based on Low (2008c).

## 9.10.1 LSS and performance functions g(x) pertaining to magnitude and rate of soft clay settlement

As in Low (2008c), the following three aspects are studied:

  i. The magnitude of the ultimate consolidation settlement $s_{cf}$;
 ii. The degree of consolidation U at time = 1 year;
iii. The consolidation settlement remaining ($s_r$) at time = 1 year, where $s_r = s_{cf} - s_{1\,yr}$.

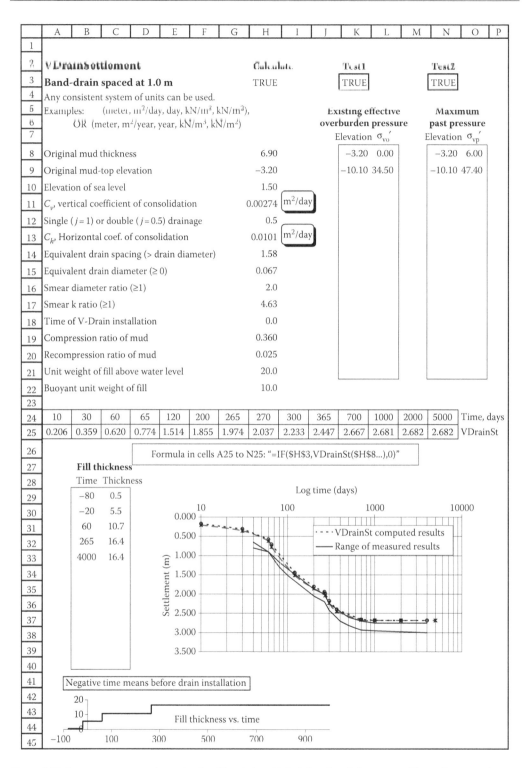

|   | A | B | C | D | E | F | G | H | I | J | K | L | M | N | O | P |
|---|---|---|---|---|---|---|---|---|---|---|---|---|---|---|---|---|
| 1 |   |   |   |   |   |   |   |   |   |   |   |   |   |   |   |   |
| 2 | VDrainSettlement |||||| Calculate || Test1 ||| Test2 |||
| 3 | Band-drain spaced at 1.0 m |||||| TRUE || TRUE ||| TRUE |||
| 4 | Any consistent system of units can be used. |||||||||||||||
| 5 | Examples: (meter, m³/day, day, kN/m³, kN/m³), |||||| | Existing effective ||| Maximum |||
| 6 |  OR (meter, m²/year, year, kN/m³, kN/m²) |||||| | overburden pressure ||| past pressure |||
| 7 |  |||||| | Elevation σ$_{vo}'$ || | Elevation σ$_{vp}'$ ||
| 8 | Original mud thickness |||||| 6.90 | | −3.20 | 0.00 | | | −3.20 | 6.00 ||
| 9 | Original mud-top elevation |||||| −3.20 | | −10.10 | 34.50 | | | −10.10 | 47.40 ||
| 10 | Elevation of sea level |||||| 1.50 |||||||||
| 11 | $C_v$, vertical coefficient of consolidation |||||| 0.00274 | m²/day ||||||||
| 12 | Single ( $j$ = 1) or double ( $j$ = 0.5) drainage |||||| 0.5 |||||||||
| 13 | $C_h$, Horizontal coef. of consolidation |||||| 0.0101 | m²/day ||||||||
| 14 | Equivalent drain spacing (> drain diameter) |||||| 1.58 |||||||||
| 15 | Equivalent drain diameter (≥ 0) |||||| 0.067 |||||||||
| 16 | Smear diameter ratio (≥1) |||||| 2.0 |||||||||
| 17 | Smear k ratio (≥1) |||||| 4.63 |||||||||
| 18 | Time of V-Drain installation |||||| 0.0 |||||||||
| 19 | Compression ratio of mud |||||| 0.360 |||||||||
| 20 | Recompression ratio of mud |||||| 0.025 |||||||||
| 21 | Unit weight of fill above water level |||||| 20.0 |||||||||
| 22 | Buoyant unit weight of fill |||||| 10.0 |||||||||
| 23 |   |   |   |   |   |   |   |   |   |   |   |   |   |   |   |   |
| 24 | 10 | 30 | 60 | 65 | 120 | 200 | 265 | 270 | 300 | 365 | 700 | 1000 | 2000 | 5000 | Time, days ||
| 25 | 0.206 | 0.359 | 0.620 | 0.774 | 1.514 | 1.855 | 1.974 | 2.037 | 2.233 | 2.447 | 2.667 | 2.681 | 2.682 | 2.682 | VDrainSt ||

Formula in cells A25 to N25: "=IF($H$3,VDrainSt($H$8...),0)"

**Fill thickness**

| Time | Thickness |
|---|---|
| −80 | 0.5 |
| −20 | 5.5 |
| 60 | 10.7 |
| 265 | 16.4 |
| 4000 | 16.4 |

Negative time means before drain installation

*Figure 9.18* Program-computed results for Check Lap Kok 1.5 m band-drain test fill quadrant, and comparison with measured settlement range. The $c_v$ and $c_h$ values of 0.00274 and 0.0101 m²/day correspond to 1 and 3.7 m²/year, respectively.

The performance functions (or limit state functions) are, respectively:

$$g_1(\mathbf{x}) = Limiting\ s_{cf} - s_{cf} \tag{9.14}$$

$$g_2(\mathbf{x}) = U - Limiting\ U \tag{9.15}$$

$$g_3(\mathbf{x}) = Limiting\ s_r - s_r \tag{9.16}$$

in which $s_{cf}$, $U$, and $s_r$ are functions of the various inputs ($\mathbf{x}$) shown in Figure 9.18, including parameters of compressibility, consolidation rate, staged loading, and stress history.

The term "limiting" connotes "acceptable" or "permissible." Positive $g(\mathbf{x})$ values correspond to the safe domain, and negative $g(\mathbf{x})$ values correspond to the unsafe domain. Hence, for $g_1(\mathbf{x})$ and $g_3(\mathbf{x})$, by virtue of "smaller settlement is safer," safe domain is indicated when $Limiting\ s_{cf} > s_{cf}$, and $Limiting\ s_r > s_r$. In contrast, for $g_2(\mathbf{x})$, by virtue of "larger degree of consolidation is safer," safe domain is indicated when $U > Limiting\ U$. These are schematically illustrated on the plane in Figures 9.19 and 9.20. Also, for all three performance functions, the parametric surface that separates safe combinations of parameters from unsafe combinations of parameters is the LSS, given by $g(\mathbf{x}) = 0$.

### 9.10.2 Distinguishing positive and negative reliability indices

In Figure 9.19, when the limiting (i.e., permissible) ultimate consolidation settlement is $s_{L1}$, the settlement evaluated at the mean-value point ($s_\mu$) is already in the unsafe zone ($>s_{L1}$). Under this circumstance, the computed reliability index $\beta$ must be regarded as negative: it is the minimum distance (in units of directional standard deviations) from the *unsafe mean-value point* to the *safe boundary* defined by limit state surface 1 (LSS1). On the other hand, if a higher permissible settlement ($s_{L2}$) is specified, the mean-value point is in the safe zone, and the reliability index $\beta$ is positive: it is the minimum distance (in units of directional standard deviations) from the *safe mean-value point* to the *unsafe boundary* defined by limit state surface 2 (LSS2).

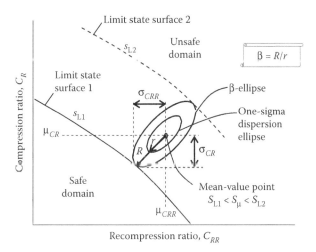

*Figure 9.19* Illustration of reliability index in the plane. With respect to LSS1, reliability index $\beta$ is negative; with respect to LSS2, $\beta$ is positive.

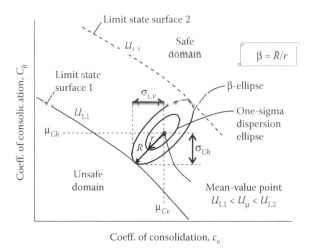

*Figure 9.20* Illustration of reliability index in the plane. With respect to LSS1, reliability index β is positive; with respect to LSS2, β is negative.

In contrast, in Figure 9.20, the reliability index β with respect to the LSS1 (for which the limiting degree of consolidation is $U_{L1}$) is positive, while that with respect to LSS2 (for which limiting $U = U_{L2}$) must be treated as negative, by virtue of $U_{L1} < U_\mu < U_{L2}$, where $U_\mu$ is the average value of the degree of consolidation evaluated using mean values ($\mu_{Cv}$, $\mu_{Ch}$) of the coefficients of consolidation $c_v$ and $c_h$.

### 9.10.3 Reliability analysis for different limiting state surfaces

The Low and Tang (2007) procedure for FORM can deal with various correlated non-normal distributions (lognormal, general beta, gamma, type 1 extreme, exponential, ...). For the case in hand, only correlated lognormals are illustrated. The values of compression ratio $C_R$, recompression ratio $C_{RR}$, and coefficients of consolidation $c_v$ and $c_h$ in Figure 9.18 are taken to be the mean values in Figure 9.21. Assumed values of standard deviations are used for illustrative purpose. Positive correlations, logical between $C_R$ and $C_{RR}$ and between $c_v$ and $c_h$, are modeled.

The deterministic setup of Figure 9.18 and the reliability analysis of Figure 9.21 are coupled easily by replacing the $C_R$, $C_{RR}$, $c_v$, and $c_h$ values (cells H19, H20, H11, and H13) of Figure 9.18 with the formulas " = Z28", " = Z29", " = Z30", and "Z31" that refer to the $x^*$ values in Figure 9.21. The performance function $g_1(\mathbf{x})$ is, by Equation 9.14, " = W34 − N25," where cell W34 has value 2.0 for this analysis. The computed β index is 1.485, treated as (−1.485) because the mean-value point is in the unsafe zone as indicated by the negative $g(\mathbf{x})$ value when the $n_i$ values were initialized to zeros prior to spreadsheet-automated reliability analysis.

By varying the $s_{\text{limit}}$ value (cell W34) between 1.2 and 4.8 at intervals of 0.2, and each time recomputing the β index, 19 values of β were obtained as shown in Figure 9.22.

### 9.10.4 Obtaining probability of failure ($P_f$) and CDF from β indices

Referring to Figure 9.19, for $s_{L1} = 2.0$ m and $β = -1.485$ from Figure 9.21, the probability of failure $P_f$ is the integration of the probability density over the entire unsafe zone ($s > s_{L1}$). A good estimate of $P_f$ can often be obtained from the established Equation 9.5,

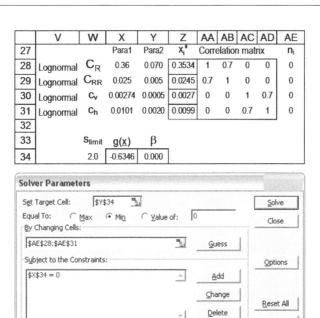

|  | V | W | X | Y | Z | AA | AB | AC | AD | AE |
|---|---|---|---|---|---|---|---|---|---|---|
| 27 |  |  | Para1 | Para2 | $x_i^*$ | Correlation matrix |  |  |  | $n_i$ |
| 28 | Lognormal | $C_R$ | 0.36 | 0.070 | 0.3534 | 1 | 0.7 | 0 | 0 | 0 |
| 29 | Lognormal | $C_{RR}$ | 0.025 | 0.005 | 0.0245 | 0.7 | 1 | 0 | 0 | 0 |
| 30 | Lognormal | $C_v$ | 0.00274 | 0.0005 | 0.0027 | 0 | 0 | 1 | 0.7 | 0 |
| 31 | Lognormal | $C_h$ | 0.0101 | 0.0020 | 0.0099 | 0 | 0 | 0.7 | 1 | 0 |
| 32 |  |  |  |  |  |  |  |  |  |  |
| 33 |  | $S_{limit}$ | g(x) | β |  |  |  |  |  |  |
| 34 |  | 2.0 | -0.6346 | 0.000 |  |  |  |  |  |  |

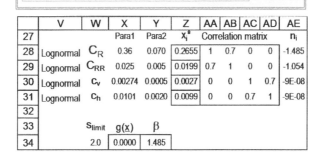

|  | V | W | X | Y | Z | AA | AB | AC | AD | AE |
|---|---|---|---|---|---|---|---|---|---|---|
| 27 |  |  | Para1 | Para2 | $x_i^*$ | Correlation matrix |  |  |  | $n_i$ |
| 28 | Lognormal | $C_R$ | 0.36 | 0.070 | 0.2655 | 1 | 0.7 | 0 | 0 | -1.485 |
| 29 | Lognormal | $C_{RR}$ | 0.025 | 0.005 | 0.0199 | 0.7 | 1 | 0 | 0 | -1.054 |
| 30 | Lognormal | $C_v$ | 0.00274 | 0.0005 | 0.0027 | 0 | 0 | 1 | 0.7 | -9E-08 |
| 31 | Lognormal | $C_h$ | 0.0101 | 0.0020 | 0.0099 | 0 | 0 | 0.7 | 1 | -9E-08 |
| 32 |  |  |  |  |  |  |  |  |  |  |
| 33 |  | $S_{limit}$ | g(x) | β |  |  |  |  |  |  |
| 34 |  | 2.0 | 0.0000 | 1.485 |  |  |  |  |  |  |

*Figure 9.21* Initially the $n_i$ values were zeros, and g($\underline{x}$) exhibits negative values. Hence, the computed β value must be treated as negative (β = −1.485).

$P_f \approx \Phi$ (−β), or 93% for the case of β = −1.485 in Figure 9.21, which means that the limiting final settlement $s_{cf}$ of 2.0 m is almost certainly to be exceeded. This is not surprising given that the computed expected $s_{cf}$ is about 2.68 m as shown in Figure 9.18. Besides estimating the probability of failure for serviceability limit states such as Equations 9.14 through 9.16 for $s_{cf}$, $U$, and $s_r$, respectively, it would also be of interest to obtain the CDF from the β indices, as follows:

$$g_1(\underline{x}) = Limiting\ s_{cf} - s_{cf}: \quad CDF = \Phi(\beta) \tag{9.17}$$

$$g_2(\underline{x}) = U - Limiting\ U: \quad CDF = \Phi(-\beta) \tag{9.18}$$

$$g_3(\underline{x}) = Limiting\ s_r - s_r: \quad CDF = \Phi(\beta) \tag{9.19}$$

The reason for Equations 9.17 and 9.19 being different from Equation 9.18 is readily appreciated if one notes that, for Equations 9.17 and 9.19, CDF = $P[s < s_{limit}]$ = $1 - P[s > s_{limit}]$ = $1 - P_f$. In contrast, for Equation 9.18, CDF = $P[U < U_{limit}]$ = $P_f$.

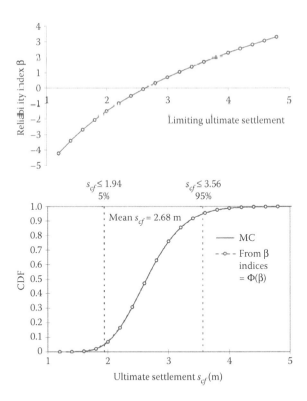

*Figure 9.22* Reliability indices for different limiting ultimate settlements, and comparison of CDF based on β indices with CDF from Monte Carlo simulations.

As shown in Figure 9.22, the CDF based on 19 values of β indices is practically indistinguishable from the CDF from 5000 realizations of MCS using the software @Risk (http://www.palisade.com). For the assumed statistical input and correlation structure, the 90% confidence interval of the ultimate consolidation settlement is (1.94, 3.56 m). The measured ultimate settlement shown in Figure 9.18 is within this interval. Other confidence intervals can also be read from Figure 9.22. Considerations such as this should be much more useful in the design stage than a deterministic analysis that yields a single ultimate settlement value with no indication at all of the effect of uncertainties (parametric, modeling, and others) on the predicted ultimate settlement.

Figures 9.23 and 9.24 show the CDF curves obtained from reliability indices (Equations 9.18 and 9.19), for the degree of consolidation (U) and the consolidation settlement remaining $(s_r)$, respectively, at time = 1 year. As in Figure 9.22, the two CDF curves in Figures 9.23 and 9.24, based on 17 and 23 values of β indices, respectively, are practically the same as the CDF curves from 5000 realizations of MCS.

It is of interest to note that the $n_i$ values of $c_v$ and $c_h$ in cells AE30:AE31 of the lower Figure 9.21 are practically zero. The implied insensitivities of the ultimate consolidation settlement to the coefficients of consolidation $(c_v$ and $c_h)$ are theoretically consistent (ultimate settlement not a function of rate parameters $c_v$ and $c_h)$ and are automatically revealed in a reliability analysis. In contrast, reliability analysis with respect to the limiting degree of consolidation at $t = 1$ year and settlement remaining at $t = 1$ year will show nonzero values for the $n_i$ values of the coefficients of consolidation $c_v$ and $c_h$, both parameters having an effect on the rate of consolidation.

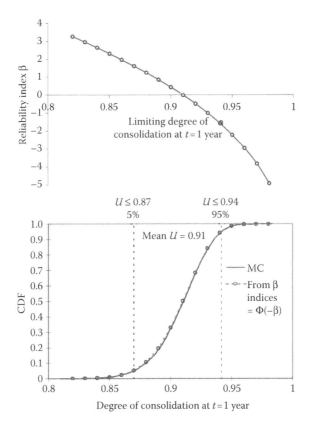

*Figure 9.23* Reliability indices for different limiting *U* at *t* = 1 year, and comparison of CDF based on β indices with CDF from Monte Carlo simulations.

## 9.10.5 Obtaining PDF curves from β index

The CDF curves shown in the lower plots of Figures 9.22 through 9.24 were obtained easily—by Equations 9.17 through 9.19—from the β values shown in the corresponding upper plots. In addition, it is simple to obtain the PDF of the respective outputs ($s_{cf}$, *U*, and $s_r$), by applying cubic spline interpolation (e.g., Kreyszig 1988) to the CDF. This is accomplished easily in the Excel spreadsheet platform, as explained below with respect to the 19 CDF values of Figure 9.22:

i. Autofill a 17-cell column vector **m** of $m_i$ (i = 2–18) with the following formula, next to the column of the 19 CDF values of $s_{cf}$:

$$m_i = \frac{3}{h}(CDF_{i+1} - CDF_{i-1}), \text{in which } h \text{ is the } s_{cf} \text{ interval of the CDF points } (=0.2\,\text{m})$$

ii. A 17 × 17 tridiagonal matrix **D** is set up, with entries $d_{i,i} = 4$, $d_{i+1,i} = d_{i,i+1} = 1$, and all other entries equal to 0.

iii. The 17 PDF values are obtained immediately and automatically upon entering " = mmult(minverse(**D**),**m**)" as a spreadsheet array formula in a 17-cell column.

The PDF curve of $s_{cf}$ thus obtained is shown in Figure 9.25. By the same procedure, 15 and 21 PDF values of the degree of consolidation (U) and of the settlement remaining (both

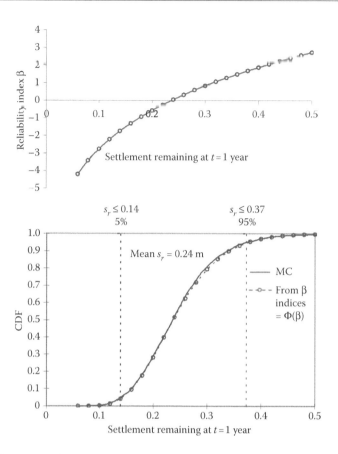

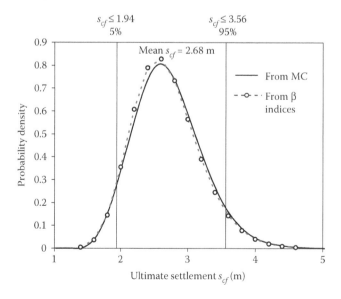

*Figure 9.24* Reliability indices for different limiting settlement remaining at $t = 1$ year, and comparison of CDF based on β indices with CDF from Monte Carlo simulations.

*Figure 9.25* PDF of ultimate primary consolidation settlement $s_{cf}$ from Monte Carlo simulations with 5000 realizations, and from 19 values of reliability index.

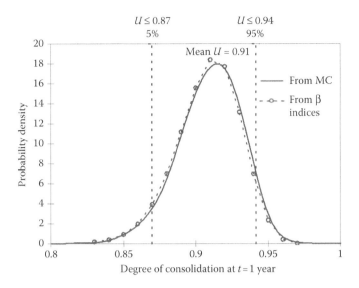

*Figure 9.26* PDF of degree of consolidation (*U*) at time = 1 year, from Monte Carlo simulations, and from 17 values of reliability index.

at time = 1 year) were obtained easily from their respective 17 and 23 CDF values of Figures 9.23 and 9.24. These two PDF curves are shown in Figures 9.26 and 9.27.

The 5000 Monte Carlo realizations performed earlier for the plots of Figures 9.22 through 9.24 can also be used to plot the outputs as PDF curves. The dashed PDF curves derived from the β indices agree remarkably well with the Monte Carlo PDF curves in Figures 9.25 through 9.27.

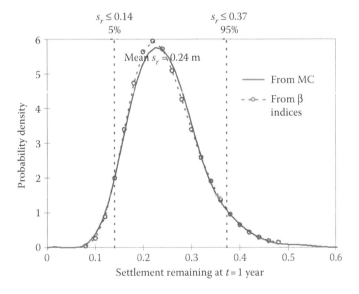

*Figure 9.27* PDF of settlement remaining (*s_r*) at time = 1 year, from Monte Carlo simulations, and from 23 values of reliability index.

## 9.11 COUPLING OF STAND-ALONE DETERMINISTIC PROGRAM AND SPREADSHEET-AUTOMATED RELIABILITY PROCEDURES VIA RESPONSE SURFACE OR SIMILAR METHODS

Programs can be written in the spreadsheet to compute the factor of safety or settlement (e.g., Low and Tang 2004, p. 87, Low et al. 2007, and the *VDrainSt* program in Low 2003b). However, there are situations where serviceability or ultimate limit states can only be evaluated using stand-alone finite element or finite difference programs, or one may already have a preferred or more accurate deterministic program in hand. In these circumstances, reliability analysis and RBD using the spreadsheet-automated FORM procedure can still be performed, provided one first obtains a response surface function (via the established response surface methodology) that closely approximates the outcome of the stand-alone finite element or finite difference programs. Once the closed-form response functions have been obtained, performing RBD for a target reliability index is straightforward and fast. Performing MCS on the closed-form approximate response surface function also takes little time. The response surface method (or other surrogate methods) was used in Li (2000) for consolidation analysis of a Singapore land reclamation project, Tandjiria et al. (2000) and Chan and Low (2012b) for laterally loaded single piles, Xu and Low (2006) for embankments on soft ground, and Lü and Low (2011) on underground rock excavations, among others.

## 9.12 SUMMARY AND CONCLUSIONS

This chapter discussed RBD/analysis for a spread footing, a rock slope, three soil slopes, a laterally loaded single pile, an anchored sheet pile wall, a rockbolt-reinforced tunnel, and soft clay consolidation accelerated by prefabricated vertical drains, using the practical FORM procedures of Low and Tang (2004, 2007) and an **x**-via-**n**-via-**u** procedure that is a very simple variation of the 2007 **x**-via-**n** procedure. The efficient spreadsheet-based FORM procedures can be coupled with stand-alone numerical packages via bridging techniques such as response surface methods and artificial neural networks.

The expanding dispersion ellipsoid perspective in the original space of the random variables was presented, as a useful alternative perspective of reliability index and the design point.

Various insights and interesting features and subtleties of RBD as revealed in the different RBD cases were discussed, testifying to the ability of RBD to locate the design point for a target risk level without presuming any partial factors and to automatically reflect parametric sensitivities and correlations from case to case. Links between RBD and EC7 design were shown, and the *complementary roles* that RBD can play to EC7 and LRFD were explained in Section 9.2.1 and further commented in the different RBD cases of subsequent sections.

RBD analysis involving spatially autocorrelated soil properties was illustrated and discussed for a Norwegian slope (Section 9.5) and for a laterally loaded single pile (Section 9.7). System FORM with multiple failure modes was discussed in Section 9.6 for a soil slope. RBD with respect to serviceability limit state was implemented for a laterally loaded pile in Section 9.7 and for consolidation settlement of soft clay accelerated by vertical drains in Section 9.10.

It was shown that for some geotechnical problems, performing SORM after FORM can improve the accuracy of the estimated failure probability, but a pragmatic stand on the adequacy of the approximate failure probability estimated from FORM $\beta$ index was also suggested and discussed in the context of the results of Figure 9.16a. This pragmatic approach does not require performing SORM and may thus reduce the hurdles confronting practitioners in using RBD.

## REFERENCES

Ang, H.S., and Tang, W.H. 1984. *Probability Concepts in Engineering Planning and Design, vol. 2—Decision, Risk, and Reliability*. John Wiley, New York.

Asadollahi, P., and Tonon, F. 2010. Definition of factor of safety for rock blocks, *International Journal of Rock Mechanics Mineral Science*, 47: 1384–1390.

Baecher, G.B., and Christian, J.T. 2003. *Reliability and Statistics in Geotechnical Engineering*. Chichester, West Sussex, England; John Wilcy, Hoboken, NJ.

Bobet, A., and Einstein, H.H. 2011. Tunnel reinforcement with rockbolts, *Tunnelling and Underground Space Technology*, 26: 100–123.

Bowles, J.E. 1996. *Foundation Analysis and Design*, 5th ed., McGraw-Hill, New York.

Brady, B.H.G., and Brown, E.T. 2006. *Rock Mechanics for Underground Mining*, 3rd ed., Kluwer, Dordrecht.

Breitung, K. 1984. Asymptotic approximations for multinormal integrals, *Journal of Engineering Mechanics*, ASCE, 110(3): 357–366.

Cai, G.Q., and Elishakoff, I. 1994. Refined second-order reliability analysis, *Structural Safety*, Elsevier, 14(4): 267–276.

Chan, C.L., and Low, B.K. 2012a. Practical second-order reliability analysis applied to foundation engineering, *International Journal for Numerical and Analytical Methods in Geomechanics*, 36(11): 1387–1409.

Chan, C.L., and Low, B.K. 2012b. Probabilistic analysis of laterally loaded piles using response surface and neural network approaches, *Computers and Geotechnics*, 43: 101–110.

Ching, J.Y., Phoon, K.K., and Hu, Y.G. 2009. Efficient evaluation of reliability for slopes with circular slip surfaces using importance sampling, *Journal of Geotechnical Geoenvironmental Engineering*, ASCE, 135(6):768–777.

Choa, V., Wong, K.S., and Low, B.K. 1990. *New Airport at Chek Lap Kok, Geotechnical Review and Assessment*. Consulting Report to Maunsell Pte Ltd., Singapore.

Ditlevsen, O. 1981. *Uncertainty Modeling: With Applications to Multidimensional Civil Engineering Systems*. McGraw-Hill, New York.

Duncan, J.M. 2000. Factors of safety and reliability in geotechnical engineering, *Journal of Geotechnical and Geoenvironmental Engineering*, 126(4): 307–316.

Duncan, J.M. 2001. Closure to discussions on factors of safety and reliability in geotechnical engineering, *Journal of Geotechnical and Geoenvironmental Engineering*, 127(8): 717–721.

Duncan, J.M., and Buchignani, A.L. 1973. Failure of underwater slope in San Francisco Bay, *Journal of Soil Mechanics and Foundation Division*, ASCE, 99(9). 687–703.

Duncan, J.M., and Wright, S.G. 2005. *Soil Strength and Slope Stability*. John Wiley & Sons, Inc., Hoboken, New Jersey, 297 pp.

EN 1997-1, *Eurocode 7: Geotechnical Design*, Part 1: *General Rules*.

Fiessler, B., Neumann, H.J., and Rackwitz, R. 1979. Quadratic limit states in structural reliability, *Journal of the Engineering Mechanics Division*, (ASCE), 105(4): 661–676.

Foott, R., Koutsoftas, D.C., and Handfelt, L.D. 1987. Test fill at Chek Lap Kok, Hong Kong, *Journal of Geotechnical Engineering*, ASCE, 113(2). 106 126.

Frank, R., Bauduin, C., Driscoll, R., Kavvadas, M., Krebs Ovesen, N., Orr, T., and Schuppener, B. 2005. *Designers' Guide to EN 1997-1 Eurocode 7: Geotechnical Design—General Rules*, Thomas Telford, London.

Haldar, A., and Mahadevan, S. 1999. *Probability, Reliability and Statistical Methods in Engineering Design*. John Wiley, New York.

Hasofer, A.M., and Lind, N.C. 1974. Exact and invariant second-moment code format, *Journal of Engineering Mechanics*, 100: 111–121, ASCE, New York.

Hoek, E. 2007. *Practical Rock Engineering*, http://www.rocscience.com/education/hoeks_corner.

Hohenbichler, M., and Rackwitz, R. 1988. Improvement of second-order estimates by importance sampling, *Journal of Engineering Mechanics*, ASCE, 114(12): 2195–2199.

Hong, H.P. 1999. Simple approximations for improving second-order reliability estimates, *Journal of Engineering Mechanics*, ASCE, 125(5): 592–595.

Koutsoftas, D.C., Foott, R., and Handfelt, L.D. 1987. Geo-technical investigations offshore Hong Kong, *Journal of Geotechnical. Engineering*, ASCE, 113(2): 87–105.

Koyluoglu, H.U., and Nielsen, S.R.K. 1994. New approximations for SORM integrals, *Structural Safety*, Elsevier, 13(4): 235–246.

Kreyszig, E. 1988. *Advanced Engineering Mathematics*, 6th ed., Wiley, New York, 972–973.

Li, D., Chen, Y., Lu, W., and Zhou, C. 2011. Stochastic response surface method for reliability analysis of rock slopes involving correlated non-normal variables, *Computers and Geotechnics*, 38(2011): 58–68.

Li, G.-J. 2000. *Soft Clay Consolidation Under Reclamation Fill and Reliability Analysis*. PhD thesis, School of CEE, Nanyang Technological University, Singapore.

Low, B.K. 2003a. Practical probabilistic slope stability analysis, *Proceedings of Soil and Rock America*, MIT, Cambridge, Massachusetts, Verlag Glückauf GmbH Essen, Vol. 2, 2777–2784 (http://www.ntu.edu.sg/home/cbklow/).

Low, B.K. 2003b. *Chapter 2: Theories, Computations, and Design Procedures Involving Vertical Drains, pp. 5–56*. (In *Soil Improvement: Prefabricated Vertical Drain Techniques*, Bo, M.W., Chu, J., Low, B.K., and Choa, V. (eds), Thomson Learning, Thomson Asia Pte Ltd., Singapore, 2003, 341pp.)

Low, B.K. 2005. Reliability-based design applied to retaining walls, *Geotechnique*, 55(1): 63–75.

Low, B.K. 2006. Practical reliability approach in geotechnical engineering, *Proceedings of GeoCongress 2006: Geotechnical Engineering in the Information Technology Age*, Atlanta, Georgia, USA, February 26–March 1, 2006, 6 pp in CD-ROM, ASCE.

Low, B.K. 2008a. Practical reliability approach using spreadsheet, Chapter 3 (pp. 134–168) of *Reliability-Based Design in Geotechnical Engineering—Computations and Applications*, Phoon, K.K. (ed.), Taylor & Francis, London.

Low, B.K. 2008b. Efficient probabilistic algorithm illustrated for a rock slope, *Rock Mechanics and Rock Engineering*, Springer-Verlag, 41(5): 715–734.

Low, B.K. 2008c. Settlement analysis of Chek Lap Kok trial embankments with probabilistic extensions, *Proceedings of the Sixth International Conference on Case Histories in Geotechnical Engineering and Symposium in Honor of Professor James K. Mitchell*, Arlington, Virginia (USA), August 11–16, 2008, 12pp.

Low, B.K. 2014. FORM, SORM, and spatial modeling in geotechnical engineering, *Structural Safety*, Elsevier, 49: 56–64.

Low, B.K., and Duncan, J.M. 2013. Testing bias and parametric uncertainty in analyses of a slope failure in San Francisco Bay mud, *Proceedings of Geo-Congress 2013*, ASCE, San Diego, March 3–6, 937–951.

Low, B.K., and Einstein, H.H. 2013. Reliability analysis of roof wedges and rockbolt forces in tunnels, *Tunnelling and Underground Space Technology*, 38: 1–10.

Low, B.K., Lacasse, S., and Nadim, F. 2007. Slope reliability analysis accounting for spatial variation, *Georisk: Assessment and Management of Risk for Engineered Systems and Geohazards*. Taylor & Francis, London, 1(4): 177–189.

Low, B.K., and Tang, Wilson H. 1997b. Reliability analysis of reinforced embankments on soft ground, *Canadian Geotechnical Journal*, 34(5): 672–685.

Low, B.K., Teh, C.I., and Tang, Wilson H. 2001. Stochastic nonlinear p–y analysis of laterally loaded piles, *Proceedings of the Eighth International Conference on Structural Safety and Reliability, ICOSSAR '01*, Newport Beach, California, June 17–21, 2001, 8pp, A.A. Balkema Publishers.

Low, B.K., and Tang, Wilson H. 1997a. Efficient reliability evaluation using spreadsheet, *Journal of Engineering Mechanics*, ASCE, 123(7): 749–752.

Low, B.K., and Tang, Wilson H. 2004. Reliability analysis using object-oriented constrained optimization, *Structural Safety*, Elsevier Science Ltd., Amsterdam, 26(1): 69–89.

Low, B.K., and Tang, Wilson H. 2007. Efficient spreadsheet algorithm for first-order reliability method, *Journal of Engineering Mechanics*, ASCE, 133(12): 1378–1387.

Low, B.K., Zhang, J., and Tang, Wilson H. 2011. Efficient system reliability analysis illustrated for a retaining wall and a soil slope, *Computers and Geotechnics*, Elsevier, 38(2): 196–204.

Lü, Q., and Low, B.K. 2011. Probabilistic analysis of underground rock excavations using response surface method and SORM, *Computers and Geotechnics*, Elsevier, 38: 1008–1021.

Madsen, H.O., Krenk, S., and Lind, N.C. 1986. *Methods of Structural Safety*. Prentice-Hall, Englewood Cliffs, NJ.

Matlock, H. 1970. Correlations for design of laterally loaded piles in soft clay, *Proceedings of the Offshore Technology Conference*, Houston, Texas, 1970, Paper OTC 1204.

Melchers, R.E. 1999. *Structural Reliability Analysis and Prediction*, 2nd ed., John Wiley, New York.

Rackwitz, R., and Fiessler, B. 1978. Structural reliability under combined random load sequences, *Computer Structures*, 9(5): 484–494.

Shinozuka, M. 1983. Basic analysis of structural safety, *Journal of Structural Engineering*, ASCE, 109(3): 721–740.

Simpson, B., and Driscoll, R. 1998. *Eurocode 7—A Commentary*. Construction Research Communications, Watford.

Sofianos, A.I., Nomikos, P., and Tsoutrelis, C.E. 1999. Stability of symmetric wedge formed in the roof of a circular tunnel: Nonhydrostatic natural stress field, *International Journal of Rock Mechanics Mineral Science*, Elsevier, 36: 687–691.

Tandjiria, V., Teh, C.I., and Low, B.K. 2000. Reliability analysis of laterally loaded piles using response surface methods, *Structural Safety*, Elsevier Science Ltd., Amsterdam, 22(4): 335–355.

Tomlinson, M.J. 1994. *Pile Design and Construction Practice*, 4th ed., E & FN Spon, London.

Tomlinson, M.J. 1995. *Foundation Design and Construction*, 6th ed., Longman Scientific, Harlow, Essex, UK.

Tvedt, L. 1983. Two second-order approximations to the failure probability. *Veritas Report RDIV/20-004083*. Det norske Veritas, Oslo.

Tvedt, L. 1988. Second-order reliability by an exact integral, *2nd IFIP Working Conference on Reliability and Optimization on Structural Systems*, Springer-Verlag, Berlin, Germany, 377–384.

Tvedt, L. 1990. Distribution of quadratic forms in normal space: Application to structural reliability, *Journal of Engineering Mechanics* (ASCE), 116(6): 1183–1197.

Veneziano, D. 1974. Contributions to second moment reliability. *Research Report No. R74-33*. Department of Civil Engineering, MIT, Cambridge, Massachusetts.

Whitman, R.V. 1984. Evaluating calculated risk in geotechnical engineering, *Journal of Geotechnical Engineering*, ASCE, 110(2): 145–188.

Xu, B., and Low, B.K. 2006. Probabilistic stability analyses of embankments based on finite-element method, *Journal of Geotechnical and Geoenvironmental Engineering*, ASCE, 132(11): 1444–1454.

Zhao, Y.-G., and Ono, T. 1999. New approximations for SORM: Part 1, *Journal of Engineering Mechanics* (ASCE), 125(1): 79–85.

Chapter 10

# Managing risk and achieving reliable geotechnical designs using Eurocode 7

*Trevor L.L. Orr*

## 10.1 INTRODUCTION

Eurocode 7, published by CEN, the European Committee for Standardization, as EN 1997 with two parts, *Part 1: General Rules* (CEN, 2004), hereafter referred to as Eurocode 7, and *Part 2: Ground Investigation and Testing* (CEN, 2007), is the new European standard for geotechnical design. Since 2010, Eurocode 7 has superseded the existing national standards for geotechnical design in all the CEN member countries. Eurocode 7 aims to achieve geotechnical designs with appropriate degrees of reliability using the limit state design method as set out in the head Eurocode, EN 1990 (CEN, 2002), which generally involves calculations with partial factors applied to characteristic parameter values. Until the introduction of Eurocode 7, there had been little experience in Europe in the use of the limit state design method for geotechnical design. Hence, its introduction has caused considerable debate and some resistance, often due to misunderstandings about how Eurocode 7 has adapted the limit state method in EN 1990 for geotechnical design. This chapter reviews how the important aspects of risk and reliability in geotechnical design are addressed in Eurocode 7. In particular, it explains how geotechnical designs with appropriate degrees of reliability are achieved by using calculations with partial factors applied to appropriately selected characteristic parameter values and quality management measures related to the different stages of a geotechnical design project which are ground investigation, design calculations, construction, and monitoring and maintenance after construction.

## 10.2 GEOTECHNICAL COMPLEXITY AND RISK

### 10.2.1 Factors affecting complexity

All civil engineering involves risk, but geotechnical designs generally involve more risk than other areas of civil engineering due to the complexity of the ground and uncertainties about its properties. Hence, Eurocode 7 provides a methodology for managing risk in geotechnical design. As the first stage in this methodology, Eurocode 7 states in §2.1(8)P that the complexity of the geotechnical design shall be identified, together with the associated risks, in order to establish the minimum requirements for the extent and content of geotechnical investigations, calculations, and construction control checks, and the expertise required of those involved in the different aspects. Note that the symbol § indicates a clause, that is, a paragraph in the standard and the letter P indicates that the clause is a Principle and hence a code requirement.

The complexity of a geotechnical design is affected by the following geotechnical factors identified in Eurocode 7, which, when they are abnormal and severe, constitute hazards that need to be taken into account in geotechnical designs:

- Ground conditions
- Groundwater situation
- Regional seismicity
- Influence of the environment

The complexity of a geotechnical design is also affected by the vulnerability of the proposed structure to its surroundings and the vulnerability of neighboring structures and services to the impact of the proposed structure. The vulnerability of the structure is a function of the nature and size of the structure itself, and its elements. Geotechnical risk is a function of two factors: the geotechnical hazards listed above, plus the vulnerability of the structure and neighboring structures, and services to the impact of the proposed structure.

## 10.2.2 Levels of risk and Geotechnical Categories

Eurocode 7 identifies different levels of risk in geotechnical designs, referring to design situations involving negligible risk, no exceptional risk, and abnormal or high risk, which are associated with the three Geotechnical Categories, GC1, GC2, and GC3, as shown in Table 10.1, adapted from Orr and Farrell (1999). GC1 corresponds to structures with low hazard and vulnerability levels and hence negligible risk, GC2 corresponds to structures with moderate hazard and vulnerability levels and no exceptional risk, and GC3 corresponds to structures with high hazard and vulnerability levels and hence abnormal or high risk. Eurocode 7 provides the following examples of typical structures with these different levels of geotechnical risk:

- Examples of structures with a low level of risk, that is GC1, are small and relatively simple structures and agricultural buildings on ground that is known, from comparable local experience, to be sufficiently straightforward and does not involve soft, loose or collapsible soil, or loose fill;
- Examples of structures with no exceptional risk, that is GC2, are conventional types of spread, pile, and raft foundations on ground that is not difficult and where the ground conditions and properties can be determined from routine investigations and tests;
- Examples of structures with abnormal or high risk, that is GC3, are very large/unusual structures on ground where the conditions are unusual or exceptionally difficult.

The use of the three Geotechnical Categories is not a Principle but is presented in Eurocode 7 as an Application Rule and hence is optional. The advantage of the Geotechnical Categories is that they provide a framework for

- Assessing the level of risk in a geotechnical design
- Selecting the appropriate geotechnical investigation and testing methods and design procedures
- Identifying the appropriate personnel to carry out the design

As noted in Section 10.2.1, the complexity of a geotechnical design situation and the associated risks affect the minimum requirements in designs to Eurocode 7 for the

*Table 10.1* Geotechnical Categories related to geotechnical complexity and risk levels

| Factors to be considered affecting design complexity | Geotechnical Categories | | |
|---|---|---|---|
| | GC1 | GC2 | GC3 |
| Hazards | Low | Moderate | High |
| Ground conditions | Known from comparable local experience to be sufficiently straightforward. Not involving soft, loose or collapsible soil, or loose fill | No difficult ground conditions. Ground conditions and properties can be determined from routine investigations and tests | Unusual or exceptionally difficult ground conditions |
| Groundwater situation | No excavation below the water table, except where comparable local experience indicates this will be straightforward | No risk of damage, without prior warning, to structures due to groundwater lowering or drainage. No exceptional water tightness requirements | High groundwater pressures and exceptional groundwater conditions, for example, multi-layered strata with variability permeability |
| Regional seismicity | Areas with no or very low earthquake hazard | Moderate earthquake hazard where seismic design code (Eurocode 8) may be used | Highly seismic areas and areas of high earthquake hazard |
| Influence of the environment | Negligible risk of problems due to surface water, subsidence, hazardous chemicals, etc. | Environmental factors covered by routine design methods. No exceptional loading conditions | Complex or difficult environmental factors requiring special design methods |
| Vulnerability | Low | Moderate | High |
| Nature and size of the structure and its elements | Small and relatively simple structures or construction. Insensitive structures in seismic areas | Conventional types of structures with no exceptional risks | Very large or unusual structures, structures involving abnormal risks, and unusual or exceptionally difficult loading conditions |
| Surroundings | Negligible risk of damage to or from neighboring structures or services and negligible risk for loss of human life | No exceptional risk of damage to neighboring structures or services due, for example, to excavation or piling and no exceptional risk for loss of human life | High risk of damage to neighboring structures or services and high risk for loss of human life |
| Risk | Negligible | Not exceptional | High |

geotechnical investigations and tests, the design procedures and construction control checks, and the expertise of the designer. These are related to the Geotechnical Category, as indicated in Table 10.2, also adapted from Orr and Farrell (1999). The Geotechnical Categories provide a framework for assessing the suitability of a site with respect to the proposed construction and the level of acceptable risks since §2.8(3) states that the Geotechnical Design Report, which is described in Section 10.3.2, should normally include a statement on this.

*Table 10.2* Geotechnical investigations, design procedures, expertise required, and examples of structures related to Geotechnical Categories

| | Geotechnical Categories | | |
|---|---|---|---|
| | *GC1* | *GC2* | *GC3* |
| Geotechnical investigations | Qualitative geotechnical investigations, including trial pits | Routine investigations involving borings, field, and laboratory tests to obtain quantitative data | Additional non-routine investigations and laboratory and field tests |
| Design procedures | Prescriptive measures and simplified design procedures, for example, designs based on experience or published presumed bearing pressures | Analyses for stability and deformations based on routine procedures for design and execution in Eurocode 7 to ensure fundamental requirements are satisfied | Alternative provisions and rules to those in Eurocode 7 |
| Expertise required | Person with appropriate comparable experience | Experienced qualified person | Experienced geotechnical specialist |
| Examples of structures | Only small and relatively simple structures:<br><br>• Simple 1 and 2 story structures and agricultural buildings having a max. column load of 250 kN and max. wall load of 100 kN/m<br>• Retaining walls and excavation supports where the ground level difference does not exceed 2 m<br>• Small excavations for drainage and pipes | Conventional types of:<br><br>• Spread, pile, and raft foundations<br>• Walls and other structures retaining or supporting soil or water<br>• Excavations<br>• Bridge piers and abutments<br>• Embankments and earthworks<br>• Ground anchors and other tie-back systems<br>• Tunnels in hard, nonfractured rock not subjected to special water tightness or other requirements | Structures or parts of structures outside the limits of GC1 or GC2:<br><br>• Very large/unusual structures or involving abnormal risks<br>• Deep excavations<br>• Tunnels in soft or highly permeable ground<br>• Areas of probable site instability or persistent ground movements requiring separate investigation or special measures |

Table 10.2 shows that GC1 structures should be designed by appropriately experienced persons, and only qualitative investigations and simplified design procedures, such as prescriptive measures, are required. Most structures will fall into GC2 and should be designed by experienced qualified persons using quantitative geotechnical data and analyses, with routine procedures for field and laboratory testing and for design and execution covered by the provisions in Eurocode 7. Very large structures or very complex ground conditions are GC3 and should be designed by experienced geotechnical specialists with additional more sophisticated tests and alternative provisions and rules to those for GC2 to ensure that the basic requirements are fulfilled.

With regard to the required level of supervision, it is noted in §4.2.1(2) that the type, quality, and frequency of supervision should be commensurate with the potential risk of failure during construction. With regard to the monitoring of structures, it is stated in §4.5(6) that for structures that may impact unfavorably on appreciable parts of the surrounding physical environment, or for which failure may involve abnormal risks to property or life, that is, GC3 structures, monitoring should be required for more than 10 years after construction is complete, or throughout the life of the structure. Eurocode 7 gives in

§12.7(2) several examples where monitoring should be applied in the case of embankments; these include

- Where the stability of an embankment acting as a dam to a large degree depends on the pore-water pressure distribution in and beneath the embankment
- Where surface erosion is a considerable risk

### 10.2.3  Risks due to adverse water pressures

Water pressures are identified in Eurocode 7 as a particular hazard that can pose significant risks in geotechnical designs and can be difficult to treat in order to achieve a consistent level of safety. As shown in Section 10.4.3, Eurocode 7 provides separate partial factors for designs against situations involving failure due to static water pressure causing uplift and hydraulic heave due to water seepage pressures. The need to consider the risks due to water pressure in the case of retaining structures is mentioned in §9.6(2)P of Eurocode 7, which refers to the example given in §9.4.1(5) of a critical limit state occurring in the form of damage to nearby structures or services due to wall movement, even though collapse of the wall may not be imminent. The treatment of water pressure in geotechnical designs, including how the characteristic value should be selected and whether or not they should be factored, is an aspect that is under discussion. The revised version of Eurocode 7, due to be published in 2020, is likely to provide more guidance on this.

### 10.2.4  Geotechnical investigations and geotechnical risks

Since soil is a natural material, not manufactured under controlled conditions like concrete and steel, its properties are not specified, as in structural design, but need to be determined. Hence, Eurocode 7 has two parts: Part 1 providing the general rules for geotechnical design and Part 2 providing the requirements for carrying out ground investigations and determining the value of soil parameter values. A geotechnical investigation with field and/or laboratory tests to establish the stratigraphy and determine the soil parameter values is the first part of the geotechnical design process. According to §2.1.1(1)P of Part 2 of Eurocode 7, geotechnical investigations shall be planned to obtain the relevant geotechnical information and data required to manage the identified and anticipated project risks. The quality and extent of the geotechnical investigations should therefore be sufficient to reduce the risk of unforeseen unfavorable ground conditions that could cause a geotechnical failure, unexpected costs, or delays to a project.

## 10.3  RELIABILITY REQUIREMENTS IN DESIGNS TO EUROCODE 7

### 10.3.1  Basic requirement

The basic requirement for a structure designed to the Eurocodes, including a geotechnical structure designed to Eurocode 7, as stated in §2.1(1)P of the head Eurocode, EN 1990: Basis of Structural Design (CEN, 2002) is that it *shall be designed and executed in such a way that it will, during its intended life, with appropriate degrees of reliability and in an economical way:*

- *Sustain all actions and influences likely to occur during execution and use, and*
- *Remain fit for the use for which it is required*

where an action is defined in EN 1990 as a set of forces (loads) or an imposed deformation or acceleration. As the above clause states, it is the basic requirement of the Eurocodes to design structures that have appropriate degrees of reliability. Reliability is defined in EN 1990 §1.5.2.17 as "*the ability of a structure or structural member to fulfill the specified requirements, including the design working life, for which it has been designed.*" Calgaro (2011), referring to the Eurocodes, notes that structural reliability covers four aspects: safety, serviceability, durability, and robustness. Also, as pointed out in *Guidance Paper L—Application and use of the Eurocodes* (EC, 2003), the word safety is encompassed in the Eurocodes in the word reliability.

## 10.3.2 Measures to achieve reliable designs

While reliability is usually expressed in probabilistic terms, it is noted in §2.2(5) of EN 1990 that achieving the appropriate degrees of reliability of a structure does not only involve using the limit state method, which is described in Section 10.4.1, but also involves adopting a combination of the following measures relating to the different aspects of the whole design and execution process:

a. Good quality investigations of the ground conditions and environmental influences
b. Preventative and protective measures
c. Measures relating to quality management
d. Measures aimed to reduce errors in design and execution, including gross human errors
e. Measures to provide adequate robustness
f. Efficient execution
g. Adequate inspection and maintenance

Further details of these measures are provided in the following paragraphs:

a. *Good quality investigations*: The importance of carrying out suitably extensive and good quality geotechnical investigations for ensuring the reliability of geotechnical designs is noted in §2.4.1(2) which states that "*Knowledge of the ground conditions depends on the extent and quality of the geotechnical investigations. Such knowledge and the control of workmanship are usually more significant to fulfilling the fundamental requirements than is precision in the calculation models and partial factors.*" Some guidance on the spacing and depth of investigation points is provided in §2.4.1.3(1)P of Part 2 of Eurocode 7.
b. *Preventative and protective measures:* Some measures provided to prevent the occurrence of a limit state or protect a structure from environmental influences when designing to Eurocode 7 are not determined using calculations and partial factors. Such measures include, for example, the provision of drains behind a retaining structure to prevent the development of pore water pressures behind the retaining structure causing failure or excessive movement, and the use of a protective sheath or the provision of sacrificial steel to protect the steel tendons in a ground anchor against corrosion.
c. *Measures relating to quality management*: Eurocode 7 provides a number of organizational measures and controls at the design, execution, use, and maintenance stages of a project. For example, Eurocode 7 includes lists of many items to be taken into account, to receive attention or be considered in the geotechnical design process. These checklists are an important quality management feature of Eurocode 7 to achieve the required reliability of geotechnical designs apart from by means of numerical calculations. Another important measure related to quality management

is the requirement in §2.8(1)P that all the assumptions, data, calculation methods, derivation of characteristic values, and the results of the ultimate and serviceability limit state verifications shall be recorded in a Geotechnical Design Report (GDR), which is required for all geotechnical designs, however, large or small. Furthermore, it is required in §3.4.1(1)P that the results of a geotechnical investigation shall be compiled in a Geotechnical Investigation Report (GIR), which shall form part of the Geotechnical Design Report. Another quality management requirement, which is given in §2.8(6), is that an extract from the Geotechnical Design Report containing the supervision, monitoring, and maintenance requirements for the completed structure shall be provided to the owner/client. Some comments on the requirements for inspection and monitoring are given in (g) below.

d. *Measures aimed to reduce errors in design and execution, including gross human errors*: Unfortunately, as many case histories and researchers have shown, for example, Sowers (1993), human error is a factor in many geotechnical and structural failures. Simpson (2011) in investigating geotechnical failures, identified the following types of errors in geotechnical design which, in his experience, were depressingly common: arithmetic errors, lack of expected basic knowledge, failures of communication, and oversight or misunderstanding of important information, for example, from geotechnical investigations. Hence, systems need to be included in the design process to reduce the occurrence of such errors and, if they occur, to identify and remedy them. The checklists and Geotechnical Design Report referred to in the previous paragraph are examples of practical measures in Eurocode 7 to reduce errors in geotechnical designs.

e. *Adequate robustness*: Robustness is the ability of a system to resist damage while maintaining its important functions. Robustness is not limited to physical structures but robustness principles can also be applied to management systems to reduce the effects of unknown risks (Calgaro, 2011). Geotechnical design calculations generally involve the consideration of a relatively small number of variables to describe a complex design situation, which, in addition to the known variability of the ground properties and actions, can also involve unforeseen variations in the actions and the geometry. Designs need to account for unforeseen variations in the actions and geometry that could occur due, for example, to the disposition of the ground strata not being precisely described or due to construction activity affecting the actions or geometry. Designs to Eurocode 7 accommodate such unforeseen secondary effects, including minor human errors, by including appropriate safety margins on the partial factors and design measures and management systems to ensure that designs are sufficiently robust.

f. *Efficient execution*: Construction work should be carried out in accordance with relevant execution standards in the case of geotechnical design situations for which such standards exist, for example, CEN has published the following execution standards, all with the prefix "Execution of Special Geotechnical Works" to their titles:

- EN 12699            Displacement piles
- EN 12715            Grouting
- EN 12716            Jet grouting
- EN 14199            Micropiles
- EN 14475            Reinforced fill
- EN 14490            Soil nailing
- EN 14679            Deep mixing
- EN 14731            Ground treatment by deep vibration
- EN 15237            Vertical drainage

- EN 1536          Bored piles
- EN 1537          Ground anchors
- EN 1538          Diaphragm walls

g. *Adequate inspection and maintenance*: Eurocode 7 states in §2.8(4)P that the Geotechnical Design Report shall include a plan of supervision and monitoring, as appropriate, and that items requiring checking during construction or maintenance afterwards shall be clearly identified and, when the required checks have been carried out during construction, they shall be included in an addendum to the Report. Supervision, monitoring, and maintenance are particularly important for ensuring the safety and reliability of geotechnical structures and hence, there are many references throughout Eurocode 7 to supervision, monitoring, and maintenance, including a whole section, Section 4, with general requirements for the supervision of construction, monitoring, and maintenance, and subsections in other parts of Eurocode 7 with specific requirements for

- Supervision in the case of pile foundations
- Supervision, monitoring, and maintenance in the case of anchors
- Monitoring in the case of overall stability
- Supervision and monitoring in the case of embankments

### 10.3.3 Design assumptions for reliable designs

Since achieving reliable geotechnical designs depends on the appropriate implementation of the measures elaborated in the previous section, the provisions in Eurocode 7 are based on a number of important assumptions regarding the competencies of those carrying out the work, the materials used and the use of the structure. It is assumed that

- Data required for design are collected, recorded, and interpreted by appropriately qualified and experienced personnel.
- Structures are designed by appropriately qualified and experienced personnel.
- Construction is carried out according to the relevant standards by personnel having the appropriate skill and experience.
- There is adequate continuity and communication between the personnel involved in data collection, design, and construction.
- Adequate supervision and maintenance are provided.
- Construction materials and products are used as specified in Eurocode 7 or other relevant specifications.
- The structure will be used for the purpose defined in the design.

To provide a framework to verify the assumptions that different aspects of the geotechnical work have been carried out by appropriately qualified personnel, some European countries have established systems for assessing and certifying the technical competencies of those involved in the design and execution of geotechnical structures. For example, in the United Kingdom, a Register of Geotechnical Engineering Professionals (RoGEP) was established in June 2011. The aim of RoGEP is to provide external stakeholders, including clients and other professionals, with a means to identify individuals who are suitably qualified and competent in ground engineering. To prevent uncertainty as to whether or not the assumptions listed above have been complied with in a geotechnical design, Eurocode 7 states in §1.3(3) that compliance with them should be documented, for example, in the Geotechnical Design Report.

## 10.4 VERIFICATION OF DESIGNS TO EUROCODE 7

### 10.4.1 Limit state design method

The limit state design method has been adopted in EN 1990, and hence in Eurocode 7, to ensure that the basic requirement stated in Section 10.3.1 regarding geotechnical structures having the appropriate degrees of reliability is fulfilled. The design requirement to be fulfilled when using the limit state method is that, for each geotechnical design situation, no relevant limit state is exceeded. Eurocode 7 states that limit states should be verified by one or a combination of the following methods:

- Use of calculations
- Adoption of prescriptive measures
- Experimental models and load tests
- An observational method

In most cases, geotechnical designs are verified by the use of calculations and, hence, Eurocode 7 and this chapter focus on that method. However, if one or a combination of the other methods is used, then the resulting design should still aim to have the appropriate degrees of reliability as achieved when using calculations. When using calculations, the reliability of the design is verified by checking that the occurrence of either an ultimate limit state (failure involving collapse or risk to life) or a serviceability limit state (excessive deformations or vibrations) will not occur for all relevant design situations when using design values of the input parameters (design geometrical data, design actions, and design material parameters or resistances). Design values are obtained by applying partial factors, combination factors, and correlation factors to characteristic, representative, or nominal parameter values, as appropriate. As explained in Section 10.5.1, the values of these factors can be determined either on the basis of calibration with previous experience or on the basis of statistical evaluation to achieve the required degrees of reliability.

When using one of the other methods on its own, the reliability of the design is verified by monitoring the performance of the structure or a component, and checking that the occurrence of either an ultimate or a serviceability limit state is sufficiently unlikely for the design conditions. The other methods, such as load tests or the observational method, are often used in combination with calculations to justify the choice of reduced partial factor values and, hence, more economical designs that still have the appropriate degrees of reliability. An example of the use of load tests to achieve a more economical design is given in Section 10.5.2. This example is the use of the reduced partial resistance factors for the design of piles in the UK National Annex to Eurocode 7, shown in Table 10.10, when the serviceability limit state is explicitly checked.

### 10.4.2 Verification by use of calculations

#### 10.4.2.1 Design equations and their components

The equations to be satisfied when designing by the use of calculations are, for ultimate limit states, to check that equilibrium is not violated; that is:

$$E_d \leq R_d \tag{10.1}$$

and, for serviceability limit states, to check that the deformations do not exceed the limiting criterion:

$$E_d \leq C_d \qquad\qquad\qquad\qquad (10.2)$$

where $E_d$ is the design value of the effect of the actions, which for ultimate limit states is the design force or moment and for serviceability limit states is the calculated design deformation, $R_d$ is the design value of the resistance, and $C_d$ is the limiting value of the deformation, that is, the relevant serviceability criterion.

The components of limit state design calculations, with examples of their constituents shown in brackets, are

- Geometrical data (ground profile, structural geometry)
- Actions (weight of soil, rock, and water, earth and water pressures, dead and imposed loads, ground movements due, for example, to mining and tunneling, temperature effects)
- Geotechnical parameters (geotechnical investigations, field and laboratory tests, derived parameters)
- Verification method (consideration of all relevant limit states, use of calculation methods, prescriptive measures and load tests, an observational method)
- Design complexity (ground conditions, nature of the structure, groundwater conditions, influence of the environmental, regional seismicity)

These components are presented in the form of the Geotechnical Design Triangle shown in Figure 10.1, based on Orr (2012), similar to Burland's (1987) Soil Mechanics Triangle, with geometrical data, the combination of actions and geotechnical parameters, and the verification method at the apexes of the triangle, each connected to the design complexity at the center of the triangle as this affects all aspects of a geotechnical design. The safety elements and safety features associated with each of these components in design calculations in order to achieve the appropriate degrees of reliability are shown in italic type in Figure 10.1, together with the linkages between them.

The components of geotechnical design calculations to Eurocode 7, their constituents, and safety elements are also shown in Table 10.3 and explained in the following paragraphs. The list in Table 10.3 of what constitute actions in Eurocode 7 is taken from §2.4.2(4) and shows that earth pressures, movements caused by mining or other caving or tunneling activities, movements due to creeping, or sliding or settling ground masses, downdrag and temperature effects should all be considered as actions. Since earth pressures and downdrag are actions that also involve the strength of the soil and since soil strength is frictional, so that the resistance is a function of actions, care is needed when calculating $E_d$ and $R_d$ to ensure that the actions and soil strengths or resistances are factored appropriately to achieve the appropriate degrees of reliability. This is commented on in Section 10.4.2.5 on design effects of actions and design resistances. Example 10.2 examines the effect of the interaction between actions and soil resistance in the case of earth resistance to uplift on the side wall of a deep basement.

### 10.4.2.2 Design geometrical data

The geometrical data involved in geotechnical design calculations include the level and slope of the ground surface, water levels, the levels of the interfaces between strata, excavation levels, and the shape of the foundation or other structural elements. Deviations and uncertainties in the geometrical data are usually small and therefore the characteristic values of

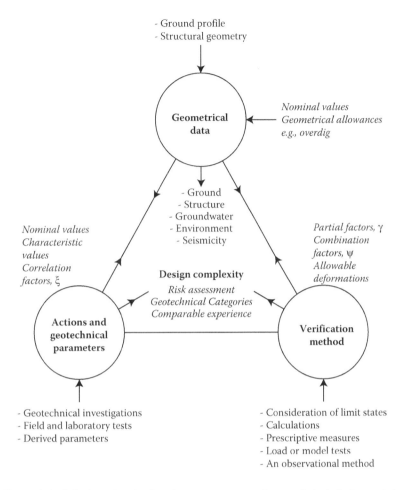

*Figure 10.1* Geotechnical design triangle showing components of geotechnical design and, in italic type, measures in Eurocode 7 to reduce risk and provide reliability.

the levels of ground and dimensions of geotechnical structures or elements should usually be nominal values, that is, values chosen on a nonstatistical basis, corresponding to values obtained from the ground investigation or specified in the design. The design value of a geometrical parameter is generally equal to the characteristic value, since the partial factors for actions and material properties include an allowance for minor variations in the geometrical data and therefore no further safety margin on the geometrical data is normally required. For situations involving water pressures and designing against limit states with severe consequences, generally ultimate limit states, §2.4.6.1(6)P states that the design values of the water pressures represented by geometrical data in the form of groundwater levels shall be the most unfavorable values that could occur during the design lifetime of the structure and, for limit states with less severe consequences, generally serviceability limit states, the design values shall be the most unfavorable values that could occur in normal circumstances.

   Where deviations in a geometrical parameter have a significant effect on the reliability of a structure, the design value of the geometrical parameter $a_d$ is assessed either directly or derived from nominal values using the following equation:

$$a_d = a_{nom} \pm \Delta a \tag{10.3}$$

*Table 10.3* Components of geotechnical designs, their constituents, and related safety elements

| Component | Constituents | Safety elements |
|---|---|---|
| Geometrical data | • Levels and slope of ground surface<br>• Water levels<br>• Levels of interfaces between strata<br>• Excavation levels<br>• Dimensions of geotechnical structure | • Nominal values<br>• Geometrical allowance, for example, change to a geometrical parameter such as the overdig allowance in front of a retaining wall |
| Actions | • Weight of soil, rock, and water<br>• Earth and water pressures<br>• Stresses in the ground<br>• Seepage forces<br>• Dead and imposed loads<br>• Surcharges<br>• Mooring forces<br>• Removal of load or excavation of ground<br>• Traffic loads<br>• Movements causing by mining or caving or other tunneling activities<br>• Swelling and shrinkage caused by vegetation, climate, and moisture changes<br>• Movements due to creeping or sliding or settling of ground masses<br>• Movements and accelerations caused by earthquakes, explosions, vibrations, and dynamic loads<br>• Temperature effects, including frost action<br>• Ice loading<br>• Imposed pre-stress in ground anchors or struts<br>• Downdrag | • Characteristic values<br>• Nominal values |
| Geotechnical parameters | • Geotechnical investigations<br>• Field and laboratory tests<br>• Derived parameters | • Characteristic values<br>• Correlation factors, $\xi$ |
| Verification method | • Consideration of all relevant limit states<br>• Use of calculation methods<br>• Prescriptive measures<br>• Load and model tests<br>• An observational method | • Combination factors, $\psi$<br>• Representative values<br>• Partial factors, $\gamma$<br>• Allowable deformations |
| Design complexity | • Ground conditions<br>• Groundwater conditions<br>• Influence of the environment<br>• Regional seismicity<br>• Nature of the structure<br>• Impact on neighboring structures | • Risk assessment<br>• Geotechnical Categories<br>• Comparable experience |

where $\Delta a$ is the change to the nominal geometrical parameter to provide the required safety margin and achieve the appropriate degrees of reliability. An example of a situation where deviation in a geometrical parameter is significant is in the design of an embedded retaining wall whose stability depends on the passive pressure of the soil in front of the wall. In this situation, Eurocode 7 specifies that the design ground surface in front of the wall should be obtained by lowering the characteristic ground surface by an amount $\Delta a$.

### 10.4.2.3 Design actions

Since Eurocode 7 is one of the harmonized sets of Eurocodes, the values of the partial factors for actions and the requirements for combining actions are given in EN 1990. The design value of an action, $F_d$ for use in verification calculations is either assessed directly or derived by multiplying the representative action, $F_{rep}$ by the appropriate partial action factor value, $\gamma_F$:

$$F_d = \gamma_F F_{rep} \tag{10.4}$$

The representative value of an action may be the characteristic value $F_k$ or an accompanying value $\psi F_k$ in the case of a variable action when there is more than one variable action and where $\psi$ is the combination factor. The characteristic value of an action is its main representative value and shall be specified as a mean value, an upper or lower value or a nominal value. The characteristic permanent action $G_k$ derived from the weights of materials, including soil and water, is normally calculated using the given nominal weight density (unit weight) of the material and the nominal geometry of the ground and the structural components, since the variability and hence the uncertainty in the weight density and geometry are not normally significant.

The design effect of the actions $E_d$ for ultimate and serviceability limit states is derived from the design value of the permanent action plus the design value of the leading variable action and the design combination values of the accompanying actions obtained using the appropriate combination rules and $\psi$ values. EN 1990 provides rules for combining permanent and variable actions in different design situations, for example, persistent or transient, accidental or seismic, using three combination factors: $\psi_0$, $\psi_1$, and $\psi_2$. The combination factor $\psi_0$ is applied to nonleading actions and takes account of the reduced probability of the simultaneous occurrence of two or more independent variable actions. The combination factor $\psi_1$ is used to obtain the frequent value of variable actions, mainly for serviceability limit state designs, while the combination factor $\psi_2$ is used to obtain the quasi-permanent value of variable actions, mainly in the assessment of long-term effects, such as creep effects in prestressed structures. Recommended values for $\psi_0$, $\psi_1$, and $\psi_2$ for different types of variable actions, including imposed loads on buildings, snow, wind, and traffic loads, are presented in tables in EN 1990. The tables are for different types of structures, for example, for buildings and different types of bridges; as an example, for imposed loads in residential buildings and offices, the recommended combination factor values are $\psi_0 = 0.7$, $\psi_1 = 0.5$, and $\psi_1 = 0.3$. The recommended $\psi_0$, $\psi_1$, and $\psi_2$ values are based on the statistical analyses and chosen to achieve structures with the required degrees of reliability for ultimate and serviceability limit states.

### 10.4.2.4 Design geotechnical parameters

The design value of a geotechnical parameter $X_d$ for use in verification calculations is either assessed directly or derived by dividing the characteristic parameter value $X_k$ by the appropriate partial material factor value $\gamma_M$:

$$X_d = X_k / \gamma_M \tag{10.5}$$

It should be noted that the partial material factor values to obtain the design values of materials are divisors and always greater than unity, unlike in some other codes, such as, for example, the Australian Standard for piling, AS 2159 (Standards Australia, 2009), which has strength reduction factors less than unity that multiply the "design geotechnical ultimate strength" to obtain the reduced (i.e., conservative) "design geotechnical strength."

The requirements in Eurocode 7 for selecting the characteristic values of geotechnical parameters and some guidance on their selection are presented in the following sections. Sometimes in geotechnical designs, there is insufficient information or it is not sufficiently precise to enable characteristic values to be selected with confidence. An example of such a situation can occur in the case of designs involving rock where the rock strength is a function of subjectively determined fuzzy parameters, such as the Geological Strength Index (GSI). In such situations, it may be appropriate to assess the design value of the geotechnical parameter directly. If this approach is adopted for an ultimate limit state design, then it is necessary to select a suitably very cautious value such that a worse value is extremely unlikely to occur. It is noted in §2.4.6.1(5) that the values of the recommended partial factors in Eurocode 7 should be used as a guide to the required level of safety when design values are selected directly.

### 10.4.2.5 Design effects of actions and design resistances

A particular feature of geotechnical design, noted already in Section 10.4.2.1, is that some actions are a function of the soil strength, for example, the active pressure on a retaining structure. Also, since soil is a frictional material, the soil strength and resistance in drained situations are functions of the normal stress and therefore the actions. Hence, the following general equations, that are functions of the actions, material properties, and dimensions, are given in Eurocode 7 for the design effect of actions and the design resistance:

$$E_d = \gamma_E E\{\gamma_F F_{rep}, X_k/\gamma_M, a_d\} \tag{10.6}$$

$$R_d = R\{\gamma_F F_{rep}, X_k/\gamma_M, a_d\}/\gamma_R \tag{10.7}$$

where $\gamma_E$ is the partial action factor and $\gamma_R$ is the partial resistance factor. It should be noted that if the partial factors $\gamma_F$ and $\gamma_M$ within the brackets are greater than unity, that is, if a material factor approach is adopted, then the partial factors $\gamma_E$ and $\gamma_R$ are equal to unity; and similarly, if $\gamma_E$ and $\gamma_R$ are greater than unity, that is, a resistance factor approach is adopted, then the partial factors $\gamma_F$ and $\gamma_M$ are equal to unity. This is so as to avoid double factoring.

## 10.4.3 Characteristic parameter values

### 10.4.3.1 Definition and selection of characteristic values

EN 1990 was originally developed for manufactured materials prepared under controlled conditions having properties, which are random variables that can be represented by particular probability distribution, generally assumed to be Gaussian. The characteristic value of these properties is defined in EN 1990 §1.5.4.1 as the *value of a material or product property having a prescribed probability of not being attained in a hypothetical unlimited test series. This value generally corresponds to a specified fractile of the assumed statistical distribution of the particular property of the material or product.* When the low value of the property is critical, EN 1990 states in §4.2(3) that *the characteristic value should be defined as the 5% fractile, unless stated otherwise stated* (author's underline). Such a situation occurs in Eurocode 7 in the case of geotechnical design where an alternative definition is given for the characteristic value of geotechnical parameters as explained below.

Due to the particular features of soil, the application of the design method in EN 1990 to geotechnical design, including the definition of the characteristic value of a material property, has caused some difficulties. The features of soil that cause geotechnical designs to differ from structural designs, and hence, Eurocode 7 to differ from the other Eurocodes, with regard to uncertainty and risk, have been identified by Orr (2012). These include the fact that soil is a natural material, nonhomogeneous, and often very variable. The properties of soil and rock vary over a very wide range compared to the properties of manufactured materials, such as steel and concrete, and have higher coefficient of variation (COV) values. However, as well as varying locally, the properties of soil and rock also vary spatially. The spatial variation maybe taken into account using the autocorrelation distance, which is the separation distance at which the correlation between the soil properties can be considered to be relatively weak, or the scale of fluctuation (SOF), which is the distance over which the properties are relatively strongly correlated. The SOF is equal to about twice the autocorrelation distance (Schneider and Schneider, 2013).

The first requirement in §2.4.5.2(1)P for selecting the characteristic values of geotechnical parameters, arising from soil being a natural material, is that they shall be based on values derived from laboratory and field tests, which shall be complemented by well-established experience. In structural design, where a large number of samples of the structural material can be tested and where the variability of the material in the section being designed is usually small, the results of individual tests are representative of the behavior of the material in that section, and hence the characteristic value is a particular fractile of the test results. In geotechnical design, however, only a few tests are carried out on a very small portion of the total volume of soil involved in a particular limit state and hence, the characteristic value of a geotechnical parameter is not defined as a particular fractile of test results. Instead, Eurocode 7 states in §2.5.4.2(2)P that the characteristic value of a soil parameter *shall be selected as a cautious estimate of the value affecting the occurrence of the limit state*. Thus, the characteristic value is normally a cautious estimate of the mean value on the failure surface, not a particular fractile of the test results.

When selecting the characteristic value of a geotechnical parameter, Eurocode 7 requires in §2.5.4.2(4)P that the following factors be taken into account:

- Geological and other background information, such as data from previous projects
- Variability of the measured property values
- Extent of the field and laboratory investigations
- Type and number of samples
- Extent of the zone governing the behavior of the structure at the limit state being considered
- Ability of the structure to transfer loads from strong to weak zones

The wording in Eurocode 7 to define and select characteristic geotechnical parameter values is a major improvement on previous geotechnical codes of practice, most of which provided no guidance on how to select soil parameter values. However, the definition of the characteristic value in Eurocode 7 has purposely been worded so as not to be prescriptive. How cautious an estimate the characteristic parameter value should be has been left to the designer to decide, taking into account the factors listed above, which include the designer's experience of the particular ground conditions, the nature of the structure being designed, and the anticipated failure mechanism. However, selection of the characteristic value of a soil parameter also needs to take into account the extent of the zone of ground involved in the limit state and prior knowledge of comparable ground behavior. This is why Eurocode 7 defines the characteristic value of a geotechnical parameter as a cautious mean value that is selected by the designer, that is, it involves subjective judgment.

Due to the subjective nature of this process, when a number of geotechnical engineers are presented with the same data, they will generally select different characteristic values, reflecting their different assessments of the design situation and interpretations of the characteristic value. Bond and Harris (2008) found this when they asked geotechnical engineers to assess the characteristic parameter values of London and Lambeth clays from the results of Standard Penetration Tests carried out in these soils. The characteristic SPT values

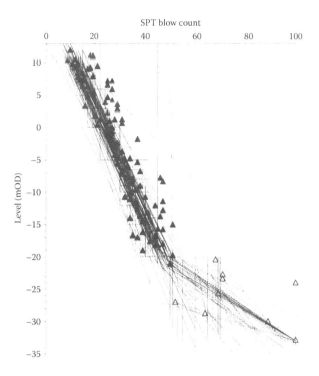

*Figure 10.2* Results of Standard Penetration Tests on London and Lambeth clays, with engineers' interpretations of the characteristic value. (After Bond A. and Harris A. 2008. *Decoding Eurocode 7*, Taylor & Francis, London.)

selected by the geotechnical engineers are presented in Figure 10.2 and show a wide spread, which indicates that geotechnical engineers are not good at selecting characteristic values, particularly when the available data are scattered. This demonstrates that there is a need for more specific guidance in Eurocode 7 on the selection of characteristic values and hence it is planned to include some specific guidance on this in the forthcoming revision of Eurocode 7, due to be published in 2020.

### 10.4.3.2 Aleatory variability and epistemic uncertainty

As noted in the previous section, one of the requirements when selecting the characteristic value of a parameter in designs to Eurocode 7 is to take account of the variability of the measured parameter. Manufactured materials prepared under controlled conditions have properties that are random variables whose inherent variability can be readily represented by a particular statistical distribution. Since a large number of tests can be carried out on such materials, the form of the distribution can be identified and the parameters to describe the distribution, for example, the mean value and COV in the case of a Gaussian distribution, can be determined. Such materials are termed aleatory and are described objectively using experimentally determined parameters. Selection of the characteristic value of such a material, defined as a specified fractile, normally 5%, of the assumed probability distribution curve, is straightforward. Aleatory variability is an inherent feature of nominally homogeneous manufactured materials. It is related to the inherent variability of a material and is not reduced by carrying out more tests. However, carrying out more tests, as in the case of an unlimited test series, will enable the variability to be known more precisely.

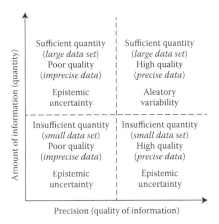

*Figure 10.3* Uncertainty and variability as functions of the quality and quantity of information. (After Bedi, A. and Harrison, J.P. 2013. *Proc. International Society for Rock Mechanics Symposium—Rock Mechanics for Resources, Energy, and Environment*, Wroclaw, Poland, 23–26 September 2013. Taylor & Francis, London.)

Since in geotechnical designs, only a few test results are usually available and because soil, being a natural material, is significantly more variable than manufactured materials, there are normally insufficient test results to identify the statistical distribution of the geotechnical property with certainty, or determine the parameters such as the standard deviation to describe it precisely, thereby taking into account the variability and selecting the characteristic value objectively. Such a situation where there is insufficient information, or where the information is not precise enough to determine the statistical distribution of a material property, is referred to as epistemic uncertainty. If, in such a situation, the material is actually inherently aleatory, then as more information is obtained the values of the material properties become more certain and the material can be modeled as being aleatory. Many soils fall into this category.

The properties of rock masses are usually more difficult to determine than the properties of soil. Some rock parameters cannot be determined precisely so that they are inherently epistemic, that is, no quantity of data would enable a precise description to be determined. Bedi and Harrison (2013) have presented a relationship between epistemic uncertainty and aleatory variability as functions of the quantity of information and the precision of the data, as shown in Figure 10.3. The epistemic nature of many rock mass properties poses difficulties in applying the limit state design method in Eurocode 7 to designs involving rock as outlined by Bedi and Orr (2014).

### 10.4.3.3 Selection of aleatory characteristic parameter values

Since, as shown in Section 10.4.3.1, it has been found that the current definition provided in Eurocode 7 for the characteristic value of a geotechnical parameter can result in very different values being selected for a particular design situation by different engineers from a set of test results, the CEN sub-committee TC250/SC7, responsible for the development of Eurocode 7, has established an evolution group to prepare more specific guidance on the selection of characteristic parameter values for the next version of Eurocode 7. The challenge facing this evolution group is to provide appropriate guidance on using statistics for the selection of characteristic values from a limited number of test results while at the same time taking account of well-established and valuable experience so that geotechnical engineers will select more consistent characteristic values.

If statistical methods are used to select the characteristic value of a geotechnical parameter, Eurocode 7 states in §2.4.5.2(11) that it should be derived such that *the calculated probability of a worse value governing the occurrence of the limit state under consideration is not greater than 5%*. A note to this clause states that *a cautious estimate of the mean value is a selection of the mean value of the limited set of geotechnical parameter values, with a confidence level of 95%; where a local failure is concerned, a cautious estimate of the low value is a 5% fractile*. It should be noted that the 5% fractile of test results on small samples normally gives too conservative of an estimate of the characteristic value.

Considering a geotechnical parameter as aleatory, if the characteristic value, $X_k$ of a parameter is the 5% fractile of an unlimited series of tests with a normal distribution, it is given by

$$X_k = X_{mean} - 1.645\sigma = X_{mean}(1 - 1.645\ COV) \tag{10.8}$$

where $X_{mean}$ is the mean value, $\sigma$ the standard deviation, COV the coefficient of variation, and 1.645 the factor defining the 5% fractile for a normal distribution, that is, the Student's t value. This equation usually gives too cautious a characteristic value of a geotechnical parameter because, as noted above, the relevant value is the 95% confidence in the mean value controlling the occurrence of a limit state, rather than the 5% fractile of the test results. Hence, Schneider (1997) proposed the following equation for selecting the characteristic geotechnical parameter value from a series of test results:

$$X_k = X_{mean}(1 - 0.5\ COV) \tag{10.9}$$

This equation aims to provide a characteristic value that is a cautious estimate of the mean value with a confidence level of 95% and results in a value that is much less conservative than the 5% fractile value given by Equation 10.8.

To account explicitly for the inherent variability and other uncertainties affecting the characteristic value, the COV in Equation 10.8 can be replaced by the following additive total COV, $COV_{total}$ proposed by Phoon and Kulhawy (1999):

$$COV_{total} = \sqrt{\Gamma_S^2 COV_{inher}^2 + COV_{meas}^2 + COV_{trans}^2 + COV_{stat}^2} \tag{10.10}$$

where
$\Gamma_S^2$ is the variance reduction function (Vanmarcke, 1977) considering the spatial extent of the governing failure mechanism
$COV_{inher}$ is the coefficient of variation of the soil's inherent variability,
$COV_{meas}$ is the coefficient of variation of the measurement errors,
$COV_{trans}$ is the coefficient of variation of the transformation errors, and
$COV_{stat}$ is the coefficient of variation of the statistical parameters.

For a 1D potential slip line in a 2D problem,

$$\Gamma_S^2 = [2L_v/\delta_v - 1 + \exp(-2L_v/\delta_v)]/(2\,L_v^2/\delta_v^2) \tag{10.11}$$

where $\delta_v$ is the SOF along the potential slip line direction, v, and $L_v$ is the length of the potential slip line in the y–z-plane. If the potential slip line is inclined at an angle $\beta$ to the horizontal,

$$\delta_v = \frac{1}{\cos(\beta)/\delta_y + \sin(\beta)/\delta_z} \tag{10.12}$$

where $\delta_y$ and $\delta_z$ are the SOFs in the horizontal and vertical directions, respectively.

For a 1D potential slip curve in a 2D problem, it is in general very difficult to determine $\Gamma_S^2$. However, with a first-order approximation, the isotropic equivalent $\delta_E$ can be computed (El-Ramly et al., 2006) from

$$\delta_E = \sqrt{\delta_y \delta_z} \tag{10.13}$$

so that $\Gamma_S^2$ can be approximated as

$$\Gamma_S^2 = [2L_v/\delta_E - 1 + \exp(-2L_v/\delta_E)]/(2L_v^2/\delta_E^2) \tag{10.14}$$

where $L_v$ is the length of the potential slip curve.

For a 2D potential slip plane in a 3D problem, for example, a slope failure on a planar surface, let the plane have extent $= L_x$ and SOF $= \delta_x$ in the direction, x along the slope perpendicular to the cross section y–z through the slope, and length $= L_{splane}$ and SOF $= \delta_{splane}$ in the cross-section. Then,

$$\Gamma_S^2 = \Gamma_x^2 \Gamma_{splane}^2 \tag{10.15}$$

where

$$\Gamma_x^2 = [2L_x/\delta_x - 1 + \exp(-2L_x/\delta_x)]/(2L_x^2/\delta_x^2) \tag{10.16}$$

$$\Gamma_{splane}^2 = [2L_{splane}/\delta_{splane} - 1 + \exp(-2L_{splane}/\delta_{splane})]/\left(2L_{splane}^2/\delta_{splane}^2\right) \tag{10.17}$$

$\delta_{splane}$ can be determined using equations similar to Equation 10.12.

For a 2D potential slip (curved) surface in a 3D problem, such as that shown in Figure 10.4, it is again in general very difficult to derive $\Gamma_S^2$, but with a first-order approximation, it can be expressed as

$$\Gamma_S^2 = \Gamma_x^2 \Gamma_{scurve}^2 \tag{10.18}$$

where $\Gamma_x^2$ is the variance reduction factor in the direction along the slope and $\Gamma_{scurve}^2$ is the variance reduction factor along the slip curve. $\Gamma_x^2$ can be determined using an equation the same as Equation 10.16, where $L_x$ is the extent of the failure surface in the x-direction along the slope perpendicular to the cross-section, as shown in Figure 10.4, and $\delta_x$ is the SOF in that direction. $\Gamma_{scurve}^2$ can be determined using an equation similar to Equation 10.14, where $\delta_E = \sqrt{\delta_y \delta_z}$ and $L_{scurve}$ is the length of the slip surface in the cross-section plane through the slope.

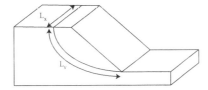

*Figure 10.4* Slope stability example showing the scales of fluctuation and autocorrelation lengths along the slope perpendicular to the slope cross-section and along the failure surface.

Ching and Phoon (2013) analyzing the mobilized strength in spatially variable soils have shown that there is a clear distinction between the spatially averaged strength along a particular failure surface and the spatial variability of the parameters within a soil mass. They found that the physical properties of a particular design situation play an important role in determining the trajectory of the failure surface so that the idea that the failure surface should seek the weakest zones within the spatially variable soil may not lead to what is actually the dominant mechanism. The selection of variance reduction factors and the effects of spatially variability is an emerging area of research.

Schneider and Schneider (2013) adopted a simplification for the variance reduction factor, which is that $\Gamma_S^2 = \Gamma_x^2 \Gamma_y^2 \Gamma_z^2$, where $\Gamma_x^2$ and $\Gamma_y^2$ are the variance reduction factors in the two horizontal directions, x and y, and $\Gamma_z^2$ is the variance reduction factor in the vertical direction, z. They have also adopted the following two equations from Vanmarcke (1983) as a simplification of Vanmarcke's (1977) Equation 10.16 for $\Gamma_i^2$ in a particular direction, i:

$$\Gamma_i^2 = \left[ \frac{\delta_i}{L_i} \left( 1 - \frac{\delta_i}{3L_i} \right) \right] \quad \text{if } L_i > \delta_i \tag{10.19}$$

and

$$\Gamma_i^2 = \left[ 1 - \frac{L_i}{3\delta_i} \right] \quad \text{if } L_i \leq \delta_i \tag{10.20}$$

where $\delta_i$ and $L_i$ are the SOF and the extent of the failure mechanism in the direction i, respectively. Comparing the $\Gamma_i^2$ values obtained using Equations 10.19 and 10.20 with those obtained using Equation 10.16 it is found that they are similar, but more cautious, particularly for $L_i$ values close to $\delta_i$.

If the measurement of the soil parameters is carried out using accurate equipment and strictly in accordance with the relevant testing standards, the measurement errors should be small so that $COV_{meas}^2 \cong 0$. Similarly, if a well-established model is used to transform the measured test results into the required parameter, then $COV_{trans}^2 \cong 0$, and if the parameters required to describe the statistical distribution are known with reasonable accuracy from experience with similar soils, then $COV_{stat}^2 \cong 0$. Hence, assuming a normal distribution and taking into account the above comments regarding the COV values, substituting the $COV_{total}$ from Equation 10.10 for COV in Equation 10.8, gives:

$$X_k = X_{mean}(1 - 1.645 \, COV_{inher} \sqrt{\Gamma_S^2}) \tag{10.21}$$

whereas for a log-normal distribution, the characteristic value is given by

$$X_k = X_{mean} 0.2 \frac{\sqrt{\ln(1 + \Gamma_S^2 COV_{inher}^2)}}{\sqrt{1 + \Gamma_S^2 COV_{inher}^2}} \tag{10.22}$$

Schneider and Schneider (2013) gave ranges of typical values for $COV_{inher}$ and recommended the values shown in Table 10.4. When $COV_{inher} < 0.3$ for a parameter, for example, $\tan \phi'$, a normal distribution may be assumed and, hence, Equation 10.21 may be used to determine the characteristic value. However, Schneider and Schneider (2013) stated that when $COV \geq 0.3$, for example, for c' and $c_u$, a log-normal distribution should be assumed

Table 10.4 COV$_{inher}$ values for soil parameters

| Soil parameter | Symbol | Range of COV values | Recommended COV$_{inher}$ |
|---|---|---|---|
| Weight density | $\gamma$ | 0.01–0.10 | 0 |
| Angle of internal friction | $\tan \phi'$ | 0.05–0.15 | 0.1 |
| Cohesion | $c'$ | 0.30–0.50 | 0.4 |
| Undrained shear strength | $c_u$ | 0.30–0.50 | 0.4 |

Source: Schneider H.R. and Schneider M.A. 2013. *Modern Geotechnical Design Codes of Practice*, eds. P. Arnold, G.A. Fenton, M.A. Hicks, T. Schweckendiek and B. Simpson, IOS Press, Amsterdam, 87–101.

and the characteristic value determined using Equation 10.22. At this point, it is safe to say that the recommendations presented in Equation 10.11 through 10.22 are crude approximations in need of more rigorous calibrations to ascertain their associated errors.

Due to the way soil is normally stratified in horizontal layers, the SOF is usually much smaller in the vertical direction than in the horizontal direction with the consequence that the extent of the failure mechanism in the vertical direction has more effect on the characteristic value than the extent in the horizontal direction. Phoon and Kulhawy (1999), from an extensive literature survey, reported mean values of 2.5 and 50.5 m for the vertical and horizontal scales of fluctuation, $\delta_v = \delta_z$ and $\delta_h = \delta_x = \delta_y$, respectively, of clay from laboratory and vane shear tests. Schneider and Fitze (2013) recommended values of 2.0 and 50.0 m for $\delta_v$ and $\delta_h$. Since the SOF is normally so different in the vertical and horizontal directions, Equations 10.21 and 10.22 may not provide appropriate $X_k$ values for some design situations as the variance reduction factor may be excessive.

Equations 10.21 and 10.22 are used to examine the effect of the extent of the failure mechanism in the vertical direction, $L_z$ on the characteristic value, $X_k$. A correlation factor $\xi = X_{mean}/X_k$ is defined for determining the characteristic soil parameter value, similar to the correlation factors in Eurocode 7 for determining the characteristic pile resistance as described in the next section. The characteristic parameter value is therefore obtained by dividing the mean value $X_{mean}$ by this correlation factor, which is greater than unity:

$$X_k = X_{mean}/\xi \tag{10.23}$$

$\xi$ is plotted against $L_z$ in Figure 10.5 for COV$_{inher} = 0.1$, assuming a normal distribution curve (Equation 10.21), and for COV$_{inher} = 0.4$, assuming a log-normal distribution (Equation 10.22), with $\delta_z = 2$ m, when the extent of the failure mechanism in the horizontal direction is equal to zero and $\Gamma_x^2 = \Gamma_y^2 = 1.0$. The graphs in Figure 10.5 show the effect of the extent of the failure mechanism in the vertical direction, that is, the $L_z$ value, on the characteristic value; it can be seen that the correlation factor, $\xi$ approaches closer to unity, that is, the characteristic value becomes closer to the mean value, as the extent of the failure zone increases.

The graphs of $\xi$, obtained from Equations 10.8 and 10.9 for $X_k$, are also plotted in Figure 10.5 for comparison. These graphs, with constant $\xi$ values as $L_z$ increases, show that Equation 10.8 for the 5% fractile gives a much too high $\xi$ value, equal to 2.92, hence a much too cautious $X_k$ value, while the Schneider (1997) Equation 10.9 provides a reasonably cautious $\xi$ value, equal to 1.25, compared to the lower $\xi$ values given by both Equations 10.21 and 10.22, except for soil parameters with COV$_{inher} \geq 0.4$ and when, for the assumed conditions, the extent of the failure mechanism is less than 17 m and Equation 10.22 gives a more conservative $\xi$ value. The graphs in Figure 10.5 demonstrate the need to take into account the extent of the failure mechanism when selecting characteristic parameter values, particularly in the case of the undrained shear strength with a high COV value.

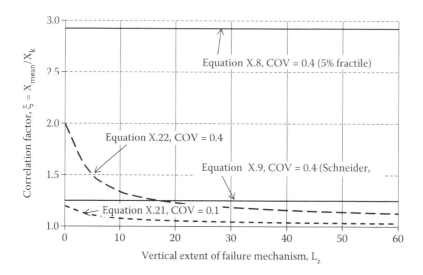

*Figure 10.5* Variation in correlation factor ξ to calculate the characteristic value with the vertical extent of failure mechanism.

It is also a Eurocode 7 requirement, as noted in Section 10.4.3.1, to take into account other background information and data from other projects when selecting characteristic parameter values. The use of a Bayesian analysis allows such background information and prior knowledge of ground conditions to be taken into account.

### 10.4.3.4 Example 10.1: Selection of characteristic parameter values

A 12-m long pile with a diameter of 0.8 m is to be installed in a homogeneous deposit of clay. From the results of field and laboratory tests on the clay, and also taking account of other information and local experience, the mean undrained shear strength was determined to be $c_{u;mean} = 40$ kPa. The characteristic undrained shear strength $c_{u;k}$ along the pile shaft and at the base of the pile are required for the design of the pile.

*Solution*
Based on the information in Table 10.4, the $COV_{inher}$ of $c_u$ is assumed to be 0.4. The extents of the failure mechanism along the pile shaft, with diameter D = 0.8 m, are $L_z = 12$ m vertically and $L_{scurve} = \pi D = \pi 0.8 = 2.5$ m circumferentially. Using Equation 10.16 with $L_z = 12$ m and $\delta_z = 2.0$ m, the reduction factor in the vertical direction $\Gamma_z^2 = 0.1528$ and using Equation 10.14 with $L_{scurve} = 2.5$ m and $\delta_{scurve} = \sqrt{\delta_x \delta_y} = \sqrt{50.0 \times 50.0} = 50.0$, the reduction factor in the horizontal direction around the pile circumference $\Gamma_{scurve}^2 = 0.9675$. Using Equation 10.18, the reduction factor for the pile shaft $\Gamma_S^2 = \Gamma_z^2 \Gamma_{scurve}^2 = 0.1528 \times 0.9675 = 0.1478$. Hence, using Equation 10.22 for a log-normal distribution and with COV = 0.4, the correlation factor $\xi = 1.29$. Using Equation 10.23, the characteristic undrained shear strength along the pile shaft is therefore

$$c_{u;k} = c_{u;mean}/\xi = 40/1.29 = 30.9 \text{ kPa}$$

Ignoring the lateral extent of the failure mechanism along the pile shaft, which is a conservative assumption, so that $\Gamma_x^2 = \Gamma_y^2 = 1.0$ and adopting Schneider and Schneider's (2103) Equation 10.19 with $L_z = 12$ m and $\delta_z = 2$ m gives $\Gamma_z^2 = 0.1574$. Using Equation 10.22 with

COV = 0.4 gives $\xi = 1.31$ along the shaft of the pile. Hence, the characteristic undrained shear strength along the pile shaft is

$$c_{u;k} = c_{u;mean}/\xi = 40/1.31 = 30.5 \text{ kPa}$$

This value is a little more conservative than the $c_{u;k}=30.9$ kPa obtained above using the Vanmarcke (1977) (Equation 10.16).

The vertical extent of the failure mechanism at the pile base is much smaller in extent than along the pile shaft. Assuming the failure mechanism beneath the pile base consists of a 45° cone and a strain fan, the vertical extent of the failure mechanism $L_z = D/\sqrt{2} = 0.8/\sqrt{2} = 0.57$ m. Again ignoring the horizontal extent of the failure mechanism, so that $\Gamma_x^2 = \Gamma_y^2 = 1.0$, and adopting Schneider and Schneider's (2013) Equation 10.20 since $L_z < \delta_z$ as $L_z = 0.57$ m and $\delta_z = 2$ m gives $\Gamma_z^2 = 0.9050$. Substituting for $\Gamma_S^2 = \Gamma_x^2\Gamma_y^2\Gamma_z^2$ in Equation 10.22 and with COV=0.4 gives $\xi$ equal to 1.93 so that the characteristic undrained shear strength for failure at the base of the pile is

$$c_{u;k} = c_{u;mean}/\xi = 40/1.93 = 20.7 \text{ kPa}.$$

This characteristic undrained shear strength value is much less than the value obtained for $c_{u;k}$ along the pile shaft and demonstrates that a more cautious characteristic value should be selected when the extent of the failure mechanism is smaller, corresponding to a local failure, for example, at the base of a pile. A less conservative $c_{u;k}$ value would be obtained if the lateral extent of the failure mechanism were also taken into account.

### 10.4.3.5 Characteristic pile compressive resistances

Four methods are presented in Eurocode 7 for calculating the compressive resistance of pile foundations from test results, which are the methods using:

- Static pile load tests
- Profiles of test results (referred to as the "model pile procedure" by Frank et al., 2007)
- Soil parameter values (referred to as the "alternative procedure" in Eurocode 7)
- Dynamic pile tests

Profiles of test results are obtained using tests such as CPT, PMT, and SPT, and these give individual test profiles that are used to calculate the pile compressive resistance at the locations of the test profiles.

The compressive resistance of a pile varies over a site due to variations in the ground properties and also due to stratification. This spatial variability needs to be taken into account when selecting the characteristic pile compressive resistance, $R_{c;k}$. Eurocode 7 provides a simple procedure involving the use of two sets of correlation factors $\xi$ to take spatial variability into account and determine $R_{c;k}$ when designing piles from static pile load tests or from profiles of test results. For piles designed from static pile load tests, the $R_{c;k}$ value is obtained by dividing the mean and lowest measured pile resistances, $(R_{c;m})_{mean}$ and $(R_{c;m})_{min}$ by correlation factors $\xi_1$ and $\xi_2$, with the recommended values given in Eurocode 7 shown in Table 10.5, and choosing the lowest value as $R_{c;k}$, that is,

$$R_{c;k} = \text{Min}\left\{\frac{(R_{c;m})_{mean}}{\xi_1}; \frac{(R_{c;m})_{min}}{\xi_2}\right\} \tag{10.24}$$

Similarly, for piles designed from profiles of test results, the $R_{c;k}$ value is obtained by dividing the mean and lowest calculated pile compressive resistances by correlation factors $\xi_3$ and

*Table 10.5* Recommended correlation factors $\xi_1$ and $\xi_2$ in Eurocode 7 to derive characteristic resistances from static pile load tests

| Number of static pile load tests (n) | 1 | 2 | 3 | 4 | ≥ 5 |
|---|---|---|---|---|---|
| $\xi_1$ applied to mean of the measured resistances | 1.40 | 1.30 | 1.20 | 1.10 | 1.00 |
| $\xi_2$ applied to lowest measured resistances | 1.40 | 1.20 | 1.05 | 1.00 | 1.00 |

$\xi_4$, with the recommended values given in Eurocode 7 shown in Table 10.6, and choosing the lowest value as $R_{c;k}$:

$$R_{c;k} = Min\left\{\frac{(R_{c;cal})_{mean}}{\xi_3}; \frac{(R_{c;cal})_{min}}{\xi_4}\right\} \tag{10.25}$$

The $\xi$ values for determining the characteristic pile resistance, taking into account the number of pile load tests and number of test profiles, are similar to the $\xi$ values in Figure 10.5 for determining the characteristic values of soil parameters, taking into account the extent of failure mechanism. As Tables 10.5 and 10.6 show, the recommended $\xi$ values decrease toward unity as the number of static pile load tests and profiles of test results increase so that the statistical uncertainty in the pile resistance is reduced and the reliability is increased. Except for when there is only one pile load test and one profile of test results, the recommended $\xi_1$ and $\xi_2$ values are all greater than the corresponding $\xi_3$ and $\xi_4$ values, reflecting the greater uncertainty in the pile resistance calculated from a profile of test results, which involves the use of a calculation model with certain assumptions about the soil behavior, compared to bearing resistances that are measured directly in static pile load tests.

The recommended $\xi$ values given in Tables 10.5 and 10.6 have been determined so as to provide characteristic pile resistance values with a confidence of 95% for soils with a range of variabilities assuming that

- When the COV of the pile resistance is less than 10%, the characteristic resistance is governed by the mean of the measured pile resistances from static pile load tests or the mean of the pile resistances calculated from profiles of tests.
- When the COV of the pile resistance is greater than 10%, the characteristic resistance is governed by the lowest measured pile resistance from static pile load tests or the lowest pile resistance calculated from profiles of tests.

Bauduin (2001) has provided some theoretical background to the $\xi$ values for the design of piles.

*Table 10.6* Recommended correlation factors $\xi_3$ and $\xi_4$ in Eurocode 7 to derive characteristic resistances from profiles of tests

| Number of profiles of tests (n) | 1 | 2 | 3 | 4 | 5 | 7 | 10 |
|---|---|---|---|---|---|---|---|
| $\xi_3$ applied to the mean of the calculated resistances | 1.40 | 1.35 | 1.33 | 1.31 | 1.29 | 1.27 | 1.25 |
| $\xi_4$ applied to lowest calculated resistance | 1.40 | 1.27 | 1.23 | 1.20 | 1.15 | 1.12 | 1.08 |

A further assumption in the derivation of the $\xi$ values is that the structure to be supported by the piles is not stiff and strong enough to transfer loads from piles where the ground is weaker to piles where the ground is stronger. If, however, the structure is stiff and strong enough to do so, then the $\xi$ values may be reduced by the factor 1.1, provided they do not become less than 1.0. This takes advantage of the Eurocode 7 requirement when selecting characteristic parameter values, noted in Section 10.4.3.1, to take account of the ability of the structure to transfer loads from strong to weak zones.

The fact that the $\xi$ values reduce as the number of pile load tests and the number of profiles of tests increase reflects the fact that obtaining more information about the ground properties results in higher, that is, less cautious, $R_{c;k}$ values and hence, more economical pile designs with the same degree of reliability. This shows the advantage of carrying out more load tests and obtaining more test profiles.

## 10.4.4 Partial factors, safety levels and reliability

### 10.4.4.1 Types of ultimate limit state and recommended partial factor values

The values of the partial factors that are applied to representative actions and characteristic resistances are chosen to obtain design values that, when used in calculation models, will provide acceptable levels of reliability against the occurrence of an ultimate limit state and a serviceability limit state. The following five different types of ultimate limit state are identified and defined in §2.4.7.1(1)P of Eurocode 7:

- EQU, which is *loss of equilibrium of the structure or the ground, considered as a rigid body, in which the strengths of structural materials and the ground are insignificant in providing resistance*
- STR, which is *internal failure or excessive deformation of the structure or structural elements in which the strength of structural materials is significant in providing resistance*
- GEO, which is *failure or excessive deformation of the ground, in which the strength of soil or rock is significant in providing resistance*
- UPL, which is *loss of equilibrium of the structure or the ground due to uplift by water pressure (buoyancy) or other vertical actions*
- HYD, which is *hydraulic heave, internal erosion and piping in the ground caused by hydraulic gradients*

Since the relative significance of, and also the uncertainties in, the actions and resistances of materials differ for these ultimate limit states, separate sets of partial factors have been established for each type of limit state. Recommended values for the partial factors are given in Eurocode 7 to achieve structures with the appropriate degrees of reliability and hence safety. Since setting the safety levels and choosing the required degrees of reliability for structures in a country is a national responsibility, the values of the partial factors to be used for geotechnical designs in a particular country are published by that country's national standards body in its national annex to Eurocode 7.

The recommended partial action, material, and resistance factors in Eurocode 7 for use in STR and GEO ultimate limit states, which are when the strengths of structural materials or the ground are significant in providing resistance, are shown in Table 10.7. In the case of pile foundations, the recommended partial resistance factors for bored piles are only given in Table 10.7; however, Eurocode 7 also provides partial resistance factors for driven and CFA piles. The partial factors for actions, soil parameters, and resistances are presented in

Table 10.7 Recommended partial factor values in Eurocode 7 for STR/GEO ultimate limit states in persistent and transient design situations

| | | | Design Approaches | | | | |
|---|---|---|---|---|---|---|---|
| | | | DA1 | | DA2 | | DA3 |
| Parameter | Factor | Set | DA1.C1 | DA1.C2 | DA2 | A1 structural actions | A2 geotechnical actions |
| **Partial factors on actions ($\gamma_F$) or the effects of actions ($\gamma_E$)** | | Set | A1 | A2 | A1 | A1 | A2 |
| Permanent unfavorable action | $\gamma_G$ | | 1.35 | 1.0 | 1.35 | 1.35 | 1.0 |
| Permanent favorable action | $\gamma_G$ | | 1.0 | 1.0 | 1.0 | 1.0 | 1.0 |
| Variable unfavorable action | $\gamma_Q$ | | 1.5 | 1.3 | 1.5 | 1.5 | 1.3 |
| Variable favorable action | $\gamma_Q$ | | 0 | 0 | 0 | 0 | 0 |
| Accidental action | $\gamma_A$ | | 1.0 | 1.0 | 1.0 | 1.0 | 1.0 |
| **Partial factors for soil parameters ($\gamma_M$)** | | Set | M1 | M2a | M1 | | M2 |
| Angle of shearing resistance (this factor is applied to $\tan\varphi'$) | $\gamma_{\tan\varphi'}$ | | 1.0 | 1.25 | 1.0 | | 1.25 |
| Effective cohesion $c'$ | $\gamma_{c'}$ | | 1.0 | 1.25 | 1.0 | | 1.25 |
| Undrained shear strength $c_u$ | $\gamma_{cu}$ | | 1.0 | 1.4 | 1.0 | | 1.4 |
| Unconfined strength $q_u$ | $\gamma_{qu}$ | | 1.0 | 1.4 | 1.0 | | 1.4 |
| Weight density of ground $\gamma$ | $\gamma_\gamma$ | | 1.0 | 1.0 | 1.0 | | 1.0 |
| **Partial resistance factors ($\gamma_R$)** | | | | | | | |
| *Spread foundations, retaining structure, and slopes* | | Set | R1 | R1 | R2 | | R3 |
| Bearing resistance | $\gamma_{R;v}$ | | 1.0 | 1.0 | 1.4 | | 1.0 |
| Sliding resistance, incl. slopes | $\gamma_{R;h}$ | | 1.0 | 1.0 | 1.1 | | 1.0 |
| Earth resistance | $\gamma_{R;h}$ | | 1.0 | 1.0 | 1.4 | | 1.0 |
| *Bored pile foundations* | | | R1 | R4 | R2 | | R3 |
| Base resistance | $\gamma_b$ | | 1.25 | 1.6 | 1.1 | | 1.0 |
| Shaft (compression) | $\gamma_s$ | | 1.0 | 1.3 | 1.1 | | 1.0 |
| Total (compression) | $\gamma_t$ | | 1.15 | 1.5 | 1.1 | | 1.0 |
| Shaft in tension | $\gamma_{t;s}$ | | 1.25 | 1.6 | 1.15 | | 1.1 |

a When calculating the resistance of piles, Set M1 is used with Set R2 in DA1.C2.

*Table 10.8* Recommended partial factor values in Eurocode 7 for EQU, UPL, and HYD ultimate limit states in persistent and transient design situations

| Parameter | Factor | EQU | UPL | HYD |
|---|---|---|---|---|
| **Partial factors on actions ($\gamma_F$) or the effects of actions ($\gamma_E$)** | | | | |
| Permanent unfavorable action | $\gamma_{G;dst}$ | 1.1 | 1.0 | 1.35 |
| Permanent favorable action | $\gamma_{G;stb}$ | 0.9 | 0.9 | 0.9 |
| Variable unfavorable action | $\gamma_{Q;dst}$ | 1.5 | 1.5 | 1.5 |
| Variable favorable action | $\gamma_{Q;stb}$ | 0 | – | – |
| **Partial factors for soil parameters ($\gamma_M$)** | | | | |
| Angle of shearing resistance (this factor is applied to $\tan\varphi'$) | $\gamma_{\tan\varphi}'$ | 1.25 | 1.25 | – |
| Effective cohesion c′ | $\gamma_c'$ | 1.25 | 1.25 | – |
| Undrained shear strength $c_u$ | $\gamma_{cu}$ | 1.4 | 1.4 | – |
| Unconfined strength $q_u$ | $\gamma_{qu}$ | 1.4 | – | – |
| Weight density of ground $\gamma$ | $\gamma_\gamma$ | 1.0 | – | – |
| Tensile pile resistance | $\gamma_{s;t}$ | – | 1.4 | – |
| Anchorage | $\gamma_R$ | – | 1.4 | – |

Eurocode 7 in Sets A1 and A2, M1 and M2, and R1, R2, R3, and R4 that are combined in different ways, referred to as Design Approaches 1, 2, and 3 (DA1, DA2, and DA3), as shown by the columns of partial factors in Table 10.7. These three Design Approaches enable partial factors to be applied either to soil parameters (DA1, Combination 2, and DA3) or to resistances (DA1, Combination 1, and DA2) for different design situations and allow partial factors of unity to be applied to permanent loads in geotechnical design (DA1, Combination 2, and DA3).

The recommended partial action, material, and resistance factors for use in EQU, UPL, and HYD ultimate limit states, which are when there is loss of equilibrium due to destabilizing actions exceeding the stabilizing actions when the strength of structural materials and the ground is insignificant, or failure due to uplift by water pressure or due to seepage pressure, are shown in Table 10.8. While in many design situations it is evident which ultimate limit state controls, in principle all limit states should be checked.

Two design examples are provided in the following sections. The first example, Example 10.2, shows the determination of the design soil resistance on the side walls of a buried structure against uplift, which is a situation where the conservative value of the earth pressure is an upper value rather than a lower value. The second example, Example 10.3, is the design of the buried structure against uplift, which is a design situation controlled by the UPL partial factors.

### 10.4.4.2 Example 10.2: Determination of the design soil resistance on walls against uplift

The buried structure shown in Figure 10.6 has a depth D = 6 m and overall horizontal dimensions of B = 8 m and L = 10 m. The groundwater level is at the surface. The soil is sand with a mean $\phi'$ value of 30° and a weight density of 18 kN/m³. The purpose of this example is to determine the design value of the frictional resistance on the side walls of the buried structure against uplift.

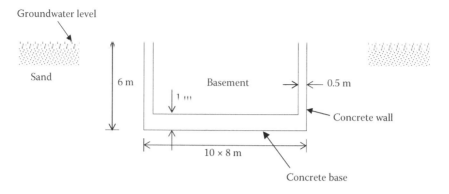

*Figure 10.6* Basement subjected to uplift pressure.

*Solution*

The first stage in solving this example is to calculate the characteristic angle of shearing resistance $\phi'_k$. It is assumed that $COV_{inher} = 0.1$, based on Table 10.4, and that the SOF values in the vertical and horizontal directions are $\delta_v = 2.0$ m and $\delta_h = 50$ m. The extents of the failure surfaces are $L_v = 6$ m vertically and $L_h = 2(8+10) = 36$ m horizontally, hence using Equations 10.16 and 10.17 together with Equation 10.13, the variance reduction factors are $\Gamma^2_v = 0.2779$ and $\Gamma^2_h = 0.6529$ so that for the 2D failure plane $\Gamma^2_\xi = \Gamma^2_v\Gamma^2_h = 0.1815$. Assuming a normal distribution and substituting for $COV_{inher}$ and $\Gamma^2_\xi$ in Equation 10.21 gives the correlation factor, $\xi = 1.075$. If the lateral extent is ignored and the $\xi$ value is obtained using Equation 10.19 or the graph in Figure 10.5, it is 1.098, which is a more conservative value. Adopting $\xi = 1.075$, the value of $\phi'_k$ is obtained as follows:

$$\phi'_k = \tan^{-1}(\tan'_{mean}/\xi) = \tan^{-1}(\tan 30/1.075) = 28.2°$$

As this is an ultimate limit state due to uplift failure, the recommended UPL $\gamma_M$ value in Table 10.8 for $\tan \phi' = 1.25$, so that the design angle of shearing resistance is

$$\phi'_d = \tan^{-1}(\tan\phi'_k/\gamma_M) = \tan^{-1}(\tan 28.2/1.25) = 23.2°$$

Assuming the angle of shearing resistance between the wall and the ground $\delta_d = 2/3\phi'_d$ and using the graph in Eurocode 7 for the horizontal component of the coefficient of active earth pressure coefficient of wall, $K_{a;d} = 0.377$ for $\phi'_d = 23.2°$. The area of the side walls $A = 6 \times (2 \times 10 + 2 \times 8) = 216$ m². Hence, the design frictional resistance on the side walls against uplift is

$$R_d = 0.5\,K_{a;d}\,\sigma'_v\,\tan\delta_d A = 0.5 \times 0.377 \times (18.0 - 9.81) \times 6 \times \tan 23.2 \times 216 = 858\,kN$$

To check the margin of safety in the wall frictional resistance provided by the above calculation, the characteristic, that is, unfactored, wall frictional resistance is calculated using $\phi'_k = 28.2°$ for which $K_{a;k} = 0.309$. Hence, the characteristic wall frictional resistance against uplift is

$$R_k = 0.5 \, K_{a;k} \, \sigma'_v \tan\delta_k A = 0.5 \times 0.309 \times (18.0 - 9.81) \times 6 \times \tan 28.2 \times 216 = 879 \, kN$$

The overall factor of safety on the wall frictional resistance against uplift is obtained by dividing the characteristic resistance by the factored resistance:

$$OFS = R_k / R_d = 879/858 = 1.02$$

Therefore in this situation factoring $\phi'_k$ and reducing it from 28.2° to 23.2° has provided virtually no margin of safety or degree of reliability on the wall frictional resistance. The reason for this is because, while applying the $\gamma_M$ to $\phi'_k$ reduces $\tan\delta_k$, at the same time it increases $K_{a;k}$ with the result that there is very little safety.

The $\phi'_d$ value above was obtained by assuming that a lower $\phi'_k$ value is conservative and dividing the inferior $\tan\phi'_k$ value by $\gamma_M$. Eurocode 7 states in §2.4.5.2(5) that a characteristic value may be an upper value, that is, greater than the mean value. If that approach is adopted in this example, the superior $\phi'_k$ is given by

$$\phi'_k = \tan^{-1}(\xi \tan\phi'_{mean}) = \tan^{-1}(1.075 \times \tan 30) = 31.8°$$

and the $\phi'_d$ value is

$$\phi'_d = \tan^{-1}(\gamma_M \tan\phi'_k) = \tan^{-1}(1.25 \times \tan 31.8) = 37.8°$$

Using this superior value for $\phi'_d$ gives $K_{a;d} = 0.203$, so that the design wall resistance is now

$$R_d = 0.5 \, K_{a;d}\sigma'_v \, \tan\delta_d A = 0.5 \times 0.203 \times (18.0 - 9.81) \times 6 \times \tan 37.8 \times 216 = 836 \, kN$$

For the upper $\phi'_k = 31.8°$, $K_{a;k} = 0.265$ so that the characteristic wall resistance is

$$R_k = 0.5 \, K_{a;k}\sigma'_v \, \tan\delta_k A = 0.5 \times 0.265 \times (18.0 - 9.81) \times 6 \times \tan 31.8 \times 216 = 872 \, kN$$

This $R_k$ is very similar to the $R_k$ calculated using the lower $\phi'_k$ value. Using these resistance values based on the superior $\phi'_k$, the overall factor of safety on the wall resistance against uplift is now

$$OFS = R_k / R_d = 872/836 = 1.04$$

The above calculations show that, for this particular example, adopting the superior $\phi'_k$ value gives a similar overall factor of safety to that obtained when using the inferior $\phi'_k$ value. This example demonstrates that in certain situations, such as when determining the wall resistance on buried structures, using factors to increase or decrease the $\phi'$ value may not provide the required level of safety and hence care is needed when determining the design parameter values. This is an example of a situation where direct selection of the design parameter value might be adopted in order to achieve the appropriate degree of reliability.

### 10.4.4.3 Example 10.3: Design of a basement against uplift

The purpose of this example is to verify the stability against uplift failure of the buried structure with the properties described in the previous example and shown in Figure 10.6. The concrete basement slab thickness $b = 1$ m and the thickness of the concrete walls is $t = 0.5$ m. The weight density of the concrete $\gamma_c = 25$ kN/m$^3$.

*Solution*
Stability against a UPL ultimate limit is verified by checking that the design destabilizing uplift force due to the water pressure, $V_{dst;d}$ does not exceed the stabilizing weight of the structure, $G_{stb;d}$ plus the additional force due to the soil frictional resistance, $R_d$ on the side walls of the buried structure:

$$V_{dst;d} \leq G_{stb;d} + R_d \tag{10.26}$$

Using the recommended UPL partial factors in Table 10.8:

$$V_{dst;d} = \gamma_{dst}\, \gamma_w\, D\, B\, L = 1.0 \times 9.81 \times 6 \times 8 \times 10 = 4709 \text{ kN}$$

$$G_{stb;d} = \gamma_{stb}\, \gamma_c\, (B\, L\, b + (D\text{-}b)(B\, L\text{-}(B\text{-}2t)(L\text{-}2t)))$$
$$= 1.0 \times 25 \times (8 \times 10 \times 1 + (6{-}1)\, (8 \times 10{-}(8{-}2 \times 0.5)(10{-}2 \times 0.5))) = 4125 \text{ kN}$$

Substituting in Equation 10.26 and using the lower, more conservative, $R_d$ value of 836 kN calculated in Example 10.2 using the superior $\phi'_d$ values gives

$$V_{dst;d} = 4709 \text{ kN} \leq G_{stb;d} + R_d = 4125 + 836 = 4961 \text{ kN}$$

Hence, UPL is satisfied and the basement has an adequate degree of reliability against the ultimate limit state of uplift failure even though the margin of safety on the $R_d$ value is low.

## 10.5 RELIABILITY LEVELS

### 10.5.1 Partial factors, uncertainty, calibration, and target reliability

According to EN 1990, the partial factors for actions, $\gamma_F$ that are used in Eurocode 7 and the other Eurocodes, take into account model uncertainty in the actions and action effects and also uncertainty in the representative values of the actions. The partial factors for material properties $\gamma_M$ take into account model uncertainty in the calculation of the resistance and also uncertainty, as opposed to variability, in the material properties. This is shown in Figure 10.7.

EN 1990 states in Annex C §C3(2) that *in principle numerical values for partial factors and ψ factors can be determined in either of two ways:*

a. *On the basis of calibration to a long experience of building tradition, or*
b. *On the basis of statistical evaluation of experimental data and field observations.*

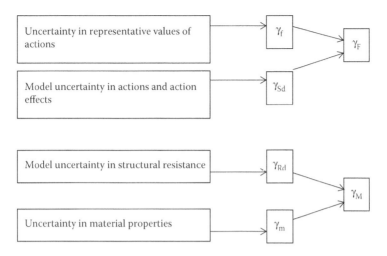

*Figure 10.7* Uncertainty and the Eurocode partial factors.

The methods that can be used to calibrate the partial factors and the relation between them are presented diagrammatically in EN 1990, as shown in Figure 10.8. The probabilistic methods are subdivided into Level I, semi-probabilistic methods, Level II, first-order reliability methods (FORM) and Level III, full probabilistic methods. Although shown in Figure 10.8, the semi-probabilistic method is not defined in EN 1990. However, in the ISO standard 2394 for the reliability of structures (ISO, 1998), the semi-probabilistic method is described as the method where design calculations are carried out, as in the Eurocodes, using consequence class categorizations (see Section 10.5.3), design situations, design values of actions, and material properties with partial factors applied to characteristic values to achieve the appropriate degree of reliability.

When a probabilistic method is used, EN 1990 states in Annex C §C3(3) that the ultimate limit states partial factors for material and actions should be calibrated such that the

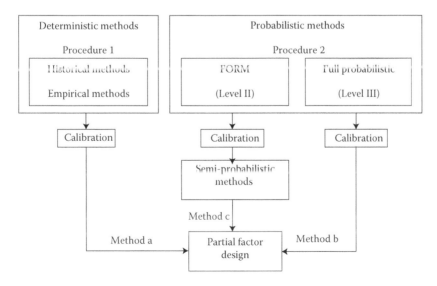

*Figure 10.8* Overview of the reliability calibration methods in EN 1990.

Table 10.9 Target reliability index, β values in EN 1990

| Limit state | Target reliability index, β | |
|---|---|---|
| | 1 year | 50 years |
| Ultimate limit state | 4.7 | 3.8 |
| Serviceability limit state (irreversible) | 2.9 | 1.5 |

reliability levels for representative structures are as close as possible to the target reliability index, β values, which are given in Table 10.9.

The values of the partial factors and combination factors determine the degrees of reliability and hence the levels of safety of structures designed using the Eurocode limit state design method. The recommended partial factor and combination factor values in the Eurocodes, including the recommended partial factor values in Eurocode 7 shown in Tables 10.7 and 10.8 and the values in the national annexes, have been chosen for the most part on the basis of a long experience of structural and geotechnical design and building tradition.

The recommended values of partial material factors for soil parameters $\gamma_M$ and partial resistance factors $\gamma_R$ given in Eurocode 7 for structural and geotechnical designs are for persistent and transient design situations for different types of ultimate limit states. For accidental design situations, Eurocode 7 states in §2.4.7.1(3) that the partial factors on actions and the effects of actions should normally be equal to 1.0, and that all partial factors on resistances should be selected according to the particular circumstances of the accidental situation. For serviceability limit states, Eurocode 7 states in §2.4.8(2) that the values of the partial factors should normally be equal to 1.0.

## 10.5.2 Partial factors in spread and pile foundations designs

The recommended partial factors given in Eurocode 7 have been chosen specifically to verify that the occurrence of just ultimate limit states is sufficiently unlikely. On account of this, the combination of partial factors in designs to Eurocode 7 can result in overall factors of safety for some design situations that are less than those used in previous practice, which were chosen to limit deformations as well as to avoid failure. An example of this is the use of the partial factor for the undrained shear strength of soil $\gamma_{cu} = 1.4$ in the design using DA1. C2 of a spread foundation loaded, for example, by a characteristic central vertical load $V_k$ consisting of 70% permanent load $G_k$ and 30% variable load $Q_k$. If the foundation rests on the ground surface, the characteristic resistance is $R_k = A\,c_{u;k}\,N_c$, where A is the base area of the foundation and $N_c = (\pi + 2)$. The overall factor of safety OFS is defined as

$$\text{OFS} = \frac{R_k}{V_k} \tag{10.27}$$

For designing to Eurocode 7 using DA1.C2, the design resistance, $R_d = A\,c_{u;d}\,N_c = A\,(c_{u;k}/\gamma_{cu})\,N_c = R_k/\gamma_{cu}$. Hence, using this equation for $R_d$ in Equation 10.1 and substituting the partial factors from Table 10.7 gives

$$E_d \leq R_d \rightarrow \left(\gamma_G G_k + \gamma_Q Q_k\right) \leq \frac{R_k}{\gamma_{cu}} \rightarrow (1.0 \times 0.7 \times V_k + 1.3 \times 0.3 \times V_k) \leq \frac{R_k}{1.4}$$

$$\rightarrow 1.09 V_k \leq \frac{R_k}{1.4}$$

Hence, the overall factor of safety for this situation is

$$\text{OFS} = \frac{R_k}{V_k} = 1.53$$

Similar calculations using the DA1.C1 partial factors give a lower OFS of 1.40 so that DA1.C2 controls the DA1 design. The OFS value of 1.53 is much less than the OFS value formerly used in traditional foundation designs, which was typically between 2 and 3 for spread foundations on fine-grained soils. Since the recommended partial factors are chosen to ensure that an ultimate limit state does not occur, Eurocode 7 has the following provisions to ensure the foundation has sufficient reliability against the occurrence of excessive settlements causing a serviceability limit state. It states in §6.6.2(16) that for conventional foundations on clays, the ratio $R_k/V_k$ should be calculated and if this ratio is less than 3, then settlement calculations should always be undertaken, and if it is less than 2, then the nonlinear stiffness effects in the ground should be taken into consideration.

In the case of pile foundations bearing in medium to dense soils and for tension piles, Eurocode 7 states in a note to §7.6.4.1(2) that *the safety requirements for the ultimate limit state design are normally sufficient to prevent a serviceability limit state in the supported structure*. However, the overall factors of safety provided by the recommended partial factors for pile design in Eurocode 7 are less than the overall factors that have been used for many years in the design of piles in most countries. For example, using the DA1.C2 partial factors in Table 10.7, including $\gamma_t = 1.5$ for the total compressive resistance, $R_k$ for a bored pile, and with a characteristic vertical load, $V_k$ consisting, as in the example above, of 70% permanent load, $G_k$ and 30% variable load, $Q_k$, substitution in Equation 10.1 gives

$$E_d \leq R_d \rightarrow (\gamma_G G_k + \gamma_Q Q_k) \leq \frac{R_k}{\gamma_t} \rightarrow (1.0 \times 0.7 \times V_k + 1.3 \times 0.3 \times V_k) \leq \frac{R_k}{1.5}$$

$$\rightarrow 1.09 V_k \leq \frac{R_k}{1.5}$$

Hence, the overall factor of safety is

$$\text{OFS} = \frac{R_k}{V_k} = 1.64$$

The overall factor of safety is lower than the value, generally between 2 and 3, that was formerly used for the design of piles. Hence, most European countries in their national annexes have adopted a method to increase the overall factor of safety in the design of piles so that it is similar to that used previously in the design. These methods include choosing partial resistance factor values for piles that are greater than those recommended in Eurocode 7, introducing a model factor to increase the resistance factor, or increasing the ξ values.

The partial resistance factors in the UK National Annex for the design of bored piles are shown in Table 10.10. These show that the R4 values, which are used with DA1.C2, are increased compared to the recommended values and that two sets are provided: one set for when the serviceability limit state is not explicitly verified and another set for when it is. When the SLS is not explicitly verified, the recommended partial resistance factor on the total pile resistance is increased by a factor of 1.54, whereas when the SLS is explicitly verified, the recommended partial resistance factor on the total pile resistance is increased by a reduced factor of 1.31. This is an example of how the use of pile load tests to verify a design

*Table 10.10* Comparison of recommended partial factor values in Eurocode 7 for bored piles with those in UK National Annex

| Parameter | Partial resistance factor | Recommended values | | | UK National Annex values | |
|---|---|---|---|---|---|---|
| Bored pile foundations | $\gamma_R$ | R1 | R4 | R1 | R4 without explicit SLS verification | R4 with explicit SLS verification |
| Base resistance | $\gamma_b$ | 1.25 | 1.6 | 1.0 | 2.0 | 1.7 |
| Shaft (compression) | $\gamma_s$ | 1.0 | 1.3 | 1.0 | 1.6 | 1.4 |
| Total (compression) | $\gamma_t$ | 1.15 | 1.5 | 1.0 | 2.0 | 1.7 |
| Shaft in tension | $\gamma_{t;s}$ | 1.25 | 1.6 | 1.0 | 2.0 | 1.7 |

may be used to justify lower partial factor values and hence, a more economical pile design. The UK National Annex states that the lower R4 values may be adopted if

a. Load tests carried out on more than 1% of the constructed piles to loads not less than 1.5 times the representative load for which they are designed.
b. Settlement is predicted by a means no less reliable than in (a).
c. Settlement at the serviceability limit state is of no concern.

It should be noted that, while the R4 values are increased in the UK National Annex, Table 10.10 shows that the R1 values are all equal to 1.0, that is, some are reduced. The reason for this is so that when designing piles to DA1 using the UK National Annex, the DA1. C2 partial factors always control the design of the length of the pile.

## 10.5.3 Reliability differentiation

EN 1990 offers the following two methods for differentiating the reliability of ULS designs:

1. The introduction of Consequences Classes (CC) and Reliability Classes (RC) established by considering the consequences of failure or malfunction of the structure and the exposure of the construction works to hazards. The bases of these classes are similar to the bases of the Eurocode 7 Geotechnical Categories, referred to in Section 10.2.2.
2. The introduction of different quality levels for the design supervision and inspection during execution related to the Reliability Class of the structure.

The Consequences Classes relate to the consequences of failure or malfunction of the structure with regard to the consequences for loss of human life as well as the economic, social, and environmental consequences. These range from low for CC1, through medium for CC2, to high for CC3, as shown in Table 10.11, adapted from Orr (2013), which includes the examples given in EN 1990. Three Reliability Classes are defined with the minimum 50-year target reliability index β values for each shown in Table 10.11. The three Reliability Classes may be associated with the three Consequences Classes as is also shown in Table 10.11. The Consequences Classes, as defined in EN 1990, differ from the Eurocode 7 Geotechnical Categories in that they only take into account the consequences of failure, that is, the vulnerability factors in Table 10.1, and do not take into account the hazards that contribute to the complexity and risk in geotechnical designs.

Eurocode 7 introduces a further method for reliability differentiation related to the duration of the action and based on the Design Situations, which are defined in §1.5.2.2 of EN

*Table 10.11* Consequences Classes, associated Reliability Classes, and β values from EN 1990

| Consequences Class | CC1 | CC2 | CC3 |
|---|---|---|---|
| Description | Low consequence for loss of human life, *or* small or negligible economic, social or environmental consequences | Medium consequence for loss of human life, or considerable economic, social or environmental consequences | High consequence for loss of human life *and* very great economic, social, or environmental consequences |
| Examples of buildings or civil engineering works | Agricultural buildings where people do not normally enter (e.g., storage buildings, greenhouses) | Residential and office buildings, public buildings where consequences of failure are medium (e.g., office buildings) | Grandstands, public buildings where consequences of failure are high (e.g., concert halls) |
| Reliability Class | RC1 | RC2 | RC3 |
| β | 3.3 | 3.8 | 4.3 |

1990 as *sets of physical conditions representing the real conditions occurring during a certain time interval for which the design will demonstrate that relevant limit states are not exceeded.* The following four Design Situations are distinguished in EN 1990:

- Transient Design Situations
- Persistent Design Situations
- Accidental Design Situations
- Seismic Design Situations

As noted above, the recommended ULS partial factor values given in Eurocode 7 are for transient and persistent Design Situations. Referring to the partial factor values, Eurocode 7 states in §2.4.7.1(4) and (5) that *More severe values than those recommended ... should be used in cases of abnormal risk or unusual or exceptionally difficult ground or loading conditions,* and that *less severe value than those recommended ... may be used for temporary structures or transient design situations where the likely consequences justify it,* that is, the values of the partial factors in Eurocode 7 should be increased or decreased, and hence the level of reliability increased or decreased, to take account of Design Situations where the level of risk may be more severe or less severe.

Examples of CEN member states that have introduced reliability differentiation involving Consequences Classes and Design Situations to introduce more severe or less severe partial factor values are Austria and Germany. Different sets of partial material factor values are given in the Austrian NA for the three Consequences Classes for the design of slopes and anchorages. These are related to the particular Design Situation (BS) as shown in Table 10.12. The highest partial factor values are for persistent Design Situations (BS 1), intermediate values are for transient Design Situations (BS 2), and lowest values are for accidental Design Situations (BS 3). In the German NA, the values of the partial action factors to be used with Eurocode 7 are related to the Design Situation for persistent, transient, and accidental Design Situations, labeled as BS-P, BS-T, and BS-A, for all types of geotechnical designs, as shown in Table 10.13. As in the case of the Austrian partial material factors, the highest partial action factors are for persistent situations, BS-P.

In France, a form of reliability differentiation has been introduced that corresponds to the second method for managing risk and differentiating reliability in geotechnical design described at the start of this section. This method does not involve changes to the partial factor values but links the Consequences Classes to the Geotechnical Categories and hence, to

*Table 10.12* Partial factor values in Austrian NA to Eurocode 7 for different Consequences Classes, and Design Situations

| Soil parameter | Symbol | Partial material factor values $\gamma_m$ for different Consequences Classes | | | | | | | | |
|---|---|---|---|---|---|---|---|---|---|---|
| | | CC 1 | | | CC 2 | | | CC 3 | | |
| | | Design Situations (BS) | | | | | | | | |
| | | BS 1 | BS 2 | BS 3 | BS 1 | BS 2 | BS 3 | BS 1 | BS 2 | BS 3 |
| Effective angle of friction | $\gamma_{\tan\phi'}$ | 1.10 | 1.05 | 1.00 | 1.15 | 1.10 | 1.05 | 1.30 | 1.20 | 1.10 |
| Effective cohesion | $\gamma_{c'}$ | 1.10 | 1.05 | 1.00 | 1.15 | 1.10 | 1.05 | 1.30 | 1.20 | 1.10 |
| Undrained shear strength | $\gamma_{cu}$ | 1.20 | 1.15 | 1.10 | 1.25 | 1.20 | 1.15 | 1.40 | 1.30 | 1.20 |
| Uniaxial compressive strength | $\gamma_{qu}$ | 1.20 | 1.15 | 1.10 | 1.25 | 1.20 | 1.15 | 1.40 | 1.30 | 1.20 |
| Weight density | $\gamma_\gamma$ | 1.00 | 1.00 | 1.00 | 1.00 | 1.00 | 1.00 | 1.00 | 1.00 | 1.00 |

*Table 10.13* Partial action factor values in German NA to Eurocode 7 for different Design Situations

| Action | Symbol | Partial action factor values $\gamma_F$ for different Design Situations | | |
|---|---|---|---|---|
| | | BS-P (persistent) | BS-T (transient) | BS-A (accidental) |
| Unfavorable permanent action | $\gamma_G$ | 1.35 | 1.20 | 1.00 |
| Favorable permanent action | $\gamma_{G, \inf}$ | 1.00 | 1.00 | 1.00 |
| Unfavorable permanent earth pressure | $\gamma_{G, EO}$ | 1.20 | 1.10 | 1.00 |
| Unfavorable variable action | $\gamma_Q$ | 1.50 | 1.30 | 1.10 |
| Favorable variable action | $\gamma_Q$ | 0 | 0 | 0 |

*Table 10.14* Linkage of Geotechnical Categories to Consequences Classes, site conditions, and basis for geotechnical design in France

| Geotechnical Category | Consequences Class | Site conditions | Basis for design |
|---|---|---|---|
| GC1 | CC1 | Simple and known | Experience and qualitative investigations |
| GC2 | CC1 | Complex | Site investigations and design |
| | CC2 | Simple or complex | |
| GC3 | GC3 | Simple or complex | Site investigations and detailed design |

the level and quality of the geotechnical investigations and design calculations, as shown in Table 10.14 from Orr (2013). This method of reliability differentiation is based on the quality management differentiation outlined in EN 1990, with three Design Supervision Levels, DSL1, DSL2, and DSL3, and three Inspection Levels during construction, IL1, IL2, and IL3, that may be related to the three Reliability Classes, as shown in Table 10.15, or chosen according to the importance of the structure and in accordance with national requirements. These are also linked in Table 10.15 with the minimum requirements in Eurocode 7 for the different Geotechnical Categories. The Design Supervision Levels are implemented through the adoption of appropriate quality management measures.

## 10.6 CONCLUSIONS

Eurocode 7 provides a broad framework to identify the complexity of geotechnical design situations and to manage and reduce the risks in geotechnical design to acceptable

*Table 10.15* Linkage of Geotechnical Categories to Design Supervision Levels and Inspection Levels and minimum requirements

| Geotechnical Category | Design Supervision Level | Inspection Level | Characteristics | Minimum recommended requirements for checking of calculations, drawings, and specifications |
|---|---|---|---|---|
| GC1 | DSL1 relating to RC1 | IL1 relating to RC1 | Normal supervision and inspection | Self-checking: checking performed by person who has prepared the design |
| GC2 | DSL2 relating to RC2 | IL2 relating to RC2 | Normal supervision and inspection | Checking by different persons than those originally responsible and in accordance with the procedure of the organization |
| GC3 | DSL3 relating to RC3 | IL3 relating to RC3 | Extended supervision and inspection | Third-party checking: checking performed by an organization different from that which has prepared the design |

levels. This framework includes measures to ensure the quality of geotechnical investigations, design calculations, the supervision of construction, and the monitoring and maintenance of a structure after completion and is based on assumptions that all of these aspects are carried out by persons with the appropriate expertise and experience. The measures include many checklists of items to be considered, taken into account, or receive attention in geotechnical design. There is also a requirement for good communication that involves the preparation of a Geotechnical Design Report and a Geotechnical Investigation Report.

The reliability of geotechnical designs to Eurocode 7 may be verified by one or a combination of four methods: calculations, prescriptive measures, model and load tests, and the observational method. When using calculations, design values are obtained using partial factors and combination factors applied appropriately to characteristic actions and characteristic material parameters or resistances. The characteristic values of parameters are selected by the designer to account for the variability in the parameters and the extent of the failure mechanism, while the partial factors and combination factors are chosen by national authorities to take account of uncertainties in the calculation models and material uncertainties as well as to provide appropriate levels of safety or degrees of reliability and robustness.

In designs to the Eurocodes, the combination of the partial factors, combination factors, and characteristic values aim to achieve designs with the appropriate target reliability index. Recognizing that statistical information about the actions and material parameters is often not available, particularly in geotechnical design, the values of the recommended partial factors in Eurocode 7 have generally been chosen on the basis of a long experience of geotechnical design and building tradition. However, as the understanding of the statistical nature of soil improves and as more research is carried out in this area, it is anticipated that statistics and probabilistic analyses will be used more frequently to select the characteristic parameter values, to calibrate the partial factors values and to assess the reliability of geotechnical designs.

## ACKNOWLEDGMENTS

Helpful comments and contributions from Professor Jianye Ching, particularly on the variance reduction factors, and from Professor K.K. Phoon are gratefully acknowledged.

# REFERENCES

Bauduin C. 2001. Design procedure according to Eurocode 7 and analysis of test results, *Proceedings Symposium on Screw Pile: Installation and Design in Stiff Clay*, Balkema, Rotterdam, 275–303.

Bedi A. and Harrison J.P. 2013. Characterisation and propagation of epistemic uncertainty in rock engineering: A slope stability example. *Proc. International Society for Rock Mechanics Symposium—Rock Mechanics for Resources, Energy and Environment*, Wroclaw, Poland, 23–26 September 2013. eds. Kwaśniewski & Łydżba, Taylor & Francis, London.

Bedi A. and Orr T.L.L. 2104. On the applicability of the Eurocode7 partial factor method for rock mechanics. *Proceedings Eurock2014*, Vigo, Spain, 27–29 May, 2014 (in print).

Bond A. and Harris A. 2008. *Decoding Eurocode 7*, Taylor & Francis, London.

Burland J.B. 1987. The teaching of soil mechanics—A personal view, Groundwater effects in geotechnical engineering. *Proceedings IX ECSMFE, Dublin*, eds. Hanrahan, E.T., Orr, T.L.L. and Widdis, T., Balkema, Rotterdam, 3:1427–1447.

Calgaro J.-A. 2011. Safety philosophy of Eurocodes, *Proceedings Geotechnical Risk and Safety. Proceedings of the 3rd International Conference on Geotechnical Safety and Risk*, Munich 2011, eds. N. Vogt, B. Schuppener, D. Straub and G. Bräu, Bundesanstadt für Wasserbau, Karlsruhe, Germany, 29–36.

CEN 2002. *EN 1990: Eurocode—Basis of Structural Design*. European Committee for Standardization, Brussels.

CEN 2004. *EN 1997-1:2004: Eurocode 7: Geotechnical Design—Part 1: General Rules*. European Committee for Standardization, Brussels.

CEN 2007. *EN 1997-2:2007: Eurocode 7—Geotechnical Design—Part 2: Ground Investigation and Testing*. European Committee for Standardization, Brussels.

Ching J. and Phoon K.K. 2013. Mobilized shear strength of spatially variable soils under simple shear stress states. *Structural Safety*, 41:20–28.

EC 2003. *Guidance Paper L (Concerning the Construction Products Directive—89/106/EEC). Application and Use of Eurocodes*, Version 27 November 2003, European Commission, Enterprise Directorate-General, Brussels.

El-Ramly H., Morgenstern N.R. and Cruden D.M. 2006. Lodalen slide: A probabilistic assessment. *Canadian Geotechnical Journal*, 43:956–968.

ISO 1998. *ISO: 2394: General Principles on Reliability for Structures*. International Standards Organization, Geneva.

Orr T.L.L. 2012. Codes and standards and their relevance. *Manual of Geotechnical Engineering*, eds. Burland J., Powrie W. and Chapman T., Institution of Civil Engineers, London, Volume 1.

Orr T.L.L. 2013. Implementing Eurocode 7 to achieve reliable geotechnical designs. *Modern Geotechnical Design Codes of Practice*, eds. P. Arnold, G.A. Fenton, M.A. Hicks, T. Schweckendiek and B. Simpson, IOS Press, Amsterdam, pp. 72–86.

Orr T.L.L. and Farrell E.R. 1999. *Geotechnical Design to Eurocode 7*, Springer, London.

Phoon K-K. and Kulhawy F.H. 1999. Characterization of geotechnical variability. *Canadian Geotechnical Journal*, Canadian Science Publishing, 36(4):612–624.

Schneider H.R. 1997. Definition and characterization of soil properties. *Proceedings XIV ICSMGE*, Hamburg, Balkema, Rotterdam.

Schneider H.R. and Fitze P. 2013. Characteristic shear strength values for EC7: Guidelines based on a statistical framework. *Proceedings XV European Conference on Soil Mechanics and Geotechnical Engineering*, September 2011, eds. A. Anagnostopoulos, M. Pachakis and C. Tsatsanifos, Athens, Greece, 4:318–324.

Schneider H.R. and Schneider M.A. 2013. Dealing with uncertainties in EC7 with emphasis on determination of characteristic soil properties. *Modern Geotechnical Design Codes of Practice*, eds. P. Arnold, G.A. Fenton, M.A. Hicks, T. Schweckendiek and B. Simpson, IOS Press, Amsterdam, 87–101.

Simpson B. 2011. Reliability in geotechnical design: Some fundamentals. *Proceedings International Symposium on Geotechnical Safety and Risk*, Munich 2010, eds. Vogt N., Schuppener B., Straub & Bräu, Bundesanstalt für Wasserbau, ISBN: 978-3-939230-01-4

Standards Australia 2009. *Piling: Design and Installation*. Standards Australia, Sydney, Australia.

Sowers G.F. 1993. Human factors in civil and geotechnical engineering failures. *Journal of Geotechnical Engineering*, 119(2):238–256.

Vanmarcke E.H. 1977. Probabilistic modeling of soil profiles. *ASCE Journal of the Geotechnical Engineering Division*, GT11, 1227–1246.

Vanmarcke E.H. 1983. *Random Fields: Analysis and Synthesis*, MIT Press, Cambridge, Massachusetts.

# Part IV

# Risk and decision

# Chapter 11

# Practical risk assessment for embankments, dams, and slopes

*Luis Altarejos-García, Francisco Silva-Tulla,*
*Ignacio Escuder-Bueno, and Adrián Morales-Torres*

## 11.1 INTRODUCTION

Geotechnical engineers can no longer ignore the benefits of performing risk assessments. Engineers from both the government and private sectors encounter the need to perform or review risk assessments for constructed facilities with increasing frequency. In the United States, the federal government is unlikely to appropriate funds for major infrastructure projects unless a risk assessment has been performed. Similarly, any major dam rehabilitation for a federal agency requires a thorough examination using risk assessment tools and techniques. In fact, the U.S. Office of Management and Budget (OMB, 2013) states:

> Project managers when developing the cost, schedule, and performance goals on developmental projects with significant risk must, therefore, provide the agency Executive Review Committee (ERC) with risk-adjusted and most likely cost, schedule, and performance goals. Without the knowledge of the risks involved managers at all levels—agency, Office of Management and Budget (OMB) and the Congress—cannot make the best decisions for the allocation of resources among the competing investments.

For all practical purposes, OMB requires an assessment of risk before major projects are considered for funding.

Practicing engineers, who in the past might have considered risk assessment as a topic of academic interest, can no longer ignore the wide acceptance of risk assessment in the business world and the engineering profession.

Risk assessments can range from very rigorous, complex, and costly analyses to pragmatic evaluations using semiempirical methods to estimate failure probabilities. This chapter presents a practical methodology for risk assessment to guide decisions for everyday projects (not unique megaprojects) or in situations where results are needed in short order. The authors expand upon the methodology presented by Silva, Lambe, and Marr (2008) and explain how to use semiempirical relationships between safety factor and probability of failure through an example of an actual engineering project. The authors address the natural (aleatory) and knowledge (epistemic) uncertainties involved in determining the level of risk and demonstrate the use of fragility curves to manage uncertainty and better understand the expected performance of a facility. The authors also stress the fact that much of the benefit from conducting a risk assessment stems from engaging the engineers to thoroughly evaluate the engineering fundamentals controlling the performance of the constructed facility.

The topics covered in this chapter include

1. How to estimate the failure probability, p(f), for slope stability, foundation stability, and soil transport problems using semiempirical charts

2. The role of fragility curves to evaluate the uncertainty in probability estimates
3. How to mathematically and numerically obtain conditional probabilities for slope stability failures related to factor of safety (FS) and fragility curves capturing natural and epistemic uncertainties
4. How to use the estimated fragility curves to obtain annualized probabilities used in the most common practices of risk analysis
5. Overall conclusions on applicability, transparency, robustness, and engineering footprint over the process

## 11.2 ESTIMATION OF CONDITIONAL PROBABILITY AS A FUNCTION OF SAFETY FACTOR

Risk assessments complement, but do not replace, other engineering analyses such as stability, flow, or deformation analyses. One of the main benefits of performing a risk assessment results from the detailed examination of engineering fundamentals required to prepare rational event trees. These frequently highlight the importance or irrelevance of mechanisms that could have been overlooked or overemphasized, respectively. Since engineering analyses shed light on which mechanisms strongly affect the safety of a facility, one could reasonably conclude that engineering analyses should precede risk assessment. However, the risk assessment process can help decide which aspects of performance, and thus which type of engineering analyses, should receive the bulk of the effort. Thus, as is so often the case in engineering, an iterative process provides the most practical alternative. In our practice, we find that starting with deterministic analyses, or what many would term "traditional engineering analyses," provides a solid base for the risk assessment, particularly if event probabilities come from expert elicitation. During the risk assessment, the need for additional or more refined deterministic or probabilistic analyses could become evident. Ultimately, the engineer needs to combine both traditional engineering analyses and risk assessment to obtain a good understanding of the structure's expected performance and risk.

### 11.2.1 FS versus p(f) charts for slope instability and soil transport

For slope stability problems, a correct slope stability analysis provides a desirable starting point for a risk assessment. The engineer should attempt to understand the likely values and variability of all relevant information. For most slope stability problems, these include

- Geometry
- Stresses
- Pore pressures
- Strengths

If the determination of one of these key parameters is not possible, the engineer should at least evaluate the likely impact of any uncertainty or knowledge gap in the results of the stability evaluation. Pitfalls to avoid when performing stability analyses include

1. Not determining the field drainage conditions—drained, undrained, or partially drained.
2. Using an incorrect strength for the stress path in the field—Lambe and Silva (2003) show examples of the influence of stress path on strength for stability analyses.
3. Confusing undrained conditions and total stresses—engineers can and do perform stability analyses for undrained conditions using strengths determined in terms of effective stresses. While this requires knowledge about the corresponding pore pressures,

not always an easy task, using a total stress analysis to avoid dealing with pore pressures does not bypass this difficulty. When using total stresses, the engineer implicitly accepts that the pore pressures developed during the tests used to determine strength apply to the field problem at hand. This is not always the case and the impact of any differences on the computed safety factor should be evaluated by the individual performing stability analyses.

Baecher and Christian (2003) present a comprehensive treatise on the various methods to estimate event probabilities. Even when engineers recognize the benefits of probabilistic risk analyses, estimating event probabilities can seem like an unsurmountable obstacle for most projects without large budgets and long durations. Empirical correlations between commonly performed stability analyses and probability of failure can make the estimation of failure probabilities less daunting. Figure 11.1 shows one correlation we have used successfully for decades with some minor adjustments to the four curves to reflect new data or improved representation of the data.

Figure 11.1 classifies earth structures into the four categories described in Table 11.1. The category of the earth structure, based on the level of engineering, ranges from Best (Category I) to Poor (Category IV). We establish the level of engineering by examining the practices followed for design, investigation, testing, analyses and documentation, construction, and operation and monitoring. The four categories correspond to the following types of facilities:

- Category I—Facilities designed, built, and operated with state-of-the-practice engineering. Generally, these facilities have high failure consequences.
- Category II—Facilities designed, built, and operated using standard engineering practice. Many ordinary facilities fall in this category.
- Category III—Facilities without site-specific design and substandard construction or operation. Temporary facilities and those with low failure consequences often fall in this category.
- Category IV—Facilities with little or no engineering.

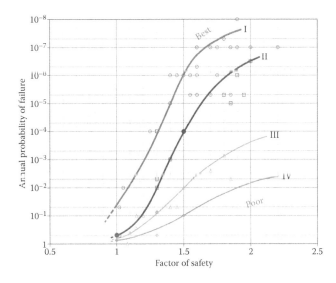

*Figure 11.1* Factor of safety versus annual probability of failure.

Table 11.1 Earth structure categories and characteristics

| Level of engineering | Design | | | Construction | Operation and monitoring |
|---|---|---|---|---|---|
| | Investigation | Testing | Analyses and documentation | | |
| I (Best) Facilities with high failure consequences | Evaluate design and performance of nearby structures<br>Analyze historic aerial photographs<br>Locate all non-uniformities (soft, wet, loose, high, or low permeability zones)<br>Determine site geologic history<br>Determine subsoil profile using continuous sampling<br>Obtain undisturbed samples for lab testing of foundation soils<br>Determine field pore pressures | Run lab tests on undisturbed specimens at field conditions<br>Run strength test along field effective and total stress paths<br>Run index field tests (e.g., field vane, cone penetrometer) to detect all soft, wet, loose, high, or low permeability zones<br>Calibrate equipment and sensors prior to testing program | Determine FS using effective stress parameters based on measured data (geometry, strength, pore pressure) for site<br>Consider field stress path in stability determination<br>Prepare flow net for instrumented sections<br>Predict pore pressures and other relevant performance parameters (e.g., stress, deformation, flow rates) for instrumented section<br>Have design report clearly document parameters and analyses used for design<br>No errors or omissions<br>Peer review | Full-time supervision by qualified engineer<br>Construction control tests by qualified engineers and technicians<br>No errors or omissions<br>Construction report clearly documents construction activities | Complete performance program including comparison between predicted and measured performance (e.g., pore pressure, strength, deformations)<br>No malfunctions (slides, cracks, artesian heads)<br>Continuous maintenance by trained crews |
| | 0.2 | 0.2 | 0.2 | 0.2 | 0.2 |

| Category | Field investigation | Laboratory testing | Analysis / FS determination | Construction supervision | Inspection and maintenance |
|---|---|---|---|---|---|
| II (Above average) Ordinary facilities | Evaluate design and performance of nearby structures Exploration program tailored to project conditions by qualified engineer  **0.4** | Run standard lab tests on undisturbed specimens Measure pore pressure in strength tests Evaluate differences between laboratory test conditions and field conditions  **0.4** | Determine FS using effective stress parameters and pore pressures Adjust for significant differences between field stress paths and stress path implied in analysis that could affect design  **0.4** | Part-time supervision by qualified engineer No errors or omissions  **0.4** | Periodic inspection by qualified engineer No uncorrected malfunctions Selected field measurements Routine maintenance  **0.4** |
| III (Average) Unimportant or temporary facilities with low failure consequences | Evaluate performance of nearby structures Estimate subsoil profile from existing data and borings  **0.6** | Index tests on samples from site  **0.6** | Rational analyses using parameters inferred from index tests  **0.6** | Informal construction supervision  **0.6** | Annual inspection by qualified engineer No field measurements Maintenance limited to emergency repairs  **0.6** |
| IV (Poor) Little or no engineering | No field investigation  **0.3** | No laboratory tests on samples obtained at the site  **0.8** | Approximate analyses using assumed parameters  **0.8** | No construction supervision by qualified engineer No construction control tests.  **0.8** | Occasional inspection by non-qualified person No field measurements  **0.8** |

Note:   The values in the dotted boxes provide a weight for each of the conditions described to facilitate the computation of the appropriate Category (or Level of Engineering) for instances where the facility includes characteristics from various categories. The Category should equal the sum of all the weights.

The family of curves in Figure 11.1 and the associated Table 11.1 with the four levels of engineering reflect the generally accepted concept that—"A larger factor of safety does not necessarily imply a smaller risk, because its effect can be negated by the presence of larger uncertainties in the design environment" (Kulhawy and Phoon, 1996; D'Andrea and Sangrey, 1982; Tavares and Serafim, 1983; Christian et al., 1994). The curves in Figure 11.1 provide a practical grouping related to the standard deviation of the safety factors calculated with four different levels of engineering.

Figure 11.1 shows data from over 75 projects spanning over four decades used to develop the relationships. The projects included zoned and homogeneous earth dams, tailings dams, natural and cut slopes, and some earth-retaining structures. The probability of failure determinations reflect quantified expert judgment as explained by Silva et al. (2008).

Figure 11.2 presents a correlation between FS and annual probability of failure (APF) similar to that in Figure 11.1, but applicable to safety with respect to internal erosion. The factor of safety against internal erosion, $FS_{ie}$, is computed as

$$FS_{ie} = \frac{i}{i_{cr}} \tag{11.1}$$

where

    i is the gradient acting at point of interest
    $i_{cr}$ is the critical gradient, or gradient required to initiate transport of soil particles in the soil at the point of interest

The critical gradient can be estimated in the laboratory as described by Southworth (1980) or Schmertmann (2000).*

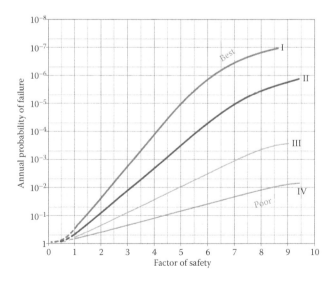

Figure 11.2  p(f) for internal erosion.

* $i_{cr}$ in Equation 11.1 corresponds to Schmertmann's $i_p$.

## 11.2.2  Example of risk assessment for an earth dam based on the empirical FS versus p(f) charts

This section presents a simplified risk analysis to evaluate the safety of an earth dam with respect to failure by internal erosion and soil transport. In addition, the risk analysis addresses the effect on risk or level of safety of proposed remedial measures for the dam. The risk assessment summarized below covers only part of the complete risk picture for the dam. Other potential failure modes exist, but failure by internal erosion/piping was considered the dominant failure mode for structure. The assessment involved the following steps:

- Estimation of failure probabilities versus peak pool elevation
- Estimation of pool elevation annual exceedance probabilities (AEPs)
- Estimation of potential loss of life versus pool elevation at time of failure
- Comparison of results with risk evaluation guidelines

### 11.2.2.1  Estimation of failure probabilities versus peak pool elevation: Example from engineering practice

The earth dam, an actual project from our engineering practice, is founded on karst and developed a sinkhole in the upstream slope during a high reservoir level event as shown schematically in Figure 11.3. The sinkhole was temporarily repaired but failure by internal

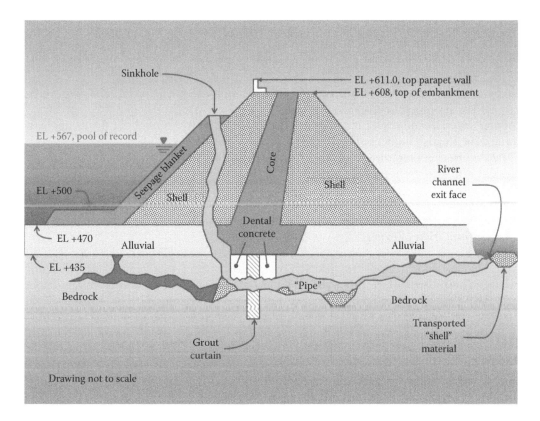

*Figure 11.3* Internal erosion/soil transport potential failure mode.

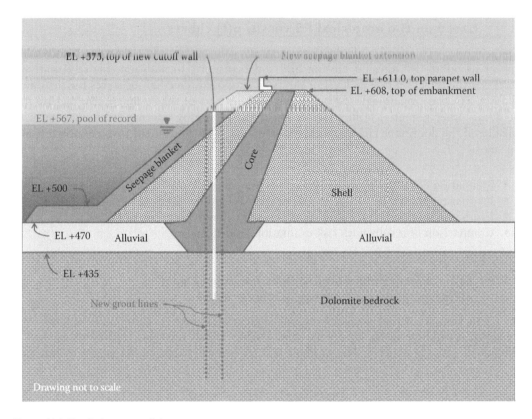

*Figure 11.4* Earth dam remedial measures.

erosion and soil transport remained the dominant failure mode. Figure 11.4 shows a schematic of the proposed "permanent" remedial measures, which included a comprehensive grouting program and construction of a concrete cutoff wall. Throughout the example, we retained the English units used in all the project documents.

The engineering literature (e.g., Morgenstern, 1995) identifies three commonly accepted ways of estimating event probabilities:

1. As result of frequent observations (historical data)
2. Derived from probability theory (mathematical modeling)
3. Quantification of expert

For the dam in the example, we used quantification of expert judgment (subjective probabilities) to estimate failure probabilities from internal erosion for a wide range of pool elevations. The set of probability estimates obtained are summarized in Table 11.2 and portrayed in Figure 11.5.

These estimated internal erosion failure probabilities are conditioned on the occurrence of the peak pool elevation. In Table 11.2, the probability of failure numbers in bold font were obtained by estimation and the numbers in regular font obtained by interpolation. A further refinement of the failure probability estimates would consider maintaining a given pool elevation for a determined time period during which a particular step in the failure mode sequence would have the opportunity to develop, instead of simply using peak values.

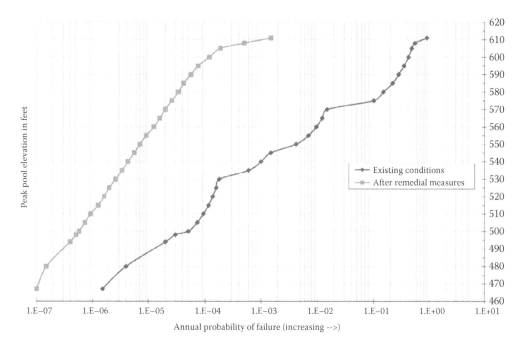

*Figure 11.5* Annual probability of failure versus earth dam peak pool elevation (internal erosion only).

### 11.2.2.2 Estimation of peak pool elevation annual exceedance probabilities

Table 11.2 shows the AEP that the peak pool elevation equals or exceeds a given level during a year. These probability estimates were based on historical reservoir pool data for 60 years of operation and reservoir operation simulation for a target pool level of +498 ft.

The annual probabilities that the pool will exceed a given elevation (PP) in Table 11.2 represent AEPs, not stage-duration or percent time exceeding. The incremental annual probability that the pool will exceed a given elevation reflects our approximate procedure for centering the pool elevation interval probabilities, that is, to put the water levels for the probabilities of failure and life loss estimates in the center of the interval rather than at the low end, for numerical integration reasons. For any given peak pool elevation (PPE$_i$), the "centered" incremental annual probability that the pool exceeds PPE$_i$ (IPP$_i$) is computed as

$$\text{IPP}_i = \frac{(\text{PP}_{i-1} - \text{PP}_{i+1})}{2} \tag{11.2}$$

The low-end approach, which does not center the pool elevations in the interval, biases downward the probabilities and annualized life loss (ALL) discussed in Section 11.2.2.4.

### 11.2.2.3 Estimation of potential loss of life versus peak pool elevation at time of failure

In our practice, we always separate potential economic consequences from potential loss of life. Potential loss of life estimates shown in Table 11.3 in terms of lives lost were derived using inundation maps and corresponding demographic data for eight pool elevations.

Table 11.2 Probability estimates

| Peak pool elevation (PPE) in feet | Probability of failure from internal erosion | | Annual probability pool elevation (PP) ≥ | Incremental annual probability pool elevation (IPP) ≥ |
| --- | --- | --- | --- | --- |
| | Existing conditions | After remedial measures | | |
| 467 (Dry reservoir—no flow) | $1.5 \times 10^{-06}$ | $1.0 \times 10^{-07}$ | 1.0000 | 0.0 |
| 480 (Operational pool) | $4.0 \times 10^{-06}$ | $1.5 \times 10^{-07}$ | 1.00000 | 0.00425 |
| 494 (Conservation pool) | $2.0 \times 10^{-05}$ | $4.0 \times 10^{-07}$ | 0.99150 | 0.02873 |
| 498 (Seasonal pool) | $3.0 \times 10^{-05}$ | $5.1 \times 10^{-07}$ | 0.94255 | 0.05854 |
| 500 | $5.1 \times 10^{-05}$ | $5.8 \times 10^{-07}$ | 0.87442 | 0.08201 |
| 505 | $7.3 \times 10^{-05}$ | $7.3 \times 10^{-07}$ | 0.77852 | 0.10164 |
| 510 | $9.4 \times 10^{-05}$ | $9.2 \times 10^{-07}$ | 0.67115 | 0.12772 |
| 515 | $1.2 \times 10^{-04}$ | $1.3 \times 10^{-06}$ | 0.52308 | 0.12383 |
| 520 | $1.4 \times 10^{-04}$ | $1.6 \times 10^{-06}$ | 0.42348 | 0.10894 |
| 525 | $1.6 \times 10^{-04}$ | $2.0 \times 10^{-06}$ | 0.30519 | 0.08123 |
| 530 (Seepage observed downstream) | $1.8 \times 10^{-04}$ | $2.6 \times 10^{-06}$ | 0.26103 | 0.06266 |
| 535 | $6.0 \times 10^{-04}$ | $3.4 \times 10^{-06}$ | 0.17987 | 0.07601 |
| 540 | $1.0 \times 10^{-03}$ | $4.3 \times 10^{-06}$ | 0.10902 | 0.05224 |
| 545 | $1.5 \times 10^{-03}$ | $5.6 \times 10^{-06}$ | 0.07539 | 0.02477 |
| 550 | $4.2 \times 10^{-03}$ | $7.0 \times 10^{-06}$ | 0.05947 | 0.01273 |
| 555 | $6.9 \times 10^{-03}$ | $9.2 \times 10^{-06}$ | 0.04993 | 0.00937 |
| 560 | $9.6 \times 10^{-03}$ | $1.3 \times 10^{-05}$ | 0.04073 | 0.01087 |
| 565 | $1.2 \times 10^{-02}$ | $1.6 \times 10^{-05}$ | 0.02819 | 0.01486 |
| 570 | $1.5 \times 10^{-02}$ | $2.0 \times 10^{-05}$ | 0.01100 | 0.01035 |
| 575 (Top of seepage blanket) | $1.0 \times 10^{-01}$ | $2.6 \times 10^{-05}$ | 0.00750 | 0.00300 |
| 580 | $1.5 \times 10^{-01}$ | $3.4 \times 10^{-05}$ | 0.00500 | 0.00240 |
| 585 | $2.2 \times 10^{-01}$ | $4.2 \times 10^{-05}$ | 0.00270 | 0.00175 |
| 590 | $2.8 \times 10^{-01}$ | $5.7 \times 10^{-05}$ | 0.00150 | 0.00094 |
| 595 | $3.5 \times 10^{-01}$ | $7.6 \times 10^{-05}$ | 0.00082 | 0.00054 |
| 600 | $4.2 \times 10^{-01}$ | $1.2 \times 10^{-04}$ | 0.00042 | 0.00030 |
| 605 | $4.8 \times 10^{-01}$ | $1.9 \times 10^{-04}$ | 0.00022 | 0.00014 |
| 608 (Top of embankment) | $5.5 \times 10^{-01}$ | $5.0 \times 10^{-04}$ | 0.00014 | 0.00006 |
| 611 (Top of parapet wall) | $9.0 \times 10^{-01}$ | $1.5 \times 10^{-03}$ | 0.00010 | 0.00007 |

Table 11.3 Pool elevation vs. potential loss of life

| Pool elevation in feet | Potential loss of life estimate |
| --- | --- |
| 494 | 23 |
| 498 | 27 |
| 500 | 35 |
| 505 | 62 |
| 525 | 100 |
| 550 | 300 |
| 575 | 339 |
| 611 | 370 |

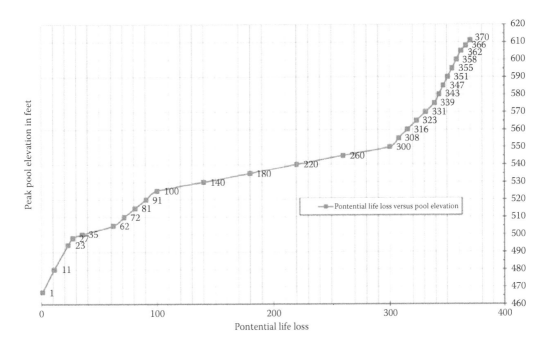

*Figure 11.6* Potential life loss versus pool elevation at failure for earth dam.

Based on these estimates, we used linear interpolation to obtain the values for all the pool elevations shown in Figure 11.6. Experience at other dams suggest that a more detailed simulation with input on evacuation effectiveness from the Emergency Management Agencies could lead to reduced potential loss of life estimates.

### 11.2.2.4 Comparison of results with risk evaluation guidelines

Figure 11.7 summarizes the U.S. Bureau of Reclamation Public Protection Guidelines for dams in a plot that shows ALL in lives/year on the sloping lines and total APF on the vertical axis. Figure 11.7 is generally known as an f–N plot and portrays "a discrete, noncumulative probability distribution in which each probability-consequence (f, N) pair is plotted" (Bowles, 2007). As mentioned earlier, in this particular case, the APF refers to failure from internal erosion/soil transport. We obtained the ALL and APF estimates shown in Table 11.4 by calculating the sums shown in Equations 11.3 and 11.4 over each of the peak pool elevation intervals shown in Table 11.2:

$$ALL = \Sigma_i \, f_i^* N_i \qquad\qquad (11.3)$$

$$APF = \Sigma_i \, f_i \qquad\qquad (11.4)$$

where
  $f_i$ is the estimated annual probability of failure for the ith peak pool elevation interval = $P_i \times p_i$
  $P_i$ is the estimated (incremental) probability of occurrence of the ith peak pool elevation during a year for the center of the elevation interval

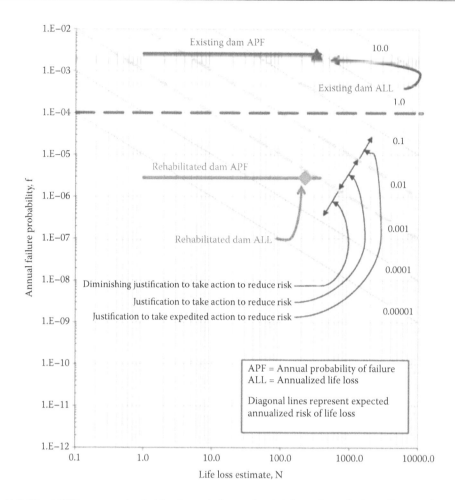

*Figure 11.7* The USBR portrayal of risks for existing and rehabilitated earth dam (From USBR. 2003. *Guidelines for Achieving Public Protection in Dam Safety Decisionmaking.* U.S. Bureau of Reclamation. Dam Safety Office, Department of the Interior, Denver, Colorado, 19 p.)

$p_i$ is the estimated conditional failure probability for the ith peak pool elevation interval
$N_i$ is the estimated life loss for failure in the ith peak pool elevation interval

The ALL estimate for existing conditions points to "justification to take expedited action to reduce risk" based on exceeding the U.S. Bureau of Reclamation Public Protection expected* annualized risk of life loss guidelines (USBR, 2003) of 0.01 $(1 \times 10^{-2})$

*Table 11.4* Annualized life loss and annual exceedance probability estimates

| Condition | ALL (lives/year) | APF (/year) |
|---|---|---|
| Existing | $8.3 \times 10^{-1}$ | $2.6 \times 10^{-3}$ |
| After remedial measures | $6.4 \times 10^{-4}$ | $2.8 \times 10^{-6}$ |

---

* USBR (2003) defines *expected value* as *estimated mean value.*

lives/year. This indicates justification for consideration of potential interim risk reduction measures. The computed ALL and APF estimates for conditions after the proposed remedial measures correspond to a risk level with "diminishing justification to take action to reduce risk" based on Reclamation's Public Protection Guidelines since ALL is estimated to be less than 0.01 ($1 \times 10^{-2}$) lives/year and APF is estimated to be less than $1 \times 10^{-4}$ per year.

In summary, these computed ALL and APF estimates indicate that the remedial measures proposed for the earth dam are justified based on Reclamation's Public Protection Guidelines and that remedial measures should be considered.

Rather than using a single "average" ALL estimate, Figure 11.8 presents the entire estimated probability distribution of life loss for earth dam for failure from internal erosion/soil transport, before and after the proposed rehabilitation. It follows the F–N chart format used by ANCOLD (2003), in which $F_{(i)}$ is plotted against $N_{(i)}$, where $F_{(i)} = \Sigma_i \, f_{(i)}$, and $f_{(i)}$ are ranked in descending order of the magnitude of life loss, $N_{(i)}$ for all n peak pool elevation intervals and (i) denotes the rank order [i.e., (i) = (1), (n)]. The resulting plot portrays a cumulative probability distribution in which "the probability–consequence (f, N) pairs are ordered in descending order of magnitude of N, and f is cumulated from largest to smallest to calculate the annual exceedance frequency. The mean of N is the area under the cumulative FN curve" (Bowles, 2007).

Figure 11.8 also shows the ANCOLD (2003) Societal Tolerable Risk Guidelines. Figure 11.8 best summarizes the impact of the proposed remedial measures of earth dam on the potential loss of life from a dam failure. While the life-loss risk before the rehabilitation falls above the limit of tolerability for existing dams, after the proposed remedial measures the life-loss risk decreases to values below the limit of tolerability for existing dams. Although not applicable to this earth dam, Figure 11.8 also shows the more stringent limit

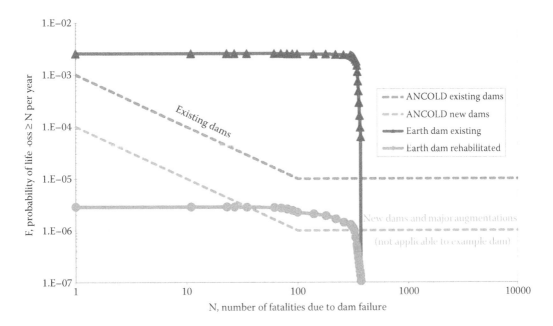

*Figure 11.8* F–N chart for existing and rehabilitated earth dam, including ANCOLD (2003) societal risk guidelines for dams.

of tolerability used for new dams or major augmentations.* The computed life-loss risk after the rehabilitation approaches this more severe limit.

ANCOLD also has an individual tolerable risk guideline of a probability of life loss of $1 \times 10^{-4}$ per year for the individual at most risk for existing dams (and $1 \times 10^{-5}$ for new dams and major augmentations). The existing earth dam does not meet this guideline, but is estimated to do so after the implementation of remedial measures.

The comparisons with the ANCOLD societal (F–N chart in Figure 11.8) and individual tolerable risk guidelines should take into account additional considerations of satisfying the ALARP (as low as reasonably practicable) principle, as explained by Bowles (2007). Nevertheless, both the ANCOLD societal and individual tolerable risk guidelines support proceeding with the proposed remedial measures at the earth dam presented in the example.

## 11.3 ROLE OF FRAGILITY CURVES TO EVALUATE THE UNCERTAINTY IN PROBABILITY ESTIMATES

### 11.3.1 Concept of uncertainty

When failures of complex structures are analyzed, evaluation of uncertainty should play an important role in the analysis of the behavior of a constructed facility. In general, two sources of uncertainty should be considered (Hartford and Baecher, 2004; Hoffman and Hammonds, 1994):

1. Natural uncertainty or randomness: Produced by the inherent variability in the natural processes. An example of this kind of uncertainty is the variability of the loads that the structure has to withstand, for instance, the variability in the earthquake intensity that can occur. Another example is the resistance's variability of the terrain where the structure is settled. This type of uncertainty, sometimes also called aleatoric uncertainty, cannot be reduced, though it can be estimated.
2. Epistemic uncertainty: Resulting from not having enough knowledge or information about the analyzed system. This lack of information can be produced by deficiency of data or because the structure's behavior is not correctly represented. The more knowledge is available about a structure, the more this type of uncertainty can be reduced. On the other hand, it is usually very difficult to estimate or quantify this uncertainty. An example of this type of uncertainty can also be found in the resistance of the terrain. The information about the foundations may be limited so the parameters used to characterize its resistance are estimated from probing and exploration. With more resources, the terrain can be better characterized and the epistemic uncertainty is reduced, although the natural variability of the terrain may still be very significant.

The distinction between natural and epistemic uncertainty takes added importance for a quantitative risk analysis in complex structures (Baraldi and Zio, 2008). In this context,

---

* From Bowles et al. (2003): According to the glossary in the draft ANCOLD (2001) guidelines, "major augmentations of existing dams … refers to modification of an existing dam involving a relatively large expenditure and creating a significant new benefit (typically, but not always, a major increase in volume of stored water), such that the economic case for marginal risk reduction would be approaching that for a new dam." McDonald (personal communication, November 17, 2002), chair of the Working Group that prepared the ANCOLD (2003) draft guidelines, states, "It is (a) subjective (guideline), since there are no clear boundaries that can be defined. Indeed, the distinction we have made can be seen to flow from the ALARP principle. If there is low marginal cost to build in additional safety, then do it if significant risk remains."

natural uncertainty is usually related to the occurrence of events that can produce the structural failure and the randomness of the structure's resistant behavior for the load produced by the events. In contrast, epistemic uncertainty is mainly focused on the lack of knowledge of the failure mechanisms, the structure's resistance parameters, and the consequences produced by the failure.

In any case, uncertainty characterization is an inherent process in any risk analysis, since it encompasses the failure probability concept. If there were no uncertainties, the load situation that produces the structural failure with 100% probability could be precisely determined.

## 11.3.2 Concept of fragility curves

Fragility curves represent a relationship between failure probability and the magnitude of the loads that produce the failure. Different empirical and analytical methodologies have been developed to obtain this kind of curves in complex structures (Altarejos-García et al., 2012; EPRI, 1994; Shinozuka et al., 2000). Fragility curves are useful in risk analysis of complex structures, since they facilitate the evaluation of the structural response for different load values.

In general, a single fragility curve is used to represent the natural uncertainty of the system response. For each loading state, this curve represents the probability of structural failure produced by the natural randomness in the structure and the terrain. In this chapter, the standard deviation of this single curve is represented with $\beta_R$.

In order to characterize the epistemic uncertainty, a probabilistic distribution is usually defined for the median of the previously defined fragility curve, which can also be called "average" fragility curve. The standard deviation of this probability distribution is represented with $\beta_U$. Variations in the fragility curve average produce a family of fragility curves that characterize both uncertainties, as shown in Figure 11.9. As can be observed, the median of the distribution for both uncertainties is normally assumed to be the same (EPRI, 1994).

In general, for a defined loading state, a complex structure can fail due to different failure modes, controlled by failure mechanisms. Commonly, fragility curves are defined for each failure mechanism separately and later combined within the framework of risk calculation by means of different techniques (i.e., common cause adjustment, SPANCOLD 2012).

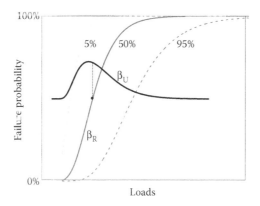

*Figure 11.9* An example of the family of fragility curves. (Adapted from EPRI. 1994. *Methodology for Developing Seismic Fragilities.* Electric Power Research Institute.)

### 11.3.3 Role of fragility curves in risk analysis

Event trees are a mathematical tool widely used to compute risk in dam engineering. They are sometimes referred to as the "risk model," as they are in fact a logical and analytical structure for risk calculation. An event tree is a graphical representation of a logical model that includes all the conceivable chains of events resulting from an initiating event that can produce the structural failure (SPANCOLD, 2012). Defining each of the probabilities along these chains of events, the overall failure probability and risk of the structure can be obtained. Figure 11.10 shows an example of an event tree.

Each node of the tree represents an event. The root node is called an initiating event. Branches that grow from an event represent the possible outcomes of their event of origin. Branches must represent mutually exclusive and collectively exhaustive events, so an event will always be reflected in a single fashion in only one branch. In this way, if a probability is assigned to each possible outcome (for every event), the addition of all probabilities of the outcomes arising from any node should be 1. Probabilities in event trees, except for the initiating event, are always conditional, that is, for any intermediate node it is assumed that all preceding events (parent nodes) have already happened.

When a structural failure is analyzed with an event tree, the initial node is usually used to introduce the full range of probability of the loads that could produce the failure of the structure. In the other nodes, the conditional probabilities of the different failure mechanisms in which the failure mode is decomposed are then introduced. These conditional probabilities are generally introduced using fragility curves, as explained in the previous section. A single value of failure probability along each branch of the tree can be obtained combining the conditional probabilities of all the failure mechanisms in the branch.

In general, when failure probability is computed using an event tree, the probabilities introduced in each node (in terms or not of fragility curves) capture the existing natural uncertainty. This uncertainty includes the probability of occurrence of different loading states (in general introduced in the first node) and the conditional probability of the different failure mechanisms for these loading states. In some event trees, epistemic and natural uncertainties are mixed (SPANCOLD, 2012), for instance, with a higher variability in the fragility curves introduced in the event tree. In general, it is advisable to analyze independently the two sources of uncertainty (Ferson and Ginzburg, 1996), to allow a better interpretation of the results.

Furthermore, event trees can be used to combine different failure modes with techniques such as common cause adjustment (SPANCOLD, 2012), which combines the conditional probabilities of different failure modes produced by the same loads.

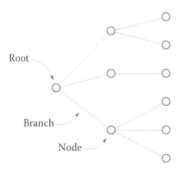

*Figure 11.10* An example of an event tree. (From SPANCOLD. 2012. Risk analysis applied to management of dam safety—Technical Guide 8 (Vol. 1), Spanish National Committee on Large Dams.)

## 11.4 MATHEMATICAL ROOTS AND NUMERICAL ESTIMATION OF FRAGILITY CURVES

### 11.4.1 Introduction

Uncertainties in geotechnical problems include parameter, system, and loading uncertainties. As an example, uplift pressure under a concrete dam, probably the most important factor influencing sliding stability (Ruggeri et al., 2004; Westberg, 2010), is a loading effect subjected to considerable uncertainty.

To calculate the probabilities of the outcomes from a single node in an event tree, two elements are needed:

- Mathematical or numerical model that simulates the physical problem
- Reliability method to be applied to the model

The selection of a mathematical or a more advanced numerical model will depend on the complexity of the problem analyzed and the quantity and quality of data available. For instance, a slope stability problem can be analyzed with a limit equilibrium model, such as Taylor, Morgersten–Price, Janbu, Spencer, or other. But it can also be analyzed using a numerical model implemented in a finite difference or finite element code.

Several reliability methods can be used. These methods include first-order second moment (FOSM) Taylor's method, point estimate method (PEM), advanced second moment (ASM) Hasofer–Lind method and Monte Carlo method.

These methods, except Monte Carlo, typically use linear approximations of the limit state function $g^*(x_1, x_2, \ldots, x_n)$. In addition, instead of considering the full probability density function of the random variables, only the mean value and standard deviation are taken into account. The result obtained using these techniques is the reliability index, $\beta$, defined as the number of standard deviations, $\sigma_{g^*}$, between the expected value of the state function, $E[g^*]$, and the value that represents the system failure, $(g^*)_{failure} = g^*(x_1, x_2, \ldots, x_n) = 0$ (Equation 11.5). This value gives a relative measure of the reliability of the system. Higher values of $\beta$ will mean higher margins of safety for the structure. On the other hand, the method does not directly provide a value of the probability of failure.

$$\beta = \frac{E[g^*] - (g^*)_{failure}}{\sigma_{g^*}} = \frac{E[g^*] - 0}{\sigma_{g^*}} = \frac{E[g^*]}{\sigma_{g^*}} \tag{11.5}$$

As $x_1, x_2, \ldots, x_n$ are random variables, the function $g^*(x_1, x_2, \ldots, x_n)$ will be a random variable as well, with a certain probability distribution, unknown in most cases. To derive a failure probability value from $\beta$, it is necessary to make an additional hypothesis on the type of probability distribution of $g^*(x_1, x_2, \ldots, x_n)$. Using a probability distribution function that is fully defined based on a mean and standard deviation of $g^*(x_1, x_2, \ldots, x_n)$, for example, normal or log-normal, the probability of failure can be derived.

The selection of the more adequate reliability method or combination of methods for a geotechnical problem depends on factors such as the quantity and quality of data available, probability distributions of the random variables, and order of magnitude of the searched probability. For linear or quasi-linear problems, with normal or quasi-normal random variables, FOSM, PEM, and ASM methods provide a good choice for the engineer. For nonlinear problems with non-Gaussian random variables and expected low probabilities of failure, the Monte Carlo method will render more accurate results.

The trade-off between accuracy and time and cost of the analysis has to be carefully considered. For instance, the use of a finite element model with the Monte Carlo method will only be justified when all the possible alternatives have proven to be insufficient to reach the objectives of the analysis.

As long as the calculated probability corresponds to a situation where outcomes of previous nodes have already happened, the calculated probability is a conditional probability. If the conditional probability of failure is calculated mapping all feasible load values, then a fragility curve is obtained.

It is important to note that the event tree approach allows for a lot of flexibility when considering the associated models to each node. Given a complex problem, the engineer can

(a)  Decompose the process to a relatively high number of nodes, or
(b)  Use an event tree with just a few nodes

In case (a), each node will represent a simpler process, which would be easier to simulate by a mathematical or numerical model. In case (b), each node will represent a group of interdependent processes, and the building of the associated simulation model will require more effort.

The engineer has to evaluate the type of model that best suits his requirements, for example, the choice between a 2D or a 3D model. In geotechnical problems where interfaces are present, another decision is to choose between continuous models with previously defined interfaces or resort to fracture mechanics models. If water is present, the interconnection with hydromechanic phenomena has to be assessed. The flow regime is another issue that has to be considered: will the flow be considered along the discontinuities or in the whole domain? If numerical models are used, a decision has to be made on which one is the more adequate constitutive model regarding the nature of the analyzed problem and the requirements of the study: elastic model, plasticity models, pore-elastic models, linear elastic fracture models, or nonlinear fracture models.

In addition, the engineer has to assess which variables will be considered as deterministic variables, subjected to none or very low uncertainty, and which variables may have values not known with precision and therefore are subjected to uncertainty (either natural or epistemic). Once this separation between deterministic and random variables is done, random variables should be characterized in a probabilistic manner. An estimation of mean values and standard deviations and a hypothesis on their probability distribution has to be made based on available data. Typical probability distributions are the uniform, normal, lognormal, triangular, and beta distributions. An important decision regarding the selection of probability distributions is related to the upper and lower truncation of unbounded distributions that theoretically extend in the $[-\infty, +\infty]$ interval. In particular, imposing a lower limit on resistance variables has a strong influence on the probability of failure.

In this context, FOSM techniques are suitable in the first stages of the analysis to identify the key variables, which contribute to the variance of the state function.

Monte Carlo analysis is also a broadly used method to compute failure probability. This analysis is made generating n groups of values for the random variables, according to their probabilistic distributions. The structural behavior is simulated for each group of values, obtaining a number of situations $n_f$ where the structure fails. Failure probability is computed dividing the number of failures ($n_f$) by the number of groups generated (n). The higher the number of groups of values, the more accurate the failure probability results. In order to characterize properly the structures with a low failure probability, a high number of simulations are required.

The two main types of uncertainty already introduced in the chapter can be addressed using a two-loop Monte Carlo analysis (Baraldi et al., 2008).

It has to be taken into account that Monte Carlo techniques require a large number of simulations and can make its practical application unfeasible, in particular, if combined with complex numerical finite-element models. The concept of limit state surface can be used to avoid this problem. The limit state or response surface, which in the general case is a hyper-surface*, sets the boundary between the safe and failure domains. This limit state surface can be approximated using the numerical model to calculate a few points on it. Then, a statistical adjustment can be performed to obtain an estimation of the surface position and shape from the calculated points. Once the limit state surface is approximated, it is no longer necessary to use the numerical model to calculate the probability of failure, since it is enough to generate samples from the probability distributions of the random variables in a Monte Carlo fashion, verifying how many of them lay in the failure domain.

## 11.4.2 Conditional probability of failure versus FS

The end nodes of an event tree typically present two possible outcomes: failure/no failure. When the system reaches the final node condition, all previous outcomes from all the nodes along the analyzed branch have already happened. At this point, it is interesting to evaluate the obtained probability of failure against the FS.

In normal practice, factors of safety are calculated and compared with reference values to assess the safety condition of a structure, thus considering the structure as "safe" if the calculated FS is higher than the reference value or "unsafe" otherwise. As it has been pointed out elsewhere (Hoek, 2007; Smith, 2003) if uncertainties are incorporated in the analysis, the FS considered as a mathematical construct, becomes another random variable, so the probability of failure is the probability $P(FS < 1)$. According to this definition, it may be expected that a higher FS will imply a lower probability of failure, but it has been shown (Silva et al., 2008; Smith, 2002) that higher factors of safety do not always correspond with lower probabilities of failure due to the uncertainties involved in the analysis.

Ching (2009) has demonstrated with his theorem of equivalence that under some circumstances, equivalence between FS and probability of failure can exist. If Z equals the uncertain variables, $\theta$ the design variables, and D is the allowable design region in the $\theta$ space, the limit state function is denoted as $g[Z,\theta]$. The nominal limit state function $g_n(\theta)$ is a positive function of $\theta$. As an example, $g_n(\theta)$ can be defined as $g[Z,\theta]$ with Z fixed at certain chosen nominal values, that is, their mean values. The system is in failure when $g[Z,\theta] < 1$. According to this, the FS is defined in Equation 11.6.

$$\frac{g_n(\theta)}{FS} \geq 1 \tag{11.6}$$

The probability of failure can be defined as in Equation 11.7.

$$P(g[Z,\theta] < 1 | \theta) = \int p(Z|\theta) \cdot I(g[Z,\theta] < 1) dZ \leq P_f^* \tag{11.7}$$

where $P_f^*$ is the desirable probability of failure and $I(g[Z,\theta] < 1)$ is a function that equals 1 when $g[Z,\theta] < 1$ and 0 otherwise. Let the normalized limit state function be defined as the limit state function divided by a nominal limit state function. If the probability distribution of the normalized limit state function is invariant over the design region, then pairs of

---

* An n-dimensional surface in a space of dimension n + 1, which represents the set of solutions to a single equation.

[FS, $P_f^*$] exist such that Equations 11.6 and 11.7 are equivalent. The functional relationship between the pair [FS, $P_f^*$] has the form expressed in Equation 11.8.

$$P\left( g[Z,\theta] - \frac{g_n(\theta)}{FS} < 0 \right) = P_f^* \tag{11.8}$$

According to its definition, the normalized limit state function, denoted by $G(Z,\theta)$, can be expressed by Equation 11.9.

$$G(Z,\theta) = \frac{g[Z,\theta]}{g_n(\theta)} \tag{11.9}$$

And Equation 11.7 can be rewritten as Equation 11.10.

$$P\left( g[Z,\theta] - \frac{g_n(\theta)}{FS} < 0 \right) = P\left( G[Z,\theta] < \frac{1}{FS} \right) = P_f^* \tag{11.10}$$

This means that 1/FS is the $(1 - P_f^*)$ percentile of the random variable G. The relationship between FS and $P_f^*$ can be determined using Monte Carlo simulation, drawing N samples from Z and $\theta$, and calculating the N values for $G^{(i)} = G(Z^{(i)},\theta^{(i)})$. For a chosen FS value, the corresponding $P_f^*$ value can be estimated with Equation 11.11.

$$P_f^* \approx \frac{1}{N} \sum I\left( G^{(i)} < \frac{1}{FS} \right) \equiv \hat{P}_f^* \tag{11.11}$$

Further details and discussion on this method can be found in the referenced paper (Ching, 2009). The premise that the distribution of $G(Z,\theta)$ is invariant over the entire allowable design region D can be achieved only if the correct nominal limit state function $g_n(\theta)$ is found, which is usually a difficult undertaking, although finding a nominal limit state function for which the premise holds approximately is relatively simple, since $g_n(\theta) = g(E(Z),\theta)$ is usually an acceptable choice.

It is important to note that for certain geotechnical problems, the FS can be defined in different ways. For instance, in shear problems (slope stability, sliding of a dam along its foundation contact), the FS can be defined as the ratio between strength and stress, or resistance, R, and loading, L. This is the case if limit equilibrium models are used (Equation 11.12).

$$FS = \frac{R}{L} \tag{11.12}$$

On the other hand, it is common that if finite-element models are used, the FS is defined as the ratio between a value of a strength parameter, $\varphi$, and the ultimate value before failure, $\varphi_{failure}$ (Equation 11.13).

$$FS = \frac{\varphi}{\varphi_{failure}} \tag{11.13}$$

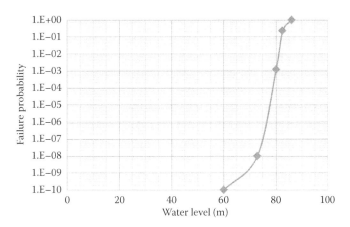

*Figure 11.11* Fragility curve for the sliding failure mode of a concrete gravity dam. (From Altarejos-García, L. et al., 2012. *Structural Safety*, 36–37, 1–13.)

This is another reason why the failure probability approach is sometimes preferred, as it provides a unique estimation of the structure safety. In both cases, it should be explicated how the values of R or φ have been derived. It is important to specify if they are best estimate values, prudent estimates, minimum values, or other. The numerical value of the FS changes dramatically depending on how the resistance and strength parameters are chosen. If failure probabilities are to be calculated, there is no doubt on what values are used, as the engineer is somehow forced to assess the whole probability distribution function of the variables if they are random, or to declare a unique value if the variable is considered deterministic, thus eliminating much confusion in the analysis.

In any case, for a specific problem, and for a certain failure mode (i.e., slope stability), it is possible to derive a relationship between FS and conditional failure probability once the fragility curve has been obtained by simply mapping the factors of safety along the loading interval. The curve thus obtained represents the equivalence between FS and probability of failure for the given problem.

Figure 11.11 shows the fragility curve derived by the authors for the problem of sliding of a concrete gravity dam along a dam foundation contact. The curve was derived with the following assumptions. The calculation model for sliding was a 2D numerical model implemented in the finite difference code FLAC (ITASCA, 2014). A cracked base analysis was performed, allowing a crack to open and progress along the dam foundation contact, with uplift updating as the crack propagates. Elastic constitutive models were used for both dam and foundation. Random variables were friction angle and cohesion at the dam–foundation contact. The reliability technique used was Monte Carlo analysis.

The FS was evaluated by a coupled progressive reduction of strength parameters, friction angle, and cohesion, from their estimated mean values to the values where sliding starts, using Equation 11.13 for the calculation for each loading level. The FS versus probability of failure curve obtained is shown in Figure 11.12.

## 11.4.3 Building fragility curves

Having already introduced all the substantial issues that concern fragility analysis in the previous points, a general flowchart to build fragility curves in the context of risk analysis is presented (Figure 11.13).

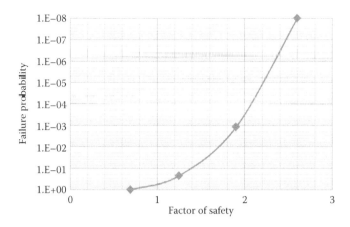

*Figure 11.12* Factor of safety–failure probability for a sliding failure mode of a concrete gravity dam. (From Altarejos-García, L. et al., 2012. *Structural Safety*, 36–37, 1–13.)

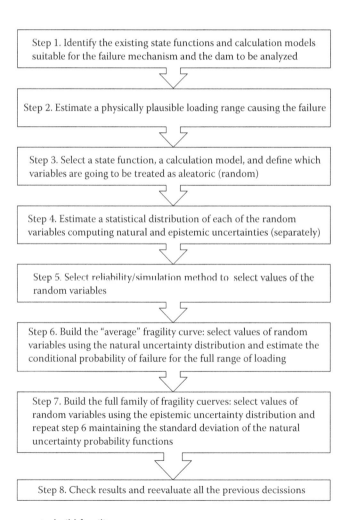

*Figure 11.13* The process to build fragility curves.

Getting more into the details of the previous chart (Figure 11.13), and the links with the concepts previously introduced in this chapter, special attention should be devoted to the following issues:

- *Step 1.* Define the mathematical or numerical model that simulates the physical problem, as defined in Section 11.4.1. The selection of a mathematical or a more advanced numerical model will depend on the complexity of the problem analyzed and the quantity and quality of data available.
- *Step 2.* Define the loading range that will make sense to estimate the fragility curve. This range will delimit the x-axis of the fragility curve. Furthermore, the number of loading cases analyzed in this range should be prechosen. The more the number of cases, the more accurate the results will be, but more time will be needed for computations.
- *Step 3.* As explained in Section 11.4.1, the engineer has to assess which variables will be considered as subjected to none or very low uncertainty, and which variables have necessarily to be treated as random.
- *Step 4.* At least two different distributions should be defined for each variable: one for natural uncertainty and one for epistemic uncertainty. Mean values, standard deviations, and probability distribution should be estimated, based on available data. Typical probability distributions are the uniform, normal, log-normal, triangular, and beta distributions. These distributions are used partly because they fit the methods in use and partly because we know how to solve the mathematics if we use them.
- *Step 5.* Select the reliability method that will be used in the model to estimate failure probability as explained in Section 11.4.1. Some examples of these methods are FOSM Taylor's method, PEM, ASM Hasofer–Lind method, and Monte Carlo method. Monte Carlo method will produce more accurate results, although computations will require more time.
- *Step 6.* For each loading case, compute the failure probability using the selected reliability method and the probability distribution defined for the natural uncertainty. The number of values of the random variables used to compute the failure probability will depend on the reliability method chosen. For each group of variables sampled, a full computation of the calculation/numerical model is required. When the failure probability is represented versus the loading range, the fragility curve capturing natural uncertainty is obtained.
- *Step 7.* First, groups of random variables are selected in the epistemic uncertainty distributions following the chosen reliability method. For each group of random variables sampled, a fragility curve is estimated using the procedure explained in the previous step but using these sampled values of random variables as new "averages" for the natural uncertainty distribution (which is assumed not to change). This family of fragility curves will then capture both the epistemic and natural uncertainty of the structure.
- *Step 8.* Check the outcomes and perform sensitivity analysis on any of the decisions previously taken. This last one is a crucial step as the engineer should never get lost in any mathematical approaches that may end up not representing sound engineering judgment.

## 11.4.4 Example of fragility analysis for stability failure mode of an earth dam

The dam to be analyzed is a homogeneous earth fill dam. Its upstream slope is 23.5° and its downstream slope is 28°, being the total height 16 m (Figure 11.14).

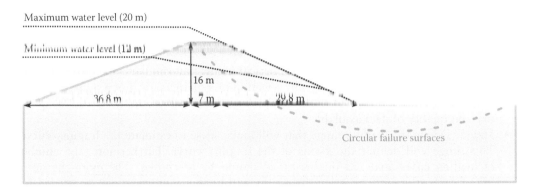

Maximum water level (20 m)

Minimum water level (12 m)

16 m

36.8 m      7 m      29.8 m

Circular failure surfaces

*Figure 11.14* Dam geometry.

The process shown in Section 11.4.3 has been followed as described below for every of the eight steps:

- *Step 1*. The calculation method selected for the stability of the embankment is the simplified Bishop method, thus making use of a Mohr–Coulomb type of failure criteria. This 2D method has been introduced in a spreadsheet to facilitate fast computations. In each computation, stability is checked for 108 different failure circles, considering that the embankment fails when the FS is lower than 1 at least for one of these circles.
- *Step 2*. Water pressure is the driving force of failure though its impact has been considered in a very simplified manner. In particular, the top flow line inside the embankment varies linearly from the water pool level in the upstream face to a fixed point in the downstream face located at 3.3 m above the downstream toe. The selected range of pool levels comprises from 12 to 20 m over the embankment base. In total, 17 water pool levels were analyzed (equally distributed every 0.5 m within the range). Beyond such levels, an existing parapet will likely fail and lead to an overtopping failure mode.
- *Step 3*. Two random variables have been considered: friction angle and cohesion.
- *Step 4*. To evaluate the natural uncertainty, the friction angle follows a truncated normal distribution with average 25°, standard deviation 2°, maximum 30°, and minimum 20°. The cohesion follows a log-normal distribution with average 10 kPa and standard deviation 3 kPa. For the epistemic uncertainty analysis, the friction angle follows a truncated normal distribution with average 25°, standard deviation 3°, maximum 30°, and minimum 20° while the cohesion follows a log-normal distribution with average 10 kPa, and standard deviation 5 kPa.
- *Step 5*. The Monte Carlo method has been chosen to sample variables in both the natural and epistemic uncertainty distributions. This method has been chosen because its results are more accurate and the state function and calculation model selected in Step 1 are simple enough to allow many stability computations to be rapidly made.
- *Step 6*. 100 pairs of values are sampled within the natural uncertainty distributions for friction angle and cohesion following the Monte Carlo method. For each water level, stability is checked for each pair of values (corresponding to the two random variables, friction angle and cohesion). Failure probability of each water level is obtained dividing the number of failure cases by the total number of cases analyzed (100). The fragility curve obtained with this process is shown in Figure 11.15 (it is important to mention that 100% probability has not been reached since for higher pool levels, overtopping is much more likely to control the failure and it does not make physical sense to increase the loading range of the stability failure mode).

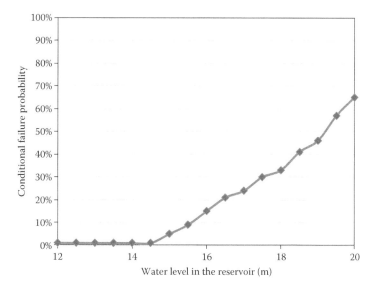

*Figure 11.15* "Average" fragility curve capturing natural variability.

- *Step* 7. 100 different pairs of values for friction angle and cohesion have been sampled from the epistemic uncertainty distributions using Monte Carlo. Each pair of values obtained has been used as new "averages" to sample 100 new pair of values from the natural uncertainty distribution, which is supposed to remain constant. In total, 10,000 pairs of values (100 × 100) of the friction angle and cohesion are thus sampled. These values have been used to compute 100 different fragility curves as explained in the previous step. Results are shown in Figure 11.16.
- *Step* 8. As it can be observed, the fragility curves obtained in this first attempt show probability ranges from 0 to 100% for almost the full range of water pool levels, which can be due to the estimated epistemic uncertainty.

When such a dispersion of results is obtained, it is worth analyzing the impact of each random variable separately, so that a thorough review of the implications of the selected

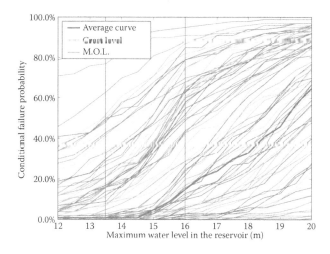

*Figure 11.16* Fragility curves obtained with friction and cohesion as random variables.

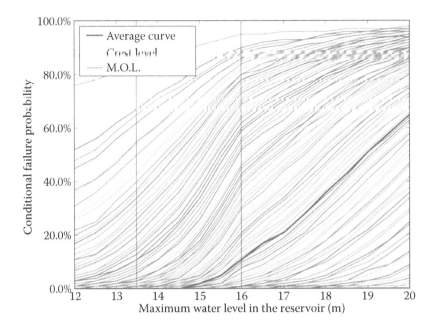

*Figure 11.17* Fragility curves obtained with only cohesion as a random variable.

probability distributions can be done. As shown in Figures 11.17 and 11.18, the full process has been repeated considering first cohesion as the only random variable (with friction angle fixed to its average value), and then friction as the only random variable (with cohesion fixed to its average value), and keeping all the other conditions unchanged.

Considering only friction or cohesion, separately, as random variables, it can be observed that the inherent uncertainty in cohesion has more influence than the friction in the dispersion of the results.

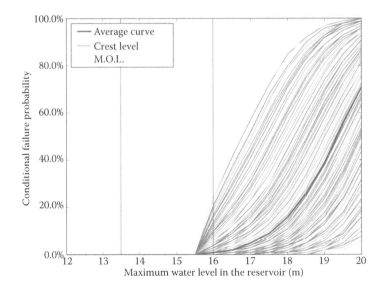

*Figure 11.18* Fragility curves obtained with only friction as a random variable.

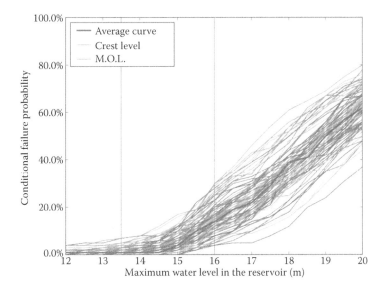

*Figure 11.19* The impact of reducing epistemic uncertainty on fragility analysis.

Moreover, accepting the fact that typically natural uncertainty cannot be reduced but epistemic can be narrowed by acquiring better knowledge (i.e., by means of improved geotechnical surveillance), the new set of fragility curves shown in Figure 11.19 has been obtained to assess the impact of reducing epistemic uncertainty to 1° for the friction angle standard deviation and 0.5 kPa for the cohesion standard deviation.

As shown in Figure 11.19, although the "average" fragility curve remains the same, the range of variation of the complete family of fragility curves has been significantly narrowed. The importance of this relies on the fact that, in principle, epistemic uncertainty can be reduced by investigations (boreholes, instrumentation records, performance evaluation, etc.).

Among the many other remaining issues, given that all the decisions taken in Step 1 through Step 7 could be critically analyzed by engineering judgment and sensitivity analysis, the optimal number of simulations to use when performing a Monte Carlo analysis is discussed. Figure 11.20 represents the comparison between "average" fragility curves obtained by 100 samples versus 10,000 samples.

Despite the similarity between the two curves in Figure 11.20, their differences can have important implications on the annual probability estimation. Significantly, with a low number of simulations (e.g., 100 or even 10,000), a single pair of sampled values resulting in failure, particularly for lower pool levels, can have a dramatic impact on the overall risk estimation, as discussed below.

## 11.5 FROM FRAGILITY CURVES TO ANNUALIZED PROBABILITY OF FAILURE COMMONLY USED IN RISK ANALYSIS

As shown and discussed in Section 11.2.2 through an example of a risk analysis for an earth dam, if results are to be provided in terms of annualized risk (which is the current state of practice in dam engineering), fragility curves should be combined with the probability of the loads that may produce the failure (Figure 11.21), which are usually characterized by their AEP (in the examples of this chapter, such load/driver of the failure is the water pool level).

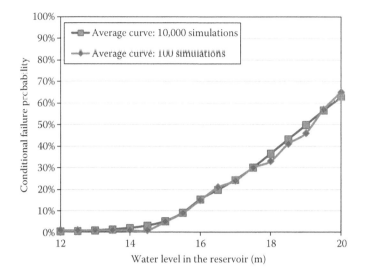

Figure 11.20 The impact of number of simulations when using Monte Carlo on fragility analysis.

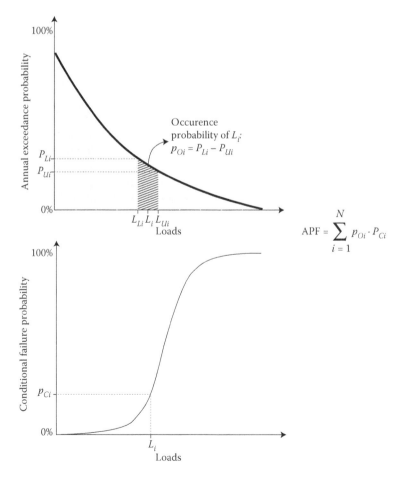

Figure 11.21 The estimation of annualized probability of failure (APF) using fragility curves.

As shown in Figure 11.21, exceedance probability curves are usually discretized in several intervals in order to obtain a representative value (usually the average) and an occurrence probability value for each interval subsequently used to compute the total failure probability. The probability of occurrence of each load interval is estimated using the probability mass function, considering pool levels as a discrete variable, as shown in Equation 11.14.

$$p_{Oi} = P_{Li} - P_{Ui} \tag{11.14}$$

where $p_{Oi}$ is the occurrence probability on interval i, $P_{Li}$ is the AEP of the lower bound of the interval i, and $P_{Ui}$ is the AEP of the upper bound of the interval i.

The failure probability for each interval is obtained multiplying the occurrence probability of each interval ($p_{Oi}$) by the conditional failure probability of the interval's representative value ($p_{Ci}$). This conditional failure probability value is obtained directly from the fragility curve. Adding the failure probability of all the intervals examined yields the total failure probability of the structure. The higher the number of intervals, the more accurate the results will be.

To obtain the annual failure probabilities, the average fragility curve from the previous section is combined with the exceedance probability of pool levels for the dam in the example. This combination has been made with a simple risk model elaborated using the iPresas Calc software (www.ipresas.com).

The AEP of maximum pool levels used for the reservoir in the example is shown in Figure 11.22. After dividing the whole plausible/meaningful range of water pool level into 20 intervals, each interval creates a branch of the event tree for this risk model. The occurrence probability of each interval is estimated as shown in Equation 11.14. Figure 11.23 shows the resulting occurrence probability results of each pool level interval.

The occurrence probability of each interval (Equation 11.14) is multiplied by the conditional probability of failure corresponding to the average water level of the interval. This conditional probability is obtained using the average fragility curve obtained in the previous section (Figure 11.15).

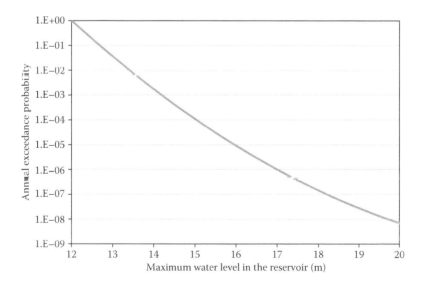

*Figure 11.22* Exceedance probability of maximum pool levels in the reservoir of the example.

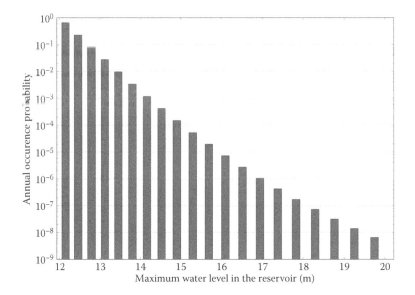

*Figure 11.23* Occurrence probability of the 20 intervals obtained with software iPresas Calc.

Figure 11.24 shows resulting annual failure probabilities from each water level interval.

Adding the annual failure probability of all the intervals yields the annual failure probability for the dam. In this example, the annual failure probability for the dam is 0.01001.

The annual failure probability obtained above is influenced by the relatively low number of simulations. When the calculations are repeated for an average fragility curve based on 10,000 simulations, the estimated APF decreases to 0.00509. This reduction suggests the need of performing at least one more set of calculations, that is, with 100,000 samples, to check if the calculated value of the annualized probability of failure fully stabilizes.

The current state of the practice in dam safety engineering consists of obtaining results in terms of APF using the average fragility curve, as done in the example above, which captures

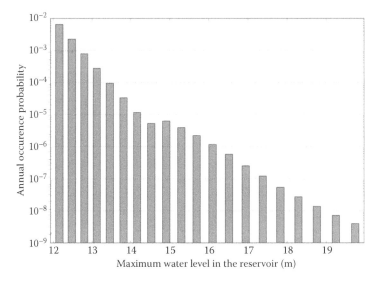

*Figure 11.24* Annual failure probability of the 20 intervals obtained with software iPresas Calc.

only the natural uncertainty. Estimating the value of the epistemic uncertainty distribution for the APF requires repeating the process for the full range of fragility curves obtained in the previous section.

## 11.6 SUMMARY OF MAIN POINTS

1. Much of the benefit from conducting a risk assessment stems from engaging the engineers to thoroughly evaluate the fundamentals controlling the performance of a constructed facility.
2. Carrying out correct engineering analyses helps engineers identify the fundamentals involved and the key parameters affecting performance.
3. Understanding the applicable fundamentals and key parameters facilitates preparation of accurate event trees and probability estimation.
4. Empirical correlations of safety factor with annual failure probability provide a practical means of estimating probabilities for a risk assessment.
5. Evaluation of uncertainty should play an important role in the analysis of the behavior of a constructed facility.
6. Fragility curves represent a relationship between failure probability and the magnitude of the loads that produce the failure.
7. An event tree is a graphical representation of a logical model that includes all the conceivable chains of events resulting from an initiating event that can produce the structural failure.
8. Probabilities in event trees, except for the initiating event, are always conditional.
9. When performing Monte Carlo simulations, using a low number of simulations (e.g., 100 or even 10,000), a single pair of sampled values resulting in failure. Always increase the number of calculations until the calculated annual probability value stabilizes.
10. Small differences in the average fragility curve can have significant influence on the calculated level of risk.

## ACKNOWLEDGMENTS

The authors gratefully acknowledge the contribution of the many colleagues that contributed with the application of risk-based decision tools for the earth dam presented in Example 11.2.2, including Donald A. Bruce, David S. Bowles, T. William Lambe, W. Allen Marr, and Steve J. Poulos.

The work presented in this chapter has been supported by the Spanish Ministry of Science and Innovation (MICINN) through the grant of the iPRESARA project (BIA 2010-17852).

## LIST OF MAIN SYMBOLS AND ACRONYMS

| | |
|---|---|
| AEP | annual exceedance probability |
| ALARP | as low as reasonably practicable principle |
| ALL | annualized life loss |
| ANCOLD | Australian National Committee on Large Dams |
| APF | annual probability of failure |
| ASM | advanced second moment |
| F | probability of life loss |

| f | estimated annual probability of failure |
|---|---|
| FOSM | first-order second method |
| FS | factor of safety |
| g | nominal limit state function |
| P | estimated incremental probability of occurrence |
| PEM | Point estimate method |
| $P_f^*$ | desirable probability of failure |
| N | estimated life loss |
| SPANCOLD | Spanish National Committee on Large Dams |
| USBR | United States Bureau of Reclamation |

## REFERENCES

Altarejos-García, L., Escuder-Bueno, I., Serrano-Lombillo, A. and de Membrillera-Ortuño, M.G. 2012. Methodology for estimating the probability of failure by sliding in concrete, *Structural Safety*, 36–37, 1–13.

ANCOLD. 2003. *Guidelines on Risk Assessment*. Australian National Committee on Large Dams, Sydney, New South Wales, Australia.

Baecher, G.B. and Christian, J.T. 2003. *Reliability and Statistics in Geotechnical Engineering*. John Wiley and Sons, Ltd., England, 605 p.

Baraldi, P. and Zio, E. 2008. A combined Monte Carlo and possibilistic approach to uncertainty propagation in event tree analysis. *Risk Analysis*, 28(5), 1309–1325.

Bowles, D.S. 2007. Tolerable risk for dams: How safe is safe enough? *Proceedings of the 2007 US Society on Dams Annual Conference*, Philadelphia, Pennsylvania, March.

Bowles, D.S., Anderson, L.R., Glover, T.F., and Chauhan, S. 2003. Dam safety decision-making: Combining engineering assessments with risk information. *Proceedings of the US Society on Dams Annual Lecture*, Charleston, South Carolina, April.

Ching, J. 2009. Equivalence between reliability and factor of safety. *Probabilistic Engineering Mechanics*, 24, 159–171.

Christian, J.T., Ladd, C.C., and Baecher, G.B. 1994. Reliability applied to slope stability analysis, *Journal of the Geotechnical Engineering Division*, ASCE, 120(12), 2180–2207.

D'Andrea, R.A. and Sangrey, D.A. 1982. Safety factors for probabilistic slope design, *Journal of the Geotechnical Engineering Division*, ASCE, 108(GT9), 1101–1118.

EPRI. 1994. *Methodology for Developing Seismic Fragilities*. Electric Power Research Institute, California.

Ferson, S. and Ginzburg, L.R. 1996. Different methods are needed to propagate ignorance and variability. *Reliability Engineering and System Safety*, 54, 133–144.

Hartford, D.N.D. and Baecher, G.B. 2004. *Risk and Uncertainty in Dam Safety*. Thomas Telford Limited, Bristol, UK.

Hoek, E. 2007. Practical Rock Engineering. On-line course.

Hoffman, F.O. and Hammonds, J.S. 1994. Propagation of uncertainty in risk assessments: The need to distinguish between uncertainty due to lack of knowledge and uncertainty due to variability. *Risk Analysis*, 14(5), 707–712.

iPresas. 2013. iPresas Calc. User Guide. [Online] iPresas Risk Analysis. Available at: www.ipresas.com.

ITASCA. 2014. FLAC. [Online] Itasca Consulting Group, Inc., An Itasca International Company. Available at: http://www.itascacg.com/software/flac.

Kulhawy, F.H. and Phoon, K.K. 1996. Engineering judgment in the evolution from deterministic to reliability-based foundation design, *Uncertainty in the Geologic Environment: From Theory to Practice, Proceedings of Uncertainty '96*, Madison, Wisconsin, July 31 to August 3, pp. 29–48.

Lambe, T.W. and Silva, F. 2003. Evaluating the stability of an earth structure. *Proceedings of the XII Pan-American Conference on Soil Mechanics and Foundation Engineering*, Cambridge, Massachusetts, U.S.A., June, pp. 2785–2790.

Morgenstern, N.R. 1995. Managing risk in geotechnical engineering. *Proc., X Panamerican Conference on Soil Mechanics and Foundation Engineering*, Vol. 4, Guadalajara, Mexico, pp. 102–126.

Office of Management and Budget. OMB. 2013. Supplement to Office of Management and Budget Circular A–11: Planning, Budgeting, and Acquisition of Capital Assets, V 3.0, July, http://www.whitehouse.gov/sites/default/files/omb/assets/a11_current_year/capital_programming_guide.pdf

Ruggeri, G., Pellegrini, R., Rubin de Célix, M., Bernsten, M., Royet, P., Bettzieche, V., Amberg, W., Gustaffson, A., Morison, T. and Zenz, G. 2004. Sliding safety of existing gravity dams. Final report. ICOLD European Club. Working group on sliding safety of existing gravity dams.

Schmertmann, J.H. 2000. The no-filter factor of safety against piping through sands. In *Judgment and Innovation, the Heritage and Future of the Geotechnical Engineering Profession*, Silva, F. and Kavazanjian, Jr. E. (eds), ASCE, Geotechnical Special Publication No. 111, Reston, Virginia, pp. 65–132.

Shinozuka, M., Feng, M.Q., Lee, J. and Naganuma, T. 2000. Statistical analysis of fragility curves. *Journal of Engineering Mechanics*, 126, Dec, 1224–1231.

Silva, F., Lambe, T.W. and Marr, W.A. 2008. Probability and risk of slope failure. *Journal of Geotechnical and Geoenvironmental Engineering*, December, 1691–1699.

Smith, M. 2002. Influence of uncertainty in the stability analysis of a dam foundation. In *Dam Maintenance and Rehabilitation: Proceedings of the International Congress on Conservation and Rehabilitation of Dams*, Madrid, 11–13 November 2002, CRC Press, pp. 151–158.

Southworth, M.J. 1980. An experimental evaluation of the Terzaghi criterion for protective filters, Thesis submitted to the Massachusetts Institute of Technology, Department of Civil Engineering, in fulfillment the fulfillment of the Master of Science degree, 122 p.

SPANCOLD. 2012. Risk analysis applied to management of dam safety—Technical Guide 8 (Vol. 1), Spanish National Committee on Large Dams.

Tavares, L.V. and Serafim, J.L. 1983. Probabilistic study on failure of large dams, *Journal of the Geotechnical Engineering Division*, ASCE, 109(11), 1483–1486.

USBR. 2003. *Guidelines for Achieving Public Protection in Dam Safety Decisionmaking*. U.S. Bureau of Reclamation. Dam Safety Office, Department of the Interior, Denver, Colorado, 19 p.

Westberg, M. 2010. Reliability-based assessment of concrete dam stability. Licenciate thesis. Division of Structural Engineering, Lund Institute of Technology, Lund University. Report TVBK-1033.

Chapter 12

# Evolution of geotechnical risk analysis in North American practice

*Gregory B. Baecher and John T. Christian*

## 12.1 INTRODUCTION

Probabilistic risk methodologies applied to geotechnical problems, in the sense that we now think of them, began to appear in the late 1960s with the work of Lumb (1966), Wu and Kraft (1967), Folayan et al. (1970), and others. Certain aspects of this direction of work can be traced back at least as far as Taylor (1948), who spoke of partial safety factors on the Coulomb strength parameters $(c, \phi)$ for slope stability, reflecting different levels of uncertainty on the two parameters (p. 414):

> It thus appears that there is no such thing as *the* factor of safety and that when a factor of safety is used its meaning should clearly defined [emphasis in in original]. [...] In order to present as general a case as possible it will first be assumed that different margins of safety are desired for the two components of shearing strength, [...].

Taylor goes on to note that the factor of safety may be defined with respect to soil properties or with respect to other factors such as the height of the embankment, and that they are not necessarily the same for a given slope. As a side observation, one often notes that the early workers in a field developed a much more subtle and nuanced understanding of a topic than the later literature evidences and that it often pays dividends to reread their original works.

Hansen (1956) spoke of this concept with respect to both load and resistance factors in the context of the limiting design of foundations (p. 385):

> Multiplying the prescribed loads with certain (partial) safety factors and limiting the corresponding maximum stress to the limit strength of the material divided by another (partial) safety factor.

Hanson also distinguished between ultimate states and service states:

> In the design of any structure two separate analyses should in principle be made: one for determining the safety against failure, and another for determining the deformations under actual working conditions.

Closely related developments in structural engineering had begun to appear in the 1950s, notably in the work of Alfred Freudenthal (1906–1977) at Columbia University (Freudenthal 1956). Emil Gumbel (1891–1966), also working at Columbia, had pioneered extreme value theory in the post-WWII period, publishing his seminal text in 1958 (Gumbel 1958). While Gumbel's work was of a generic nature, extreme value theory quickly found applications in civil engineering, particularly in hydrology and flood frequency analysis.

This brief chapter looks at some of the milestones in the development of probabilistic methods and risk analysis in geotechnical engineering, at some of the trends in practice that drove these developments, and at some of the formative concepts that derived from particular applications. Our apologies *ex anti* that we will surely overlook some of the formative concepts of geotechnical risk and reliability that are personal favorites to particular readers. Let it be a take-home exercise for the reader to add his or her own favorites to our list, and as a favor to communicate them to the authors.

The story starts in 1971 and charts a few application areas that stimulated what the authors consider to be among the most interesting and unique developments: mining, offshore, environmental remediation, and dam safety. There were others areas of application that led to important insights—for example, risk analysis of regional landslides, seismic hazard analysis, and risk in underground works—but these will have to wait another venue. The chapter ends by taking stock of current directions in the development of geotechnical risk and reliability analysis.

## 12.2 BEGINNINGS

With the beginning of the 1970s, serious work on geotechnical reliability began to appear. The first ICOSSAR conference had been held at Columbia in 1969, but contained little of geotechnical interest. This was followed by the first ICASP, held at the University of Hong Kong in 1971 and organized by Peter Lumb, which was the initial gathering of people working on geotechnical reliability (Lumb 1971). Slowly, journal papers began to appear. The data of Figure 12.1 show the number of papers in the ASCE Database having the keyword, *risk*, starting in 1950. Using this as a surrogate for publications on probabilistic

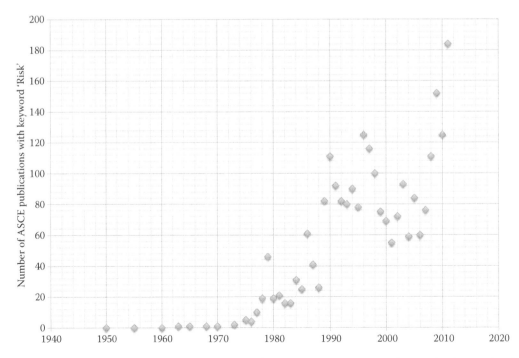

*Figure 12.1* Numbers of publications in the ASCE Civil Engineering Database having the keyword risk 1950–2011.

approaches, one sees an emerging number in the 1970s followed by a rapid increase in the 1980s and 1990s.

Like most things in research, interest in geotechnical risk and reliability has followed economic trends and national programs. Looking back in history, there is a background phase of interest in geotechnical reliability (as opposed to risk analysis) starting about 1971 and continuing until the mid-1990s. Against this background, various economic or strategic interests appeared in the 1980s and 1990s, which built on the existing fundamental work in geotechnical reliability. These included, for example, interest in the nuclear fuel cycle, offshore energy production, environmental remediation, and dam safety.

We start with the seminal period 1971–1996. This period saw the creation of a geotechnical reliability literature, and the emergence of many of the insights about risk and reliability in geotechnical practice, which now inform the literature. The second half of the period, starting about 1980, saw tailored development in a number of economic sectors: mining, offshore, environmental remediation, and dam safety, among others. Each of these sectors developed somewhat independent of the others as workers specialized in applications of their choosings, but each of these led to interesting and different insights. We overview a selection of these but unfortunately ignore others due to space limitations.

By the early twenty-first century, interest had begun shifting to risks associated with engineered systems. Perhaps, no development was more critical to this development than Hurricane Katrina and its aftermath for planning the reconstruction of New Orleans, Louisiana. Many of the long-time contributors to geotechnical reliability in North America and to some extent in Europe contributed to the response. A small library of reports and papers resulted from this effort.

While all this development has been going on, parallel activities have been afoot around the world. Asia has been a particular hotbed for the development of geotechnical risk and reliability in the current century. But this chapter has to have bounds; so, it focuses on North America. We leave the task to others to provide a global review.

## 12.3 GEOTECHNICAL RELIABILITY (1971–1996)

The period 1971 through 1996 reflects an era during which the focus of probabilistic methods in geotechnical engineering was principally on the uncertainty in soil properties and on reliability. Geotechnical reliability concerns itself with the probability that foundations, earth structures, underground works, and other geotechnical designs perform as required. The beginning date is that of the *First International Conference on Applications of Statistics and Probability to Soil and Structural Engineering* hosted by Professor Peter Lumb in 1971. The ending date is that of the ASCE conference in 1996 on *Uncertainty in the Geological Environment* at which modern risk analysis papers appear in increasing numbers. Toward the end of this period, the (US) National Research Council report on *Probabilistic Methods in Geotechnical Engineering* chaired by Wilson Tang appeared (NRC 1995).

### 12.3.1 Probabilistic veneer on deterministic models

This was a formative time. In the early years of this period, attention focused on how uncertainty in engineering models and parameters affected the probability of satisfactory performance, or its complement, "failure." In essence, much of this work mostly placed a probabilistic veneer on traditional deterministic modeling, propagating uncertainties in loads, soil-engineering parameters, and other factors through design equations, while being blissfully inattentive to measurement bias and noise, spatial variation, the separation of aleatory from epistemic uncertainties, and other subtleties that prominently figure in the later epochs.

The focus was on lumped-parameter models in which spatial variation and the taxonomy of sources of uncertainty was more or less ignored. As the period evolved, the texture of consideration in geotechnical reliability became even more complex. As the period progressed, the principal contributors to the literature transformed from researchers who had primarily matriculated as old-fashioned engineering mechanics types to people who had fundamental training in stochastic theory.

Typical work of the period looked at the probability of bearing capacity failures of shallow foundations or the probability of slope failures of embankments when soil properties or empirical factors were described by probability distributions. In the early years of this period, rock mechanics and mining applications of reliability were as common, maybe more common, as those to soil mechanics (Einstein et al. 1976; Priest and Hudson 1976). Among the important contributors in these early years—in addition to those referenced earlier— were (alphabetically) Harr, Schultze, Tang, Vanmarke, and Veneziano, among many others. The early textbook by Cornell (Benjamin and Cornell 1970) as well as later books by Tang (Ang and Tang 1975) and Harr (1987) also contributed to the rising interests in reliability analysis for geotechnical problems.

## 12.3.2 Variability of soil-engineering properties

In retrospect, this was a period in which many naïve assumptions were made in geotechnical modeling that now in hindsight we recognize as simplistic. The multiattributed sources of uncertainty in geotechnical data were widely ignored (Figure 12.2), as was the distributed nature of soil properties (i.e., soil properties are spatially variable). This sometimes led to wild overestimates of the probabilities of adverse performance, since measurement noise, spatial variation, model uncertainty, and other sources of uncertainty were undifferentiated from one another.

This period saw growing interest among senior names in traditional soil mechanics in probabilistic approaches and formal methods of uncertainty analysis. Casagrande's Terzaghi Lecture, while ignoring formal probability analysis, caught the attention of mainline practice in grappling with uncertainty (Casagrande 1965). Notable contributions to this strategic level of thinking about uncertainty followed from Whitman (1984, 1996), Duncan (2000), Kulhawy and Phoon (1996), and others.

Much of the work of this period began with lumped-parameter analysis of traditional geotechnical models. Soil-engineering parameters were modeled as uncertain but uniform within strata. A good deal of attention was invested in empirical data and the variation of soil properties that those data suggested. Many early applications made the (retrospectively) innocent assumption that the variability observed in empirical data was representative of the uncertainty that should be imputed to engineering parameters in the design. This sometimes led to significant overestimates of uncertainty and therefore of probability of failure

| Soil property uncertainty | | | |
|---|---|---|---|
| Natural variation (aleatory) | | Bias error (epistemic) | |
| Spatial variation | Temporal variation | Parameter error | Model bias |

Figure 12.2 Sources of error in soil-engineering properties. (Adapted from Baecher, G., and J. Christian. 2003. *Reliability and Statistics in Geotechnical Engineering*. 1st ed. Wiley.)

because random errors in soil testing and spatial variability of geological properties were assumed to be of the operative soil properties themselves.

### 12.3.3 Slope stability analysis

Slope stability analysis was the mainstay of this period of geotechnical reliability research, and to some extent remains so today. The slope stability problem is well formulated, relies on a limited number of uncertain inputs, and has a simple output, namely the factor of safety against instability. Among the early contributors to this work were Wu and Kraft (1970), Matsuo (1976), Alonzo (1976), Vanmarcke (1977), and Harr (1987).

Much of the early work on slope reliability simply replaced deterministic soil-engineering parameters with corresponding probability distributions and generated probability distributions of margin of safety or factor of safety against instability as output (Figure 12.3). The probability of failure of the slope was taken to be the probability of $MS \leq 0$ or $FS \leq 1$ (Figure 12.4). Cornell in his Hong Kong paper (Cornell 1971) had introduced the concept of the nonparametric *reliability index* ($\beta$), and this caught on across geotechnical practice:

$$\beta = \frac{E(FS) - 1.0}{SD(FS)} = \frac{E(MS) - 0.0}{SD(MS)} \tag{12.1}$$

in which $\beta$ is the reliability index, $E(.)$ is the expected value, and $SD(.)$ is the standard deviation. The reliability index measures the distance from the best estimate of a performance index to its limiting state (failure) in units of the standard deviation.

The first-order second-moment (FOSM) reliability index proved to be a highly useful index property even though it had limitations. The most serious limitation was that the FOSM reliability index is not invariant to changes in the definition of the index property and the limiting condition. For example, the reliability index expressed in $MS$ may and often does differ from that expressed in $FS$. The alternative Hasofer–Lind measure of reliability largely replaced the FOSM measure in structural engineering by about 1980 (Hasofer and Lind 1974), but Cornell's FOSM $\beta$ remains common in soil mechanics applications—with their less exact models and greater parameter uncertainty—even currently.

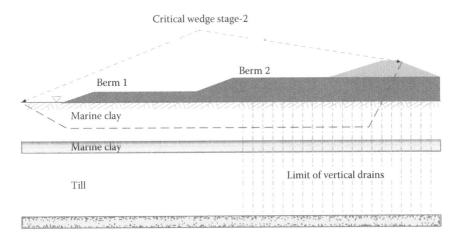

*Figure 12.3* A two-dimensional slope reliability analysis typical of ca. the 1980s practice. James Bay Dikes of HydroQuébec.

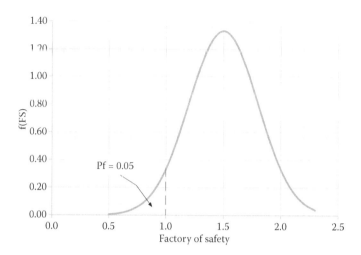

*Figure 12.4* Probability density function of factor of safety for slope reliability with mean *E[FS]* = 1.5 and standard deviation *S[FS]* = 0.3.

A stochastic finite element (SFE) or finite difference methods permit the analyst to model the spatial details of a physical process instead of relying on a few variables to represent the complexities of nature. They are particularly useful for dealing with variables that are spatially correlated. Fenton and Griffiths (2008) present a detailed treatment of SFEs applied to geotechnical problems.

It is relatively easy to assign properties randomly to each element in a mesh as long as the properties are not spatially correlated. When spatial correlation must be addressed, the analytical procedure must assign to each element values of the random properties that both satisfy the underlying probability distributions and retain the correlation structure. There are several methods for doing this, including the moving average (MA), discrete Fourier transform (DFT), covariance matrix decomposition, fast Fourier transform (FFT), turning bands (TBM), and local average subdivision (LAS). Each of these approaches has advantages and disadvantages. At this time, there is no consensus on the best method to use.

### 12.3.4 Lumped versus distributed parameter models

This lumped-parameter analysis emphasized the centrality of derived distributions in modeling geotechnical reliability. A derived distribution is that which results from propagating uncertainty in input parameters through a mathematical model to obtain corresponding uncertainty in output parameters. A variety of tools were developed to achieve these derived distributions. Cornell's first-order second-moment methods suited this purpose (Cornell 1971); Rosenblueth introduced point-estimate methods (Rosenblueth 1975). Today, most people use Monte Carlo simulation to achieve the same end, with a great deal more power and less thinking. Harr (1977) was an early proponent of Monte Carlo methods, but most researchers at the time looked upon stochastic simulation with suspicion. Today, the error is the reverse, if it is an error at all.

A significant advancement during this period was the realization that soil and rock mass properties are spatially variable and exhibit a spatial correlation structure (Lumb 1966). This has several important implications. The first is that the variability among the average properties of large soil volumes is considerably less than that among the properties of small volumes, such as boring samples or *in situ* tests. Matheron in the related context of mineral deposit statistics had made this observation earlier, and his work mainly influenced the emerging field of soil reliability (Matheron 1965). Using empirically observed variances among test specimens grossly exaggerates the variability of the soil properties within the mobilized zones that control field performance. The second implication is that spatial variability leads to size effects, seen most clearly in long, linear structures such as levees (Vanmarcke 1983). Size effects are important manifestations of spatial variation and, even today, they are often ignored.

### 12.3.5  Aleatory versus epistemic uncertainty

Another significant realization during this period was the duality of uncertainty between aleatory and epistemic types and the corresponding centrality of subjective (degree-of-belief) probability in geotechnical risk (Vick 2002). This realization led to the understanding that much or maybe most of the uncertainty geotechnical engineers face has to do with information and knowledge rather than randomness in time and space. This opened the door to Bayesian methods, which have come to dominate geotechnical reliability analysis. A frequent joke among workers in the field is that hydrologists are invariably frequentist statisticians, geotechnical engineers are invariably Bayesians, and structural engineers do not care as long as things align with building codes.

### 12.4  MINING ENGINEERING (1969–1980)

Mining engineering, especially surface mining, was an early venue for reliability and decision-theoretic methods. The nominal starting date for this period, 1969, was the year of the 11th U.S. National Symposium on Rock Mechanics in Berkeley, at which early probabilistic work in rock mechanics began to make its appearance. The closing date, 1980, brackets the period during which much of the similar work had appeared.

The issues in mining, especially surface mines, were unique in geotechnical reliability in that they were strongly driven by financial considerations of optimal slope angles and amounts of excavation, and were highly dependent on the pervasive fracturing of natural rock masses, that is, the so-called jointing systems. As the later decades of the century rolled out, these applications of geotechnical risk were quickly extended to underground works, but these are ignored in this chapter.

Joints are pervasive natural fracturing systems of rock masses induced by geological forces. They are semiparallel, usually finite-size separations, often organized into suborthogonal sets. They introduce planes of weakness and high transmissivity in a rock mass. As a result, they are important determinants of rock mass strength, permeability, and other physical properties.

An issue with rock mass jointing is that it is characterized based on statistical samples. The engineer or geologist goes into the field and observes the surface manifestation of jointing as a population of fracture traces upon the outcrop or excavation wall. These are lines of intersection of the joint sets with various semiplanar sampling windows. From measurements of these lines of intersection, the three-dimensional (3D) stochastic geometry of the joints needs to be inferred. This raises interesting questions of geometric probability.

Among the early contributors to this geometric sampling problem were Snow (1970), Barton (1975), Priest and Hudson (1976), Cruden (1977), and Baecher et al. (1977). The early work on this statistical problem addressed joint spacing or intensity as separate from orientation and trace length. Later, modeling attempted to integrate the 3D geometry of jointing (Dershowitz and Einstein 1988).

The interesting and original aspect of this body of work was that statistical thinking and geometrical probability led to nonintuitive inferences about geological and engineering properties of the subsurface. Sampling for soil properties had been, and remains, straightforward. Sampling for geometric properties raised issues of truncation, censoring, orientation bias, nonplanar support of probability distributions, and a host of other complications

The purpose behind so much interest in joint systems at the time was analyzing rock slope stability, especially in surface mines. For economic reasons, these slopes are designed with low factors of safety; yet, it is undesirable for them to fail prior to the completion of mining operations. Their stability depends both on the engineering properties of the intact rock and on the fabric of the rock mass including the joint systems. The combination of economic incentive, significant uncertainties about *in situ* conditions of the rock mass, and the need to balance costs against safety made this an obvious candidate for statistical decision theory. McMahon was among the early contributors to this direction of work (McMahon 1975), and the Ministry of Natural Resources, Canada, developed guidance along these lines (Coates 1977). Einstein et al. (2010) surveyed this early work.

## 12.5 OFFSHORE RELIABILITY (1974–1990)

Following the oil crisis of 1973–1974, oil exploration and production became an international priority. Offshore exploitation of oil had started in California in the late 1800s and in the Gulf of Mexico before World War II, but accelerated following the war. By the 1970s, offshore production was routine and widespread (National Commission on the BP Deepwater Horizon Oil Spill and Offshore Drilling, 2010). The industry had experienced a series of serious failures over the years. In 1969, a blow out at a Union Oil Company rig in the Santa Barbara Channel produced an 800 square miles oil slick on the coast, which was followed in the next 2 years by three other blowouts and a major fire on rigs in American waters. By 1980, these were followed by failures in the Persian Gulf, Niger Delta, North Sea, and Mexico. In 1988, the Piper Alpha tower in the North Sea exploded killing 167.

Geotechnical reliability developments in the offshore realm are closely integrated with those of structural reliability. Since earlier generations of offshore platforms were typically founded on driven piles, a great deal of research was devoted to piles. Some of this early work is summarized in Høeg and Tang (1978) and in Wu et al. (1989). Developments on the reliability of offshore piles are summarized by Lacasse and Goulois ( 1989) and by Tang et al. (1990); a recent overview is provided by Gilbert et al. (2013). A major contribution to this body of work was the introduction of pile resistance calibrations against historical data and the relation of such calibration to codes.

A closely related area of code development driven by offshore applications was the introduction of load and resistance factor design (LRFD) to foundation engineering. Motivated by the AISC structural steel code, which was an early adopter of LRFD, both the offshore industry and the Federal Highway Administration (FHWA) became early proponents of LRFD in the geotechnical design. These efforts were closely integrated with structural reliability. Benchmark contributions include Lloyd and Karsan (1988) and Moses and Larrabee (1988). In transportation engineering, corresponding contributions include Paikowsky et al. (2004, 2010).

In more recent years, the offshore industry has moved to updating and calibrating reliability-based methods in light of field observations in a series of serious hurricanes: Andrew (1992), Ivan (2004), Katrina/Rita (2005), Ike (2008), and others.

## 12.6 ENVIRONMENTAL REMEDIATION (1980–1995)

The beginning date of 1980 is the year Superfund was passed by the U.S. Congress (P.L. 96-510, 42 U.S.C., December 11, 1980). *Superfund*—more formally known as the Comprehensive Environmental Response, Compensation, and Liability Act of 1980 (CERCLA)—authorized U.S. Government agencies to recover natural resource damages caused by releases of hazardous substances. This led to a spate of site investigations and plans for site remediation across the country and to an equally large number of lawsuits and enforcement actions. The ending date, 1995, is the year of expiration of the Superfund Tax on oil and chemical companies for cleaning up abandoned sites.

The Superfund era led to at least two innovative developments in geotechnical risk analysis: a focus on spatial sampling methodologies for characterizing subsurface soil properties, especially those properties associated with chemicals rather than engineering mechanics, and the creation of a literature on probabilistic geohydrology models by which to forecast fate and transport through groundwater and to characterize the uncertainty in these forecasts.

Spatial sampling methodologies received considerable advancement during the environmental remediation (ER) period especially under the support of the U.S. Environmental Protection Agency (USEPA). The agency mandates health-risk analysis for engineering options based on risk standards, for example, for municipal landfills and for Superfund cleanups. This created a need for statistically valid assessments of site properties and to guidance on sampling plans and statistical data analysis (USEPA 1991, 2002). A number of textbooks appeared during this period on statistical methods of environmental site characterizations (Ott 1995). This period also saw a dramatic increase in the use of "geostatistics" for soils sampling.

Geohydrology was another important area of development in the ER era, leading to the evolution of a suite of probabilistic tools for modeling subsurface flows, along with several textbooks on stochastic approaches (Marsily 1986; Kitanidis 1997; Rubin 2003).

## 12.7 DAM SAFETY (1986–ONGOING)

Prior to the failure of the Teton Dam in 1976, little attention was paid to probabilistic characterizations of embankment reliability or to quantitative risk analysis of dam safety. With the passage of the *Reclamation Safety of Dams Act* (P.L. 95-578) in 1978, and *Executive Order* 12148 requiring U.S. federal agencies to implement federal guidelines for dam safety, the US Bureau of Reclamation (USBR) began a concerted effort to address dam safety, and eventually, through the lens of risk analysis. The act was amended in 1984, 2000, 2002, and 2004; the program continues to operate. The starting date for this period is the year that USBR began publicly reporting on its emerging risk analysis approach.

In the mid-1980s (Parrett 1986), the USBR risk cadre pioneered the use of risk analysis in dam safety in the United States, championing a number of methodological approaches that would come to characterize dam safety evaluations: event tree analysis, subjective probability assignment, quantified loss-of-life modeling, and tolerable risk criteria, among others. Publication of the *Public Protection Guidelines* in 1997 was a benchmark in this progression (USBR 1997). BCHydro and Ontario Power Generation working through the Dam Safety Interest Group with USBR were also early promoters of risk analysis for dam safety (Salmon and Hartford 1995), as was ANCOLD that created a working group on risk analysis in 1987 (McDonald 1995), and the Dam Safety Committee of the Government of New South Wales promulgated risk analysis approaches starting in the 1990s.

Dam safety continues to be a test bed for developments in geotechnical risk analysis. Following Hurricane Katrina in 2005, the U.S. Army Corps of Engineers (USACE) converted to risk-informed planning, creating its Risk Management Center (RMC) as a platform for the transformation in the programs for dam and levee safety. The RMC in recent years has championed risk-based analysis of internal erosion in dams, dam and levee-screening criteria, and international standards for tolerable risk.

A number of important developments have come out of this work on dam safety risk, especially that done by USBR and private sector hydropower operators. In recent years, the International Committee on Large Dams (ICOLD) has begun promulgating guidance on dam safety risk analysis methods (ICOLD 2005). Scott (2011) has presented a detailed overview of the major contributions of this body of work. Among other things, he cites the following contributions as important, and it is hard to disagree with his assessment:

- Potential failure mode analysis
- Hazard analysis of annual exceedance probabilities for seismic, flood, and reservoir loading
- Event tree analysis
- Subjective probability and expert elicitation methods
- Fragility curves
- Consequence (loss of life) evaluation

## 12.8 SYSTEMS RISK ASSESSMENT (2005–ONGOING)

A major evolution in the practice of geotechnical risk analysis arrived in 2005 with the landfall of Hurricane Katrina in New Orleans, Louisiana, and the Delta Risk Management Program in California. In each case, the cognizant authority elected to employ systematic and large-scale risk assessments to understand the respective hazards, vulnerabilities, and consequences of failure of an extensive levee network (350 miles in New Orleans and 1100 miles in California) and to develop risk management plans.

### 12.8.1 New Orleans

A primary contributor to the flooding of New Orleans during Hurricane Katrina was the failure of levees and floodwalls that make up the hurricane protection system and in some parts of the system their overtopping by storm surge (the levee system and its various components is now called the *Hurricane and Storm Damage Risk Reduction System*, HSDRRS).

The HSDRRS had been designed to provide protection from storm-induced surges and waves in an attempt to control naturally occurring conditions. The HSDRRS was designed to perform this function without imposing unacceptable risks to public safety, property, and welfare; however, some level of risk always remains. Even with the reconstruction and strengthening of the hurricane protection system (HPS) in the future, still some risk will remain.

The term *risk* was used to define hazards, losses, and potential outcomes, defined by *Risk = Hazard probability × Vulnerability × Consequences of failure*, in which hazard probability is the rate or uncertainty of occurrence of the causal event; vulnerability is the reliability with which the constructed system withstands the loads or other demands caused by the hazard; and consequences of failure are the costs in lives and dollars accruing in the event of a failure (IPET8 2008). This is sometimes called the hazard–vulnerability–consequence (HVC) model (or, in security risk, TVC, for "threat").

The application of the probabilistic risk analysis to the HSDRRS of New Orleans was challenging, because the system is a complex of levees, floodwalls, pumping stations, and other components that serve a large geographical region and our capability to model hurricanes

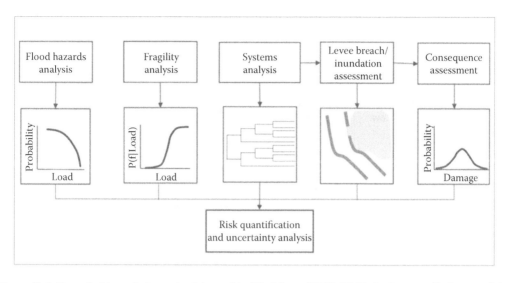

*Figure 12.5* Overall risk analysis methodology. (Modified from IPET8. 2008. *Performance Evaluation of the New Orleans and Southeast Louisiana Hurricane Protection System, Final Report, v.8— Engineering and Operational Risk and Reliability Analysis.* Washington, DC: Interagency Performance Evaluation Taskforce.)

and their effects is limited. Hurricane models can predict winds, waves, and surges only with limited accuracy, and the reliability models used to predict levee performance when subjected to hurricane forces are similarly limited. Hence, the risk profiles of hurricane-induced flooding cannot be established with certainty. Risk analysis, therefore, had to include not just a best estimate of risk, but also an estimate of the uncertainty in that best estimate. By identifying the sources of uncertainty in the analysis, measures, such as gathering additional data, could be taken to reduce the uncertainty and improve the risk estimates.

The analysis examined risks associated with the performance of the hurricane protection system (HSDRRS). Probabilistic risk analysis used to develop the overall risk analysis methodology of the hurricane protection system is presented in Figure 12.5. The calculations of risk were affected by large event-tree analyses. Much of this approach was informed by earlier work on dam safety risk analysis, as discussed above.

## 12.8.2 California delta

The Delta Risk Management Strategy study (DRMS) in California undertook a similar risk analysis of the historical and fragile levee system in the San Francisco Bay Delta, which is subject to seismic hazard. The project was authorized by the Department of Water Resources to perform a risk analysis of the Delta and Suisun Marsh and to develop a set of improvement strategies to manage these risks (URS/JBA 2007)

The hazards considered in the DRMS study were (a) earthquakes that cause levees or their foundations to fail, (b) high storm runoff that can overtop levees or increase seepage and cause them to fail, (c) "sunny day events" caused by undetected flaws that fail levees during non-flood-flow periods, and (d) wind waves and erosion that can weaken levees. The analysis included the frequency that events of different magnitudes occur: Smaller earthquakes and floods occur less often than the more extreme events, but all of them pose some risk to the levee system. Small events may fail only one levee, and larger events may fail multiple levees.

| | Negligible | Low | Moderate | High | Extreme |
|---|---|---|---|---|---|
| **High** | Low risk | Medium risk | Medium risk | High risk | High risk |
| **Moderate** | Low risk | Low risk | Medium risk | Medium risk | High risk |
| **Low** | Low risk | Low risk | Medium risk | Medium risk | High risk |
| **Very low** | Low risk | Low risk | Low risk | Medium risk | High risk |

Probability (vertical axis), Consequence (horizontal axis)

*Figure 12.6* An example of a risk matrix or "heat" diagram.

### 12.8.3 Risk registers

One of the advances to be reinforced in New Orleans and California was the use of risk registers. Risk registers are not new: Anyone who has ever written down a checklist of things that could go wrong in a project has created a risk register. But modern systems risk assessments take the risk register to a new level of detail and comprehensiveness. Such registers are now common on large geotechnical projects informed by risk management, particularly tunnels, to aid in informed risk management.

The risk register typically structures things that might go wrong on a project as rows in a spreadsheet. These are usually organized by the asset type, or by phases of the project, or in some other orderly way. The columns in the spreadsheet provide a description of the risk item to a level of detail that two engineers reading the same description take away the same understanding of the event, an appraisal of the probability and consequence should the risk occur, and usually an assignment of responsibility to someone within the organization for monitoring the risk item. Sometimes, the risk register also lists response and remediation action plans for each significant risk item.

It is increasingly common to color code the risk items in the register on a simple ordinal scale—low–medium–high—as shown in Figure 12.6. This has proved to be a useful way for grouping risks and tracking those with the greatest expected consequence and for communicating risks to project stakeholders. However, this practice is not without its critics and suffers limitations for which account should be taken (Cox 2008, 2009). The issue with risk matrices is that the rankings are ordinal scales and thus do not admit the mathematical operations of addition and multiplication; thus, weighted sums of these measures or even comparisons along diagonals through the matrix are not meaningful, despite that many people overlook this limitation.

## 12.9 EMERGING APPROACHES: SYSTEM SIMULATION, STRESS TESTING, AND SCENARIO APPRAISALS

While geotechnical risk analysis has matured over the four decades since 1970, it is still evolving. An argument could be plausibly made that the rate of advance is greater now than ever before. Certainly, its reception in mainstream geotechnical practice is now an accomplished fact.

There are many ways in which the current geotechnical risk analysis is evolving. Three ways that may eventually prove of importance are system simulation, stress or scenario testing, and dynamic risk analysis and management. Certainly, there are other trends afoot, but the authors are less familiar with them.

## 12.9.1 Systems simulation methods

Systems simulation methods arise from the observation that the reliable performance of a system such as a dam, reservoir, diversion structure, and so on depends on the time-varying demands placed upon it by geology, hydrology, operating rules, the interactions among a cascade of reservoirs, and vagaries of operator interventions and natural disturbances. In managing such systems, the goal is to bound the likelihood of events that might lead to a loss of control over stored-water volume and releases. The function of such systems is to retain water volumes and to pass flows in a controlled way (Figure 12.7).

Adverse performance typically arises from unanticipated combinations of loadings and responses within the normal operating ranges of a dam's performance. That is, failures often do not occur under the extreme loadings considered in the design, but under the uncommon combination of relatively common things happening. For example, instrumentation misperforms; so, pool levels are allowed to rise, simultaneously ice has formed on the spillway gates that then cannot be opened sufficiently to accommodate necessary discharges, and finally the operator misperceives the danger and an accident follows. Failures often occur because several and sometimes not unusual deviations from normal conditions occur simultaneously.

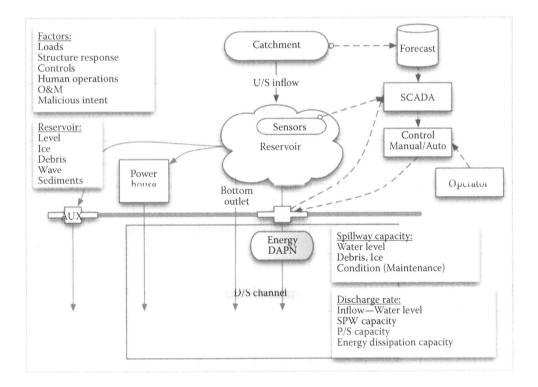

*Figure 12.7* System of flow-control and discharge functions for a single dam. (From Baecher, G.B. et al. 2014. *Second-Generation Risk Analysis for Dam Safety*. Toronto.)

A new approach to analyzing flow-control systems is needed. The awareness of this need is not unique to dams, hydropower, and large water management structures but has been recognized with respect to many safety critical technical systems. The simulation framework involves four parts: (1) *Simulation* by which stochastic reservoir inflows are generated and propagated through the river–reservoir–spillway–outflow system; (2) *physics of failure modeling* to infer the impact of spillway heads and discharges on the hydraulic structures accommodating outflows; (3) *component reliability analysis* to ascertain the performance of individual components of the outflow works; and (4) *systems reliability assessment* through which demands of the river system and performance of flow-control systems components coupled with interactions of humans are convolved into annual exceedance probabilities of adverse performance.

## 12.9.2 Stress testing and scenario analysis

Stress testing and scenario analysis is a concept that has arisen out of system failures in other arenas, for example, in seismic performance of buildings or in the response of the financial system to disturbances. It remains an evolving concept in geotechnical applications but is receiving attention in natural hazards assessments (Short 2013), particularly in cases involving low-probability high-consequence events (Figure 12.8).

Stress tests attempt to determine the load under which a system as a whole fails and how it fails, and whether cascading failures are important to systems performance. The overall system is hypothetically loaded by a catastrophic hazard, and as the loading continues, the response of the individual system components and their interactions is tracked. The goal is to understand the systems behavior during an extreme event and the modalities of its degradation (Lacasse and Nadim 2011).

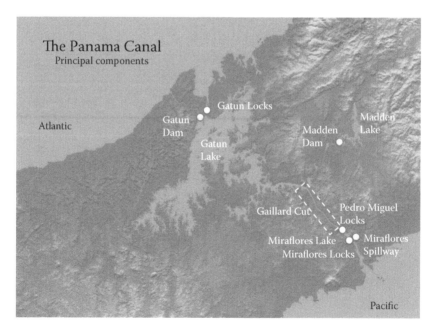

*Figure 12.8* The principal infrastructure assets of the Panama Canal Authority. (From Alfaro, L. 2013. *The Expansion of the Panama Canal _in the Context of the Canal's History*, February 22. West Lafayette: Purdue University.)

### 12.9.3 Dynamic risk analysis and management

The emergence of real-time sensing combined with web-based data communication and visualization has led to risk-monitoring capabilities to enhance risk management that has great potential and, at one time, was only speculated about. Modern sensor systems have allowed geotechnical performance data to be measured in real time and communicated to central databases where even more powerful computer systems are used to model field conditions as they change. As noted by Marr (2011), who terms the approach *Active Risk Management*™, this is principally an extension of Peck's concept of the *observational method* (Peck 1969) by using modern tools, of which risk modeling and decision science are central components. Real-time monitoring is used during construction and operation to identify emerging risks as soon as possible so that steps can be taken to reduce the probability of the risk event, reduce the consequence, or do both.

### 12.10 TEN UNRESOLVED QUESTIONS

While a great deal of progress has been made, certain questions still remain that deserve continued thought. The authors recently suggested a short list of 10 unresolved questions in geotechnical risk and reliability analysis and that list may bear repeating for those without easy access to ASCE conference proceedings (Christian and Baecher 2011). The 10 issues range from technical mechanics problems to matters of communication. The list is certainly not exhaustive, but we propose it to challenge the reliability community. It will be interesting to return to these issues 10 years hence to see what progress we have made.

1. *Why are failures less frequent than predicted?* Typical coefficients of variation for soil-engineering properties are reported to be in the order of 20–30%. Presuming a mean factor of safety of 1.5, corresponding reliability indices ($\beta$) are about 1.67, implying probabilities of failure of about 0.05. These are an order of magnitude larger than the observed frequency of adverse performance and two orders of magnitude larger than the frequency of all-modes failures of earth dams. Why the difference?

2. *What is the actual variability of soil and rock properties?* Several studies have been published on the variability of soil properties, but more work needs to be done, especially on how to improve estimates at a particular site. Variations in soil-engineering data involve at least two things: (1) actual variability from one point to another, and (2) noise. Some of that noise arises because the measurements are index properties and only loosely correlated to engineering properties (e.g., the field vane with its Bjerrum correction factor) introduced by methods of measurement. In addition, however, there are at least two bias errors that creep into assessments: (3) statistical error due to limited numbers of observations, and (4) model error due to the approximate nature of our mathematical descriptions of soil behavior.

3. *What are the effects of spatial correlation?* Geological materials arrive at their present configurations by a geologic process that follows physical principles. Therefore, their physical properties exhibit spatial correlation. While there have been successes in describing spatial correlation statistically and in modeling spatially correlated variables, the techniques for dealing with spatial correlation are difficult to implement, and they are poorly understood in practice; thus, their consequences are often ignored.

4. *How should we account for scale effects?* Much of geotechnical engineering involves scaling properties from laboratory or field tests on limited volumes. We know that

scaling to field-sized properties is a statistical problem, but the profession has not assimilated it.

5. *Can we develop reliability models for internal erosion?* About one-third of the failures of modern earth dams occur because of internal erosion. However, practical design methods are deterministic, and we do not have adequate methods to detect piping in existing structures. The major problem is that we lack a physical model for predicting internal erosion from first principles. It is a problem for which the basic physics of failure has not been described adequately because they are not well understood. It is also a problem dominated by local conditions of soil gradation, density, strength, and permeability to such an extent that a realistic model is difficult to formulate with meaningful parameters.

6. *Can we more strongly connect the observational method to quantitative Bayesian updating?* In a paper near the end of his life, Terzaghi wrote, "Soil engineering projects [...] require a vast amount of effort and labor securing only roughly approximate values for the physical constants that appear in the equations. The results of the computations are not more than working hypotheses, subject to confirmation or modification during construction. [...]. The elements of this method are 'learn-as-you-go:' Base the design on whatever information can be secured. Make a detailed inventory of all the possible differences between reality and the assumptions. Then compute, on the basis of the original assumptions, various quantities that can be measured in the field. On the basis of the results of such measurements, gradually close the gaps in knowledge, and if necessary modify the design during construction." This is Bayesian logic.

7. *We need better ways of communicating risk to stakeholders.* There are actually two types of risk communication involved: communication with owners and communication with the public. In both cases, the problem is that the engineer must convert technical knowledge into language that can be understood by stakeholders not conversant with the technology. An underlying issue is that it is difficult for the public, owners, and even engineers to deal with and plan for extremely rare events that have severe consequences. How should resources be allocated? Which of many dire scenarios should be taken seriously?

8. *Can we improve and validate the application of LRFD to geotechnical problems?* LRFD was developed as a rational way to quantify the contributions of different loadings in the analysis of steel and concrete structures. Resistance was and remains a single variable. The factors are primarily applied to the loads. LRFD has been applied with some success to design procedures for piles, which are essentially structural members embedded in the ground. The approach has been less satisfactory when applied to retaining structures, and even then effects such as strength dependent on normal stress and varying water pressures are ignored.

9. *Can we develop practical guidance on statistical considerations in exploration strategies?* There is a large literature on sampling strategy, but it has had little impact on site characterization. Part of the problem is that the results are not very useful unless they are combined with Bayesian thinking and subjective probability. There is a need for serious study of what can be done in a realistic environment for communicating these insights to practitioners.

10. *Can we improve on rules for multiple-failure modes?* Calculating the failure probability of a system from the computed failure probabilities of the individual modes often leads to unreasonably conservative results. The current practice is to assume the extreme cases that the behaviors of the individual failure modes are either perfectly correlated or perfectly uncorrelated. The probability of system failure, presuming all the modes are mutually independent, is an upper bound. The lower bound on the

probability of system failure is given by the case of perfect dependence among all the probabilities; that is, for the case in which, if one mode fails, all of them fail. The so-called De Morgan bounds have been suggested for this problem, but they are too broad.

## 12.11 CONCLUDING THOUGHTS

Recent years have seen a chronicling of the important insights and open questions about risk analysis, in general (Greenberg et al. 2012), and about risk in geotechnical practice, in particular (Christian and Baecher 2011).

What are the transformative insights about geotechnical risk and reliability that have shaped our current understanding? There are several, but we will not attempt a tentative list here, that await another opportunity to do so; it would be with the understanding that everyone working in the field will have his or her own list, and that not everyone will agree with our list.

Among the insights that we consider the most transformative (in no particular order) are (1) the understanding that geotechnical properties are spatially variable, and this variability controls engineering performance in subtle ways; (2) the distinction between aleatory and epistemic uncertainty is fundamental to practice, and this necessitates a conclusion that geotechnical reliability is fundamentally Bayesian; and (3) the observation that variations in geotechnical data arise from a host of sources, and the source of the variation dictates how the corresponding uncertainty affects our predictions. One might also draw conclusions about models, methods, and approaches, but they, too, will wait for another opportunity.

## ACKNOWLEDGMENTS

The authors acknowledge the valued advice and suggestions of their colleagues Robert Patev of the USACE on the risk analysis of New Orleans post-Katrina, Robert Gilbert of the University of Texas on developments in offshore engineering, Herbert Einstein of MIT on developments in rock mechanics and mining, and W. Allen Marr of Geocomp Corporation on active risk management.

## REFERENCES

Alfaro, L. 2013. *The Expansion of the Panama Canal _in the Context of the Canal's History*, February 22. West Lafayette: Purdue University.

Alonzo, E.E. 1976. Risk analysis of slopes and its application to slopes in Canadian sensitive clays. *Geotechnique* 26(3): 453–72.

Ang, A.H. S., and W.II. Tang. 1975. *Probability Concepts in Engineering Planning and Design*. New York: Wiley.

Baecher, G., and J. Christian. 2003. *Reliability and Statistics in Geotechnical Engineering*. 1st ed. Chichester, England: Wiley.

Baecher, G.B., D.N.D. Hartford, R.C. Patev, K. Rytters, and P.A. Zielinski. 2014. *Second-Generation Risk Analysis for Dam Safety*. Toronto.

Baecher, G.B., N.A. Lanney, and H.H. Einstein. 1977. Statistical description of rock properties and sampling. In *Proceedings of the 18th US Symposium on Rock Mechanics*, 5C1-1 to 5C1-8, Golden, CO.

Barton, N. 1975. Suggested methods for the descriptions of rock masses, joints and discontinuities, 2nd draft. IRSM Working Party, categories I9, I10.

Benjamin, J.R., and C.A. Cornell. 1970. *Probability, Statistics, and Decision for Civil Engineers*. New York: McGraw-Hill.

Casagrande, A. 1965. The role of the calculated risk in earthwork and foundation engineering. *Journal of the Soil Mechanics and Foundations Division, ASCE* 91 (SM4): 1–40.

Christian, J., and G. Baecher. 2011. Unresolved problems in geotechnical risk and reliability. In *Geo-Risk 2011*, 50–63. American Society of Civil Engineers. http://dx.doi.org/10.1061/41183(418)3.

Coates, D.F. 1977. *CANMET Pit Slope Manual*. Ottawa: Canada Centre for Mineral and Energy Technology.

Cornell, C.A. 1971. First-order uncertainly analysis of soils deformation and stability. In *International Conference on Applications of Statistics and Probability to Structural and Geotechnical Engineering*. Honk Kong: University of Hong Kong Press.

Cox, L.A. 2008. What's wrong with risk matrices? *Risk Analysis* 28(2): 497–512. doi:10.1111/j.1539-6924.2008.01030.x.

Cox, L.A. 2009. What's wrong with hazard-ranking systems? An expository note. *Risk Analysis* 29(7): 940–8. doi:10.1111/j.1539-6924.2009.01209.x.

Cruden, D.M. 1977. Describing the size of discontinuities. *International Journal of Rock Mechanics and Mining Sciences and Geomechanics Abstracts* 14(3): 133–7. doi:10.1016/0148-9062(77)90004-3.

Dershowitz, W.S., and H.H. Einstein. 1988. Characterizing rock joint geometry with joint system models. *Rock Mechanics and Rock Engineering* 21(1): 21–51. doi:10.1007/BF01019674.

Duncan, J. 2000. Factors of safety and reliability in geotechnical engineering. *Journal of Geotechnical and Geoenvironmental Engineering* 126(4): 307–16. doi:10.1061/(ASCE)1090-0241(2000) 126:4(307).

Einstein, H.H., D.A. Labreche, M.J. Markow, and G.B. Baecher. 1976. Decision analysis applied to rock tunnel exploration. *Engineering Geology* 12(2): 143–61.

Einstein, H.H., Sousa, R., Karam, K., Manzella, I., and Kveldsvik, V. 2010. Rock slopes from mechanics to decision making. In *Rock Mechanics in Civil and Environmental Engineering*. London: CRC Press.

Fenton, G.A., and D.V. Griffiths. 2008. *Risk Assessment in Geotechnical Engineering*. Hoboken, NJ: John Wiley & Sons.

Folayan, J., K. Høeg, and J. Benjamin. 1970. Decision theory applied to settlement predictions. *Journal Soil Mechanics and Foundation Engineering* 96(4): 1127–41.

Freudenthal, A.M. 1956. Safety and the probability of structural failure. *American Society of Civil Engineers Transactions* 121: 1337–97.

Gilbert, R.B., S. Lacasse, and F. Nadim. 2013. Advances in geotechnical risk and reliability for offshore applications. In *Geotechnical Safety and Risk IV*, 29–42. Boca Raton, FL: CRC Press.

Greenberg, M., C. Haas, A. Cox, K. Lowrie, K. McComas, and W. North. 2012. Ten most important accomplishments in risk analysis, 1980–2010: Ten most important accomplishments in risk analysis. *Risk Analysis* 32(5): 771–81. doi:10.1111/j.1539-6924.2012.01817.x.

Gumbel, E.J. 1958. *Statistics of Extremes*. New York: Columbia University Press.

Hansen, J.B. 1956. *Limit Design and Safety Factors in Soil Mechanics*. Bulletin No. 1. Copenhagen: Danish Geotechnical Institute.

Harr, M.E. 1977. *Mechanics of Particulate Media: A Probabilistic Approach*. New York: McGraw-Hill.

Harr, M.E. 1987. *Reliability-Based Design in Civil Engineering*. New York: McGraw-Hill Book Company.

Hasofer, A.M., and N. Lind. 1974. An exact and invariant first-order reliability format. *Journal of Engineering Mechanics, ASCE* 100(EM1): 111–21.

Høeg, K., and W.H. Tang. 1978. *Probabilistic Considerations in the Foundation Engineering for Offshore Structures*. Publication No. 120. Oslo: Norwegian Geotechnical Institute.

ICOLD. 2005. *Risk Assessment in Dam Safety Management*, Bulletin 130. Paris: International Commission on Large Dams.

IPET8. 2008. *Performance Evaluation of the New Orleans and Southeast Louisiana Hurricane Protection System, Final Report, v.8—Engineering and Operational Risk and Reliability Analysis*. Washington, DC: Interagency Performance Evaluation Taskforce.

Kitanidis, P.K. 1997. *Introduction to Geostatistics: Applications to Hydrogeology*. New York: Cambridge University Press.

Kulhawy, F., and K.K. Phoon. 1996. Engineering judgment in the evolution from deterministic to reliability-based foundation design. In *Uncertainty in the Geologic Environment*. ASCE Press, Madison, WN.

Lacasse, S., and A.M. Goulois. 1989. Uncertainty in API parameters for prediction of axial capacity of driven piles in sand. In *OTC Paper 6001*. Houston: Society of Petroleum Engineers.

Lacasse, S., and F. Nadim. 2011. Learning to live with geohazards: From research to practice. In *Memorias de La Conferencia GeoRisk*, 64–116, ASCE Press, Reston, VA.

Lloyd, J.R., and D.I. Karsan. 1988. Development of a reliability-based alternative to API RP2A. In *OTC Paper 5882*. Houston: Society of Petroleum Engineers.

Lumb, P. 1966. The variability of natural soils. *Canadian Geotechnical Journal* 3: 74–97. doi:* Book Section.

Lumb, P. 1971. *Proceedings, First International Conference on Applications of Statistical and Probability to Soil and Structural Engineering*. Hong Kong: Hong Kong University Press.

Marr, W.A. 2011. Active risk management in geotechnical engineering. In *GeoRisk 2011*, 894–901. Atlanta: ASCE.

de Marsily, G. 1986. *Quantitative Hydrogeology: Groundwater Hydrology for Engineers*. Orlando, FL: Academic Press.

Matheron, G. 1965. *Les Variables Régionalisées et Leur Estimation, Une Application de La Théorie de Fonctions Aléatoires Aux Sciences de La Nature*. Paris: Masson et Cie.

Matsuo, M. 1976. *Reliability in Embankment Design*. PhD, Cambridge: MIT.

McDonald, L. 1995. ANCOLD risk assessment guidelines. In *Acceptable Risks for Major Infrastructure*, 105–21. Rotterdam: Balkema.

McMahon, B.K. 1975. A statistical method for the design of rock slopes. In *Proceedings of the Australia–New Zealand Conference on Geomechanics*. Lindfield, N.S.W: Australian Rock Engineering Consultants.

Moses, F., and R.D. Larrabee. 1988. Calibration of the draft RP2A-LRFD for fixed platforms. In *OTC Paper 5699*. Houston: Society of Petroleum Engineers.

National Commission on the BP Deepwater Horizon Oil Spill and Offshore Drilling. 2010. *A Brief History of Offshore Oil Drilling*. Staff Working Paper No. 11. Washington, DC: National Commission on the BP Deepwater Horizon Oil Spill and Offshore Drilling.

NRC. 1995. *Probabilistic Methods in Geotechnical Engineering*. Washington, DC: The National Academies Press.

Ott, W.R. 1995. *Environmental Statistics and Data Analysis*. Boca Raton, FL: Lewis Publishers.

Paikowsky, S.G., National Cooperative Highway Research Program, National Research Council (U.S.), and Transportation Research Board. 2004. *Load and Resistance Factor Design (LRFD) for Deep Foundations*. Washington, DC: Transportation Research Board.

Paikowsky, S.G., National Research Council (U.S.), Transportation Research Board, National Cooperative Highway Research Program, American Association of State Highway and Transportation Officials, United States, and Federal Highway Administration. 2010. *LRFD Design and Construction of Shallow Foundations for Highway Bridge Structures*. Washington, DC: Transportation Research Board. http://books.google.com/books?id=vgUnAQAAMAAJ.

Parrett, N. 1986. U.S. Bureau of reclamation use of risk analysis. In *Decision Making in Water Resources*, 154–71. Santa Barbara: The Engineering Foundation.

Peck, R.B. 1969. Advantages and limitations of the observation method in applied soil mechanics. *Geotechnique* 19(2): 171–87.

Priest, S.D., and J.A. Hudson. 1976. Discontinuity spacings in rock. *International Journal of Rock Mechanics and Mining Science* 13: 135–48.

Rosenblueth, E. 1975. Point estimates for probability moments. *Proceedings, National Academy of Science* 72(10): 3812–4.

Rubin, Y. 2003. *Applied Stochastic Hydrogeology*. Oxford, New York: Oxford University Press.

Salmon, G., and D. Hartford. 1995. Lessons from the application of risk assessment to dam safety. In *ANCOLD/NXSOLD Conference on Dams*, 54–67. Christchurch: ANCOLD Bulletin 101.

Scott, G.A. 2011. The practical application of risk assessment to dam safety. In *Proceedings of GeoRisk*. http://ascelibrary.org/doi/abs/10.1061/41183(418)6.

Short, J.R. 2013. Stress testing the USA: Public policy and reaction to disaster events, NY: Macmillian.

Snow, D.T. 1970. The frequency and apertures of fractures in rock. *International Journal of Rock Mechanics and Mining Sciences and Geomechanics Abstracts* 7(1): 23–40. doi:10.1016/0148-9062(70)90025-2.

Tang, W.H., D.L. Woodford, and J.H. Pelletier. 1990. Performance reliability of offshore piles. In *OTC Paper 6379*. Houston: Society of Petroleum Engineers.

Taylor, D.W. 1948. *Fundamentals of Soil Mechanics*. New York: John Wiley & Sons.

URS/JBA. 2007. *Delta Risk Management Strategy (DRMS) Phase 1 Risk Analysis Report*. Sacramento: Prepared by URS Corporation/Jack R. Benjamin & Associates, Inc. for the California Department of Water Resources.

USBR. 1997. *Guidelines for Achieving Public Protection in Dam Safety Decision Making*. Interim. Denver: U.S. Bureau of Reclamation.

USEPA. 1991. *Removal Program Representative Sampling Guidance, Volume 1: Soil*. Washington, DC: Office of Emergency and Remedial Response.

USEPA. 2002. *Guidance on Choosing a Sampling Design for Environmental Data Collection for Use in Developing a Quality Assurance Project Plan*. EPA QA/G-5S. Washington, DC: U.S. Environmental Protection Agency.

Vanmarcke, E. 1983. *Random Fields, Analysis and Synthesis*. Cambridge, MA: MIT Press.

Vanmarcke, E.H. 1977. Reliability of earth slopes. *Journal of Geotechnical Engineering Division, ASCE* 103 (GT11): 1247–65.

Vick, S.G. 2002. *Degrees of Belief: Subjective Probability and Engineering Judgment*. Reston: ASCE.

Whitman, R.V. 1984. Evaluating the calculated risk in geotechnical engineering. *Journal of the Geotechnical Engineering Division, ASCE* 110(2): 145–88.

Whitman, R.V. 1996. Organizing and evaluating uncertainty in geotechnical engineering. In *Uncertainty in the Geologic Environment*, 1–28. Madison: ASCE.

Wu, T.H., and L.M. Kraft. 1967. The probability of foundation safety. *Journal of the Soil Mechanics and Foundations Division, ASCE* 93 (SM5): 213–31.

Wu, T.H., and L.M. Kraft. 1970. Safety analysis of slopes. *Journal of the Soil Mechanics and Foundations Division* 96: 609–30.

Wu, T.H., W.H. Tang, D.A. Sangrey, and G.B. Baecher. 1989. Reliability of offshore foundations: State-of-the-art. *Journal of Geotechnical Engineering Division, ASCE* 115(2): 157–78.

Chapter 13

# Assessing the value of information to design site investigation and construction quality assurance programs

*Robert B. Gilbert and Mahdi Habibi*

## 13.1 INTRODUCTION

The potential to improve the types and quantities of data that are collected in the practice of geotechnical engineering is enormous. There are many situations where additional information would be valuable in developing designs and making decisions. Likewise, there are also many situations where data that are obtained have little or no value in developing designs or making decisions.

The goal of this chapter is to describe a framework and provide practical tools and insights for assessing the value of information to design site investigation and construction quality assurance programs. A decision framework for assessing the value of information is presented, the role and practical application of Bayes' Theorem is explored, and illustrative examples and case histories are provided throughout to demonstrate and elucidate the theory.

## 13.2 VALUE OF INFORMATION FRAMEWORK

Assessing the value of information involves considering how the information might be used to guide decision-making. This section describes the basic framework for decision analysis and provides illustrative examples of applying this framework.

### 13.2.1 Decision analysis

The generic decision tree in Figure 13.1 shows the role of information in making a decision, see Benjamin and Cornell (1970), Ang and Tang (1984), and Gilbert et al. (2008) for details on decision trees. The decision here is between two alternatives, Plan A versus Plan B. The uncertainty in the decision is the consequence that will be realized when one of the plans is implemented, which is represented by a probability distribution, $p_C(C)$, for the different possibilities. The preferred alternative is that with the greatest expected consequence,[*] $E(C) = \sum_{all\ c} c \times p_C(c)$ where the expected consequence once the decision is made is the maximum of the possible expected consequence values

$$E(C\ for\ Decision) = max[E(C_{Alternative\ A}), (C_{Alternative\ B})] \tag{13.1}$$

---

[*] Consequence is used here assuming that larger values are preferred to smaller values, such as net profit with positive values being a gain and negative values being a loss or nondimensional utility with a range between the least—preferred consequence of zero and the most preferred consequence of one.

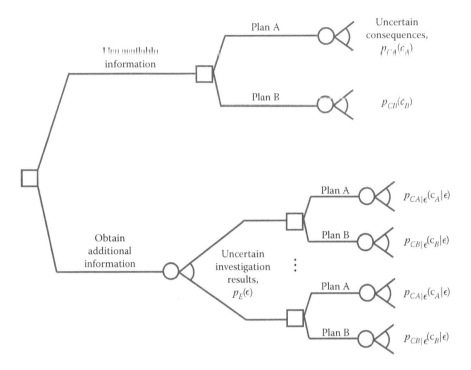

*Figure 13.1* Decision tree framework for assessing value of information.

The value of obtaining additional information depends on what that information might be, which is represented by a probability distribution, $p_E(\epsilon)$, and how that information affects the probability distribution for the possible consequences, $p_{C|\epsilon}(c|\epsilon)$. For each possible outcome of information, $\epsilon$, the decision between the two alternatives is evaluated on the basis of the expected consequence, $E(C|\epsilon) = \sum_{all\ c} c \times p_{C|\epsilon}(c|\epsilon)$, where the expected consequence for the decision given a possible outcome of information is

$$E(C\ for\ Decision\ with\ New\ Information\|\epsilon) = max[\ E(C_{Alternative\ A}|\epsilon), (C_{Alternative\ B}\|\epsilon)] \qquad (13.2)$$

The expected consequence associated with obtaining the new information is then obtained from the Theorem of Total Probability

$$E\big(C\ for\ Decision\ with\ New\ Information\big)$$
$$= \sum_{all\ c} E\big(C\ for\ Decision\ with\ New\ Information|\epsilon\big) \times p_E(\epsilon) \qquad (13.3)$$

The value of information is defined as the maximum cost (negative consequence), $\Delta c_{New\ Information}$, the decision maker would be willing to spend in obtaining that information:

$$E(C\ for\ Decision\ with\ New\ Information\ Including\ \Delta c_{New\ Information}) = E(C\ for\ Decision)$$
$$(13.4)$$

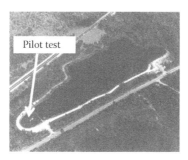

*Figure 13.2* Aerial photograph of contaminated lagoon.

## 13.2.2 Illustrative example: Remediation of contaminated lagoon

An example illustrating the value of information framework is for the remediation of a contaminated lagoon (Figure 13.2). Two approaches are considered to remediate this site: Alternative A is a conventional approach to pump and treat the contaminated groundwater on-site, effectively containing the contamination to prevent it from migrating off-site; Alternative B is an innovative approach to remove the source of contamination by breaking down the volatile organic compound contaminants in the lagoon into benign by-products with *in situ* bioremediation. The decision tree for these two approaches based on existing information is shown in Figure 13.3; this step is referred to as a *Prior* decision since it is analyzed prior to obtaining any additional information. The probabilities for different outcomes are based on the available information, including historical information about operation of the site, geologic and hydrogeologic models, and site investigation and monitoring data. On the basis of the available information, the preferred alternative is the conventional approach, Alternative A (Figure 13.3).

An alternative to consider before deciding the best approach for remediating the site is to obtain additional information about the innovative bioremediation approach. Specifically, a pilot test of the *in situ* bioremediation process could be performed in one corner of the lagoon. Since this pilot test is implementing the approach at a smaller scale and shorter duration than it will be implemented in the lagoon itself, the results are not perfectly reliable. Three test outcomes are possible: a successful result (*S*), a questionable result (*Q*)

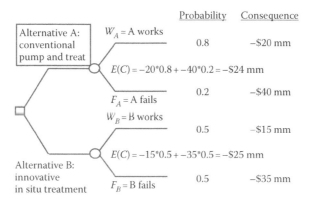

*Figure 13.3* Decision tree to select preferred alternative to remediate contaminated lagoon based on existing information.

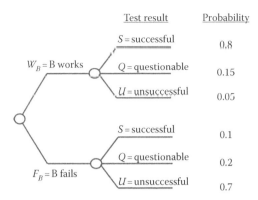

*Figure 13.4* Event tree showing conditional probabilities of possible test outcomes for contaminated lagoon.

(e.g., concentrations of contaminants decrease but at a slow rate relative to the duration of the pilot test), and an unsuccessful result ($U$). An event tree showing how the probabilities for these test outcomes are related to whether or not the in situ bioremediation will work at the full scale is shown in Figure 13.4. An important feature of these conditional probabilities is that they are assessed for a defined set of physical conditions, for example, $P(S|W_B)$ is the probability of a successful test result if the geologic, chemical, and biological conditions at the site are such that the full-scale bioremediation will work. These conditional probabilities can be assessed by analyzing data from other sites where both pilot-scale and full-scale implementations were used or more plausibly, since this technology is new, by simulating pilot-scale and full-scale results over a range of possible site conditions.

The updated probabilities in the decision tree with additional information are obtained using Bayes' Theorem

$$P_{C|\epsilon}(c|\epsilon) = \frac{P(E = \epsilon|c)P_C(c)}{\sum_{all\ c} P(E = \epsilon|c)P_C(c)} \tag{13.5}$$

For example, the probability that the bioremediation alternative will work given a successful pilot test is obtained as

$$P_{C_B|S}(C_B = -\$15\ MM + \Delta c_{test}|S) = P(W_B|S)$$
$$= \frac{P(S|W_B)P(W_B)}{P(S|W_B)P(W_B) + P(S|F_B)P(F_B)} = \frac{0.8 \times 0.5}{0.8 \times 0.5 + 0.1 \times 0.5} \tag{13.6}$$
$$= \frac{0.400}{0.400 + 0.050} = \frac{0.400}{0.450} = 0.889$$

The resulting decision trees for the three possible test outcomes are shown in Figure 13.5; this step is referred to a *Posterior* decision since it is made after additional information is obtained. The pilot test will change the preferred alternative to the *in situ* bioremediation approach if a successful result is obtained from the pilot test, while it will not change the preferred alternative if the other possible test outcomes occur.

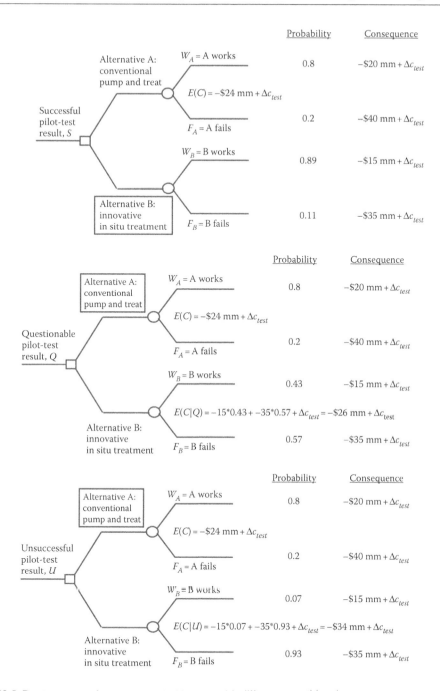

Figure 13.5 Decision trees for contaminated lagoon with different possible pilot-test results.

The *Value of Information* for the pilot test can then be assessed using the decision tree in Figure 13.6; this step is referred to as a *Preposterior* decision since it is made before additional information is obtained by considering all possible outcomes for the additional information. The value of information for this pilot test is the value of $\Delta c_{test}$ such that $-\$21\ \text{MM} + \Delta c_{test} = -\$24\ \text{MM}$, or $\Delta c_{test} = -\$3\ MM$. This value provides guidance on

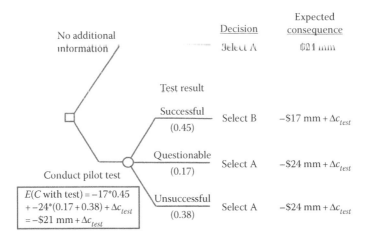

*Figure 13.6* Decision tree to assess value of information for pilot test in contaminated lagoon.

whether or not to conduct a pilot test; if a pilot test costs <$3 MM, then pursuing it is preferred compared to not conducting the test (Figure 13.6).

Bounds on the value of information can be established. A lower bound is *No Information* in which the updated probabilities for the decision outcomes are not changed by the additional information. In this example, the case of *No Information* corresponds to $P(S|W_B) = P(S|F_B)$, $P(Q|W_B) = P(Q|F_B)$, and $P(U|W_B) = P(U|F_B)$. At this limit, no possible outcome for the additional information will change the preferred decision alternative, giving a value of information equal to zero (i.e., $\Delta c_{test} = 0$). An upper bound is *Perfect Information* in which the information is perfectly reliable at indicating which consequence will be realized. A decision tree to assess the value of perfect information about the performance of the *in situ* bioremediation approach is shown in Figure 13.7. The value of perfect information about the performance of the *in situ* bioremediation is $\Delta c_{test} = -\$4.5$ MM. Therefore, no matter how reliable the pilot test is, it would not be preferred to pursue it if the cost is >$4.5 MM. This value of perfect information provides practical guidance because it is straightforward to obtain from the original decision tree without needing to account for the reliability of a test.

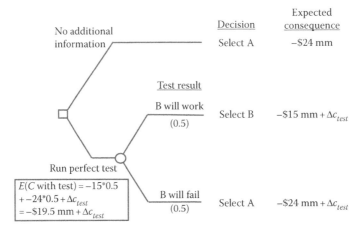

*Figure 13.7* Decision tree to assess value of information for pilot test in contaminated lagoon.

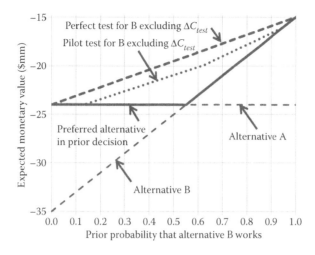

*Figure 13.8* Sensitivity of expected monetary values to prior probability that in situ bioremediation will work for contaminated lagoon.

A sensitivity analysis for the value of information is shown in Figures 13.8 and 13.9. The expected monetary value for Alternatives A and B is shown in Figure 13.8 versus the prior probability that Alternative B will work, $P(W_B)$ in Figure 13.3. The preferred alternative is Alternative A at $P(W_B) < 0.55$ and Alternative B at $P(W_B) > 0.55$ (Figure 13.8); the alternatives are equally preferable at $P(W_B) = 0.55$. The expected monetary values for the pilot test and the perfect test without including the cost of the test are also shown in Figure 13.8; the value of information is the distance between the expected monetary value for a test without including the cost of the test and the expected monetary value for the preferred alternative in the prior decision. The value of information versus the prior probability that Alternative B will work is shown in Figure 13.9. The value of information is a maximum when there is the greatest potential to change the decision, which is at the indifference point of $P(W_B) = 0.55$ (Figure 13.9). The value of information for the pilot test is zero when the prior probability that Alternative B will work is small [$P(W_B)$ less than about 0.15] or large [$P(W_B)$ greater than about 0.95]; at these extremes, there is no possible outcome from the pilot test that will change the decision about whether or not to implement Alternative B.

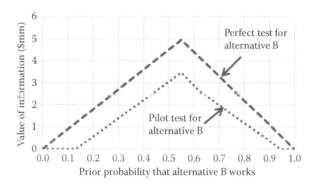

*Figure 13.9* Sensitivity of value of information to prior probability that in situ bioremediation will work for contaminated lagoon.

This example of the contaminated lagoon is loosely based on an actual case history. In the case history, the pilot test produced a successful result and *in situ* bioremediation was selected and implemented. While this approach was effective in essentially removing the source of contamination from the lagoon, it was not successful in removing all of the contamination in the ground at the site because pools of non-aqueous-phase-contaminant liquids had migrated from the lagoon. Consequently, the conventional pump-and-treat approach was still required in the long term to contain groundwater contamination on site, resulting in a significantly greater cost (i.e., the event $F_B$ occurred when Alternative B was selected in Figure 13.3). Therefore, this case history underscores the important role of uncertainty. A preferred alternative may have the greatest expected value, but the actual outcome when the decision is implemented may not be preferred. In addition, the value of information is an expected value and not necessarily the actual value that will be realized by obtaining additional information.

## 13.3 INSIGHTS FROM BAYES' THEOREM

Bayes' Theorem (Equation 13.5) plays an integral role in the value of information. Bayes' Theorem can be expressed as follows:

$$
\begin{aligned}
&P(Decision\ Consequence | Information) \\
&= P(Information | Decision\ Consequence)P(Decision\ Consequence) \\
&\Bigg/ \left[ \sum_{\substack{all \\ Consequences}} P(Information | Decision\ Consequence)P(Decision\ Consequence) \right]
\end{aligned}
\tag{13.7}
$$

where $P(Decision\ Consequence | Information)$ is the *updated probability* for the possible decision consequence given the information, $P(Decision\ consequence)$ is the *prior probability* for the consequence, and $P(Information | Decision\ Consequence)$ is the *likelihood function* relating the probability of obtaining the information if that decision consequence is realized. The two key components in Bayes' Theorem are the *prior probability* and the *likelihood function*; the set of prior probabilities provides the starting point and the likelihood function serves to filter the starting point based on the information.

### 13.3.1 Prior probabilities

The prior probabilities in Bayes' Theorem are important because the updated probability for a decision consequence is proportional to the prior probability for that consequence (Equation 13.7). The importance of the prior probabilities is underscored by considering the case where the prior probability for a particular decision consequence is zero; no matter how strong the likelihood function is in amplifying the probability of that consequence, the updated probability will always be zero.

The prior probabilities establish the initial uncertainty in the decision consequences. A common misconception is that additional information will reduce uncertainty; uncertainty can decrease, remain unchanged, or increase with additional information. Consider the simple example shown in Figure 13.10. While the prior probability distribution is heavily weighted toward a consequence of $10 MM with little uncertainty (Figure 13.10a), the

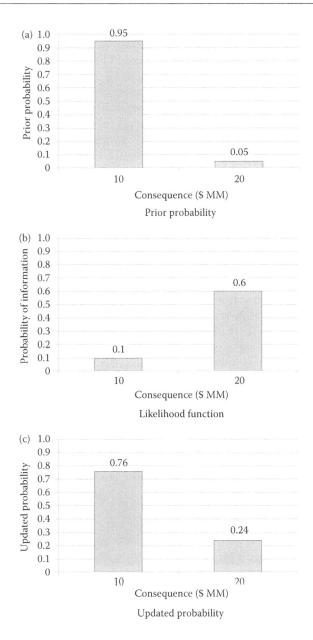

*Figure 13.10* Illustrative example showing that uncertainty can increase with additional information.

likelihood function (Figure 13.10b) amplifies the probability of a \$20 MM consequence, reduces the probability of a \$10 MM consequence, and produces an updated probability distribution with more uncertainty in the consequence (Figure 13.10c).

The magnitude of uncertainty in a probability distribution is rationally and conveniently measured with the theory of information entropy (e.g., Shannon 1948; Jaynes 1957; Tribus 1969):

$$Entropy\ of\ Information = -\sum_{all\ c_i} ln\big[p_C(c_i)\big] \times p_C(c_i) \tag{13.8}$$

Analogous to entropy in thermodynamics, the entropy of the information reflects the relative frequencies of possible decision consequences (states) that could be realized if the decision was implemented a large number of times. The entropy of information is maximized when the probabilities are equal for all possible consequences and it is minimized when only one consequence is possible. For the example in Figure 13.10, the entropy of the prior probability distribution (Figure 13.10a), $-[ln(0.95)] \times 0.95 - [ln(0.05)] \times 0.05 = 0.20$, is smaller than the entropy of the updated prior probability distribution (Figure 13.10c), $-[ln(0.76)] \times 0.765 - [ln(0.24)] \times 0.24 = 0.55$. This simple example therefore demonstrates that uncertainty can increase as additional information is obtained. In other words, *sometimes we learn we know less than we thought we did.*

The importance of the prior probabilities in Bayes' Theorem has motivated the development of several approaches for establishing prior probabilities.[*] The most commonly applied approach in practice is to rely on subjective information or experience to establish a reasonable starting point (e.g., Luce and Raiffa 1957). The challenge with this approach is that it is difficult to account for possibilities beyond our range of experience (e.g., "Black Swan" events as coined by Taleb 2007). Another approach is to maximize the entropy of the prior probability distribution in Bayes' Theorem (e.g., Jaynes 1968; Tribus 1969). For example, the prior probability distribution with maximum entropy for the example in Figure 13.10 would be a uniform distribution with equal probabilities of 0.5 for the two possible consequences. The challenge with this approach is that it is not necessarily rational or consistent. Maximizing the entropy of the probability distribution for variables in a decision is not rational because different variables affect a decision in different ways. Journel and Deutsch (1993) demonstrate this lack of rationality with an example of spatial variability in the permeability of an oil reservoir. They show that large entropy in the permeability field (i.e., little structure or spatial correlation between high and low permeability values) produces small entropy in the probability distribution for well production, while small entropy in the permeability field (e.g., high-permeability channels and low-permeability barriers) produces large entropy in the probability distribution for well production. This approach is also not consistent because the definition of an input variable is ambiguous. For example, a uniform probability distribution for permeability does not give a uniform probability distribution for the logarithm of permeability.

Practical guidance in establishing prior probabilities is to consider the sensitivity of the preferred decision alternative and the value of information to these prior probabilities. Figures 13.8 and 13.9 demonstrate an example of implementing this guidance. A proposed approach to formally implement this guidance is described later in this chapter in the case-history application of test wells for an energy resource.

## 13.3.2 Likelihood functions

The likelihood function, $P(Information|Decision\ Consequence)$ in Equation 13.7 is important because it filters the prior probabilities for decision consequences, potentially amplifying the probabilities for some possibilities and reducing the probabilities for other possibilities. The *no information* and *perfect information* bounds are controlled by the shape of the likelihood function. A likelihood function that has the same probabilities for the information

---

[*] To be fair, one perspective is to not use Bayes' Theorem in practice. Classical statisticians such as Fisher (1935) entirely focus on the likelihood function because of the difficulty they perceive in defensibly establishing prior probabilities. Keynes (1921) also supports this perspective. While avoiding difficulties with prior probabilities, this approach is of little practical value because it does not allow for the use of probabilities in making decisions with uncertain outcomes.

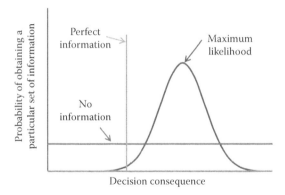

Figure 13.11 Illustrative example showing that uncertainty can increase with additional information.

for all possible decision consequences is noninformative (Figure 13.11). A likelihood function that has zero probabilities for all but one possible decision consequence is perfectly informative (Figure 13.11). In general, for a continuous likelihood function, a measure of its "informativeness" is its curvature or sharpness at the point of maximum likelihood (Figure 13.11); zero curvature is noninformative and infinite absolute curvature is perfectly informative.* The sharpness of the likelihood function is affected by attributes of the information, such as the quantity of information (e.g., the number of measurements), relationships between pieces of information (e.g., correlations between measurements), and the clarity of the information (e.g., errors in measurements).

In general, the likelihood function is formulated in terms of variables (model parameters) that affect the consequences of a decision, but not explicitly in terms of the consequences themselves:

$$P(\Theta_1, ..., \Theta_n | \epsilon) = \frac{P(E = \epsilon | \theta_1, ..., \theta_n)P(\Theta_1, ..., \Theta_n)}{\sum_{\theta_1, ..., \theta_n} {}_{all} P(E = \epsilon | \theta_1, ..., \theta_n)P(\Theta_1, ..., \Theta_n)}$$    (13.9)

where $P(\Theta_1, ..., \Theta_n | \epsilon)$ is the updated probability distribution for the set of model parameters given the observed information, $\epsilon$; $P(E = \epsilon | \theta_1, ..., \theta_n)$ is the likelihood function for the information in terms of the model parameters; and $P(\Theta_1, ..., \Theta_n)$ is the prior probability distribution for the set of model parameters. The prior and posterior probabilities for each decision consequence are then obtained from the Theorem of Total Probability:

$$P(Decision\ Consequence) = \sum_{all\ \theta_1, ..., \theta_n} P(Consequence | \theta_1, ..., \theta_n)P(\Theta_1, ..., \Theta_n)$$    (13.10)

and

---

* It is sometimes convenient to work with the logarithm of the likelihood function due to its values ranging over orders of magnitude and due to its exponential form with normally distributed data. Since the logarithm is a single-valued increasing function of the argument, the maximum of the logarithm of the likelihood function occurs at the same location as the maximum of the likelihood function and the curvature of the logarithm of the likelihood function is as useful as the curvature of the likelihood function in indicating "informativeness."

$$P(Decision\ Consequence|Information) = \sum_{all\ \theta_1,\ldots,\theta_n} P(Consequence|\theta_1,\ldots,\theta_n)P(\Theta_1,\ldots,\Theta_n|\epsilon)$$

$$(13.11)$$

When working with model parameters, it can be insightful to express the likelihood function in terms of the decision consequences (e.g., Figure 13.11) since it is the relationship between decision consequences and information that governs the value of information:

$$P(Information|Decision\ Consequence)$$

$$= \sum_{all\ \theta_1,\ldots,\theta_n} P(Consequence|\theta_1,\ldots,\theta_n)P(\Theta_1,\ldots,\Theta_n|\epsilon) \times \sum_{\substack{all\\ \theta_1,\ldots,\theta_n}} P(E = \epsilon|\theta_1,\ldots,\theta_n)P(\Theta_1,\ldots,\Theta_n)$$

$$/ \left[ \sum_{all\ \theta_1,\ldots,\theta_n} P(Consequence|\theta_1,\ldots,\theta_n)P(\Theta_1,\ldots,\Theta_n) \right]$$

$$(13.12)$$

### 13.3.3 Illustrative example: Design of pile foundation

An example illustrating insights from Bayes' Theorem is for the design of a pile foundation for axial loading (Figure 13.12). The decision is between two different pile lengths. Spatial variability in axial pile capacities across the site is modeled with a normal distribution,

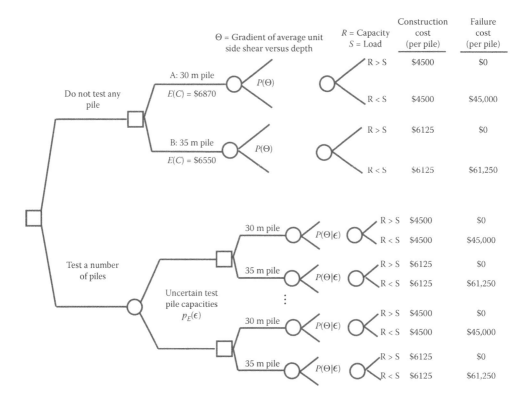

*Figure 13.12* Decision tree to assess value of information for load tests in pile foundation design.

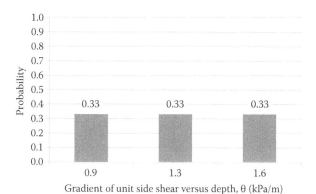

Figure 13.13 Prior probability distribution for model parameter in pile foundation design.

$R = Axial\,Pile\,Capacity \sim N(M_R, \sigma_R = 0.2 \times M_R)$, where the mean capacity is a function of an uncertain gradient of unit side shear versus depth, $M_R = (1/2 \times c \times l^2) \times \Theta$ where $c$ is the pile circumference of 1.2 m, $l$ is the pile length, and $\Theta$ is the gradient of unit side shear versus depth. Figure 13.13 shows the prior probability distribution for the gradient of unit side shear versus depth, $P(\Theta = \theta)$, and Figure 13.14 shows the possible conditional probability distributions for axial pile capacity given the possible values for the model parameter (i.e., gradient of unit side shear vs. depth). The axial loads on the piles are also uncertain and modeled with a normal distribution: $S = Axial\,Pile\,Load = N(\mu_S = 400\,kN, \sigma_S = 60\,kN)$. The expected consequence for a given pile length is obtained as a function of the cost of construction for that length, plus the expected cost of failure:

$$E(Consequence\ for\ Pile\ Length\ l) = Cost(l) + \left[10 \times Cost(l)\right] \times P(S > R|l) \qquad (13.13)$$

where $Cost(l)$ is the construction cost, the cost of failure is 10 times the construction cost, and $P(S > R|l)$ is the probability of failure (load exceeds capacity) for that pile length

$$P(S > R|l) = \sum_{all\ \theta}\left[\Phi\left(-\frac{\mu_{R|\theta} - \mu_S}{\sqrt{\sigma_{R|\theta}^2 + \sigma_S^2}}\right)P(\Theta = \theta)\right] \qquad (13.14)$$

where $\Phi(.)$ is the standard normal cumulative distribution function, $\mu_{R|\theta} = (1/2 \times c \times l^2) \times \theta$, and $\sigma_{R|\theta} = 0.2 \times \mu_{R|\theta}$. The resulting prior probability distributions for the consequences obtained from Equation 13.11 are shown in Figure 13.15. For the prior decision, the longer pile is preferred with an expected cost of $6550 per pile versus the shorter pile with an expected cost of $6870 per pile (Figure 13.15).

A possible alternative is to perform pile load tests before deciding between the shorter or longer pile lengths (Figure 13.12). First, consider the possibility of a single-load test. The probability of measuring a single capacity, $\epsilon_1$, is given by

$$P(\epsilon_1|\theta) = P(r = \epsilon_1) = \phi\left(\frac{\epsilon_1 - \mu_{R|\theta}}{\sigma_{R|\theta}}\right)d\epsilon_1 \qquad (13.15)$$

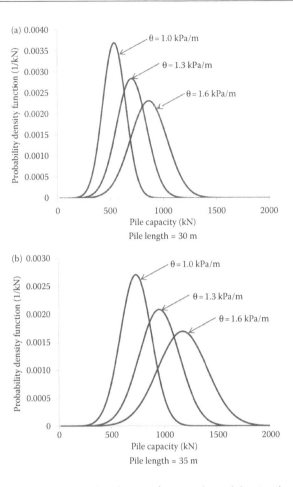

*Figure 13.14* Conditional prior probability distributions for spatial variability in pile capacity given different model parameter values for alternative pile lengths in pile foundation design.

where $\phi(.)$ is the standard normal probability density function and $d\epsilon_i$ is a negligibly small interval about the measurement result (note that the magnitude of this interval is not significant since it does not affect the relative likelihood). This likelihood function is shown in Figure 13.16 for an example test result. Note that since it is the relative likelihood that controls the "informativeness" of the function via filtering, the likelihood values are shown relative to the maximum likelihood. The resulting updated probability distribution for the model parameter obtained from Equation 13.9 is shown in Figure 13.17, and the resulting updated probability distributions for the consequences obtained from Equations 13.10 and 13.11 are shown in Figure 13.18. The preferred pile length based on this load test result is now the shorter pile (Figure 13.18).

Next, consider the possibility of performing multiple pile load tests. If $n$ tests are performed and the results are statistically independent between piles, then the likelihood function is given by

$$P(\epsilon_1, \ldots, \epsilon_n | \theta) = \prod_{all\ \epsilon_i} P(r = \epsilon_i) = \prod_{all\ \epsilon_i} \phi\left(\frac{\epsilon_i - \mu_{R|\theta}}{\sigma_{R|\theta}}\right) d\epsilon_i \qquad (13.16)$$

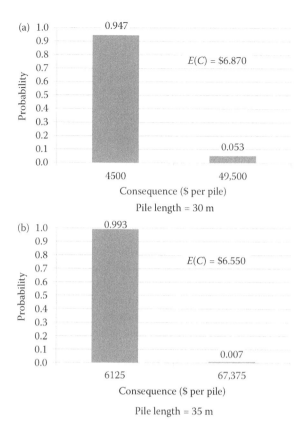

*Figure 13.15* Prior probability distributions for consequences for alternative pile lengths in pile foundation design.

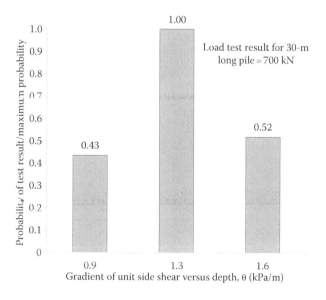

*Figure 13.16* Likelihood function for model parameter given one pile load test result in pile foundation design.

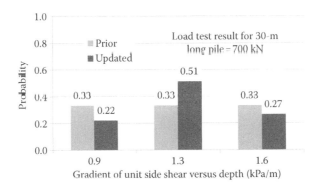

*Figure 13.17* Prior and updated probability distributions for model parameter given one pile load test result in pile foundation design.

The likelihood function in this case is a function of the sample mean of the individual load test results, $m_t$. This likelihood function is shown in Figure 13.19 as a function of the number of independent tests for an example set of test results; the likelihood function becomes significantly sharper as the number of independent tests increases.

The test results may not be independent; there may be a systematic variation due to spatial proximity or measurement error as well as an independent variation. If the total variation

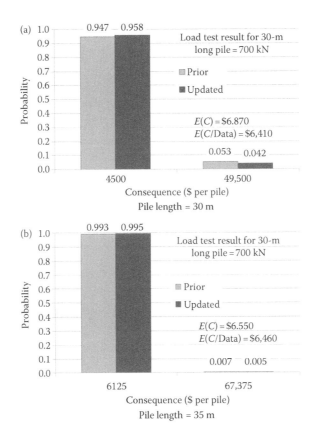

*Figure 13.18* Prior and updated probability distributions for consequences given one pile load test result for alternative pile lengths in pile foundation design.

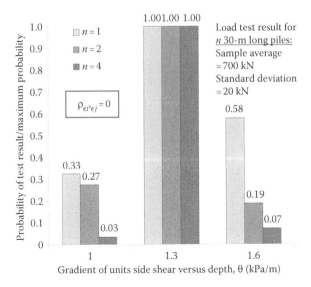

*Figure 13.19* Likelihood functions for model parameter given an average load test result from n independent pile load tests in pile foundation design.

in pile capacity between locations is divided into systematic and random components, $\sigma_{R|\theta}^2 = \sigma_s^2 + \sigma_r^2$, where $\sigma_s$ is the standard deviation of the systematic component and $\sigma_r$ is the standard deviation of the random component, then the correlation coefficient between two test results $\epsilon_i$ and $\epsilon_j$ is the percentage contribution of the systematic variance to the total variance, $\rho_{\epsilon_i,\epsilon_j} = \dfrac{\sigma_s^2}{\sigma_s^2 + \sigma_r^2} = \dfrac{\sigma_s^2}{\sigma_{R|\theta}^2}$. The probability of measuring a set of correlated capacities, $\bar{\epsilon} = \epsilon_1, ..., \epsilon_n$, is obtained from the multivariate normal distribution

$$P(\epsilon_1, ..., \epsilon_n | \theta) = \frac{1}{(2\pi)^{n/2} \left| C_{R|\theta} \right|^{1/2}} \times exp\left[ -\frac{1}{2} (\bar{\epsilon} - \bar{\mu}_{R|\theta})^T C_{R|\theta}^{-1} (\bar{\epsilon} - \bar{\mu}_{R|\theta}) \right] d\epsilon_n \, ... \, d\epsilon_1 \qquad (13.17)$$

where $\bar{\mu}_{R|\theta}$ is an $n \times 1$ vector of mean pile capacities and $C_{R|\theta}$ is an $n \times n$ covariance matrix with diagonal terms equal to $\sigma_{R|\theta}^2$ and off-diagonal terms equal to $\rho_{\epsilon_i,\epsilon_j}\sigma_{R|\theta}^2$. The likelihood function in this case is a function of the sample mean and the sample standard deviation of the individual load test results, $m_\epsilon$ and $s_\epsilon$, respectively. This likelihood function is shown in Figure 13.20 as a function of the correlation coefficient $\rho_{\epsilon_i,\epsilon_j}$ for an example set of test results; the sharpness of the likelihood function is reduced as the systematic variation increases relative to the total variation between test results. There are two limiting cases for systematic variations: if the variation error is zero, then $\rho_{\epsilon_i,\epsilon_j} = 0$ and the likelihood function reduces to the case of independent test results (Equation 13.16); if there is only systematic variation, then $\rho_{\epsilon_i,\epsilon_j} = 1$ and the likelihood function reduces to the case of a single-test result (Equation 13.15), since $\epsilon_1 = \epsilon_2 = \cdots = \epsilon_n$.

Finally, consider the possibility of performing proof load tests where the test result is either that the test pile does or does not hold a proof load. The probability of measuring a set of independent proof test results is given by

$$P(\epsilon_1, ..., \epsilon_n | \theta) = \prod_{all \, \epsilon_i} \left\{ P(r > \epsilon_{proof})^{n_{successes}} \times \left[ 1 - P(r > \epsilon_{proof}) \right]^{n - n_{successes}} \right\} \qquad (13.18)$$

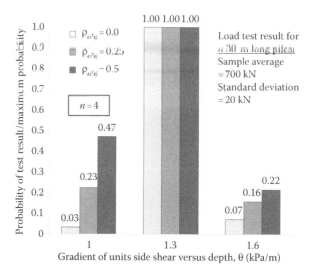

*Figure 13.20* Likelihood functions for model parameter given an average load test result from a set of correlated pile load tests in pile foundation design.

where $\epsilon_{proof}$ is the proof load, $n_{successes}$ is the number of successful test results in $n$ tests, and

$$P(r > \epsilon_{proof}) = 1 - \Phi\left(\frac{\epsilon_{proof} - \mu_{R|\theta}}{\sigma_{R|\theta}}\right) \qquad (13.19)$$

This likelihood function is shown in Figure 13.21 as a function of the proof load, $\epsilon_{proof}$, for an example set of test results. The likelihood function becomes sharper as the proof load increases since the result (three passing tests out of three proof tests) can be readily

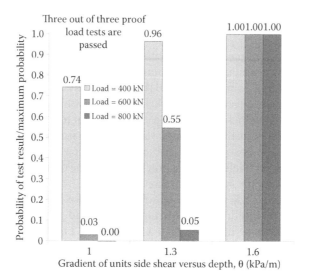

*Figure 13.21* Likelihood functions for model parameter given a result for a set of proof load tests with different proof loads in pile foundation design.

explained by all possible models with small proof loads but can be best explained by only one model at the higher proof loads (Figure 13.14).

## 13.4 IMPLEMENTATION OF VALUE OF INFORMATION ASSESSMENT

There are useful tools that facilitate the implementation of the framework for assessing the value of information. Both analytical and numerical methods are described in the following sections.

### 13.4.1 Analytical methods

A useful set of analytical methods for implementing Bayes' Theorem is available for select cases of prior and posterior distributions for model parameters and likelihood functions that are updated with random sampling (i.e., statistical independence between individual measurements or observations). These cases are referred to as conjugate pairs since the form for the prior distribution fits nicely with the form of the likelihood function in deriving an analytical result for the posterior distribution, which has the same form as the prior distribution. These results are widely published (e.g., Benjamin and Cornell 1970; Ang and Tang 2007); a summary of commonly used cases is presented in Table 13.1.

When analytical results are available for implementing Bayes' Theorem, it may be possible with simple decision problems to develop analytical results for assessing the value of information.

### 13.4.2 Illustrative example: Design quality control program for compacted fill

To illustrate the derivation of analytical tools, consider an example of designing a quality assurance/quality control (QA/QC) program for a compacted fill. The quality of the fill to serve as a foundation for future construction will be indicated by its undrained shear strength. The decision to be made is how many tests of undrained shear strength to conduct during the construction of the fill.

Assume the undrained shear strength, $X$, is normally distributed with a mean value, $M_X$, that is uncertain and an inherent standard deviation due to spatial variations and testing error, $\sigma_X$, that is equal to 200 psf. Model uncertainty in the mean strength is represented by a normal distribution with a mean value, $\mu_\mu = 1000$ psf, and a standard deviation, $\sigma_\mu = 150$ psf.

The cost of implementing a foundation design depends on the design shear strength that is used, which is denoted as $x^*$. The smaller the design shear strength, the greater the cost of the foundation: Implementation cost = \$20(1000−$x^*$), where $x^*$ is in pounds per square foot. If the actual strength, $x$, is less than the design strength, then the foundation will settle excessively and there is an associated failure cost of \$100,000.

The testing program will consist of n-independent measurements of the undrained shear strength and it has a cost that is proportional to the number of tests: Test cost = \$100n.

The decision tree for selecting the optimal design strength is shown in Figure 13.22. The design strength is obtained by minimizing the expected cost:

$$E(C|x^*) = C_{Implementation} + C_{Failure} \times P(Failure) = (20 - 0.4x^*) + 100 \times \Phi\left(\frac{x^* - \mu'_\mu}{\sqrt{\sigma_X^2 + \sigma_\mu'^2}}\right) \quad (13.20)$$

Table 13.1 Analytical results for Bayesian updating with random sampling conjugate pair distributions

| Data | Unknown parameter | Prior and posterior | Likelihood function | Conjugate density | Posterior parameter |
|---|---|---|---|---|---|
| x Occurrences in n trials from Bernoulli Sequence | p (frequency) | Beta $\alpha, \beta$ | $L(p) = P(X = x, n\|p)$ $= \binom{n}{x} p^x (1-p)^{n-x}$ | $f_p(p) = \dfrac{\Gamma(\alpha,\beta)}{\Gamma(\alpha)\Gamma(\beta)} \times p^{\alpha-1}(1-p)^{\beta-1}$ | $\alpha'' = \alpha' + x$ $\beta'' = \beta' + (n-x)$ |
| x Occurrences within interval t from Poisson Process | $\lambda$ (rate) | Gamma $\alpha, \beta$ | $L(\lambda) = P(X = x, t\|\lambda) = \dfrac{e^{-\lambda t}(\lambda t)^x}{x!}$ | $f_\lambda(\lambda) = \dfrac{1}{\Gamma(\alpha)}\beta^\alpha \times \lambda^{\alpha-1} e^{-\lambda\beta}$ | $\alpha'' = \alpha' + x$ $\beta'' = \beta' + 1$ |
| n Observations with sample mean $\bar{x}$ from normal variate | $\mu$ (mean) | Normal $\mu_\mu, \sigma_\mu$ | $L(\mu) = P(\bar{x}, n\|\mu)$ $= \prod_{i=1}^{n} \dfrac{1}{\sqrt{2\pi\sigma^2}}$ $\times \exp\left\{\dfrac{-1}{2}\left(\dfrac{x_i - \mu}{\sigma}\right)^2\right\}$ $= K \times \phi_\mu(\bar{x}, \sigma/\sqrt{n})$ | $f_\mu(\mu) = \phi_\mu(\mu_\mu, \sigma_\mu)$ $= \dfrac{1}{\sqrt{2\pi}\sigma_\mu}$ $\times \exp\left\{\dfrac{-1}{2}\left(\dfrac{\mu - \mu_\mu}{\sigma_\mu}\right)^2\right\}$ | $\mu''_\mu = \dfrac{\mu'_\mu(\sigma^2/n) + \bar{x}(\sigma_L^2)}{(\sigma^2/n) + (\sigma_\mu^2)}$ $\sigma''^2_\mu = \dfrac{(\sigma^2/n) \times (\sigma_\mu^2)}{(\sigma^2/n) + (\sigma_\mu^2)}$ |

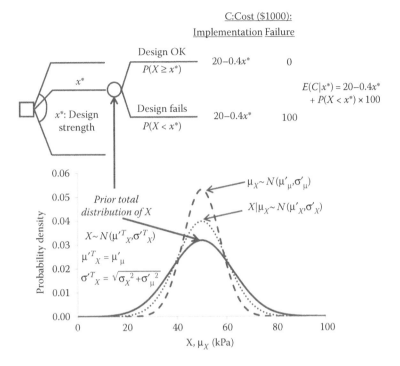

*Figure 13.22* Prior decision tree to select design shear strength for compacted fill.

Taking the derivative of $E(C|x*)$ with respect to $x*$ and setting it equal to zero to find the minimum produces the following result for the optimal value of $x*$:

$$x^*_{Opt} = \mu'_\mu - \left(\sqrt{\sigma^2_X + \sigma'^2_\mu}\right) \times \sqrt{\ln\left(\frac{100^2}{2\pi \times 0.4^2}\right) - \ln(\sigma^2_X + \sigma'^2_\mu)} \qquad (13.21)$$

The minimum cost that corresponds to the optimal $x*$ is given by

$$E(C)_{Min} = E(C|x^*_{Opt}) = (20 - 0.4 \times x^*_{Opt}) + 100 \times \Phi\left(-\sqrt{\ln\left(\frac{100^2}{2\pi \times 0.4^2}\right) - \ln(\sigma^2_X + \sigma'^2_\mu)}\right)$$

$$(13.22)$$

For example, the optimum value for $x*$ is shown in Figure 13.23.

The posterior decision tree to select the design strength for the compacted fill based on QA/QC test results is shown in Figure 13.24. Since this case with a normal prior distribution for an uncertain mean value and a likelihood function corresponding to random sampling from a normal distribution constitutes a conjugate pair (Table 13.1), the posterior distribution is also normal with the following parameters:

$$\mu''_\mu = \frac{\mu'_\mu(\sigma^2_X/n) + \bar{x}(\sigma'^2_\mu)}{(\sigma^2_X/n) + (\sigma'^2_\mu)} \qquad (13.23)$$

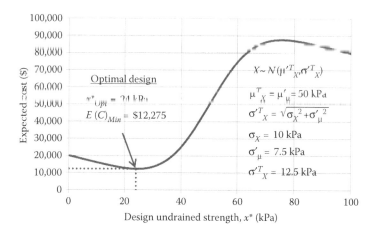

*Figure 13.23* Expected cost versus design strength for compacted fill.

$$\sigma_\mu''^2 = \frac{(\sigma_X^2/n) \times (\sigma_\mu'^2)}{(\sigma_X^2/n) + (\sigma_\mu'^2)} \tag{13.24}$$

where $\mu_\mu'$ is the prior mean value for the mean, 1000 psf, and $\sigma_\mu'$ is the prior standard deviation for the mean, 150 psf, and $\bar{x}$ is the sample mean of the undrained shear strength measured in n QA/QC tests. Therefore, the information from the test will affect the design

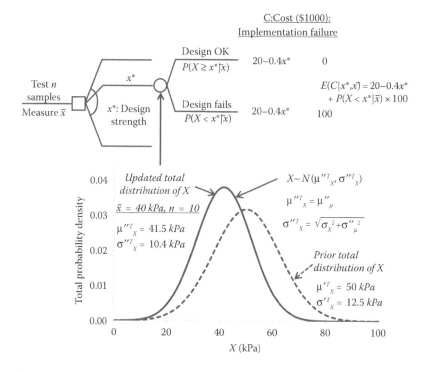

*Figure 13.24* Posterior decision tree to select design strength based on test data for compacted fill.

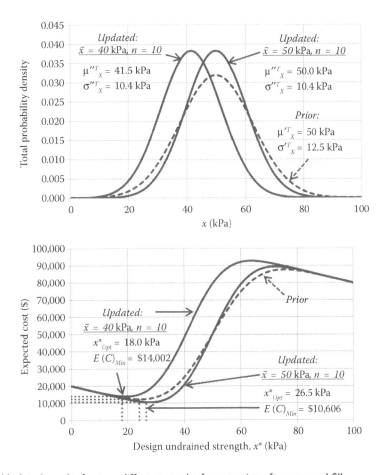

*Figure 13.25* Updated results for two different results from testing of compacted fill.

strength, $x^*$, and the expected total cost in two ways. First, the mean value for $X$ will be updated from $\bar{x}$. Second, the total standard deviation for $X$ will be reduced with the magnitude of the reduction increasing with increasing $n$.[*]

The optimal design value in a posterior analysis is shown in Figure 13.25 for two different possible QA/QC test results. Note that the updated optimal design value increases and the expected cost decreases from the prior analysis if the measured sample mean, $\bar{x}$, is equal to the prior mean of 1000 psf. This result occurs because we have reduced uncertainty in $X$ by obtaining 10 measurements, meaning that a less-conservative design is possible. Conversely, the updated optimal design value decreases and the expected cost increases from the prior analysis if the measured sample mean is 200 psf lower than the prior mean (800 vs. 1000 psf). In this case, while we are more certain about $X$, the updated mean value for $X$ is lower; so, a more conservative and costly design is required.

---

[*] Note that this result where the uncertainty in the updated distribution for the model parameter always decreases with added information is not general; the uncertainty in the updated distribution for a parameter can increase with added information (e.g., Figure 13.10).

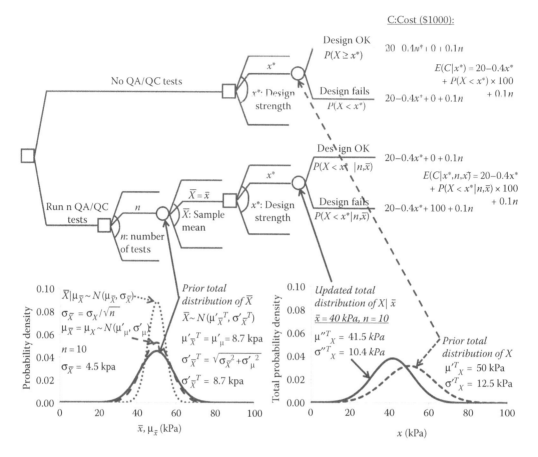

*Figure 13.26* Preposterior decision tree to design QA/QC program for testing compacted fill.

The preposterior decision tree to design the QA/QC program for the compacted fill is shown in Figure 13.26. The expected cost for a given value of n can then be obtained as follows:

$$E(C|n) = \int_{\bar{X}=-\infty}^{\bar{X}=+\infty} E(C|n, \bar{x})_{Min} \times f_{\bar{X}}(\bar{x}) d\bar{x} \tag{13.25}$$

where

$$E(C|n, \bar{x})_{Min} = [20 - 0.4 \times (x^*_{Opt}|n, \bar{x})] + 100 \times \Phi\left(-\sqrt{\ln\left(\frac{100^2}{2\pi \times 0.4^2}\right) - \ln(\sigma_X^2 + \sigma_\mu''^2)}\right) + 0.1n \tag{13.26}$$

and

$$x^*_{Opt}|n, \bar{x} = \mu_\mu'' - \left(\sqrt{\sigma_X^2 + \sigma_\mu''^2}\right) \times \sqrt{\ln\left(\frac{100^2}{2\pi \times 0.4^2}\right) - \ln(\sigma_X^2 + \sigma_\mu''^2)} \tag{13.27}$$

and the equation for $x_{opt}^*$ and $\mu_\mu''$ and $\sigma_\mu''$ is given above. For the case of $n = 0$, the pile is designed without obtaining additional data (the first step in the decision tree becomes the "Select Design Strength" decision node). The result for the expected cost for a given value of n can then be analytically obtained through integration:

$$E(C|n) = \int_{\bar{X}=-\infty}^{\bar{X}=+\infty} \left\{ [20 - 0.4 \times (x_{Opt}^*|n, \bar{x})] + 100 \times \Phi\left( -\sqrt{\ln\left(\frac{100^2}{2\pi \times 0.4^2}\right) - \ln(\sigma_{\bar{X}}^2 + \sigma_\mu''^2)} \right) + 0.1n \right\}$$
$$\times f_{\bar{X}}(\bar{x})d\bar{x} \tag{13.28}$$

$$E(C|n) = 20 - 0.4 \times 50 + 0.4 \times \left( \sqrt{10^2 + \frac{\left(\frac{10^2}{n}\right) \times (7.5^2)}{(10^2/n) + (7.5^2)}} \right)$$

$$\times \sqrt{\ln\left(\frac{100^2}{2\pi \times 0.4^2}\right) - \ln\left( 10^2 + \frac{\left(\frac{10^2}{n}\right) \times (7.5^2)}{(10^2/n) + (7.5^2)} \right)} + 100$$

$$\times \Phi\left( -\sqrt{\ln\left(\frac{100^2}{2\pi \times 0.4^2}\right) - \ln\left( 10^2 + \frac{\left(\frac{10^2}{n}\right) \times (7.5^2)}{(10^2/n) + (7.5^2)} \right)} \right) + 0.1n \tag{13.29}$$

Results are presented in Figure 13.27; the optimum number of tests is 4. In addition, results for sensitivity analyses are presented in Figure 13.28. The optimum number of tests increases as the benefit of being less conservative with the design strength increases, as the cost of testing decreases, as the spatial variability in the shear strength increases, and as the prior uncertainty in the mean value for the undrained shear strength increases.

## 13.4.3 Numerical methods

In general, for realistic problems, it is not possible to derive analytical methods to assess the value of information. In these cases, numerical methods such as the Monte Carlo simulation can be used. The basic flowchart for approximating the value of information numerically is shown in Figure 13.29; this approach of simulating model parameters and then information as a function of these parameters is commonly referred to as "bootstrapping." It is important to calculate and report confidence bounds on the approximate value of information, which will require performing multiple sets of N realizations to estimate the variability from set to set. In some cases, the efficiency of the Monte Carlo simulation can be improved using variance reduction techniques, such as Latin Hypercube Sampling when the expected consequences tend to be dominated by the expected values for the model parameters or Importance Sampling when the expected consequences tend to be dominated by small probability but high-consequence events (e.g., failure of a system). Also, simulation

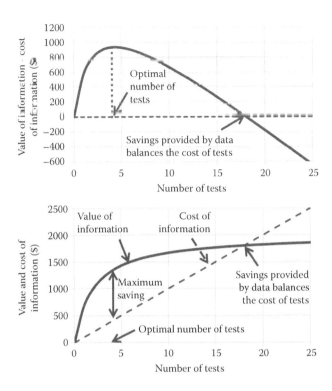

*Figure 13.27* Optimal number of tests obtained by value of information analysis for QA/QC of compacted fill.

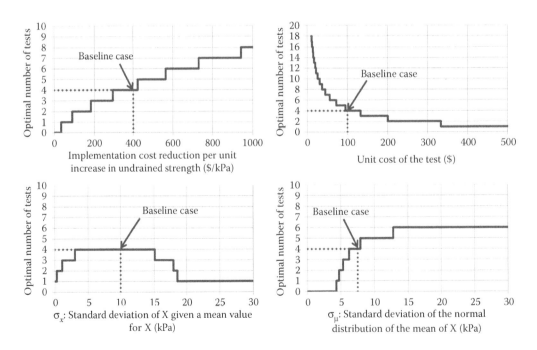

*Figure 13.28* Relationship between optimal number of tests and input variables for QA/QC of compacted fill.

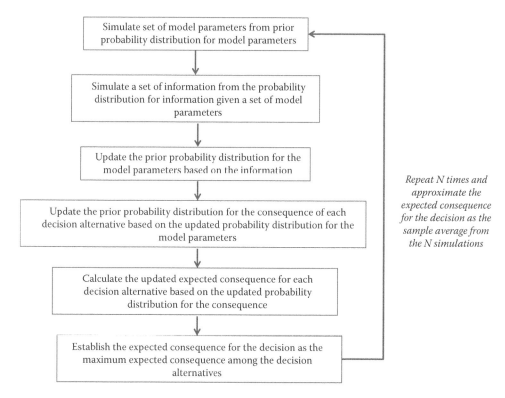

Figure 13.29 Conceptual flow chart to approximate value of information numerically.

of the possible sets of information, which generally will involve sampling from multivariate probability distributions, can be facilitated using Markov Chain Monte Carlo methods (e.g., Hastings 1970).

### 13.4.4 Illustrative example: Pile foundation load tests

The pile foundation design problem described earlier (Figures 13.12 through 13.21) provides an example to illustrate the use of numerical methods in assessing the value of information from pile load tests. An excerpt from an Excel® spreadsheet used to approximate the value of information is shown in Figure 13.30.

The value of information is shown for load tests to failure in Figure 13.31. For each point on these graphs, 10 sets of 10,000 realizations were used producing 95% confidence bounds on the estimated value for the value of information that are all smaller than ±1% of the estimate. There are diminishing returns in increasing the value of information by increasing the number of load tests. For independent test results, the value of perfect information is approached with 10 load tests. The value of information from multiple tests decreases as the correlation between test results increases (i.e., the contribution of systematic variations to the total variations increases). Also, the number of tests required to achieve perfect information increases as the correlation between test results increases; at the limit of perfect correlation, it is not possible to achieve perfect information no matter how many load tests are performed (Figure 13.31).

The value of information is shown for proof load tests in Figure 13.32. The value of information is greatest when the proof load is at 800 kN, which is near the center of the possible

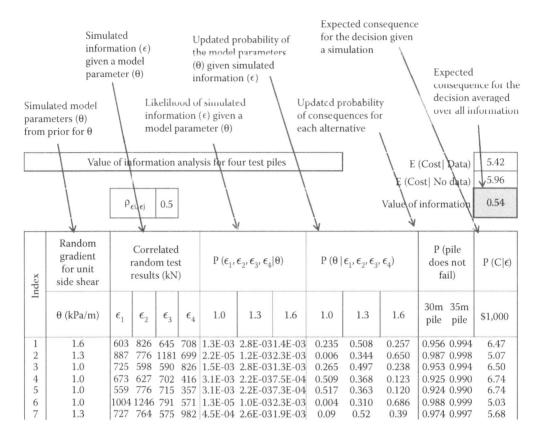

Simulated information ($\epsilon$) given a model parameter ($\theta$)

Updated probability of the model parameters ($\theta$) given simulated information ($\epsilon$)

Expected consequence for the decision given a simulation

Expected consequence for the decision averaged over all information

Simulated model parameters ($\theta$) from prior for $\theta$

Likelihood of simulated information ($\epsilon$) given a model parameter ($\theta$)

Updated probability of consequences for each alternative

| Value of information analysis for four test piles | | E (Cost\| Data) | 5.42 |
| | | E (Cost\| No data) | 5.96 |
| $\rho_{\epsilon i, \epsilon j}$ 0.5 | | Value of information | 0.54 |

| Index | Random gradient for unit side shear | Correlated random test results (kN) | | | | $P(\epsilon_1, \epsilon_2, \epsilon_3, \epsilon_4 \| \theta)$ | | | $P(\theta \| \epsilon_1, \epsilon_2, \epsilon_3, \epsilon_4)$ | | | P (pile does not fail) | | P (C\|$\epsilon$) |
|---|---|---|---|---|---|---|---|---|---|---|---|---|---|---|
| | $\theta$ (kPa/m) | $\epsilon_1$ | $\epsilon_2$ | $\epsilon_3$ | $\epsilon_4$ | 1.0 | 1.3 | 1.6 | 1.0 | 1.3 | 1.6 | 30m pile | 35m pile | $1,000 |
| 1 | 1.6 | 603 | 826 | 645 | 708 | 1.3E-03 | 2.8E-03 | 1.4E-03 | 0.235 | 0.508 | 0.257 | 0.956 | 0.994 | 6.47 |
| 2 | 1.3 | 887 | 776 | 1181 | 699 | 2.2E-05 | 1.2E-03 | 2.3E-03 | 0.006 | 0.344 | 0.650 | 0.987 | 0.998 | 5.07 |
| 3 | 1.0 | 725 | 598 | 590 | 826 | 1.5E-03 | 2.8E-03 | 1.3E-03 | 0.265 | 0.497 | 0.238 | 0.953 | 0.994 | 6.50 |
| 4 | 1.0 | 673 | 627 | 702 | 416 | 3.1E-03 | 2.2E-03 | 7.5E-04 | 0.509 | 0.368 | 0.123 | 0.925 | 0.990 | 6.74 |
| 5 | 1.0 | 559 | 776 | 715 | 357 | 3.1E-03 | 2.2E-03 | 7.3E-04 | 0.517 | 0.363 | 0.120 | 0.924 | 0.990 | 6.74 |
| 6 | 1.0 | 1004 | 1246 | 791 | 571 | 1.3E-05 | 1.0E-03 | 2.3E-03 | 0.004 | 0.310 | 0.686 | 0.988 | 0.999 | 5.03 |
| 7 | 1.3 | 727 | 764 | 575 | 982 | 4.5E-04 | 2.6E-03 | 1.9E-03 | 0.09 | 0.52 | 0.39 | 0.974 | 0.997 | 5.68 |

*Figure 13.30* Excerpt from Excel® spreadsheet illustrating value of information assessment with Monte Carlo simulation.

probability distributions for pile capacity (Figure 13.14). For small proof loads, the information is less valuable because it is unlikely that a pile will be loaded to failure, meaning that the results will not be able to discriminate between the different models of pile capacity. Likewise, for large proof loads, the information is less valuable because it is unlikely that a pile will survive the proof load.

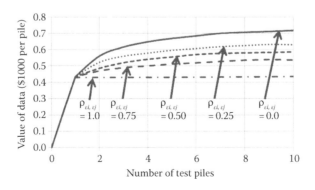

*Figure 13.31* Value of information versus number of load tests for pile foundation design.

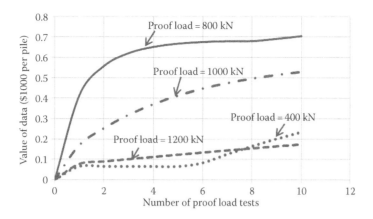

Figure 13.32 Value of information verus number of proof load tests for pile foundation design.

## 13.5 CASE-HISTORY APPLICATIONS

Four case-history applications are presented to illustrate the use of a value of information assessment in practice.

### 13.5.1 Site investigation for foundation design

A common need for a value of information assessment in practice is to decide how many soil borings to drill for the design of a foundation. Figure 13.33 shows an offshore structure with steel pipe pile foundations. Random variable models have been developed based on the existing geotechnical data for offshore fields in a variety of different geologic settings (e.g., Gilbert et al. 1999a, b, 2008; Gambino and Gilbert 1999; Cheon and Gilbert 2014). Example models for the side capacity of steel pipe pile are shown in Figure 13.34. The geological setting

Figure 13.33 Steel pipe piles being inserted into and then driven through legs of steel jacket into seafloor for offshore oil platform.

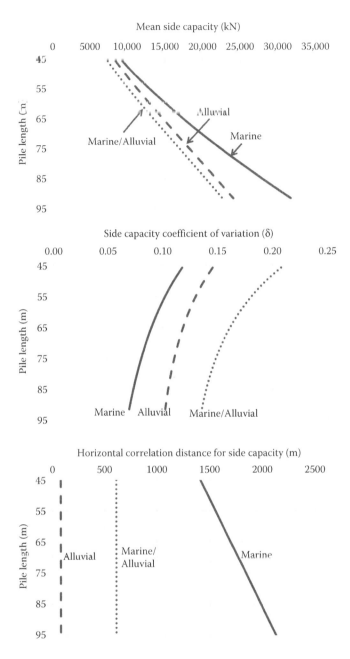

*Figure 13.34* Models for mean, coefficient of variation, and horizontal correlation distance for spatial variability in axial pile capacities due to side shear in different offshore fields.

here corresponds to a normally consolidated marine clay that is relatively homogeneous, an
alluvial deposit with layers of clay and sand that is relatively heterogeneous, and an alluvial
deposit that is incised with channels filled with marine clay. The marine clay has the smallest
spatial variability and the largest horizontal correlation distance,[*] while the mixed setting

---

[*] Correlation distance is defined here for an exponential correlation model as the distance between locations
where the correlation coefficient in capacity is $e^{-1}$.

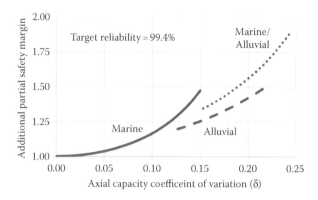

Figure 13.35 Relationship between additional partial safety margin required to achieve a target reliabilty and coefficient of variation due to spatial variability in axial pile capacity in different offshore fields.

with alluvial and marine deposits has the greatest variability and smallest horizontal correlation distance (Figure 13.34).

It is common in offshore practice to not have soil borings exactly at the location of structures because locations and configurations for facilities often change during the design due to other considerations (e.g., reservoir engineering) and because new facilities are added in the future. The consequence of not having a site-specific boring to design a pile foundation is increased uncertainty due to spatial variability in pile capacity. If a target reliability is desired, then this reliability can be achieved when there is additional uncertainty by increasing the factor of safety with an additional partial safety factor or decreasing the resistance factor with an additional partial resistance factor (e.g., Gilbert et al. 1999a, b). Figure 13.35 shows how this additional partial safety factor increases with an increasing coefficient of variation in the axial capacity due to spatial variability.

The value of information from drilling a new soil boring quantifies the expected benefit of this boring in reducing uncertainty in the foundation design. This value of information is shown in Figure 13.36, represented as an expected savings in pile length, versus the distance to an existing boring. The greatest value from a new soil boring is in the most variable geologic setting (i.e., "Marine/Alluvial" in Figure 13.36). The value is smallest when the

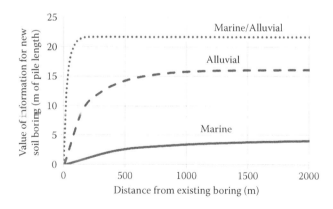

Figure 13.36 Value of information for new boring versus distance from existing boring in designing axially loaded pile in different offshore fields.

new boring is very close to the existing boring, the value increases as the distance increases, and the value eventually reaches an asymptote once the spatial correlation between the new boring and the existing boring becomes negligible. This asymptote is reached at the smallest distance in the most heterogeneous geologic setting (i.e., "Marine/Alluvial" in Figure 13.36) and at the largest distance in the most homogeneous geologic setting (i.e., "Marine" in Figure 13.36). These results demonstrate how the value of information from site investigation data heavily depends on the geologic setting.

## 13.5.2 Remedial investigation for a contaminated site

Figure 13.37 shows an oil refinery with potential contamination that may require remediation (see Gilbert 2002 for more details). One contaminant of concern was benzene in the soil below storage tanks and processing facilities. Information was available about historical land use and documented spills in various areas throughout the facility. In addition, several investigations had been performed over the years to measure benzene concentrations in soil samples.

The metric of interest in this analysis was the fraction of soil, H, with a benzene concentration greater than the regulatory limit of 50 mg/kg. A prior probability distribution for H was established based on historical land use and spill information. A β-distribution was used to model H since it can be bounded between zero and one while taking on a variety of shapes within the bounds and since it provides a mathematically convenient conjugate pair for random sampling. Example distributions for the fraction of the contaminated samples in different subareas are shown in Figure 13.38. If there was no available information about benzene contamination in a subarea, then a uniform distribution (one possible shape for a β-distribution) was assumed. This prior β-distribution for H was then updated through Bayes' Theorem with all available benzene concentration measurements for that subarea. The sampling results were assumed to be independent within a subarea, meaning that the likelihood function for the number of "hits" (i.e., samples with benzene concentrations

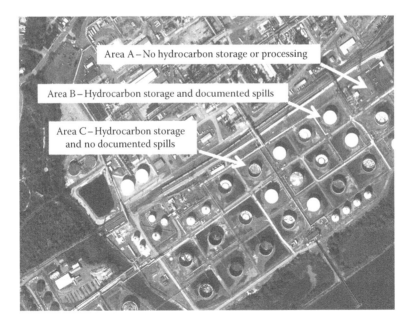

*Figure 13.37* Oil refinery with potential need for remediation.

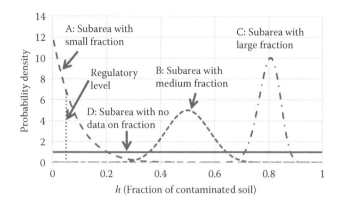

*Figure 13.38* Prior probability distributions for fraction of contaminated soil based on historical land-use in subareas of oil refinery (see Figure 13.37 for subareas).

above the regulatory limit) in a set of samples followed a binomial distribution. Therefore, the updated distribution for the fraction of contaminated samples in a subarea would also follow a β-distribution (Table 13.1).

A framework to make decisions on whether or not to remediate a subarea was then formulated. Guidance on the relative cost of failing to remediate a subarea that was contaminated was established for remediation based on input from the regulators. If an area was suspected to have benzene contamination, the regulators would generally require that 20 samples be taken and support a decision to not remediate the area if all 20 samples were clean. Mathematically, if there is nothing known about an area (i.e., the prior probability distribution of the fraction exceeding 50 mg/kg is uniform between 0 and 1), then obtaining 20 clean samples is equivalent to achieving a probability of 66% that the fraction is <1/20. Therefore, the relative cost of failing to remediate a contaminated site was set to be three times greater than the cost of remediation, meaning that the decision to remediate would be preferred if the probability that the contaminated fraction was <1/20 was smaller than 66% (Figure 13.39). On the basis of prior information, the probability was assessed for each subarea that the contaminated fraction was smaller than 0.05 (Figure 13.40). The decision to remediate a subarea could be made if there was a large probability of high benzene concentrations in the soil or if little was known about it based on the prior information.

Finally, an assessment of the value of information from additional soil samples was conducted to design a remedial investigation program. An example of the preposterior decision tree for this assessment in one particular subarea is shown in Figure 13.41. Results from

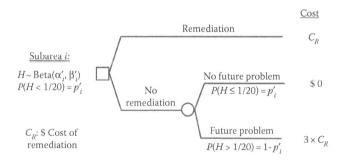

*Figure 13.39* Prior decistion tree to analyze the remediation decision.

*Figure 13.40* Prior probability that fraction of contaminated soil is less than 0.05 in subareas of oil refinery.

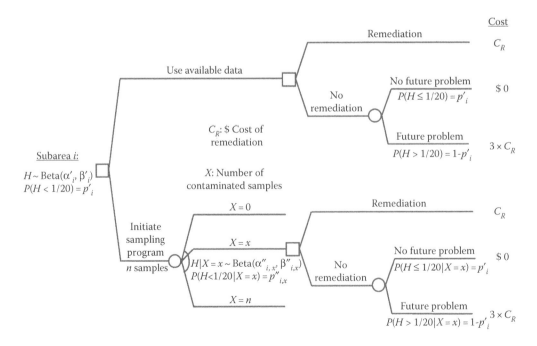

*Figure 13.41* Preposterior decision tree to assess value of information in obtaining additional soil samples from a subarea of oil refinery.

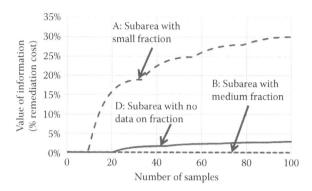

Figure 13.42 Value of information versus number of additional soil samples for different subareas of oil refinery.

this value of information assessment are shown in Figure 13.42; the value of information is expressed as a percentage of the cost to remediate the subarea. For a subarea where it was likely that there is benzene contamination, the value of information from additional soil samples was relatively small because it is unlikely that the additional information will change the decision to remediate the subarea. On the other hand, the value of information was relatively high for subareas where there was the greatest potential to change the prior decision to remediate (i.e., those subareas in Figure 13.40 with prior probabilities near 66% that the contaminated fraction was <1/20).

## 13.5.3 Exploration program for resources

The development of unconventional oil and gas resources is an interesting application because the technologies are in their infancy and there is little "prior" information that is necessarily relevant to a new field. Figure 13.43 shows a preposterior decision tree to assess the value of information from test wells in a new play. The consequence is related to the frequency of good wells in the new play, $\theta$. If the frequency is greater than a breakeven value,

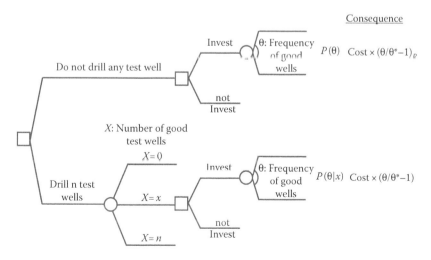

Figure 13.43 Preposterior decistion tree to assess value of information in drilling test wells for an unconventional gas reservoir.

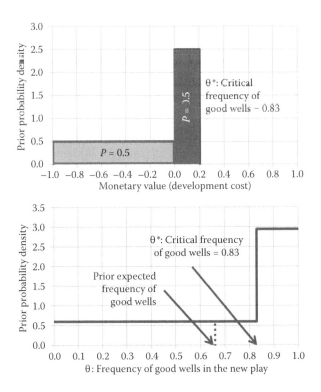

*Figure 13.44* Prior probability distributions for decision consequences and for frequency of good wells in developing an unconventional gas reservoir.

$\theta^*$, then the play will be profitable (i.e., consequence greater than zero). Likewise, if the frequency is less than $\theta^*$, then the play will not be profitable. The frequency of good wells in the new play is uncertain.

The prior probability distribution for the frequency of good wells in the new play is intended to reflect the case of no information or maximum uncertainty. A theory called Decision Entropy Theory (Gilbert et al. 2012; Habibi et al. 2014) is currently under development to establish a "non-informative" prior probability distribution in the context of decision making. The principle of this theory is to establish prior probabilities for variables in a decision by maximizing the entropy of the outcomes of the decision and what might or might not be learned about these outcomes. Figure 13.44 shows the product of this theory in this application; the entropy of the decision is maximized such that there is a 50–50 probability that developing the play will be preferred (i.e., that it will be profitable if implemented) and there are equal probabilities for the possible profits if development is preferred and possible losses if development is not preferred (Figure 13.44). Note that this play is rather risky in that a high frequency of good wells is required for it to be profitable, $\theta^* = 0.83$. The noninformative prior probability distribution for the decision consequence is then mapped into a noninformative prior probability distribution for the uncertain model parameter, the frequency of good wells in the new play (Figure 13.44).

Information will generally be available before making an investment decision on a new play. Specifically, the performance of similar wells in similar geologic conditions, or analog fields, will be used in making the decision to develop the new play. Consider the case where an analog field had 20 good wells out of 100 total wells. If this information is assumed to apply directly to the new play, then we would choose not to develop it because the frequency

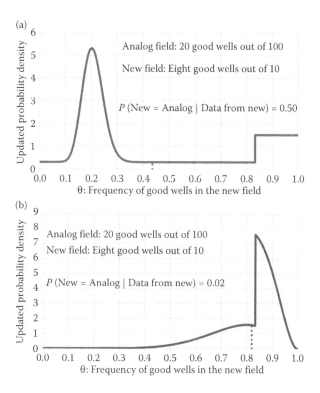

(a)

Analog field: 20 good wells out of 100

New field: Eight good wells out of 10

$P$ (New = Analog | Data from new) = 0.50

Updated probability density

θ: Frequency of good wells in the new field

(b)

Analog field: 20 good wells out of 100

New field: Eight good wells out of 10

$P$ (New = Analog | Data from new) = 0.02

Updated probability density

θ: Frequency of good wells in the new field

*Figure 13.45* Example updated probability distributions for frequency of good wells in developing an unconventional gas reservoir.

of good wells based on a large sample of data is 0.20 while the frequency of good wells in the new field needs to be >0.83 to be profitable. However, it is possible that information from the analog field is not directly relevant to the performance of the new field since the development technology is advancing rapidly and the reservoir conditions may not be identical (i.e., the new field could be a "Black Swan" in the words of Taleb 2007).

The principle of Decision Entropy Theory to maximize the entropy of what might or might not be learned about the decision outcomes suggests starting with a 50–50 probability that the data from the analog field are relevant to the performance of the new field (Habibi et al. 2014). Figure 13.45 shows the updated distribution for the frequency of good wells in the new field based on the analog field and test wells drilled in the new field. If there are no test wells in the new field, the prior probability distribution reflects the data from the analog field and the prior probability distribution for the frequency of good wells in the new field (Figure 13.45). As data from test wells in the new field become available indicating a much higher frequency of good wells than that in the analog field, the probability that the analog field is relevant to the new field decreases and the updated probability distribution for the frequency of good wells in the new field more strongly reflects the data from the new field than the analog field (Figure 13.45).

The value of information versus the number of test wells in the new field is shown in Figure 13.46. For small numbers of test wells, there is no value to the information because there will not be enough information to justify developing the play even if all of the wells are good. The value of information from test wells in the new field assuming that the analog field is the same as the new field is also shown for comparison in Figure 13.46; there is no value to this information even with 50 test wells. A simple explanation for this result is to consider the

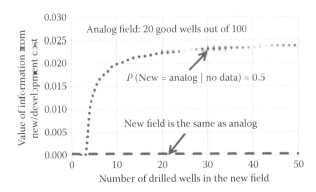

*Figure 13.46* Value of information versus number of test wells in developing an unconventional gas reservoir.

case where all 50 test wells in the new field are good. Since the analog field with only 20 good wells out of 100 is directly relevant, the combined data give 70 good wells out of 150 total wells, which are still well below the break-even frequency of 0.83 in the new play.

This case-history application illustrates the following key points:

1. It is not reasonable to assume that prior probability distributions can be established based entirely on experience because that precludes the possibility of events beyond our experience.
2. The value of test wells in this risky play is enhanced when leaving open the possibility for profit even though available information from analog fields are not encouraging.
3. There is a balance between relying entirely on experience (i.e., historical data from analog fields) versus not relying on it at all.

### 13.5.4 QA/QC testing

Figure 13.47 shows a geomembrane liner being installed for waste containment; the geo-membrane panels are welded together to form seams. QA/QC testing for the installation involves taking destructive samples of the seams and measuring their strength. For destructive samples that fail the strength test, additional destructive samples are taken to bound and then repair the length of the defective seam (Figure 13.48).

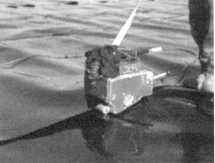

*Figure 13.47* Installing geomebrane liner by seaming panels together in the field.

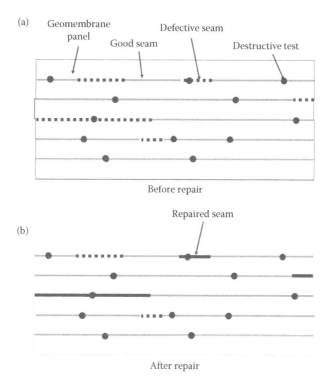

(a)

Geomembrane panel

Good seam

Defective seam

Destructive test

Before repair

(b)

Repaired seam

After repair

*Figure 13.48* Coneptual model of seam quality and destructive testing for QA/QC of geomembrane liner.

To assess the optimal frequency for destructive tests, a two-state Poisson model was developed (Gilbert 1997). This model is described by two model parameters: the mean fraction of defective seam length and the mean length of defective seam segments. Data from geomembrane installations were used to develop typical values for these model parameters. Monte Carlo simulation was then used with Bayes' Theorem to determine the updated percentage of defective seam length repaired as a function of the sampling interval relative to the mean length for a defective segment (Figure 13.49); for small sampling intervals close to 100%, the defective seam is detected and repaired. However, there is a cost to small sampling intervals beyond the cost of testing since destructively sampling the membrane requires cutting holes in it.

A simplified analysis was performed to assess the value of information from destructive sampling. The cost of having a defective seam in the final installation (after QA/QC) was assumed to be proportional to the length of the defective seam, and the cost of destructive testing was assumed to be proportional to the number of samples. Results are shown in Figure 13.50 for the expected net benefit from destructive testing, where the expected net benefit is the expected cost with QA/QC less the expected cost without QA/QC. The optimal sampling interval corresponds to about 50 m. The practical significance of this result is that typical sampling intervals today are in the order of 100–150 m. Therefore, there is significant potential to improve the value of the information from destructive testing by reducing the sampling interval.

## 13.6 SUMMARY

This chapter has described a framework and provided practical tools and insights for assessing the value of information to design site investigation and construction quality assurance

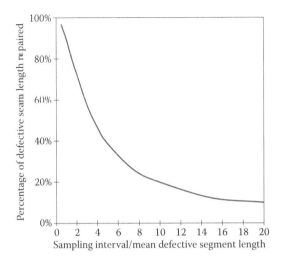

*Figure 13.49* Expected percentage of defective seam length that is repaired versus sampling interval of destructive testing for QA/QC of geoembrane liner.

programs. The intent has been to provide illustrative examples with enough detail that readers can recreate the analyses, to demonstrate with actual case histories of how these tools can benefit the practice, and to offer practical insights into the value of information.

The following important concepts have been illustrated and emphasized:

1. The value of information depends on how useful that information will be in making decisions. The value of information is enhanced when leaving open the possibility that decisions may change based on the information.
2. Additional information does not necessarily decrease uncertainty; uncertainty can increase, remain unchanged, or decrease with additional information.
3. Relationships (such as correlations) between pieces of information can significantly affect the value of the information. While assuming pieces of information are

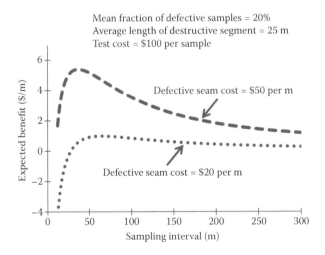

*Figure 13.50* Expected benefit versus destructive testing interval for QA/QC of geoembrane liner.

independent can be mathematically convenient, this assumption is not necessarily realistic and it can overstate the value of the information.
4. The value of information from site investigation data will generally depend strongly on the geologic setting.
5. It is not reasonable to assume that probabilities can be established based entirely on experience because that approach precludes the possibility of events beyond our experience. There is an important balance between relying entirely on experience versus not relying on it at all.

There is ample room to improve the practicality in assessing the value of information to design investigation and testing programs and to establish simple guidance and insights without requiring detailed analyses.

## ACKNOWLEDGMENTS

The contents of this chapter were developed through a variety of research and consulting projects supported by the United States National Science Foundation, the United States Department of Interior, the State of Texas, the Center for Petroleum Asset Risk Management at The University of Texas at Austin, and industry.

## REFERENCES

Ang, A. H.-S. and Tang, W. H. 1984, *Probability Concepts in Engineering Planning and Design: Volume 2, Decision, Risk, and Reliability*, John Wiley & Sons, New York.

Ang, A. H.-S. and Tang, W. H. 2007, *Probability Concepts in Engineering: Emphasis on Applications in Civil and Environmental Engineering*, 2nd edition, John Wiley & Sons, New York.

Benjamin, J. R. and Cornell, C. A. 1970, *Probability, Statistics, and Decision for Civil Engineers*, McGraw-Hill Book Company, New York.

Cheon, J. Y. and Gilbert, R. B. 2014, Modeling spatial variability in offshore geotechnical data for reliability-based foundation design, *Structural Safety*, 49, 18–26.

Fisher, R. A. 1935, *The Design of Experiments*, The MacMillan and Co. Ltd., London, UK.

Gambino, S. J. and Gilbert, R. B. 1999, Modeling spatial variability in pile capacity for reliability-based design, In *Analysis, Design, Construction and Testing of Deep Foundations*, Roesset Ed., ASCE Geotechnical Special Publication No. 88, Reston, VA, 135–149.

Gilbert, R. B. 1997, Model to design QA/QC programs for geomembrane liner seams, *Journal of Engineering Mechanics, ASCE*, 123(6), 586–595.

Gilbert, R. B. 2002, Questions for the future of risk assessment in environmental geotechnics, *Proceedings, Fourth International Congress on Environmental Geotechnics*, Rio de Janeiro, Brazil, Vol. 2, 1007–1012.

Gilbert, R. B., Gambino, S. J. and Dupin, R. M. 1999a, Reliability-based approach for foundation design without site-specific soil borings, *Proceedings, Offshore Technology Conference*, Houston, Texas, OTC 10927, 631–640.

Gilbert, R. B., Habibi, M. and Min, N. 2012, Achieving reliability in the face of extreme uncertainty, Keynote Paper, *Proceedings, Asian Pacific Symposium on Structural Reliability and Its Applications*, Singapore.

Gilbert, R. B., Najjar, S. S., Choi, Y. J. and Gambino, S. J. 2008, Practical application of reliability-based design in decision making, In *Reliability-Based Design in Geotechnical Engineering: Computations and Applications*, Phoon Ed., Taylor & Francis Books Ltd., London.

Gilbert, R. B., Stong, T. J., Lang, J. T., Albrecht, R. S. and Dupin, R. M. 1999b, Optimizing investigation programs for offshore platform foundations—Effect of geology on axial pile capacity, *Proceedings, 2nd International Conference on Seabed Geotechnics*, IBC Ltd., London, UK.

Habibi, M., Gilbert, R. B., Lake, L. W. and McIntosh, P. 2014, Assessing value of test wells in developing an unconventional play—Decision entropy theory, SPE *Hydrocarbon Economics and Evaluation Symposium*, Society of Petroleum Engineers, Houston, TX, May, pp. 19–20.

Hastings, W. 1970, Monte Carlo sampling methods using Markov Chains and their application, *Biometrika*, 57, 97–109.

Jaynes, E. T. 1957, Information theory and statistical mechanics, *Physical Review*, 106(4), 620–630.

Jaynes, E. T. 1968, Prior probabilities, *IEEE Transactions on System Science and Cybernetics*, 4(3), 227–241.

Journel, A. G. and Deutsch, C. V. 1993, Entropy and spatial disorder, *Mathematical Geology*, 25(3), 329–355.

Keynes, J. M. 1921, *A Treatise on Probability*, The MacMillan and Co. Ltd., London, UK.

Luce, R. D. and Raiffa, H. 1957, *Games and Decisions*, John Wiley & Sons, New York.

Shannon, C. E. 1948, A mathematical theory of communication, *Bell System Technical Journal*, 27, 379–423 and 623–656, July and October.

Taleb, N. N. 2007, *The Black Swan: The Impact of the Highly Improbable*, Random House, Inc., New York.

Tribus, M. 1969, *Rational Descriptions, Decisions and Designs*, Pergamon Press, New York.

# Chapter 14

# Verification of geotechnical reliability using load tests and integrity tests

*Limin Zhang*

## 14.1 INTRODUCTION

Pile tests are an important means to cope with uncertainties in the design and construction of pile foundations and avoid the use of excessively conservative designs. Both quantitative tests, such as static or dynamic load tests, and qualitative integrity tests, such as coring examination and cross-hole sonic logging, are routinely employed. Pile tests serve several purposes. Some of their functions include verification of design parameters (this is especially important if the geotechnical data are uncertain), establishment of the effects of construction methods on foundation capacities, meeting regulatory requirements, and provision of data for the improvement of design methodologies in use and for research purposes (Zhang 2004). In addition to these functions, savings may be derived from pile tests. With pile tests, lower factors of safety (FOSs) or higher-strength parameters may be used. Hence, there are cost savings even when no changes in the design are made after the tests. For example, the U.S. Army Corps of Engineers (USACE) (1993), the American Association of State Highway and Transportation Officials (AASHTO) (1997), and the Geotechnical Engineering Office (2006) recommend the use of different FOSs depending on whether load tests are carried out or not to verify the design. The FOSs recommended by the USACE are shown in Table 14.1. The FOS under the usual load combination may be reduced from 3.0 for designs based on theoretical or empirical predictions to 2.0 for designs based on the same predictions that are verified by a sufficient number of proof load tests.

Engineers have used proof pile tests in the allowable stress design (ASD) as follows. The piles are sized using a design method suitable for the soil conditions and the construction procedure for the project considered. In designing the piles, an FOS of 2.0 is often used when proof load tests are scheduled. Preliminary piles or early test piles are then constructed, necessary design modifications made based on construction control, and proof tests on the test piles or some working piles conducted. If the test piles do not reach a prescribed failure criterion at the maximum test load (i.e., twice the design load), then the design is validated; otherwise, the design load for the piles must be reduced or additional piles or pile lengths must be installed.

Integrity tests are also performed on selected piles to detect defects such as toe debris, short piles, voids, honeycombing, cracks, cavities, and necking. If the detected defect sizes are larger than the tolerable defect sizes, the defects need to be repaired or the defective piles need to be replaced. Whether actions are taken or not after the tests, the information obtained from these tests will reduce the uncertainties involved in the design variables, such as occurrence probability of defect and defect size. As the distributions of these variables affecting the reliability of the pile foundations are updated using the additional information from the tests, the reliability of the pile foundations can be updated.

*Table 14.1* Factor of safety for pile capacity

| Method of determining capacity | Loading condition | Minimum factor of safety | |
|---|---|---|---|
| | | Compression | Tension |
| Theoretical or empirical prediction to be verified by pile load test | Usual | 2.0 | 2.0 |
| | Unusual | 1.5 | 1.5 |
| | Extreme | 1.15 | 1.15 |
| Theoretical or empirical prediction to be verified by pile driving analyzer | Usual | 2.5 | 3.0 |
| | Unusual | 1.9 | 2.25 |
| | Extreme | 1.4 | 1.7 |
| Theoretical or empirical prediction not verified by load test | Usual | 3.0 | 3.0 |
| | Unusual | 2.25 | 2.25 |
| | Extreme | 1.7 | 1.7 |

Source: After US Army Corps of Engineers (USACE). 1993. *Design of Pile Foundations,* New York: ASCE Press.

In a reliability-based design (RBD), the value of pile tests can be further maximized. In an RBD incorporating the Bayesian approach (e.g., Ang and Tang 2007), the same load test results reveal more information. Predicted pile capacity using theoretical or empirical methods can be very uncertain. Results from load tests not only suggest a more realistic pile capacity value, but also greatly reduce the uncertainty of the pile capacity since the error associated with load-test measurements is much smaller than that associated with predictions. In other words, the reliability of a design for a test site can be updated by synthesizing existing knowledge of pile design and site-specific information from pile tests.

This chapter presents systematic methods to incorporate results of quantitative proof load tests and qualitative integrity tests into foundation design. Two examples are worked out to evaluate the reliability of piles after verification by proof load tests and integrity tests. Reliability and FOS are used together in this chapter since they provide complementary measures of an acceptable design (Duncan 2000) as well as a better understanding of the effect of pile tests on the costs and reliability of pile foundations. This chapter will focus only on ultimate limit states.

## 14.2 WITHIN-SITE VARIABILITY OF PILE CAPACITY

In addition to uncertainties with site investigation, laboratory testing, and design models, the capacity values of supposedly identical piles within one site also vary. Suppose several "identical" test piles are constructed at a seemingly uniform site and are load tested following an "identical" procedure. The measured values of the ultimate capacity of the piles would usually be different due to the so-called "within-site" variability following Baecher and Rackwitz (1982). For example, Evangelista et al. (1977) tested 22 "identical" bored piles in a sand–gravel site. The piles were "all" 0.8 m in diameter and 20 m in length, and construction of the piles was assisted with bentonite slurry. The load tests revealed that the coefficient of variation (COV) of the settlement of these "identical" piles at the intended working load was 0.21, and the COV of the applied loads at the mean settlement corresponding to the intended load was 0.13.

Evangelista et al. (1977) and Zhang et al. (2004) described several sources of the within-site variability of pile capacity: inherent variability of properties of the soil in the influence zone of each pile, construction effects, variability of pile geometry (length and diameter), variability of properties of the pile concrete, and soil disturbance caused by pile driving and afterward setup. Construction effects and setup effects are worth a particular mention because these effects can introduce many mechanisms that cause a large scatter of data. One example of construction effects is the influence of drilling fluids on the capacity of drilled shafts. According to O'Neill and Reese (1999), improper handling of bentonite slurry alone could reduce the beta factor for pile shaft resistance from a common range of 0.4–1.2 to below 0.1. Setup effects refer to the phenomenon that the pile capacity increases with time following pile installation. Chow et al. (1998) and many others revealed that the capacity of driven piles in sand could increase from a few percent to over 200% after the end of initial driving. Such effects make the pile capacity from a load test a "nominal" value.

The variability of properties of the soil and the pile concrete is affected by the space over which the properties are estimated (e.g., Fenton and Griffiths 2003). The issue of spatial variability of soils is the subject in several chapters of this book. Note that, after including various effects, the variability of the soils at a site may not be the same as the variability of the pile capacity at the same site. Zhang and Dasaka (2010) investigated the spatial variability of a weathered ground using random field theory and geostatistics. The scale of fluctuation of the depth of competent rock (moderately decomposed rock) beneath a building block is approximately 30 m, whereas the fluctuation scale of the as-build depth of the piles that provide the same nominal capacity is only approximately 10 m.

Table 14.2 lists the values of the COV of the capacity of driven piles from load tests in nine sites. These values range from 0.12 to 0.28. Note that the variability reported in the table is among test piles at one site; the variability among production piles and among different sites may be larger (Baecher and Rackwitz 1982). In this chapter, a mean value of COV = 0.20 is adopted for analysis. This assumes that the standard deviation is proportional to the mean pile capacity.

Kay (1976) noted that the possible values of the pile capacity within a site favor a log-normal distribution. This may be tested with the data from 16 load tests in a site in southern Italy reported by Evangelista et al. (1977). Figure 14.1 shows the measured and assumed normal and log-normal distributions of the pile capacity in the site. Using the Kolmogorov–Smirnov test, both normal and log-normal theoretical curves do not appear to fit the observed cumulative curve very well. However, it is acceptable to adopt a log-normal distribution for mathematical convenience, since the maximum deviation in Figure 14.1 is still smaller than the 5% critical value. The tail part of the cumulative distribution of the pile capacity is of more interest to designers since the pile capacity in that region is smaller. In Figure 14.1, it can be seen that the assumed log-normal distribution underestimates the pile capacity at a

*Table 14.2* Within-site viability of the capacity of driven piles in eight sites

| Site | Number of piles | Diameter (m) | Length (m) | Soil | COV | Reference |
|---|---|---|---|---|---|---|
| Ashdod, Israel | 12 | – | – | Sand | 0.22 | Kay (1976) |
| Bremerhaven, Germany | 9 | – | – | Sand | 0.28 | Kay (1976) |
| San Francisco, USA | 5 | – | – | Sand and clay | 0.27 | Kay (1976) |
| Southern Italy | 12 | 0.40 | 8.0 | Sand and gravel | 0.25 | Evangelista et al. (1977) |
| Southern Italy | 4 | 0.40 | 12.0 | Sand and gravel | 0.12 | Evangelista et al. (1977) |
| Southern Italy | 17 | 0.52 | 18.0 | Sand and gravel | 0.19 | Evangelista et al. (1977) |
| Southern Italy | 3 | 0.36 | 7.3 | Sand and gravel | 0.12 | Evangelista et al. (1977) |
| Southern Italy | 4 | 0.46 | 7.0 | Sand and gravel | 0.14 | Evangelista et al. (1977) |
| Southern Italy | 16 | 0.50 | 15.0 | Clay, sand, and gravel | 0.20 | Evangelista et al. (1977) |

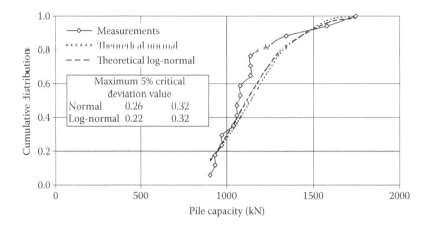

*Figure 14.1* Observed and theoretical distributions of within-site pile capacity.

particular cumulative percentage near the tail. Therefore, the assumed distribution will lead to results on the conservative side.

From the definition, the within-site variability is inherent in a particular geological setting and a geotechnical construction procedure at a specific site. The within-site variability represents the minimum variability for a construction procedure at a site, which cannot be reduced using load tests.

## 14.3 UPDATING PILE CAPACITY WITH PROOF LOAD TESTS

### 14.3.1 Proof load tests that pass

For proof load tests that are not carried out to failure, the pile capacity values are not known although they are greater than the maximum test loads. Define the variate, $x$, as the ratio of the measured pile capacity to the predicted pile capacity (called the "bearing capacity ratio" hereafter). At a particular site, $x$ can be assumed to follow a log-normal distribution (Whitman 1984; Barker et al. 1991) with the mean and standard deviation of $x$ being $\mu$ and $\sigma$ and those of $\ln(x)$ being $\eta$ and $\xi$, where $\sigma$ or $\xi$ describes the within-site variability of the pile capacity. Suppose the specified maximum test load corresponds to a value of $x = x_T$. For example, if the maximum test load is twice the design load and an FOS of 2.0 is used, then the value of $x_T$ is 1.0. Using the log-normal probability density function (PDF), the probability that the test pile does not fail at the maximum test load (i.e., $x > x_T$) is

$$p(x \geq x_T) = \int_{x_T}^{\infty} \frac{1}{\sqrt{2\pi}\xi x} \exp\left\{ -\frac{1}{2}\left( \frac{\ln(x) - \eta}{\xi} \right)^2 \right\} dx \tag{14.1}$$

Noting that $y = (\ln(x) - \eta)/\xi$, Equation 14.1 becomes

$$p(x \geq x_T) = \int_{\frac{\ln(x_T) - \eta}{\xi}}^{\infty} \frac{1}{\sqrt{2\pi}} \exp\left( -\frac{1}{2}y^2 \right) dy = 1 - \Phi\left( \frac{\ln(x_T) - \eta}{\xi} \right) = \Phi\left( -\frac{\ln(x_T) - \eta}{\xi} \right) \tag{14.2}$$

where $\Phi$ is the cumulative distribution function of the standard normal distribution. Suppose that $n$ proof tests are conducted and none of the test piles fails at $x_T$. If the standard deviation, $\xi$, of $\ln(x)$ is known but its mean, $\mu$ or $\eta$, is a variable, then the probability that all of the $n$ test piles do not fail at $x_T$ is

$$L(\mu) = \prod_{i=1}^{n} p_X(x \geq x_T)\big|_{\mu} = \Phi^n\left(-\frac{\ln(x_T) - \eta}{\xi}\right) \tag{14.3}$$

$L(\mu)$ is also called "likelihood function" of $\mu$. Given $L(\mu)$, the updated distribution of the mean of the bearing capacity ratio is (e.g., Ang and Tang 2007)

$$f''(\mu) = \alpha\Phi^n\left(-\frac{\ln(x_T) - \eta}{\xi}\right)f'(\mu) \tag{14.4}$$

where $f'(\mu)$ is the prior distribution of $\mu$, which can be constructed based on the empirical log-normal distribution of $x$ and the within-site variability information (Zhang 2004), and $\alpha$ is a normalizing constant:

$$\alpha = \left[\int_{-\infty}^{\infty} \Phi^n\left(-\frac{\ln(x_T) - \eta}{\xi}\right)f'(\mu)\,d\mu\right]^{-1} \tag{14.5}$$

Given an empirical distribution $N(\mu_X, \sigma_X)$, the prior distribution can also be assumed as a normal distribution with the following parameters:

$$\mu' = \mu_X \tag{14.6}$$

$$\sigma' = \sqrt{\sigma_X^2 - \sigma^2} \tag{14.7}$$

The updated distribution of the bearing capacity ratio, $x$, is thus (e.g., Ang and Tang 2007)

$$f_X(x) = \int_{-\infty}^{\infty} f_X(x|\mu)f''(\mu)\,d\mu \tag{14.8}$$

where $f_X(x|\mu)$ is the distribution of $x$ given the distribution of its mean. This distribution is assumed to be log-normal as mentioned earlier.

## 14.3.2 Proof load tests that do not pass

More generally, suppose only m out of n test piles do not fail at $x = x_T$. The probability that this event occurs is

$$L(\mu) = \binom{n}{m}\left[p(x \geq x_T)\big|_{\mu}\right]^m\left[1 - p(x \geq x_T)\big|_{\mu}\right]^{(n-m)} \tag{14.9}$$

and the normalizing constant, $\alpha$, is

$$\alpha = \left[ \int_{-\infty}^{\infty} \binom{n}{m} \left[ p(x \geq x_T)\big|_\mu \right]^m \left[ 1 - p(x \geq x_T)\big|_\mu \right]^{(n-m)} f'(\mu)\,d\mu \right]^{-1} \tag{14.10}$$

Accordingly, the updated distribution of the mean of the bearing capacity ratio is

$$f''(\mu) = \alpha \binom{n}{m} \left[ p(x \geq x_T)\big|_\mu \right]^m \left[ 1 - p(x \geq x_T)\big|_\mu \right]^{(n-m)} f'(\mu) \tag{14.11}$$

Equation 14.11 has two special cases. When all tests pass (i.e., $m = n$), Equation 14.11 reduces to Equation 14.4. If none of the tests passes (i.e., $m = 0$), a dispersive posterior distribution will result. The updated distribution of the bearing capacity ratio can be obtained using Equation 14.8.

### 14.3.3 Proof load tests conducted to failure

Let us now consider the proof load tests carried out to failure, that is, cases the pile capacity values are known. To start with, the log-normally distributed pile capacity assumed previously is now transformed into a normal variate. If the test outcome is a set of $n$ observed values representing a random sample following a normal distribution with a known standard deviation $\sigma$ and a mean of $\bar{x}$, then the distribution of the pile capacity can be updated directly using Bayesian sampling theory. Given a known prior normal distribution, $N_\mu\,(\mu', \sigma')$, the posterior density function of the bearing capacity ratio, $f_X''(x)$ is also normal (e.g., Ang and Tang 2007):

$$f_X''(x) = N_X(\mu_X'', \sigma_X'') \tag{14.12}$$

where

$$\mu_X'' = \frac{\bar{x}(\sigma')^2 + \mu'(\sigma^2/n)}{(\sigma')^2 + (\sigma^2/n)} \tag{14.13}$$

$$\sigma_X'' = \sigma\sqrt{1 + \frac{(\sigma'^2/n)}{(\sigma')^2 + (\sigma^2/n)}} \tag{14.14}$$

### 14.3.4 Multiple types of tests

In many cases, multiple test methods are used for construction quality assurance at a single site. The construction control of driven H-piles in Hong Kong (Zhang et al. 2006b) and the seven-step construction control process of Bell et al. (2002) are two examples. In the Hong Kong practice, both dynamic formula and static analysis are adopted for determining the pile length. Early test piles or preliminary test piles are required at the early stage of construction to verify the construction workmanship and design parameters. In addition, not fewer than 10% of working piles may be tested using high-strain dynamic tests, for example, using a pile driving analyzer (PDA). One-half of these PDA tests should be analyzed using a wave equation analysis such as the CAse Pile Wave Analysis Program (CAPWAP). Finally, at the end of construction, 1% of working piles should be proof tested using static loading tests. With these construction control measures in mind, a low FOS is commonly used. Zhang et al. (2006b) described the piling practice in the language of Bayesian updating. A preliminary pile length is first set based primarily on a dynamic formula. The pile performance is then verified by PDA tests during the final setting tests. The combination of the dynamic formula (prior) and the PDA tests will result in a posterior distribution. Taking the posterior distribution after the PDA tests as a prior, the pile performance is further updated based on the outcome of the CAPWAP analysis, and finally updated based on the outcome of the proof load tests. This exercise can be repeated if more indirect or direct verification tests are employed.

## 14.4 UPDATING PILE CAPACITY WITH INTEGRITY TESTS

### 14.4.1 Reliability updating based on integrity tests

Integrity tests do not offer pile capacity values directly. When using reliability theory, however, the outcome of qualitative integrity tests is routinely used as the basis for verifying the reliability of piles. Testing of toe debris in bored piles is used in this chapter to illustrate the use of qualitative testing information for the verification of pile reliability. Toe debris is the accumulation of loose or undesirable foreign materials at the position of pile rock socket. It is detrimental to piles that are designed to found on rock, as the quality of the interface between pile concrete and bedrock greatly affects the mobilization of the pile base resistance. There are three categories of defective toes, namely soil inclusions, unbounded concrete aggregate, and their combinations (HKCA 2003). Causes of toe debris include poor workmanship during concreting, soil dislodgement from unlined portion of excavation, substandard bentonite slurry, percolating groundwater, and objects falling from the ground surface. In order to ensure a satisfactory pile toe condition, interface coring, which is the formation of access to the pile shaft by drilling through a pile toe interface, is often conducted after pile construction following guidelines such as those specified by Hong Kong's Buildings Department (2004).

Consider the case of single-bored piles with toe debris along with other types of imperfections such as cracks, necking, and voids. If thick debris exists, then the pile capacity or settlement will be adversely affected. The toe debris, if present, is assumed to be uniformly distributed over the pile cross section for simplicity. The pile performance is still uncertain even when the pile is free from toe debris due to the presence of other types of imperfections as well as many other sources of uncertainty such as spatial variability of soils, construction effects, and load effects. Based on the total probability theorem (e.g., Ang and Tang 2007), the probability of unsatisfactory performance of the pile, $p_f$, can be calculated as follows:

$$p_f = P(F|\bar{E})(1 - p_d) + P(F|E)p_d \tag{14.15}$$

where $p_d$ is the occurrence probability of toe debris, $F$ is the event of pile failure, $E$ and $\bar{E}$ are the events of toe debris presence and toe debris absence, respectively, and $P(F|\bar{E})$ and $P(F|E)$ are the conditional probabilities of unsatisfactory performance of the pile given the absence and presence of toe debris, respectively.

For a particular pile, the toe debris may be described by its thickness $x$. If $x$ is taken to be a discrete variable, the conditional probability of unsatisfactory performance of the pile is

$$P(F|E) = \sum_{i=1}^{n} P(F|x_i)P(x_i|E) \tag{14.16}$$

in which $P(x_i|E)$ is the occurrence probability of a toe debris with thickness $x_i$ and $P(F|x_i)$ is the probability of unsatisfactory performance of the pile with the given toe debris. Similarly, if toe debris does exist and $x$ is taken to be a continuous variable, its probability distribution can be expressed as $f(x|t)$ in which the parameter, $t$, is the mean of $x$ and treated as another random variable with a probability distribution $f(t)$. The conditional probability of unsatisfactory performance of the pile can be further given by

$$P(F|E) = \int_{x_L}^{x_U} P(F|x) \left[ \int_{t_L}^{t_U} f(x|t)f(t)\,dt \right] dx \tag{14.17}$$

where $x_L$ and $x_U$ are the lower and upper bounds of $x$ and $t_L$ and $t_U$ are the lower and upper bounds of $t$, respectively.

For a pile that is constructed without any integrity testing, the empirical occurrence probability of toe debris can be considered as a prior distribution of $p_d$. The empirical perception of toe debris thickness can also be used to establish the prior distribution of toe debris thickness. A prior $p_f$ can then be calculated using Equation 14.15. If field measurements from interface coring are available, the distributions of $p_d$ and toe debris thickness can be updated with the additional measurement information using the Bayesian approach (e.g., Ang and Tang 2007), and $P(F|E)$ of the pile with toe debris can be calculated using Equation 14.16 or Equation 14.17. Once the updated distribution of $p_d$ and the updated conditional reliability of the pile with toe debris are available, the updated $p_f$ can be calculated using Equation 14.15.

## 14.4.2 Updating occurrence probability of toe debris

The Bayesian approach provides a method to combine empirical information and field observations. Take the occurrence probability of toe debris, $p_d$, as an example. An initial PDF of $p_d$, $f'(p_d)$, can be established based on theoretical models or expert judgments. When field observations of $p_d$ through on-site interface coring at a particular site are available, $f'(p_d)$ can be updated by combining the initial PDF with the field observations. The updated distribution can be regarded as a weighted average of the initial information and the additional observations. More specifically, the updated PDF of $p_d$, $f''(p_d)$, can be expressed as

$$f''(p_d) = KL(R|p_d)f'(p_d) \tag{14.18}$$

where $K$ is a normalization constant and $L(R|p_d)$ is the likelihood of observing the outcome $R$. Suppose $n$ piles are selected randomly from a site for inspection, and $m$ of them are found to contain toe debris. Since a thick toe debris, if present in a pile, can be detected with certainty in most cases by interface coring, it is reasonable to assume that the detection probability of interface coring is equal to one. If statistical independence among pile inspections is further assumed, then the likelihood that the assumed event occurs can be expressed as a binomial function

$$L(R|p_d) = \binom{n}{m} p_d^m (1 - p_d)^{n-m} \tag{14.19}$$

$f'(p_d)$ can be formulated based on available test data on toe debris. Since $p_d$ is bounded between 0 and 1, it may be described by a beta distribution

$$f'(p_d) = \frac{1}{B(q,r)} p_d^{(q-1)} (1 - p_d)^{(r-1)} \tag{14.20}$$

in which $q$ and $r$ are the parameters of beta distribution and $B()$ is the beta function. The beta distribution is a conjugate distribution of the binomial distribution. That is, if the prior $p_d$ is a beta distribution, the updated $p_d$ is still a beta distribution (Raiffa and Schlaifer 2000). In this case, the updated PDF of $p_d$ is a beta distribution with parameters $q'' = q + m$ and $r'' = r + n - m$.

If no information on $p_d$ is available, then a diffuse prior for $f'(p_d)$ may be used. In this case, one can obtain

$$f''(p_d) = \frac{p_d{}^m (1 - p_d)^{n-m}}{\int_0^1 p_d{}^m (1 - p_d)^{n-m} \, \mathrm{d}p_d} \tag{14.21}$$

### 14.4.3 Updating mean thickness of toe debris

The thickness of toe debris, $x$, is also assumed to follow an exponential distribution with the mean thickness $t$ as a parameter

$$f(x|t) = \frac{1}{t} \exp\left(-\frac{x}{t}\right), \quad x \geq 0 \tag{14.22}$$

For mathematical convenience and simplicity, in the absence of any information, the inverted gamma distribution, which is the conjugate distribution of the exponential distribution, may be taken as the prior distribution of $t$ (Raiffa and Schlaifer 2000)

$$\dot{f}'(t) = \frac{v^k}{\Gamma(\kappa)}\left(\frac{1}{t}\right)^{\kappa+1} \exp\left(-\frac{v}{t}\right), \quad t \geq 0 \tag{14.23}$$

in which $\Gamma()$ is the gamma function and $\kappa$ and $v$ are the parameters of the distribution. When results from on-site interface coring are available, the PDF of $t$ can be updated using the Bayesian approach. The posterior PDF of $t$ is also an inverted gamma distribution with parameters $\kappa'' = \kappa + m$ and $v''v + \Sigma_{i=1}^{m} x_{ii}$ (Raiffa and Schlaifer 2000) in which $m$ is the number of piles found to be defective and $x_i$ is the observed toe debris thickness in the $i$th defective pile.

### 14.4.4 Cases of test outcome

Three cases of test outcome are considered in this chapter: (1) no toe debris is detected, (2) toe debris is detected without repair, and (3) toe debris is detected and repaired.

Suppose $n$ piles are selected randomly from a site for interface coring. If the construction quality at the site is excellent and no toe debris is found in the sampled piles, then no actions need to be taken after the tests. In this case, the occurrence probability of toe debris will be updated after the tests, and the updated occurrence probability will be smaller than the estimated prior value. Since no information on the thickness of toe debris is obtained, the distribution of thickness of toe debris cannot be updated. According to Equation 14.15, the updated probability of unsatisfactory performance of the pile will be smaller than that before the integrity tests. The reliability of the piles will be improved although no remedial actions are taken.

If $n$ piles are selected randomly from a site for interface coring and $m$ out of these are found to contain toe debris, then a decision on immediate repair or replacement must be made. Let us assume the detected amount of toe debris is tolerable so that no repair actions are necessary. In this case, on-site information on both the occurrence probability and the toe debris thickness can be obtained from the tests. Accordingly, the distributions of occurrence probability and toe debris thickness can both be updated. Thus, the $p_f$ of the pile can be updated. In particular, if the observed occurrence probability is smaller than the prior occurrence probability, and the observed toe debris thickness is less than the prior thickness, then the updated $p_f$ after the tests will be smaller than that before the tests. The reliability of the piles will then be changed even though no remedial actions are taken.

If some toe debris is found in a random sampling at a site as in case 2 and the detected toe debris is deemed intolerable, then the toe debris needs to be repaired (e.g., by pressure grouting) or the defective piles need to be replaced. Upon detecting serious toe debris in the sampled piles, further inspection on other piles at this site may be carried out. If it is further assumed that all toe debris present will be detected and repaired, the occurrence probability of toe debris becomes zero. Obviously, the updated $p_f$ will be significantly smaller than that before the tests and repair.

## 14.5 RELIABILITY OF PILES VERIFIED BY PROOF LOAD TESTS

### 14.5.1 Calculation of reliability index

In an RBD, the safety of a pile can be described by a reliability index $\beta$. To be consistent with current efforts in code development, such as AASHTO's LRFD Bridge Design Specifications or the National Building Code of Canada (Barker et al. 1991; NRC 1995; Becker 1996; AASHTO 1997; Witham et al. 2001; Phoon et al. 2003), the first-order reliability method is used for calculating the reliability index. If both resistance and load effects are log-normal variates, then the reliability index for a linear performance function can be written as

$$\beta = \frac{\ln\left((\bar{R}/\bar{Q})\sqrt{(1 + COV_Q^2)/(1 + COV_R^2)}\right)}{\sqrt{\ln[(1 + COV_R^2)(1 + COV_Q^2)]}} \tag{14.24}$$

where $\bar{Q}$ and $\bar{R}$ are the mean values of load effect and resistance, respectively, and $COV_Q$ and $COV_R$ are the COV for the load effect and resistance, respectively. If the only load effects to be considered are dead and live loads, Barker et al. (1991), Becker (1996), and Witham et al. (2001) have shown that

$$\frac{\bar{R}}{\bar{Q}} = \frac{\lambda_R FOS((Q_D/Q_L) + 1)}{\lambda_{QD}(Q_D/Q_L) + \lambda_{QL}} \tag{14.25}$$

and

$$COV_Q^2 = COV_{QD}^2 + COV_{QL}^2 \tag{14.26}$$

where $Q_D$ and $Q_L$ are the nominal values of dead and live loads, respectively; $\lambda_R$, $\lambda_{QD}$, and $\lambda_{QL}$ are the bias factors for the resistance, dead load, and live load, respectively, with the bias factor referring to the ratio of the mean value to the nominal value; $COV_{QD}$, and $COV_{QL}$ are the coefficients of variation for the dead load and live load, respectively; and

FOS is the factor of safety in the traditional ASD. Najjar and Gilbert (2009) show that a more realistic quantification of reliability can be achieved by incorporating a lower bound in the distribution of foundation capacity.

If the load statistics are prescribed, Equations 14.24 and 14.25 indicate that the reliability of a pile foundation designed with an FOS is a function of $\lambda_R$ and $COV_R$. As will be shown later, if a few pile tests are conducted at a site, the values of $\lambda_R$ and $COV_R$ associated with the design analysis can be updated using the test results. If the test results are favorable, then the updated $COV_R$ will decrease but the updated $\lambda_R$ will increase. The reliability level or the $\beta$ value will therefore increase. Conversely, if a target reliability, $\beta_T$, is specified, the FOS required to achieve the $\beta_T$ can be calculated using Equations 14.24 through 14.26 and the costs of construction can be reduced if the pile test results are favorable.

## 14.5.2 Example: Design based on SPT and verified by proof load tests

For illustration purposes, a standard penetration test (SPT) method proposed by Meyerhof (1976) for pile design is considered. This method uses the average corrected SPT blow count near the pile toe to estimate the toe resistance and the average uncorrected blow count along the pile shaft to estimate the shaft resistance. According to statistical studies conducted by Orchant et al. (1988), the bias factor and COV of the pile capacity from the SPT method are $\lambda_R = 1.30$ and $COV_R = 0.50$, respectively. The within-site variability of the pile capacity is assumed to be $COV = 0.20$ (Zhang 2004), which is smaller than the $COV_R$ of the design analysis. This is because the $COV_R$ of the prediction includes more sources of errors such as model errors and differences in construction effects between the site where the model is applied and the sites from which the information was extracted to formulate the model.

Calculations for updating the probability distribution can be conducted using an Excel spreadsheet. Figure 14.2 shows the empirical distribution of the bearing capacity ratio, $x$, based on the given $\lambda_R$ and $COV_R$ values and the updated distributions after verification by proof tests. The translated mean $\eta$ and standard deviation $\xi$ of $\ln(x)$, calculated based on the given $\lambda_R$ and $COV_R$, are used to define the empirical log-normal distribution. In Figure 14.2a, all proof tests are positive (i.e., the test piles do not fail at twice the design load); the mean value of the updated pile capacity after verification by the proof tests increases with the number of tests while the COV value decreases. Specifically, the updated mean increases from 1.30 for the empirical distribution to 1.74 after the design has been verified by three positive tests. In Figure 14.2b, the cases in which no test, one test, two tests, and all three tests are positive are considered. As expected, the updated mean decreases significantly when the number of positive tests decreases. The updated mean and COV values with different outcomes from the proof tests are summarized in Table 14.3. The updated distributions in Figure 14.2 may be approximated by the log-normal distribution for the convenience of reliability calculations using Equation 14.15. Because of the differences in the distributions updated by proof tests of different outcomes, the COV values in Table 14.3 do not change in a descending or ascending manner as the number of positive tests decreases.

The following typical load statistics in the LRFD Bridge Design Specifications (AASHTO 1997) are adopted for illustrative reliability analyses: $\lambda_{QD} = 1.08$, $\lambda_{QL} = 1.15$, $COV_{QD} = 0.13$, and $COV_{QL} = 0.18$. The dead-to-live load ratio, $Q_D/Q_L$, is structure-specific. Investigations by Barker et al. (1991) and McVay et al. (2000) show that $\beta$ is relatively insensitive to this ratio. For the SPT method considered, the calculated $\beta$ values are 1.92 and 1.90 for distinct $Q_D/Q_L$ values of 3.69 (75 m span length) and 1.58 (27 m span length), respectively, if an

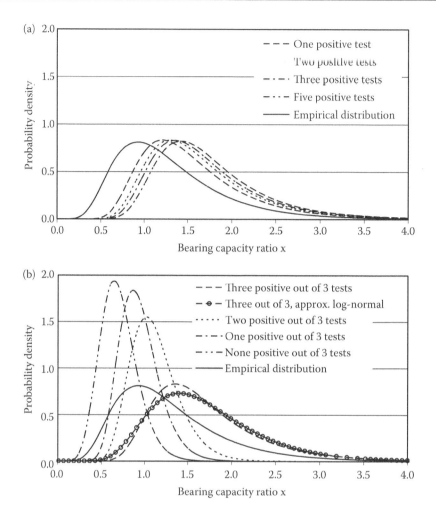

*Figure 14.2* Distributions of the pile capacity ratio after verification by (a) proof tests that pass at twice the design load and (b) proof tests in which some piles fail.

FOS of 2.5 is used. The difference between the two β values is indeed small. In the following analyses, the larger value of $Q_D/Q_L = 3.69$, which corresponds to a structure on which the dead loads dominate, is adopted. This $Q_D/Q_L$ value was used by Barker et al. (1991) and Zhang (2004) and would lead to FOSs on the conservative side.

Figure 14.3 shows the reliability index β values for single piles designed with an FOS of 2.0 and verified by several proof tests. Each of the curves in this figure represents the reliability of the piles when they have been verified by n proof tests out of which r tests are positive (see the likelihood function Equation 14.9). The β value corresponding to the empirical distribution (see Figure 14.2) and an FOS of 2.0 is 1.49. If one conventional proof test is conducted to verify the design and the test is positive, then the β value will be updated to 2.23. The updated β value will continue to increase if more proof tests are conducted and if all the tests are positive. In the cases when not all tests are positive, the reliability of the piles will decrease with the number of tests that are not positive. For instance, the β value of the piles verified by three positive tests is 2.58. If one, two, or all of the three test piles fail, the β values decrease to 2.21, 1.70, and 0.67, respectively. If multiple proof tests are conducted,

Table 14.3  Updated mean and COV of the bearing capacity ratio based on SPT (Meyerhof 1976) and verified by proof tests

| Total number of tests | Number of positive tests | Mean | COV | β at FOS = 2.0 |
|---|---|---|---|---|
| 4 | 4 | 1.77 | 0.38 | 2.67 |
|   | 3 | 1.19 | 0.24 | 2.38 |
|   | 2 | 1.04 | 0.24 | 2.00 |
|   | 1 | 0.91 | 0.24 | 1.57 |
|   | 0 | 0.69 | 0.30 | 0.60 |
| 3 | 3 | 1.74 | 0.38 | 2.58 |
|   | 2 | 1.14 | 0.25 | 2.21 |
|   | 1 | 0.96 | 0.25 | 1.70 |
|   | 0 | 0.72 | 0.31 | 0.67 |
| 2 | 2 | 1.68 | 0.39 | 2.45 |
|   | 1 | 1.06 | 0.26 | 1.91 |
|   | 0 | 0.75 | 0.32 | 0.77 |
| 1 | 1 | 1.59 | 0.41 | 2.23 |
|   | 0 | 0.82 | 0.34 | 0.95 |

the target reliability marked by the shaded zone can still be satisfied even if some of the tests are not positive.

The FOS that will result in a sufficient level of reliability is of interest to engineers. Figure 14.4 shows the calculated FOS values required to achieve a β value of 2.0 for the pile foundation verified by proof tests with different outcomes. It can be seen that an FOS of 2.0 is sufficient for piles that are verified by one or more consecutive positive tests, or by three or four tests in which no more than one test is not positive. However, larger FOS values should be used if the only proof test or one out of two tests is not positive.

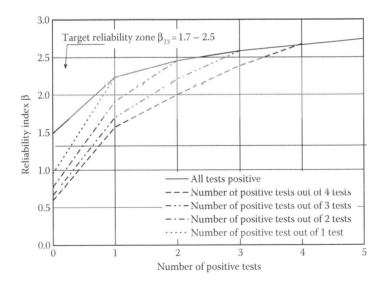

Figure 14.3  The reliability of single driven piles designed with an FOS of 2.0 and verified by conventional proof tests of different outcomes.

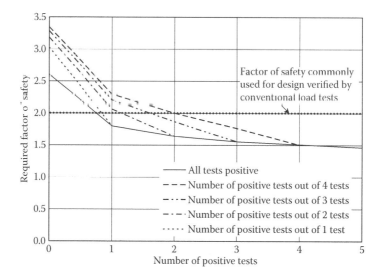

*Figure 14.4* Factor of safety for single driven piles that are required to achieve a target reliability index of 2.0 after verification by proof load tests.

The updating effects of the majority of the proof tests conducted to twice the design load and the tests in which the piles have a mean measured capacity of twice the design load are different. Table 14.4 compares the updated statistics of the two types of tests. For the proof tests that are not conducted to failure, the updated mean bearing capacity ratio and $\beta$ values significantly increase with the number of tests. For the tests that are conducted to failure at $\bar{x} = 1.0$, the updated mean value will approach 1.0 and the updated COV value will approach the assumed within-site COV value of 0.20 as expected when the number of tests is sufficiently large. This is because test outcomes carry a larger weight than that of the prior information as described in Equations 14.13 and 14.14.

## 14.5.3 Accuracy effect of design methods

The example in the previous section focuses on a single design method. One may ask if the observations from the example would apply to other design methods. Table 14.5 presents

*Table 14.4* Comparison of the updating effects of proof tests that pass at $x_T = 1.0$ and tests in which the test piles fail at a mean bearing capacity ratio of $x_T = 1.0$

| Number of tests | Tests that pass at $x_T = 1.0$ | | | Tests that fail at $x_T = 1.0$ | | |
|---|---|---|---|---|---|---|
| | Mean | COV | $\beta$ | Mean | COV | $\beta$ |
| 0 | 1.30 | 0.50 | 1.49 | 1.30 | 0.50 | 1.49 |
| 1 | 1.59 | 0.41 | 2.23 | 1.06 | 0.27 | 1.89 |
| 2 | 1.68 | 0.39 | 2.45 | 1.04 | 0.24 | 1.98 |
| 3 | 1.74 | 0.38 | 2.58 | 1.04 | 0.23 | 2.02 |
| 4 | 1.77 | 0.38 | 2.67 | 1.03 | 0.22 | 2.04 |
| 5 | 1.80 | 0.37 | 2.74 | 1.03 | 0.22 | 2.05 |
| 10 | 1.89 | 0.36 | 2.94 | 1.02 | 0.21 | 2.09 |

Note: Mean and COV of the bearing capacity ratio *x*.

Table 14.5 Reliability statistics of single driven piles

| Method category | Prediction method | Number of cases | Bias factor $\lambda_R$ | Coefficient of variation $COV_R$ | ASD factor of safety FOS | Reliability index $\beta$ | Reference |
|---|---|---|---|---|---|---|---|
| Dynamic methods with measurements (Construction stage) | CAPWAP (EOD, Florida database) | 44 | 1.60 | 0.35 | 2.5 | 3.11 | McVay et al. (2000) |
| | CAPWAP (BOR, Florida database) | 79 | 1.26 | 0.35 | 2.5 | 2.55 | McVay et al. (2000) |
| | CAPWAP (EOD, national database) | 125 | 1.63 | 0.49 | 2.5 | 2.39 | Paikowsky and Stenersen (2000) |
| | CAPWAP (BOR, national database) | 162 | 1.16 | 0.34 | 2.5 | 2.38 | Paikowsky and Stenersen (2000) |
| | Energy approach (EOD, Florida database) | 27 | 1.11 | 0.34 | 2.5 | 2.29 | McVay et al. (2000) |
| | Energy approach (BOR, Florida database) | 72 | 0.84 | 0.36 | 3.25 | 2.11 | McVay et al. (2000) |
| | Energy approach (EOD, national database) | 128 | 1.08 | 0.40 | 2.5 | 1.94 | Paikowsky and Stenersen (2000) |
| | Energy approach (BOR, national database) | 153 | 0.79 | 0.37 | 3.25 | 1.92 | Paikowsky and Stenersen (2000) |
| Static methods (Design stage) | Alpha method, clay type I | – | 1.10 | 0.21 | 2.5 | 3.08 | Sidi (1985), Barker et al. (1991) |
| | Alpha method, clay type II | – | 2.34 | 0.57 | 2.5 | 2.72 | Sidi (1985), Barker et al. (1991) |
| | Beta method | – | 1.03 | 0.21 | 2.5 | 2.82 | Sidi (1985), Barker et al. (1991) |
| | CPT method | – | 1.03 | 0.36 | 2.5 | 1.98 | Orchant et al. (1988), Barker et al. (1991) |
| | Lambda method, clay type I | – | 1.02 | 0.41 | 2.5 | 1.74 | Sidi (1985), Barker et al. (1991) |
| | Meyerhof's SPT method | – | 1.30 | 0.50 | 2.5 | 1.92 | Orchant et al. (1988), Barker et al. (1991) |

Source: Zhang, L. M., Tang, W. H. and Ng, C. W. W. 2001. Journal of Geotechnical and Geoenvironmental Engineering ASCE, 127, 1051–1060.

Note: Type I refers to soils with undrained shear strength $S_u < 50$ kPa; Type II refers to soils with $S_u > 50$ kPa.

statistics of a number of commonly used methods for design and construction of driven piles (Zhang et al. 2001). In this table, failure of piles is defined by the Davisson criterion. For each design method in the table, the driven pile cases are put together regardless of ground conditions or types of pile response (i.e., end-bearing or frictional). Ideally, the cases should be organized into several subsets according to their ground conditions and types of pile response. The statistics in the table are intended to be used only to illustrate the proposed methodology. The ASD approach results in designs with levels of safety that are rather uneven from one method to another (i.e., $\beta = 1.74$–$3.11$). If an FOS of 2.0 is used for all these methods and each design analysis is verified by two positive proof tests conducted to twice the design load (i.e., $x_T > 1.0$), the statistics of these methods can be updated as shown in Table 14.6. The updated bias factors are greater, but the updated $COV_R$ values are smaller than those in Table 14.5. The updated $\beta$ values of all these methods fall into a narrow range (i.e., $\beta = 2.22$–$2.89$) with a mean of approximately 2.5.

Now consider the case when an FOS of 2.0 is applied to all these methods in Table 14.5 and the design is verified by two proof tests conducted to failure at an average load of twice the design load (i.e., $x_T = 1.0$). The corresponding updated statistics using the Bayesian sampling theory (Equations 14.12 through 14.14) are also shown in Table 14.6. Both the updated bias factors ($\lambda_R = 0.97$–$1.12$) and the updated $COV_R$ values ($COV_R = 0.21$–$0.24$) fall into narrow bands. Accordingly, the updated $\beta$ values also fall into a narrow band (i.e., $\beta = 1.78$–$2.31$) with a mean of approximately 2.0.

The results in Table 14.6 indicate that the safety level of a design verified by proof tests is less influenced by the accuracy of the design method. This is logical in the context of Bayesian statistical theory in which the information of the empirical distribution will play a smaller role when more measured data at the site become available. This is consistent with foundation engineering practice. In the past, reliable designs were achieved by subjecting design analyses of varying accuracies to proof tests and other quality control measures (Hannigan et al. 1997; O'Neill and Reese 1999). This also shows the effectiveness of the observational method (Peck 1969), with which uncertainties can be managed and acceptable safety levels can be maintained by acquiring additional information during construction. However, the importance of the accuracy of a design method in sizing the pile should be emphasized. A more accurate design method has a smaller $COV_R$ and utilizes a larger percentage of the actual pile capacity (McVay et al. 2000). Hence, the required safety level can be achieved more economically.

## 14.6 RELIABILITY OF PILES VERIFIED BY INTEGRITY TESTS

### 14.6.1 Worked example

The impact of interface coring on the reliability of large-diameter bored piles with toe debris is illustrated in this worked example. The same procedure can be applied to study the impact of other integrity tests on the reliability of piles with various imperfections. If a pile contains toe debris, the pile capacity will be adversely affected. A pile capacity reduction factor, $R_F$, is used to measure the effect of toe debris, which is defined as the ratio of the capacity of the pile with toe debris to the capacity of the pile without toe debris.

Strictly speaking, $R_F$ should be treated as a random variable. Owing to lack of sufficient statistical data for $R_F$, $R_F$ is treated as a deterministic quantity in this study, but its value will depend on the debris thickness, and the length and diameter of the pile. Once $R_F$ is obtained, the bias factor of the capacity of the pile with toe debris, $\lambda_{RD}$, is given by

$$\lambda_{RD} = R_F \lambda_R \tag{14.27}$$

Table 14.6 Reliability statistics of single driven piles designed using an FOS of 2.0 and verified by two load tests

| Prediction method | Verified by two positive tests, $x_T > 1.0$ | | | Verified by two tests with a mean $x_T = 1.0$ | | |
|---|---|---|---|---|---|---|
| | Bias factor $\lambda_R$ | Coefficient of variation $COV_R$ | Reliability index $\beta$ | Bias factor $\lambda_R$ | Coefficient of variation $COV_R$ | Reliability index $\beta$ |
| CAPWAP (EOD, Florida database) | 1.71 | 0.32 | 2.89 | 1.12 | 0.24 | 2.21 |
| CAPWAP (BOR, Florida database) | 1.46 | 0.30 | 2.62 | 1.07 | 0.24 | 2.07 |
| CAPWAP (EOD, national database) | 1.89 | 0.42 | 2.58 | 1.07 | 0.24 | 2.05 |
| CAPWAP (BOR, national database) | 1.38 | 0.29 | 2.54 | 1.05 | 0.24 | 2.02 |
| Energy approach (EOD, Florida database) | 1.34 | 0.28 | 2.50 | 1.04 | 0.24 | 1.99 |
| Energy approach (BOR, Florida database) | 1.22 | 0.28 | 2.26 | 0.98 | 0.24 | 1.81 |
| Energy approach (EOD, national database) | 1.42 | 0.31 | 2.46 | 1.03 | 0.24 | 1.95 |
| Energy approach (BOR, national database) | 1.20 | 0.28 | 2.22 | 0.97 | 0.24 | 1.78 |
| Alpha method, clay type I | 1.12 | 0.21 | 2.40 | 1.09 | 0.21 | 2.31 |
| Alpha method, clay type II | 2.54 | 0.52 | 2.69 | 1.09 | 0.24 | 2.09 |
| Beta method | 1.07 | 0.21 | 2.22 | 1.03 | 0.21 | 2.10 |
| CPT method | 1.33 | 0.29 | 2.43 | 1.02 | 0.24 | 1.93 |
| Lambda method, clay type I | 1.40 | 0.32 | 2.42 | 1.02 | 0.24 | 1.92 |
| Meyerhof's SPT method | 1.68 | 0.39 | 2.45 | 1.04 | 0.24 | 1.98 |

Note: (1) Type I refers to soils with undrained shear strength $S_u < 50$ kPa; Type II refers to soils with $S_u > 50$ kPa.
(2) $x_T$ = Ratio of the maximum test load to the predicted pile capacity.

Subsequently, the reliability of the defective pile can be calculated using Equations 14.24 through 14.26 by replacing $\lambda_R$ with $\lambda_{RD}$. For simplicity, $COV_R$ for a pile with toe debris is assumed to be the same as that for a pile free of toe debris.

## 14.6.2 Survey of toe debris

To obtain the prior information on occurrence probability and thickness of toe debris in Hong Kong, a practice survey was conducted in the form of questionnaires to solicit views from practicing engineers on bored pile quality (Zhang et al. 2006a). The results for toe debris are shown in Figure 14.5a. The perceived occurrence probability of toe debris is generally smaller than 5%. A histogram of the occurrence probability of toe debris is plotted in Figure 14.5b. Through the Chi-square goodness-of-fit test at the 5% significance level, a beta distribution as shown in Equation 14.20 with parameters $q = 2.0$ and $r = 34.2$ is supported as an adequate model. The corresponding mean and standard deviation of the occurrence probability are 5.5 and 3.7%, respectively.

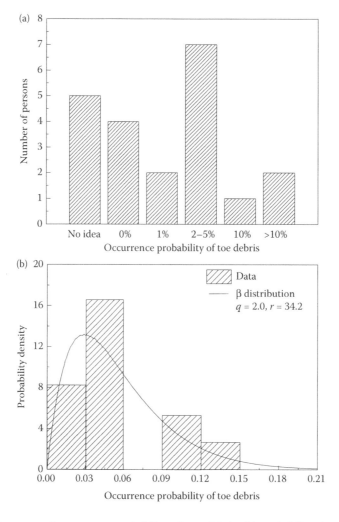

Figure 14.5 (a) Histogram of occurrence probability of toe debris; (b) probability density of occurrence probability of toe debris.

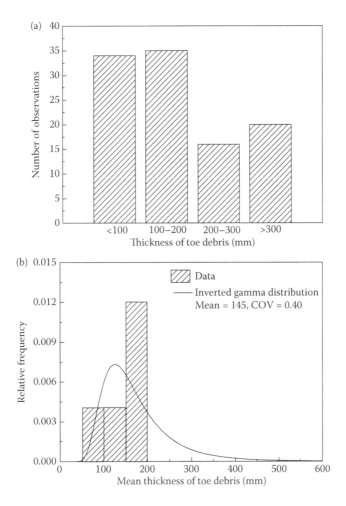

*Figure 14.6* (a) Histogram of thickness of toe debris; (b) probability density of mean thickness of toe debris.

To obtain the information on toe debris thickness, as-built information of 263 bored piles constructed at five different sites in Hong Kong in 2000 is adopted for analysis. The data were reported by the Hong Kong Construction Association (HKCA 2003) and the study was conducted in response to recent bored pile cases where confirmatory interface coring revealed pile toe irregularities. The results are summarized in Figure 14.6a. About 70% of toe debris found is less than 200-mm thick. The thickness of toe debris in 20 piles is larger than 300 mm. Based on these data, a probability density of the mean thickness is plotted in Figure 14.6b. Through the Kolmogorov–Smirnov goodness-of-fit test at a significance level of 5%, an inverted gamma distribution as shown in Equation 14.23 with parameters $\kappa = 6.3$ and $v = 906.3$ is confirmed to fit the frequency diagram of the mean thickness. The corresponding mean and standard deviation of the inverted beta distribution are 145 and 58 mm, respectively.

## 14.6.3 Updating the priors based on interface coring tests

After some coring tests have been conducted, information on the occurrence probability and thickness of toe debris can be obtained. In a field study, 67 bored piles 2.5 m in diameter were randomly selected from a site and were interface cored. Two piles were found to contain toe debris, whose thicknesses were $x_1 = 100$ mm and $x_2 = 50$ mm, respectively. This observation corresponds to case 2 described earlier. For comparison purposes, imaginary scenarios corresponding to case 1 (i.e., no toe debris is found in the 67 piles) and case 3 (i.e., all the toe debris at the site are detected and repaired) are also studied.

The prior distributions of the occurrence probability and the toe debris thickness are updated for the three cases. Figure 14.7 compares the updated PDFs of occurrence probability of toe debris for cases 1 and 2 for a diffuse prior and a prior beta distribution. The four updated PDFs in Figure 14.7 are sharper than the respective prior PDFs. The updated standard deviations of the occurrence probability, calculated by $\sqrt{q''r''/(q'' + r'' + 1)}/(q'' + r'')$ based on the beta prior distribution, are 1.3 and 1.9% for cases 1 and 2, respectively. All these updated standard deviations are significantly smaller than the prior standard deviation of 3.7% found earlier. The uncertainties in the occurrence probability are substantially reduced through the tests. For the prior beta distribution, the Bayesian estimators of occurrence probability, calculated by $(q + m)/(q + r + n)$, for the three cases are 1.9, 4.0, and 0%, respectively. For the diffuse prior distribution, the Bayesian estimators, calculated by $(m + 1)/(n + 2)$, for the three cases are 1.5, 4.3, and 0%, respectively, which will be used for $p_d$ in Equation 14.15.

Similarly, the mean thickness of toe debris can also be updated based on the test results. Figure 14.8 shows the prior and updated PDFs of mean toe debris thickness for case 2. The updated PDF is sharper than the prior PDF. The corresponding standard deviation of the mean thickness of toe debris after the tests, calculated by $v''/\sqrt{(\kappa'')^3}$, is 45 mm, which is smaller than the prior standard deviation of 58 mm found in the previous section. Hence, the uncertainty in the mean thickness is substantially reduced through the coring tests. The Bayesian estimators of the mean thickness, calculated by $\hat{t}'' = (v + \Sigma_{i=1}^{m} x_i)/(\kappa + m)$, for the three cases are 145, 128, and 0 mm, respectively.

There is no observation of toe debris thickness in case 1 and all toe debris, if any, is repaired in case 3. Therefore, no updating exercise for toe debris thickness is conducted for the two cases.

Table 14.7 summarizes the assumptions in the three cases and whether the PDFs of $f(p_d)$ and $f(t)$ are updated or not in each case. Note that all piles would be tested and repaired in case 3 and hence no toe debris would be present.

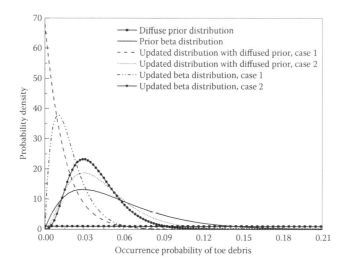

*Figure 14.7* A comparison between updated and prior PDFs of occurrence probability of toe debris.

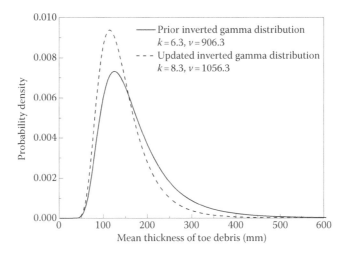

*Figure 14.8* A comparison between updated and prior distributions of mean toe debris thickness.

## 14.6.4 Updating reliability of piles based on interface coring tests

Having obtained the updated distributions of the occurrence probability and thickness of toe debris based on the tests, the reliability of the piles can be updated. The following load statistics are adopted: $\lambda_{QD} = 1.08$, $\lambda_{QL} = 1.15$, $\gamma_D = 1.25$, $\gamma_L = 1.75$, $\text{COV}_{QD} = 0.13$, $\text{COV}_{QL} = 0.18$, and $Q_D/Q_L = 2.0$. These statistics are similar to those in Section 14.5.2. When the soil shaft resistance, the rock shaft resistance, and the end bearing are all considered, the bias factor and COV of the pile capacity are assumed as $\lambda_R = 1.15$ and $\text{COV}_R = 0.17$ following McVay et al. (1998). The test piles in the calibration exercise are assumed to have been constructed properly. If excessive toe debris exists, then $\lambda_R$ will decrease due to the reduction in the pile capacity.

Ideally, the effect of toe debris on the behavior of axially loaded, large-diameter bored piles founded on rock should be studied by systematic field tests. Such tests will be costly, and numerical modeling is often resorted to instead. Zhang et al. (2006a) reported the behavior of large-diameter bored piles with different lengths and toe debris of varying thickness using a nonlinear finite-element program. The relationships between the applied load and the pile-head settlement for piles of three lengths (25, 50, and 75 m) and three diameters (1.2, 1.5, and 2.0 m) were obtained. Based on these relationships, the pile capacity reduction factors defined by Equation 14.27 can be obtained. Figure 14.9 shows the reduction factors plotted against the toe debris thickness. As expected, the reduction factor decreases with the toe debris thickness. The toe debris has a significant effect on the capacity of the 25 m long piles, but a minor effect on the 75 m long piles. Obviously, when a pile is very long, the

*Table 14.7* Summary of updated items in three cases

| Case | Assumptions | PDF of occurrence probability of toe debris, $f(p_d)$ | PDF of toe debris thickness, $f(t)$ |
|---|---|---|---|
| Case 1 | No toe debris detected | Updated | Not updated |
| Case 2 | Toe debris detected but not repaired | Updated | Updated |
| Case 3 | Toe debris detected and repaired | $p_d = 0, x = 0$ | Not relevant |

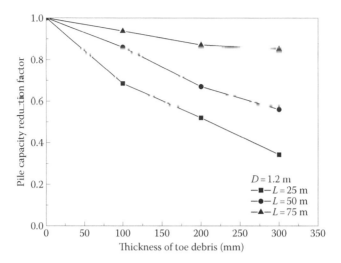

*Figure 14.9* The relationships between pile capacity reduction factor and toe debris thickness.

mobilized toe resistance only consists of a small fraction of the pile capacity. Therefore, the pile capacity is less affected by the presence of toe debris.

After obtaining the reduction factor, the resulting bias factor $\lambda_{RD}$ for a pile with toe debris can be calculated using Equation 14.28. Accordingly, the reliability index of the defective pile can be calculated using Equation 14.24 and the probability of unsatisfactory performance of the pile with a given toe debris, $P(F|x)$, can be calculated based on the calculated reliability index. Figure 14.10 shows the relationships between the reliability index and the toe debris thickness for three pile lengths. The reliability index decreases significantly due to the effect of toe debris when the bedrock levels are not extremely deep (say, smaller than 50 m). For instance, for a 1.2 m diameter, 25 m long pile, the reliability index decreases from 3.5 when no toe debris is present to 2.11 when a 100-mm-thick toe debris is present. The reliability index at zero toe debris thickness, 3.50, and the corresponding probability of failure, $P(F|\bar{E}) = 2.33 \times 10^{-4}$, account for other types of imperfections besides toe debris as well as other sources of uncertainty. The conditional probability of unsatisfactory performance given the presence of toe debris, $P(F|E)$, can be calculated using Equation 14.17. To do so, three functions, $P(F|x)$, $f(x|t)$, and $f(t)$ must be available. $P(F|x)$ can be obtained by curve fitting based on the data in Figure 14.10; $f(x|t)$ is expressed by Equation 14.22; and $f(t)$ is expressed by the inverted gamma distribution shown in Figure 14.8 and Equation 14.23. The calculations with Equation 14.17 are conducted using MATLAB®. The corresponding prior $P(F|E)$ values are 0.077 for $L = 25$ m, 0.039 for $L = 50$ m, and 0.00088 for $L = 75$ m, and the updated values are 0.07 for $L = 25$ m, 0.034 for $L = 50$ m, and 0.00087 for $L = 75$ m.

Having determined $P(F|E)$, $P(F|\bar{E})$, and the Bayesian estimators of $p_d$, the reliability of the piles at a site before and after the tests can be calculated by Equation 14.15 for each of the three cases of test outcome and follow-up actions described earlier. Let us examine the scenario of a diffuse prior $p_d$ first. The updated PDFs of the occurrence probability of toe debris have been shown in Figure 14.7. Figure 14.11 shows the reliability of the 1.2-m diameter piles of different lengths before and after tests. The reliability after the tests is significantly higher than that before the tests, particularly for relatively short piles. The effectiveness of the tests decreases with the pile length. For the 75-m long piles in case 2, the reliability index only increases from 3.26 before the tests to 3.47 after the tests. The outcome of the tests also affects the updated reliability as expected. For example, for the 25-m long piles, the

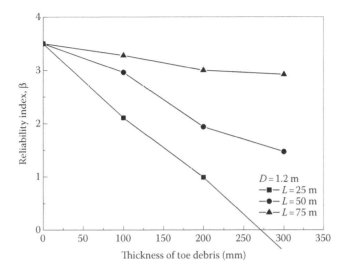

*Figure 14.10* The reliability of piles given toe debris of varying thickness.

updated reliability indexes for cases 1, 2, and 3 of different outcomes are 3.00, 2.72, and 3.50, respectively. Toe debris is found in case 2 but no remedial actions are taken; therefore, the updated reliability is lower than other two cases. The reliability index for case 3 is the highest among the three cases because the assumption that all piles would be tested and repaired has been made for this case.

## 14.7 SUMMARY

It has been shown that pile tests, either load tests or integrity tests, are routinely conducted during and after pile construction to ensure the safety of pile foundations and reduce costs. Uncertainties in the key factors affecting the reliability of piles and the estimated reliability

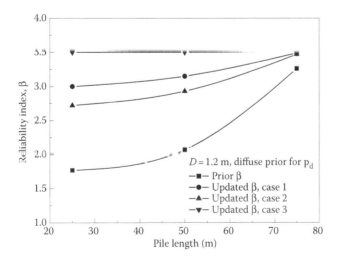

*Figure 14.11* A comparison between updated and prior reliability indexes of piles using a diffuse prior for $p_d$.

of the piles will be changed after these tests. This chapter describes procedures for quantitatively evaluating the impact of routine load tests and integrity tests on the reliability of piles using the Bayesian approach and illustrates the proposed procedures by considering the effect of proof load tests and interface coring on the reliability of piles.

In the proof load test example, whether or not a load test is carried out to failure, the test outcome can be used to ensure that the acceptance reliability is met using the methods described in this chapter. Thus, contributions of load tests can be included in a design in a systematic manner. Although various analysis methods could arrive at considerably different designs, the reliability of the designs associated with these analysis methods is rather uniform if the designs adopt the same FOS of 2.0 and are verified by consecutive positive proof tests.

In the example of reliability analysis of bored piles with defective toes, the occurrence probability and mean thickness of toe debris are updated with the information from the interface coring tests. The test information can improve the reliability of pile foundations significantly in the example. The degree of reliability improvement depends on outcomes of the tests, actions taken after the tests, the pile length, and the pile diameter.

## ACKNOWLEDGMENT

The research described in this chapter was substantially supported by the Research Grants Council of the Hong Kong Special Administrative Region (Project No. HKUST6126/03E).

## LIST OF SYMBOLS

| | |
|---|---|
| $\beta$ | reliability index |
| $\beta_T$ | target reliability index |
| $\xi$ | standard deviation of logarithm of a random variable |
| $\eta$ | mean of logarithm of a random variable |
| $\kappa$ and $\nu$ | parameters for a gamma distribution |
| $\lambda_R$ | bias factor for resistance |
| $\lambda_{QD}$ | bias factor for dead load |
| $\lambda_{QL}$ | bias factor for live load |
| $\mu$ | mean value of random variable |
| $\mu'$ | prior mean of the mean value when the mean value is also a variable |
| $\mu''$ | posterior mean of a random variable |
| $\sigma$ | standard deviation of a random variable |
| $\sigma'$ | prior standard deviation of a random variable |
| $\sigma''$ | posterior standard deviation of a random variable |
| $\Phi()$ | cumulative distribution function |
| $COV_Q$ | coefficient of variation for load effect |
| $COV_R$ | coefficient of variation for resistance |
| $D$ | pile diameter |
| $L$ | pile length |
| $k$ | ratio of the maximum load in a load test to the design load |
| $m$ | number of pile tests with unsatisfactory test outcome |
| $n$ | total number of pile tests |
| $p_d$ | occurrence probability of toe debris |
| $p_f$ | probability of unsatisfactory performance |
| $q$ and $r$ | parameters for a beta distribution |

$Q_D$    dead load
$Q_L$    live load
$\overline{Q}$    mean value of load effect
$\overline{R}$    mean value of resistance
$R_F$    a bearing capacity reduction factor
$S_u$    undrained shear strength
$t$    mean thickness of toe debris
$x_T$    ratio of the maximum load in a load test to the estimated pile capacity

## REFERENCES

AASHTO 1997. *LRFD Bridge Design Specifications*, Washington, DC: American Association of State Highway and Transportation Officials.

Ang, A. H.-S. and Tang, W. H. 2007. *Probability Concepts in Engineering—Emphasis on Applications to Civil and Environmental Engineering*, 2nd Edition, New York: John Wiley and Sons.

Baecher, G. R. and Rackwitz, R. 1982. Factors of safety and pile load tests, *International Journal for Numerical and Analytical Methods in Geomechanics*, 6, 409–424.

Barker, R. M., Duncan, J. M., Rojiani, K. B., Ooi, P. S. K., Tan, C. K. and Kim, S. G. 1991. *Manuals for the Design of Bridge Foundations*, NCHRP Report 343, Washington, DC: Transportation Research Board, National Research Council.

Becker, D. E. 1996. Eighteenth Canadian geotechnical colloquium: Limit states design for foundations. Part I. An overview of the foundation design process, *Canadian Geotechnical Journal*, 33, 956–983.

Bell, K. R., Davie, J. R., Clemente, J. L. and Likins, G. 2002. Proven success for driven pile foundations, *Proceedings of the International Deep Foundations Congress, Geotechnical Special Publication No. 116*, O'Neill, M. W. and Townsend, F. C. (eds), ASCE, Reston, VA, 1029–1037.

Buildings Department 2004. *Code of Practice for Foundations*, Hong Kong: Buildings Department.

Chow F. C., Jardine R. J., Brucy F. and Nauroy J. F. 1998. Effect of time on the capacity of pipe piles in dense marine sand, *Journal of Geotechnical and Geoenvironmental Engineering, ASCE*, 124(3), 254–264.

Duncan, J. M. 2000. Factors of safety and reliability in geotechnical engineering, *Journal of Geotechnical and Geoenvironmental Engineering ASCE*, 126, 307–316.

Evangelista, A., Pellegrino, A. and Viggiani, C. 1977. Variability among piles of the same foundation, *Proceedings of the 9th International Conference on Soil Mechanics and Foundation Engineering*. Rotterdam: A. A. Balkema, 493–500.

Fenton, G. A. and Griffiths, D. V. 2003. Bearing-capacity prediction of spatially random c φ soils, *Canadian Geotechnical Journal*, 40(1), 54–65.

Geotechnical Engineering Office 2006. *Foundation Design and Construction*, GEO Publication 1/2006, Hong Kong: Geotechnical Engineering Office.

Hannigan, P. J., Goble, G. G., Thendean, G., Likins, G. E. and Rausche, F. 1997. *Design and Construction of Driven Pile Foundations, Workshop Manual, Vol. 1, Publication No. FHWA-HI-97-014*, Washington, DC: Federal Highway Administration.

Hong Kong Construction Association Ltd (HKCA). 2003. *Study for Bored Pile Interface Acceptance Criteria*. Hong Kong: Prepared by Ove Arup and Partners Hong Kong Ltd.

Kay, J. N. 1976. Safety factor evaluation for single piles in sand, *Journal of Geotechnical Engineering, ASCE*, 102, 1093–1108.

McVay, M. C., Birgisson, B., Zhang, L. M., Perez, A. and Putcha, S. 2000. Load and resistance factor design (LRFD) for driven piles using dynamic methods—A Florida perspective, *Geotechnical Testing Journal ASTM*, 23, 55–66.

McVay, M. C., Kuo, C. L. and Singletary, W. A. 1998. *Calibrating Resistance Factors in the Load and Resistance Factor Design for Florida Foundations*. Final Report, Department of Civil Engineering, University of Florida, Gainesville.

Meyerhof, G. G. 1976. Bearing capacity and settlement of pile foundations, *Journal of Geotechnical Engineering, ASCE*, 102, 195–228.

Najjar, S. S. and Gilbert, R. B. 2009. Importance of lower-bound capacities in the design of deep foundations, *Journal of Geotechnical and Geoenvironmental Engineering, ASCE*, 135(7), 890–900.

NRC 1995. *National Building Code of Canada*, Ottawa: National Research Council Canada.

O'Neill, M. W. and Reese, L. C. 1999. *Drilled Shafts: Construction Procedures and Design Methods, Publication No. FHWA-IF-99-025*, Washington, DC: Federal Highway Administration.

Orchant, C. J., Kulhawy, F. H. and Trautmann, C. H. 1988. *Reliability-Based Foundation Design for Transmission Line Structures, Vol. 2: Critical Evaluation of In-Situ Test Methods*, EL-5507 Final Report, Palo Alto: Electrical Power Institute.

Paikowsky, S. G. and Stenersen, K. L. 2000. The performance of the dynamic methods, their controlling parameters and deep foundation specifications, *Proceedings of the Sixth International Conference on the Application of Stress-Wave Theory to Piles*. Rotterdam: A. A. Balkema, 281–304.

Peck, R. B. 1969. Advantages and limitations of the observational method in applied soil mechanics, *Geotechnique*, 19, 171–187.

Phoon, K. K., Kulhawy, F. H. and Grigoriu, M. D. 2003. Development of a reliability-based design framework for transmission line structure foundations, *Journal of Geotechnical and Geoenvironmental Engineering, ASCE*, 129(9), 798–806.

Raiffa, H. and Schlaifer, R. 2000. *Applied Statistical Decision Theory*, New York: Wiley.

Sidi, I. D. 1985. Probabilistic prediction of friction pile capacities, PhD thesis, Urbana-Champaign: University of Illinois.

US Army Corps of Engineers (USACE) 1993. *Design of Pile Foundations*, New York: ASCE Press.

Whitman, R. V. 1984. Evaluating calculated risk in geotechnical engineering, *Journal of Geotechnical Engineering, ASCE*, 110, 145–188.

Withiam, J. L., Voytko, E. P., Barker, R. M., Duncan, J. M., Kelly, B. C., Musser, S. C. and Elias, V. 2001. *Load and Resistance Factor Design (LRFD) for Highway Bridge Substructures*, Report No. FHWA HI-98-032, Washington, DC: Federal Highway Administration.

Zhang, L. M. 2004. Reliability verification using proof pile load tests, *Journal of Geotechnical and Geoenvironmental Engineering, ASCE*, 130, 1203–1213.

Zhang, L. M. and Dasaka, S. M. 2010. Uncertainties in geologic profiles vs. variability in pile founding depths. *Journal of Geotechnical and Geoenvironmental Engineering, ASCE*, 136(11), 1475–1488.

Zhang, L. M., Li, D. Q. and Tang, W. H. 2006a. Impact of routine quality assurance on reliability of bored piles. *Journal of Geotechnical and Geoenvironmental Engineering, ASCE*, 132(5), 622–630.

Zhang, L. M., Li, D. Q. and Tang, W. H. 2006b. Level of construction control and safety of driven piles, *Soils and Foundations*, 46(4), 415–425.

Zhang, L. M., Tang, W. H. and Ng, C. W. W. 2001. Reliability of axially loaded driven pile groups, *Journal of Geotechnical and Geoenvironmental Engineering ASCE*, 127, 1051–1060.

Zhang, L. M., Tang, W. H., Zhang, L. L. and Zheng, J. G. 2004. Reducing uncertainty of prediction from empirical correlations, *Journal of Geotechnical and Geoenvironmental Engineering, ASCE*, 130, 526–534.

Part V

# Spatial variability

# Chapter 15

# Application of the subset simulation approach to spatially varying soils

*Ashraf Ahmed and Abdul-Hamid Soubra*

## 15.1 INTRODUCTION

The probabilistic analysis of geotechnical structures presenting spatial variability in the soil properties is generally performed using Monte Carlo simulation (MCS) methodology. This method is not suitable for the computation of the failure probabilities encountered in practice (especially when using a computationally expensive finite-element/finite-difference deterministic model) due to the large number of simulations required to calculate a small failure probability. For this reason, only the mean value and the standard deviation of the system response were extensively investigated in literature when using this method.

As an alternative to MCS methodology, Au and Beck (2001) proposed the subset simulation (SS) approach to calculate the small failure probabilities. The SS method was mainly applied in the literature to problems where the uncertain parameters are modeled by random variables. In this chapter, the SS method is employed in the case of a spatially varying soil. The Karhunen–Loève (K–L) expansion is used to discretize the random field (i.e., to transform the random field into a finite number of random variables) and to generate random field realizations that respect a prescribed autocorrelation function.

After a brief overview of the K–L expansion method, both the classical SS approach (involving the case where the uncertain parameters are modeled by random variables) and its extension to the case of a spatially varying soil are presented in the form of a step-by-step procedure for practical use. Three application examples are provided in this chapter. They aim at showing the practical implementation of (i) the method of generation of a random field by K–L expansion, (ii) the SS approach in case where the uncertain parameters are modeled by random variables, and (iii) the SS approach in case where the uncertain soil properties are modeled by random fields.

## 15.2 KARHUNEN–LOÈVE EXPANSION METHODOLOGY FOR THE DISCRETIZATION OF A RANDOM FIELD

A random field is typically described by (i) a probability density function (PDF) and (ii) a covariance function (or an autocorrelation function, which is the covariance function divided by the variance of the random field). There are several types of autocorrelation functions (e.g., white noise, linear, exponential, squared exponential, and power autocorrelation functions). For more details, one can refer to Baecher and Christian (2003). In the present chapter, an exponential covariance function was used to represent the correlation structure of the random field. For a two-dimensional (2D) Gaussian random field, this function is given as follows:

$$C[(x_1,y_1),(x_2,y_2)] = \sigma^2 \exp\left(-\frac{|x_1 - x_2|}{l_x} - \frac{|y_1 - y_2|}{l_y}\right) \tag{15.1}$$

in which $\sigma$ is the standard deviation of the random field, and $l_x$ and $l_y$ are, respectively, the horizontal and vertical autocorrelation lengths. An autocorrelation length is the distance within which the values of the uncertain property are significantly correlated. It is the distance required for the autocorrelation function to decay from 1 to $e^{-1}$ (i.e., 0.3679). Finally, $(x_1, y_1)$ and $(x_2, y_2)$ are the coordinates of two arbitrary points in the space.

Let $R(x, y, \theta)$ be a Gaussian random field where $x$ and $y$ denote, respectively, the horizontal and vertical coordinates and $\theta$ indicates the stochastic character of the random field. The random field $R$ can be approximated by the K–L expansion as follows (Spanos and Ghanem 1989; Ghanem and Spanos 1991):

$$R(x, y, \theta) \approx \mu + \sum_{i=1}^{M} \sqrt{\lambda_i}\,\phi_i(x, y)\xi_i(\theta) \tag{15.2}$$

where $\mu$ is the mean value of the random field, $M$ is the size of the series expansion, $\lambda_i$ and $\phi_i$ are the eigenvalues and eigenfunctions of the covariance function, respectively, and $\xi_i(\theta)$ (for $i = 1, ..., M$) is a vector of standard uncorrelated random variables. For the exponential covariance function used in this chapter, Ghanem and Spanos (1991) provided an analytical solution for the eigenvalues and eigenfunctions as follows:

a. For a one-dimensional (1D) horizontal random field generated in the interval $[-a_x, a_x]$, the eigenvalues can be calculated as follows:

$$\lambda_i = \frac{2c}{\omega_i^2 + c^2} \tag{15.3}$$

where

$$c = \frac{1}{l_x} \tag{15.4}$$

and

$$\begin{cases} c - \omega_i \tan(\omega_i a_x) = 0 & \text{for } i \text{ odd} \\ \text{and} \\ \omega_i + c \tan(\omega_i a_x) = 0 & \text{for } i \text{ even} \end{cases} \tag{15.5}$$

The eigenfunctions are calculated as follows:

$$\phi_i = \frac{\cos(\omega_i x)}{\sqrt{a_x + (\sin(2\omega_i a_x)/2\omega_i)}} \quad \text{for } i \text{ odd} \tag{15.6}$$

$$\phi_i = \frac{\sin(\omega_i x)}{\sqrt{a_x - (\sin(2\omega_i a_x)/2\omega_i)}} \quad \text{for } i \text{ even} \tag{15.7}$$

b. For a one-dimensional vertical random field generated in the interval $[-a_y, a_y]$, the eigenvalues and eigenfunctions are calculated using the same equations after replacing the horizontal coordinate $(x)$, the horizontal half-width $(a_x)$ of the domain and the horizontal autocorrelation length $(l_x)$, respectively, by the vertical coordinate $(y)$, the vertical half-depth $(a_y)$ of the domain, and the vertical autocorrelation length $(l_y)$.

c. In the case of a 2D random field, the eigenvalues are calculated as the product of all possible combinations of eigenvalues of the 1D random fields of each direction as follows:

$$\lambda_i^{2D} = \lambda_j^x \lambda_k^y \tag{15.8}$$

where $\lambda_i^{2D}$ are the eigenvalues of the 2D random field, $\lambda_j^x \{j = 1, \ldots, M\}$ are the eigenvalues of the horizontal direction, and $\lambda_k^y \{k = 1, \ldots, M\}$ are the eigenvalues of the vertical direction. Similarly

$$\phi_i^{2D}(x, y) = \phi_j(x)\phi_k(y) \tag{15.9}$$

in which $\phi_i^{2D}(x, y)$ are the eigenfunctions of the 2D random field, $\phi_j(x)\{j = 1, \ldots, M\}$ are the eigenfunctions of the horizontal direction, and $\phi_k(y)\{k = 1, \ldots, M\}$ are the eigenfunctions of the vertical direction.

Notice that the eigenvalues (and the corresponding eigenfunctions) of the 2D random field retained in the analysis are the highest $M$ ones in the list of values obtained after arranging these eigenvalues in a decreasing order. It should be emphasized that the choice of the number $M$ of terms retained in the K–L expansion (cf. Equation 15.2) depends on the desired accuracy of the problem being treated. In the case of a Gaussian random field, the error estimate of the K–L expansion with $M$ terms can be calculated as follows (Sudret and Berveiller 2008):

$$err(x, y) = 1 - (1/\sigma)\sum_{i=1}^{M} \lambda_i \phi_i^2(x, y) \tag{15.10}$$

in which $\sigma$ is the standard deviation of the Gaussian random field. In the case of a log-normal random field, one should use $\sigma_{ln}$ instead of $\sigma$ in Equation 15.10, where $\sigma_{ln}$ is the standard deviation of the underlying normal random field. It should be mentioned that in this case, the K–L expansion given in Equation 15.2 becomes (Cho and Park 2010)

$$R(x, y, \theta) \approx \exp\left[\mu_{ln} + \sum_{i=1}^{M} \sqrt{\lambda_i}\phi_i(x, y)\xi_i(\theta)\right] \tag{15.11}$$

where $\mu_{ln}$ is the mean value of the underlying normal random field. Notice that $\sigma_{ln}$ and $\mu_{ln}$ can be computed using the following equations:

$$\sigma_{ln} = \sqrt{ln(1 + (\sigma/\mu)^2)} \tag{15.12}$$

$$\mu_{ln} = ln\mu - 0.5\sigma_{ln}^2 \tag{15.13}$$

Finally, it should be mentioned that in the case of a log-normal random field, $l_x$ and $l_y$ in Equation 15.1 should be replaced by $l_{ln\,x}$ and $l_{ln\,y}$, where $l_{ln\,x}$ and $l_{ln\,y}$ are, respectively, the horizontal and vertical autocorrelation lengths of the underlying normal random field. They are the lengths over which the values of $ln[R(x, y, \theta)]$ are highly correlated.

## 15.3 BRIEF OVERVIEW OF THE SUBSET SIMULATION APPROACH

SS approach was proposed by Au and Beck (2001) as an alternative to MCS methodology to compute small failure probabilities. The basic idea of the SS approach is that the small failure probability can be expressed as a product of larger conditional failure probabilities. In this section, one presents a brief description of the steps of SS approach in the case of two random variables (for more details, the reader may refer to Chapters 1 and 4 by Phoon 2008). A quasi-similar procedure will be used later in this chapter for the case of a spatially random field where a finite number of random variables are involved in the analysis. The steps of the SS approach in the case of two random variables $(V_1, V_2)$ can be described as follows:

1. Generate a vector of two random variables $(V_1, V_2)$ according to a target PDF using direct MCS.
2. Using the deterministic model, calculate the system response corresponding to $(V_1, V_2)$.
3. Repeat steps 1 and 2 until obtaining a prescribed number $N_s$ of vectors of random variables and the corresponding values of the system response.
4. Determine the value of the performance function corresponding to each value of the system response and then arrange the values of the performance function in an increasing order within a vector $G_0$, where $G_0 = \{G_0^1, ..., G_0^k, ..., G_0^{Ns}\}$. Notice that the subscripts "0" refer to the first level (level 0) of the SS approach.
5. Prescribe a constant intermediate conditional failure probability $p_0$ for the failure regions $F_j \{j = 1, 2, ..., m - 1\}$ and evaluate the first failure threshold $C_1$, which corresponds to the first level of the SS approach (see Figure 15.1). The failure threshold $C_1$ is equal to the $[(N_s \times p_0) + 1]$th value in the increasing list of elements of the vector $G_0$.

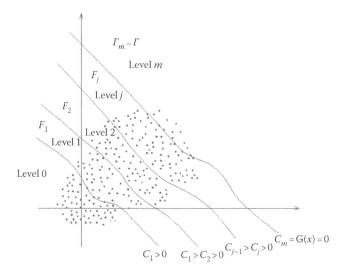

Figure 15.1 Nested failure domain.

This means that the value of the conditional failure probability of the first level $P(F_1)$ will be equal to the prescribed $p_0$ value.

6. Among the $N_s$ vectors of random variables, there are $[N_s \times p_0]$ ones whose values of the performance function are less than $C_1$ (i.e., they are located in the failure region $F_1$). These vectors are used as "mother vectors" to generate $N_s$ new vectors of random variables (according to a proposal PDF $P_p$) using the Markov chain method based on the modified Metropolis–Hastings (M–H) algorithm by Santoso et al. (2011). This algorithm is presented in Appendix 15A.

7. Using the deterministic model, calculate the values of the system response corresponding to the new vectors of random variables (which are located in level 1). Then, calculate the corresponding values of the performance function. Finally, gather the values of the performance function in an increasing order within a vector $G_1$, where $G_1 = \{G_1^1, ..., G_1^k, ..., G_1^{Ns}\}$.

8. Evaluate the second failure threshold $C_2$ as the $[(N_s \times p_0) + 1]$th value in the increasing list of elements of the vector $G_1$.

9. Repeat steps 6–8 to evaluate the failure thresholds $C_3$, $C_4$, ..., $C_m$ corresponding to the failure regions $F_3$, $F_4$, ..., $F_m$. Notice that contrary to all other thresholds, the last failure threshold $C_m$ is negative. Thus, $C_m$ is set to zero and the conditional failure probability of the last level $P(F_m|F_{m-1})$ is calculated as

$$P(F_m|F_{m-1}) = \frac{1}{N_s} \sum_{k=1}^{Ns} I_{F_m}(s_k) \qquad (15.14)$$

where $I_{F_m} = 1$ if the performance function $G(s_k)$ is negative and $I_{F_m} = 0$ otherwise. Finally, the failure probability $P(F)$ is evaluated as follows:

$$P(F) = P(F_1) \prod_{j=2}^{m} P(F_j|F_{j-1}) \qquad (15.15)$$

It should be mentioned that a normal PDF was used herein as a target PDF $P_t$. However, a uniform PDF was used as a proposal PDF $P_p$ (for more details, refer to Appendix 15A).

## 15.4 METHOD OF COMPUTATION OF THE FAILURE PROBABILITY BY THE SS APPROACH IN THE CASE OF A SPATIALLY VARYING SOIL PROPERTY

This section aims at employing the SS methodology for the computation of the failure probability in the case of a spatially varying soil property modeled by a random field. The random field was discretized in this chapter using the K–L expansion. In order to calculate the failure probability, a link between the SS approach and the K–L expansion was performed.

In fact, the K–L expansion (cf. Equation 15.2) includes two types of parameters (deterministic and stochastic). The deterministic parameters are the eigenvalues $\lambda_i$ and eigenfunctions $\phi_i$ of the covariance function. The role of these parameters is to ensure the correlation between the values of the random field at different points in the space. On the other hand, the stochastic parameters are represented by the vector of the standard normal random variables $\{\xi_i\}_{i=1, ..., M}$. The role of these parameters is to ensure the random nature of the uncertain property. The link between the SS approach and the K–L expansion was performed through the vector $\{\xi_i\}_{i=1, \ldots, M}$. This ensures that the SS technique does not affect the correlation structure of the random field.

The basic idea of the link is that for a given random field realization obtained by K–L expansion, the vector $\{\xi_i\}_{i=1, \ldots, M}$ represents a sample of the SS method for which the system response is calculated in two steps. The first step is to substitute the vector $\{\xi_i\}_{i=1, \ldots, M}$ in the K–L expansion to calculate the values of the random field at the centers of the different elements of the deterministic mesh according to their coordinates. The second step is to use the deterministic model to calculate the corresponding system response.

The algorithm of the SS approach in the case of a spatially varying soil property can be described as follows:

1. Choose the number $M$ of terms of K–L expansion. This number must be sufficient to accurately represent the target random field.
2. Generate a vector of $(M)$ standard normal random variables $\{\xi_1, \ldots, \xi_i, \ldots, \xi_M\}$ by direct MCS.
3. Substitute the vector $\{\xi_1, \ldots, \xi_i, \ldots, \xi_M\}$ in the K–L expansion to obtain the first realization of the random field. Then, use the deterministic model to calculate the corresponding system response.
4. Repeat steps 2 and 3 until obtaining a prescribed number $N_s$ of realizations of the random field and their corresponding values of the system response. Then, evaluate the corresponding values of the performance function to obtain the vector $G_0$, where $G_0 = \{G_0^1, \ldots, G_0^k, \ldots, G_0^{Ns}\}$. Notice that the values of the performance function of the different realizations are arranged in an increasing order in the vector $G_0$. Notice also that the subscripts "0" refer to the first level (level 0) of the SS approach.
5. Prescribe a constant intermediate conditional failure probability $p_0$ for the failure regions $F_j$ ($j = 1, 2, \ldots, m-1$) and evaluate the first failure threshold $C_1$, which corresponds to the failure region $F_1$, where $C_1$ is equal to the $[(N_s \times p_0) + 1]$th value in the increasing list of elements of the vector $G_0$. This ensures that the value of $P(F_1)$ will be equal to the prescribed $p_0$ value.
6. Among the $N_s$ realizations, there are $[N_s \times p_0]$ ones whose values of the performance function are less than $C_1$ (i.e., they are located in the failure region $F_1$). The corresponding vectors of standard normal random variables are used as "mother vectors" to generate $N_s$ new vectors of standard normal random variables using Markov chain method based on the modified M–H algorithm by Santoso et al. (2011). These new vectors are substituted in the K–L expansion to obtain the random field realizations of level 1.
7. The values of the performance function corresponding to the realizations of level 1 are gathered in an increasing order within a vector of performance function values $G_1 = \{G_1^1, \ldots, G_1^k, \ldots, G_1^{Ns}\}$
8. Evaluate the second failure threshold $C_2$ as the $[(N_s \times p_0) + 1]$th value in the increasing list of elements of the vector $G_1$.
9. Repeat steps 6–8 to evaluate the failure thresholds $C_3, C_4, \ldots, C_m$ corresponding to the failure regions $F_3, F_4, \ldots, F_m$. Notice that contrary to all other thresholds, the last failure threshold $C_m$ is negative. Thus, $C_m$ is set to zero and the conditional failure probability of the last level, that is, $P(F_m|F_m-1)$ is calculated as

$$P(F_m|F_{m-1}) = \frac{1}{N_s} \sum_{k=1}^{Ns} I_{F_m}(s_k) \qquad (15.16)$$

where $I_{F_m} = 1$, if the performance function $G(s_k)$ is negative and $I_{F_m} = 0$ otherwise.

Finally, the failure probability $P(F)$ is evaluated as follows:

$$P(F) = P(F_1) \prod_{j=2}^{m} P(F_j | F_{j-1}) \tag{15.17}$$

It should be emphasized that a normal PDF must be used herein as a target PDF $P_t$. However, a uniform PDF can be used as a proposal PDF $P_p$.

## 15.5 EXAMPLE APPLICATIONS

The next sections are devoted to three application examples. The first example involves the generation of a random field using K–L expansion. The second example involves the application of SS methodology to a case where the uncertain parameters are modeled by random variables. This example aims at computing the failure probability against bearing capacity failure of a strip footing resting on a $(c, \varphi)$ soil and subjected to a service vertical load $P_s$, where $c$ and $\varphi$ are considered as two random variables. Finally, the third example involves the application of the SS approach to a case of a spatially varying soil. It aims at computing the probability of failure against exceeding a tolerable vertical displacement of a strip footing resting on a spatially varying soil, where the soil Young's modulus was considered as a random field.

### 15.5.1 Example 1: Generation of a random field by K–L expansion

This example aims at illustrating the practical implementation of some theoretical concepts presented before for the generation of a random field. In this example, the soil Young's modulus was considered as a 2D log-normal random field. The statistical parameters used in the analysis are $\mu = 60$ MPa, COV = 15%, and $l_{ln\,x} = l_{ln\,y} = 1$ m. A small number of K–L terms ($M = 2$) is considered for illustrative purposes. The half-widths in both the $x$- and $y$-directions were taken as follows: $a_x = 7.5$ m and $a_y = 3$ m.

In order to generate a 2D random field, 1D random fields were first considered to calculate the eigenvalues and eigenfunctions corresponding to the horizontal and vertical directions according to Equations 15.3 through 15.7. They are shown in Table 15.1. Notice that in this table, the 1D eigenfunctions were computed based on the arbitrary values $x = 0.5$ m and $y = 3$ m.

After the computation of the eigenvalues and eigenfunctions of the 1D random fields, the product of all possible combinations was carried out according to Equations 15.8 and 15.9. The results of this process are shown in Table 15.2.

Table 15.1 Eigenvalues and eigenfunctions of 1D random fields in the $x$ and $y$ directions, where $\mu = 60$ MPa, COV = 15%, $l_{ln\,x} = l_{ln\,y} = 1$ m, $a_x = 7.5$ m, $a_y = 3$ m, and $M = 2$

| | 1D random field in the x-direction | | | | 1D random field in the y-direction | | |
|---|---|---|---|---|---|---|---|
| $j$ | $\omega_j^x$ | $\lambda_j^x$ | $\phi_j^x\ (x = 0.5\text{ m})$ | $k$ | $\omega_k^y$ | $\lambda_k^y$ | $\phi_k^y\ (y = 3\text{ m})$ |
| 1 | 0.18505 | 1.93378 | 0.34220 | 1 | 0.39740 | 1.72722 | 0.18802 |
| 2 | 0.37146 | 1.75749 | 0.06380 | 2 | 0.81861 | 1.19752 | 0.33382 |

Table 15.2 Eigenvalues and eigenvectors of a 2D random field, where $\mu = 60$ MPa, COV = 15%, $l_{ln\ x} = l_{ln\ y} = 1$ m, $a_x = 7.5$ m, $a_y = 3$ m, and $M = 2$

| J | k | $\lambda_j^x$ | $\lambda_k^y$ | $\lambda_i^{1D} = \lambda_j^x \lambda_k^y$ | $\phi_j^x$ $(x = 0.5$ m) | $\phi_k^y$ $(y = 3$ m) | $\phi_i^{2D} = \phi_j^x \phi_k^y$ |
|---|---|---|---|---|---|---|---|
| 1 | 1 | 1.93378 | 1.72722 | 3.34008 | 0.34220 | 0.18802 | 0.06434 |
| 1 | 2 | 1.93378 | 1.19752 | 2.31575 | 0.34220 | 0.33382 | 0.11423 |
| 2 | 1 | 1.75749 | 1.72722 | 3.03557 | 0.06380 | 0.18802 | 0.01199 |
| 2 | 2 | 1.75749 | 1.19752 | 2.10465 | 0.06380 | 0.33382 | 0.02130 |

Table 15.3 Retained eigenvalues and eigenvectors of a 2D random field where $\mu = 60$ MPa, COV = 15%, $l_{ln\ x} = l_{ln\ y} = 1$ m, $a_x = 7.5$ m, $a_y = 3$ m, and $M = 2$

| i | $\lambda_i^{2D}$ | $\phi_i^{2D}$ |
|---|---|---|
| 1 | 3.34008 | 0.06434 |
| 2 | 3.03557 | 0.01199 |

The rows of Table 15.2 should now be arranged according to a decreasing order of the eigenvalues $\lambda_i^{2D}$. Finally, the highest $M$ eigenvalues and the corresponding eigenfunctions are retained (Table 15.3). These values are then substituted in the K–L expansion to calculate the value of the 2D random field at a given point. For a given realization (say $\xi_1 = 0.815$, $\xi_2 = 0.906$), the value of the Young's modulus at the arbitrary point ($x = 0.5$ m and $y = 3$ m) as given by Equation 15.11 is equal to E = 60,360,433 Pa.

The accuracy of the discretized random field depends on the size of the K–L expansion (i.e., the number of terms $M$). Figure 15.2 presents the error estimate of the discretized 2D random field versus the number of eigenmodes (i.e., the number of terms of K–L expansion) for two values of the autocorrelation distance of an isotropic soil ($l_{ln\ x} = l_{ln\ y} = 1$ m and $l_{ln\ x} = l_{ln\ y} = 3$ m). For a prescribed accuracy, a greater number of terms is required for a smaller value of the autocorrelation distance. Figure 15.2 also indicates (as expected) that a more accurate representation of the target random field is obtained for a greater number

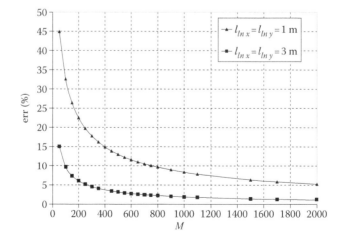

Figure 15.2 Error estimate versus the number of eigenmodes for two values of the autocorrelation distance of an isotropic soil.

of terms. For instance, when $l_{ln\,x} = l_{ln\,y} = 1$ m, the error is equal to 45% when $M = 50$ and it decreases to 12% when $M = 600$.

A computer program was developed that allows one to compute the values of a random field (for a given realization) at different points of the soil mass. These points (as will be seen in Example 3) are located on the centers of gravity of the different elements of a deterministic mesh. The obtained values of the random field (for a given realization) will be used to compute the system response (displacement at a given point of the structure, ultimate load of a footing, safety factor, etc.).

### 15.5.2 Example 2: Computation of the failure probability by SS approach in the case of random variables

In order to illustrate the algorithm of SS methodology in a simple way, a numerical example is provided herein. In this example, the SS approach was used to calculate the failure probability $P_f$ against bearing capacity failure of a strip footing of breadth B. The footing rests on a $(c, \varphi)$ soil and it is subjected to a service vertical load $P_s$. The soil cohesion $c$ and the soil angle of internal friction $\varphi$ were considered as random variables. The following formula was used for the computation of the ultimate bearing capacity:

$$q_u = \frac{\gamma B}{2} N_\gamma + c N_c + q N_q \tag{15.18}$$

in which

$$N_\gamma = 2(N_q - 1)\tan\varphi \tag{15.19a}$$

$$N_q = e^{\pi\tan\varphi} \cdot \tan^2\left(\frac{\pi}{4} + \frac{\varphi}{2}\right) \tag{15.19b}$$

$$N_c = \frac{N_q - 1}{\tan\varphi} \tag{15.19c}$$

where $N_\gamma$, $N_q$, and $N_c$ are the bearing capacity factors due to soil weight, surcharge loading, and cohesion, respectively. These coefficients are functions of the soil friction angle. On the other hand, $\gamma$ is the soil unit weight and $q$ is the surcharge loading. The performance function used in the analysis is

$$G = \frac{P_u}{P_s} - 1 \tag{15.20}$$

where $P_u$ is the ultimate footing load and $P_s$ is the footing applied load. As mentioned previously, only the soil cohesion and friction angle were considered as random variables. All the other parameters were considered as deterministic. These parameters are given in Table 15.4.

In this example, the intermediate failure probability $p_0$ of a given level $j$ ($j = 1, 2, ..., m - 1$) was arbitrarily chosen equal to 0.2. A small number of samples per level ($N_s = 10$ samples) were used to facilitate the illustration.

Table 15.5 presents (i) the values of $c$ and $\varphi$ of each sample for the successive levels, (ii) the corresponding values of the performance function, and (iii) the values of the failure thresholds $C_j$ for the different levels. Notice that only the first two levels and the last level for which

*Table 15.4* Data used for the probabilistic analysis of a strip footing against bearing capacity failure

| Parameter | Type of parameter | Mean and coefficient of variation of the parameter |
|---|---|---|
| Breadth $B$ | Deterministic | 2 m |
| Surcharge loading $q$ | Deterministic | 10 kPa |
| Soil unit weight $\gamma$ | Deterministic | 20 kN/m³ |
| Service vertical load $P_s$ | Deterministic | 1000 kN/m |
| Cohesion $c$ | Random normal variable | $\mu_c = 20$ kPa |
|  |  | $COV_c = 0.3$ |
| Friction angle $\varphi$ | Random normal variable | $\mu_\varphi = 30°$ |
|  |  | $COV_\varphi = 0.1$ |

*Table 15.5* Results of SS algorithm when $N_s = 10$ and $p_0 = 0.2$

| Level's number j | Cohesion c (kPa) | Angle of internal friction $\varphi$ (deg) | Performance function | Failure threshold $C_j$ |
|---|---|---|---|---|
| 1 | 23.23 | 26.0 | 0.926 | 1.4875 |
|  | 31.00 | 39.1 | 1.119 |  |
|  | 6.45 | 32.2 | 1.488 |  |
|  | 25.17 | 29.9 | 1.560 |  |
|  | 21.91 | 32.1 | 2.002 |  |
|  | 12.15 | 29.4 | 2.662 |  |
|  | 17.40 | 29.6 | 3.860 |  |
|  | 22.06 | 34.5 | 4.989 |  |
|  | 41.47 | 34.3 | 5.391 |  |
|  | 36.62 | 34.3 | 9.191 |  |
| 2 | 25.83 | 26.5 | 0.859 | 0.9740 |
|  | 27.29 | 26.2 | 0.941 |  |
|  | 25.32 | 26.8 | 0.974 |  |
|  | 23.98 | 25.4 | 1.056 |  |
|  | 25.92 | 26.0 | 1.084 |  |
|  | 14.86 | 28.0 | 1.151 |  |
|  | 14.14 | 28.8 | 1.159 |  |
|  | 12.27 | 28.8 | 1.175 |  |
|  | 13.14 | 29.4 | 1.193 |  |
|  | 11.80 | 30.0 | 1.236 |  |
| 6 | 15.17 | 22.9 | −0.160 | −0.0936 |
|  | 14.88 | 23.0 | −0.100 |  |
|  | 14.88 | 23.0 | −0.094 |  |
|  | 14.56 | 22.5 | 0.046 |  |
|  | 14.56 | 22.5 | 0.072 |  |
|  | 15.84 | 22.5 | 0.072 |  |
|  | 16.36 | 21.5 | 0.116 |  |
|  | 14.53 | 20.7 | 0.142 |  |
|  | 12.89 | 20.6 | 0.148 |  |
|  | 15.43 | 20.3 | 0.148 |  |
| $P_f$ | $9.6 \times 10^{-5}$ |  |  |  |

the failure threshold becomes negative were provided herein for illustration. Table 15.5 indicates that the failure threshold decreases with the successive levels until reaching a negative value at the last level. This means that the samples generated by the SS successfully progress toward the limit state surface $G = 0$. In order to select the failure threshold of a given level, the calculated values of the performance function of this level were arranged in an increasing order as shown in Table 15.5. Then, the failure threshold was selected as the $[(N_s \times p_0) + 1]$ th value of the arranged values of the performance function. Since $N_s = 10$ and $p_0 = 0.2$, the failure threshold is equal to the third value of the arranged values of the performance function. The SS computation continues until reaching a negative value (or a value of zero) of the failure threshold. In this example, the negative value was reached in the sixth level (where $C_6 = -0.0936$ as shown in Table 15.5. Theoretically, the last failure threshold should be equal to zero. For this reason, $C_6$ was set to zero. This means that the last conditional failure probability $P(F_6|F_5)$ is not equal to $p_0$. In this case, the last conditional failure probability $P(F_6|F_5)$ is calculated as the ratio between the number of samples for which the performance function is negative and the chosen number $N_s$ of samples (i.e., 10). According to Table 15.5, $P(F_6|F_5)$ is equal to $3/10 = 0.3$. Thus, the failure probability of the footing under consideration is equal to $0.2^5 \times 0.3 = 9.6 \times 10^{-5}$. It should be emphasized that the failure probability calculated in Table 15.5 is not accurate due to the small value of $N_s$. For an accurate computation of the failure probability, $N_s$ should be increased. This number should be greater than 100 to provide a small bias in the calculated $P_f$ value (see Chapter 4 by Honjo in Phoon 2008).

In order to determine the optimal number of samples $N_s$ to be used per level, different values of $N_s$ were considered to calculate $P_f$ and its coefficient of variation $COV_{Pf}$ as shown in Table 15.6. The thresholds corresponding to each $N_s$ value were calculated and shown in this table. Table 15.6 indicates (as was shown before when $N_s = 10$) that for the different values of $N_s$, the failure threshold decreases with the successive levels until reaching a negative value at the last level.

Figure 15.3a shows the effect of $N_s$ on the failure probability. It shows that for small values of $N_s$, the failure probability largely changes with $N_s$. However, for high values of $N_s$, the failure probability converges to an almost constant value. Figure 15.3a indicates that 2200 samples per level are required to accurately calculate the failure probability. This is because (i) the $C_j$ values corresponding to $N_s = 2200$ and 2400 samples are quasi-similar as it may be seen from Table 15.6 and (ii) the corresponding final $P_f$ values are too close (they are respectively equal to $2.60 \times 10^{-3}$ and $2.63 \times 10^{-3}$).

*Table 15.6* Evolution of the failure threshold $C_j$ with the different levels $j$ and with the number of realizations $N_s$ when $p_0 = 0.2$

| Failure threshold $C_j$ for each level $j$ | Number of samples $N_s$ per level | | | | | | |
|---|---|---|---|---|---|---|---|
| | 10 | 100 | 200 | 1000 | 2000 | 2200 | 2400 |
| $C_1$ | 1.4875 | 0.9397 | 1.0071 | 1.0638 | 1.0532 | 1.0466 | 1.0803 |
| $C_2$ | 0.9740 | 0.4157 | 0.3969 | 0.4916 | 0.4467 | 0.4466 | 0.4942 |
| $C_3$ | 0.7391 | 0.1011 | 0.1016 | 0.1513 | 0.1434 | 0.1347 | 0.1549 |
| $C_4$ | 0.4007 | −0.0491 | −0.0437 | −0.0307 | −0.0616 | −0.0536 | −0.0564 |
| $C_5$ | 0.1573 | − | − | − | − | − | − |
| $C_6$ | −0.0937 | − | − | − | − | − | − |
| $p_f$ | $9.60 \times 10^{-5}$ | $2.80 \times 10^{-3}$ | $2.72 \times 10^{-3}$ | $2.20 \times 10^{-3}$ | $2.80 \times 10^{-3}$ | $2.60 \times 10^{-3}$ | $2.63 \times 10^{-3}$ |
| $COV_{P_f}$ (%) | 221.4 | 57.9 | 42.1 | 18.7 | 13.3 | 12.8 | 12.4 |

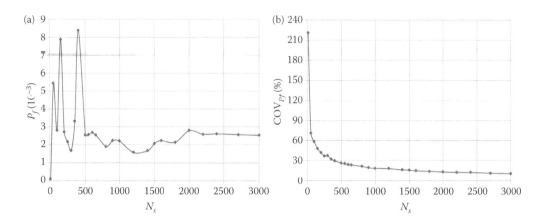

*Figure 15.3* $P_f$ and COV($P_f$) versus the number of realizations $N_s$.

Figure 15.3b shows the effect of $N_s$ on the coefficient of variation of the failure probability COV$_{Pf}$. As expected, COV$_{Pf}$ decreases with increasing $N_s$. Notice that the values of COV$_{Pf}$ for $N_s = 2200$ and 2400 samples are equal to 12.8 and 12.4%, respectively, which indicates (as expected) that the COV$_{Pf}$ decreases with the increase in the number of realizations.

It should be mentioned here that for $p_0 = 0.2$, four levels of SS were found necessary to reach the limit state surface $G = 0$ as may be seen from Table 15.6. Therefore, when $N_s = 2200$ samples, a total number of $N_t = 2200 \times 4 = 8800$ samples were required to calculate the final $P_f$ value. Remember that in this case, the COV of $P_f$ was equal to 12.8%. Notice that if the same value of COV (i.e., 12.8%) is desired by MCS to calculate $P_f$, the number of samples would be equal to 20,000. This means that, for the same accuracy, the SS approach reduces the number of realizations by 56%. On the other hand, if one uses MCS with the same number of samples (i.e., 8800 realizations), the value of COV of $P_f$ would be equal to 19.6%. This means that for the same computational effort, the SS approach provides a smaller value of COV($P_f$) than MCS.

### 15.5.3 Example 3: Computation of the failure probability by an SS approach in the case of random fields

This section presents a probabilistic analysis at the serviceability limit state (SLS) of a strip footing resting on a spatially varying soil using the SS approach. The objective is the computation of the probability $P_e$ of exceeding a tolerable vertical displacement under a prescribed footing load. Only one soil variability ($l_{ln\,x} = 10$ m and $l_{ln\,y} = 1$ m) is considered in this section. An extensive probabilistic parametric study on the same problem may be found in Ahmed and Soubra (2012).

A footing of breadth $b = 2$ m that is subjected to a central vertical load $P_s = 1000$ kN/m (i.e., an applied uniform vertical pressure $q_s = 500$ kN/m²) was considered in the analysis. As in Example 1, the Young's modulus was modeled by a random field and it was assumed to follow a log-normal PDF. The mean value and the coefficient of variation of the Young's modulus were, respectively, $\mu_E = 60$ MPa and COV$_E = 15\%$. An exponential covariance function was used to represent the correlation structure of the random field. The random field was discretized using K–L expansion. The performance function used to calculate the probability $P_e$ of exceeding a tolerable vertical displacement was defined as follows:

$$G = \delta v_{max} - \delta v \qquad (15.21)$$

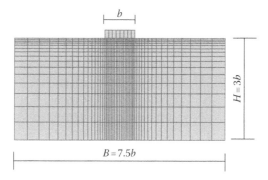

*Figure 15.4* Soil domain and mesh used in the numerical simulations.

where $\delta v_{max}$ is a tolerable vertical displacement of the footing center and $\delta v$ is the vertical displacement of the footing center due to the applied pressure $q_s$. In this chapter, it was assumed that $\delta v_{max}/b = 2 \times 10^{-2}$.

The deterministic model used to calculate the footing vertical displacement $\delta v$ was based on the commercial numerical code FLAC³ᴰ. For this calculation, a footing of breadth $b$ that rests on a soil domain of width $B$ and depth $H$ was considered in the analysis (Figure 15.4).

An optimal nonuniform but symmetrical mesh composed of 750 zones was employed. Although an SLS analysis is considered herein, the soil behavior was modeled by a conventional elastic–perfectly plastic model based on Mohr–Coulomb failure criterion in order to take into account the possible plastification that may occur near the edges of the foundation even under the service load. On the other hand, the strip footing was modeled by a linear elastic model. It is connected to the soil via interface elements that have the same shear strength parameters as the soil. The values of the shear strength and elastic properties of the soil, footing, and interface are given in Table 15.7. The soil unit weight is equal to 18 kN/m³.

In order to calculate the footing vertical displacement for a given random field realization, (i) the vertical and horizontal coordinates of the center of each element of the mesh were calculated; then the K–L expansion was used to calculate the value of the Young's modulus at the center of each element using Equation 15.11, (ii) geostatic stresses were applied to the soil, (iii) the obtained displacements were set to zero in order to obtain the footing displacement due to only the footing applied pressure, and finally (iv) the uniform vertical pressure was applied to the footing and the vertical displacement at the footing center due to this pressure was calculated.

*Table 15.7* Shear strength and elastic properties of soil, footing, and interface

| Variable | Soil | Footing | Interface |
|---|---|---|---|
| $c$ | 20 kPa | N/A | 20 kPa |
| $\varphi$ | 30° | N/A | 30° |
| $\psi = 2/3\varphi$ | 20° | N/A | 20° |
| $E$ | 60 MPa | 25 GPa | N/A |
| $\nu$ | 0.3 | 0.4 | N/A |
| $K_n$ | N/A | N/A | 1 GPa |
| $K_s$ | N/A | N/A | 1 GPa |

Table 15.8 Evolution of the failure threshold $C_j$ with different levels $j$ of the SS approach and with the number of realizations $N_s$ per level

| Failure threshold $C_j$ for each level $j$ | Number of realizations $N_s$ per level | | | | |
|---|---|---|---|---|---|
| | 50 | 100 | 150 | 200 | 250 |
| $C_1$ | 0.0086 | 0.0077 | 0.0080 | 0.0076 | 0.0076 |
| $C_2$ | 0.0058 | 0.0048 | 0.0050 | 0.0041 | 0.0040 |
| $C_3$ | 0.0044 | 0.0015 | 0.0019 | 0.0011 | 0.0011 |
| $C_4$ | 0.0017 | −0.0019 | −0.0007 | −0.0020 | −0.0018 |
| $C_5$ | −0.0015 | − | − | − | − |

Table 15.9 Values of $P_e$ and $COV_{Pe}$ versus the number $N_s$ of realizations per level

| | Number of realizations $N_s$ per level | | | | |
|---|---|---|---|---|---|
| | 50 | 100 | 150 | 200 | 250 |
| $P_e$ | $0.34 \times 10^{-4}$ | $4.60 \times 10^{-4}$ | $2.07 \times 10^{-4}$ | $3.78 \times 10^{-4}$ | $3.77 \times 10^{-4}$ |
| $COV_{Pe}$ (%) | 92 | 71 | 60 | 51 | 38 |

As mentioned before, a random field with $l_{lnx} = 10$ m and $l_{lny} = 1$ m was considered herein. The number $M$ of terms of K–L expansion adopted in the analysis is equal to $M = 100$. This corresponds to an error estimate that is smaller than 13%. The intermediate failure probability $p_0$ of a given level $j$ ($j = 1, 2, ..., m − 1$) was chosen equal to 0.1.

The failure thresholds $C_j$ of the different levels of the SS were calculated and presented in Table 15.8 for different values of $N_s$. Table 15.9 presents the $P_e$ values and the corresponding values of the coefficient of variation for the different numbers of realizations $N_s$. As expected, the coefficient of variation of $P_e$ decreases with the increase in the number of realizations $N_s$.

From Tables 15.8 and 15.9, it was found that $P_e$ converges when $N_s = 200$ realizations. This is because (i) the $C_j$ values corresponding to $N_s = 200$ and 250 realizations are quasi-similar and (ii) the corresponding final $P_e$ values are too close (they are equal to $3.78 \times 10^{-4}$ and $3.77 \times 10^{-4}$, respectively). Notice that the values of $COV_{Pe}$ for $N_s = 200$ and 250 realizations are equal to 51 and 38%, respectively, which indicates (as expected) that the $COV_{Pe}$ decreases with the increase in the number of realizations. It should be mentioned here that for $p_0$ equal to 0.1, four levels of SS were found necessary to reach the limit state surface $G = 0$ as may be seen from Table 15.8. Therefore, when $N_s = 200$ realizations, a total number of $N_t = 200 \times 4 = 800$ realizations were required to calculate the final $P_e$ value. Remember that in this case, the COV of $P_e$ was equal to 51%. Notice that if the same value of COV (i.e., 51%) is desired by MCS to calculate $P_e$, the number of realizations would be equal to 12,000 (see Ahmed and Soubra 2012). This means that, for the same accuracy, the SS approach reduces the number of realizations by 93.3%. On the other hand, if one uses MCS with the same number of realizations (i.e., 800 realizations), the value of COV of $P_e$ would be equal to 189% (see Ahmed and Soubra 2012). This means that for the same computational effort, the SS approach provides a smaller value of $COV_{Pe}$ than MCS.

## 15.6 CONCLUSION

The probabilistic analysis of shallow foundations resting on a spatially varying soil was generally performed in the literature using MCS methodology. The mean value and the standard deviation of the system response were extensively investigated. This was not the case for the failure probability because MCS methodology requires a large number of calls of the deterministic model to accurately calculate a small failure probability. This chapter fills this gap. An extension of the SS to the case of a spatially varying soil (where the soil property was modeled by a random field) was presented. The random field was discretized using K–L expansion methodology.

Three example applications were provided. They aim at showing the practical implementation of (i) the method of generation of random field by K–L expansion, (ii) the computation of the failure probability against bearing capacity failure (using the SS approach) of a strip footing resting on a $(c, \varphi)$ soil and subjected to a service vertical load $P_s$ where $c$ and $\varphi$ are considered as two random variables, and (iii) the probability of failure against exceeding a tolerable vertical displacement (using the SS approach) of a strip footing resting on a spatially varying soil where the soil Young's modulus was considered as a random field.

It was found that for a prescribed accuracy, the SS approach significantly reduces the number of realizations as compared to MCS methodology (the reduction was found to be equal to 93.3% in the present chapter). In other words, for the same computational effort, the SS approach provides a smaller value of the coefficient of variation of $P_e$ than MCS.

Finally, it should be mentioned that the MATLAB® codes used for three example applications are provided in http://www.univ-nantes.fr/soubra-ah for practical use.

## APPENDIX 15A: MODIFIED M–H ALGORITHM

The M–H algorithm is a Markov chain Monte Carlo (MCMC) method. It is used to generate a sequence of new realizations from existing realizations (that follow a target PDF called "$P_t$"). Refer to Figure 15.1 and let $s_k \in F_j$ be a current realization that follows a target PDF "$P_t$". Using a proposal PDF "$P_p$", a next realization $s_{k+1} \in F_j$ that follows the target PDF "$P_t$" can be simulated from the current realization $s_k$ as follows:

a. A candidate realization $\hat{s}$ is generated using the proposal PDF ($P_p$). The candidate realization $\hat{s}$ is centered at the current realization $s_k$.
b. Using the deterministic model, evaluate the value of the performance function $G(\hat{s})$ corresponding to the candidate realization $\hat{s}$. If $G(\hat{s}) < C_j$ (i.e., $\hat{s}$ is located in the failure region $F_j$), set $s_{k+1} = \hat{s}$; otherwise, reject $\hat{s}$ and set $s_{k+1} = s_k$ (i.e., the current realization $s_k$ is repeated).
c. If $G(\hat{s}) < C_j$ in the preceding step, calculate the ratio $r_1 = P_t(\hat{s})/P_t(s_k)$ and the ratio $r_2 = P_p(s_k|\hat{s})/P_p(\hat{s}|s_k)$, and then compute the value $r = r_1 r_2$.
d. If $r \geq 1$ (i.e., $\hat{s}$ is distributed according to the $P_t$), one continues to retain the realization $s_{k+1}$ obtained in step b; otherwise, reject $\hat{s}$ and set $s_{k+1} = s_k$ (i.e., the current realization $s_k$ is repeated).

Notice that in step b, if the candidate realization $\hat{s}$ does not satisfy the condition $G(\hat{s}) < C_j$, it is rejected and the current realization $s_k$ is repeated. Also in step d, if the candidate realization $\hat{s}$ does not satisfy the condition $r \geq 1$ (i.e., $\hat{s}$ is not distributed according to the $P_t$), it is rejected and the current realization $s_k$ is repeated. The presence of several repeated realizations is not desired as it leads to high probability that the chain of realizations remains in the current state. This means that there is high probability that the next failure threshold $C_{j+1}$ is equal to the current failure threshold $C_j$. This decreases the efficiency of the SS approach.

To overcome this inconvenience, Santoso et al. (2011) proposed to modify the classical M–H algorithm as follows:

a. A candidate realization $\hat{s}$ is generated using the proposal PDF ($P_p$). The candidate realization $\hat{s}$ is centered at the current realization $s_k$.

b. Calculate the ratio $r_1 = P_t(\hat{s})/P_t(s_k)$ and the ratio $r_2 = P_p(s_k|\hat{s})/P_p(\hat{s}|s_k)$, and then compute the value $r = r_1 r_2$.

c. If $r \geq 1$, set $s_{k+1} = \hat{s}$; otherwise, another candidate realization is generated. Candidate realizations are generated randomly until the condition $r \geq 1$ is satisfied.

d. Using the deterministic model, evaluate the value of the performance function $G(s_{k+1})$ of the candidate realization that satisfies the condition $r \geq 1$. If $G(s_{k+1}) < C_j$ (i.e., $s_{k+1}$ is located in the failure region $F_j$), one continues to retain the realization $s_{k+1}$ obtained in step c; otherwise, reject $\hat{s}$ and set $s_{k+1} = s_k$ (i.e., the current realization $s_k$ is repeated).

These modifications reduce the repeated realizations and allow one to avoid the computation of the system response of the rejected realizations. This becomes of great importance when the time cost for the computation of the system response is expensive (i.e., for the finite element or finite difference models).

## LIST OF SYMBOLS

| | |
|---|---|
| $x$ | horizontal coordinate |
| $y$ | vertical coordinate |
| $\mu$ | mean value of a random variable or random field |
| $\sigma$ | standard deviation of a random variable or random field |
| $\mu_{ln}$ | mean value of an underlying normal random variable or random field |
| $\sigma_{ln}$ | standard deviation of an underlying normal random variable or random field |
| $l_x$ | horizontal autocorrelation length of a random field |
| $l_y$ | vertical autocorrelation length of a random field |
| $l_{ln\,x}$ | horizontal autocorrelation length of an underlying normal random field |
| $l_{ln\,y}$ | vertical autocorrelation length of an underlying normal random field |
| $M$ | number of terms in K–L expansion |
| $\lambda$ | eigenvalue of the covariance function |
| $\phi$ | eigenfunction of the covariance function |
| $\xi$ | vector of standard normal uncorrelated random variables |
| $p_0$ | intermediate conditional failure probability |
| $N_s$ | number of simulations in each level of subset simulation approach |
| $b$ | footing width |
| $c$ | soil cohesion |
| $\varphi$ | angle of internal friction of the soil |
| $\gamma$ | soil unit weight |
| $N_c, N_q,$ and $N_\gamma$ | bearing capacity factors |
| $q_u$ | ultimate bearing capacity |
| $q_s$ | footing vertical pressure |
| $m$ | number of levels of subset simulation approach |
| $P_t$ | target probability density function |
| $P_p$ | proposal probability density function |
| $P_e$ | probability of exceeding a tolerable settlement |

| | |
|---|---|
| $P_s$ | applied footing load |
| $G$ | performance function |
| $\delta v$ | vertical displacement of the footing center |
| $\delta v_{max}$ | tolerable vertical displacement of the footing center |

## REFERENCES

Ahmed, A. and Soubra, A.-H. 2012. Probabilistic analysis of strip footings resting on a spatially random soil using subset simulation approach. *Georisk: Assessment and Management of Risk for Engineered Systems and Geohazards*, 6(3), 188–201.

Au, S.K. and Beck, J.L. 2001. Estimation of small failure probabilities in high dimensions by subset simulation. *Journal of Probabilistic Engineering Mechanics*, 16, 263–277.

Baecher, G. and Christian, J.T. 2003. *Reliability and Statistics in Geotechnical Engineering*. England: Wiley, 605 pp.

Cho, S.E. and Park, H.C. 2010. Effect of spatial variability of cross-correlated soil properties on bearing capacity of strip footing. *International Journal for Numerical and Analytical Methods in Geomechanics*, 34, 1–26.

Ghanem, R. and Spanos, P. 1991. *Stochastic Finite Elements—A Spectral Approach*. New York: Springer.

Phoon, K.K. 2008. Numerical recipes for reliability analysis—A primer. In: Phoon, K.K., ed. *Reliability-Based Design in Geotechnical Engineering: Computations and Applications*. London: Taylor & Francis, 1–75.

Santoso, A.M., Phoon, K.K. and Quek, S.T. 2011. Modified Metropolis–Hastings algorithm with reduced chain-correlation for efficient subset simulation. *Probabilistic Engineering Mechanics*, 26(2), 331–341.

Spanos, P.D. and Ghanem, R. 1989. Stochastic finite element expansion for random media. *Journal of Engineering Mechanics, ASCE*, 115(5), 1035–1053.

Sudret, B. and Berveiller, M. 2008. Stochastic finite element methods in geotechnical engineering. In: Phoon, K.K., ed. *Reliability-Based Design in Geotechnical Engineering: Computations and Applications*. London: Taylor & Francis, 260–297.

# Index